Michell Structures

Tomasz Lewiński · Tomasz Sokół
Cezary Graczykowski

Michell Structures

Springer

Tomasz Lewiński
Faculty of Civil Engineering
Warsaw University of Technology
Warsaw, Poland

Tomasz Sokół
Faculty of Civil Engineering
Warsaw University of Technology
Warsaw, Poland

Cezary Graczykowski
Institute of Fundamental Technological Research
Polish Academy of Sciences
Warsaw, Poland

ISBN 978-3-030-06988-9 ISBN 978-3-319-95180-5 (eBook)
https://doi.org/10.1007/978-3-319-95180-5

Softcover re-print of the Hardcover 1st edition 2019

This Springer imprint is published by the registered company Springer International Publishing AG
The registered company address is: Gewerbestrasse 11, 6330 Cham, Switzerland

In memory of
Professor George Rozvany

Preface

The publication written by Anthony George Maldon Michell (1904) was far ahead of its time. This remarkable paper has posed and solved selected optimum design problems in which a structure as a whole is treated as the design variable, contrary to the traditional setting in which only some selected dimensions or parameters may be chosen to play this role. Michell has raised the question of a safe and economic way of transmitting the given loading to the prescribed zone of the support, without imposing *a priori* assumptions on the layout of the bars (i.e. on the structural topology concerning the position of nodes and members connections) designed to make this load transmission possible. The adopted safety conditions require keeping the uniaxial stress in members between the lower bound $-\sigma_C$ and the upper bound σ_T. The cost of the structure is assumed as proportional to its volume.

Forty years later, Leonid Kantorovich (1942) has shed a new light on the mass transport problem of Gaspard Monge. Only in 2001, Guy Bouchitté and Giuseppe Buttazzo noted essential links between the Monge–Kantorovich—and the Michell problems. Although these two problems do not coincide, both share the same theoretical background. To pose the problems correctly, one should construct the same kind of abstract variational settings expressed in terms of signed Radon measures.

Thus, a correct setting of the Michell problem necessitates an abstract reformulation. Michell's paper starts from referring to James C. Maxwell's results on truss design. Michell adopted Maxwell's concept of the embedding of a structure in a field $\mathbf{v}$ representing virtual displacements. The values of this field at the nodes are interpreted as their virtual displacements. Since the field $\mathbf{v}$ fills up the whole design domain, frequently being the whole space, the concept of introducing such a field is set in an abstract manner; the virtual displacements at nodes are in no connection with nodal displacements caused by the given load. Michell discovered that for the lightest, fully stressed structure, one can indicate a virtual displacement field $\mathbf{v}$ in which this structure is embedded such that: (i) this field makes the value of the virtual work of the load maximal, and (ii) this field generates virtual relative elongations of truss members lying within the bounds: $\frac{-\sigma_0}{\sigma_C}, \frac{\sigma_0}{\sigma_T}$; the referential stress σ_0 may be assumed arbitrarily.

It is really surprising that the problem of the volume minimization leads to the problem of maximization of the virtual work. Moreover, the problem set originally for trusses is rearranged to a problem of optimum design of continuum bodies. The continuum setting is the result of the phenomenon of the decrease in the volume along with increasing the number of possible truss members. This process has no its natural end, thus leading to the optimum structures being continuum bodies of fibrous microstructure. Indeed, if the number of possible members increases, the truss members can be formed according to a better layout contributing to a smaller value of the volume. Since no lower limit on the areas of the member cross sections is imposed, the members may be thinner and thinner thus allowing the total volume being smaller and smaller. The optimal Michell structures assume the form of discrete-continuum structures, thus extending the initial class of truss structures.

The next theoretical step is dualization of the mentioned problem of maximization of the virtual work leading to the problem of minimization of a certain measure of the stress field equilibrating the given loading and transmitting this loading to the prescribed support zone of the continuum body considered; minimization is taken over all statically admissible stress fields. This measure of stress fields was not known to Michell; it was proposed much later by George Rozvany (1976) in the context of flexural problems, and then rediscovered (for the case of the permissible stresses in tension and compression being equal) by Gilbert Strang and Robert Kohn (1983). A complete set of mathematical theorems, including both the maximization and minimization problems, can be found in the seminal paper by Guy Bouchitté, Wilfrid Gangbo and Pierre Seppecher (2008), where, probably for the first time, not only Michell's theory but even the linear theory of statics of trusses have been formulated there from scratch in terms of a fully mathematical language. Thus, the Michell problem reduces to the two, mutually dual problems whose solutions are linked by the optimality conditions. The main property of the optimum design is a pointwise locking of the virtual strains which naturally links this theory with William Prager's (1957) theory of deformations of bodies made of materials with locking.

The theory of Michell's structures would be only a forgotten artefact if the theory were not augmented and illustrated by carefully chosen examples, representing the exact and clear illustrative solutions to sufficiently rich class of fundamental problems of structural analysis. Spectacular is admitting the point loads, impossible to handle properly in elasticity. Indeed, the elasticity theory does not encompass the material concentrations on surfaces and curves, thus not allowing for the presence of the point loads as well the loads applied along the curves. The available Michell's structures, i.e. the exact solutions to Michell's optimum design problems, have been an inspiration for structural and aircraft engineers, architects and machine designers for the last decade and found applications even in the modern areas like mechatronics and biotechnology. One of the most important problems is to find the Michell structure carrying three self-equilibrated forces in the plane. The main family of the exact solutions to this problem, for the case of equal permissible stresses in tension and compression, has been constructed by Henry Chan (1966), but this report has been concealed for 40 years, being not mentioned even in the fundamental book by Walter Hemp (1973). In 2006, Prof. Henry Chan has generously sent this report to one of the authors of the present book, which has paved the way towards constructing new exact solutions composed of the substructures being the solutions to the three-force problems.

The exact solutions to Michell's problems are both educating and inspiring. In particular, they show that the edges should be strengthened by ribs; but they disclose also the necessity of using the ribs inside the structure. The optimal structures show also very fine methods of surrounding the point loads by fibrous substructures to minimize displacements of the nodes where the point loads are applied. The exact solutions constitute certain families of layouts; an exact solution may change its topology from one layout to the other, depending on the data. Small changes of the data may cause drastic differences in the optimum layouts. This phenomenon resembles the theory of attractors in the non-linear dynamics. Due to these subtle properties of the exact solutions, a precise numerical method for predicting new solutions or verifying the old solutions is pending. Let it be stressed here that all the exact solutions to Michell's problems reported in the present book have been checked numerically by the highly precise version of the ground structure method. This method has been proposed by W. S. Dorn, Ralf E. Gomory and Harvey J. Greenberg (1964), Matthew Gilbert and Andy Tyas (2003) and then improved by Tomasz Sokół (2011).

An anti-plane counterpart of the Michell problem in the plane is the optimum design of grillages of rigid joints and systems of beams, as set up by William Prager and George Rozvany, see Rozvany (1976) for the rich bibliography. Although the theory of optimal layout of grillages looks similar to Michell's theory, this similarity is only apparent. The properties of the exact solutions are different; in particular, the fibres in the optimum grillages are never curved. The mathematical theory of grillage optimization is being elaborated by Karol Bołbotowski; the results presented in this book are published by his courtesy.

The theories of Michell's structures and Prager–Rozvany grillages are prototypes of the contemporary field of optimization, called optimization of structural topology, or, simply topology optimization. This field, well represented in the books by Martin Bendsøe (1995) and Martin Bendsøe and Ole Sigmund (2003), develops rapidly, mostly by the activity of the Society of Structural and Multidisciplinary Optimization (ISSMO), the main organizer of the series of the biennial World Congresses on Structural and Multidisciplinary Optimization (WCSMO) and the patron of the journal: Structural and Multidisciplinary Optimization, edited by Springer-Verlag since 1989.

The book includes some material published by permissions generously given to us. This courtesy is highly acknowledged.

The book was prepared within the Research Grant no 2013/11/B/ST8/04436 financed by the National Science Centre (Poland), entitled: "Topology optimization of engineering structures. An approach synthesizing the methods of: free material design, composite design and Michell-like trusses." The text editing has been financed by the statutory grant of the Faculty of Civil Engineering, Warsaw University of Technology. The authors would like to express their thanks to Mrs Iwona Malicka for her dedicated and highly professional work of word processing, editing the whole manuscript and drawing the figures.

Moreover, we express our thanks to the editors, and especially to Mrs. Silvia Schilgerius, Senior Editor of Applied Sciences, Springer-Verlag, for the invaluable editorial remarks and understanding, when a deadline had gone. Last but not least, the authors express their thanks to Sindhu Sundararajan and her team for the final, highly precise work on the manuscript.

Warszawa–Radom
May 2018

Tomasz Lewiński
Tomasz Sokół
Cezary Graczykowski

References

Bendsøe, M. P. (1995). *Optimization of Structural Topology, Shape and Material*. Springer: Berlin.

Bendsøe, M. P., & Sigmund, O. (2003). *Topology Optimization, Theory, Methods and Applications*. Springer: Berlin.

Bouchitté G., Gangbo, W., & Seppecher, P. (2008). Michell trusses and lines of principal action. *Mathematical Models and Methods in Applied Sciences* **18**, 1571–1603.

Chan, H. S. Y. (1966). Minimum weight cantilever frames with specified reactions. University of Oxford. Department of Engineering Science Laboratory, Parks Road, Oxford. June 1966, No 1,010.66, p 11 with 4 figures.

Dorn, W. S., Gomory, R. E., & Greenberg, H. J. (1964). Automatic design of optimal structures. *Journal de Mécanique*, *3*, 25–52.

Gilbert, M. & Tyas, A. (2003). Layout optimization of large-scale pin-jointed frames. *Engineering Computations* **20**, 1044–1064.

Hemp, W. S. (1973). *Optimum Structures*. Clarendon Press: Oxford.

Kantorovich, L. V. (1942). On the transfer of masses. Dokl.Akad.Nauk SSSR **37**, 227–229 (in Russian).

Michell, A. G. M. (1904). The limits of economy of material in frame structures. *Philosophical Magazine Series 6*, **8** (47), 589–597.

Prager, W. (1957). On ideal locking materials, Trans. *Society of Rheology*, **1**, 169–175.

Rozvany, G. I. N. (1976). *Optimal Design of Flexural Systems*. Pergamon Press: London.

Sokół, T. (2011). A 99 line code for discretized Michell truss optimization written in Mathematica. *Structural and Multidisciplinary Optimization* **43**, 181–190.

Strang, G., & Kohn, R. V. (1983). Hencky-Prandtl nets and constrained Michell trusses. *Computer Methods in Applied Mechanics and Engineering*, *36*, 207–222.

Contents

Chapter 1
Selected Problems of Statics

1.1 Introduction

In its original formulation, Michell (1904) problem reads as follows: find the framework of least volume in which state of stress $\boldsymbol{\sigma}$ satisfies the conditions $-\sigma_C \leq \min \lambda_i(\boldsymbol{\sigma}) \leq \max \lambda_i(\boldsymbol{\sigma}) \leq \sigma_T$ and which a given load transmits to a given segment of the boundary of the design domain; $\lambda_i(\boldsymbol{A})$ represents ith eigenvalue of a symmetric matrix $\boldsymbol{A}$. The solution to this problem depends essentially on the ratio $\kappa = \frac{\sigma_T}{\sigma_C}$ but does not depend on other material data. Therefore, this problem belongs to the class called the plastic optimum design.

The point of departure of Michell's theory is the theory of small deformations of bar structures of rigid joints. The material in these structures is fully stressed only if the bending is eliminated. The simplest method to eliminate the bending is to choose the trusses instead of frames. Just optimization of trusses allows for the most economical, fully stressed designs and leads to the least volume solutions.

Although at the starting point, the Michell's approach concerns the bar structures, the condition of the least volume extends this class of structures to the discrete-continuous structures—usually reinforced along some parts of their boundary and filled up densely with fibres. In 3D problems, the boundaries can be covered by fibrous membranes interacting with the fibrous interior. That is why the discussion of this theory requires elements from the theory of curves bars and from the theory of membrane shells.

If the load acts orthogonally to the plane being the design domain then in a natural way, a problem of optimization of grillages arises. Since all the bars are bent, the condition of constant stress should be replaced with the condition of constant maximal through the thickness axial stress component, which leads to the condition of constant change of curvature. The theory of such optimal grillages has been proposed by William Prager and George Rozvany, see Rozvany (1976) and Save and Prager (1985). In order to discuss this theory later, we consider in this chapter the statics problems of grillages and systems of beams.

T. Lewiński et al., *Michell Structures*,
https://doi.org/10.1007/978-3-319-95180-5_1

In case of $\sigma_T = \sigma_C$, the Michell problem has the same solutions as the problem of compliance minimization of linearly elastic trusses under the condition of the volume being fixed, see Achtziger (1996, 1997). A similar analogy can be shown for the grillage optimization. That is why it is thought appropriate to put forward here a complete theory of statics of linearly elastic bar structures with an emphasis put on the Castigliano's theorem which links many models of bars structures in one mathematical setting.

This chapter includes a complete of information on statics of thin bar structures, thin plates in bending as well as membrane shells. This unified approach paves the way towards optimum design problems to be discussed in the forthcoming chapters.

1.2 The Equilibrium Problems of Elastic Frames, Thin Plates in Bending and Membrane Shells

The subject of Sects. 1.2.1–1.2.9 is the statics of frames made of straight and prismatic thin bars formulated in terms of the simplest theory of small elastic deformations with the additional assumption of the transverse shear deformation being neglected (this is called the *assumption of thinness*). In each theory of structures, and also in the theory of thin bars, the *internal forces* are introduced. They are usually called stress and couple resultants, but the term internal forces sounds better and will be used in the sequel. The internal forces are conjugated with strains, specifically defined for each theory. Each product of the internal force and the conjugated to this internal force the strain measure, contributes to the density of energy of the structure. Apart from the internal forces, the theory of thin bars involves the transverse shear forces which represent reactions due to the constraints of the transverse shear deformation being zero. The transverse shear forces do not contribute to the elastic energy, hence they are not classified as the internal forces.

The problem of statics will be reduced to the three groups of algebraic equations:

- linking the internal forces and the transverse shear forces with the loads;
- defining the strain measures by the displacements;
- interrelating the internal forces with the strain measures.

These equations will be called as equilibrium equations, kinematic equations and the constitutive equations, respectively.

The following statics problems of bar structures will be considered: a single bar, a plane frame, a plane frame made of inextensional bars, systems of beams, grillages of rigid joints, spatial frames and trusses. Then we turn to statics of thin elastic plates in bending (or plates transversely loaded) and to membrane shells. The discussion concerns small deformations, hence in all cases, the equilibrium equations refer to the undeformed configuration. The equilibrium equations are not affected by the deformation state.

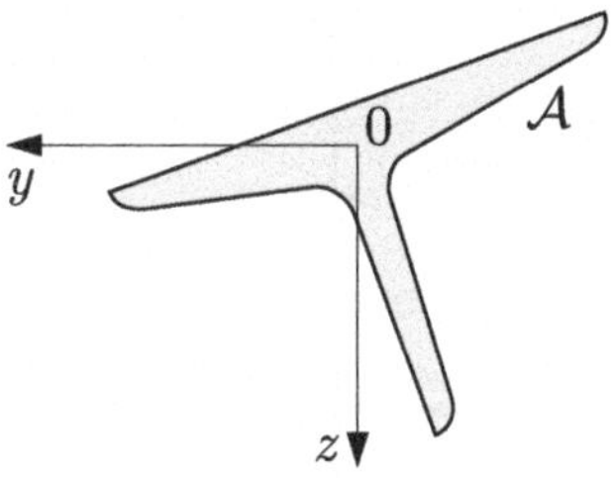

Fig. 1.1 The cross section $\mathcal{A}$ and its principal axes y, z

1.2.1 A Single Straight and Prismatic Bar

We consider a straight bar of length l of given cross section $\mathcal{A}$ of the principal axes y, z. The bar is made of an isotropic material of Young's modulus E and Poisson's ratio ν; the shear modulus equals $\mu = \dfrac{E}{2(1+\nu)}$.

We assume that the cross section $\mathcal{A}$ does not vary along the axis x, which means that the bar is prismatic (Fig. 1.1).

Let us introduce the operators

$$\varepsilon(f) = \frac{df}{dx}, \qquad \varkappa(f) = -\frac{d^2 f}{dx^2} \tag{1.1}$$

These definitions (as well as similar definitions later) are introduced for future convenience: for showing analogies between theories of bars and plates. Thin straight bars undergo three modes of deformation: the axial (longitudinal), flexural (bending in two planes) and torsional. Thus, the deformation of the bar will be determined by the displacements: $u(x)$, $v(x)$, $w(x)$ along the axes: x, y, z and by the angle of rotation $\phi(x)$ about the x axis. These fields determine the state of deformation characterized by the deformation measures:

$$\begin{aligned} \varepsilon_x = \varepsilon(u), \qquad \varkappa_M &= \varkappa(w), \qquad \varkappa_L = \varkappa(v), \qquad \tau = \varepsilon(\phi), \\ \beta_M &= -\frac{dw}{dx}, \qquad \beta_L = -\frac{dv}{dx} \end{aligned} \tag{1.2}$$

The internal forces and the transverse shear forces at the sections $x = \text{const}$, $x + dx = \text{const}$ are shown in Fig. 1.2. They are called:

- the axial force[1] $T(x)$;
- the transverse shear forces $Q_L(x)$, $Q_M(x)$;
- the bending moments $M(x)$, $L(x)$;
- the torsional moment $\mathfrak{M}(x)$.

The quantities $T(x)$, $M(x)$, $L(x)$, $\mathfrak{M}(x)$ represent the internal forces of the theory. The load applied to the bar is replaced by the axial load of intensity $p_x(x)$, the

[1]The non-typical notation T for the axial force is taken from Hemp (1973).

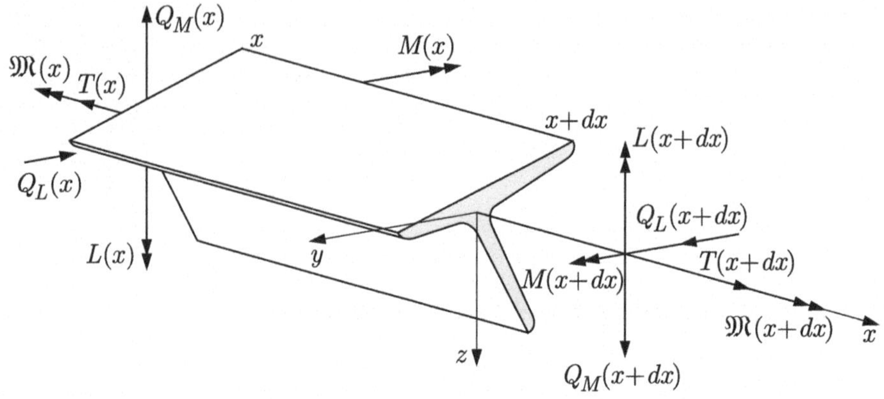

Fig. 1.2 The internal forces, the transverse shear forces and their sign conventions

transverse and loads in directions y, z of intensities $p_y(x)$, $p_z(x)$; the loads are referred to the unit length of the bar. The distributed torsional load is neglected.

In the theory of thin bars, the internal forces and the transverse shear forces are linked with the loads by the equilibrium equations:

$$\frac{dT}{dx} + p_x = 0, \qquad \frac{dQ_M}{dx} + p_z = 0, \qquad \frac{dQ_L}{dx} + p_y = 0,$$
$$Q_M = \frac{dM}{dx}, \qquad Q_L = \frac{dL}{dx}, \qquad \frac{d\mathfrak{M}}{dx} = 0 \tag{1.3}$$

These equations are equivalent to a single variational equation

$$\int_0^l \left(T\varepsilon(\bar{u}) + M\varkappa(\bar{w}) + L\varkappa(\bar{v}) + \mathfrak{M}\varepsilon(\bar{\phi})\right)dx = \mathcal{L}(\bar{u}, \bar{v}, \bar{w}, \bar{\phi})$$
$$\text{for each choice of } \bar{u}, \bar{v}, \bar{w}, \bar{\phi} \text{ on } [0, l] \tag{1.4}$$

where the right-hand side is expressed as below

$$\mathcal{L}(\bar{u}, \bar{v}, \bar{w}, \bar{\phi}) = \int_0^l (p_x\bar{u} + p_y\bar{v} + p_z\bar{w})dx + \underline{\left(T\bar{u} + M\bar{\beta}_M + L\bar{\beta}_L + \mathfrak{M}\bar{\phi} + Q_L\bar{v} + Q_M\bar{w}\right)\Big|_0^l} \tag{1.5}$$

The equivalence of (1.3) and (1.4) requires introducing appropriate regularity assumptions, but this mathematical subtlety will not be discussed. Equation (1.4) is the variational equilibrium equation of a single bar. If the bar is additionally loaded by concentrated forces, the right-hand side of (1.4) should include the additional terms representing the virtual work of these loads.

If the bar is supported at both the ends and is not loaded at the ends, then the assumption of the fields $\bar{u}$, $\bar{v}$, $\bar{w}$, $\bar{\phi}$ being kinematically admissible implies vanishing of the term underlined in (1.5).

Let A be the area of the cross section $\mathcal{A}$, let $J^{(M)}$; $J^{(L)}$ be moments of inertia with respect to the axes y, z ; let $J^{(\mathfrak{M})}$ be the torsional characteristic of torsion. The constitutive equations are assumed in the form

$$T = EA\varepsilon_x, \quad \mathfrak{M} = \mu J^{(\mathfrak{M})}\tau, \quad M = EJ^{(M)}\varkappa_M, \quad M = EJ^{(L)}\varkappa_L \tag{1.6}$$

They link the internal forces with the strains. The quantities EA, $EJ^{(M)}$, $EJ^{(L)}$, $\mu J^{(\mathfrak{M})}$ are called the stiffnesses due to tension/compression, bending in the plane (x, z), bending in the plane (x, y), and due to torsion, respectively.

Substitution of (1.6), (1.2) into (1.4) leads to a governing variational equation linking the unknown displacements and the loads; it is, in general, posed correctly, which directly follows from the celebrated F. Riesz representation theorem.

Let $\mathcal{R}$ be the set of the fields $\bar{u}$, $\bar{v}$, $\bar{w}$, $\bar{\phi}$ such that

$$\varepsilon(\bar{u}) = 0, \quad \varkappa(\bar{v}) = 0, \quad \varkappa(\bar{w}) = 0, \quad \varepsilon(\bar{\phi}) = 0 \tag{1.7}$$

Thus, the set $\mathcal{R}$ is the set of the fields $\bar{u}$, $\bar{v}$, $\bar{w}$, $\bar{\phi}$ in the form

$$\bar{u} = c_1, \quad \bar{v} = c_2 + c_3x, \quad \bar{w} = c_4 + c_5x, \quad \bar{\phi} = c_6 \tag{1.8}$$

where c_i are constants.

If the bar is not supported, then the condition of well posedness of the static problem has the form

$$\mathcal{L}(\bar{u}, \bar{v}, \bar{w}, \bar{\phi}) = 0 \;\; \forall \;\; (\bar{u}, \bar{v}, \bar{w}, \bar{\phi}) \in \mathcal{R} \tag{1.9}$$

The substitution of (1.8) into (1.9) leads to the well-known conditions of the loads being self-equilibrated: the resultant of the load and the couple resultant must vanish.

1.2.2 Plane Frames Made of Compressible Bars

Assume now that the cross sections $\mathcal{A}$ of the bars have symmetry axes: $y = 0$; the bending stiffnesses are $EJ^{(M)} = EJ$. We introduce the local coordinate systems x_k, y_k, z_k of the bars: $k = 1, \ldots, e$; the axes y_k are all parallel while the axes x_k, z_k determine a plane π. The frame is subject to the load p_x, p_z which causes the state of deformation characterized by the displacement fields u, w, the angle of rotation β_M, changes of curvature $\varkappa(w)$ and the relative elongations $\varepsilon(u)$; the internal forces are T, M. The equilibrium of the frame is governed by the variational equation

$$\int (T\varepsilon(\bar{u}) + M\varkappa(\bar{w}))\,dx = \mathcal{L}(\bar{u}, \bar{w}) \tag{1.10a}$$

to be fulfiled for each choice of the fields $\bar{u}$, $\bar{w}$. The virtual work $\mathcal{L}(\bar{u}, \bar{v})$ represents the virtual work of the loads p_x, p_z referred to the local coordinate systems x_k, z_k, other loads acting in the plane Π and the work of reactions. The integral in (1.10a) refers to all bars of the frame.

Let us substitute the geometric relations (1.2) in the constitutive equations (1.6)

$$T = EA\varepsilon(u), \qquad M = EJ\varkappa(w) \tag{1.10b}$$

Substituting (1.10b) into the equilibrium equation (1.10a) results in the governing equation for the unknowns u, w; this equation will be discussed in Sect. 1.2.8. Its solution u, w is determined up to rigid body motions in the plane Π.

The aim of this section is reducing Eqs. (1.10a), (1.10b) to the algebraic form, without any approximations. This is possible because the problem is described by the linear ordinary differential equations and the unknowns are functions of one argument. In order to obtain the algebraic form of the equations, we choose the trial fields $(\bar{u}, \bar{v})$ in a specific form such that they represent displacements of the frame loaded at the nodes. Then the fields $(\bar{u}, \bar{v})$ are determined by the nodal displacements and nodal angles of rotation. Thus, for each bar, the fields $(\bar{u}, \bar{v})$ are expressed by the displacements and nodal angles of rotation of both the ends of a bar, each of the ends being a node of the frame. Thus, the field $\bar{u}$ is barwise a polynomial of order 1 and the field $\bar{w}$ is barwise a polynomial of order 3. Let $q_1, q_2, \ldots, q_s$ be the displacements and angles of rotation (whose vectors are directed along the axes y) of the nodes of the frame. For the sake of simplicity assume that the nodes at the supports are not subject to displacements. Thus, all the quantities $q_1, q_2, \ldots, q_s$ are the unknowns, see Fig. 1.3.

All the bars must be numbered ($k = 1, \ldots, e$) and the local coordinate systems x_k, z_k should be determined for each bar. These axes determine the left and the right end of the bar as well as the sign rules of the quantities T, M, u, w, $\beta_M = -\frac{dw}{dx}$, $\varkappa(w)$, $\varepsilon(u)$, p_x, p_z, see Fig. 1.4.

For the kth bar, we define the left and right quantities

$$\begin{aligned} {}^{\bullet}u_k &= u(0), & {}^{\bullet}w_k &= w(0), & {}^{\bullet}\beta_k &= -\frac{dw}{dx}(0), \\ u_k^{\bullet} &= u(l_k), & w_k^{\bullet} &= w(l_k), & \beta_k^{\bullet} &= -\frac{dw}{dx}(l_k) \end{aligned} \tag{1.11}$$

These quantities are linear combinations of the unknowns $q_1, q_2, \ldots, q_s$, forming a column vector $\boldsymbol{q} \in \mathbb{R}^s$. For example, the left end displacements of the bar no 2 of Fig. 1.3 are expressed by

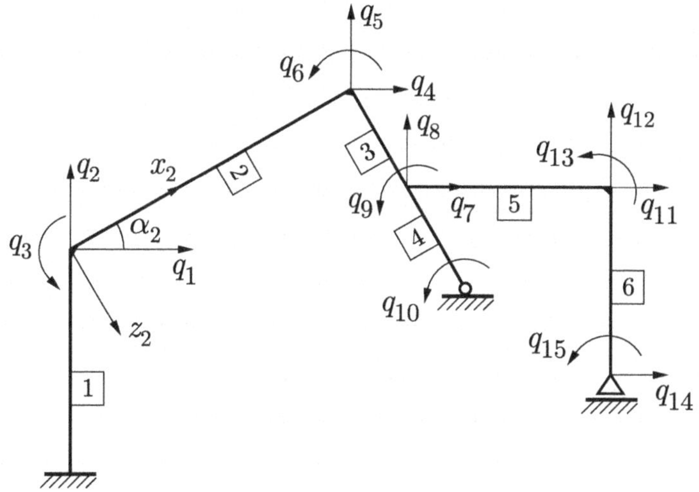

Fig. 1.3 An illustration of the choice of the unknowns $q_1, q_2, \ldots, q_s$, $s = 15$ for a plane frame

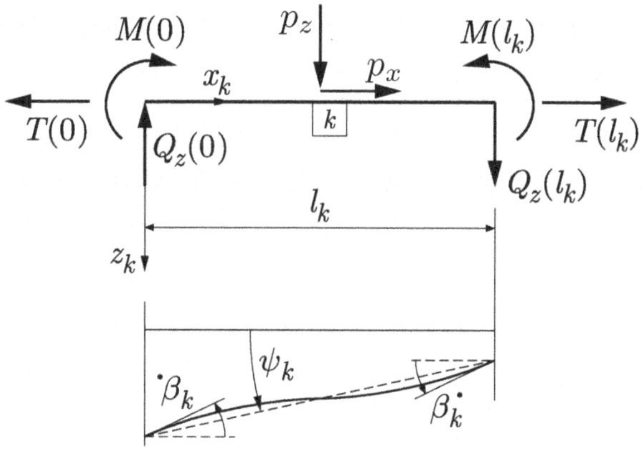

Fig. 1.4 A single bar of the frame. The sign convention

$$
\begin{aligned}
{}^{\bullet}u_2 &= q_1 \cos \alpha_2 + q_2 \sin \alpha_2 \\
{}^{\bullet}w_2 &= q_1 \sin \alpha_2 - q_2 \cos \alpha_2 \\
{}^{\bullet}\beta_2 &= q_3
\end{aligned}
\tag{1.12}
$$

and the right end displacements can be written in a similar manner. In general, one can express all the left and right end displacements and angles of rotation in terms of the components of the unknown column vector $\boldsymbol{q}$, as follows:

$$\begin{aligned}
{}^{\bullet}u_k &= \sum_{j=1}^{s} {}^{\bullet}A_{kj}^{(u)} q_j, \qquad u_k^{\bullet} = \sum_{j=1}^{s} A_{kj}^{\bullet(u)} q_j \\
{}^{\bullet}w_k &= \sum_{j=1}^{s} {}^{\bullet}A_{kj}^{(w)} q_j, \qquad w_k^{\bullet} = \sum_{j=1}^{s} A_{kj}^{\bullet(w)} q_j \\
{}^{\bullet}\beta_k &= \sum_{j=1}^{s} {}^{\bullet}A_{kj}^{(\beta)} q_j, \qquad \beta_k^{\bullet} = \sum_{j=1}^{s} A_{kj}^{\bullet(\beta)} q_j
\end{aligned} \tag{1.13}$$

which are called allocation equations. As it was mentioned above, the virtual fields $\bar{u}(x_k)$, $\bar{w}(x_k)$ are polynomials of 1st and 3rd orders, respectively. Consequently, the axial deformation $\bar{\varepsilon}_k$ is constant while the graph of the change of curvature $\bar{\varkappa}_k$ is linear. Let us define for the kth bar the kinematic quantities

$$\begin{aligned}
\Delta_k &= u_k^{\bullet} - {}^{\bullet}u_k, \qquad \psi_k = \frac{1}{l_k}({}^{\bullet}w_k - w_k^{\bullet}) \\
{}^{\bullet}\chi_k &= -{}^{\bullet}\beta_k + \psi_k, \qquad \chi_k^{\bullet} = \beta_k^{\bullet} - \psi_k
\end{aligned} \tag{1.14}$$

We note that Δ_k represents the bar's elongation; ψ_k is the angle of rotation of the bar's chord; ${}^{\bullet}\chi_k$, $\chi_k^{\bullet}$ are the angles of rotations of the left and right ends, measured with respect to the bar's chord. One can prove that the trial deformation fields are expressed by

$$\bar{\varepsilon}_k = \frac{1}{l_k}\bar{\Delta}_k, \qquad \bar{\varkappa}_k = \frac{1}{l_k}\,{}^{\bullet}\bar{\chi}_k\,{}^{\bullet}f\left(\frac{x_k}{l_k}\right) + \frac{1}{l_k}\bar{\chi}_k{}^{\bullet}\,f^{\bullet}\left(\frac{x_k}{l_k}\right) \tag{1.15}$$

where

$${}^{\bullet}f(\xi) = 4 - 6\xi, \qquad f^{\bullet}(\xi) = -2 + 6\xi \tag{1.16}$$

Substitution of (1.13) into (1.14) leads to the following linear equations:

$$\Delta_k = \sum_{j=1}^{s} B_{kj} q_j, \qquad {}^{\bullet}\chi_k = \sum_{j=1}^{s} {}^{\bullet}B_{kj} q_j, \qquad \chi_k^{\bullet} = \sum_{j=1}^{s} B_{kj}^{\bullet} q_j \tag{1.17}$$

or

$$\boldsymbol{\Delta} = \boldsymbol{B}\boldsymbol{q}, \qquad {}^{\bullet}\boldsymbol{\chi} = {}^{\bullet}\boldsymbol{B}\boldsymbol{q}, \qquad \boldsymbol{\chi}^{\bullet} = \boldsymbol{B}^{\bullet}\boldsymbol{q} \tag{1.18}$$

where $\boldsymbol{\Delta} = [\Delta_1, \ldots, \Delta_e]^{\top}$ and the column vectors ${}^{\bullet}\boldsymbol{\chi}$, $\boldsymbol{\chi}^{\bullet}$ are defined similarly. The matrices $\boldsymbol{B}$, ${}^{\bullet}\boldsymbol{B}$, $\boldsymbol{B}^{\bullet}$ can be easily expressed by means of the allocation matrices involved in (1.13).

Let us rewrite (1.10a) in the form

$$\sum_{k=1}^{e} \int_0^{l_k} (T\varepsilon(\bar{u}) + M\varkappa(\bar{w}))dx_k = \mathcal{L}(\bar{u}, \bar{w}) \tag{1.19}$$

Assume now that the fields $(\bar{u}, \bar{w})$ determine the displacements associated with the displacements and rotations of the nodes. Then the trial deformation fields $\varepsilon(\bar{u})$, $\varkappa(\bar{w})$, are given by (1.15). Let us define the average axial force and bending moments

$$\begin{aligned} T_k &= \int_0^1 T(l_k\xi)d\xi \\ {}^{\bullet}M_k &= \int_0^1 {}^{\bullet}f(\xi)M(l_k\xi)d\xi \\ M_k^{\bullet} &= \int_0^1 f^{\bullet}(\xi)M(l_k\xi)d\xi \end{aligned} \tag{1.20}$$

where $T(x_k)$, $M(x_k)$, $x_k = l_k\xi$ determine the fields T, M in the k th bar. Substitution of (1.15) into (1.19) gives

$$\sum_{k=1}^{e} \left(T_k\bar{\Delta}_k + {}^{\bullet}M_k\,{}^{\bullet}\bar{\chi}_k + M_k^{\bullet}\bar{\chi}_k^{\bullet}\right) = \mathcal{L}(\bar{u}, \bar{w}) \tag{1.21}$$

where $(\bar{u}, \bar{w})$ has an assumed form, compatible with (1.15). Substitution of (1.17) for $\boldsymbol{q} = \bar{\boldsymbol{q}}$ into (1.21) reduces the left-hand side of (1.21) to the form

$$\bar{\boldsymbol{q}}^{\top}\left[\boldsymbol{B}^{\top}\boldsymbol{T} + ({}^{\bullet}\boldsymbol{B})^{\top}\,{}^{\bullet}\boldsymbol{M} + (\boldsymbol{B}^{\bullet})^{\top}\boldsymbol{M}^{\bullet}\right] \tag{1.22}$$

where $\boldsymbol{T} = [T_1, T_2, \ldots, T_e]^{\top}$ and the column vectors ${}^{\bullet}\boldsymbol{M}, \boldsymbol{M}^{\bullet}$ are defined in a similar manner. The right-hand side of (1.21) can always be expressed by the scalar product $\boldsymbol{Q} \cdot \bar{\boldsymbol{q}} = (\bar{\boldsymbol{q}})^{\top}\boldsymbol{Q}$, where $Q_1, Q_2, \ldots, Q_s$ are the effective loads forming a column vector $\boldsymbol{Q} \in \mathbb{R}^s$; the component Q_i is the value of the virtual work done by the load on the field $(\bar{u}, \bar{w})$ generated by the set of displacement $\bar{q}_j = \delta_{ij}$, $j = 1, \ldots, s$. Due to arbitrariness of $\bar{q}_1, \ldots, \bar{q}_s$, one obtains

$$\boldsymbol{B}^{\top}\boldsymbol{T} + ({}^{\bullet}\boldsymbol{B})^{\top}\,{}^{\bullet}\boldsymbol{M} + (\boldsymbol{B}^{\bullet})^{\top}\boldsymbol{M}^{\bullet} = \boldsymbol{Q} \tag{1.23}$$

Thus, we have obtained the algebraic equation of equilibrium which fulfils the condition (1.10a) under arbitrary choice of the trial fields $(\bar{u}, \bar{w})$ which reflect a virtual deformation of the frame caused by arbitrary loads concentrated at the nodes.

We shall prove now that the average internal forces (1.20) are linked with the deformation measures by the constitutive equations

$$\begin{aligned} T_k &= \frac{EA_k}{l_k}\Delta_k, \\ {}^{\bullet}M_k &= \frac{2EJ_k}{l_k}(2\,{}^{\bullet}\chi_k - \chi_k^{\bullet}) \\ M_k^{\bullet} &= \frac{2EJ_k}{l_k}(-{}^{\bullet}\chi_k + 2\chi_k^{\bullet}) \end{aligned} \tag{1.24}$$

where EA_k, $EJ_k = EJ_k^{(M)}$ are the stiffnesses of the kth bar. Prior to derive the formula, let us note that they do not involve free terms depending on the span load (p_x, p_z).

Irrespectively of the type of the load, one can express the internal forces T, M by the formulae

$$\begin{aligned} T(x_k) &= EA_k\varepsilon_k + T_k^0(x_k) \\ M(x_k) &= EJ_k\varkappa_k(x) + M_k^0(x_k) \end{aligned} \tag{1.25}$$

where ε_k, $\varkappa_k$ are given by Eqs. (1.15), (1.14), (1.13) in which all the quantities are unknown (not virtual). The functions T_k^0, M_k^0 stand for the axial and bending moment functions in the bar being fixed at both the ends and subject to the given span load. Let us note that the following identities hold:

$$\begin{aligned} &\int_0^1 T_k^0(l_k\xi)d\xi = 0, \\ &\int_0^1 {}^{\bullet}f(\xi)M_k^0(l_k\xi)d\xi = 0, \\ &\int_0^1 f^{\bullet}(\xi)M_k^0(l_k\xi)d\xi = 0 \end{aligned} \tag{1.26}$$

for each choice of the span load. Indeed, the first equality holds, since the elongation of the bar fixed at both the ends vanishes. The second and the third equalities hold, since their left-hand sides stand for the angles of rotations of the fixed supports, while they are not rotated. Substituting (1.15) into (1.25) and performing integration in (1.26) leads to (1.24), which ends the derivation.

Let us introduce now the diagonal matrices of dimensions $e \times e$:

$$\begin{aligned} \boldsymbol{E} &= \operatorname{diag}\left\{\frac{EA_k}{l_k}\right\} \\ \boldsymbol{D} &= \operatorname{diag}\left\{\frac{2EJ_k}{l_k}\right\} \end{aligned} \tag{1.27}$$

and rewrite (1.24) in the form

$$\boldsymbol{T} = \boldsymbol{E}\boldsymbol{\Delta}, \qquad {}^{\bullet}\boldsymbol{M} = \boldsymbol{D}(2\,{}^{\bullet}\boldsymbol{\chi} - \boldsymbol{\chi}^{\bullet}), \qquad \boldsymbol{M}^{\bullet} = \boldsymbol{D}(-{}^{\bullet}\boldsymbol{\chi} + 2\boldsymbol{\chi}^{\bullet}) \tag{1.28}$$

Now the complete system of the algebraic equations describing deformation and stress states in the static problems of plane frames is constructed. It consists of three groups of equations:

- equation of equilibrium (1.23),
- kinematic equations (1.18),
- constitutive equations (1.28).

This system is balanced, i.e. the number of equations is equal to the number of the unknowns. To show it, let us insert (1.18) into (1.28) and then substitute the result into (1.23). We get

$$\boldsymbol{K}\boldsymbol{q} = \boldsymbol{Q} \tag{1.29}$$

where

$$\boldsymbol{K} = \boldsymbol{K}_T + \boldsymbol{K}_M \tag{1.30}$$

and

$$\boldsymbol{K}_T = \boldsymbol{B}^\top \boldsymbol{E}\boldsymbol{B} \tag{1.31}$$

$$\boldsymbol{K}_M = 2{}^\bullet\boldsymbol{B}^\top \boldsymbol{D}^\bullet \boldsymbol{B} - \left({}^\bullet\boldsymbol{B}^\top \boldsymbol{D}\boldsymbol{B}^\bullet + ({}^\bullet\boldsymbol{B}^\top \boldsymbol{D}\boldsymbol{B}^\bullet)^\top\right) + 2(\boldsymbol{B}^\bullet)^\top \boldsymbol{D}\boldsymbol{B}^\bullet \tag{1.32}$$

The matrix $\boldsymbol{K}$ is of dimensions $s \times s$, is symmetric and positive semi-definite. The system (1.29) is uniquely solvable if this matrix is positive definite.

The quantities ${}^\bullet M_k$, $M_k^\bullet$ are not, in general, the left and right bending moments. To compute the latter, one should add the values of the initial end moments caused by the span load under the boundary conditions of both clamped ends.

A peculiar class of frames is the statically determinate frames. By definition of statical determinacy, the equilibrium equations (1.23) are then solvable uniquely (without any reference to the remaining equations of statics). The necessary condition of the statical determinacy is the equality: $3e - s = 0$. Note that solvability of the equilibrium equations implies the solvability of the kinematic equations (1.17). Thus, having the forces $\boldsymbol{T}$, ${}^\bullet\boldsymbol{M}$, $\boldsymbol{M}^\bullet$, one can compute the strains $\boldsymbol{\Delta}$, ${}^\bullet\boldsymbol{\chi}$, $\boldsymbol{\chi}^\bullet$ and then compute $\boldsymbol{q}$ from (1.17).

If $s > 3e$, the frame is geometrically variable. The equilibrium equations are not contradictory if the vector $\boldsymbol{Q}$ is orthogonal to the vectors $\boldsymbol{p} \in \mathcal{R}_0$ where $\mathcal{R}_0$ is the set of vectors $\boldsymbol{p} \in \mathbb{R}^s$ satisfying

$$\boldsymbol{B}\boldsymbol{p} = \boldsymbol{0}, \qquad {}^\bullet\boldsymbol{B}\boldsymbol{p} = \boldsymbol{0}, \qquad \boldsymbol{B}^\bullet \boldsymbol{p} = \boldsymbol{0} \tag{1.33}$$

1.2.3 Plane Frames of Incompressible Bars

The frames designed to resist bending deformation are inevitably overstiffened with respect to the axial deformation. Consequently, in such frames, the axial deformation is negligible and it is rational to introduce this constraint into the analysis to reduce the number of unknowns in the static analysis and to reduce the dimensions of the optimum design problems. Thus, we set $\varepsilon(\bar{u}) = 0$ in the variational equation of equilibrium (1.10a) and arrive at

$$\int M\varkappa(\bar{w})dx = \mathcal{L}(\bar{w}) \tag{1.34}$$

to hold for the trial fields $(\bar{u}, \bar{w})$ compatible with the constraints $\varepsilon(\bar{u}) = 0$. Imposing these constraints means that the nodal displacements are linked by some linear equations. Thus, it is convenient to redefine the unknowns: instead of the nodal displacements, we shall deal with sways, characterizing specific motions of the whole frame or of segments of bars of the frame. The unknowns will be still denoted by $q_1, \ldots, q_s$; some of these unknowns are still angles of rotation of nodes and some of them are characteristic angles of sways; keeping the same notation as before should not lead to misunderstandings.

Proceeding as in Sect. 1.2.2, we arrive at the equilibrium equation

$$({}^{\bullet}\boldsymbol{B})^{\top}\,{}^{\bullet}\boldsymbol{M} + (\boldsymbol{B}^{\bullet})^{\top}\boldsymbol{M}^{\bullet} = \boldsymbol{Q} \tag{1.35}$$

while the constitutive equations assume the form of the second and third equations of (1.28); the kinematic equations are given by the second and third equations of (1.18). The governing equation is still (1.29), but the stiffness matrix is now given by $\boldsymbol{K} = \boldsymbol{K}_M$. The column vector $\boldsymbol{Q}$ has now a completely different meaning. If q_i is an angle of a sway, then Q_i represents the value of the virtual work of the load done on the deformation state: $q_i = 1$ and $q_j = 0$ if $j \neq i$. While construction of the diagram of the bending moment is standard, computing the axial forces is not always possible, although never contradictions occur. In most cases, the system of equilibrium equations (1.23) to find the axial forces is overdetermined, so they can only be found up to some arbitrary constants.

1.2.4 Grillages of Rigid Joints

The grillage is a plane frame but loaded out of the plane. Thus, the load is orthogonal to the plane formed by the bars. Such structures are subject to bending and torsion; as always the bending is accompanied by the appearance of the transverse forces. Remind that they are not internal forces, but they are treated as reactions to the constraints of the shear deformations being neglected. Due to the specific kind of the load, the axial forces are zero. The only internal forces are the bending moments M (to be consequent, the bending moments should be denoted by L yet we still use the notation M, which should not lead to misunderstandings) and the torsional moments $\mathfrak{M}$. The equation of equilibrium reads

$$\int \left(M\varkappa(\bar{w}) + \mathfrak{M}\varepsilon(\bar{\phi})\right) dx = \mathcal{L}(\bar{w}, \bar{\phi}) \tag{1.36}$$

for each choice of $(\bar{w}, \bar{\phi})$. In general, the right-hand side represents the virtual work of the loads and reactions. If the fields $(\bar{w}, \bar{\phi})$ are kinematically admissible, the virtual work $\mathcal{L}(\bar{w}, \bar{\phi})$ represents only the work of the load.

The equation of equilibrium (1.36), the constitutive equations

$$M = EJ\varkappa_M, \qquad \mathfrak{M} = \mu J^{(\mathfrak{M})}\tau \tag{1.37}$$

and the definitions of the strain measures

$$\varkappa_M = \varkappa(w), \qquad \tau = \varepsilon(\phi) \tag{1.38a}$$

constitute a system of equations of statics of grillages. Now, we note a mathematical analogy between the equations of statics of the plane frames: (1.10a), (1.6)—the first and third equations; (1.2)—the two first equations and the above equations of grillages. We conclude that the algebraic system of equations of grillages has the same mathematical structure as the system: (1.23), (1.28) and (1.18).

Let us introduce the constitutive matrix involving stiffnesses due to torsion

$$\boldsymbol{G} = \mathrm{diag}\left\{\frac{\mu J_k^{(\mathfrak{M})}}{l_k}\right\} \tag{1.38b}$$

Let us define the angles of axial rotations of the bars

$$\theta_k = \phi_k^{\bullet} - {}^{\bullet}\phi_k \tag{1.39}$$

where ${}^{\bullet}\phi_k$, $\phi_k^{\bullet}$ are the angles of axial rotation at the left and right end of the bar. The measure of torsion is assumed in the form

$$\tau_k = \frac{1}{l_k}\theta_k \tag{1.40}$$

For each bar, we introduce the average moment of torsion

$$\mathfrak{M}_k = \int_0^1 \mathfrak{M}(l_k\xi)d\xi \tag{1.41}$$

and we form the column vector $\mathfrak{M} = [\mathfrak{M}_1, \ldots, \mathfrak{M}_e]^\top$. The measures of torsion depend linearly on the unknowns $q_1, \ldots, q_s$, which is written as follows:

$$\tau_k = \sum_{j=1}^{s} \hat{B}_{kj} q_j \tag{1.42}$$

The quantities q_j are displacements of nodes (orthogonal to the grillage plane) and the components of the vectors of angles of rotation lying in this plane, see Fig. 1.5.

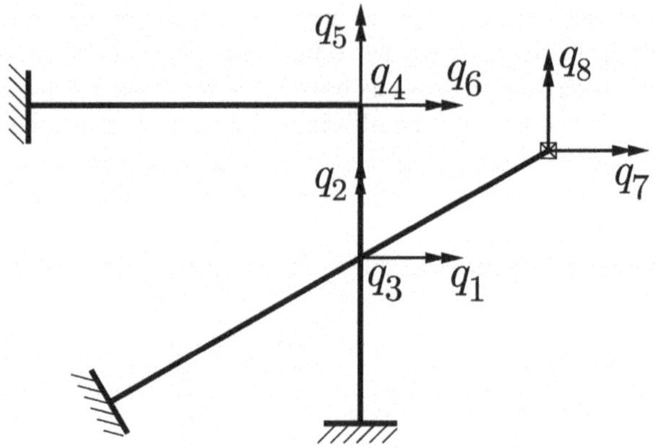

Fig. 1.5 Unknowns in a grillage (of rigid joints)

Adopting this notation, we write down the equations of statics of grillages

$$\hat{\boldsymbol{B}}^{\top}\mathfrak{M} + ({}^{\bullet}\boldsymbol{B})^{\top}\,{}^{\bullet}\boldsymbol{M} + (\boldsymbol{B}^{\bullet})^{\top}\boldsymbol{M}^{\bullet} = \boldsymbol{Q} \tag{1.43}$$

$$\mathfrak{M} = \boldsymbol{G}\boldsymbol{\theta} \tag{1.44}$$

$${}^{\bullet}\boldsymbol{M} = \boldsymbol{D}(2\,{}^{\bullet}\chi - \chi^{\bullet}), \qquad \boldsymbol{M}^{\bullet} = \boldsymbol{D}(-{}^{\bullet}\chi + 2\chi^{\bullet}) \tag{1.45}$$

$$\boldsymbol{\theta} = \hat{\boldsymbol{B}}\boldsymbol{q}, \qquad {}^{\bullet}\chi = {}^{\bullet}\boldsymbol{B}\boldsymbol{q}, \qquad \chi^{\bullet} = \boldsymbol{B}^{\bullet}\boldsymbol{q} \tag{1.46}$$

The governing equation has still the form (1.29) where $\boldsymbol{K} = \boldsymbol{K}_{\mathfrak{M}} + \boldsymbol{K}_M$, the matrix $\boldsymbol{K}_M$ being given by (1.32) and

$$\boldsymbol{K}_{\mathfrak{M}} = \hat{\boldsymbol{B}}^{\top}\boldsymbol{G}\hat{\boldsymbol{B}} \tag{1.47}$$

Remark 1.2.1 In the practice of civil engineering, the profiles of the bars are frequently thin-walled profiles of open cross sections; the bending stiffnesses of such bars are much greater than the stiffnesses due to torsion: $\mu J^{\mathfrak{M}} \ll EJ$. Then one can omit the first term in Eq. (1.43). The bending stiffness will be given by $\boldsymbol{K} = \boldsymbol{K}_M$. The assumption of small torsional stiffness does not lead to contradictions. The problem of optimum design of such grillages will be discussed in Sects. 2.5.2 and 2.5.3.

1.2.5 Systems of Beams

Assume that the bars of the grillage are connected in such a manner, that only the out of plane deflection at the nodes are kept the same, with no restraints on the nodal rotations of the connected beams; this is cleared up in Fig. 1.6.

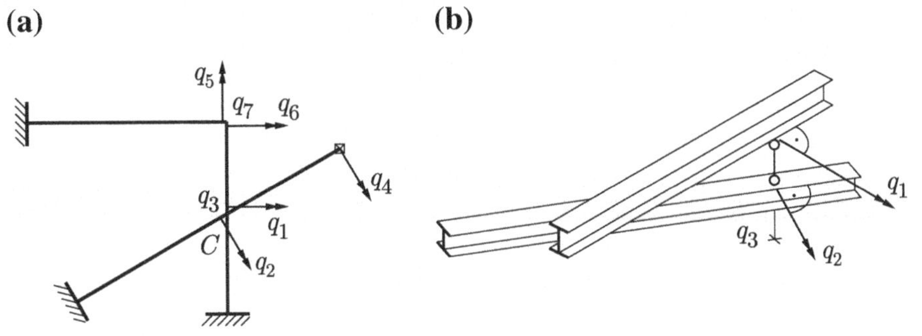

Fig. 1.6 System of beams: **a** choice of unknowns, **b** unknowns around the node C; q_3 is the common vertical displacement, while the angles of rotation: q_1, q_2 are independent

Let the load be applied out of the plane (e.g. the transverse load—orthogonal to the plane). Then the bars are not subject to torsion; the bars are subjected to bending; this is not an approximation but the result of the assumptions. Under this system of connections, the vector of the unknowns $\boldsymbol{q}$ encompasses the displacements orthogonal to the grillage plane as well as two independent angles of rotation of vectors orthogonal to the bars, see Fig. 1.6. The system of equilibrium equations has the formal structure known from the theory of grillages of weak torsional stiffness (Remark 1.2.1) but now the vector of the unknowns $\boldsymbol{q}$ and the load vector $\boldsymbol{Q}$ have different meanings. The equilibrium equation has the form (1.35), while the matrices ${}^{\bullet}\boldsymbol{B}$ and $\boldsymbol{B}^{\bullet}$ have the meaning like in (1.46).

1.2.6 *The Spatial Frames*

The deformation of a spatial frame may be decomposed into the in-plane deformation (as in the plane frames) and the out of the plane deformation (as in grillages). Having found the equations of statics of plane frames and grillages, one can write the spatial frames equations directly without an additional derivation.

At each bar, we introduce the local coordinate system (x_k, y_k, z_k). The quantities

$$
\begin{gathered}
{}^{\bullet}M_k,\quad M_k^{\bullet},\quad {}^{\bullet}\chi_k,\quad \chi_k^{\bullet},\quad {}^{\bullet}B_{kj},\quad B_{kj}^{\bullet},\quad EJ_k^{(M)},\\
\boldsymbol{D}_M = \operatorname{diag}\left\{\frac{2EJ_k^{(M)}}{l_k}\right\}
\end{gathered}
\tag{1.48}
$$

refer to the bending deformation in the plane (x_k, z_k), while their counterparts for the bending in the plane (x_k, y_k) are denoted by

$$
{}^{\bullet}L_k, \quad L_k^{\bullet}, \quad {}^{\bullet}\rho_k, \quad \rho_k^{\bullet}, \quad {}^{\bullet}C_{kj}, \quad C_{kj}^{\bullet}, \quad EJ_k^{(L)},
$$

$$
\boldsymbol{D}_L = \mathrm{diag}\left\{\frac{2EJ_k^{(L)}}{l_k}\right\} \tag{1.49}
$$

The equilibrium equations are formed by merging the plane frames equations and the grillages equations, hence they read

$$
\boldsymbol{B}^{\top}\boldsymbol{T} + \hat{\boldsymbol{B}}^{\top}\mathfrak{M} + ({}^{\bullet}\boldsymbol{B})^{\top}{}^{\bullet}\boldsymbol{M} + (\boldsymbol{B}^{\bullet})^{\top}\boldsymbol{M}^{\bullet} + ({}^{\bullet}\boldsymbol{C})^{\top}{}^{\bullet}\boldsymbol{L} + (\boldsymbol{C}^{\bullet})^{\top}\boldsymbol{L}^{\bullet} = \boldsymbol{Q} \tag{1.50}
$$

The constitutive equations are collected below

$$
\begin{aligned}
\boldsymbol{T} &= \boldsymbol{E}\boldsymbol{\Delta}, & {}^{\bullet}\boldsymbol{M} &= \boldsymbol{D}_M(2{}^{\bullet}\boldsymbol{\chi} - \boldsymbol{\chi}^{\bullet}), & {}^{\bullet}\boldsymbol{L} &= \boldsymbol{D}_L(2{}^{\bullet}\boldsymbol{\rho} - \boldsymbol{\rho}^{\bullet}) \\
\mathfrak{M} &= \boldsymbol{G}\boldsymbol{\theta}, & \boldsymbol{M}^{\bullet} &= \boldsymbol{D}_M(-{}^{\bullet}\boldsymbol{\chi} + 2\boldsymbol{\chi}^{\bullet}), & \boldsymbol{L}^{\bullet} &= \boldsymbol{D}_L(-{}^{\bullet}\boldsymbol{\rho} + 2\boldsymbol{\rho}^{\bullet})
\end{aligned} \tag{1.51}
$$

The deformation measures are expressed by the unknowns $(q_1, q_2, \ldots, q_s)$ as follows:

$$
\begin{aligned}
\boldsymbol{\Delta} &= \boldsymbol{B}\boldsymbol{q}, \quad \boldsymbol{\theta} = \hat{\boldsymbol{B}}\boldsymbol{q}, \quad {}^{\bullet}\boldsymbol{\chi} = {}^{\bullet}\boldsymbol{B}\boldsymbol{q}, \quad \boldsymbol{\chi}^{\bullet} = \boldsymbol{B}^{\bullet}\boldsymbol{q} \\
{}^{\bullet}\boldsymbol{\rho} &= {}^{\bullet}\boldsymbol{C}\boldsymbol{q}, \quad \boldsymbol{\rho}^{\bullet} = \boldsymbol{C}^{\bullet}\boldsymbol{q}
\end{aligned} \tag{1.52}
$$

The stiffness matrix $\boldsymbol{K}$ is expressed by

$$
\boldsymbol{K} = \boldsymbol{K}_T + \boldsymbol{K}_{\mathfrak{M}} + \boldsymbol{K}_M + \boldsymbol{K}_L
$$

where $\boldsymbol{K}_T$ is given by (1.31), $\boldsymbol{K}_{\mathfrak{M}}$ is given by (1.47), while $\boldsymbol{K}_M$, $\boldsymbol{K}_L$ are given by (1.32) with $\boldsymbol{D} = \boldsymbol{D}_M$ and $\boldsymbol{D} = \boldsymbol{D}_L$, respectively.

Equations (1.50)–(1.52) can be rearranged to the form in which all the constitutive equations are decoupled, which results in the diagonal form of the constitutive matrices. To this end, let us introduce the decompositions

$$
\begin{aligned}
{}^{\bullet}\boldsymbol{\chi} &= \boldsymbol{\chi}_s + \boldsymbol{\chi}_a, & \boldsymbol{\chi}^{\bullet} &= \boldsymbol{\chi}_s - \boldsymbol{\chi}_a \\
{}^{\bullet}\boldsymbol{\rho} &= \boldsymbol{\rho}_s + \boldsymbol{\rho}_a, & \boldsymbol{\rho}^{\bullet} &= \boldsymbol{\rho}_s - \boldsymbol{\rho}_a \\
{}^{\bullet}\boldsymbol{M} &= \boldsymbol{M}_s + \boldsymbol{M}_a, & \boldsymbol{M}^{\bullet} &= \boldsymbol{M}_s - \boldsymbol{M}_a \\
{}^{\bullet}\boldsymbol{L} &= \boldsymbol{L}_s + \boldsymbol{L}_a, & \boldsymbol{L}^{\bullet} &= \boldsymbol{L}_s - \boldsymbol{L}_a
\end{aligned} \tag{1.53}
$$

Now we express the deformations $\boldsymbol{\chi}_s$, $\boldsymbol{\chi}_a$, $\boldsymbol{\rho}_s$, $\boldsymbol{\rho}_a$, as linear combinations of the unknowns $(q_1, q_2, \ldots, q_s)$, or

$$
\begin{aligned}
\boldsymbol{\chi}_s &= \boldsymbol{B}_s\boldsymbol{q}, & \boldsymbol{\chi}_a &= \boldsymbol{B}_a\boldsymbol{q}, \\
\boldsymbol{\rho}_s &= \boldsymbol{C}_s\boldsymbol{q}, & \boldsymbol{\rho}_a &= \boldsymbol{C}_a\boldsymbol{q}
\end{aligned} \tag{1.54a}
$$

where

$$\begin{array}{ll} \boldsymbol{B}_s + \boldsymbol{B}_a = {}^\bullet\boldsymbol{B}, & \boldsymbol{B}_s - \boldsymbol{B}_a = \boldsymbol{B}^\bullet, \\ \boldsymbol{C}_s + \boldsymbol{C}_a = {}^\bullet\boldsymbol{C}, & \boldsymbol{C}_s - \boldsymbol{C}_a = \boldsymbol{C}^\bullet \end{array} \tag{1.54b}$$

The constitutive equations assume the diagonal form, or

$$\begin{array}{ll} \boldsymbol{T} = \boldsymbol{E}\boldsymbol{\Delta}, & \mathfrak{M} = \boldsymbol{G}\boldsymbol{\theta} \\ \boldsymbol{M}_s = \boldsymbol{D}_M \boldsymbol{\chi}_s, & \boldsymbol{M}_a = 3\boldsymbol{D}_M \boldsymbol{\chi}_a \\ \boldsymbol{L}_s = \boldsymbol{D}_L \boldsymbol{\rho}_s, & \boldsymbol{L}_a = 3\boldsymbol{D}_L \boldsymbol{\rho}_a \end{array} \tag{1.55}$$

The equation of equilibrium (1.50) assumes the form

$$\boldsymbol{B}^\top \boldsymbol{T} + \hat{\boldsymbol{B}}^\top \mathfrak{M} + 2\left[(\boldsymbol{B}_s)^\top \boldsymbol{M}_s + (\boldsymbol{B}_a)^\top \boldsymbol{M}_a + (\boldsymbol{C}_s)^\top \boldsymbol{L}_s + (\boldsymbol{C}_a)^\top \boldsymbol{L}_a\right] = \boldsymbol{Q} \tag{1.56}$$

The stiffness matrix reads

$$\begin{aligned} \boldsymbol{K} = \boldsymbol{B}^\top \boldsymbol{E}\boldsymbol{B} + \hat{\boldsymbol{B}}^\top \boldsymbol{G}\hat{\boldsymbol{B}} + 6\left[(\boldsymbol{B}_a)^\top \boldsymbol{D}_M \boldsymbol{B}_a + (\boldsymbol{C}_a)^\top \boldsymbol{D}_L \boldsymbol{C}_a\right] + \\ 2\left[(\boldsymbol{B}_s)^\top \boldsymbol{D}_M \boldsymbol{B}_s + (\boldsymbol{C}_s)^\top \boldsymbol{D}_L \boldsymbol{C}_s\right] \end{aligned} \tag{1.57}$$

Let us introduce the column vectors

$$\begin{aligned} \boldsymbol{\sigma} &= \left[\boldsymbol{T}^\top, \mathfrak{M}^\top, \sqrt{2}\boldsymbol{M}_s^\top, \sqrt{2}\boldsymbol{L}_s^\top, \sqrt{2}\boldsymbol{M}_a^\top, \sqrt{2}\boldsymbol{L}_a^\top\right]^\top \\ \boldsymbol{\varepsilon} &= \left[\boldsymbol{\Delta}^\top, \boldsymbol{\theta}^\top, \sqrt{2}\boldsymbol{\chi}_s^\top, \sqrt{2}\boldsymbol{\rho}_s^\top, \sqrt{2}\boldsymbol{\chi}_a^\top, \sqrt{2}\boldsymbol{\rho}_a^\top\right]^\top \end{aligned} \tag{1.58}$$

of dimension $m \times 1$, $m = 6e$, respectively, and the block matrices

$$\begin{aligned} \mathcal{B}_{m\times s} &= \mathrm{col}\left[\boldsymbol{B}, \hat{\boldsymbol{B}}, \sqrt{2}\boldsymbol{B}_s, \sqrt{2}\boldsymbol{C}_s, \sqrt{2}\boldsymbol{B}_a, \sqrt{2}\boldsymbol{C}_a\right] \\ \mathcal{E}_{m\times m} &= \mathrm{diag}\left[\boldsymbol{E}, \boldsymbol{G}, \boldsymbol{D}_M, \boldsymbol{D}_L, 3\boldsymbol{D}_M, 3\boldsymbol{D}_L\right] \end{aligned} \tag{1.59}$$

the latter being a diagonal matrix of positive diagonal components.

The system of statics equations assumes the form similar to that of the truss theory:

$$\mathcal{B}^\top \boldsymbol{\sigma} = \boldsymbol{Q}, \qquad \boldsymbol{\sigma} = \mathcal{E}\boldsymbol{\varepsilon}, \qquad \boldsymbol{\varepsilon} = \mathcal{B}\boldsymbol{q} \tag{1.60}$$

Let us note that the transpose of the matrix which determines the deformation measures is the operator of the equilibrium equation. This is called Clebsch's theorem. This property extends to the main problems of mathematical physics, where it is expressed in terms of functional analysis and is of vital importance for their well posedness, see Strang (1988).

1.2.7 Trusses

A frame is a truss if its bars are straight and connected by hinges; the load acts at the nodes in the form of the concentrated forces. Then, the bars are not subject to torsion, to bending and the transverse shearing. The unknowns $(q_1, q_2, \ldots, q_s)$ represent nodal displacements, hence will be denoted by u_i.

Assume that the forces at the nodes P_j, $j = 1, 2, \ldots, s$ are directed along the unknown displacements u_j. Thus, the virtual work of the load is expressed by $\bar{\boldsymbol{u}}^\top \boldsymbol{P} = \boldsymbol{P} \cdot \boldsymbol{u}$. The variational equation of equilibrium (1.21) reduces to

$$\sum_{k=1}^{e} T_k \bar{\Delta}_k = \sum_{j=1}^{s} P_j \bar{u}_j \tag{1.61}$$

which holds for arbitrary $u_1, \ldots, \bar{u}_s$; moreover $\bar{\Delta}_k = \Delta_k(\bar{\boldsymbol{u}})$ and according to (1.17), we have

$$\Delta_k(\bar{\boldsymbol{u}}) = \sum_{j=1}^{s} B_{kj} \bar{u}_j \tag{1.62}$$

Substitution of (1.62) into (1.61) and change of the sequence of summation lead to

$$\sum_{j=1}^{s} R_j \bar{u}_j = 0, \qquad R_j = \sum_{k=1}^{e} B_{kj} T_k - P_j \tag{1.63}$$

The assumption $\bar{u}_j = \delta_{ij}$ implies $R_i = 0$, hence the equilibrium equations assume the form

$$\sum_{k=1}^{e} (\boldsymbol{B}^\top)_{jk} T_k = P_j \qquad \text{or} \quad \boldsymbol{B}^\top \boldsymbol{T} = \boldsymbol{P} \tag{1.64}$$

the matrix $\boldsymbol{B}$ has dimensions $e \times s$ and is called the geometric matrix (or the matrix of directional cosines).

The system of equations of statics of trusses can be formally inferred from the equations of frames: (1.50)–(1.52). It has the form

$$\boldsymbol{B}^\top \boldsymbol{T} = \boldsymbol{P}, \qquad \boldsymbol{T} = \boldsymbol{E}\boldsymbol{\Delta}, \qquad \boldsymbol{\Delta} = \boldsymbol{B}\boldsymbol{u} \tag{1.65}$$

which reflects Clebsch's theorem.

If $e = s$ and if $\boldsymbol{B}^\top$ is invertible, the column vector $\boldsymbol{T}$ can be determined from the equilibrium equation (the first of (1.65)). Such trusses are called statically determinate. Having $\boldsymbol{T}$ one can compute $\boldsymbol{\Delta}$ and then one can compute $\boldsymbol{u}$, since both the matrices: $\boldsymbol{B}^\top$ and $\boldsymbol{B}$ are invertible. Thus, in the statically determinate trusses, one can compute the vectors: $\boldsymbol{T}$, $\boldsymbol{\Delta}$, $\boldsymbol{u}$ subsequently, without any need of forming the matrix $\boldsymbol{K} = \boldsymbol{B}^\top \boldsymbol{E} \boldsymbol{B}$ and solving Eq. (1.29).

The theory of trusses is the prototype of the theory of frames, which is reflected in the equations of frames (1.60) including all the terms concerning the axial deformation.

The condition of solvability of equilibrium equations (first of (1.65)) is

$$\boldsymbol{P} \in \text{range}\,(\boldsymbol{B}^\top) \tag{1.66}$$

where range $(\boldsymbol{A})$ is the range of the linear operator given by the matrix $\boldsymbol{A}$. Since

$$\text{range}\,(\boldsymbol{B}^\top) = (\ker(\boldsymbol{B}))^\perp \tag{1.67}$$

where $\ker(\boldsymbol{B})$ is the kernel of $\boldsymbol{B}$ or

$$\ker(\boldsymbol{B}) = \left\{\bar{\boldsymbol{u}} \in \mathbb{R}^s \mid \boldsymbol{B}\bar{\boldsymbol{u}} = \boldsymbol{0}\right\} \tag{1.68}$$

and the sign $\perp$ stands for the orthogonal complement. Thus, ker $(\boldsymbol{B})$ is the set of the inextensional (or rigid) motions of the truss, i.e. such that $\bar{\boldsymbol{\Delta}} = \boldsymbol{0}$. The condition (1.66) is equivalent to

$$\boldsymbol{P} \cdot \bar{\boldsymbol{u}} = 0 \quad \forall \quad \bar{\boldsymbol{u}} \text{ such that } \Delta_k(\bar{\boldsymbol{u}}) = 0, \quad k = 1, \ldots, e \tag{1.69}$$

which expresses the virtual work on the inextensional motions being zero. Note, that the condition (1.69) is a direct consequence of the variational equation (1.61), since if there exist motions such that $\bar{\boldsymbol{\Delta}} = \boldsymbol{0}$ then Eq. (1.61) imposes the condition (1.69) on the load applied, to make Eq. (1.61) non-contradictory.

1.2.8 The Variational Continuum Approach to Statics of Plane Frames, Systems of Beams and Trusses: The Theorems on Minimum of the Elastic Potential and of the Minimum Complementary Energy

Consider first the problem of statics of a frame made of extensional bars, see Sect. 1.2.2. Substitution of Eq. (1.10b) or

$$T = EA\varepsilon(u), \qquad M = EJ\varkappa(w) \tag{1.70}$$

into (1.10a) leads to the variational equation

$$a(u, w, \bar{u}, \bar{w}) = \mathcal{L}(\bar{u}, \bar{w}) \quad \forall \quad (\bar{u}, \bar{w}) \in \mathcal{V} \tag{1.71}$$

where $\mathcal{V}$ is the set of kinematically admissible displacements $\bar{u}$, $\bar{w}$ of the frame, while the left-hand side (1.71) is expressed by the bilinear form

$$a(u, w, \bar{u}, \bar{w}) = \int (EA\varepsilon(u)\varepsilon(\bar{u}) + EJ\varkappa(w)\varkappa(\bar{w}))\, dx \tag{1.72}$$

and the integral concerns all the bars of the frame. By endowing the set $\mathcal{V}$ with the structure of a Hilbert space and then treating the equality (1.71) as the Riesz' equation of the theorem on representation, one notes that the problem (1.71) is in general well posed; its solution (u, w) exists, provided that the known conditions of the mentioned theorem are fulfiled. It is remarkable, that the problem (1.71) has the same mathematical setting as the 3D linear elasticity problem, see Ciarlet (1997); only here there is no need to precede the analysis by proving Korn-like inequalities. Let us consider the problem

$$\mathcal{J} = \max \left\{ \mathcal{L}(\bar{u}, \bar{w}) - \frac{1}{2} a(\bar{u}, \bar{w}, \bar{u}, \bar{w}) \mid (\bar{u}, \bar{w}) \in \mathcal{V} \right\} \tag{1.73}$$

By assuming appropriate regularity conditions, one can prove that the maximizer (u, w) is the solution to the problem (1.71) of statics of the frame. The problem (1.73) can be easily rearranged to a minimization problem and then it is called the theorem on minimization of the elastic potential. Yet the formulation (1.73) is preferred in the present book, since the result of the maximization is a positive number and this result is just proportional to the minimum value of the complementary energy discussed below.

Let Σ be the set of fields $(\bar{T}, \bar{M})$ given on the bars of the frame satisfying the variational equation of equilibrium (1.10a). Then these fields satisfy the local equations of equilibrium, see (1.3)

$$\frac{d\bar{T}}{dx} + p_x = 0, \qquad \frac{d^2\bar{M}}{dx^2} + p_z = 0 \tag{1.74}$$

and the statical boundary conditions. Let

$$\mathcal{C}(\bar{T}, \bar{M}) = \int \left(\frac{\bar{T}^2}{EA} + \frac{\bar{M}^2}{EJ} \right) dx \tag{1.75}$$

Consider the problem

$$\wp = \min \left\{ \mathcal{C}(\bar{T}, \bar{M}) \mid (\bar{T}, \bar{M}) \in \Sigma \right\} \tag{1.76}$$

One can prove that the minimizer (T, M) of the problem (1.76) is linked with the maximizer (u, w) of the problem (1.73) by the equations:

$$T = EA\varepsilon(u), \qquad M = EJ\varkappa(w).$$

Moreover, $\wp = 2J$. This is the thesis of Castigliano's theorem.

Let us show that (T, M) determined by solving (1.71) is the minimizer of (1.75) over the set Σ.

The pairs (T, M) and $(\bar{T}, \bar{M})$ are statically admissible, hence they satisfy (1.10a). Let us set $\bar{u} = u$, $\bar{w} = w$, where (u, w) is the solution to (1.71). Hence

$$\begin{aligned} &\int (T\varepsilon(u) + M\varkappa(w))\,dx = \mathcal{L}(u, w) \\ &\int \left(\bar{T}\varepsilon(u) + \bar{M}\varkappa(w)\right)dx = \mathcal{L}(u, w) \end{aligned} \tag{1.77}$$

By subtracting these equations, we have

$$\int \left[(T - \bar{T})\varepsilon(u) + (M - \bar{M})\varkappa(w)\right]dx = 0 \tag{1.78}$$

Let us make use of the constitutive equations and rearrange the above equality to the form

$$\int \left[(T - \bar{T})\frac{T}{EA} + (M - \bar{M})\frac{M}{EJ}\right]dx = 0 \tag{1.79}$$

Let us rearrange (1.75) by using the identity

$$\mathcal{C}(\bar{T}, \bar{M}) = \mathcal{C}\left((\bar{T} - T) + T, (\bar{M} - M) + M\right)$$

to obtain

$$\begin{aligned} \mathcal{C}(\bar{T}, \bar{M}) = \mathcal{C}(T, M) + \underline{\mathcal{C}(\bar{T} - T, \bar{M} - M)} + \\ 2\int \left[(\bar{T} - T)\frac{T}{EA} + (\bar{M} - M)\frac{M}{EJ}\right]dx \end{aligned} \tag{1.80}$$

The underlined term is non-negative, while the last term vanishes due to (1.79). Thus

$$\mathcal{C}(\bar{T}, \bar{M}) \geq \mathcal{C}(T, M) \tag{1.81}$$

and the equality is attained if $\bar{T} = T$, $\bar{M} = M$. This proves that the pair (T, M) solving (1.71) is the minimizer of (1.76).

Consider now the problem of statics of frames of inextensional bars. Substitution of the second constitutive equation in (1.70) to the equilibrium equation (1.34) leads to the variational problem of finding w in $\mathcal{V}$ such that

$$a(w, \bar{w}) = \mathcal{L}(\bar{w}) \quad \forall\, \bar{w} \in \mathcal{V} \tag{1.82}$$

where $\mathcal{V}$ stands for the set of kinematical admissible transverse displacements (or deflections) of bars of the frame; the bilinear form of the left-hand side of (1.82) reads

$$a(w, \bar{w}) = \int EJ\varkappa(w)\varkappa(\bar{w})dx \tag{1.83}$$

which corresponds to flexural deformations. The maximizer of the problem

$$\mathcal{J} = \max\left\{\mathcal{L}(\bar{w}) - \frac{1}{2}a(\bar{w}, \bar{w}) \mid \bar{w} \in \mathcal{V}\right\} \tag{1.84a}$$

is also the solution to the problem (1.82).

Let Σ be the set of fields M satisfying the equilibrium equation (1.34). Let

$$\mathcal{C}(\bar{M}) = \int \frac{\bar{M}^2}{EJ}dx \tag{1.84b}$$

One can prove that the minimizer M of problem

$$\wp = \min\left\{\mathcal{C}(\bar{M}) \mid \bar{M} \in \Sigma\right\} \tag{1.84c}$$

is linked with the maximizer w of the problem (1.84b) by the constitutive equation: $M = EJ\varkappa(w)$ and $\wp = 2\mathcal{J}$. This is the thesis of the Castigliano theorem referred to frames of inextensional bars.

Equations (1.82)–(1.84c) refer also to the systems of beams, cf. Sect. 1.2.5. Then $\bar{w}$ is the deflection field of the system of beams and the set $\mathcal{V}$ is composed of the deflection fields compatible with the system of connections of the beams and satisfying appropriate kinematic boundary conditions. In particular, Eqs. (1.82)–(1.84c) refer to a single beam and then the fields M, w, $\varkappa$ are determined on the interval $[0, l]$, where l is the length of the beam.

The equations of plane frames (1.70)–(1.76) refer also to plane trusses (in which $M = 0$, $\varkappa = 0$). Thus, these equations reduce to the equation of equilibrium

$$\int T\varepsilon(\bar{u})dx = \mathcal{L}(\bar{u}) \tag{1.85}$$

where $\mathcal{V}$ is the set of kinematical admissible axial displacements of the truss members; $\mathcal{L}(\bar{u})$ represents the virtual work of the applied forces concentrated at the nodes. The set of fields T satisfying (1.85) is denoted by Σ. The constitutive equation has the form

$$T = EA\varepsilon(u) \tag{1.86}$$

Substitution of (1.86) into (1.85) leads to the variational equation of the form (1.82); this equation can be easily set in a mathematically correct form, compatible with Riesz' theorem of representation. The solution u of this equation determines T by (1.86) and this function is the minimizer of the problem

$$\mathcal{C}(T) = \min\left\{\mathcal{C}(\bar{T}) \mid \bar{T} \in \Sigma\right\} \tag{1.87a}$$

with

$$\mathcal{C}(\bar{T}) = \int \frac{\bar{T}^2}{EA} dx \tag{1.87b}$$

and the integral runs over the bars of the truss.

The Castigliano's theorem referred to the continuum formulation of the statics problem applies also for spatial frames and grillages; the details will not be reported.

1.2.9 The Spatial Frames. The Discretized Versions of the Theorems on the Minimum of the Elastic Potential and of the Minimum of Complementary Energy

The point of departure is the set of Eqs. (1.60). Let us write down the column vectors involved in (1.60) as follows:

$$\begin{aligned} \boldsymbol{q} &= [q_1\, q_2\, \ldots\, q_s]^\top \\ \boldsymbol{\sigma} &= [\sigma_1\, \sigma_2\, \ldots\, \sigma_m]^\top, \qquad m = 6e \\ \boldsymbol{\varepsilon} &= [\varepsilon_1\, \varepsilon_2\, \ldots\, \varepsilon_m]^\top \\ \boldsymbol{\mathcal{Q}} &= [\mathcal{Q}_1\, \mathcal{Q}_2\, \ldots\, \mathcal{Q}_s]^\top \end{aligned} \tag{1.88}$$

and

$$\begin{aligned} \boldsymbol{\mathcal{B}} &= [\mathcal{B}_{kj}]_{k=1,\ldots,m;\quad j=1,\ldots,s} \\ \boldsymbol{\mathcal{E}} &= \mathrm{diag}[\mathcal{E}_{11}\, \ldots\, \mathcal{E}_{mm}] \end{aligned} \tag{1.89}$$

where $\mathcal{E}_{ii}$ $i > 0, i = 1, \ldots, m$. Equations (1.60) imply

$$\boldsymbol{\mathcal{K}} \boldsymbol{q} = \boldsymbol{\mathcal{Q}}, \qquad \boldsymbol{\mathcal{K}} = \boldsymbol{\mathcal{B}}^\top \boldsymbol{\mathcal{E}} \boldsymbol{\mathcal{B}} \tag{1.90}$$

where $\boldsymbol{\mathcal{K}} = [K_{ij}]$, $i, j = 1, \ldots, s$. Let us consider the elastic energy stored in the frame

$$W = \frac{1}{2} \sum_{i,j=1}^{s} q_i \mathcal{K}_{ij} q_j \tag{1.91}$$

or, equivalently

$$W = \frac{1}{2} \sum_{k,l=1}^{m} \varepsilon_l \mathcal{E}_{kl} \varepsilon_k = \frac{1}{2} \sum_{k=1}^{m} \mathcal{E}_{kk} (\varepsilon_k)^2 \tag{1.92}$$

Since the diagonal components of the matrix $\boldsymbol{\mathcal{E}}$ are positive, we see that W is non-negative and the stiffness matrix $\boldsymbol{\mathcal{K}}$ is positive semi-definite.

Assume that the frame is geometrically invariable, or there exists no displacement state $\tilde{\boldsymbol{q}} = [\tilde{q}_1, \ldots, \tilde{q}_s]^\top$ such that all the components $\tilde{\varepsilon}_k = \sum_{j=1}^{s} \mathcal{B}_{kj} \tilde{q}_j$ vanish. Thus, then $W = 0$ implies $\boldsymbol{q} = \boldsymbol{0}$ or the matrix $\mathcal{K}$ is then positive definite, hence is non-singular. Let us define the function

$$\mathcal{J}(\boldsymbol{p}) = \boldsymbol{p}^\top \mathcal{Q} - \frac{1}{2}\boldsymbol{p}^\top \mathcal{K}\boldsymbol{p} \tag{1.93}$$

The quantity $-J(\boldsymbol{q})$ is called the elastic potential of the frame. We shall show that

$$\mathcal{J}(\boldsymbol{q}) \geq \mathcal{J}(\bar{\boldsymbol{p}}) \quad \forall \quad \bar{\boldsymbol{p}} \in \mathbb{R}^s \tag{1.94}$$

To prove it, let us consider the expression $-J(\boldsymbol{p}+\boldsymbol{q}) + J(\boldsymbol{q})$. An elementary computation gives

$$\mathcal{J}(\boldsymbol{q}) - \mathcal{J}(\boldsymbol{p}+\boldsymbol{q}) = \frac{1}{2}\left(\underline{\boldsymbol{p}^\top \mathcal{K}\boldsymbol{q}} + \underline{\boldsymbol{q}^\top \mathcal{K}\boldsymbol{p}} + \boldsymbol{p}^\top \mathcal{K}\boldsymbol{p}\right) - \boldsymbol{p}^\top \mathcal{Q} \tag{1.95}$$

Due to symmetry of the matrix $\mathcal{K}$, the underlined terms are equal. Thus

$$\mathcal{J}(\boldsymbol{q}) - \mathcal{J}(\boldsymbol{p}+\boldsymbol{q}) = \boldsymbol{p}^\top(\mathcal{K}\boldsymbol{q} - \mathcal{Q}) + \frac{1}{2}\boldsymbol{p}^\top \mathcal{K}\boldsymbol{p} \geq 0 \tag{1.96}$$

since the first term of the r.h.s vanishes while the second is non-negative. Let $\bar{\boldsymbol{p}} = \boldsymbol{p} + \boldsymbol{q}$. Since $\boldsymbol{p}$ is arbitrary, the vector $\bar{\boldsymbol{p}}$ runs over the whole set $\mathbb{R}^s$ and we arrive at the thesis (1.94). Since there exists only one $\boldsymbol{q}$ such that ((1.90), the first equation) holds, hence the solution to the problem $\max\{\mathcal{J}(\boldsymbol{p}) \mid \boldsymbol{p} \in \mathbb{R}^s\}$ coincides with the solution of the problem (1.90).

Let us note that $\mathcal{J}(\boldsymbol{q}) = \frac{1}{2}\boldsymbol{q}^\top \mathcal{Q} = \frac{1}{2}\boldsymbol{q}^\top \mathcal{K}\,\boldsymbol{q} > 0$.

Let us formulate now for the spatial frames the Castigliano theorem in the discretized version. Let us introduce the set of vectors in $\mathbb{R}^m$

$$\Sigma = \left\{\tilde{\boldsymbol{\sigma}} \in \mathbb{R}^m \mid \mathcal{B}^\top \tilde{\boldsymbol{\sigma}} = \mathcal{Q}\right\} \tag{1.97}$$

or the set of statically admissible internal forces. Let us define the function of arguments in $\mathbb{R}^m$ and the values in $\mathbb{R}$

$$\tilde{\boldsymbol{\sigma}} \to \mathcal{C}(\tilde{\boldsymbol{\sigma}}) = \tilde{\boldsymbol{\sigma}}^\top \mathcal{E}^{-1} \tilde{\boldsymbol{\sigma}} \tag{1.98}$$

or

$$\mathcal{C}(\tilde{\boldsymbol{\sigma}}) = \sum_{l=1}^{m} \frac{(\tilde{\sigma}_l)^2}{\mathcal{E}_{ll}} \tag{1.99}$$

Let $\boldsymbol{\sigma}$ be the solution to the problem (1.60) then

$$\mathcal{C}(\boldsymbol{\sigma}) \leq \mathcal{C}(\tilde{\boldsymbol{\sigma}}) \quad \forall \quad \tilde{\boldsymbol{\sigma}} \in \Sigma \tag{1.100}$$

The simplest proof of this inequality runs as follows. If $\tilde{\boldsymbol{\sigma}}, \boldsymbol{\sigma} \in \Sigma$ then

$$\begin{aligned}\sum_{l=1}^{m} \tilde{\sigma}_l \bar{\varepsilon}_l &= \sum_{i=1}^{s} \mathcal{Q}_i \bar{q}_i \quad \forall \;\; \bar{\boldsymbol{q}} \in \mathbb{R}^s \\ \sum_{l=1}^{m} \sigma_l \bar{\varepsilon}_l &= \sum_{i=1}^{s} \mathcal{Q}_i \bar{q}_i \quad \forall \;\; \bar{\boldsymbol{q}} \in \mathbb{R}^s\end{aligned} \tag{1.101}$$

where $\bar{\varepsilon} = \mathcal{B}\bar{\boldsymbol{q}}$. In both the equations, the same trial vectors are assumed. Subtracting these equations by sides gives

$$\sum_{l=1}^{m} (\tilde{\sigma}_l - \sigma_l)\bar{\varepsilon}_l = 0 \;\; \forall \;\; \bar{\boldsymbol{q}} \in \mathbb{R}^s \tag{1.102}$$

Let us insert $\tilde{\sigma}_l = \tilde{\sigma}_l - \sigma_l + \sigma_l$ into (1.99). An elementary computation gives

$$\mathcal{C}(\tilde{\boldsymbol{\sigma}}) = \mathcal{C}(\boldsymbol{\sigma}) + \underline{\sum_{l=1}^{m} \frac{(\tilde{\sigma}_l - \sigma_l)^2}{\mathcal{E}_{ll}}} + R \tag{1.103}$$

with

$$R = 2\sum_{l=1}^{m} \frac{\sigma_l}{\mathcal{E}_{ll}}(\tilde{\sigma}_l - \sigma_l) \tag{1.104}$$

Let us assume $\bar{\boldsymbol{q}} = \boldsymbol{q}$ in (1.102). Then $\bar{\varepsilon}_l = \varepsilon_l$ and $\varepsilon_l = \sigma_l/\mathcal{E}_{ll}$. Therefore $R = 0$. Since the underlined term in (1.103) is non-negative, the result (1.103) leads to (1.100). We know that $\boldsymbol{\sigma}$ is unique and that is why it is the only minimizer in Castigliano's problem: $\min\{\mathcal{C}(\tilde{\boldsymbol{\sigma}}) \mid \tilde{\boldsymbol{\sigma}} \in \Sigma)\}$. From the mathematical viewpoint, this is the problem dual to $\max\{\mathcal{J}(\boldsymbol{p})$ over $\boldsymbol{p} \in \mathbb{R}^s\}$. Thus, the following equality is proved:

$$\min\left\{\frac{1}{2}\tilde{\boldsymbol{\sigma}}^\top \mathcal{E}^{-1}\tilde{\boldsymbol{\sigma}} \mid \tilde{\boldsymbol{\sigma}} \in \Sigma\right\} = \max\left\{\boldsymbol{p}^\top \boldsymbol{\mathcal{Q}} - \frac{1}{2}\boldsymbol{p}^\top \mathcal{K}\boldsymbol{p} \mid \boldsymbol{p} \in \mathbb{R}^s\right\} \tag{1.105}$$

For $\tilde{\boldsymbol{\sigma}} = \boldsymbol{\sigma}$ and $\boldsymbol{p} = \boldsymbol{q}$, we have

$$\frac{1}{2}\boldsymbol{\sigma}^\top \boldsymbol{\mathcal{E}}^{-1}\boldsymbol{\sigma} = \boldsymbol{q}^\top \boldsymbol{\mathcal{Q}} - \frac{1}{2}\boldsymbol{q}^\top \mathcal{K}\boldsymbol{q} \tag{1.106}$$

which clears up why the left-hand side is called the complementary energy. The right-hand side of (1.106) is equal to the energy W stored in the frame and is equal to $\frac{1}{2}\boldsymbol{q}^\top \boldsymbol{\mathcal{Q}}$. The quantity $\boldsymbol{q}^\top \boldsymbol{\mathcal{Q}}$ will be called the compliance of the frame. It is equal to $\wp(\boldsymbol{\sigma})$.

The formulated theorems extend to all the specific bar structures discussed in the present section.

Lastly, let us write down the discretized version of Castigliano theorem for trusses. Let $(\boldsymbol{u}, \boldsymbol{\Delta}, \boldsymbol{T})$ solve the problem of statics, see (1.65). Then $\boldsymbol{u}$ is the maximizer of $J(\boldsymbol{p})$ given by (1.93) with the matrix $\boldsymbol{K} = \boldsymbol{B}^{\top}\boldsymbol{E}\boldsymbol{B}$. The forces in the bars: $T_1, \ldots, T_e$ are minimizers of the problem

$$\wp = \min\left\{ \sum_{k=1}^{e} \frac{(\bar{T}_k)^2 l_k}{EA_k} \;\middle|\; \text{over } \bar{\boldsymbol{T}} \text{ such that } \boldsymbol{B}^{\top}\bar{\boldsymbol{T}} = \boldsymbol{P} \right\} \tag{1.107}$$

The above quantity is equal to $\boldsymbol{u}^{\top}\boldsymbol{P}$ and is called the compliance of the truss. In the sequel, the quantity $\wp$ will be treated as a function of arguments $EA_1, \ldots, EA_e$ while the position of nodes and the load vector $\boldsymbol{P}$ will be fixed (viz. do not undergo the process of optimum design). In a similar manner, the discretized versions of the Castigliano theorem may be formulated for systems of beams and for grillages.

1.2.10 Thin Plates in Bending

Consider a transversely homogeneous thin plate of constant thickness h, its middle plane Ω being parameterized by a Cartesian system x_1, x_2; the axis x_3 is orthogonal to the middle plane. The domain of the plate is determined by the conditions $x = (x_1, x_2) \in \Omega$, $|x_3| < h/2$. We assume that along the lateral surface $\Gamma_2 = \gamma_2 \times (-h/2, h/2)$, where γ_2 is a segment of the contour of the boundary $\partial\Omega$ of Ω, the plate is clamped while on the remaining part $\Gamma_1 = \gamma_1 \times (-h/2, h/2)$ of the lateral surface the plate is transversely loaded or subject to bending loads. The contour $\partial\Omega = \gamma_1 \cup \gamma_2$ is parameterized by a natural parameter s, see Fig. 1.7. Let us assume that the contour γ_1 is piecewise smooth, composed of segments of vertices V_k, $k = 1, \ldots, r$ of coordinates s_k, see Fig. 1.7, where $r = 3$. If $\boldsymbol{n} = (n_1, n_2)$ is a versor (i.e. a vector of unit length) outward normal to γ, while $\boldsymbol{\tau} = (\tau_1, \tau_2)$ is the versor tangent to the contour γ, then

$$\|\boldsymbol{n}\| = 1, \qquad \|\boldsymbol{\tau}\| = 1, \qquad \boldsymbol{n} \cdot \boldsymbol{\tau} = 0$$

and these versors undergo jumps at the vertices V_k. If $g = g(s)$ is a function given on the contour γ then its jump at V_k is defined by

$$[\![g]\!]_k = g(s_k + 0) - g(s_k - 0) \tag{1.108}$$

The response of the transversely loaded thin plate is determined by its deflection $w = w(x)$ which satisfies the kinematic conditions

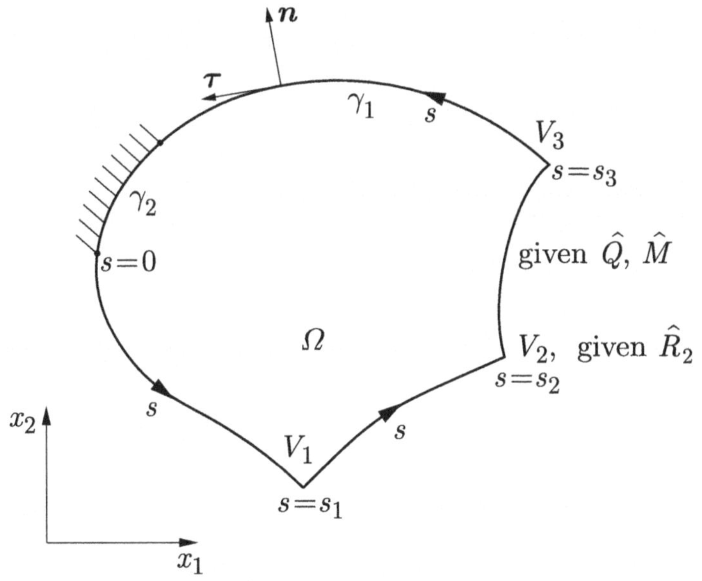

Fig. 1.7 Parameterization of the middle plane and its contour

$$w = 0, \qquad \frac{\partial w}{\partial \boldsymbol{n}} = 0 \quad \text{on} \quad \gamma_2 \tag{1.109}$$

where $\frac{\partial w}{\partial \boldsymbol{n}} = \frac{\partial w}{\partial x_1} n_1 + \frac{\partial w}{\partial x_2} n_2$ is the directional derivative in the $\boldsymbol{n}$ direction. Such field w forms a space $\mathcal{V}(\Omega)$ of kinematically admissible deflections.

In the present section, the summation convention is adopted; the sign of summation will be omitted if the expression summed involves repeating indices, e.g.:

$$A^i B_i = A^1 B_1 + \cdots + A^n B_n$$

where $i = 1, \ldots, n$. In this section $n = 2$.

Let $\bar{w} = v \in \mathcal{V}(\Omega)$. The virtual work of the transverse surface load acting on the plate faces and on the boundary is expressed by the value of a linear form $\mathcal{L}(\cdot)$ on the virtual deflection field v:

$$\mathcal{L}(v) = \int_{\Omega} q v dx + \int_{\gamma_1} \left(\hat{Q} v + \hat{M} \left(-\frac{\partial v}{\partial \boldsymbol{n}} \right) \right) ds + \sum_k \hat{R}_k v(s_k) \tag{1.110}$$

where q is the effective intensity of the transverse surface load, $\hat{Q}$ is the intensity of the transverse load distributed along the contour γ_1, $\hat{M}$ is the external moment bending the plate edge along γ_1 and $\hat{R}_k$ are concentrated forces acting orthogonally to the plate at the vertices V_k. Let us introduce the tensor of changes of curvature

$$\varkappa_{ij}(v) = -\frac{\partial^2 v}{\partial x_i \partial x_j} \tag{1.111}$$

The applied load causes bending; the bending moments M^{ij} within the plate form a tensor field $\boldsymbol{M} = (M^{ij})$ on Ω. The field $w = w(x)$ represents the plate deflection. The deformations of the plate are given by $\varkappa_{ij}(w)$. The bending moments and the deformations are linked by the constitutive equation (the summation convention is now adopted)

$$M^{ij} = D^{ijkl}\varkappa_{kl}(w) \tag{1.112}$$

where $\boldsymbol{D} = (D^{ijkl})$ is the tensor of bending stiffnesses satisfying the following symmetry conditions:

$$D^{ijkl} = D^{klij}, \qquad D^{ijkl} = D^{jikl} \tag{1.113}$$

The formulae (1.112) will be written shortly as $\boldsymbol{M} = \boldsymbol{D}\boldsymbol{\varkappa}(w)$.

Let the tensor $\boldsymbol{D}$ satisfy the condition of positive definiteness

$$\varkappa_{ij} D^{ijkl}\varkappa_{kl} \geq \alpha_0 ||\boldsymbol{\varkappa}||^2 \tag{1.114}$$

where $\alpha_0 > 0$ and $||\boldsymbol{\varkappa}|| = \sqrt{\varkappa_{ij}\varkappa_{ij}}$ is the Euclidean norm of tensor $\boldsymbol{\varkappa}$. Locally $\boldsymbol{M}$ and $\boldsymbol{\varkappa}$ are elements of the set $\mathbb{E}_s^2$ of symmetric 2nd rank tensors. The scalar product in this set is defined by

$$\boldsymbol{M} \cdot \boldsymbol{\varkappa} = M^{ij}\varkappa_{ij} \tag{1.115}$$

The field $\boldsymbol{M}$ is said to be statically admissible if

$$\int_\Omega \boldsymbol{M} \cdot \boldsymbol{\varkappa}(v)dx = \mathcal{L}(v) \quad \forall \quad v \in \mathcal{V}(\Omega) \tag{1.116}$$

The above equation implies

- the local equilibrium equations

$$\frac{\partial^2 M^{ij}}{\partial x_i \partial x_j} + q = 0 \quad \text{in } \Omega \tag{1.117}$$

- the natural boundary conditions along γ_1

$$M = \hat{M}, \quad Q = \hat{Q}, \tag{1.118}$$

and $R_k = \hat{R}_k$ at the vertices V_k of the contour γ_1 where

$$M = M^{ij} n_i n_j \tag{1.119}$$

is the bending moment,

$$Q = \frac{\partial M^{ij}}{\partial x_j} n_i + \frac{\partial M_\tau}{\partial s} \tag{1.120}$$

is the effective transverse shear force (or Kirchhoff's transverse force), while

$$M_\tau = M^{ij} n_i \tau_j \tag{1.121}$$

stands for the torsion component of the moment. Moreover,

$$R_k = [\![M_\tau]\!]_k \tag{1.122}$$

or R_k is equal to the jump of M_τ at the vertices.

The set of fields $\boldsymbol{M}$ satisfying (1.116), or, locally satisfying (1.117) and the appropriate natural boundary conditions along γ_1 is denoted by $\Sigma(\Omega)$. This set is linear affine; note that $\mathbf{0} \notin \Sigma(\Omega)$. Usually, this set is non-empty.

The formulation of the problem of statics of a thin transversely loaded plate is: find $w \in \mathcal{V}(\Omega)$ such that the field $\boldsymbol{M}$ determined by

$$\boldsymbol{M} = \boldsymbol{D}\boldsymbol{\varkappa}(w)$$

belongs to the set $\Sigma(\Omega)$.

Thus, the scalar field $w \in \mathcal{V}(\Omega)$ satisfies the variational equation

$$a(w, v) = \mathcal{L}(v) \quad \forall \quad v \in \mathcal{V}(\Omega) \tag{1.123}$$

where

$$a(w, v) = \int_\Omega \boldsymbol{\varkappa}(v) \cdot (\boldsymbol{D}\boldsymbol{\varkappa}(w)) dx \tag{1.124}$$

The field w determined this manner is the maximizer of the problem

$$\mathcal{J} = \max_{v \in \mathcal{V}(\Omega)} \left\{ \mathcal{L}(v) - \frac{1}{2} a(v, v) \right\} \tag{1.125}$$

while $\boldsymbol{M} = \boldsymbol{D}\boldsymbol{\varkappa}(w)$ is the minimizer of

$$\wp = \min_{\bar{\boldsymbol{M}} \in \Sigma(\Omega)} \int_\Omega \bar{\boldsymbol{M}} \cdot (\boldsymbol{D}^{-1} \bar{\boldsymbol{M}}) dx \tag{1.126}$$

while the values of (1.125) and (1.126) are linked by $2\mathcal{J} = \wp$.

Due to the properties (1.113), (1.114) of tensor $\boldsymbol{D}$, the bilinear form $a(\cdot, \cdot)$ has all the properties of the scalar product. Appropriate endowing the space $\mathcal{V}(\Omega)$ with the structure of the Hilbert space with the scalar product given by $a(\cdot, \cdot)$ paves the way towards the theorems on existence of the solution to problem (1.123), on the basis of the Riesz' representation theorem.

1.2.11 Membrane Shells

1.2.11.1 Arbitrary Parameterization of the Middle Surface

Let the domain $\Omega \in \mathbb{R}^2$ be a domain parameterized by the Cartesian system ξ^1, ξ^2. Let $\boldsymbol{r} : \Omega \to S \in \mathbb{R}^3$ be the mapping of class $C^3(\Omega)$. Here S is a surface in $\mathbb{R}^3$ determined by $x_i = x_i(\xi^1, \xi^2)$, $i = 1, 2, 3$, $\xi = (\xi^1, \xi^2) \in \Omega$ or $\boldsymbol{r}(\xi) = \sum_{i=1}^{3} x_i(\xi)\hat{\boldsymbol{e}}_i$, where $\hat{\boldsymbol{e}}_i$ are versors of the Cartesian system $\{x_i\}$ of the origin at O. The vectors tangent to the parametric lines are given by

$$\boldsymbol{g}_i = \frac{\partial \boldsymbol{r}}{\partial \xi^i}, \qquad i = 1, 2 \tag{1.127}$$

Similarly as in Sect. 1.2.10, the summation convention is applied for repeating indices. Moreover, the partial differentiation will be denoted by a comma, i.e.

$$\frac{\partial f}{\partial x_i} = f_{,i} \qquad \frac{\partial^2 f}{\partial x_i \partial x_j} = f_{,ij} \tag{1.128}$$

Thus, Eq. (1.127) will be abbreviated to the form $\boldsymbol{g}_i = \boldsymbol{r}_{,i}$. Let

$$\boldsymbol{g}_3 = \frac{\boldsymbol{g}_1 \times \boldsymbol{g}_2}{||\boldsymbol{g}_1 \times \boldsymbol{g}_2||} \tag{1.129}$$

be the versor normal to the surface. The metric tensor on S is defined by

$$g_{ij} = \boldsymbol{g}_i \cdot \boldsymbol{g}_j \tag{1.130}$$

while (g^{ij}) constitute the matrix inverse to (g_{ij}); $i, j = 1, 2$. Let us introduce the co-basis by

$$\boldsymbol{g}^i = g^{ij}\boldsymbol{g}_j, \qquad i, j = 1, 2 \tag{1.131}$$

and $\boldsymbol{g}^3 = \boldsymbol{g}_3$. The curvature tensor is defined by

$$b_{ij} = -\boldsymbol{g}_i \cdot \boldsymbol{g}_{3,j} = \boldsymbol{g}_3 \cdot \boldsymbol{g}_{i,j} \tag{1.132}$$

Let us introduce Christoffel's symbols on S by

$$\Gamma^i_{jk} = g^{il}\Gamma_{jkl} \qquad \Gamma_{ijk} = \frac{1}{2}(g_{ik,j} + g_{jk,i} - g_{ij,k}) \tag{1.133}$$

where $i, j, k = 1, 2, 3$. Consider a vector $\boldsymbol{u} = u_i \boldsymbol{g}^i$ tangent to S; $i = 1, 2$. The covariant derivative of a vector $\boldsymbol{u}$ is defined by

$$u_{i|j} = u_{i,j} - \Gamma_{ij}^k u_k \tag{1.134}$$

Let $g = \det(g_{ij})$. The elementary area of S is

$$dS = \sqrt{g} d\xi^1 d\xi^2 \tag{1.135}$$

A vector $\boldsymbol{v}$ can be locally referred to the basis $\boldsymbol{g}^i$, $\boldsymbol{g}^3$ and we write: $\boldsymbol{v} = v_i \boldsymbol{g}^i + v_3 \boldsymbol{g}^3$; note that $v_3 = v^3$, since $\boldsymbol{g}^3 = \boldsymbol{g}_3$. Let us introduce the operator

$$\varepsilon_{ij}(\boldsymbol{v}) = \frac{1}{2}(v_{i|j} + v_{j|i}) - b_{ij} v_3 \tag{1.136}$$

defining the strain tensor $\boldsymbol{\varepsilon} = \varepsilon_{ij} \boldsymbol{g}^i \otimes \boldsymbol{g}^j$, $i, j = 1, 2$. Let $\boldsymbol{n}, \boldsymbol{t}$ be the unit vectors: outward and tangent to ∂S and lying in the tangent plane of ∂S. Let us define

$$v_n = \boldsymbol{v} \cdot \boldsymbol{n}, \qquad v_t = \boldsymbol{v} \cdot \boldsymbol{t} \tag{1.137}$$

The membrane theory of shells admits the following kinematic boundary conditions: on Γ_2 being a part of ∂S given are v_n, v_t; only v_n; only v_t; both v_n, v_t are arbitrary at the free edge.

The surface S will be the middle surface of the shell. Assume that the shell is subject to the surfacial loads of intensity $\boldsymbol{p} = \tilde{p}_i \boldsymbol{g}^i + \tilde{p}_3 \boldsymbol{g}^3$ and to the edge load of intensity $\hat{\boldsymbol{p}} = \hat{p}_t \boldsymbol{t} + \hat{p}_n \boldsymbol{n}$; note that the edge load cannot be orthogonal to the middle surface.

Let $\mathcal{V}(S)$ be the set of kinematically admissible displacements $\boldsymbol{v}$, appropriately regular on S and satisfying the kinematic boundary conditions on Γ_2, a part of ∂S. In the case of a clamped edge both v_n and v_t vanish. The virtual work of the load is expressed by

$$\mathcal{L}(\boldsymbol{v}) = \int_S \boldsymbol{p} \cdot \boldsymbol{v} dS + \int_{\Gamma_1} \hat{\boldsymbol{p}} \cdot \boldsymbol{v} ds \tag{1.138}$$

The field $\boldsymbol{N} = \tilde{N}^{ij} \boldsymbol{g}_i \otimes \boldsymbol{g}_j$, $i, j = 1, 2$, of stress resultants satisfies the variational equation of equilibrium

$$\int_S \boldsymbol{N} \cdot \boldsymbol{\varepsilon}(\boldsymbol{v}) dS = \mathcal{L}(\boldsymbol{v}) \quad \forall \quad \boldsymbol{v} \in \mathcal{V}(S) \tag{1.139}$$

where $\boldsymbol{N} \cdot \boldsymbol{\varepsilon} = \tilde{N}^{ij} \varepsilon_{ij}$. This equation implies the local equations of equilibrium

$$\tilde{N}^{ij}|_j + \tilde{p}^i = 0, \qquad b_{ij} \tilde{N}^{ij} + \tilde{p}^3 = 0 \tag{1.140}$$

and the natural boundary conditions on Γ_1

$$\tilde{N}^{ij} n_i n_j = \hat{p}_n \qquad \tilde{N}^{ij} n_j t_i = \hat{p}_t \tag{1.141}$$

The fields $\boldsymbol{N}$ on S satisfying (1.139) or (1.140) and (1.141) constitute a set $\Sigma_p(S)$. The constitutive equations of linear elasticity are assumed in the simplest form

$$\tilde{N}^{ij} = A^{ijkl} \varepsilon_{kl}(\boldsymbol{u}) \tag{1.142}$$

where $\boldsymbol{A} = (A^{ijkl})$ represents the tensor of membrane stiffnesses; this tensor is characterized by usual symmetries and is positive definite. Let us introduce the bilinear form

$$a(\boldsymbol{u}, \boldsymbol{v}) = \int_S A^{ijkl} \varepsilon_{ij}(\boldsymbol{u}) \varepsilon_{kl}(\boldsymbol{v}) dS \tag{1.143}$$

Let

$$\mathcal{R} = \{\boldsymbol{v} \mid \varepsilon_{11}(\boldsymbol{v}) = 0, \quad \varepsilon_{22}(\boldsymbol{v}) = 0, \quad \varepsilon_{12}(\boldsymbol{v}) = 0\} \tag{1.144}$$

The set $\mathcal{R}$ encompasses the isometric deformations of the shell, or the deformations characterized by the membrane strains being zero. Due to positive definiteness of the tensor $\boldsymbol{A}$, the set $\mathcal{R}$ is given by $\boldsymbol{v}$ such that $a(\boldsymbol{v}, \boldsymbol{v}) = 0$.

The problem of statics reads: find $\boldsymbol{u} \in \mathcal{V}(S)$ such that

$$a(\boldsymbol{u}, \boldsymbol{v}) = \mathcal{L}(\boldsymbol{v}) \quad \forall\, \boldsymbol{v} \in \mathcal{V}(S) \tag{1.145}$$

The necessary condition of existence of $\boldsymbol{u}$ is compatibility of the load with the set $\mathcal{R}$ or

$$\mathcal{L}(\boldsymbol{v}) = 0 \quad \forall \quad \boldsymbol{v} \in \mathcal{V}(S) \cap \mathcal{R} \tag{1.146}$$

If $\boldsymbol{v}$ satisfies the conditions $v_n = 0$, $v_t = 0$ along ∂S and if the tensor $\boldsymbol{b} = (b_{ij})$ is positive definite on S, then $\mathcal{V}(S) \cap \mathcal{R} = \emptyset$ and then the condition (1.146) is satisfied for each choice of the surface loads, see Ciarlet and Sanchez-Palencia (1996).

1.2.11.2 The Membrane Shells of Revolution

The middle surface S of the shell of revolution is constructed by rotation of the meridian around the axis x_3, see Figs. 1.8 and 1.9.

Position of a point A is given by x_3, or, alternatively by the angle φ directed as in Fig. 1.8; here $\varphi_U \leq \varphi \leq \varphi_L$, where φ_U, φ_L refer to the upper and lower edges. The shape of the meridian is given by $|A_0 A| = R_0(\varphi)$. Point A_2 is the centre of the curvature of the meridian, hence $|AA_2| = R_2(\varphi)$ represents the radius of the curvature of the meridian. Note that the radius of a curvature of the S in the circumferential direction equals $|AA_1| = R_1(\varphi)$ and $R_0(\varphi) = \sin(\varphi) R_1(\varphi)$. The radii R_0, R_1, R_2 are linked by

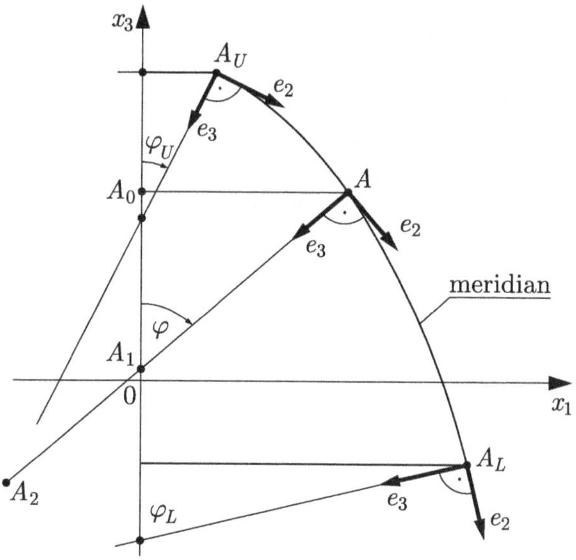

Fig. 1.8 Geometry of the meridian

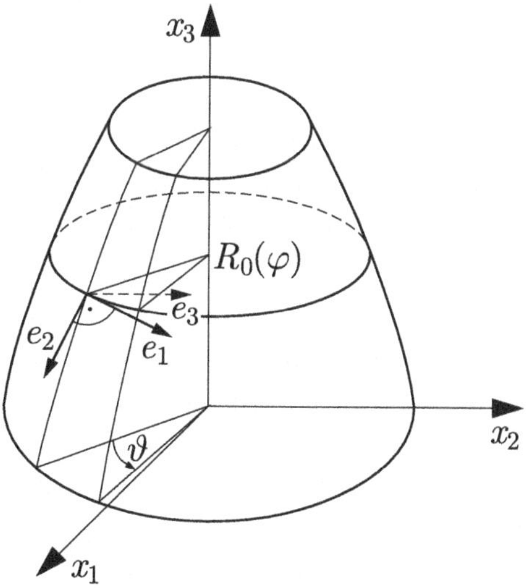

Fig. 1.9 Parameterization of the shell

$$\frac{dR_0}{d\varphi} = \cos\varphi R_2(\varphi) \qquad \frac{dR_1}{d\varphi} = \cot\varphi[R_2(\varphi) - R_1(\varphi)] \tag{1.147}$$

which follows from Codazzi–Mainardi identities, see Mazurkiewicz (2004).

We assume the geographic parameterization: $\xi^1 = \vartheta$ determines the parallels, while $\xi^2 = \varphi$ determines the meridians. We define the versors $\boldsymbol{e}_i = \frac{\boldsymbol{g}_i}{\|\boldsymbol{g}_i\|}$; the versor $\boldsymbol{e}_1$ is tangent to the parallels; the versor $\boldsymbol{e}_2$ is tangent to the meridian; the versor $\boldsymbol{e}_3$ is orthogonal to S, directed as shown in Fig. 1.9. The surface load $\boldsymbol{p}$ is referred to the basis $\boldsymbol{e}_i$: $p_i = \boldsymbol{p} \cdot \boldsymbol{e}_i$; the edges $\varphi = \varphi_U$, $\varphi = \varphi_L$ are subject to the tractions $\hat{\boldsymbol{p}}_U(\vartheta)$, $\hat{\boldsymbol{p}}_L(\vartheta)$.

The displacement vector $\boldsymbol{u}$ is decomposed in the basis $\boldsymbol{e}_i$; let us denote: $u = \boldsymbol{u} \cdot \boldsymbol{e}_1$, $v = \boldsymbol{u} \cdot \boldsymbol{e}_2$, $w = \boldsymbol{u} \cdot \boldsymbol{e}_3$. The fields $u(\vartheta, \varphi)$, $v(\vartheta, \varphi)$, $w(\vartheta, \varphi)$ are displacements in the circumferential, meridian and normal to S directions, respectively.

The physical components of the tensor ε_{ij} (referred to the basis $\boldsymbol{e}_i$) are denoted by ϵ_{ij}; they are given by the formulae below

$$
\begin{aligned}
\epsilon_{11}(\boldsymbol{u}) &= \frac{1}{R_0}\frac{\partial u}{\partial \vartheta} + \frac{1}{R_1}(v \cot \varphi - w) \\
\epsilon_{22}(\boldsymbol{u}) &= \frac{1}{R_2}\left(\frac{\partial v}{\partial \varphi} - w\right) \\
2\epsilon_{12}(\boldsymbol{u}) &= \frac{1}{R_0}\frac{\partial v}{\partial \vartheta} + \frac{1}{R_2}\frac{\partial u}{\partial \varphi} - \frac{\cos \varphi}{R_0} u
\end{aligned}
\tag{1.148}
$$

The virtual work of the load is expressed by

$$
\begin{aligned}
\mathcal{L}(\boldsymbol{v}) = \int_S \boldsymbol{p} \cdot \boldsymbol{v} dS + \int_0^{2\pi} \hat{\boldsymbol{p}}_U(\vartheta) \cdot \boldsymbol{v}(\vartheta, \varphi_U) R_0(\varphi_U) d\vartheta + \\
\int_0^{2\pi} \hat{\boldsymbol{p}}_L(\vartheta) \cdot \boldsymbol{v}(\vartheta, \varphi_L) R_0(\varphi_L) d\vartheta
\end{aligned}
\tag{1.149}
$$

where $dS = R_0(\varphi) R_2(\varphi) d\vartheta d\varphi$.

The variational equation of equilibrium has the form (1.139), where

$$
\boldsymbol{N} \cdot \boldsymbol{\varepsilon} = N^{11}\epsilon_{11} + 2N^{12}\epsilon_{12} + N^{22}\epsilon_{22}
\tag{1.150}
$$

and N^{ij} are physical components of tensor $\boldsymbol{N}$ in the basis $\boldsymbol{e}_i$. These quantities satisfy the local equations of equilibrium in the form, see (1.140)

$$
\begin{aligned}
R_2 \frac{\partial N^{11}}{\partial \vartheta} + \frac{\partial}{\partial \varphi}(R_0 N^{12}) + R_2 N^{12} \cos \varphi + R_0 R_2 p_1 &= 0 \\
R_2 \frac{\partial N^{12}}{\partial \vartheta} + \frac{\partial}{\partial \varphi}(R_0 N^{22}) - R_2 N^{11} \cos \varphi + R_0 R_2 p_2 &= 0 \\
R_2 N^{11} + R_1 N^{22} + R_1 R_2 p_3 &= 0
\end{aligned}
\tag{1.151}
$$

In case of shells of revolution, the set $\mathcal{R}$ of isometric deformations does not coincide with the set of translations and rigid rotations, but is more rich; its elements describe complex deformation modes, usually generated from the folding of the edges. In paper by Love (1888), one can find the following solutions to the set of equations: $\varepsilon_{ij}(\boldsymbol{v}) = 0$, see (1.144) referring to the spherical shells

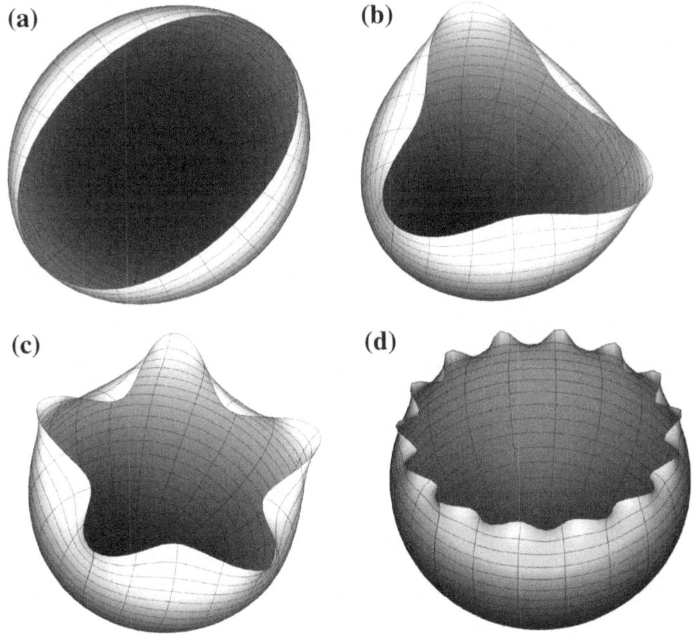

Fig. 1.10 The isometric deformations of a segment of a spherical shell for the subsequent values of the parameter n: **a** $n = 2$; **b** $n = 3$; **c** $n = 5$; **d** $n = 15$

$$
\begin{aligned}
u &= \sin\varphi\left[A(\tan(\varphi/2))^n - B(\cot(\varphi/2))^n\right]\sin(n\vartheta) \\
v &= \sin\varphi\left[A(\tan(\varphi/2))^n + B(\cot(\varphi/2))^n\right]\cos(n\vartheta) \qquad (1.152) \\
w &= \sin\varphi\left[A(n+\cos\varphi)(\tan(\varphi/2))^n + \right. \qquad (1.153) \\
&\quad \left. B(-n+\cos\varphi)(\cot(\varphi/2))^n\right]\cos(n\vartheta)
\end{aligned}
$$

where $n = 1, 2, \ldots,\ A, B \in \mathbb{R}$. These deformations are excluded, if the spherical shell forms the full sphere, since then the functions (1.152), (1.153) become singular at both the poles ($\varphi = 0$ and $\varphi = \pi$).

Selected deformations expressed by (1.152), (1.153) given in Fig. 1.10 show characteristic folding of the edges. To satisfy the condition (1.146), the applied load should do the zero virtual work on such deformations. Since this is very difficult to fulfil, to make the equilibrium problem well posed, one should impose appropriate kinematic conditions to prevent from the appearance of the isometric deformations.

1.2.11.3 Pure Torsion of Shells of Revolution

Consider the problem of torsion of a shell of revolution subject to the circumferential load at both the edges, see Fig. 1.11, the intensities of the edge loads being

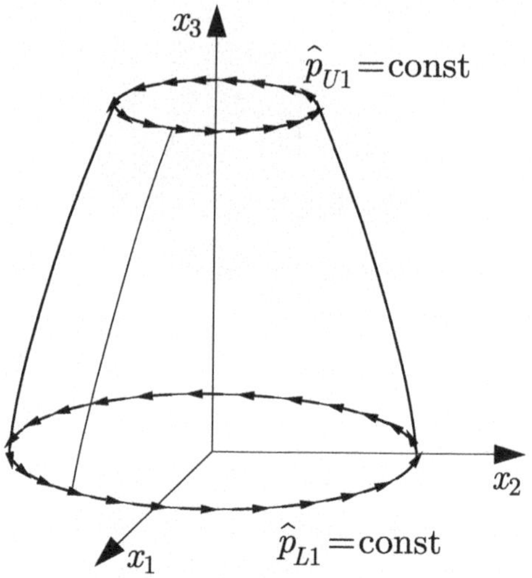

Fig. 1.11 A shell of revolution subject to a torsion load

$$p_1 = p_1(\varphi), \quad \hat{p}_{U1} = \text{const}, \quad \hat{p}_{L1} = \text{const}, \quad p_2 = p_3 = 0, \\ \hat{p}_{Ui} = 0, \quad \hat{p}_{Li} = 0, \quad i = 2, 3 \tag{1.154}$$

The shell is not supported, hence the condition (1.146) must be satisfied. The set $\mathcal{R}$ has now the form

$$\mathcal{R} = \{\boldsymbol{u} = (u, v, w) \mid u = \psi R_0(\varphi), \quad v = 0, \quad w = 0, \quad \psi \in \mathbb{R}\} \tag{1.155}$$

The condition (1.146) reduces to

$$\int_S p_1(\varphi) R_0(\varphi) dS + 2\pi \left[(R_0(\varphi_U))^2 \hat{p}_{U1} + (R_0(\varphi_L))^2 \hat{p}_{L1} \right] = 0 \tag{1.156}$$

If $p_1 = 0$, then $\hat{p}_{U_1}$, $\hat{p}_{L_1}$, are linked by

$$\hat{p}_{U1} = \frac{M}{2\pi[R_0(\varphi_U)]^2}, \qquad \hat{p}_{L1} = -\frac{M}{2\pi[R_0(\varphi_L)]^2} \tag{1.157}$$

where M and $-M$ are the resultants of the loads; M is called a torque.

In the problem discussed the state of stress resultants reduces to:

$$N^{11} = 0, \qquad N^{22} = 0, \qquad N^{12} = N^{12}(\varphi) \tag{1.158}$$

The first of Eqs. (1.151) gives

$$\frac{d}{d\varphi}(R_0 N^{12}) + R_2 \cos\varphi N^{12} = 0 \tag{1.159}$$

By using (1.147), we rearrange this equation to the form

$$\frac{d}{d\varphi}[(R_0(\varphi))^2 N^{12}(\varphi)] = 0 \tag{1.160}$$

hence

$$N^{12}(\varphi) = \frac{C}{(R_0(\varphi))^2} \tag{1.161}$$

The boundary conditions

$$N^{12}(\varphi_L) = -\hat{p}_{L1}, \qquad N^{12}(\varphi_U) = \hat{p}_{U1} \tag{1.162}$$

give the final results

$$N^{12}(\varphi) = \frac{M}{2\pi[R_0(\varphi)]^2} \tag{1.163}$$

1.2.12 Bibliographical Comments

The linear theory of bars and systems of bars has its origin in the works by Jacques and Jean Bernoulli, see Timoshenko (1953). The contemporary version of this theory may be found in the book by Hartmann (1985). Less known aspect of this theory is proper determination of the torsional stiffness of bars $\mu J^{(\mathfrak{M})}$, see Arutyunyan and Abramyan (1963). The theory of bars is included in the books on optimum design of engineering structures and on plasticity and shakedown problems of bar structures, cf. Borkowski (1985) and König (1987). This theory can also be learned from the books on numerical methods: from those classical, like Błaszkowiak and Kączkowski (1966) and those written in terms of the finite element method, like Argyris and Mlejnek (1986).

The main feature of the linear theory of bar structures is an analogy between geometric equations defining the strain measures and the equilibrium equations linking internal forces with the loads. In the discretized setting (or the algebraic setting) of the static problems, one notes that the geometric matrix is transpose to the matrix of equilibrium equations. This remarkable property implies that there is no necessity to construct the equilibrium equations once the equations which determine deformations are already constructed. This property of the system of statics equations is called Clebsch theorem (see the famous book by Alfred Clebsch 1862) and this theorem has wide, really far-reaching implications in the most important problems of mathematical physics, see Strang (1988).

The correctness of the statics equations (1.4)–(1.6) of spatial frames is conditioned by its reducibility to the functional analysis equation of the type:

$$(u, v) = f(v) \quad \text{for each} \quad v \in \mathcal{V} \tag{1.164}$$

where $(\cdot, \cdot)$ is a scalar product and $f(\cdot)$ is a linear continuous form. The space $\mathcal{V}$ can be usually endowed with a structure of a Hilbert space. This reduction sets the problems of linear statics in the form of the identity known from the celebrated F.Riesz' representation theorem, see Ch. III in the book by Yosida (1980). There is no necessity here to refer to a more general Lax–Milgram lemma, see the same chapter in Yosida's (1980) book.

The linear problem of statics of plane frames can be rearranged to the set of algebraic equations (1.18), (1.23) and (1.28), as derived in Lewiński (2001), which makes it possible to decompose the stiffness matrix into terms referring to the axial and bending deformation, see (1.30). This approach can be easily extended to the case of frames made of extensible as well as inextensible bars. The approach applies also to grillages; their equations form a similar set of Eqs. (1.43)–(1.46), where the terms referred to torsion replace the terms referred to tension/compression.

The system of equations of statics of spatial frames can be put in the form fully similar to that known from statics of trusses; this has been shown in Sect. 1.2.6 following Lewiński (2001). In this setting, the constitutive equations are diagonal (see (1.55)) and the stiffness matrix is clearly decomposed into terms reflecting subsequent modes of the frame deformations. In this setting, Clebsch theorem is built in and seen clearly, like in the theory of trusses.

The theorem of Castigliano (called also the theorem on minimum of complementary energy) plays the central role in the whole book; its formulation is taken from Nečas and Hlávaček (1981). From the mathematical viewpoint, no special theorem is needed, since the thesis of this theorem is just a direct result of a passage to the dual formulation. Nevertheless, we use this name for the dual setting of the problems of statics to put emphasis on its vital role in the theory of compliance minimization. The derivations shown in Sect. 1.2.8 (for plane frames) and in Sect. 1.2.9 (for spatial frames) follow the lines of the proof concerning the linear elasticity problem given by Nečas and Hlávaček (1981). On the other hand, the proof of the inequality (1.94) concerning the elastic potential follows the lines of the proof given by Ciarlet (1997) for linear elasticity.

The Castigliano theorem is crucial for the whole book, since it makes it possible to express the compliance of a bar structure directly in terms of stiffnesses of bars. The compliance is subject to minimization in order to find the stiffest design. The compliance may be expressed in several ways. Its definition is very simple; it is the value of the work of the load done on the state of displacements caused by this load. This is the very definition of the compliance which shows that this is also the doubled elastic energy stored in the structure. However, the latter definition conceals the way it depends on the stiffness; this definition is not only inconvenient but highly misleading. To have a clear insight how the stiffnesses affect the compliance, we shall make use of the Castigliano theorem. The weak point of this approach is that it cannot be extended to non-linear problems; indeed, this book does not go beyond the linear settings of statics.

The Michell structures have their origin in the theory of trusses, the latter being a part of the theory of frames. The theory of statics of trusses is given in the book by Hemp (1973) and in the papers by Achtziger (1996, 1997) on truss optimization.

A variational approach to the boundary value problems of thin plates was already put forward by Kirchhoff (1876). A contemporary setting can be found in Ciarlet (1997).

The theory of thin elastic shells has been put forward by Love (1888), but these equations involved some mathematical inconsistencies, removed only in the middle of 1960s. The first mathematically correct theory was proposed by Novozhilov in 1950s (see his book of 1970). The drawback of this approach was that this theory was written only for the curvilinear parameterization coinciding with the lines of principal curvatures of the middle surface. Moreover, this theory could not be easily extended to a general parameterization. To have a universal theory whose equations are valid for all admissible curvilinear parameterizations, one should construct the theory in a slightly different manner; this discussion can be found in Naghdi (1963), see also Ciarlet (1998). Unfortunately, the seminal text by Naghdi (1963) was published one year too early, since only in 1964 Budiansky and Sanders proposed the final version of the theory of thin shells, free of any drawbacks of Love's theory and preserving all the good properties of Novozhilov's theory, see Sec. 16.3 in Lewiński and Telega (2000).

The theory of thin shells encompasses two phenomena: the membrane deformation and the bending deformation. Consequently, this theory involves stiffnesses of various units, which is the reason of the difficulties in deriving this theory from 3D elasticity by a natural asymptotic method, without artificial scalings of the moduli or loads. By omitting the bending stiffnesses, we come across the membrane theory, well known from nineteenth century, see Flügge (1960), Gol'denveizer (1976), Vlasov (1949). Unfortunately, the membrane theory is not well posed in general: substitution of the constitutive equations into the equilibrium equations does not lead to an equation of the form (1.164), since the left-hand side obtained is not a scalar product; more precisely, the condition $(v, v) = 0$ does not imply $v = 0$ but leads to a rich family of isometric deformations. Such deformations for spherical shells have been constructed by Rayleigh, see Love (1888) and Flügge (1960). The isometric deformations were also the subject of the works by Vekua (1982). If the tensor of curvature of the middle surface is positive definite at each point of the middle surface and if the edge of the shell is fixed, then the left-hand side of (1.164) becomes a scalar product and the problem of membrane shells may be set correctly, see Ciarlet and Sanchez-Palencia (1996).

In many problems of membrane shells, the state of stress resultants is uniquely determined by the load, while the state of displacements is not unique. Shells of revolution constitute an exceptional family for which numerous analytical solutions are available, see Vlasov (1949), Martin (1949), Gol'denveizer (1976), Flügge (1960). One of the simplest is the problem of the response of the shell to the uniform torsional load applied along the circumferences of the upper and lower edges; the solution reported in Sec. 2.11.3 is taken after Mazurkiewicz (2004) and Lewiński (2004).

The notion: statical determinacy refers mainly to bar structures. A bar structure is statically determinate, if the internal forces and, consequently, all the stress resultants can be uniquely determined from the equilibrium equations. In other words, for such structures, the set of statically admissible stress resultants is one element set. This notion refers only to the linear theories, since only in these theories, the equations of equilibrium refer to the undeformed configuration and do not involve the displacements fields; for the non-linear settings this notion is irrelevant. In the statically determinate structures, the stress resultants are determined by the geometry of the bar layout and by the load. Hence, the bar cross sections may be chosen to make the stresses in the sections uniform and this design is done in one shot. The notion of statical determinacy may be extended to elasticity, but the family of statically determinate problems in the continuum setting is very narrow. A good example (and probably, the only one) is pure torsion of an annular plate loaded uniformly along both the circumferences. Among the 3D continuum problems and among the problems of bending of plates, no one is statically determinate. Yet the notion of statical determinacy appears in the theory of membrane shells. Although the success is only partial, since this theory is wrongly posed in general (as emphasized in Sect. 1.2.11.1), this property may be used for solving selected optimum design problems of membrane shells. In the membrane theory, all the three stress resultants are involved in the three equations of equilibrium, thus making this system balanced. Moreover, the last equation is algebraic, see Eq. (1.140), which simplifies essentially the equilibrium analysis. Just due to the middle surface being curved in two directions the equilibrium of the forces in the direction normal to the surface can be satisfied, despite omitting the transverse shear forces. These exceptional properties of the membrane theory pave the way towards posing and solving important problems of optimum design of shells, see Flügge (1960), Novozhilov (1970), Krużelecki and Życzkowski (1985).

The problems of large deformations of thin shells, see Chróścielewski et al. (2004), lie beyond the scope of this book.

References

Achtziger, W. (1996). Topology optimization of discrete structures: An introductory course using convex and nonsmooth analysis. Advanced School, International Centre for Mechanical Sciences (CISM), Udine, 24–28 June 1996.

Achtziger, W. (1997). Topology optimization of discrete structures: An introduction in view of computational and nonsmooth aspects. In G. I. N. Rozvany (Ed.), *Topology optimization in structural mechanics* (Vol. 374, pp. 57–100). CISM courses and lectures. Wien: Springer.

Argyris, J. H., & Mlejnek, H.-P. (1986). Die Methode der Finiten Elementen in der elementaren Strukturmechanik, Band I, Verschiebungsmethode in der Statik, Friedr. Vieweg u.Sohn, Braunschweig/Wiesbaden.

Arutyunyan, N. Kh., & Abramyan, B. L. (1963). *Torsion of elastic bodies*. Moscow: Fizmatgiz (in Russian).

Błaszkowiak, S., & Kączkowski, Z. (1966). *Iterative methods in structural analysis*. Warsaw-London: Pergamon Press and Polish Scientific Publishers.

Borkowski, A. (1985). *Statical analysis of bar systems in the elastic and plastic range*. Warszawa-Poznań (in Polish): Polish Scientific Publishers.

Chróścielewski, J., Makowski, J., & Pietraszkiewicz, W. (2004). Statics and dynamics of multifolded shells. The nonlinear theory and the finite element method. IPPT PAN, Warsaw (in Polish).

Ciarlet, Ph. (1997). *Mathematical elasticity. Vol. II: Elastic plates*. Amsterdam: Elsevier.

Ciarlet, Ph. (1998). *Introduction to linear shell theory*. Paris: Gauthier-Villars.
Ciarlet, Ph., & Sanchez-Palencia, E. (1996). An existence and uniqueness theorem for the two-dimensional linear membrane shell equations. *Journal de Mathématiques Pures et Appliquées, 75*, 51–67.
Clebsch, A. (1862). *Theorie der Elastizität fester Korper*. Leipzig: Teubner.
Flügge, W. (1960). *Stresses in shells*. Berlin: Springer.
Gol'denveizer, A. L. (1976). *Theory of elastic thin shells* (2nd ed.). Moscow: Nauka (in Russian).
Hartmann, F. (1985). *The mathematical foundation of structural mechanics*. Berlin: Springer.
Hemp, W. S. (1973). *Optimum structures*. Oxford: Clarendon Press.
Kirchhoff, G. (1876). Vorlesungen uber mathematische Physik. Bd 1, Mechanik, Teubner. Leipzig.
König, A. (1987). *Shakedown of elastic-plastic structures*. Warsaw: Elsevier.
Krużelecki, J., & Życzkowski, M. (1985). Optimal structural design of shells - a survey. *SM Archives, 10*, 101–170.
Lewiński, T. (2001). On algebraic equations of elastic trusses, frames and grillages. *Journal of Theoretical and Applied Mechanics, 39*, 307–322.
Lewiński, T. (2004). Michell structures formed on surfaces of revolution. *Structural and Multidisciplinary Optimization, 28*(2004), 20–30.
Lewiński, T., & Telega, J. J. (2000). *Plates, laminates and shells. Asymptotic analysis and homogenization* (Vol. 52). Series on advances in mathematics for applied sciences. Singapore: World Scientific Publishing Co.
Love, A. E. H. (1888). The small free vibrations and deformation of a thin elastic shell. *Philosophical Transactions of the Royal Society of London A, 179*, 491–546.
Martin, F. (1949). Die Membran-Kugelschale unter Einzellasten. *Ingenieur-Archiv, 17*, 167–186.
Mazurkiewicz, Z. E. (2004). *Thin elastic shells. Linear theory*. Warszawa (in Polish): Oficyna Wydawnicza PW.
Michell, A. G. M. (1904). The limits of economy of material in frame structures. *Philosophical Magazine. Series 6, 8*(47), 589–597.
Naghdi, P. M. (1963). Foundation of elastic shell theory. In I. N. Sneddon & R. Hill (Eds.), *Progress in solid mechanics* (pp. 2–90). Amsterdam: North Holland.
Nečas, J., & Hlávaček, I. (1981). *Mathematical theory of elastic and elasto-plastic bodies: An introduction*. Rotterdam: Elsevier.
Novozhilov, V. V. (1970). *Thin shell theory*. Groningen: Walters-Nordhoff.
Rozvany, G. I. N. (1976). *Optimal design of flexural systems*. London: Pergamon Press.
Save, M., & Prager, W. (Eds.). (1985). *Structural optimization. Vol. 1. Optimality criteria*. New York and London: Plenum Press.
Strang, G. (1988). A framework for equilibrium equations. *SIAM Review, 30*, 283–297.
Timoshenko, S. P. (1953). *History of strength of materials*. New York: Mc Graw Hill Book Co.
Vekua, I. N. (1982). *Some general methods of constructing selected variants of the theories of shells*. Moscow (in Russian): Nauka.
Vlasov, V. Z. (1949). *General theory of shells and its applications in engineering*. Moscow-Leningrad (in Russian): Gostekhizdat.
Yosida, K. (1980). *Functional analysis*. Berlin: Springer.

Chapter 2
Optimum Design of Structures of Finite Number of Bars. Single Load Variant

The central problem of the present chapter is the minimization of the compliance of bar structures. It turns out that the optimum design of trusses of minimal compliance leads to an auxiliary problem, which naturally appears in the problem of minimization of truss volume under the condition of stresses in bars being bounded from both sides. This auxiliary problem is a problem of linear programming, which makes it possible to construct optimal structures of huge number of bars. The numerical algorithm is based on the ground structure method. These optimum designs of huge number of bars are approximants of Michell's structures to be discussed in the next chapter. The optimal designs of grillages (Sect. 2.5) are governed by more complicated equations. Only additional approximations imposed pave the way towards the linear programming problems thus delivering approximate solutions of Prager–Rozvany grillages, being flexural counterparts of Michell's structures.

2.1 Trusses of Minimal Compliance and Fixed Volume

The optimal trusses will be formed from straight bars connecting nodes whose positions are a priori fixed in a design domain. Let e be the maximal number of bars connecting the nodes. The loads are concentrated at the nodes. The subject of the consideration is construction of the trusses of the least compliance and a given cost whose nodes are chosen from the given set and the bars linking selected nodes lie within the given design domain. The cost of the design is assumed in the form

$$\sum_{k=1}^{e} E A_k l_k = \Lambda \tag{2.1}$$

T. Lewiński et al., *Michell Structures*,
https://doi.org/10.1007/978-3-319-95180-5_2

where A_k are areas of the cross sections; l_k represent lengths of the bars; E is Young's modulus. Thus, $\Lambda/E = V$ represents the volume of the material used to form a given truss.

The given forces $P_1, \dots, P_s$ applied at nodes cause nodal displacements $u_1, \dots, \dots, u_s$ directed along the given forces. The positions of the loads are not subject to optimization; they are fixed. Similarly, the positions of supports are fixed, yet not all the nodes on the support zones must be used. The areas of the cross sections A_k are the design variables and are subject to the condition: $A_k \geq 0$; the case of $A_l = 0$ means that this bar does not exist. Therefore, the number of bars in the optimal truss is not imposed.

The equations of equilibrium of a truss are encompassed by the variational equation called the virtual work equation, see (1.61)

$$\boldsymbol{T} \cdot \boldsymbol{\Delta}(\bar{\boldsymbol{u}}) = \boldsymbol{P} \cdot \bar{\boldsymbol{u}} \quad \forall\, \bar{\boldsymbol{u}} \in \mathbb{R}^s \tag{2.2}$$

where the column vector of virtual elongations of bars is determined by the vector of virtual displacements by the third of equations (1.65) or $\boldsymbol{\Delta}(\bar{\boldsymbol{u}}) = \boldsymbol{B}\bar{\boldsymbol{u}}$, and $\boldsymbol{B}$ is the geometric matrix or the matrix of directional cosines. The right-hand side of (2.2) represents the virtual work of the forces P_i on virtual displacements u_i; s is the number of independent unknown components of nodal displacements. Equation (2.2) links the axial forces in bars T_k, $k = 1, \dots, e$ with the nodal loads P_i, $i = 1, \dots, s$.

The compliance of the truss (1.107) will be treated as a function of arguments EA_k, $k = 1, \dots, e$:

$$\wp\,(\{EA_k\}) = \min\left\{\sum_k \frac{(T_k)^2 l_k}{EA_k} \mid T_k \ \ \text{satisfying (2.2)}\right\} \tag{2.3}$$

The overbars present in (1.107) are omitted above to simplify the notation; also, the range of summation is not indicated. Consider the minimization problem

$$\breve{\wp} = \min\left\{\wp\,(\{EA_k\}) \mid EA_k \geq 0, \ \ \sum_k EA_k l_k = \Lambda\right\} \tag{2.4}$$

in which the position of nodes is fixed, while the stiffnesses EA_k are design variables. We shall show that

(i)

$$\breve{\wp} \geq \frac{1}{\Lambda}(Z_1)^2 \tag{2.5}$$

with

$$Z_1 = \max\{\boldsymbol{P} \cdot \bar{\boldsymbol{u}} \mid |\Delta_k(\bar{\boldsymbol{u}})| \leq l_k, \ \ \bar{\boldsymbol{u}} \text{ kinematically admissible}\} \tag{2.6}$$

(ii)

$$\wp \leq \frac{1}{\Lambda}(Z_2)^2 \tag{2.7}$$

with

$$Z_2 = \min\left\{\sum_k |T_k| l_k \quad | \; T_k \text{ satisfying (2.2)}\right\} \tag{2.8}$$

(iii)

$$Z_1 = Z_2 \tag{2.9}$$

The proof of (i)
Let us write down the Eq. (2.2) in the form

$$\boldsymbol{P} \cdot \bar{\boldsymbol{u}} = \sum_k \left(\frac{T_k\sqrt{l_k}}{\sqrt{EA_k}}\right)\left(\frac{\sqrt{EA_k}\Delta_k(\bar{\boldsymbol{u}})}{\sqrt{l_k}}\right) \tag{2.10}$$

and let us estimate the right-hand side by using the Cauchy–Schwarz inequality

$$\boldsymbol{P} \cdot \bar{\boldsymbol{u}} \leq \sqrt{\sum_k \frac{(T_k)^2 l_k}{EA_k}} \sqrt{\sum_k EA_k \left(\frac{\Delta_k(\bar{\boldsymbol{u}})}{l_k}\right)^2 l_k} \tag{2.11}$$

Let us choose the vector $\bar{\boldsymbol{u}}$ such that

$$\left|\frac{\Delta_k(\bar{\boldsymbol{u}})}{l_k}\right| \leq 1 \qquad k = 1, 2, \ldots \tag{2.12}$$

The estimate (2.11) simplifies to the form

$$\boldsymbol{P} \cdot \bar{\boldsymbol{u}} \leq \sqrt{\sum_k \frac{(T_k)^2 l_k}{EA_k}} \sqrt{\sum_k EA_k l_k} \tag{2.13}$$

Let us now impose the cost condition (2.1) and rearrange (2.13) to the form

$$\sum_k \frac{(T_k)^2 l_k}{EA_k} \geq \frac{1}{\Lambda}(\boldsymbol{P} \cdot \bar{\boldsymbol{u}})^2 \tag{2.14}$$

where $\bar{\boldsymbol{u}}$ satisfies (2.12) while the stiffnesses EA_k satisfy (2.1). The axial forces T_k must be made statically compatible with the load P_i, $i = 1, \ldots, s$.

Now, we perform minimization of the left-hand side of (2.14) over EA_k and T_k, and maximization of the right-hand side over $\bar{u}_i$. The inequality (2.14) leads to the estimate

$$\min\left\{\sum_k \frac{(T_k)^2 l_k}{EA_k} \mid EA_k \geq 0, \sum_k EA_k l_k = \Lambda, \quad T_k \text{ satisfying (2.2)}\right\} \geq \\ \geq \frac{1}{\Lambda}\left(\max\left\{\boldsymbol{P}\cdot\bar{\boldsymbol{u}} \mid \quad \left|\frac{\Delta_k(\bar{\boldsymbol{u}})}{l_k}\right| \leq 1\right\}\right)^2 \tag{2.15}$$

which proves the inequality (2.5).

The proof of (ii)
Let

$$Y(\boldsymbol{T}) = \min\left\{\sum_k \frac{(T_k)^2 l_k}{EA_k} \mid EA_k \geq 0, \sum_k EA_k l_k = \Lambda\right\} \tag{2.16}$$

Now, the problem (2.4) can be expressed as below

$$\breve{\wp} = \min\left\{Y(\boldsymbol{T}) \mid \{T_k\} \text{ satisfying (2.2)}\right\} \tag{2.17}$$

Thus, $\breve{\wp} \leq Y(\boldsymbol{T})$ if T_k are arbitrary. The minimization in (2.16) can be performed analytically. We shall make use of

Lemma 2.1.1 *Let $F_k \geq 0$, $k = 1, \ldots, e$ and $\Lambda > 0$. Then*

$$\min\left\{\sum_{k=1}^{e} \frac{F_k}{x_k} \mid x_k \geq 0, \sum_{k=1}^{e} x_k = \Lambda\right\} = \frac{1}{\Lambda}\left(\sum_{k=1}^{e}\sqrt{F_k}\right)^2 \tag{2.18}$$

and the minimum is attained for

$$\breve{x}_k = \Lambda\frac{\sqrt{F_k}}{\sum_{k=1}^{e}\sqrt{F_k}}, \qquad k = 1, \ldots, e \tag{2.19}$$

The proof.
Let $F_k = 0$ if $k \in I_F$; I_F being a set of indices of non-active bars. Once again, the Cauchy–Schwarz inequality will play the crucial role; we estimate

$$\sum_k \sqrt{F_k} = \sum_{k\notin I_F}\sqrt{F_k} = \sum_{k\notin I_F}\sqrt{\frac{F_k}{x_k}}\sqrt{x_k} \leq \sqrt{\sum_{k\notin I_F}\frac{F_k}{x_k}}\sqrt{\sum_{k\notin I_F} x_k} \leq \\ \sqrt{\sum_{k\notin I_F}\frac{F_k}{x_k}}\sqrt{\Lambda} \tag{2.20}$$

Hence,

$$\sum_{k \notin I_F} \frac{F_k}{x_k} \geq \frac{1}{\Lambda} \left(\sum_{k=1}^{e} \sqrt{F_k} \right)^2 \tag{2.21}$$

Note that the equality is attained if one chooses $x_k = \breve{x}_k$ according to (2.19), for $k \notin I_F$, which ends the proof.

Let us perform in (2.16) minimization over EA_k. Let

$$x_k = EA_k l_k, \qquad F_k = (T_k l_k)^2.$$

We apply Lemma 2.1.1 and arrive at

$$Y(\boldsymbol{T}) = \frac{1}{\Lambda} \left(\sum_k |T_k| l_k \right)^2 \tag{2.22}$$

the minimum being attained at

$$E\breve{A}_m l_m = \Lambda \frac{|T_m| l_m}{\sum_k |T_k| l_k} \tag{2.23}$$

hence, we conclude that the absolute values of stresses in the bars of the optimum truss

$$|\sigma_m| = \frac{|T_m|}{\breve{A}_m} \tag{2.24}$$

are expressed by

$$|\sigma_m| = \sigma_0, \qquad \sigma_0 = \frac{E}{\Lambda} \sum_k |T_k| l_k \tag{2.25}$$

Thus, the stresses are equal to σ_0 in all the bars.

We conclude that the optimization over areas cross sections aimed at the minimization of the compliance leads to making the stresses in bars uniform.
We have proved the estimate

$$\breve{\wp} \leq \frac{1}{\Lambda} \left(\sum_k |T_k| l_k \right)^2 \tag{2.26}$$

This estimate can be made as sharp as possible by taking minimum over the statically admissible axial forces T_k, which leads to (2.7).

The proof of (iii)
We shall make use of the following representation of the modulus of a real number

$$|a| = \min\left\{a' + a'' \mid a' \geq 0,\ a'' \geq 0,\ a' - a'' = a\right\} \tag{2.27}$$

The minimization operation selects $a' = a^{(+)}$, $a'' = a^{(-)}$, where $a^{(+)}$ represents the positive part of the number a

$$a^{(+)} = \begin{cases} a & \text{if } a > 0 \\ 0 & \text{if } a < 0 \end{cases} \tag{2.28}$$

while $a^{(-)}$ is the negative part of the number a

$$a^{(-)} = \begin{cases} 0 & \text{if } a > 0 \\ -a & \text{if } a < 0 \end{cases} \tag{2.29}$$

Let us note that $a^{(+)}a^{(-)} = 0$, since one of these numbers is zero.
The representation (2.27) will be applied in problem (2.8) or

$$Z_2 = \min\left\{\min\left\{\sum_{k=1}^{e}(T_k' + T_k'')l_k \mid T_k' \geq 0,\ T_k'' \geq 0,\ T_k' - T_k'' = T_k\right\} \mid \boldsymbol{B}^\top \boldsymbol{T} = \boldsymbol{P}\right\} \tag{2.30}$$

where $\boldsymbol{T} = (T_1, \ldots, T_e)^\top$. The problem above may be written as below

$$Z_2 = \min\left\{\sum_{k=1}^{e}(T_k' + T_k'')l_k \mid T_k' \geq 0,\ T_k'' \geq 0,\ \boldsymbol{B}^\top(\boldsymbol{T}' - \boldsymbol{T}'') = \boldsymbol{P}\right\} \tag{2.31}$$

Upon solving this problem and finding $(\breve{T}_k', \breve{T}_k'')$ one can compute $\breve{T}_k$ by $\breve{T}_k = \breve{T}_k' - \breve{T}_k''$. Note that the minimization operations in (2.31) will determine the quantities $(\breve{T}_k', \breve{T}_k'')$ such that they will be the positive and negative parts of $\breve{T}_k$, respectively.

$$\breve{T}_k' = \breve{T}_k^{(+)}, \qquad \breve{T}_k'' = \breve{T}_k^{(-)} \tag{2.32}$$

Thus, upon solving (2.31) and finding the minimizer $(\breve{T}_k', \breve{T}_k'')$ the axial forces in the bars become

$$\breve{T}_k = \begin{cases} \breve{T}_k' & \text{if } \breve{T}_k' > 0 \text{ (then } \breve{T}_k'' = 0) \\ -\breve{T}_k'' & \text{if } \breve{T}_k'' > 0 \text{ (then } \breve{T}_k' = 0) \end{cases} \tag{2.33}$$

Let us rewrite (2.6) in the form

$$Z_1 = \max \left\{ \boldsymbol{P}^\top \bar{\boldsymbol{u}} \mid \sum_{j=1}^{s} B_{kj}\bar{u}_j \le l_k, \; -\sum_{j=1}^{s} B_{kj}\bar{u}_j \le l_k \right\} \tag{2.34}$$

introduce the matrices

$$\boldsymbol{x} = \begin{vmatrix} \boldsymbol{T}' \\ \boldsymbol{T}'' \end{vmatrix} \qquad \boldsymbol{A} = [\boldsymbol{B}^\top \;\; -\boldsymbol{B}^\top] \qquad A^\top = \begin{bmatrix} \boldsymbol{B} \\ -\boldsymbol{B} \end{bmatrix} \tag{2.35}$$

and

$$\boldsymbol{c} = \begin{vmatrix} \boldsymbol{l} \\ \boldsymbol{l} \end{vmatrix}, \qquad \boldsymbol{c}^\top = [\boldsymbol{l}^\top \, \boldsymbol{l}^\top] \qquad \boldsymbol{l} = (l_1, \ldots, l_e)^\top \tag{2.36}$$

The problem (2.31) may be written as follows

$$Z_2 = \min \left\{ \boldsymbol{c}^\top \boldsymbol{x} \mid \boldsymbol{A}\boldsymbol{x} = \boldsymbol{P}, \; \boldsymbol{x} \ge \boldsymbol{0} \right\} \tag{2.37}$$

and the problem (2.34), in the form

$$Z_1 = \max \left\{ \boldsymbol{P}^\top \bar{\boldsymbol{u}} \mid \boldsymbol{A}^\top \bar{\boldsymbol{u}} \le \boldsymbol{c} \right\} \tag{2.38}$$

The above problems of linear programming are mutually dual and the equality $Z_1 = Z_2$ holds. Concluding, the minimal compliance of a truss of given volume $V = \Lambda/E$ equals

$$\breve{\wp} = \frac{1}{\Lambda} Z^2 \tag{2.39}$$

where Z may be expressed in two ways

(a) $Z = Z_1$, Z_1 given by (2.6)
(b) $Z = Z_2$, Z_2 given by (2.8)

The solutions $\breve{\boldsymbol{T}}'$, $\breve{\boldsymbol{T}}''$ and $\breve{\bar{\boldsymbol{u}}}$ are linked as explained below.

If $\breve{T}_k' > 0$ and $\breve{T}_k'' = 0$, then $\breve{T}_k > 0$ and (2.24) reduces to the form $\breve{T}_k/\breve{A}_k = \sigma_0$. Moreover, if $\breve{\bar{\boldsymbol{u}}}$ is the solution to (2.38), then $\breve{\bar{\Delta}}_k/l_k = 1$, where $\breve{\bar{\Delta}}_k = (\boldsymbol{B}\breve{\bar{\boldsymbol{u}}})_k$ is the virtual elongation of the kth bar by (2.38).

If $\breve{T}_k' = 0$ and $\breve{T}_k'' > 0$, then $\breve{T}_k < 0$ and (2.24) holds assuming the form $\breve{T}_k/\breve{A}_k = -\sigma_0$. Moreover, $\breve{\bar{\Delta}}_k/l_k = -1$.

If $\breve{T}_k' = 0$ and $\breve{T}_k'' = 0$, then $\breve{T}_k = 0$ and $\breve{A}_k = 0$ The following inequalities hold: $-1 \le \breve{\bar{\Delta}}_k/l_k \le 1$.

Note, there is no rule: if kth bar has a zero cross section, then $\breve{\bar{\Delta}}_k/l_k \in (-1, 1)$.

Moreover, among the optimal trusses (i.e. those having the same optimal compliance) there exists one which is statically determinate.

The proof of these rules will be given in Sect. 2.2 in a more general context.

2.2 The Least Volume Trusses of Bounded Stress Level

The subject of the discussion will be the optimum design of trusses of minimal volume with the stresses in their bars being bounded from both sides:

$$-\sigma_C \leq \sigma \leq \sigma_T \tag{2.40}$$

where σ_C, σ_T are permissible stresses in compression and tension, respectively.

The loads are applied at fixed nodes, as in Sect. 2.1. The support conditions are also fixed, yet the places where the structure is supported are not fixed: the nodes, where the structure is supported are chosen from the given set of nodes belonging to the zone of a possible support. The design variables are the areas A_k of bars' cross sections; we assume that $A_k \geq 0$. Thus, the case of $A_k = 0$ is admitted, which means that the main aim of the optimization is structural topology. No assumptions concerning the constitutive equations of the material are imposed. Thus, the problem has the form

$$V_{\min} = \min \left\{ \sum_{k=1}^{e} A_k l_k \mid \sum_{k=1}^{e} T_k \Delta_k(\boldsymbol{v}) = \sum_j P_j v_j \quad \forall \quad \boldsymbol{v} \in \mathbb{R}^s \right.$$

$$\left. \Delta_k(\boldsymbol{v}) = \sum_{j=1}^{s} B_{kj} v_j, \; T_k \leq A_k \sigma_T, \; -T_k \leq A_k \sigma_C \right\} \tag{2.41}$$

2.2.1 Case of $\sigma_T = \sigma_C = \sigma_0$

We shall prove that the problem (2.41) is equivalent to the problem (2.31) which determines the quantity Z_2, see (2.8). In problem (2.41), the unknowns A_k, T_k are involved. These unknowns will be replaced by T_k', T_k'' by the rules (cf. Achtziger 1997)

$$A_k = \frac{1}{\sigma_0}(T_k' + T_k''), \qquad T_k = T_k' - T_k'' \tag{2.42}$$

The conditions

$$-A_k \sigma_0 \leq T_k \leq A_k \sigma_0 \tag{2.43}$$

will be expressed by the unknowns T_k', T_k''. We obtain

$$-T_k' - T_k'' \leq T_k' - T_k'' \leq T_k' + T_k'' \tag{2.44}$$

or $T_k' \geq 0$, $T_k'' \geq 0$. We conclude that the problem (2.41) reduces to

$$V_{\min} = \frac{1}{\sigma_0} Z_2 \tag{2.45}$$

where Z_2 is given by the solution to the problem (2.31), or by the solution to the problem (2.8).

2.2.2 Case of $\sigma_T \neq \sigma_C$

Let $\sigma_T > 0$, $\sigma_C > 0$. We introduce new unknowns: T_k', T_k''. by the rules

$$A_k = \frac{1}{\sigma_T} T_k' + \frac{1}{\sigma_C} T_k'', \qquad T_k = T_k' - T_k'' \tag{2.46}$$

The conditions

$$-A_k \sigma_C \leq T_k \leq A_k \sigma_T \tag{2.47}$$

assume the form

$$-\frac{\sigma_C}{\sigma_T} T_k' - T_k'' \leq T_k' - T_k'' \leq T_k' + \frac{\sigma_T}{\sigma_C} T_k'' \tag{2.48}$$

hence

$$\left(1 + \frac{\sigma_C}{\sigma_T}\right) T_k' \geq 0 \qquad \left(1 + \frac{\sigma_T}{\sigma_C}\right) T_k'' \geq 0 \tag{2.49}$$

Thus, once again $T_k' \geq 0$, $T_k'' \geq 0$ and the problem (2.41) assumes the form

$$V_{\min} = \frac{1}{\sigma_0} \tilde{Z}_2 \tag{2.50}$$

where

$$\tilde{Z}_2 = \min \left\{ \sum_{k=1}^{e} \left(\frac{\sigma_0}{\sigma_T} T_k' + \frac{\sigma_0}{\sigma_C} T_k'' \right) l_k \mid T_k' \geq 0,\ T_k'' \geq 0,\ \boldsymbol{B}^\top (\boldsymbol{T}' - \boldsymbol{T}'') = \boldsymbol{P} \right\} \tag{2.51}$$

and the parameter $\sigma_0 > 0$ is here arbitrary. Usually, we shall assume $\sigma_0 = \sigma_T$.

We conclude that the considered problem of optimal design of least volume fully stressed trusses reduces to the problem of linear programming (2.51). From the discussion in Sect. 2.1 we know that T_k' stands for the positive part of the axial force T_k, while T_k'' represents its negative part. Hence, one of the numbers T_k', T_k'' vanishes.

The problem dual to (2.31) has the form (2.6). Till now, the problem dual to (2.51) has not been constructed; this will be the subject of the present consideration. We generalize the setting (2.36) by introducing a new formula for $\boldsymbol{c}$

$$\boldsymbol{c}^\top = \left[\frac{\sigma_0}{\sigma_T} \boldsymbol{l}^\top, \ \frac{\sigma_0}{\sigma_C} \boldsymbol{l}^\top \right] \tag{2.52}$$

Problem (2.51) has now the form (2.37), where $\boldsymbol{c}$ is given by (2.52). The problem dual to (2.37) has the form (2.38). Its explicit formulation reads

$$\tilde{Z}_1 = \max \left\{ \boldsymbol{P}^\top \bar{\boldsymbol{u}} \mid -\frac{\sigma_0}{\sigma_C} \leq \frac{\Delta_k(\bar{\boldsymbol{u}})}{l_k} \leq \frac{\sigma_0}{\sigma_T} \right\} \tag{2.53}$$

where $\Delta_k(\bar{\boldsymbol{u}})$ is given by (1.62). The solutions $(\breve{\boldsymbol{T}}', \breve{\boldsymbol{T}}'', \breve{\bar{\boldsymbol{u}}})$ to the problems (2.51) and (2.53) are linked as follows.

If $\breve{T}_k' > 0$, $\breve{T}_k'' = 0$, then $\breve{T}_k > 0$ and the equality: $\breve{T}_k = \breve{A}_k \sigma_T$ holds. Moreover, $\breve{\Delta}_k / l_k = \sigma_0 / \sigma_T$.

If $\breve{T}_k' = 0$, $\breve{T}_k'' > 0$, then $\breve{T}_k < 0$, $\breve{T}_k = -\breve{A}_k \sigma_C$, and $\breve{\Delta}_k / l_k = -\sigma_0 / \sigma_C$.

If $\breve{T}_k' = 0$, $\breve{T}_k'' = 0$, then $\breve{T}_k = 0$. Then $\breve{A}_k = 0$. Moreover, $-\frac{\sigma_0}{\sigma_C} \leq \breve{\Delta}_k / l_k \leq \frac{\sigma_0}{\sigma_T}$.

The numerical methods for solving the problem (2.41) will be based on the formulations (2.51), (2.53).

2.2.3 Maxwell's Theorem

The aim of this section is to provide an important property of the trusses equilibrating a given set of point loads. This property was discovered by Maxwell (1872), and then utilized by Michell (1904) in deriving optimality criteria of least-weight trusses.

Let us consider a truss equilibrating a system of forces $\boldsymbol{P}_i$ applied at points with position vectors $\boldsymbol{r}_i$, $i = 1, 2, \ldots, n_p$. Here, $n_p \geq 2$ denotes the number of applied forces. Without the loss of generality, we can assume that all forces are non-zero, i.e. $|\boldsymbol{P}_i| > 0$. Moreover, we assume that the trusses under consideration are not supported at all or are supported without redundant reactions. Consequently, all reactions can be determined from the equilibrium conditions and treated as prescribed external forces. Such structures are said to be externally statically determinate.

Thus, consider a truss transmitting a self-equilibrating system of forces, i.e. satisfying:

$$\sum_{i=1}^{n_p} \boldsymbol{P}_i = \boldsymbol{0} \quad \text{and} \quad \sum_{i=1}^{n_p} \boldsymbol{P}_i \times \boldsymbol{r}_i = \boldsymbol{0} \tag{2.54}$$

For 3D problems parameterized by a Cartesian coordinate system $Oxyz$, we assume the following representations: $\boldsymbol{P}_i = (P_{x,i}, P_{y,i}, P_{z,i})^\top$ and $\boldsymbol{r}_i = (x_i, y_i, z_i)^\top$. Note that plane problems can be treated as a specific case of 3D problem with all $P_{z,i} = 0$ and $z_i = 0$.

Let $\boldsymbol{S}_p$ be a set of points $\boldsymbol{r}_i$, and $\boldsymbol{S}_n$ be a set of points $\boldsymbol{r}_j$ defining positions of nodes of a given truss, $j = 1, 2, \ldots, n_n$. Obviously, for non-zero forces $\boldsymbol{P}_i$ we shall assume $\boldsymbol{S}_n \supseteq \boldsymbol{S}_p$ and $n_n \geq n_p$. In other words, a set of nodes of a truss considered here includes all points $\boldsymbol{r}_i$ but can also include some additional nodes not loaded by external forces.

Assume that a truss transmitting a set of forces $\boldsymbol{P}_i$ is composed of n_e bars. Let T_k be a member axial force, l_k a length, and A_k a cross-sectional area of kth bar; $k = 1, 2, \ldots, n_e$. Recall the principle of virtual work, (cf. Eq. (1.61))

$$\sum_k T_k \bar{\Delta}_k = \sum_i \boldsymbol{P}_i \cdot \bar{\boldsymbol{u}}_i \tag{2.55}$$

where $\bar{\boldsymbol{u}}_i$ and $\bar{\Delta}_k$ stand for the vector of virtual displacement of ith node and the elongation of kth bar, respectively. Note that due to the absence of supports any set of vectors $\bar{\boldsymbol{u}}_i$ is kinematically admissible. This set determines the elongations $\bar{\Delta}_k$ of bars (cf. Eq. (1.62)). Now, let us consider the continuous virtual displacement field proposed by Maxwell, representing a uniform dilatation of the full space:

$$\bar{\boldsymbol{u}}(\boldsymbol{r}) = \bar{\varepsilon}\boldsymbol{r} \tag{2.56}$$

where $\bar{\varepsilon}$ is a fixed small positive or negative constant. Thus, the distance between two points A and B upon deformation is equal to

$$\begin{aligned} l_{AB} + \bar{\Delta}_{AB} = {} & |(\boldsymbol{r}_B + \bar{\boldsymbol{u}}_B) - (\boldsymbol{r}_A + \bar{\boldsymbol{u}}_A)| = |(\boldsymbol{r}_B + \bar{\varepsilon}\boldsymbol{r}_B) - (\boldsymbol{r}_A + \bar{\varepsilon}\boldsymbol{r}_A)| = \\ & |\boldsymbol{r}_B(1+\bar{\varepsilon}) - \boldsymbol{r}_A(1+\bar{\varepsilon})| = |\boldsymbol{r}_B - \boldsymbol{r}_A|(1+\bar{\varepsilon}) = \\ & l_{AB}(1+\bar{\varepsilon}) \end{aligned} \tag{2.57}$$

and $\bar{\Delta}_{AB} = \bar{\varepsilon} l_{AB}$. Thus, the elongation of any bar of any truss immersed in this virtual displacement field is proportional to its length

$$\bar{\Delta}_k = \bar{\varepsilon} l_k \quad \text{for} \quad k = 1, 2, \ldots, n_e \tag{2.58}$$

and does not depend on the orientation of the bar in the space. For this particular displacement field, the elongations of bars can be evaluated independently of the internal structure of a truss and its geometric matrix.

Applying (2.56) and (2.58) in (2.55) we find

$$\bar{\varepsilon} \sum_k T_k l_k = \bar{\varepsilon} \sum_i \boldsymbol{P}_i \cdot \boldsymbol{r}_i \tag{2.59}$$

The above equation, after cancelling $\bar{\varepsilon}$ and splitting bars into ones in tension and compression can be written as

$$\sum_t T_t^+ l_t - \sum_c T_c^- l_c = \sum_i \boldsymbol{P}_i \cdot \boldsymbol{r}_i \tag{2.60}$$

where the first sum is taken for all the ties and the second for all the struts; $t = 1, 2, \ldots, n_t, c = 1, 2, \ldots, n_c$ and $n_t + n_c = n_e$. Note that the sum of scalar products on the right-hand side is independent of the form of the truss and independent of the assumed coordinate system. The right-hand side of (2.60) is called Maxwell's number and will be denoted by $\mathcal{M}$. The number $\mathcal{M}$ is an invariant for the given set of point loads. As an example, consider a new Cartesian coordinate system $\tilde{O}\tilde{x}\tilde{y}\tilde{z}$ with rotated axes and translated origin, so the new coordinates are related to original ones by

$$\tilde{\boldsymbol{r}} = \tilde{\boldsymbol{r}}_o + \boldsymbol{C}\boldsymbol{r} \tag{2.61}$$

where $\tilde{\boldsymbol{r}}_O$ is a translation vector describing position of the origin O and $\boldsymbol{C}$ is an orthogonal matrix representing a rotation (i.e. we have: $\boldsymbol{C}^\top \boldsymbol{C} = \boldsymbol{I}$). Note that we use here a simplified vector notation in which bold letters denote arrays of coordinates rather than strict vectors as in tensor notation. Thus, the coordinates of loads $\boldsymbol{P}_i$ in the new system are defined by

$$\tilde{\boldsymbol{P}}_i = \boldsymbol{C}\boldsymbol{P}_i \tag{2.62}$$

After taking into account (2.60), (2.62) and $(2.54)_1$, it is easy to show that

$$\sum_i \tilde{\boldsymbol{P}}_i^\top \tilde{\boldsymbol{r}}_i = \sum_i \boldsymbol{P}_i^\top \boldsymbol{C}^\top (\tilde{\boldsymbol{r}}_o + \boldsymbol{C}\boldsymbol{r}_i) =$$

$$\left(\sum_i \boldsymbol{P}_i^\top\right) \boldsymbol{C}^\top \tilde{\boldsymbol{r}}_o + \sum_i \boldsymbol{P}_i^\top \boldsymbol{C}^\top \boldsymbol{C}\boldsymbol{r}_i = \sum_i \boldsymbol{P}_i^\top \boldsymbol{r}, \tag{2.63}$$

which confirms the invariance of $\mathcal{M}$ under translations and rotations.

For fully stressed trusses, all bars are stressed to the limits σ_T and σ_C for tension and compression, respectively. Thus, $T_t^+ = \sigma_T A_t$ and $T_c^- = \sigma_C A_c$, where A_t and A_c denote cross-sectional areas of bars in tension and compression. Consequently,

$$\sum_t T_t^+ l_t - \sum_c T_c^- l_c = \sigma_T \sum_t A_t l_t - \sigma_C \sum_c A_c l_c =$$

$$\sigma_T \sum_t V_t - \sigma_C \sum_c V_c = \sigma_T V_T - \sigma_C V_C, \tag{2.64}$$

where V_T is the volume of all the ties and V_C that of all the struts. Having in mind the definition of the Maxwell number, we conclude that the difference

$$\sigma_T V_T - \sigma_C V_C = \mathcal{M} \tag{2.65}$$

is a constant for any framework designed to equilibrate a given set of point loads in which the stresses in the ties equal σ_T and the stresses in the struts equal σ_C.

The total volume of the truss is given by

$$V = V_T + V_C \tag{2.66}$$

and using (2.65) it can be written alternatively as

$$V = V_T\left(1 + \frac{\sigma_T}{\sigma_C}\right) - \frac{\mathcal{M}}{\sigma_C} = V_C\left(1 + \frac{\sigma_C}{\sigma_T}\right) + \frac{\mathcal{M}}{\sigma_T} \tag{2.67}$$

The above formula implies that of all the trusses of bars being fully stressed and equilibrating a given system of forces $\boldsymbol{P}_i$ the truss of minimal weight is characterized by the smallest volume of the bars in tension and the smallest volume of the bars in compression. In particular, if there exists a truss equilibrating the given system of loads such that all its bars are in tension up to a given limit σ_T (or in compression, with stresses equal $-\sigma_C$), then this truss has the smallest volume. Usually, for a given system of loads one can construct a great number of such trusses (see Cox (1965)), but all of them have the same volume; in case of trusses in tension, this volume equals $V_T = \mathcal{M}/\sigma_T$; the least volume trusses in compression have the volume equal to $V_C = -\mathcal{M}/\sigma_C$.

Last, let us consider the sets of forces for which $\mathcal{M} = 0$. Then, $\sigma_T V_T = \sigma_C V_C$, and in addition, if simultaneously $\sigma_T = \sigma_C$, then the volumes of the bars in tension and compression are the same—$V_T = V_C$. This property may be used while computing the volumes of the optimal trusses or their parts, cf. Sect. 4.5.5, Eq. (4.258).

Let us come back to the problem (2.51) related to minimization of the volume for the case of $\sigma_T \neq \sigma_C$. Let us note that

$$\sum_{k=1}^{e}\left(\frac{\sigma_0}{\sigma_T}T_k' + \frac{\sigma_0}{\sigma_C}T_k''\right)l_k = \frac{1}{2}\left(\frac{\sigma_0}{\sigma_T} + \frac{\sigma_0}{\sigma_C}\right)\sum_{k=1}^{e}(T_k' + T_k'')l_k + \frac{1}{2}\left(\frac{\sigma_0}{\sigma_T} - \frac{\sigma_0}{\sigma_C}\right)\mathcal{M} \tag{2.68}$$

If the loads applied are known the Maxwell number $\mathcal{M}$ is independent of the design. In this case, the quantity (2.51) may be expressed by

$$\tilde{Z}_2 = \frac{1}{2}\left(\frac{\sigma_0}{\sigma_T} + \frac{\sigma_0}{\sigma_C}\right)Z_2 + \frac{1}{2}\left(\frac{\sigma_0}{\sigma_T} - \frac{\sigma_0}{\sigma_C}\right)\mathcal{M} \tag{2.69}$$

where Z_2 is given by (2.31), which corresponds to the case of $\sigma_T = \sigma_C$. Therefore, the Maxwell theorem provides a shortcut for deriving the optimal volume of Michell structures equilibrating a given system of forces, for which the Maxwell number $\mathcal{M}$ is an invariant of the design. In the first step, one computes the volume for the case of $\sigma_T = \sigma_C$ and arrives at the result of the form $V = \bar{V} Pl \frac{2}{\sigma_0}$. In the second step, one computes the volume of the optimal truss corresponding to the unequal permissible stresses by the formula

$$V = \bar{V} Pl \left(\frac{1}{\sigma_T} + \frac{1}{\sigma_C} \right) + \frac{1}{2} \left(\frac{1}{\sigma_T} - \frac{1}{\sigma_C} \right) \mathcal{M} \tag{2.70}$$

In case of $\mathcal{M} = 0$, the formula above simplifies appropriately and both the volumes differ in a scaling factor.

Let us stress that the formula (2.70) holds only if all the forces applied (i.e. including reactions) are given; such trusses are said to be externally statically determinate. In particular, this formula is wrong for all the cantilever problems, including the original Michell's structure supported on a circle, to be discussed in Sect. 4.6. Thus, the formulae of Sect. 4.6 confined to the case of $\sigma_T = \sigma_C$ cannot be easily extended to the case of unequal permissible stresses in tension and compression.

2.3 On Geometric Properties of the Least Volume Symmetric Cantilever Trusses of Prescribed Number of Nodes

In the optimum design problems of trusses considered in the previous sections, the positions of nodes have been fixed, while the stiffnesses EA_k were the only design variables. In specific problems, the optimum design methods discussed can be enhanced by the graphic statics methods, thus making it possible to tackle the problem of optimal position of nodes.

Consider the problem of minimization of the volume of the 8-bar truss of Fig. 2.1a, discussed in Prager (1977). The permissible stresses in tension and compression are equal: $\sigma_T = \sigma_C = \sigma_0$. The positions of pin support at B and B' as well as of the node A, where the force $2P$ is applied, are fixed. Assume that the mid-perpendicular to BB' is a symmetry axis of the truss; the position of the nodes D and C are determined by the angles α_1, β, α_2.

Let us construct the *force diagram*. We start from the construction of the triangle of forces: $\boldsymbol{R}_B$, $\boldsymbol{R}_{B'}$, $2\boldsymbol{P}$ of vertices b, b', a, cf. Fig. 2.1b. From the point a, we draw a line parallel to BC; then from the point b, we draw a line parallel to BD. We find the point f at the intersection. Then, we draw a line crossing f and parallel to DC. We construct the point e. Thus, we build the *force diagram* of nodes: b, b', f', a, f, e. The sides of this diagram are proportional to the magnitudes of the axial forces in the bars of the truss of Fig. 2.1a.

The nodes of the *force diagram* may be treated as nodes of a truss subject to a horizontal force at a and pin supported at b, b'. This truss will be called an *adjoint*

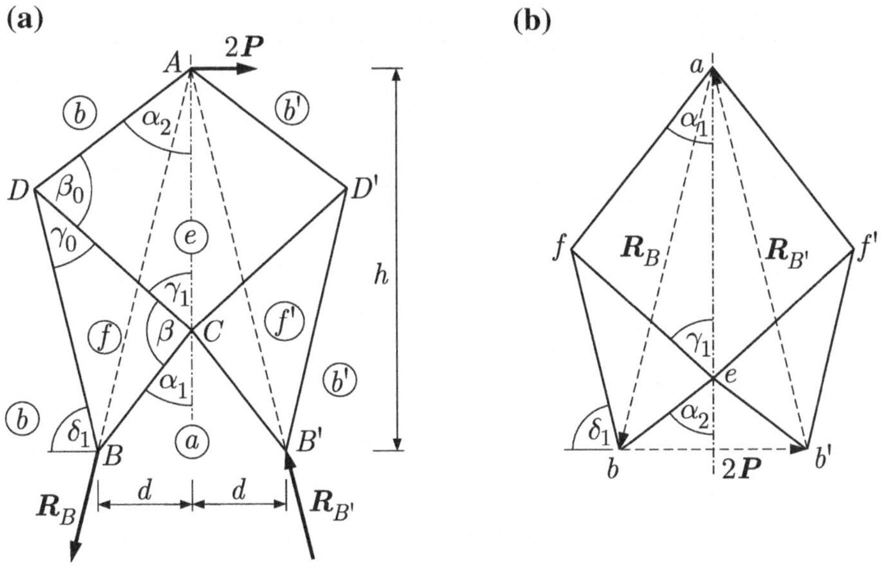

Fig. 2.1 Problem formulation and notation of the graphic statics method (**a**); the *force diagram* (**b**)

truss for the truss in Fig. 2.1a. For the adjoint truss, the force diagram assumes the shape of the original truss.

In general, the force diagram does not resemble the given truss and no such a comparison is possible, see Beghini et al. (2014). The class of cantilevers discussed here is an exception discovered by Prager (1977).

In case of $\alpha_1 = \alpha_2 = \alpha$, the given truss and its force diagram have the same shapes. Such a truss is said to be *self-adjoint*. We conjecture that the optimal truss is characterized by this property. The condition of self-adjointness reduces the number of design variables; the volume of the fully stressed truss becomes a function $V = V(\alpha, \beta)$. The conditions

$$\frac{\partial V}{\partial \alpha} = 0, \qquad \frac{\partial V}{\partial \beta} = 0 \tag{2.71}$$

lead to $\beta = \pi/2$ and to an optimal value of α satisfying the condition

$$\frac{\cot 2\alpha}{\sin^2 \alpha} + \cot \alpha = \frac{h}{d} \tag{2.72}$$

Thus, $\beta = \pi/2$ independently of the ratio h/d. Hence, $\beta_0 = \pi/2$. Moreover, the optimal value of $\gamma_0 = \angle BDC$ equals $\gamma_0 = 2\alpha$, α being given by (2.72). The values of angle α corresponding to various ratios d/h are set up in Table 2.1.

The above results found by Prager (1977) have made it possible to develop numerical codes producing optimal cantilever trusses of finite number of nodes, see Fig. 2.2 showing Mazurek et al. (2011) result.

Table 2.1 The values of angle α in terms of d/h ratio

d/h	0.1	0.2	0.3	0.4	0.5	0.6	0.7	0.8	0.9	1.0
α	21.887°	27.885°	32.012°	35.152°	37.646°	39.677°	41.359°	42.772°	43.971°	45°

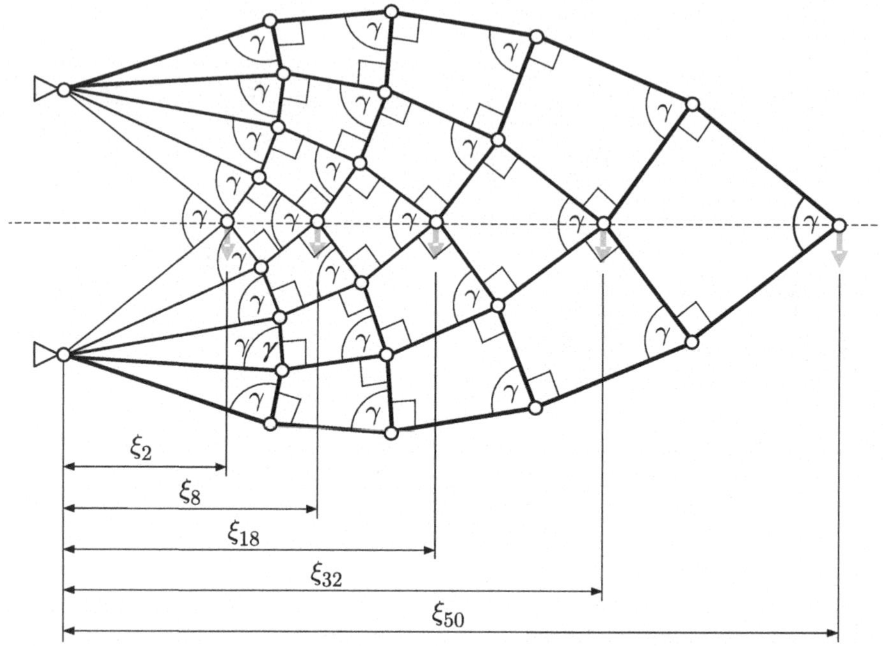

Fig. 2.2 Optimum solutions for 2, 8, 18, 32, 50-element structures (by the courtesy of SMO; Mazurek et al. (2011))

Although it is straightforward to obtain the optimum structure for given value of angle γ (see Fig. 2.2), the process of finding the value of an angle γ which gives a cantilever of a given length ξ requires an iterative procedure. The solutions with 2, 8, 18, 32 and 50—bar trusses corresponding to the cantilevers of subsequent lengths are depicted in Fig. 2.2. The total number of members in these solutions equals $2(n_0)^2$, where n_0 is the number of elements connected to one of the supports.

The same method can be extended to solve the three-force problem for any set of three balanced forces, see Mazurek (2012).

2.4 The Least Compliant Beams and Systems of Beams

This section concerns the minimization of compliance of systems of beams (see Sect. 1.5) with the cost condition expressed by

$$\int EJ(x)dx = \Lambda \tag{2.73}$$

where the integral is taken over all the bars of the structure; $EJ(x) = EJ_k(x)$ for the kth bar. Thus, the bending stiffnesses may vary along the bars' axes. If one assumes that the heights of the bars h_k are constant and the widths $b_k(x)$ may vary, then $J_k(x) = \alpha b_k(x)$, α being a proportionality coefficient. Then, the areas of the cross sections $A_k(x)$ are proportional to $b_k(x)$, hence the volume of the k th bar is proportional to the integral of $b_k(x)$ along the given bar. Then, the condition (2.73) determines the volume of the structure.

2.4.1 The Continuum Formulation of the Optimum Design Problem

The problem of statics of systems of beams has been the subject of Sect. 1.5. Assume now that the load is concentrated at the nodes. Then, the equilibrium equation of the system of beams has the form

$$\int M\varkappa(\bar{w})dx = \sum_{j=1}^{s} Q_j \bar{q}_j \quad \forall\, \bar{w} \text{ kinematically admissible} \tag{2.74}$$

where the integral is taken over all bars; the right-hand side is the virtual work of the load. The component Q_j of the load represents either a concentrated moment of a vector orthogonal to a beam of the structure (and then $\bar{q}_j$ is a virtual angle of rotation of the beam cross section along this direction), or Q_j stands for the point load at a given node, orthogonal to the plane (and then $\bar{q}_j$ represents the virtual deflection of this node). The quantities $\bar{q}_j$ correspond to arbitrary, kinematically admissible virtual deflections $\bar{w}$ of the beams of the structure. We remember that a function $\bar{w}$ is smooth beam-wise. The set of bending moments M satisfying the equilibrium equation (2.74) is denoted by Σ_Q.

According to Castigliano's theorem (see Sect. 1.8), the compliance of the system of beams is expressed by

$$\wp(EJ) = \min_{M\in\Sigma_Q} \int \frac{M^2}{EJ} dx \tag{2.75}$$

We consider the problem of designing the least compliant system of beams with the cost condition (2.73). This problem reads

$$\breve{\wp} = \min\left\{\wp(EJ) \mid EJ \geq 0 \quad \text{such that} \quad \int EJdx = \Lambda\right\} \tag{2.76}$$

Substitution of (2.75) gives

$$\breve{\wp} = \min_{M \in \Sigma_Q} Y(M) \tag{2.77}$$

with

$$Y(M) = \min\left\{ \int \frac{M^2}{EJ} dx \mid \int EJdx = \Lambda, \quad EJ \geq 0 \right\} \tag{2.78}$$

To solve the problem above we shall make use of

Lemma 2.4.1 *Let Ω be an open set in $\mathbb{R}^n$ and $F(x) \geq 0$ in Ω. The solution to the problem*

$$\mathcal{J} = \min\left\{ \int_\Omega \frac{F(x)}{u(x)} dx \mid u \geq 0, \quad \int_\Omega udx = \Lambda \right\} \tag{2.79}$$

has the form

$$\hat{u}(x) = \Lambda \frac{\sqrt{F(x)}}{\displaystyle\int_\Omega \sqrt{F(x)}dx} \tag{2.80}$$

and the minimal value of (2.79) is expressed by

$$\mathcal{J} = \frac{1}{\Lambda}(Z_F)^2 \tag{2.81}$$

where

$$Z_F = \int_\Omega \sqrt{F(x)}dx \tag{2.82}$$

This lemma will be used several times, that is why it is appropriate to show its proof. The lemma is a continuum counterpart of Lemma 2.1.1.

The proof.
Let us write down the quantity $Z = Z_F$ in the form

$$Z = \int_\Omega \sqrt{\frac{F(x)}{u(x)}} \sqrt{u(x)}dx \tag{2.83}$$

and assume that $u(x) > 0$. Let us estimate the integral (2.83) by using Cauchy–Schwarz inequality

$$Z \leq \sqrt{\int_\Omega \frac{F(x)}{u(x)} dx} \sqrt{\int_\Omega u(x)dx} \tag{2.84}$$

Let us assume now that u satisfies the condition of its integral being equal Λ. Hence

$$\int_\Omega \frac{F(x)}{u(x)} dx \geq \frac{1}{\Lambda} Z^2 \tag{2.85}$$

Note now that the equality can be attained by taking $u = \hat{u}$ given by (2.80). This ends the proof.
Lemma 2.4.1 applies to (2.78) since $F = M^2 \geq 0$ and $u = EJ \geq 0$. Thus

$$Y(M) = \frac{1}{\Lambda}(Z_M)^2 \tag{2.86}$$

where

$$Z_M = \int |M| dx \tag{2.87}$$

and the integral runs over all the beams.

Let us substitute the results (2.86), (2.87) into (2.77) to get

$$\breve{\wp} = \frac{1}{\Lambda} \min_{M \in \Sigma_Q} \left(\int |M| dx \right)^2 \tag{2.88}$$

Since the function $x \to x^2$ is an increasing function and the values of Z_M are non-negative, the optimal compliance can be expressed by

$$\breve{\wp} = \frac{1}{\Lambda} Z^2 \tag{2.89}$$

where $Z = Z_2$ and

$$Z_2 = \min \left\{ \int |M| dx \mid M \in \Sigma_Q \right\} \tag{2.90}$$

Thus, we have arrived at the problem above, which has been the subject of intense studies in the past, since 1950's, see Rozvany (1976) and Prager (1985), where a rich bibliography may be found.

The problem (2.90) should be considered parallel with its dual. The condition $M \in \Sigma_Q$ is equivalent to (2.74), hence (2.90) may be rewritten as

$$Z = \min \left\{ \int |M| dx + \max_{\substack{\bar{w} \text{ kinematically} \\ \text{admissible}}} \left\{ \sum_{j=1}^{s} Q_j \bar{q}_j - \int M \varkappa(\bar{w}) dx \right\} \right\} \tag{2.91}$$

In the formulation above, the field M does not undergo the condition of being statically admissible. Now, we change the order of the min and max operations (which can be justified) and arrive at

$$Z = \max \left\{ \sum_{j=1}^{s} Q_j \bar{q}_j + R(\bar{w}) \mid \begin{array}{c} \bar{w} \text{ kinematically} \\ \text{admissible} \end{array} \right\} \tag{2.92}$$

where

$$R(\bar{w}) = \min \left\{ \int [|M| - M\varkappa(\bar{w})]\, dx \mid \begin{array}{c} M \text{ defined on the} \\ \text{system of beams} \end{array} \right\} \tag{2.93}$$

Let $M(x) = m(x)\alpha(x)$, $\alpha(x) = |M(x)|$, $|m(x)| = 1$. We write (2.93) in the form

$$-\mathcal{R}(\bar{w}) = \max \left\{ \int \alpha(x)[m(x)\varkappa(\bar{w}) - 1]dx \mid \alpha \text{ arbitrary and } m(x) \in \{-1, 1\} \right\} \tag{2.94}$$

The above quantity vanishes if $|\varkappa(\bar{w}(x))| \leq 1$ with $\bar{w}$ being smooth almost everywhere on the beams, and is $+\infty$ otherwise. Thus, $Z = Z_1$, where

$$Z_1 = \max \left\{ \sum_{j=1}^{s} Q_j \bar{q}_j \mid \begin{array}{c} \bar{w} \text{ kinematically admissible,} \\ |\varkappa(\bar{w}(x))| \leq 1 \text{ a.e.} \end{array} \right\} \tag{2.95}$$

The equality $Z_1 = Z_2$ means vanishing the duality gap. Let the solutions to the problems (2.90) and (2.95) be denoted by: $M = \breve{M}$, $\bar{w} = \breve{w}$, $\bar{q} = \breve{q}$. The optimality condition $|\varkappa(\breve{w})| = 1$ is satisfied there, where $\breve{M} \neq 0$.

2.4.2 *Construction of the Optimum Solution and Its Properties*

The problem of minimum compliance (2.76) reduces to the auxiliary problems (2.90) and (2.95). The optimal layout of stiffnesses $E\breve{J}$ is given by (2.80) or

$$E\breve{J}(x) = \Lambda \frac{|\breve{M}(x)|}{\int |\breve{M}(x)|dx} \tag{2.96}$$

where the integral is taken over all the beams. We conclude that the maximal axial stresses (or the σ_x stresses in fibres near the lower and upper faces of the beams) will be made uniform in the whole system of beams. Let us substitute the result (2.96) into (2.75); the optimum compliance is expressed by

$$\breve{\wp} = \left[\min_{M \in \Sigma_Q} \int \frac{M^2}{|\breve{M}|} dx \right] \frac{1}{\Lambda} \int |\breve{M}(x)|dx \tag{2.97}$$

We shall show now that $M = \breve{M}$ is the minimizer in the problem above. To this end, by using Cauchy–Schwarz inequality, we estimate

$$\int |M|dx = \int \frac{|M|}{\sqrt{|\breve{M}|}}\sqrt{|\breve{M}|}dx \leq \sqrt{\int \frac{M^2}{|\breve{M}|}dx}\sqrt{\int |\breve{M}|dx} \tag{2.98}$$

hence

$$\int \frac{M^2}{|\breve{M}|}dx \geq \frac{1}{\int |\breve{M}|dx}\left(\int |M|dx\right)^2 \tag{2.99}$$

Since $\breve{M}$ is the minimizer of (2.90), we have

$$\int |\breve{M}|dx \leq \int |M|dx \tag{2.100}$$

Therefore,

$$\int \frac{M^2}{|\breve{M}|}dx \geq \int |\breve{M}|dx \tag{2.101}$$

and this equality is attained for $M = \breve{M}$. This means that $M = \breve{M}$ is the minimizer of (2.97). We conclude that the diagram of the bending moments in the optimal system of beams coincides with the diagram of bending moments being the solution to the auxiliary problem (2.90), or to the problem (2.115) derived later. This means that according to (2.96), there holds: $|\breve{\varkappa}| = \text{const}$ in the optimum system of beams. The deflection $\breve{w}$ of the optimum structure is composed of polynomials of degree 2. Let us note that at the points where $\breve{M}$ vanishes, the optimal stiffness $E\breve{J}$ vanishes as well, hence these points of contraflexure may be interpreted as hinges; these points are also points of inflection of the deflection curve. Just the appearance of hinges makes a chance for statical determinacy of the optimum structure. Bołbotowski (2018) proved recently that at least one optimal system of beams is statically determinate.

Remark 2.4.1 If the cross sections of beams are rectangular of width $b(x)$ and height $h(x)$, then $J = bh^3/12$. Thus, the condition (2.73) may be interpreted as the condition of a fixed volume and the problem (2.76) may be interpreted as the problem of designing the width of beams under the condition of its height being fixed.

On the other hand, if $h = h(x)$ is the design variable and the width b is kept fixed, then the condition (2.73) cannot be interpreted as the condition which fixes the volume of the structure.

Let us try to replace (2.73) by

$$b\int h(x) = V_0$$

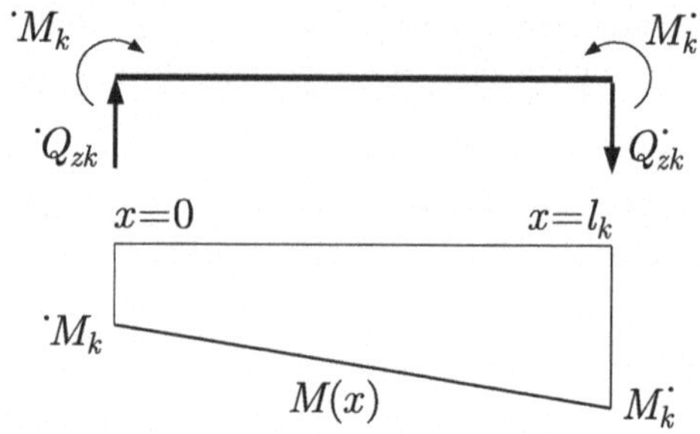

Fig. 2.3 The sign convention for the bending moments and transverse forces

where V_0 is a given volume and reformulate the problem (2.76) as below

$$\min\left\{\min_{M\in\Sigma_Q}\int\frac{M^2}{h^3}dx \mid h\geq 0 \text{ such that } \int hdx=\frac{V_0}{b}\right\}$$

Unfortunately, the problem above is badly posed, as noted already in 1960's, see Kozłowski and Mróz (1969). This is a warning: not all rationally posed engineering problems may be solved.

2.4.3 *The Algebraic Approach*

Assume, as previously, that the load is concentrated at the nodes. Then, the field M at each beam is expressed by a polynomial of degree 1. On each beam, the diagram of M has the form

$$M(x)=\left(1-\frac{x}{l_k}\right){}^{\bullet}M_k+\frac{x}{l_k}M_k^{\bullet} \tag{2.102}$$

where x parameterizes the neutral axis of a beam; $0\leq x\leq l_k$ and

$${}^{\bullet}M_k=M(x=0),\qquad M_k^{\bullet}=M(x=l_k) \tag{2.103}$$

The positive ordinates of M lie at the lower side of beams, see Fig. 2.3.

The bending moments acting at both the ends are accompanied by the transverse forces at both the ends:

$${}^{\bullet}Q_{zk}=Q_z(x=0),\qquad Q_{zk}^{\bullet}=Q_z(x=l_k) \tag{2.104}$$

We remember that the span load is absent.

Let us define a two-argument function

$$\varrho(x_1, x_2) = \begin{cases} \frac{1}{2}(|x_1| + |x_2|) & \text{if } x_1 x_2 \geq 0 \\ \frac{1}{2}\frac{(x_1)^2 + (x_2)^2}{|x_1| + |x_2|} & \text{otherwise} \end{cases} \tag{2.105}$$

If the diagram M on the kth beam is of the form (2.102), then

$$\int_0^{l_k} |M| dx_k = l_k \varrho({}^{\bullet}M_k, M_k^{\bullet}) \tag{2.106}$$

Let us note that the function $\boldsymbol{x} \to \varrho(\boldsymbol{x}) = \varrho(x_1, x_2)$ has all the properties of a norm in $\mathbb{R}^2$, or

$$\begin{aligned} &\text{(i)} \quad \varrho(\lambda \boldsymbol{x}) = |\lambda| \varrho(\boldsymbol{x}), \quad \lambda \in \mathbb{R} \\ &\text{(ii)} \quad \varrho(\boldsymbol{x}) \geq 0 \text{ and } \varrho(\boldsymbol{x}) = 0 \Rightarrow \boldsymbol{x} = (0, 0) \\ &\text{(iii)} \quad \varrho(\boldsymbol{x} + \boldsymbol{y}) \leq \varrho(\boldsymbol{x}) + \varrho(\boldsymbol{y}) \quad \boldsymbol{x}, \boldsymbol{y} \in \mathbb{R}^2 \end{aligned} \tag{2.107}$$

Let us define the norm dual to ϱ by

$$\varrho^{\star}(\boldsymbol{y}) = \max_{\boldsymbol{x} \neq 0,\ \boldsymbol{x} \in \mathbb{R}^2} \frac{\boldsymbol{y} \cdot \boldsymbol{x}}{\varrho(\boldsymbol{x})} \tag{2.108}$$

or

$$\varrho^{\star}(\boldsymbol{y}) = \max_{\varrho(\boldsymbol{x})=1,\ \boldsymbol{x} \in \mathbb{R}^2} \boldsymbol{y} \cdot \boldsymbol{x} \tag{2.109}$$

Bołbotowski (2018) derived the explicit form of the latter norm

$$\varrho^{\star}(\boldsymbol{y}) = |y_1 - y_2| + \sqrt{(y_1 - y_2)^2 + (y_1 + y_2)^2} \tag{2.110}$$

Indeed, the function $\boldsymbol{y} \to \varrho^{\star}(\boldsymbol{y})$ has all the properties (2.107) of a norm in $\mathbb{R}^2$. Moreover,

$$\boldsymbol{x} \cdot \boldsymbol{y} \leq \varrho(\boldsymbol{x}) \varrho^{\star}(\boldsymbol{y}) \quad \forall\, \boldsymbol{x}, \boldsymbol{y} \in \mathbb{R}^2 \tag{2.111}$$

It is now worth showing the contour graphs : $\varrho(x_1, x_2) = \text{const}$ and: $\varrho^{\star}(x_1, x_2) = \text{const}$, see Fig. 2.4.

If the span load is absent, the problem (2.90) reduces to the minimization problem over a set of finite number of elements. Instead of the unknown field M, we deal now with the values ${}^{\bullet}M_k$, $M_k^{\bullet}$ at each beam. Drawing on the knowledge from Sect. 1.2.2, we note that the equilibrium equation (2.74) reduces to the algebraic equations

$$\sum_k \left({}^{\bullet}M_k\, {}^{\bullet}\bar{\chi}_k + M_k^{\bullet} \bar{\chi}_k^{\bullet}\right) = \sum_{j=1}^{s} Q_j \bar{q}_j \qquad \forall\ \bar{\boldsymbol{q}} \in \mathbb{R}^s \tag{2.112}$$

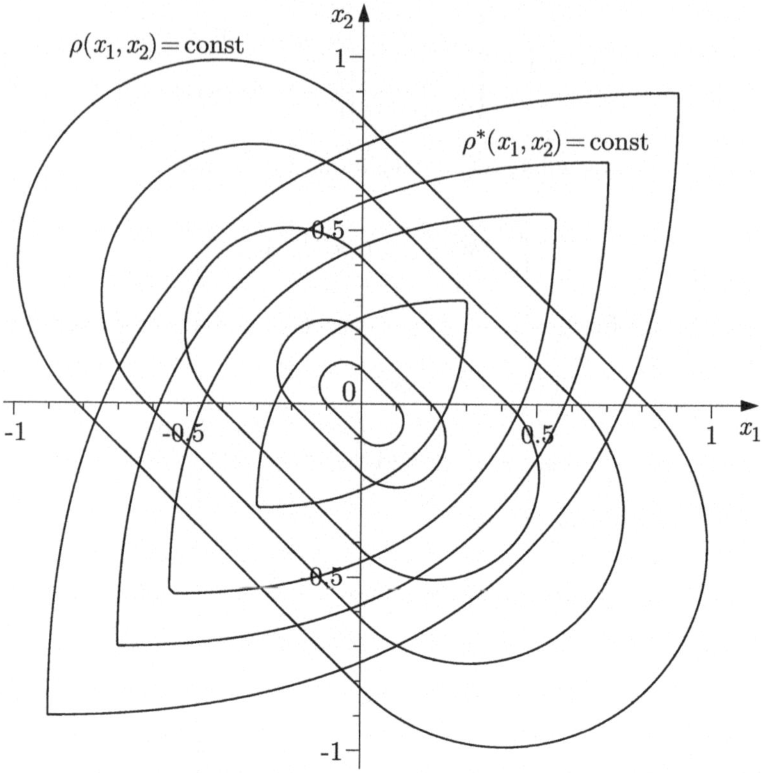

Fig. 2.4 Juxtaposition of the contours: $\varrho(x_1, x_2) = \text{const}$, $\varrho^\star(x_1, x_2) = \text{const}$

where, see (1.18)

$$\dot{}\bar{\chi} = \dot{}\boldsymbol{B}\bar{\boldsymbol{q}} \qquad \bar{\chi}^{\cdot} = \boldsymbol{B}^{\cdot}\bar{\boldsymbol{q}} \tag{2.113}$$

Equations (2.112), (2.113) imply

$$(\dot{}\boldsymbol{B})^{\top}\dot{}\boldsymbol{M} + (\boldsymbol{B}^{\cdot})^{\top}\boldsymbol{M}^{\cdot} = \boldsymbol{q} \tag{2.114}$$

see (1.23). Thus, the condition $M \in \Sigma_Q$ is equivalent to (2.114). By using (2.106), we write (2.90) in the form $Z = Z_2$, where

$$Z_2 = \min\left\{\sum_{k=1}^{e} l_k \varrho(\dot{}\boldsymbol{M}_k, \boldsymbol{M}_k^{\cdot}) \mid (\dot{}\boldsymbol{M}, \boldsymbol{M}^{\cdot}) \text{ satisfying (2.114)}\right\} \tag{2.115}$$

and in this manner the problem (2.90) has been changed into an algebraic problem.

Now, we shall form the problem dual to (2.115). Due to the equivalence of problems (2.112) and (2.114), let us express the problem (2.115) in the form (the bar over $\boldsymbol{q}$ being omitted)

$$Z_2 = \min\left\{\sum_{k=1}^{e} l_k \varrho({}^{\bullet}M_k, M_k^{\bullet}) + \max_{q\in\mathbb{R}^s}\left\{\sum_{j=1}^{s} Q_j q_j - \sum_{k=1}^{e} l_k({}^{\bullet}M_k {}^{\bullet}\varkappa_k(q) + M_k^{\bullet}\varkappa_k^{\bullet}(q))\right\}\right\} \tag{2.116}$$

with

$$^{\bullet}\varkappa_k(\boldsymbol{q}) = \frac{1}{l_k}{}^{\bullet}\chi_k(\boldsymbol{q}) \qquad \varkappa_k^{\bullet}(\boldsymbol{q}) = \frac{1}{l_k}\chi_k^{\bullet}(\boldsymbol{q}) \tag{2.117}$$

and the relations ${}^{\bullet}\chi_k(\boldsymbol{q})$, $\chi_k^{\bullet}(\boldsymbol{q})$ are given by (1.17). Let us interchange the sequence of operations : min and max, according to the arguments of Sect. 2.1

$$Z_2 = \max_{q\in\mathbb{R}^s}\left\{\sum_{j=1}^{s} Q_j q_j + \mathcal{R}(\boldsymbol{q})\right\} \tag{2.118}$$

where

$$\mathcal{R}(\boldsymbol{q}) = \min\left\{\sum_{k=1}^{e} l_k\left[\varrho({}^{\bullet}M_k, M_k^{\bullet}) - ({}^{\bullet}\varkappa_k(q){}^{\bullet}M_k + \varkappa_k^{\bullet}(q)M_k^{\bullet})\right] \mid ({}^{\bullet}M_k, M_k^{\bullet}) \in \mathbb{R}^2\right\} \tag{2.119}$$

We shall show that

$$\mathcal{R}(\boldsymbol{q}) = \begin{cases} 0 & \text{if } \varrho^{\star}({}^{\bullet}\varkappa_k(\boldsymbol{q}), \varkappa_k^{\bullet}(\boldsymbol{q})) \le 1 \quad k = 1, \ldots, e \\ -\infty & \text{otherwise} \end{cases} \tag{2.120}$$

and this result reduces the problem (2.118) to

$$Z_2 = Z_1 \tag{2.121}$$

where

$$Z_1 = \max_{q\in\mathbb{R}^s}\left\{\sum_{j=1}^{2} Q_j q_j \mid \varrho^{\star}({}^{\bullet}\varkappa_k(\boldsymbol{q}), \varkappa_k^{\bullet}(\boldsymbol{q})) \le 1 \quad k = 1, \ldots, e\right\} \tag{2.122}$$

with

$$ {}^{\bullet}\varkappa_k(\boldsymbol{q}) = \frac{1}{l_k} \sum_{j=1}^{s} {}^{\bullet}B_{kj} q_j \qquad \varkappa_k^{\bullet}(\boldsymbol{q}) = \frac{1}{l_k} \sum_{j=1}^{s} B_{kj}^{\bullet} q_j \tag{2.123} $$

The problems (2.115), (2.122) are similar to problems (2.8), (2.6) in the theory of truss optimization. Yet, now these settings do not reduce to the linear programming problems.

Let us proceed to prove the result (2.120). Let

$$ ({}^{\bullet}M_k, M_k^{\bullet}) = \boldsymbol{x}^{(k)}, \qquad ({}^{\bullet}\varkappa_k, \varkappa_k^{\bullet}) = \boldsymbol{y}^{(k)} \tag{2.124} $$

Let us rewrite (2.119) in the form

$$ \mathcal{R} = \min_{\boldsymbol{x}^{(k)} \in \mathbb{R}^2,\ k=1,\dots,e} \sum_{k=1}^{e} l_k \left[\varrho(\boldsymbol{x}^{(k)}) - \boldsymbol{y}^{(k)} \cdot \boldsymbol{x}^{(k)} \right] \tag{2.125} $$

Let us represent $\boldsymbol{x}^k$ in the form

$$ \boldsymbol{x}^{(k)} = \gamma_k \boldsymbol{e}^{(k)}, \qquad \varrho(\boldsymbol{e}^{(k)}) = 1, \qquad \gamma_k = \varrho(\boldsymbol{x}^{(k)}) \tag{2.126} $$

and consider the problem

$$ -\mathcal{R} = \max_{\gamma_k \geq 0,\ \boldsymbol{e}^{(k)} \in \mathbb{R}^2,\ \varrho(\boldsymbol{e}^{(k)})=1} \sum_{k=1}^{e} l_k \gamma_k \left(\boldsymbol{y}^{(k)} \cdot \boldsymbol{e}^{(k)} - 1 \right) \tag{2.127} $$

Thus,

$$ -\mathcal{R} = \max_{\gamma_k \geq 0} \sum_{k=1}^{e} l_k \gamma_k \left[\max_{\boldsymbol{e}_k \in \mathbb{R}^2,\ \varrho(\boldsymbol{e}_k)=1} (\boldsymbol{y}^{(k)} \cdot \boldsymbol{e}^{(k)}) - 1 \right] \tag{2.128} $$

and, taking into account (2.110) we may write

$$ -\mathcal{R} = \max_{\gamma_k \geq 0} \sum_{k=1}^{e} l_k \gamma_k \left[\varrho^{\star}(\boldsymbol{y}^{(k)}) - 1 \right] \tag{2.129} $$

It is seen that $-\mathcal{R} = \infty$, if $\varrho^{\star}(\boldsymbol{y}^k) > 1$ and $-\mathcal{R} = 0$ if $\varrho^{\star}(\boldsymbol{y}^k) - 1 \leq 0$, which confirms (2.120).

Let $({}^{\bullet}\breve{\boldsymbol{M}}, \breve{\boldsymbol{M}}^{\bullet})$ be the minimizer of (2.115). In the considered case of the loads being concentrated at nodes, the optimal distribution of the bending moment on the kth bar has the form

$$ \breve{M}(x) = \left(1 - \frac{x}{l_k}\right) {}^{\bullet}\breve{M}_k + \frac{x}{l_k} \breve{M}_k^{\bullet} \tag{2.130} $$

The diagram of bending moments thus constructed determines the optimal distribution of the bending stiffness $E\breve{J}$ according to (2.96). If a span load is present then

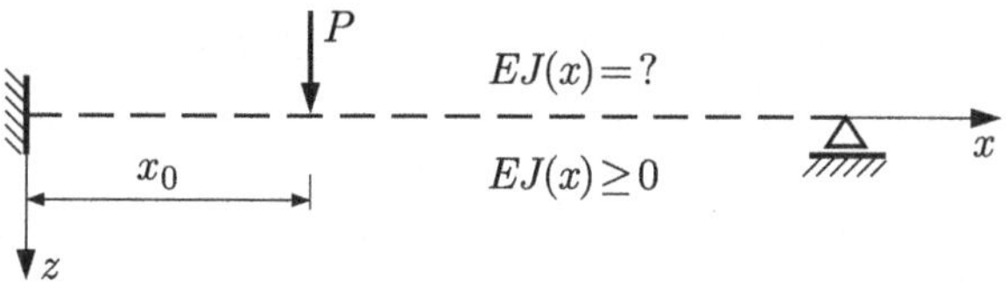

Fig. 2.5 The problem for the given point load construct the stiffest beam of prescribed boundary conditions

the formula (2.130) should be extended to include a term representing the bending moment at the given beam clamped at both the ends.

2.4.4 *Examples of Optimal Beams*

In the simplest case, a system of beams consists of a single beam, which will be the subject of the present section. We shall make use of the formulations (2.90), (2.95). The results of Sect. 2.4.3 will not be used.

2.4.4.1 A Propped Cantilever

Consider a beam clamped at its left end and simply supported at the right end, subject to a vertical point load at point of ordinate x_0, see Fig. 2.5.

The aim is to construct the stiffest beam capable of transmitting the given load to the given supports (or to the left support only), under the cost condition expressed by (2.73). For illustrative reasons, we assume that the stiffness EJ is proportional to the beam's width $t(x)$, while its height is constant. First, we solve the auxiliary problem (2.95). We know that the maximum of the virtual work is attained on $\bar{w}$ being piecewise continuously stitched polynomials of degree 2. The simplest smooth (i.e whose first derivative is continuous) function of this kind and satisfying the prescribed kinematic boundary conditions is $\bar{w}(x) = \breve{w}(x)$ given by

$$\breve{w}(x) = \begin{cases} \frac{1}{2}x^2 & 0 \le x \le \breve{a} \\ -\frac{1}{2}x^2 + 2\breve{a}x - \breve{a}^2 & \breve{a} \le x \le l \end{cases} \tag{2.131}$$

where

$$\breve{a} = \left(1 - \frac{1}{\sqrt{2}}\right) l \tag{2.132}$$

see Fig. 2.6.

The function (2.131) has one inflection point $x = \breve{a}$. It is seen that $|\varkappa(\breve{w})| = 1$, where $\varkappa$ is defined by (1.1). Moreover, $\frac{d\breve{w}}{dx}$ is continuous at $x = \breve{a}$, hence $\breve{w}$ is a function from the space $C^1[0, l]$ and fulfills the kinematic boundary conditions.

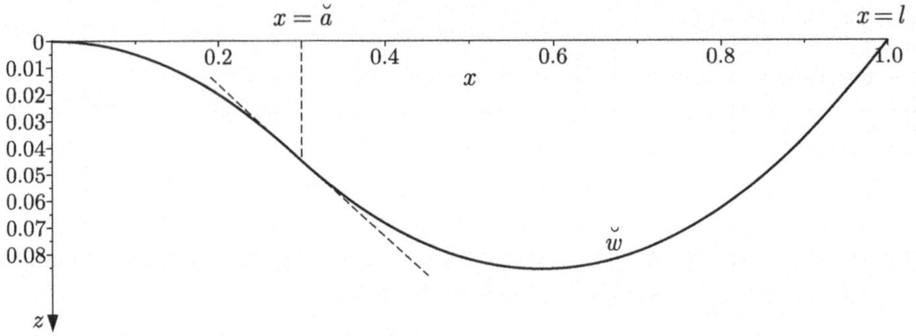

Fig. 2.6 The plot of beam's deflection described by function (2.131)

One can assume that $\bar{w}$ has two inflection points at $x = a$ and $x = b$, $b > a$. Such virtual deflection is determined up to one parameter (say, the parameter a), which will be written as $\bar{w} = \tilde{w}_a(x)$

$$\tilde{w}_a(x) = \begin{cases} \frac{1}{2}x^2 & 0 \leq x \leq a \\ -\frac{1}{2}x^2 + 2ax - a^2 & a \leq x \leq b \\ \frac{1}{2}x^2 - 2(b-a)x + (b^2 - a^2) & b \leq x \leq l \end{cases} \tag{2.133}$$

where

$$\frac{b}{l} = 1 - \sqrt{\left(1 - \frac{a}{l}\right)^2 - \frac{1}{2}} \tag{2.134}$$

The formula (2.133) refers to the case of $a \leq \breve{a}$, $\breve{a}$ given by (2.132). If $a > \breve{a}$ then the second inflection point cannot appear for $x = b \leq l$. A selected function $\tilde{w}_a$ is shown in Fig. 2.7.

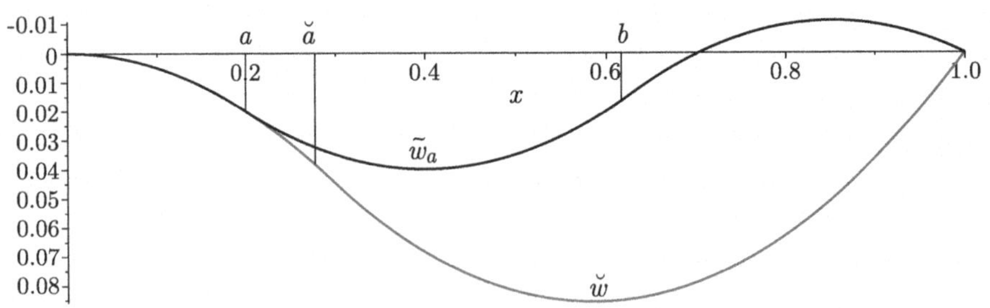

Fig. 2.7 Juxtaposition of $\tilde{w}_a$ (for $a = 0.2$) and $\breve{w}$

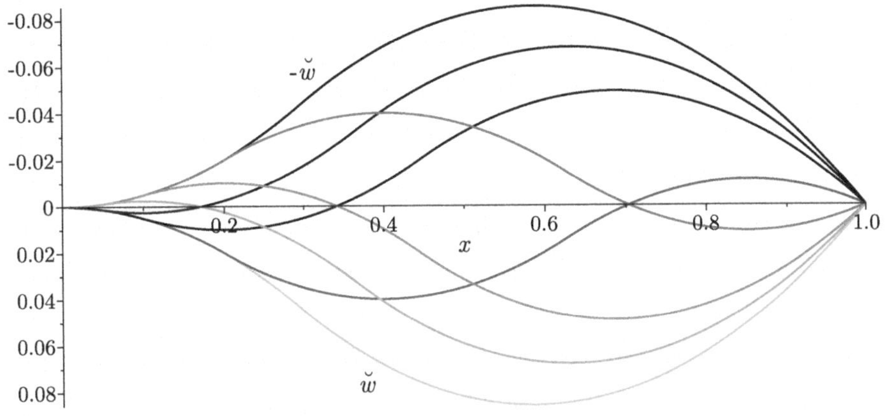

Fig. 2.8 Selected virtual deflections: of one inflection point ($\pm\breve{w}$) and of two inflection points $\tilde{w}_a$ (for $a = 0.05, 0.1, 0.2$)

It is seen that $\tilde{w}_a \in C^1[0, l]$ are kinematically admissible and $|\varkappa(\tilde{w}_a)| = 1$. Moreover, the parameter a varies between 0 and $\breve{a}$. One can prove that all the functions $\tilde{w}_a(x)$, $a \in [0, \breve{a}]$ lie between $\pm\breve{w}(x)$, see Fig. 2.8.

In a similar manner, one can construct a two-parameter family of virtual deflections $\bar{w}$ of class $C^1[0, l]$, satisfying $|\varkappa(\bar{w})| = 1$ and having three inflection points. One can note that such functions lie between $\pm\breve{w}(x)$.

One can prove that the plots of all the kinematically admissible deflections $w(x)$ of class $C^1[0, l]$ satisfying $|\varkappa(w)| = 1$ in $(0, l)$ lie between $\pm\breve{w}(x)$. Thus, the virtual work attains a maximal value on $w = \breve{w}$ given by (2.131). We conclude that the solution to problem (2.95), where $s = 1$, $Q_1 = P$, $\bar{q}_1 = \bar{w}(x_0)$ is $\bar{w} = \breve{w}$. Thus,

$$Z_1 = P\breve{w}(x_0) \tag{2.135}$$

This result can be confirmed by the construction of the solution to the dual problem (2.90). We guess that the minimizer $M = \breve{M}$ is the bending moment field in the beam supported as in Fig. 2.5, having a hinge at the inflection point $x = \breve{a}$, see Fig. 2.9. One can prove this property and show that other statically admissible bending moment fields M cause greater values of the integral (2.87).

If $x_0 < \breve{a}$, the effective domain of the field $\breve{M}$ is the interval $[0, x_0]$. According to (2.96), the optimal bending stiffness $E\breve{J}$ is proportional to $|\breve{M}|$. The optimal width of the beam is shown in Fig. 2.9c. On the interval $x > \breve{a}$, the structure is not necessary. If $x_0 > \breve{a}$, the effective domain of $\breve{M}$ is the whole span of the beam. The optimal thickness $t(x)$ is shown in Fig. 2.9e.

If $x_0 < \breve{a}$, then

$$Z_2 = \int_0^l |\breve{M}(x)|dx = \frac{1}{2}P(x_0)^2 \tag{2.136}$$

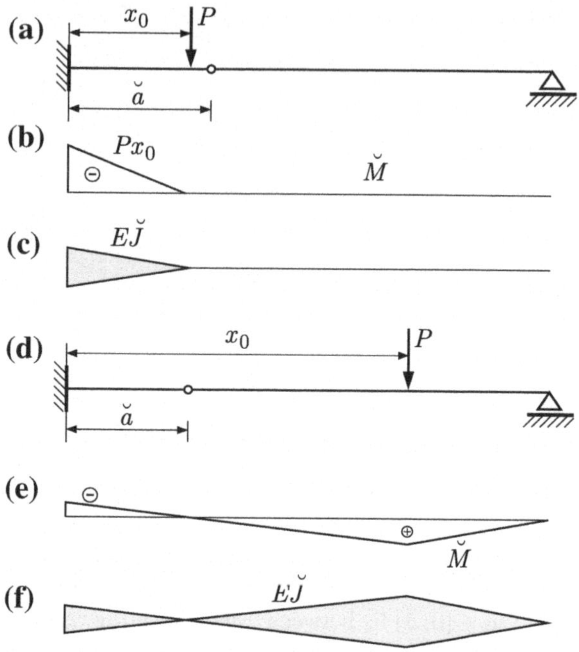

Fig. 2.9 Problem setting (**a**), the solutions for the case: $x_0 < \breve{a}$, see (**b**, **c**), and for the case $x_0 > \breve{a}$, see (**d**, **e**)

and this result coincides with the value of the virtual work of the force P on the displacement $\breve{w}(x_0)$, see (2.131). If $x_0 > \breve{a}$, the computation is more complicated, but also leads to the result of the integral of $|\breve{M}|$ being equal to $P\breve{w}(x_0)$. Thus, we prove the equality $Z_1 = Z_2$, which proves that both solutions are correct: the field $\breve{w}$ is the maximizer of the virtual work and $\breve{M}$ is the minimizer of problem (2.90). This confirms that the optimal $E\breve{J}$ has been constructed correctly.

Let us now consider the deformation of the optimal beam. According to (2.96) $E\breve{J}(x) = \alpha|\breve{M}(x)|$, where $\alpha = \Lambda/Z_2$, Z_2 being given by (2.90), or Z_2 is given by the integral of $|\breve{M}|$ where $\breve{M}$ is the bending moment in the beam of the statical scheme of Fig. 2.9. The displacement at point x_1 is computed by the Maxwell–Mohr rule

$$\bar{1}w(x_1) = \int \bar{M}\varkappa dx$$

where $\varkappa = \breve{M}/E\breve{J}$, hence $|\varkappa| = 1/\alpha$. We note that the sign of $\varkappa$ coincides with the sign of $\bar{M}$.

Let us note that the function thus constructed is a second rank polynomial with respect to x_1, and due to the given boundary condition the field w is proportional to $\breve{w}$, the latter being given by (2.131), see Fig. 2.6. We conclude that the virtual deflection $\breve{w}$ determines the shape of the optimal beam (or its fragment close to the clamped edge, if $x_0 < \breve{a}$. Therefore, the position of the hinge in the optimal beam (loaded in the transverse direction) is independent of the load distribution along its span.

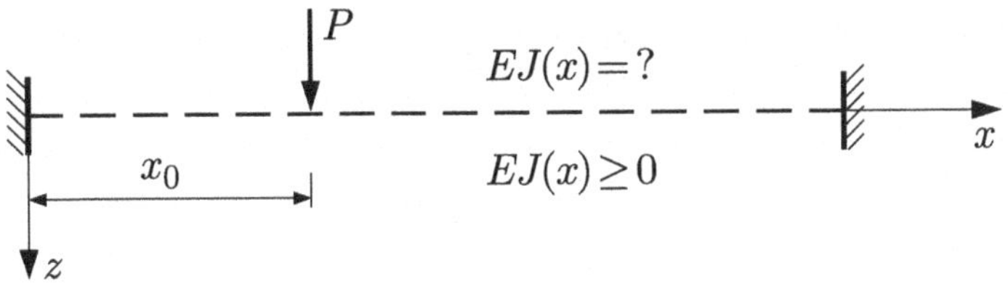

Fig. 2.10 Problem setting

Let us remind that a beam is deflected in the shape of a parabola only if is not subject to shear. Thus, an optimal beam deform the same manner as a beam working in the state of the pure bending.

2.4.4.2 The Beam Possibly Clamped at Both the Ends, Subject to a Vertical Force

The aim is the construction of the stiffest beam clamped at both the ends (or at one end, if a part of beam will turn out to be unnecessary), see Fig. 2.10.

We start from solving problem (2.95). The simplest function w satisfying the given kinematic boundary conditions, the smoothness conditions and the condition $|\varkappa(w)| = 1$ is symmetric with respect to the mid-perpendicular, the formula for $x < l/2$ being

$$\breve{w}(x) = \begin{cases} \frac{1}{2}x^2 & x \in \left(0, \frac{l}{4}\right) \\ -\frac{1}{2}\left(x - \frac{l}{2}\right)^2 + \frac{1}{16}l^2 & x \in \left(\frac{l}{4}, \frac{l}{2}\right), \end{cases} \tag{2.137}$$

see Fig. 2.11.

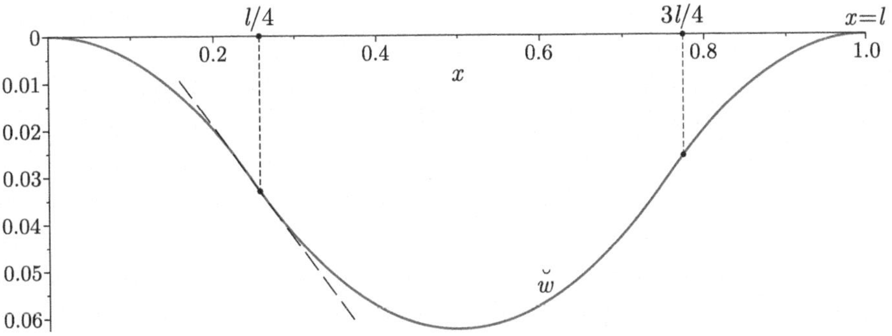

Fig. 2.11 The plot of the optimal deflection $\breve{w}$

Fig. 2.12 Construction of optimal design of a beam of two clamped ends

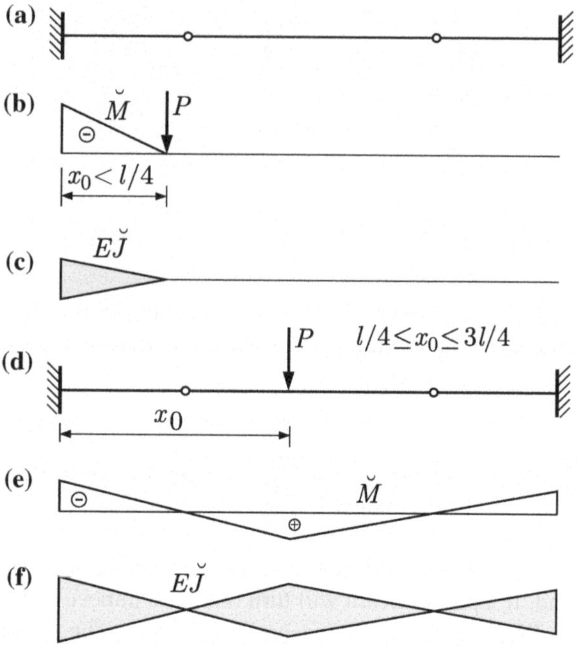

The function $\breve{w}$ has two inflection points ($x = l/4, x = 3l/4$). One can construct a one-parameter family of functions describing the virtual deflections having three inflection points, and, in general, two-parameter functions of required properties having n inflection points. As in the previous example, all these functions lie between $\pm\breve{w}(x)$. Thus, the virtual work $Pw(x_0)$ assumes a maximal value if one chooses $w = \breve{w}$. Then, $Z_1 = P\breve{w}(x_0)$, where $\breve{w}$ is given by (2.137). This result can be confirmed by the statical consideration. The inflection points of the curves of Fig. 2.11 determine the position of the hinges; introduction of these hinges leads to a simply supported beam between the inflection points, cf. Fig. 2.12.

Consider the case of $x_0 < l/4$ (or $x_0 > 3l/4$), then $\breve{M} \neq 0$ only for $x < x_0$ (or $x > x_0$). If $x_0 < l/4$, at the right-hand side of the point $x = x_0$ the material is not necessary, see Fig. 2.12c, where the width of the optimal beam is shown. If $l/4 < x_0 < 3l/4$, the bending moment $\breve{M}$ is non-zero along the whole beam span, see Fig. 2.12d. The optimal width is proportional to $|\breve{M}|$, see Fig. 2.12e. We omit the computational details leading to the equality

$$\int |\breve{M}| dx = P\breve{w}(x_0) \tag{2.138}$$

equivalent to $Z_1 = Z_2$, hence confirming that both the solutions to the problems (2.90), (2.95) have been correctly constructed.

Remark 2.4.4.1 The examples above refer to single point loads, yet the functions $\breve{w}$ constructed (cf. Figs. 2.6 and 2.11) apply to other vertical loads: the concentrated loads $P_1, \ldots, P_n$ applied at points $x_1, \ldots, x_n$ and to the distributed load $p_z(x)$ directed downwards, i.e. $P_i > 0$, $p_z(x) > 0$. For such a system of loads, we have

$$Z_1 = \sum_{j=1}^{n} P_j \breve{w}(x_j) + \int_0^l p_z(x)\breve{w}(x)dx \tag{2.139}$$

while the diagrams $\breve{M}$ appear in the optimal beams of Figs. 2.9 and 2.12. The optimal bending stiffness $E\breve{J}(x)$ is proportional to $\breve{M}(x)$.

2.4.4.3 The Beam Subject to the Load of Non-equal Sign

If the load is of non-equal sign, like in Fig. 2.13, the problem (2.95) changes its character. The maximizer $\breve{w}$ should be searched among the functions $\bar{w}$ satisfying the conditions of clamping at both the ends and constructed by parabolas of the form: $\pm\frac{1}{2}x^2 + ax + b$ smoothly stitched at the inflection points x_i, $i = 1, \ldots, n$. The number n is not known a priori.

Alternatively, one can attack the problem (2.90) in which the minimizer $\breve{M}$ is searched among statically admissible fields of the bending moments. The problem involves only two design variables which can be chosen as values of $\breve{M}$ at the beam's ends. To find these two parameters, any available programs of nonlinear optimization can be applied.

The solution to the problem of Fig. 2.13 is parameterized by the ratio: P/ql. If $P = 0$ or if $q = 0$, the maximizer of the problem (2.95) is given by $\breve{w}$ defined by (2.137), see Fig. 2.11. Otherwise the maximizer of (2.95) may be characterized by the number of inflection points greater than $n = 2$; see the results in Fig. 2.14 a, b, c, d corresponding to the case of P/ql being: 1/8, 1/4, 1/2, 1, respectively.

In all the cases $n = 3$. These inflection points fix the position of hinges. Consequently, all the optimal beams are geometrically variable. Their dimensions and stiffnesses are ideally suited for the given load only. The optimal bending stiffness $E\breve{J}$ is proportional to $|\breve{M}|$ and the corresponding curvature function $\breve{\varkappa}$ is piecewise constant and assumes values ± 1. The function $\breve{w}$ can be readily recovered by twice integrating $\breve{\varkappa} = \varkappa(\breve{w})$, i.e.

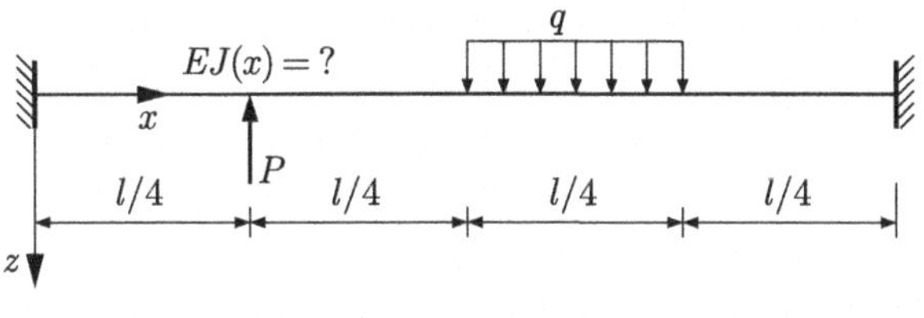

Fig. 2.13 The problem of constructing the stiffest beam under the cost condition (2.73). The load is of non-equal sign

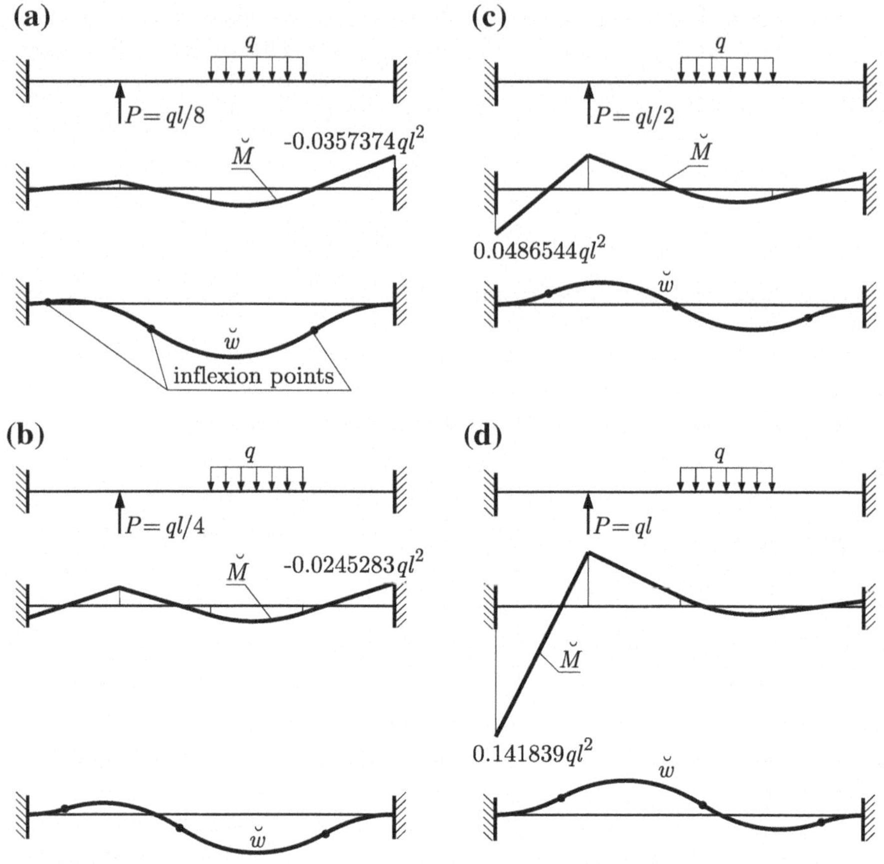

Fig. 2.14 The solutions to the problem of Fig. 2.13: the minimizer $\breve{M}$ of (2.90), the maximizer $\breve{w}$ of (2.95), for the cases of P/ql equal to: $1/8$ (**a**);$1/4$ (**b**), $1/2$ (**c**), 1 (**d**)

$$\breve{w}(x) = -\int_0^x \left(\int_0^\xi \breve{\varkappa}(\eta) d\eta \right) d\xi$$

This function satisfies the conditions of clamping at the left end of the beam. The conditions of clamping at the right end are satisfied approximately, with accuracy corresponding to the accuracy of computing the ordinates of the inflection points. The stiffness vanishes at the hinges. To rationalize the results, one should consider multiple load formulations, see Sect. 2.5.

2.4.5 Least Compliant Systems of Beams. A Simplified Approach for Dense Layouts of Bars

The problems (2.115) and (2.122) of optimum design of systems of beams are not problems of linear programming, which confines the range of their applications to small number of beams. Here, a simplified approach will be outlined, the solution of which reduces to a linear programming problem. This approach refers also to grillages, provided that the torsion effects can be neglected.

Let us come back to the problem (2.76) of minimum compliance of systems of beams with the cost condition (2.73). This problem reduces to problem (2.90), which may be rearranged to the algebraic setting (2.115). The minimizer of problem (2.115) determines the optimal bending stiffnesses according to (2.96).

If a grillage is composed of a great number of bars, then, in majority of bars, the bending moments have equal sign, which simplifies (2.106) to the form

$$\int_0^{l_k} |M| dx_k = l_k d({}^{\cdot}M_k, M_k^{\cdot}) \tag{2.140}$$

with

$$d(x, y) = \frac{1}{2}(|x| + |y|), \qquad x, y \in \mathbb{R} \tag{2.141}$$

cf. (2.105). Thus, the problem (2.115) simplifies to the form

$$Z_2 = \min \left\{ \sum_{k=1}^{e} l_k d({}^{\cdot}\boldsymbol{M}_k, \boldsymbol{M}_k^{\cdot}) \mid ({}^{\cdot}\boldsymbol{M}, \boldsymbol{M}^{\cdot}) \text{ satisfying (2.114)} \right\} \tag{2.142}$$

Let us introduce the aggregated matrices

$$\boldsymbol{\sigma} = \begin{bmatrix} {}^{\cdot}\boldsymbol{M} \\ \boldsymbol{M}^{\cdot} \end{bmatrix}, \qquad \boldsymbol{\mathcal{B}} = \begin{bmatrix} {}^{\cdot}\boldsymbol{B} \\ \boldsymbol{B}^{\cdot} \end{bmatrix} \tag{2.143}$$

and rewrite (2.142) as

$$Z_2 = \min \left\{ \frac{1}{2} \sum_{k=1}^{2e} l_k |\sigma_k| \mid \boldsymbol{\mathcal{B}}^{\top} \boldsymbol{\sigma} = \boldsymbol{Q} \right\} \tag{2.144}$$

where $l_{e+i} = l_i$, $i = 1, \ldots, e$.

This problem has the same mathematical structure as (2.8), hence can also be put in the form similar to (2.31), and then to the form (2.37).

Let us find the norm dual to $d(x, y)$. According to (2.108), we find

$$d^{\star}(a, b) = 2 \max\{|a|, |b|\}, \quad a, b \in \mathbb{R} \tag{2.145}$$

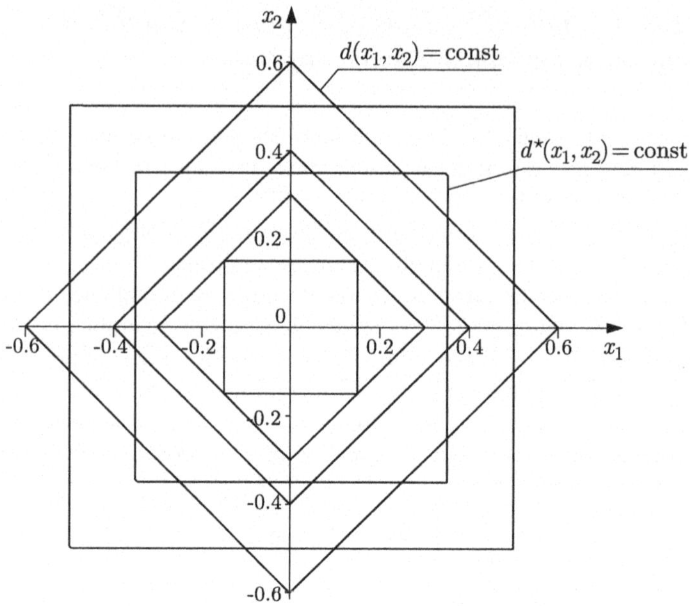

Fig. 2.15 Juxtaposition of the contours of the norms (2.141), (2.145)

The contours of the norms (2.141), (2.145) are shown in Fig. 2.15.

To construct the problem dual to (2.144), we proceed as in Sect. 2.4.2. We obtain the equality $Z_1 = Z_2$ where

$$Z_1 = \max_{q \in \mathbb{R}^s} \left\{ \sum_{j=1}^{s} Q_j q_j \mid d^\star({}^{\bullet}\varkappa_k(\boldsymbol{q}), \varkappa_k^{\bullet}(\boldsymbol{q})) \le 1 \right\} \tag{2.146}$$

and ${}^{\bullet}\varkappa_k, \varkappa^{\bullet}{}_k$ are given by (2.123). The condition in (2.146) can be written as below

$$\max \left\{ |{}^{\bullet}\chi_k(\boldsymbol{q})|, |\chi_k^{\bullet}(\boldsymbol{q})| \right\} \le \frac{1}{2} l_k \tag{2.147}$$

where ${}^{\bullet}\chi_k, \chi^{\bullet}{}_k$ are given by (1.14). The above condition bounds the values of the curvature at both the ends of the bar. By using (2.113), we write the condition (2.147), in terms of $\boldsymbol{q}$

$$\max \left\{ \left| \sum_{j=1}^{s} {}^{\bullet}B_{kj} q_j \right|, \left| \sum_{j=1}^{s} B_{kj}^{\bullet} q_j \right| \right\} \le \frac{1}{2} l_k \tag{2.148}$$

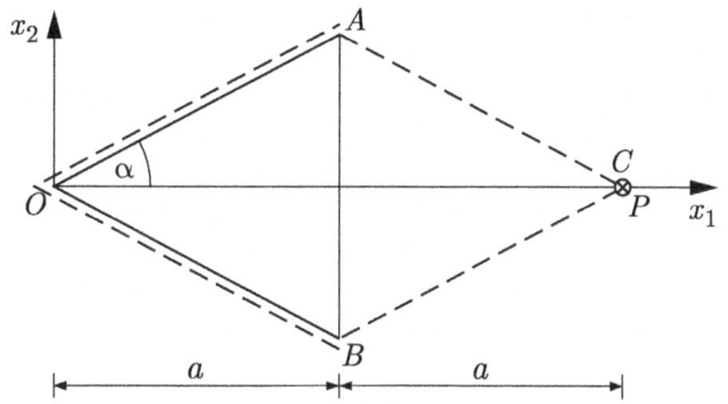

Fig. 2.16 Prager's problem: find the stiffest system of beams transmitting the force at C to the simply supported edges OA and OB; the edges AC and BC are free

which may be written as

$$-\frac{1}{2}l_k \leq \sum_{j=1}^{s} {}^{\bullet}B_{kj}q_j \leq \frac{1}{2}l_k, \qquad -\frac{1}{2}l_k \leq \sum_{j=1}^{s} B^{\bullet}_{kj}q_j \leq \frac{1}{2}l_k \tag{2.149}$$

The problem (2.146) with the condition (2.149) is the linear programming problem. Thus, problems of great number of unknowns may be effectively solved by using the ground structure algorithm, see Gilbert and Tyas (2003), Sokół (2011a, b).

Such a ground structure based program has been written by Bołbotowski et al. (2018). This program makes it possible to construct the stiffest system of beams for a wide range of design domains, boundary conditions and loads. For illustration, let us consider the problem of transmitting a point load to two sides: OA, OB of a rhombic design domain, see Fig. 2.16.

This problem has been the subject of theoretical consideration by Prager (1985), which will be discussed in Sect. 5.4 with using the tools of the layout theory of continuum grillages. In this section, we show the numerical solutions for a subsequent nets of beams covering the design domain $OACB$; we take $|AB| = a$, hence $\tan\alpha = 1/2$.

We introduce the grid: $x_1 = j\frac{a}{n}$, $x_2 = \frac{a}{2n}$, $j = 1, \ldots, 2n$ and assume that the beams can link all the nodes of the grid. The ground structure method delivers the optimal system of beams for given division n and the corresponding value of $Z_2(n)$, see (2.142).

For the case of $\alpha < \pi/4$ the sequence $Z_2(n)$ is decreasing and tends to the result of Prager (1985):

$$\lim_{n\to\infty} Z_2(n) = \left(\frac{1}{2} + \frac{3}{2}(\tan\alpha)^2\right) Pa^2 \tag{2.150}$$

For the case of $\tan\alpha = 1/2$ the optimal layouts for the subsequent divisions: $n = 4, 8, 16, 32, 64, 128$ are shown in Fig. 2.17, while the values of $Z_2(n)$ are set up in Table 2.2.

The results have been found by K. Bołbotowski; they are published here by his courtesy.

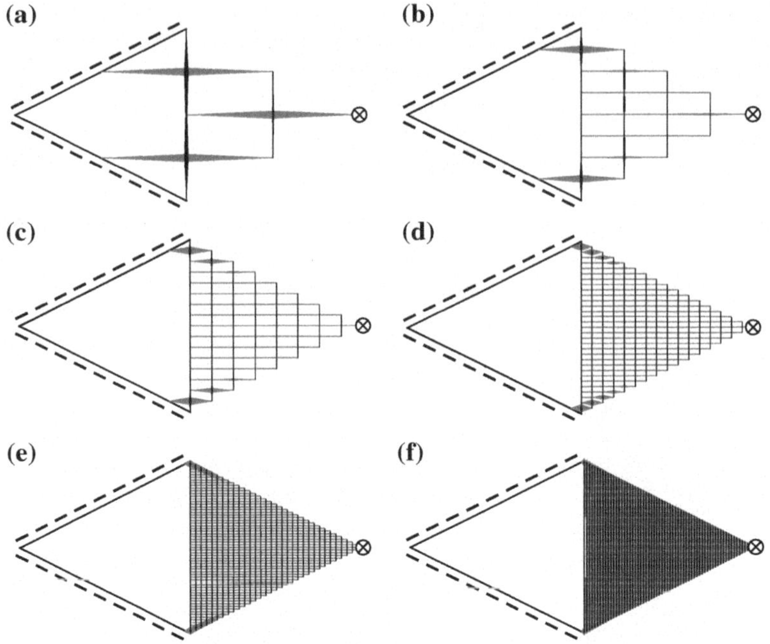

Fig. 2.17 The sequence of suboptimal systems of beams for the problem of Fig. 2.16 for the case of $\alpha = \arctan(1/2)$; the division index n is equal to: 4 (**a**), 8 (**b**),16 (**c**), 32 (**d**), 64 (**e**), 128 (**f**); published by the courtesy of K. Bołbotowski

Table 2.2 Sequence of $Z_2(n)$ for $\alpha = \arctan(1/2)$

n	4	8	16	32	64	128	∞
$\frac{Z_2(n)}{Pa^2}$	1.0625	0.96875	0.921875	0.898438	0.886719	0.880859	$\frac{7}{8}$

We note that the volume of beams within the domain OAB decreases if n is bigger and bigger. We guess that the optimal structure will have all beams outside the interior of this triangle.

Moreover, if n increases, there appears a strengthening of the free edges AC and BC. In the limit, these boundary ribs play the role of bars capable of transmitting torsion; they are called beam-weaves. These special bars have been discovered by Prager (1977) and their appearance has been recently confirmed by the high precision numerical methods, see Bołbotowski (2018). This topic will be continued in Sect. 5.4.

2.5 The Stiffest Grillages of Rigid Joints

2.5.1 Standard Approach

The problem of statics of grillages of rigid joints (called further grillages, in contrast to systems of beams) is governed by the equilibrium equation (1.36) and the constitutive equations (1.37), in which the strain measures are defined by (1.38a). Upon discretization, the problem reduces to the equations of linear algebra: (1.43)–(1.46) which reduce to the governing equation (1.29) with the stiffness matrix $\boldsymbol{K} = \boldsymbol{K}_M + \boldsymbol{K}_{\mathfrak{M}}$, where $\boldsymbol{K}_M$ determines the elastic bending energy and $\boldsymbol{K}_{\mathfrak{M}}$ – the elastic energy of torsion, see (1.32), (1.47).

A natural extension of the isoperimetric condition (2.73) used in the optimum design problem of systems of beams is assuming now this condition as below

$$\int \left[EJ(x) + \alpha\mu J^{(\mathfrak{M})}(x)\right] dx = \Lambda \tag{2.151}$$

where $\alpha > 0$ is a fixed parameter. The compliance of the grillage is expressed by

$$\wp(EJ, \mu J^{(\mathfrak{M})}) = \min_{(M,\mathfrak{M})\in\Sigma_Q} \int \left(\frac{M^2}{EJ} + \frac{\mathfrak{M}^2}{\mu J^{(\mathfrak{M})}}\right) dx \tag{2.152}$$

where Σ_Q is the set of fields $M, \mathfrak{M}$ satisfying the equilibrium condition (1.36).

Consider the problem of designing the stiffest grillage satisfying the cost condition (2.151)

$$\breve{\wp} = \min\left\{\wp(EJ, \mu J^{(\mathfrak{M})}) \mid EJ \geq 0,\ \mu J^{(\mathfrak{M})} \geq 0,\ \int (EJ + \alpha\mu J^{(\mathfrak{M})})dx = \Lambda\right\} \tag{2.153}$$

The stiffnesses EJ, $\mu J^{(\mathfrak{M})}$ are treated as design variables, linked by the cost condition (2.151). The optimization process may lead to the degenerated designs as follows:

(a) $EJ > 0, \mu J^{(\mathfrak{M})} = 0$;
(b) $EJ = 0, \mu J^{(\mathfrak{M})} > 0$

Case (a) refers to the grillages composed of thin-walled bars of open profiles whose torsional stiffness is much smaller than the bending stiffness. Case (b) refers to the case of a grillage made of an isotropic material whose bulk modulus is much smaller than the shear modulus, and then also the Young modulus is much smaller than the shear modulus. Although the problem (2.153) may lead to some degeneracies, it is thought appropriate to keep this formulation due to its mathematical correctness. This problem can be written in the form

$$\breve{\wp} = \min_{(M,\mathfrak{M})\in\Sigma_Q} Y(M, \mathfrak{M}) \tag{2.154}$$

where

$$Y(M, \mathfrak{M}) = \min \left\{ \int \left(\frac{M^2}{EJ} + \frac{\mathfrak{M}^2}{\mu J^{(\mathfrak{M})}} \right) dx \mid EJ \geq 0,\ \mu J^{(\mathfrak{M})} \geq 0, \right.$$
$$\left. \int (EJ + \alpha\mu J^{(\mathfrak{M})}) dx = \Lambda \right\} \tag{2.155}$$

Let us change the design variables

$$E_1 = EJ + \alpha\mu J^{(\mathfrak{M})}, \quad E_2 = EJ - \alpha\mu J^{(\mathfrak{M})}, \quad A = 2M^2, \quad B = 2\alpha\mathfrak{M}^2 \tag{2.156}$$

Note that

$$E_1 \geq 0, \qquad |E_2| \leq E_1$$

The problem (2.155) is now rearranged to the form

$$Y = \min \left\{ \int \left(\frac{A}{E_1 + E_2} + \frac{B}{E_1 - E_2} \right) dx \mid E_1 \geq 0,\ |E_2| \leq E_1,\ \int E_1 dx = \Lambda \right\} \tag{2.157}$$

The minimization over E_2 may be performed analytically by using the following result from calculus

$$\min_{|E_2| \leq E_1} \left(\frac{A}{E_1 + E_2} + \frac{B}{E_1 - E_2} \right) = \frac{(\sqrt{A} + \sqrt{B})^2}{2E_1} \tag{2.158}$$

valid for $A > 0$, $B > 0$, $E_1 > 0$, the minimum being attained for $E_2 = \breve{E}_2$ given by

$$\breve{E}_2 = \frac{\sqrt{A} - \sqrt{B}}{\sqrt{A} + \sqrt{B}} E_1 \tag{2.159}$$

The problem (2.157) reduces to the form

$$Y = \min \left\{ \int \frac{F}{E_1} dx \mid E_1 \geq 0,\ \int E_1 dx = \Lambda \right\} \tag{2.160}$$

where

$$F = \frac{1}{2}(\sqrt{A} + \sqrt{B})^2 \tag{2.161}$$

The above problem can be solved with using the Lemma 2.4.1. According to (2.80)–(2.82) one finds

$$Y = \frac{1}{\Lambda}(Z_F)^2 \tag{2.162}$$

where

$$Z_F = \int \sqrt{F} dx \tag{2.163}$$

The optimal modulus $E_1 = \breve{E}_1$ equals

$$\breve{E}_1(x) = \Lambda \frac{\sqrt{F(x)}}{\int \sqrt{F(x)} dx} \tag{2.164}$$

We come back now to (2.154). The optimal compliance is expressed by

$$\breve{\wp} = \frac{1}{\Lambda} Z^2 \tag{2.165}$$

with $Z = Z_2$ and

$$Z_2 = \min \left\{ \int (|M| + \sqrt{\alpha}|\mathfrak{M}|) dx \mid (M, \mathfrak{M}) \in \Sigma_Q \right\} \tag{2.166}$$

Let the pair $M = \breve{M}_\alpha$, $\mathfrak{M} = \breve{\mathfrak{M}}_\alpha$ be the minimizer of (2.166). By (2.164), we find

$$\breve{E}_1(x) = \frac{\Lambda}{Z_2} \left(|\breve{M}_\alpha(x)| + \sqrt{\alpha} |\breve{\mathfrak{M}}_\alpha(x)| \right) \tag{2.167}$$

It is seen that the isoperimetric condition in (2.160) is fulfilled. The optimal modulus $E = \breve{E}_2$ is given by (2.159), (2.167). Thus, the optimal stiffnesses are expressed by

$$E\breve{J} = \frac{\Lambda}{Z_2} |\breve{M}_\alpha|, \qquad \mu \breve{J}^{(\mathfrak{M})} = \frac{\Lambda}{\sqrt{\alpha} Z_2} |\breve{\mathfrak{M}}_\alpha| \tag{2.168}$$

We note that in the optimal grillage the absolute values of the strains (1.38a) are constant in all the bars. Consequently, the maximal (in each cross section) values of normal axial stresses caused by bending and the maximal tangent stresses due to torsion are uniformly distributed in the whole grillage.

The coefficient α indicates how much the cost of making the bar appropriately stiff with respect to torsion is greater than the cost of assuring the desired stiffness due to bending. If the cost of restraining torsion is high, then the optimal torsion stiffness will be small, see the second equation in (2.168).

By using the technique of Sect. 2.4 one can construct the problem dual to (2.166). Let $\mathcal{B}$ be a ball in $\mathbb{R}^2$

$$\mathcal{B} = \left\{ (y_1, y_2) \mid |y_1| \le 1, \quad |y_2| \le \sqrt{\alpha} \right\} \tag{2.169}$$

One can prove that $Z_1 = Z_2$ with

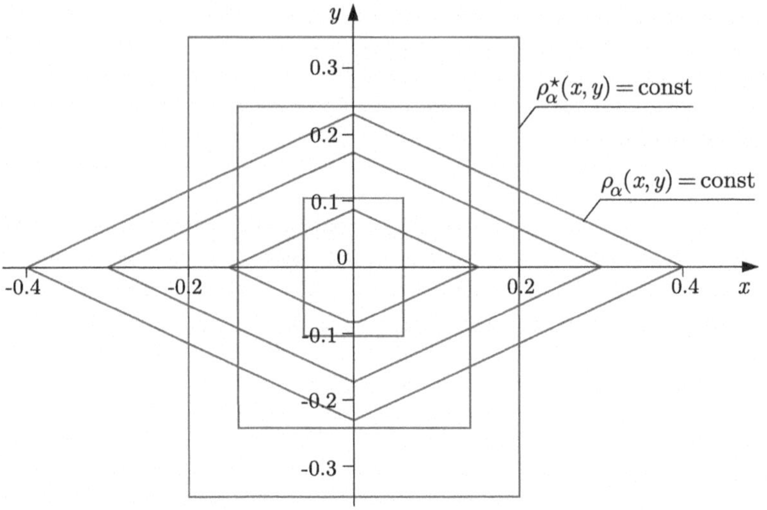

Fig. 2.18 The contour graphs: $\varrho_\alpha(x, y) = \text{const}$ and $\varrho^{\star}_\alpha(x, y) = \text{const}$. Case of $\alpha = 3$

$$Z_1 = \max \left\{ \mathcal{L}(\bar{w}, \bar{\phi}) \mid (\varkappa(\bar{w}), \varepsilon(\bar{\phi})) \in \mathcal{B}, \quad \bar{w},\ \bar{\phi} \text{ kinematically admissible} \right\} \tag{2.170}$$

Let $\bar{w} = \breve{w}, \bar{\phi} = \breve{\phi}$ be a maximizer of this problem. To discuss relations between the solutions to the problems (2.166) and (2.170), we introduce the norms in $\mathbb{R}^2$

$$\varrho_\alpha(x_1, x_2) = |x_1| + \sqrt{\alpha}|x_2| \tag{2.171}$$

$$\varrho^{\star}_\alpha(y_1, y_2) = \max \left\{ |y_1|, \frac{1}{\sqrt{\alpha}}|y_2| \right\} \tag{2.172}$$

These norms are mutually dual (see Fig. 2.18), or the second of them can be constructed by performing maximization operation in (2.108). The following estimate holds

$$\boldsymbol{x} \cdot \boldsymbol{y} \le \varrho_\alpha(\boldsymbol{x}) \varrho^{\star}_\alpha(\boldsymbol{y}) \quad \forall\, \boldsymbol{x}, \boldsymbol{y} \in \mathbb{R}^2 \tag{2.173}$$

and the equality can be achieved, for a given $\boldsymbol{x} = (x_1, x_2)$ one can find $\boldsymbol{y} = (y_1, y_2)$ such that

$$\begin{array}{lll}
\text{(i)}\ x_1 x_2 > 0, & \operatorname{sgn}(y_1) = \operatorname{sgn}(x_1), & y_2 = \sqrt{\alpha} y_1 \\
\text{(ii)}\ x_1 x_2 < 0, & \operatorname{sgn}(y_1) = \operatorname{sgn}(x_1), & y_2 = -\sqrt{\alpha} y_1 \\
\text{(iii)}\ x_1 = 0,\ x_2 \ne 0, & \sqrt{\alpha}|y_1| \le |y_2|, & \operatorname{sgn}(x_2) y_2 \ge 0 \\
\text{(iv)}\ x_1 \ne 0,\ x_2 = 0, & |y_2| \le \sqrt{\alpha}|y_1|, & \operatorname{sgn}(x_1) y_1 \ge 0 \\
\text{(v)}\ x_1 = 0,\ x_2 = 0, & y_1 \in \mathbb{R}, & y_2 \in \mathbb{R}
\end{array} \tag{2.174}$$

The problem (2.166) can be expressed by using the norm (2.171)

$$Z_2 = \min\left\{\int \varrho_\alpha(M, \mathfrak{M})dx \mid (M, \mathfrak{M}) \in \Sigma_Q\right\} \tag{2.175}$$

and the problem (2.170), can be rewritten as

$$\begin{aligned} Z_1 = \max\{&\mathcal{L}(\bar{w}, \bar{\phi}) \mid (\bar{w}, \bar{\phi}) \text{ kinematically admissible} \\ &\text{and } \varrho^\star_\alpha(\varkappa(\bar{w}), \varepsilon(\bar{\phi})) \leq 1\} \end{aligned} \tag{2.176}$$

Let us note that the equality $Z_1 = Z_2$ implies

$$\varrho^\star_\alpha(\varkappa(\breve{w}), \varepsilon(\breve{\phi})) = 1 \tag{2.177}$$

Indeed, the pair $(\breve{w}, \breve{\phi})$ is kinematically admissible, hence

$$\mathcal{L}(\breve{w}, \breve{\phi}) = \int \left(\breve{M}\varkappa(\breve{w}) + \breve{\mathfrak{M}}\varepsilon(\breve{\phi})\right) dx, \tag{2.178}$$

and using (2.173) we can estimate

$$\mathcal{L}(\breve{w}, \breve{\phi}) \leq \int \varrho_\alpha(\breve{M}, \breve{\mathfrak{M}})\varrho^\star_\alpha\left(\varkappa(\breve{w}), \varepsilon(\breve{\phi})\right) dx \tag{2.179}$$

and taking into account that $Z_1 = Z_2$, we find

$$\mathcal{L}(\breve{w}, \breve{\phi}) = \int \varrho_\alpha(\breve{M}, \breve{\mathfrak{M}})dx \tag{2.180}$$

which can be true only if (2.177) holds, where both the fields $M, \mathfrak{M}$ are non-zero. Moreover, by (2.174) we note that

$$\begin{aligned} &\text{(i)}\ \ \breve{M} \neq 0,\ \breve{\mathfrak{M}} \neq 0 \Rightarrow |\varepsilon(\breve{\phi})| = \sqrt{\alpha}|\varkappa(\breve{w})| \\ &\text{(ii)}\ \ \breve{M} = 0 \Rightarrow \sqrt{\alpha}|\varkappa(\breve{w})| \leq |\varepsilon(\breve{\phi})| \\ &\text{(iii)}\ \ \breve{\mathfrak{M}} = 0 \Rightarrow |\varepsilon(\breve{\phi})| \leq \sqrt{\alpha}|\varkappa(\breve{w})| \end{aligned} \tag{2.181}$$

Moreover,

$$|\varkappa(\breve{w})| \leq 1, \qquad |\varepsilon(\breve{\phi})| \leq \sqrt{\alpha} \tag{2.182}$$

and (2.177) holds.

Remark 2.5.1 Problem (2.166) may be put in an algebraic form by using the methods of Sects. 1.2.4 and 2.4.2. According to (2.106) and due to $\mathfrak{M} = \text{const}$ bar-wise, we have

$$Z_2 = \min \left\{ \sum_{k=1}^{e} l_k \left[\varrho(^{\boldsymbol{\cdot}}M_k, M_k^{\boldsymbol{\cdot}}) + \sqrt{\alpha}|\mathfrak{M}_k| \right] \,\middle|\, \hat{\boldsymbol{B}}^{\top}\mathfrak{M} + (^{\boldsymbol{\cdot}}\boldsymbol{B})^{\top}\,^{\boldsymbol{\cdot}}\boldsymbol{M} + (\boldsymbol{B}^{\boldsymbol{\cdot}})^{\top}\boldsymbol{M}^{\boldsymbol{\cdot}} = \boldsymbol{Q} \right\} \tag{2.183}$$

where $\varrho(\cdot, \cdot)$ is given by (2.105). The construction of the dual problem is left to the reader.

Remark 2.5.2 Consider the optimum design of stiffnesses EA, EJ of a plane frame with the cost condition

$$\int (EJ + \alpha EA)dx = \Lambda \tag{2.184}$$

The aim is to minimize the compliance

$$\wp(EJ, EA) = \min_{(M,T)\in\Sigma_Q} \int \left(\frac{T^2}{EA} + \frac{M^2}{EJ} \right) dx \tag{2.185}$$

where Σ_Q is the set of fields T, M satisfying the equilibrium equations (1.10a). Thus, the problem assumes the form

$$\breve{\wp} = \min_{(M,T)\in\Sigma_Q} \left\{ \int \wp(EJ, EA) \mid EJ \geq 0,\ EA \geq 0, \int (EJ + \alpha EA)dx = \Lambda \right\} \tag{2.186}$$

similar to (2.153) concerning grillages. The main difference lies in the units of stiffnesses, now the stiffnesses EA, EJ have different units; consequently, the parameter α has dimension of $[\mathrm{m}^2]$. Due to the mentioned analogy, the problem (2.186) reduces to

$$Z_2 = \min \left\{ \int (|M| + \sqrt{\alpha}|T|)dx \mid (M, T) \in \Sigma_Q \right\} \tag{2.187}$$

the solution $\breve{M}_\alpha$, $\breve{T}_\alpha$ of which determines the optimal stiffnesses by the formulae

$$E\breve{J} = \frac{\Lambda}{Z_2}|\breve{M}_\alpha|, \qquad E\breve{A} = \frac{\Lambda}{\sqrt{\alpha}Z_2}|\breve{T}_\alpha| \tag{2.188}$$

These formulae show that the strains $\breve{\varkappa}$, $\breve{\varepsilon}$ in the optimal frame are distributed uniformly. Hence, the maximal (over cross sections) stresses (per section) are also constant in the frame.

The problem dual to (2.187) has the form (2.176) where one should replace $\bar{\phi}$ with $\bar{u}$. The bars appear, where $\varrho_{\alpha(M,T)} \neq 0$ and there

$$\varrho^{\star}_{\alpha}(\varkappa(\breve{w}), \varepsilon(\breve{u})) = 1 \tag{2.189}$$

Discretization replaces the problem (2.187) with

$$Z_2 = \min\left\{ \sum_{k=1}^{e} l_k \left[\varrho({}^{\bullet}M_k, M_k^{\bullet}) + \sqrt{\alpha}|T_k|\right] \;\middle|\; \boldsymbol{B}^{\top}\boldsymbol{T} + ({}^{\bullet}\boldsymbol{B})^{\top}\,{}^{\bullet}\boldsymbol{M} + (\boldsymbol{B}^{\bullet})^{\top}\boldsymbol{M}^{\bullet} = \boldsymbol{Q} \right\} \tag{2.190}$$

where $\boldsymbol{B}$ is the matrix linking the elongations of bars Δ_k with nodal unknowns $q_1, \ldots, q_s$, see (1.18), (1.23).

The solution of (2.188) admits extreme cases as follows:

(a) $E\breve{J} > 0$, $E\breve{A} = 0$,
(b) $E\breve{J} = 0$, $E\breve{A} > 0$.

These extreme cases cannot be attained since in both the conditions the same material constant is involved. In order to prevent such extreme results from appearing in the optimum design, one should consider more load variants that one usually excludes the degeneracies of the design, see Sect. 2.5.

Remark 2.5.3 In the problem (2.153) of optimum design of a grillage, the stiffnesses EJ, $\mu J^{(\mathfrak{M})}$ are viewed as independent design variables. In the problem (2.186) of optimum design of a plane frame, the independent design variables are EA, EJ. To avoid the problem of recovery of the cross section of the bars on the basis of these data one can assume the following:

(a) for the design of the grillage: $EJ(x) = b(x)EJ_0$, $\mu J^{(\mathfrak{M})} = b(x)\mu J_0^{(\mathfrak{M})}$
(b) for the design of the frame: $EA(x) = b(x)EA_0$, $EJ(x) = b(x)EJ_0$

where $b(x)$ is the only design variable subject to the condition

$$\int b(x)dx = \Lambda_0 \tag{2.191}$$

where Λ_0 is a given constant. In case of rectangular sections, $b(x)$ represents the width of the bar. The problem of the grillage optimization assumes the form

$$\breve{\wp} = \min\left\{\wp(bEJ_0, b\mu J_0^{(\mathfrak{M})}) \mid b \geq 0, \int b(x)dx = \Lambda_0\right\} \tag{2.192}$$

where $\wp$ is given by (2.152). Performing minimization over b results in

$$\breve{\wp} = \frac{1}{\Lambda_0} Z^2 \tag{2.193}$$

where $Z = Z_2$ and

$$Z_2 = \min\left\{ \int \varrho_0(M, \mathfrak{M})dx \mid (M, \mathfrak{M}) \in \Sigma_Q \right\} \tag{2.194}$$

with

$$\varrho_0(x_1, x_2) = \sqrt{\frac{1}{EJ_0}(x_1)^2 + \frac{1}{\mu J_0^{(\mathfrak{M})}}(x_2)^2} \tag{2.195}$$

The norm dual to this norm has the form

$$\varrho_0^\star(y_1, y_2) = \sqrt{EJ_0(y_1)^2 + \mu J_0^{(\mathfrak{M})}(y_2)^2} \tag{2.196}$$

and the problem dual to (2.194) reads

$$Z_1 = \max\{\mathcal{L}(\bar{w}, \bar{\phi}) \mid (\bar{w}, \bar{\phi}) \text{ kinematically admissible and } \varrho_0^\star(\varkappa(\bar{w}), \varepsilon(\bar{\phi})) \le 1\} \tag{2.197}$$

The contours of the norms (2.195), (2.196) are shown in Fig. 2.19.

Let $(\breve{M}, \breve{\mathfrak{M}})$ be the minimizer of (2.194). The optimal stiffnesses are determined by the optimal width $\breve{b}$ given by

$$\breve{b}(x) = \frac{\Lambda_0}{Z_2}\varrho_0(\breve{M}(x), \breve{\mathfrak{M}}(x)) \tag{2.198}$$

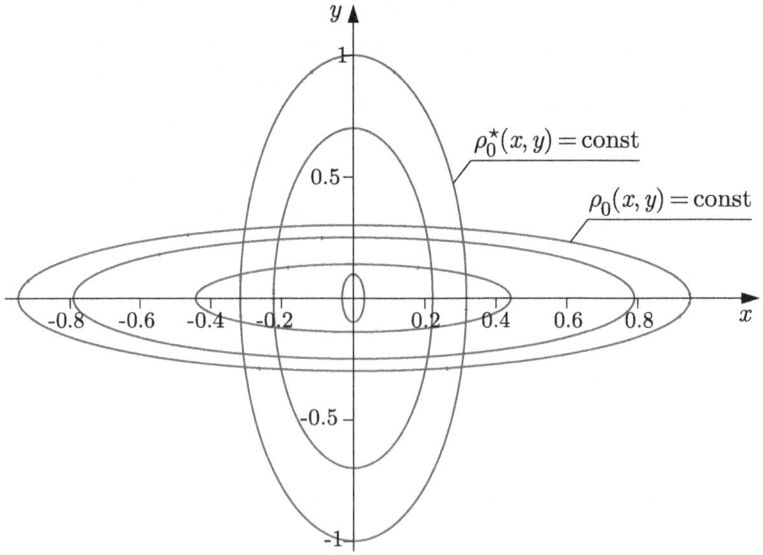

Fig. 2.19 The contour graphs: $\varrho_0(x, y) = \text{const}$ and $\varrho_0^\star(x, y) = \text{const}$

Thus, $\breve{b} = 0$ only where $\breve{M}$, $\breve{\mathfrak{M}}$ vanish simultaneously. The problem of optimum design of plane frames with b being the design variable can be reformulated in the same manner.

2.5.2 Case of Negligible Torsional Stiffness

Consider a grillage of rigid joints. At each node, we assign an unknown transverse displacement and an in-plane vector representing rotation. These quantities constitute the vector of unknowns $\boldsymbol{q}$.

We put $\boldsymbol{G} = \boldsymbol{0}$ in Eqs. (1.43)–(1.46), which reduces this set to the set of equations known from Sect. 1.2.5, but now the vector $\boldsymbol{q}$ has a different meaning and the matrices ${}^{\bullet}\boldsymbol{B}$, $\boldsymbol{B}^{\bullet}$, $\boldsymbol{Q}$ are constructed in a different manner. Thus, consider the problem of minimum compliance of a grillage of rigid joints in which torsion is neglected. The mathematical continuum description coincides with (2.76), (2.75); the term Σ_Q refers to a grillage of rigid joints. The problem reduces to the formulae (2.89)–(2.90). Upon discretization, this problem simplifies to (2.115), or to the dual form (2.122) in which $q_1, \ldots, q_s$ represent now the transverse displacements and angles of rotation of subsequent nodes.

2.5.3 A Simplified Approach for Dense Grillages

The formulation of the optimum design problem of Sect. 2.5.2 can be further simplified by the method of Sect. 6.6 assuming that in majority of bars the diagram of bending moment does not change the sign. The minimum compliance problem reduces to the auxiliary problems (2.144) or (2.146). The latter problem can be written as

$$\max_{q} \boldsymbol{Q}^{\top}\boldsymbol{q} \tag{2.199}$$

with $\boldsymbol{q}$ such that

$$-\frac{1}{2}\boldsymbol{l} \le {}^{\bullet}\boldsymbol{B}\boldsymbol{q} \le \frac{1}{2}\boldsymbol{l}, \qquad -\frac{1}{2}\boldsymbol{l} \le \boldsymbol{B}^{\bullet}\boldsymbol{q} \le \frac{1}{2}\boldsymbol{l} \tag{2.200}$$

where the inequality signs in (2.200) refer to the subsequent components of the vectors involved. The problem (2.199)–(2.200) coincides with the kinematic formulation of the minimum volume problem of a grillage discussed in Bolbotowski, He and Gilbert (2018). The static problem (2.144) and the kinematic problem (2.199)–(2.200) are the starting points of the ground structure method applied to grillages. The ground structure method may be applied to very dense grids of bars, which enables verification of numerous available exact solutions (e.g. those constructed by W. Prager and G. Rozvany) concerning continuum grillages. This problem will be discussed in Sect. 5.4.

2.6 Bibliographical Comments

The problem of minimum compliance in the form (2.4) has been the subject of numerous papers, see the book by Bendsøe (1995) and the references cited therein. The results (2.9) as well (2.6) and (2.8) can be found in Achtziger (1997) with complete proofs, yet the derivation of Sect. 2.1 seems new.

The problem of minimization of the weight (volume) of a truss is reduced to the form (2.51). This formulation can be found in Hemp (1973) and is used in the contemporary papers by Gilbert and Tyas (2003), Sokół (2011a, b). The theorems of Sect. 2.2.3 can be found in Achtziger (1997).

The ground structure method has been proposed by Dorn et al. (1964) and then was improved for large-scale problems by Gilbert and Tyas (2003), Sokół (2011b, 2013), Rozvany and Sokół (2014), Zegard and Paulino (2013, 2014, 2015).

The Lemma 2.4.1 was formulated in Czarnecki and Lewiński (2013), while its proof given in Sect. 2.4 follows the lines of the proof given in Czarnecki and Lewiński (2017a).

The problem (2.90) of optimum design of beams has been formulated by Heyman (1959) and Rozvany (1976), see the papers cited therein. However, the problem (2.90) was not obtained there as an auxiliary problem of the minimum compliance design, but was put forward as the problem expressing the optimum design of minimum reinforcement of beams in bending. The algebraic setting given in Sect. 2.4.3 has been recently proposed by Bołbotowski (2018). The main new results—the norm (2.110) dual to the norm (2.105) and the formulation (2.122) are due to Bołbotowski.

The optimum beams of Sects. 2.4.4.1, 2.4.4.2 can be found in Rozvany (1976, 1989).

The reduction of the problem of optimum design of grillages (2.155) to the problem (2.166) has been done by using the method developed within the theories of material design—the isotropic material design (IMD) and the Young modulus design (YMD), see Czarnecki and Lewiński (2017a, b).

References

Achtziger, W. (1997). Topology optimization of discrete structures: An introduction in view of computational and nonsmooth aspects. In G. I. N. Rozvany (Ed.), *Topology optimization in structural mechanics* (Vol. 374, pp. 57–100), CISM courses and lectures. Wien New York: Springer.

Beghini, A., & Baker, W. F. (2015). On the layout of a least weight multiple span structure with uniform load. *Structural and Multidisciplinary Optimization*, *52*, 447–457.

Bendsøe, M. P. (1995). *Optimization of structural topology, shape and material*. Berlin: Springer.

Bojczuk, D. (1999). *Sensitivity analysis and optimization of bar structures*. Monographs, Studies, Treatises, Kielce University of Technology, Kielce (in Polish).

Bojczuk, D., & Mróz, Z. (1998). On optimal design of supports in beam and frame structures. *Structural Optimization*, *16*, 47–57.

Bołbotowski, K. (2018). Theory of grillage optimization – a discrete setting (In preparation).

Bołbotowski, K., He, L., & Gilbert, M. (2018). Design of optimum grillages using layout optimization. *Structural and Multidisciplinary Optimization, 58*, 851–868.

Cox, H. L. (1965). *The design of structures of least weight*. Oxford: Pergamon Press.

Czarnecki, S., & Lewiński, T. (2013). On minimum compliance problems of thin elastic plates of varying thickness. *Structural and Multidisciplinary Optimization, 48*(1), 17–31.

Czarnecki, S., & Lewiński, T. (2017a). On material design by the optimal choice of Young's modulus distribution. *International Journal of Solids and Structures, 110–111*, 315–331.

Czarnecki, S., & Lewiński, T. (2017b). Pareto optimal design of non-homogeneous isotropic material properties for the multiple loading conditions. *Physica Status Solidi B: Basic Solids State Physics, 254*, 1600821.

Dorn, W., Gomory, R., & Greenberg, M. (1964). Automatic design of optimal structures. *Journal de Mécanique, 3*, 25–52.

Hemp, W. S. (1973). *Optimum structures*. Oxford: Clarendon Press.

Heyman, J. (1959). On the absolute minimum weight design of framed structures. *The Quarterly Journal of Mechanics and Applied Mathematics, 12*, 314–324.

Gilbert, M., & Tyas, A. (2003). Layout optimization of large-scale pin-jointed frames. *Engineering Computations, 20*, 1044–1064.

Kozłowski, W., & Mróz, Z. (1969). Optimal design of solid plates. *International Journal of Solids and Structures, 5*, 781–794.

Lowe, P. G., & Melchers, R. E. (1972). On the theory of optimal constant thickness. Fibrereinforcement plates. *International Journal of Mechanical Sciences, 14*, 311–324.

Lowe, P. G., & Melchers, R. E. (1973). On the theory of optimal constant thickness, fibrereinforcement plates. II. *International Journal of Mechanical Sciences, 15*, 157–170.

Martinez, P., Marti, P., & Querin, Q. M. (2007). Growth method for size, topology, and geometry optimization of truss structures. *Structural and Multidisciplinary Optimization, 33*, 13–26.

Maxwell, J. C. (1872). On reciprocal figures, frames and diagrams of forces. *Transactions of the Royal Society of Edinburgh, 26*, 1–40.

Mazurek, A. (2012). Geometrical aspects of optimum truss like structures for three-force problem. *Structural and Multidisciplinary Optimization, 45*, 21–32.

Mazurek, A., Baker, W., & Tort, C. (2011). Geometrical aspects of optimum truss like structures. *Structural and Multidisciplinary Optimization, 43*, 231–242.

Morley, C. T. (1966). The minimum reinforcement of concrete slabs. *International Journal of Mechanical Sciences, 8*, 305–319.

Prager, W. (1977). Optimal lauout of contilever trusses. *The Journal of Optimization Theory and Applications, 23*, 111–117.

Prager, W. (1978a). Optimal layout of trusses of finite number of joints. *Journal of the Mechanics and Physics of Solids, 26*, 241–250.

Prager, W. (1978b). Nearly optimal design of trusses. *Computers and Structures, 8*, 451–454.

Prager, W. (1985). Optimal design of grillages. In M. Save & W. Prager (Eds.), *Structural optimization. Vol 1. Optimality criteria* (pp. 153–200). New York and London: Plenum Press.

Prager, W., & Rozvany, G. I. N. (1977). Optimal layout of grillages. *Journal of Structural Mechanics, 5*, 1–18.

Rozvany, G. I. N. (1972). Grillages of maximum strength and maximum stiffness. *International Journal of Mechanical Sciences, 14*, 651–666.

Rozvany, G. I. N. (1976). *Optimal design of flexural systems*. London: Pergamon Press.

Rozvany, G. I. N. (1989). *Structural design via optimality criteria*. The Netherlands: Kluwer Academic Publishers Dordrecht.

Rozvany, G. I. N., & Sokół, T. (2014). Validation of numerical methods by analytical benchmarks, and verification of exact solutions by numerical methods. In G.I.N. Rozvany & T. Lewiński (Eds.), *Topology Optimization in Structural and Continuum Mechanics* (Vol. 549), CISM International Centre for Mechanical Sciences. Vienna: Springer.

Sokół, T. (2011a). A 99 line code for discretized Michell truss optimization written in Mathematica. *Structural and Multidisciplinary Optimization, 43*, 181–190.

Sokół, T. (2011b). Topology optimization of large-scale trusses using ground structure approach with selective subsets of active bars. In A. Borkowski, T. Lewiński & G. Dzierżanowski (Eds.), *19th International Conference on Computer Methods in Mechanics CMM-2011* (pp. 457–458), Warsaw, 9–12 May 2011.

Sokół, T. (2013). Numerical approximations of exact Michell solutions using the adaptive ground structure approach. In S. Jemioło & M. Lutomirska (Eds.), *Mechanics and Materials, Sec. VI* (pp. 87–98). Warsaw University of Technology.

Steven, G., Querin, O., & Xie, M. (2000). Evolutionary structural optimization (ESO) for combined topology and size optimization of discrete structures. *Computer Methods in Applied Mechanics and Engineering, 188*, 743–754.

Stolpe, M. (2010). On some fundamental properties of structural topology optimization problems. *Structural and Multidisciplinary Optimization, 41*, 661–670.

Strang, G., & Kohn, R. V. (1983). Hencky-Prandtl nets and constrained Michell trusses. *Computer Methods in Applied Mechanics and Engineering, 36*, 207–222.

Zegard, T., & Paulino, G. H. (2013). Truss layout optimization within a continuum. *Structural and Multidisciplinary Optimization, 48*, 1–16.

Zegard, T., & Paulino, G. H. (2014). GRAND-Ground structure based topology optimization for arbitrary 2D domains using MatLab. *Structural and Multidisciplinary Optimization, 50*, 861–882.

Zegard, T., & Paulino, G. H. (2015). GRAND3 - Ground structure based topology optimization for arbitrary 3D domains using MATLAB. *Structural and Multidisciplinary Optimization, 52*, 1161–1184.

Chapter 3
Theory of Michell Structures. Single Load Case

The Michell structures are solutions to the problems of optimum design put forward and discussed in Sects. 2.1 and 2.2 provided that the nodes of the trusses may be placed at arbitrary point of the design domain being a subdomain of the plane or of the Euclidean space. The present chapter introduces the reader to the main topic of the book.

3.1 Kinematic Formulation. Case of Equal Permissible Stresses in Tension and Compression

Let us come back to the main problem of optimum design of Sect. 2.1. We know now that the problem of minimum compliance of a truss of nodes placed at points selected from a given set of points within the design domain reduces to two mutually dual problems (2.6) and (2.8). The latter problem can be rearranged to the form (2.31) known from the linear programming.

The numerical experiments teach us that the suboptimal trusses constructed by the ground structure method with a bigger and bigger number of possible nodes form sequences of structures whose compliance is smaller and smaller while the numbers of their bars are bigger and bigger. This tendency is illustrated in Fig. 3.1. The same property characterizes the suboptimal trusses for the least volume problem, see Eq. (2.45). By adding new bars, we extend the design space; the structure can be designed better, and since there is no lower bound for the areas of the cross sections, the bars become thinner and thinner.

This process has no natural end, and consequently, a sequence of suboptimal trusses tends to a discrete-continuous structure of fibrous microstructure, possibly reinforced by bars of finite cross sections (i.e. of finite axial stiffness $EA > 0$. Such a solution is called a Michell structure or a Michell truss to honour A.G.M. Michell

T. Lewiński et al., *Michell Structures*,
https://doi.org/10.1007/978-3-319-95180-5_3

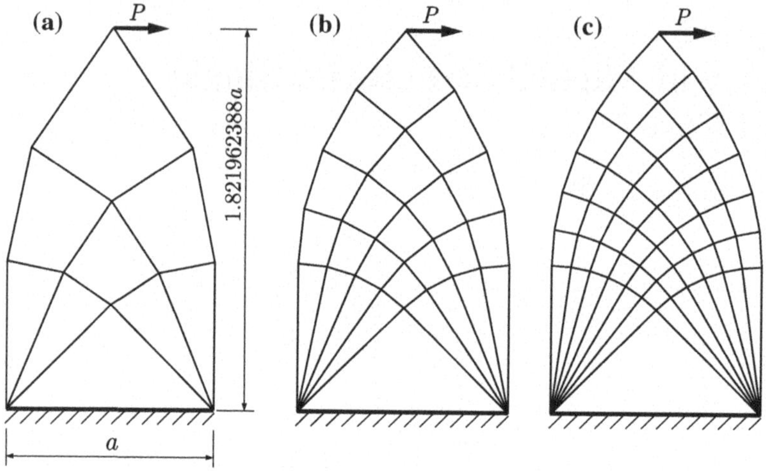

Fig. 3.1 Three suboptimal designs of the least volume trusses transmitting a given force to the straight rigid support of length a. The volumes form a decreasing sequence: $\bar{V} = 6.1605$ (**a**); $\bar{V} = 6.0953$ (**b**); $\bar{V} = 6.0835$ (**c**), where $\bar{V} = V/V_0$, $V_0 = Pa/\sigma_0$, σ_0 being the permissible stress

who discovered these structures in 1904. The aim of the present section is to put forward the kinematic formulation of the problem of constructing Michell trusses corresponding to the equal values of permissible stresses in tension and compression.

The simplest passage from the truss optimization problem to Michell's problem takes the formulation (2.6) as the point of departure and is based on the appropriate extension of the virtual kinematic field $\bar{\boldsymbol{u}}$.

Consider a vector field $\boldsymbol{v}$ given in the design domain included within $\mathbb{R}^n$, $n = 2$ or $n = 3$, which assumes at given nodes $\boldsymbol{x}_i$ of a given truss the values coinciding with the displacements $\bar{\boldsymbol{u}}(\boldsymbol{x}_i)$ of this truss; for the case of $n = 3$ we require

$$\boldsymbol{v}(\boldsymbol{x}_i) = \bar{u}_{i_1}\boldsymbol{e}_1 + \bar{u}_{i_2}\boldsymbol{e}_2 + \bar{u}_{i_3}\boldsymbol{e}_3 \tag{3.1}$$

where i_1, i_2, i_3 are the corresponding numbers of degrees of freedom at the node i. Assume that the bar of number k links the nodes at the points $\boldsymbol{x}_k$ and $\boldsymbol{y}_k$. The length of this bar is $l_k = ||\boldsymbol{x}_k - \boldsymbol{y}_k||$. The displacements along the axis of the bar are

$$u_{\text{left}} = \boldsymbol{v}(\boldsymbol{x}_k) \cdot \frac{\boldsymbol{y}_k - \boldsymbol{x}_k}{||\boldsymbol{x}_k - \boldsymbol{y}_k||} \qquad u_{\text{right}} = \boldsymbol{v}(\boldsymbol{y}_k) \cdot \frac{\boldsymbol{y}_k - \boldsymbol{x}_k}{||\boldsymbol{x}_k - \boldsymbol{y}_k||} \tag{3.2}$$

The elongation of this bar equals

$$\Delta_k = \left[\boldsymbol{v}(\boldsymbol{y}_k) - \boldsymbol{v}(\boldsymbol{x}_k)\right] \cdot \frac{\boldsymbol{y}_k - \boldsymbol{x}_k}{||\boldsymbol{x}_k - \boldsymbol{y}_k||} \tag{3.3}$$

The condition (2.12) may be expressed as below

$$\left|[\boldsymbol{v}(\boldsymbol{x}_k) - \boldsymbol{v}(\boldsymbol{y}_k)] \cdot (\boldsymbol{x}_k - \boldsymbol{y}_k)\right| \leq ||\boldsymbol{x}_k - \boldsymbol{y}_k||^2 \tag{3.4}$$

Now, let us assume that the above condition applies to all pairs of points of the design domain Ω. For simplicity, we introduce convexity property to assure that all these connections may form bars. Thus, the field $\boldsymbol{v}$ satisfies

$$|[\boldsymbol{v}(\boldsymbol{x}) - \boldsymbol{v}(\boldsymbol{y})] \cdot (\boldsymbol{x} - \boldsymbol{y})| \leq ||\boldsymbol{x} - \boldsymbol{y}||^2 \quad \forall\, (\boldsymbol{x}, \boldsymbol{y}) \in \Omega \times \Omega \tag{3.5}$$

For the subdomains where this vector field is differentiable, we introduce the tensor

$$\varepsilon_{ij}(\boldsymbol{v}) = \frac{1}{2}\left(\frac{\partial v_i}{\partial x_j} + \frac{\partial v_j}{\partial x_i}\right) \tag{3.6}$$

or the symmetric part of the gradient of the field $\boldsymbol{v}$. For a given Cartesian parameterization, locally, this tensor forms a matrix from the set $\mathbb{E}^2_s$ of symmetric matrices of rank 2. At each point of the domain, we determine the spectral norm of this matrix:

$$||\boldsymbol{\varepsilon}||_{\text{spec}} = \max_{i=1,\ldots,n} \{|\lambda_i(\boldsymbol{\varepsilon})|\} \tag{3.7}$$

Upon ordering, the eigenvalues of this tensor are: $\varepsilon_I \geq \varepsilon_{II} \geq \varepsilon_{III}$, if $n = 3$ or $\varepsilon_I \geq \varepsilon_{II}$, if $n = 2$. The spectral norm of the strain tensor can be expressed as

$$||\boldsymbol{\varepsilon}||_{\text{spec}} = |\varepsilon_I| \tag{3.8}$$

One can prove that the two-point condition (3.5) implies the pointwise condition

$$||\boldsymbol{\varepsilon}(\boldsymbol{v})||_{\text{spec}} \leq 1 \qquad \text{a.e. in } \Omega \tag{3.9}$$

if the domain Ω is convex. Let us express the relationship between (3.5) and (3.9) in a mathematical language. Let $\mathcal{D}(\mathbb{R}^n, \mathbb{R}^n)$ be Schwartz space of C^∞ functions with compact support. Let $\mathbb{E}^2$ be the space of matrices of dimensions $n \times n$, let $z \in \mathbb{E}^2$ and z^{sym} be the symmetric part of z. Define

$$\varrho(z) = ||z^{sym}||_{\text{spec}}, \quad z \in \mathbb{E}^2 \qquad z^{sym} = \frac{1}{2}(z + z^\top) \tag{3.10}$$

and introduce the set

$$\mathcal{W} = \left\{\boldsymbol{u} \in \mathcal{D}(\mathbb{R}^n, \mathbb{R}^n) \mid \varrho(\nabla \boldsymbol{u}) \leq 1\right\} \tag{3.11}$$

and its closure $\mathcal{U}_1(\mathbb{R}^n)$ in $C(\mathbb{R}^n, \mathbb{R}^n)$. One can prove that the conditions: $\boldsymbol{v} \in \mathcal{U}_1(\mathbb{R}^n)$ and (3.5) for $\Omega = \mathbb{R}^n$ are equivalent. Let us define the norm for the vector fields $\boldsymbol{v}$ defined on the domain Ω:

$$|||\boldsymbol{v}|||_{\Omega} = \sup \left\{ \frac{|[\boldsymbol{v}(\boldsymbol{x}) - \boldsymbol{v}(\boldsymbol{y})] \cdot (\boldsymbol{x} - \boldsymbol{y})|}{||\boldsymbol{x} - \boldsymbol{y}||^2} \,|\, \boldsymbol{x} \neq \boldsymbol{y}, \quad \boldsymbol{x} \in \Omega, \quad \boldsymbol{y} \in \Omega \right\} \tag{3.12}$$

This norm is inspired by the condition (3.5). If the domain Ω is convex, the following conditions are equivalent:

$$\text{(i)} \quad |||\boldsymbol{v}|||_{\Omega} \leq 1 \tag{3.13}$$

$$\begin{aligned} \text{(ii)} \quad & \boldsymbol{v} \in C(\bar{\Omega}, \mathbb{R}^n) \\ & \varepsilon(\boldsymbol{v}) \in L^{\infty}(\Omega, \mathbb{E}_s^2) \\ & \varrho(\nabla \boldsymbol{v}) \leq 1 \quad \text{a.e. in } \Omega \end{aligned} \tag{3.14}$$

where the last condition is equivalent to $|\varepsilon_I| \leq 1$, see Bouchitté et al. (2008). It is seen that the condition (i) has strong consequences; it implies continuity of the vector field and introduces the local upper bound on the principal strains.

We conjecture that in case of $n = 2$, the condition (i) implies continuity of the rigid rotations

$$\omega = \frac{1}{2} \left(\frac{\partial v_2}{\partial x_1} - \frac{\partial v_1}{\partial x_2} \right)$$

in the whole design domain. This property will be used frequently in the analytical constructions of the virtual fields satisfying (3.14).

Let us recall that the condition (3.13) can be interpreted as the condition bounding the relative elongation Δ_k / l_k of each potential bar linking two points (nodes) of the design domain Ω, i.e.: $|\Delta_k / l_k| \leq 1$. Thus, the condition (3.13) imposes conditions on the bar structure based on selected nodes within the design domain. On the other hand, the condition (3.9) (or the third condition of (3.14)) refers to the discrete-continuous structure. We note that the theorem on equivalence of (i) and (ii) paves the way from the theory of bars to the theory of discrete-continuous Michell's structures.

The condition of convexity of the design domain is not essential. The Michell structures may be constructed within non-convex domains.

Let us come back to the problem (2.6). It involves the scalar product of the force vector and the vector of virtual displacements of the nodes. This scalar product is replaced here by the value of a linear form $\mathcal{L}$ on the field $\boldsymbol{v}$, or by $\mathcal{L}(\boldsymbol{v})$. Assume that the load is self-equilibrated, or

$$\mathcal{L}(\boldsymbol{v}) = 0 \qquad \forall\, \boldsymbol{v} = \boldsymbol{v}^0 + \boldsymbol{\omega} \times \boldsymbol{x} \tag{3.15}$$

where $\boldsymbol{\omega}$ and $\boldsymbol{v}_0$ are vectors from $\mathbb{R}^n$. In case of $n = 3$, the fields of (3.15) are of the form

$$\begin{aligned} v_1 &= v_1^0 - \omega_3 x_2 + \omega_2 x_3 \\ v_2 &= v_2^0 + \omega_3 x_1 - \omega_1 x_3 \\ v_3 &= v_3^0 - \omega_2 x_1 + \omega_1 x_2 \end{aligned} \tag{3.16}$$

If $n = 2$, the field $\boldsymbol{v}$ in (3.15) has the form

$$v_1 = v_1^0 - \omega_3 x_2, \qquad v_2 = v_2^0 + \omega_3 x_1 \tag{3.17}$$

The condition $\mathcal{L}(\boldsymbol{v}_0) = 0$ means that the resulting vector of the load is a zero vector. The condition $\mathcal{L}(\boldsymbol{\omega} \times \boldsymbol{x}) = 0$ means that the resulting moment of the loads vanishes. If the load is self-equilibrated, the natural extension of problem (2.6) to encompass the bar structures with nodes arbitrarily selected from the domain Ω reads

$$\mathcal{Z}_1 = \sup \{\mathcal{L}(\boldsymbol{v}) \mid \boldsymbol{v} \in \mathcal{U}_1(\Omega)\} \tag{3.18}$$

Let us note that the set $\mathcal{U}_1(\Omega)$ is non-empty and compact, while the form $\mathcal{L}(\cdot)$ is linear and continuous. We conclude that the problem (3.18) is well posed. In general, the solution is non-unique.

Let us write the problem (3.18) more explicitly

$$\begin{aligned} \mathcal{Z}_1 = \sup \big\{ & \mathcal{L}(\boldsymbol{v}) \mid \boldsymbol{v} \in C(\bar{\Omega}, \mathbb{R}^n), \\ & \boldsymbol{\varepsilon}(\boldsymbol{v}) \in L^\infty(\Omega, \mathbb{E}_s^2), \quad ||\boldsymbol{\varepsilon}(\boldsymbol{v})||_{\text{spec}} \leq 1 \text{ a.e. in } \Omega \big\} \end{aligned} \tag{3.19}$$

The problem (3.19) may be extended to the case of the load being non-self-equilibrated. Then, the field $\boldsymbol{v}$ should satisfy appropriate kinematic conditions: this field should vanish on a segment Γ_2 of the boundary of the domain Ω; the problem (3.19) extends to the form:

$$\begin{aligned} \mathcal{Z}_1 = \sup \big\{ & \mathcal{L}(\boldsymbol{v}) \mid \boldsymbol{v} \in C(\bar{\Omega}, \mathbb{R}^n), \quad \boldsymbol{v} = \boldsymbol{0} \text{ on } \Gamma_2, \\ & \boldsymbol{\varepsilon}(\boldsymbol{v}) \in L^\infty(\Omega, \mathbb{E}_s^2), \quad ||\boldsymbol{\varepsilon}(\boldsymbol{v})||_{\text{spec}} \leq 1 \text{ a.e. on } \Omega \big\} \end{aligned} \tag{3.20}$$

According to (2.50), the volume of the optimal Michell truss equals

$$V_{\min} = \frac{1}{\sigma_0} \mathcal{Z} \tag{3.21}$$

where $\mathcal{Z} = \mathcal{Z}_1$.

In the 2D setting, the problem (3.20) should be completed by the condition concerning boundedness of the virtual strain ε_Γ of the reinforcing bars along some line Γ_r. Thus, the conditions involved in (3.20) should be completed by

$$|\varepsilon_\Gamma(s)| \leq 1 \text{ on } \Gamma_r \tag{3.22}$$

The formula for the strain measure ε_Γ involves geometric characteristics of the curve Γ_r; this formula will be given in Sect. 4.2, see Remark 4.16.1. In the 3D setting, there can appear also membrane-like reinforcements, then additional conditions on the membrane strains will be added, see Chap. 5.

In the 3D setting, the load can be distributed within the design domain as a body force, applied as traction on a segment Γ_2 of the boundary of the given design domain, but also can be concentrated along curves or at points. In the 2D setting, the load can be body forces continuously distributed in the given plane domain, can be an in-plane traction applied along a given segment Γ_2 of the contour or can be an in-plane point load. The value $\mathcal{L}(\boldsymbol{v})$ is the work done by the given load on the virtual displacement $\boldsymbol{v}$. This work can be viewed as a dual product $\langle \boldsymbol{F}, \boldsymbol{v} \rangle$, where $\boldsymbol{F}$ is an element of the space dual to $\mathcal{U}_1(\Omega)$. Just this notation is usually used in mathematical papers to specify the quantity $\mathcal{L}(\boldsymbol{v})$.

3.2 Static Formulation. Case of $\sigma_C = \sigma_T$

In the preceding section, the kinematic formulation (2.6) concerning bar structures has been extended to the formulation (3.18) referring to Michell's structures by using James Clerk Maxwell's concept of an embedding of the bar structure into a virtual displacement field, reflected in Eq. (3.1). The setting (3.18)–(3.19) has been put forward only in 2008, yet it is consistent with Michell's (1904) formulation of the optimum design of fully stressed frames of least volume.

It is much more difficult to extend the stress-based formulation (2.8) to its stress-based counterpart in the theory of Michell structures. This extension has not been given by Michell (1904). The present section is just aimed at extending the setting (2.8) to a stress-based problem being dual to the kinematic problem (3.18) derived in the previous section. The procedure is not obvious from the point of view of structural mechanics since there is no direct link between forces in bars T_k and components of a stress field replacing somehow the stress resultants T_k.

The problem (3.18) involves the spectral norm of the strain field, see (3.7). In the problem dual to (3.18), there will be involved a norm dual to the spectral norm. Let us recall now the definition of a norm dual to a given norm $||\cdot||_0$ in $\mathbb{R}^n$. Let $\boldsymbol{x} \in \mathbb{R}^n$. The dual norm of $\boldsymbol{x}$ is defined by

$$||\boldsymbol{x}||_0^{\star} = \max_{\boldsymbol{y} \in \mathbb{R}^n \setminus \{\boldsymbol{0}\}} \frac{\boldsymbol{x} \cdot \boldsymbol{y}}{||\boldsymbol{y}||_0} \tag{3.23}$$

If $||\cdot||_0 = ||\cdot||$, or $||\cdot||_0$ is the Euclidean norm, then $||\cdot||_0^{\star} = ||\cdot||$. Thus, the norm dual to the Euclidean norm is the Euclidean norm. In this case, the duality operation (3.23) does not change the definition of the norm. In other cases, the duality operation changes the norm. For instance, assume that a matrix $\boldsymbol{A}$ of dimensions $n \times n$ is symmetric and positive definite. Such a matrix defines the norm in $\mathbb{R}^n$ by

$$||\boldsymbol{x}||_A = \sqrt{\boldsymbol{x}^{\top} \boldsymbol{A} \boldsymbol{x}} \tag{3.24}$$

One can easily prove that the norm dual to the norm above is expressed by

$$||\boldsymbol{x}||_A^\star = \sqrt{\boldsymbol{x}^\top \boldsymbol{A}^{-1}\boldsymbol{x}} \tag{3.25}$$

hence involves the matrix inverse to $\boldsymbol{A}$.

The problem dual to (3.18) involves the norm dual to the spectral norm (3.7). Let $\tau \in \mathbb{E}_s^2$. We define

$$||\boldsymbol{\tau}||_{\text{spec}}^\star = \max_{\varepsilon \in \mathbb{E}_s^2 \setminus \{\boldsymbol{0}\}} \frac{\boldsymbol{\tau}\cdot\boldsymbol{\varepsilon}}{||\boldsymbol{\varepsilon}||_{\text{spec}}} \tag{3.26}$$

The result of the maximization operation in (3.26) is known; it reads

$$||\boldsymbol{\tau}||_{\text{spec}}^\star = \sum_{i=1}^{n} |\lambda_i(\boldsymbol{\tau})| \tag{3.27}$$

see Rockafellar (1970, Sec. 15) and Bouchitté et al. (2008, Sec. 2). Moreover, the norm dual to (3.27) coincides with (3.7). From the definition (3.26), we obtain the upper bound

$$\boldsymbol{\tau}\cdot\boldsymbol{\varepsilon} \le ||\boldsymbol{\tau}||_{\text{spec}}^\star ||\boldsymbol{\varepsilon}||_{\text{spec}} \tag{3.28}$$

and one of the conditions of making this bound sharp is the collinearity of tensors $\boldsymbol{\tau}$ and $\boldsymbol{\varepsilon}$: these tensors should have a common set of the eigenvectors. Let $n = 3$ and $\text{diag}(\tau_1, \tau_2, \tau_3)$, $\text{diag}(\varepsilon_1, \varepsilon_2, \varepsilon_3)$ are representations of these tensors in the common eigenbasis. The equality in (3.28) implies

$$\tau_1\varepsilon_1 + \tau_2\varepsilon_2 + \tau_3\varepsilon_3 = (|\tau_1| + |\tau_2| + |\tau_3|) \max_j \left\{|\varepsilon_j|\right\} \tag{3.29a}$$

and this is possible if the tensor $\boldsymbol{\varepsilon}$ is appropriately chosen for the given tensor $\boldsymbol{\tau}$, namely (do not sum over i)

$$\tau_i\varepsilon_i = |\tau_i| \max_j \left\{|\varepsilon_j|\right\} \qquad i = 1, 2, 3 \tag{3.29b}$$

or

$$\varepsilon_i = \text{sgn}\,(\tau_i) \max_j \left\{|\varepsilon_j|\right\} \qquad i = 1, 2, 3 \qquad \text{if } \tau_i \neq 0; \tag{3.30}$$

otherwise ε_i is arbitrary.

Consider now the problem of optimum design of a fibrous structure (i.e. having a truss-like microstructure) loaded by tractions of intensity $\boldsymbol{p}$ on part Γ_1 of the boundary and supported on the part Γ_2 of the boundary. The load causes the state of stress within the body expressed by the tensor field $\boldsymbol{\tau} = (\tau_{ij})$ within the domain Ω. The equations of equilibrium of the interior and of the boundary layer near Γ_1 are encompassed by the variational equation linking the stress field with the load:

$$\int \boldsymbol{\tau}\cdot\boldsymbol{\varepsilon}(\boldsymbol{v})dx = \mathcal{L}(\boldsymbol{v}) \quad \forall \quad \boldsymbol{v} \in \mathcal{V}(\Omega) \tag{3.31}$$

where $\mathcal{V}(\Omega)$ is the set of kinematically admissible displacement field $\boldsymbol{v}$. The regularity conditions will not be reported; the operation $\boldsymbol{\varepsilon}(\boldsymbol{v})$ is given by (3.6), i.e. it defines the symmetric part of the gradient of $\boldsymbol{v}$. The right-hand side of (3.31) represents the virtual work of the traction load $\boldsymbol{p}$ applied on Γ_1, or

$$\mathcal{L}(\boldsymbol{v}) = \int_{\Gamma_1} \boldsymbol{p} \cdot \boldsymbol{v} d\Gamma \tag{3.32}$$

The set of stress fields $\boldsymbol{\tau}$ in Ω satisfying (3.31), and hence also satisfying indispensable regularity assumptions, is denoted by $\Sigma_p(\Omega)$ and is called the set of statically admissible stress fields. In contrary, a set of stress fields defined in the same domain, being locally tensors of the set $\mathbb{E}_s^2$, satisfying the same regularity assumptions, will be denoted by $\mathcal{S}(\Omega, \mathbb{E}_s^2)$. Thus $\Sigma_p(\Omega) \subset \mathcal{S}(\Omega, \mathbb{E}_s^2)$.

Consider the optimum design problem

$$\mathcal{Z}_2 = \inf \left\{ \int ||\boldsymbol{\tau}||^{\star}_{\text{spec}} dx \mid \boldsymbol{\tau} \in \Sigma_p(\Omega) \right\} \tag{3.33}$$

Putting now apart its meaning, let us rearrange this problem to its dual setting following the methods known from the linear programming. The problem (3.33) is equivalent to

$$\mathcal{Z}_2 = \inf_{\boldsymbol{\tau} \in \mathcal{S}(\Omega, \mathbb{E}_s^2)} \left\{ \int ||\boldsymbol{\tau}||^{\star}_{\text{spec}} dx + \sup_{\boldsymbol{v} \in \mathcal{V}(\Omega)} \left\{ \mathcal{L}(\boldsymbol{v}) - \int_{\Omega} \boldsymbol{\tau} \cdot \boldsymbol{\varepsilon}(\boldsymbol{v}) dx \right\} \right\} \tag{3.34}$$

Assume now that the operations: *inf* and *sup* may be interchanged, which gives

$$\mathcal{Z}_2 = \sup \left\{ \mathcal{L}(\boldsymbol{v}) + g(\boldsymbol{v}) \mid \boldsymbol{v} \in \mathcal{V}(\Omega) \right\} \tag{3.35}$$

where

$$g(\boldsymbol{v}) = -\sup \left\{ \int_{\Omega} [\boldsymbol{\tau} \cdot \boldsymbol{\varepsilon}(\boldsymbol{v}) - ||\boldsymbol{\tau}||^{\star}_{\text{spec}}] dx \mid \boldsymbol{\tau} \in \mathcal{S}(\Omega, \mathbb{E}_s^2) \right\} \tag{3.36}$$

Let

$$\boldsymbol{\tau} = \rho \hat{\boldsymbol{\tau}}, \qquad \rho = ||\boldsymbol{\tau}||^{\star}_{\text{spec}}, \qquad ||\hat{\boldsymbol{\tau}}||^{\star}_{\text{spec}} = 1$$

Now we rewrite (3.36) in the form

$$g(\boldsymbol{v}) = -\sup \left\{ \int_{\Omega} [\boldsymbol{\varepsilon}(\boldsymbol{v}) \cdot \hat{\boldsymbol{\tau}} - 1] \rho dx \mid \rho \geq 0, \quad ||\hat{\boldsymbol{\tau}}||^{\star}_{\text{spec}} = 1 \quad \text{a.e. in } \Omega \right\} \tag{3.37}$$

The conditions involved in (3.37) are algebraic, hence

$$g(\boldsymbol{v}) = -\sup \left\{ \int_{\Omega} \max_{||\hat{\boldsymbol{\tau}}||^{\star}_{\text{spec}} = 1} [\boldsymbol{\varepsilon}(\boldsymbol{v}) \cdot \hat{\boldsymbol{\tau}} - 1] \rho dx \mid \rho \geq 0 \text{ in } \Omega \right\} \tag{3.38}$$

Thus, taking into account that $(||\cdot||^{\star}_{\text{spec}})^{\star} = ||\cdot||_{\text{spec}}$, we write the above formulation as

$$g(\boldsymbol{v}) = -\sup\left\{\int_{\Omega}[||\varepsilon(\boldsymbol{v})||_{\text{spec}} - 1]\rho dx \mid \rho \geq 0 \text{ in } \Omega\right\} \tag{3.39}$$

hence

$$g(\boldsymbol{v}) = \begin{cases} 0 & \text{if } ||\varepsilon(\boldsymbol{v})||_{\text{spec}} \leq 1 \\ -\infty & \text{otherwise} \end{cases} \tag{3.40}$$

We come back to (3.35) and reduce it to the form

$$\mathcal{Z}_2 = \sup\left\{\mathcal{L}(\boldsymbol{v}) \mid \boldsymbol{v} \in \mathcal{V}(\Omega) \text{ and } ||\varepsilon(\boldsymbol{v})||_{\text{spec}} \leq 1 \text{ a.e. in } \Omega\right\} \tag{3.41}$$

or to the problem (3.18), where the regularity assumptions are put forward.

The above passage from (3.33) to (3.41) (or to (3.18)) suggests that the problem (3.33), upon making the regularity assumptions more precise, is dual to (3.18).

Let $\check{\boldsymbol{v}}$ be a solution to the problem (3.18) while $\check{\boldsymbol{\tau}}$ be a solution to problem (3.33). The equality $\mathcal{Z}_1 = \mathcal{Z}_2$ holds under the condition (3.30). Indeed, we require that

$$\mathcal{Z}_2 = \int_{\Omega} ||\check{\boldsymbol{\tau}}||^{\star}_{\text{spec}} dx = \mathcal{Z}_1 = \mathcal{L}(\check{\boldsymbol{v}}) \tag{3.42}$$

but

$$\mathcal{L}(\check{\boldsymbol{v}}) = \int_{\Omega} \boldsymbol{\tau} \cdot \varepsilon(\check{\boldsymbol{v}}) dx \quad \forall \quad \boldsymbol{\tau} \in \Sigma_p(\Omega) \tag{3.43}$$

Let us take $\boldsymbol{\tau} = \check{\boldsymbol{\tau}}$; then (3.42) results in

$$\int_{\Omega} ||\check{\boldsymbol{\tau}}||^{\star}_{\text{spec}} dx = \int_{\Omega} \check{\boldsymbol{\tau}} \cdot \varepsilon(\check{\boldsymbol{v}}) dx \tag{3.44}$$

while taking into account (3.28), we have

$$\int_{\Omega} ||\check{\boldsymbol{\tau}}||^{\star}_{\text{spec}} dx \leq \int_{\Omega} ||\check{\boldsymbol{\tau}}||^{\star}_{\text{spec}} ||\varepsilon(\check{\boldsymbol{v}})||_{\text{spec}} dx \tag{3.45}$$

and the equality is possible only if

$$||\varepsilon(\check{\boldsymbol{v}})||_{\text{spec}} = 1 \tag{3.46}$$

there, where $||\check{\boldsymbol{\tau}}||^{\star} \neq 0$. Moreover, the equality in (3.45) means

$$\check{\boldsymbol{\tau}} \cdot \varepsilon(\check{\boldsymbol{v}}) = ||\check{\boldsymbol{\tau}}||^{\star}_{\text{spec}} \cdot ||\varepsilon(\check{\boldsymbol{v}})||_{\text{spec}} \tag{3.47}$$

and this equality determines the values $\lambda_i(\varepsilon(\boldsymbol{v}))$ on the basis of the stress field $\check{\boldsymbol{\tau}}$ according to (3.30), or

$$\lambda_i\left(\varepsilon(\breve{v})\right) = \operatorname{sgn}\left(\lambda_i(\breve{\tau})\right) \max_j \left\{|\lambda_j\left(\varepsilon(\breve{v})\right)|\right\} \tag{3.48}$$

if $\lambda_i(\breve{\tau}) \neq 0$. Otherwise, if $\lambda_i(\breve{\tau}) = 0$, the only condition for the ith eigenvalue of $\varepsilon(\breve{v})$ is

$$|\lambda_i(\varepsilon(\breve{v}))| \leq 1 \tag{3.49}$$

Let us imagine that the body has a fibrous microstructure and the load on Γ_1 is transmitted to the support Γ_2 along the trajectories of the stress field $\breve{\tau}$ which is the minimizer of (3.33). The field $\breve{\varepsilon} = \varepsilon(\breve{v})$, where $\breve{v}$ being a maximizer of problem (3.18) must be suited to the field $\breve{\tau}$ such that the equalities (3.9_a) and (3.46) hold, or

$$\sum_{j=1}^{n} \lambda_j(\breve{\tau})\lambda_j(\breve{\varepsilon}) = \sum_{j=1}^{n} |\lambda_j(\breve{\tau})| \tag{3.50}$$

and this condition is satisfied if (3.48), (3.49) are fulfilled. Thus

$$\begin{aligned} \lambda_i(\breve{\tau}) > 0 \quad &\Rightarrow \quad \lambda_i(\breve{\varepsilon}) = 1 \\ \lambda_i(\breve{\tau}) = 0 \quad &\Rightarrow \quad |\lambda_i(\breve{\varepsilon})| \leq 1 \\ \lambda_i(\breve{\tau}) < 0 \quad &\Rightarrow \quad \lambda_i(\breve{\varepsilon}) = -1 \end{aligned} \tag{3.51a}$$

for $i = 1, \ldots, n$. The cases $n = 2$ and $n = 3$ will be discussed independently. Let $n = 2$. Let

$$\breve{\tau}_I = \max_{1\leq i\leq 2} \{\lambda_i(\breve{\tau})\}, \qquad \breve{\tau}_{II} = \min_{1\leq i\leq 2} \{\lambda_i(\breve{\tau})\} \tag{3.51b}$$

In a similar manner the quantities $\breve{\varepsilon}_I \geq \breve{\varepsilon}_{II}$ are defined. The following cases are possible:

$$\begin{aligned}
&\text{a)}\ \breve{\tau}_I > 0, \quad \breve{\tau}_{II} > 0 \\
&\quad \text{then} \quad \breve{\varepsilon}_I = \breve{\varepsilon}_{II} = 1 \\
&\text{b)}\ \breve{\tau}_I > 0, \quad \breve{\tau}_{II} < 0 \\
&\quad \text{then} \quad \breve{\varepsilon}_I = 1, \quad \breve{\varepsilon}_{II} = -1 \\
&\text{c)}\ \breve{\tau}_I > 0, \quad \breve{\tau}_{II} = 0 \\
&\quad \text{then} \quad \breve{\varepsilon}_I = 1, \quad |\breve{\varepsilon}_{II}| \leq 1 \\
&\text{d)}\ \breve{\tau}_I < 0, \quad \breve{\tau}_{II} < 0 \\
&\quad \text{then} \quad \breve{\varepsilon}_I = -1, \quad \breve{\varepsilon}_{II} = -1 \\
&\text{e)}\ \breve{\tau}_I = 0, \quad \breve{\tau}_{II} = 0 \\
&\quad \text{then} \quad |\breve{\varepsilon}_I| \leq 1, \quad |\breve{\varepsilon}_{II}| \leq 1 \\
&\text{f)}\ \breve{\tau}_I = 0, \quad \breve{\tau}_{II} < 0 \\
&\quad \text{then} \quad |\breve{\varepsilon}_I| \leq 1, \quad \breve{\varepsilon}_{II} = -1
\end{aligned} \tag{3.52}$$

In the theory of Michell structures, the cases (b), (c), (f) are the most important.

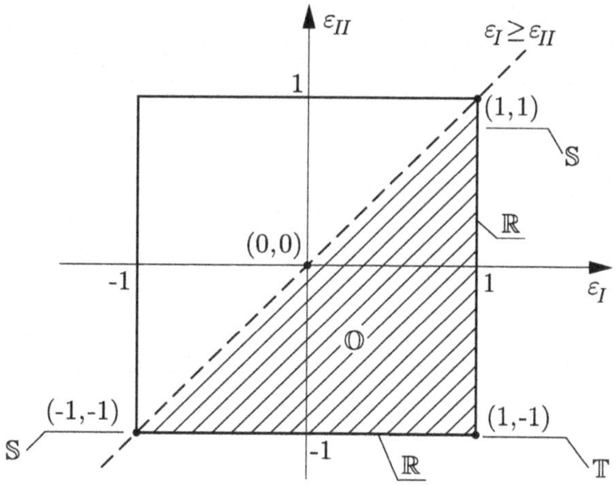

Fig. 3.2 The locking locus for the case of $\sigma_C = \sigma_T$

In case (b), the body is composed of two mutually orthogonal families of fibres. At each point of the body, the two conditions: $\breve{\varepsilon}_I = 1$ and $\breve{\varepsilon}_{II} = -1$ are simultaneously fulfilled and they impose strong geometric constraints on such families of curves. Such lines are known from the theory of plasticity as Hencky nets. Their properties will be discussed in the sequel.

In cases (c) and (f), the load is transmitted by a single family of fibres; in many cases, these fibres form circular fans.

In cases (a) and (d), the microstructure is not unique.

In case (e), a subdomain of the design domain becomes empty. Then in many examples, $\breve{\varepsilon}_I = 0$ and $\breve{\varepsilon}_{II} = 0$ and the virtual field $\breve{\boldsymbol{v}}$ corresponds to a small rigid rotation.

The set (3.9) is called a locking locus. If $n = 2$, in the plane ε_I and ε_{II}, the locking locus is the interior of the square (in fact, a half of the square, since $\varepsilon_I \geq \varepsilon_{II}$, see Fig. 3.2.

The domain Ω is divided into subdomains corresponding to the vertices, the sides or the interior of the locking locus. Let us introduce the notation used in papers by Prager and Rozvany:

$$\begin{aligned}
\mathbb{O}: &\quad -1 < \varepsilon_I < 1, \quad \varepsilon_{II} > -1, \quad \varepsilon_I \geq \varepsilon_{II} \\
\mathbb{S}: &\quad \text{vertices } (-1,-1),\ (1,1) \\
\mathbb{T}: &\quad \text{the vertex } (1,-1) \\
\mathbb{R}: &\quad \text{the sides } \varepsilon_I = 1, \quad -1 < \varepsilon_{II} < 1 \\
&\quad \text{and } \varepsilon_{II} = -1, \quad -1 < \varepsilon_I < 1
\end{aligned}$$

In the problem of optimum design of trusses discussed in Sect. 2.2, the objective function was the volume of the truss, see (2.41). Admitting fibrous structures extends the formula for the volume to the form (3.21) where $\mathcal{Z} = \mathcal{Z}_1$ and according to (3.42), we have $\mathcal{Z} = \mathcal{Z}_2$, where $\mathcal{Z}_2$ is given by (3.33). If the field $\breve{\boldsymbol{\tau}}$ is the minimizer of (3.33), then the formula for the optimum volume may be expressed as below

$$V_{\min} = \int_{\Omega} \upsilon dx \tag{3.53}$$

where

$$\upsilon = \frac{1}{\sigma_0} ||\breve{\boldsymbol{\tau}}||^{\star}_{\text{spec}} \tag{3.54}$$

represents the volume fraction of the material of fibrous microstructure, υ being a nondimensional quantity describing the volume fraction of fibres. Let us note that in the optimum truss problem for the case of $\sigma_C = \sigma_T = \sigma_0$ the counterpart of the formula (3.53) has the form

$$V_{\min} = \sum_{k=1}^{e} \breve{V}_k \tag{3.55}$$

where

$$\breve{V}_k = \frac{1}{\sigma_0} |\breve{T}_k| l_k \tag{3.56}$$

is the volume of the kth bar and $\breve{T}_k$ is the axial force in this bar.

Remark 3.2.1 If $n = 2$, the problem (3.33) corresponds to the result on optimal distribution of a small amount of an isotropic and homogeneous material to minimize the compliance of the structure to be formed from this material. The formulation (3.33) appears in the relaxed formulation performed via homogenization, as shown in Allaire (2002, Sect. 4.2.3) and in Lewiński and Telega (2000, pp. 686–687). In Sect. 3.3, we shall show that the problem (3.33) arises in the problem of optimal design of Young's modulus distribution of a material of specific constitutive properties, sensitive to the sign of det $\boldsymbol{\sigma}$.

Remark 3.2.2 The formulation (3.33) does not encompass the case of accumulating the axial stiffnesses *EA* along curves Γ_r along which the reinforcing bars are placed. Such bars appear to transmit the applied point loads thus avoiding stress singularities. Usually these bars lie along the boundary, but this is not the rule. The stress resultants in these bars transmit tension (then $F = F_T$) or compression (then $F = F_C$). The forces F along with the stress field $\boldsymbol{\tau}$ satisfy the equilibrium equations, which will be denoted by: $(\boldsymbol{\tau}, F) \in \Sigma_p(\Omega, \Gamma_r)$. The problem (3.33) for $n = 2$ should be written, in general, as below

$$\mathcal{Z}_2 = \inf \left\{ \int_{\Omega} ||\boldsymbol{\tau}||^{\star}_{\text{spec}} dx + \int_{\Gamma_r} |F| ds \mid (\boldsymbol{\tau}, F) \in \Sigma_p(\Omega, \Gamma_r) \right\} \tag{3.57}$$

to take into account the occurrence of the reinforcing bars.

Remark 3.2.3 If $n = 3$ the fields τ_{ij} are stresses inside the body; their units are [Pa]. If $n = 2$, these fields represent stress resultants in a plate loaded in its plane. The units are then [N/m]. In the papers on shells, the membrane stress resultants are usually denoted by $N^{\alpha\beta}$, see Naghdi (1963). Yet in the present book, we do not introduce a new notation for stress resultants in plates loaded in plane, the notation τ_{ij} will be still used, despite different meaning and even different units than in the 3D case.

Remark 3.2.4 In the spatial setting, only a small number of exact solutions to Michell's problem are at our disposal. Our experience in constructing such solutions is very modest. The 3D solutions strongly depend on the type of the load: whether the load is a point load, a load along a curve or a traction on a surface.

In case of the point loads and the loads along curves we expect that one of principal stresses $\lambda_i(\breve{\boldsymbol{\tau}})$ vanishes. Then the optimal body will be composed of two families of fibres forming a surface and the whole solution will be composed of many surfacial segments, see Chap. 5.

Remark 3.2.5 In the 3D setting, the problem (3.33) does not correspond to the solution to the problem of minimum compliance of a porous body. If relaxed by homogenization, the problem is rearranged to a problem which for the case of a small amount of the material tends to the generated problem of the form, see Allaire (2002), p. 339

$$\mathcal{Z}_2^H = \inf\left\{\int_\Omega \rho_{3D}(\boldsymbol{\tau})dx \mid \boldsymbol{\tau} \in \Sigma_p(\Omega)\right\} \tag{3.58}$$

the integrand being

$$\rho_{3D}(\boldsymbol{\tau}) = \begin{cases} |\tau_I| + |\tau_{II}| + |\tau_{III}| & \text{if } \ |\tau_I| \le |\tau_{II}| + |\tau_{III}| \\ \sqrt{(|\tau_{II}| + |\tau_{III}|)^2 + |\tau_I|^2} & \text{otherwise} \end{cases} \tag{3.59}$$

Here, $\tau_I \geq \tau_{II} \geq \tau_{III}$ are ordered principal stresses.

Having in mind the discrepancy between the formulations (3.58) and (3.33), we state that the 3D Michell structures are not solutions to the problem of minimum compliance of a porous body of a small mass. It is easy to note that

$$\rho_{3D}(\boldsymbol{\tau}) \le ||\boldsymbol{\tau}||^{\star}_{\text{spec}} \tag{3.60}$$

since for $a > 0$, $b > 0$, $c > 0$ we have

$$\sqrt{(a+b)^2 + c^2} < (a+b) + c. \tag{3.61}$$

Thus we can expect that the quantity $\mathcal{Z}_2^H$ is smaller than $\mathcal{Z}_2$. Thus, according to (2.39), the compliance of the optimal spatial body of a small mass is smaller that the compliance of the corresponding Michell's structure. This justifies the numerical results in Sigmund et al. (2016). Thus, we may conclude that the truss-based point of departure in the spatially formulated optimum design problem is a stronger constraint than in the planar optimum design.

3.3 Reconstruction of the Optimal Continuum Body of Least Compliance Corresponding to Michell's Problem

The problem of compliance minimization of trusses (viz. structures composed of finite number of bars) leads to two mutually dual problems: (2.6) and (2.8). It has been shown in Sects. 3.1 and 3.2 that the problems (2.6) and (2.8) can be extended to encompass fibrous solutions without a reference to the initial problem (2.4) of minimum compliance. A question arises, whether problem (2.4) corresponds to the problem of minimum compliance of a solid body.

Consider a body made from a material whose density energy has the form

$$j^{\star}(\boldsymbol{\sigma}) = \frac{1}{2E}\left(||\boldsymbol{\sigma}||^{\star}_{\text{spec}}\right)^{2} \tag{3.62}$$

for $\boldsymbol{\sigma} \in \mathbb{E}^2_s$; here E is a reference modulus of units: [Pa]. Assume that the problem of statics is expressed by the counterpart of Castigliano's theorem

$$\wp(E) = \inf\left\{\int_{\Omega} 2j^{\star}(\boldsymbol{\tau})dx \mid \boldsymbol{\tau} \in \Sigma_p(\Omega)\right\} \tag{3.63}$$

Let $E = E(x) \geq 0$ be the design variable subject to the cost condition

$$\int_{\Omega} Edx = \Lambda \tag{3.64}$$

Consider the problem of optimum material design concerning the optimal layout of the modulus E within the given design domain Ω

$$\breve{\wp} = \inf\left\{\wp(E) \mid E \geq 0, \quad \int_{\Omega} Edx = \Lambda\right\} \tag{3.65}$$

According to Lemma 4.1 and the results of Sects. 3.1 and 3.2, we find

$$\breve{\wp} = \frac{1}{\Lambda}Z^2 \tag{3.66}$$

where $Z = \mathcal{Z}_1$, $\mathcal{Z}_1$ being given by (3.20), or $Z = \mathcal{Z}_2$, $\mathcal{Z}_2$ given by (3.33). According to (2.80), the optimal modulus $\breve{E}$ is given by

$$\breve{E}(x) = \Lambda\frac{||\breve{\boldsymbol{\tau}}(x)||^{\star}_{\text{spec}}}{\displaystyle\int_{\Omega}||\breve{\boldsymbol{\tau}}||^{\star}_{\text{spec}}dx} \tag{3.67}$$

where $\breve{\boldsymbol{\tau}}$ is the minimizer of (3.33). The modulus $\breve{E}$ is determined in the subdomain, where $\breve{\boldsymbol{\tau}} \neq 0$, or in the effective domain of $\breve{\boldsymbol{\tau}}$ which determines the optimal shape of the body. The problem (3.65) solves simultaneously two problems: the problem of shape optimization and the problem of optimal distribution of the modulus $\check{E}$.

If $n = 2$, the elastic potential can be expressed as below

$$2Ej^{\star}(\boldsymbol{\sigma}) = \begin{cases} (\sigma_I)^2 + (\sigma_{II})^2 + 2\sigma_I\sigma_{II} & \text{if} \quad \sigma_I\sigma_{II} \geq 0 \\ (\sigma_I)^2 + (\sigma_{II})^2 - 2\sigma_I\sigma_{II} & \text{if} \quad \sigma_I\sigma_{II} \leq 0 \end{cases} \tag{3.68}$$

where $\sigma_I = \lambda_1(\sigma)$, $\sigma_{II} = \lambda_2(\sigma)$, $\sigma_I \geq \sigma_{II}$.

Equivalently, we can write

$$Ej^{\star}(\boldsymbol{\sigma}) = \begin{cases} \frac{1}{2}(\operatorname{tr}\boldsymbol{\sigma})^2 & \text{if} \quad \det\boldsymbol{\sigma} \geq 0 \\ ||\operatorname{dev}\boldsymbol{\sigma}||^2 & \text{if} \quad \det\boldsymbol{\sigma} \leq 0 \end{cases} \tag{3.69}$$

where $\operatorname{tr}\boldsymbol{\sigma} = \sigma_I + \sigma_{II}$, $\operatorname{dev}\boldsymbol{\sigma} = \boldsymbol{\sigma} - \frac{1}{2}\operatorname{tr}\boldsymbol{\sigma}\boldsymbol{I}$, $\boldsymbol{I}$ being a unit tensor in $\mathbb{E}^2_s$; $||\cdot||$ is the Euclidean norm.

Let us recall the expression for the complementary energy of the isotropic body (in the 2D setting)

$$j^{\star}_{iso}(\boldsymbol{\sigma}) = \frac{1}{4k}\left(\frac{\operatorname{tr}\boldsymbol{\sigma}}{\sqrt{2}}\right)^2 + \frac{1}{4\mu}||\operatorname{dev}\boldsymbol{\sigma}||^2 \tag{3.70}$$

where k, μ are the bulk and shear moduli defined within the 2D setting. Let

$$k = \begin{cases} \frac{1}{2}E & \text{if} \quad \det\boldsymbol{\sigma} \geq 0 \\ +\infty & \text{otherwise} \end{cases} \tag{3.71}$$

$$\mu = \begin{cases} +\infty & \text{if} \quad \det\boldsymbol{\sigma} \geq 0 \\ \frac{1}{4}E & \text{otherwise} \end{cases} \tag{3.72}$$

Under this interpretation of the moduli k, μ, the complementary energy density (3.62) may be interpreted as referring to an isotropic body whose bulk modulus tends to infinity if $\det\boldsymbol{\sigma} \leq 0$ while the shear modulus tends to infinity if $\det\boldsymbol{\sigma} \geq 0$. The Young modulus E_m of the isotropic body is given by

$$\frac{2}{E_m} = \frac{1}{2k} + \frac{1}{2\mu} \tag{3.73}$$

or

$$E_m = \begin{cases} 2E & \text{if} \quad \det\boldsymbol{\sigma} \geq 0 \\ E & \text{otherwise} \end{cases} \tag{3.74}$$

The Poisson coefficient ν_m for the 2D setting is given by

$$\nu_m = \frac{k-\mu}{k+\mu} \tag{3.75}$$

Thus, the Poisson ratio of the body of energy density (3.69) is given by

$$\nu_m = \begin{cases} -1 & \text{if} \quad \det \boldsymbol{\sigma} \geq 0 \\ 1 & \text{otherwise} \end{cases} \tag{3.76}$$

We note that the moduli E, ν suffer jumps if $\det \boldsymbol{\sigma}$ changes its sign. Note that the quantities $\det \boldsymbol{\sigma}$ and $\sigma_{11}\sigma_{22}$ have the same signs independently of the observation frame x_1, x_2. Despite the jumps of the moduli E, ν the function (3.69) is continuous, since (3.69) may be expressed as below

$$Ej^{\star}(\boldsymbol{\sigma}) = \begin{cases} \frac{1}{2}(\operatorname{tr} \boldsymbol{\sigma})^2 & \text{if} \quad \det \boldsymbol{\sigma} \geq 0 \\ \frac{1}{2}(\operatorname{tr} \boldsymbol{\sigma})^2 - 2\det \boldsymbol{\sigma} & \text{otherwise} \end{cases} \tag{3.77}$$

The above interpretation adds a new meaning to problem (3.65). It is a formation of an optimal body whose response is dependent on the sign of $\det \boldsymbol{\sigma}$. The problem (3.65) is similar to the problem of minimum compliance of bodies the constitutive laws of which are dissymmetric in tension–compression, see Desmorat and Duvaut (2003).

3.4 Kinematic Formulation for the 2D and 3D Settings. Case of $\sigma_C \neq \sigma_T$

In case of $\sigma_C = \sigma_T$, the Michell problem arises as an auxiliary problem appearing within the theory of the minimum compliance problem, see Sect. 3.3. In case of $\sigma_C \neq \sigma_T$, the Michell problem is not linked with the problem of minimum compliance but is associated with the problem of minimum weight of trusses made of bars whose cross sections can be infinitely thin.

Let us extend the problem (2.53) to the case of trusses of an infinite number of bars. Assume that the vector $\bar{\boldsymbol{u}} = (\bar{u}_1, \ldots, \bar{u}_i, \ldots)$ is linked with the vector field $\boldsymbol{v}$ defined on $\Omega \subset \mathbb{R}^n$ by Eqs. (3.1). These equations state that the nodes $\boldsymbol{x}_i$ of a truss displace according to the values of the field $\boldsymbol{v}$ at these nodes. If the nodes $\boldsymbol{x}_i$ fill up the set Ω densely, then the conditions

$$-\frac{\sigma_0}{\sigma_C} \leq \bar{\varepsilon}_k = \frac{1}{l_k}\Delta_k(\bar{\boldsymbol{u}}) \leq \frac{\sigma_0}{\sigma_T} \tag{3.78}$$

involved in (2.53) must be satisfied along each direction around a given node. The relative elongations ε_n of fibres of origin at $\boldsymbol{x}$ along the direction $\boldsymbol{n}$ are expressed by projecting the tensor $\boldsymbol{\varepsilon} = \varepsilon_{ij}(\boldsymbol{v}(\boldsymbol{x}))$ on the direction $\boldsymbol{n}$ according to the rule

$$\varepsilon_n(\boldsymbol{v}(x)) = \varepsilon_{ij}(\boldsymbol{v}(x))n_i n_j \tag{3.79}$$

where $n_i n_i = 1$; the summation over i, j runs over $1, \ldots, n$; $n = 2$ or $n = 3$. A natural extension of the conditions (3.78) is the estimate

$$-\frac{\sigma_0}{\sigma_C} \leq \varepsilon_n(\boldsymbol{v}(x)) \leq \frac{\sigma_0}{\sigma_T} \tag{3.80}$$

for arbitrary unit vector $\boldsymbol{n}$, at almost each point of the design domain.

Similarly as in Sect. 3.1, we replace the scalar product $\boldsymbol{P}\boldsymbol{\cdot}\bar{\boldsymbol{u}}$ with $\mathcal{L}(\boldsymbol{v})$ representing the virtual work of the load done on the displacement $\boldsymbol{v}$. Thus, the volume of the optimal structure equals, see (2.50)

$$V_{\min} = \frac{1}{\sigma_0}\tilde{\mathcal{Z}}_1 \tag{3.81}$$

where

$$\begin{aligned}\tilde{\mathcal{Z}}_1 = \sup\{&\mathcal{L}(\boldsymbol{v}) \mid \boldsymbol{v} \text{ kinematically admissible in } \Omega \text{ such that}\\ &\left.\max_{\|\boldsymbol{n}\|=1} \varepsilon_n(\boldsymbol{v}(x)) \leq \frac{\sigma_0}{\sigma_T} \quad \min_{\|\boldsymbol{n}\|=1} \varepsilon_n(\boldsymbol{v}(x)) \geq -\frac{\sigma_0}{\sigma_C} \quad \text{a.e. in } \Omega\right\}\end{aligned} \tag{3.82}$$

Let $\lambda_i(\varepsilon)$ represent the eigenvalues of ε. Let us note that

$$\max_{1\leq i\leq n} \lambda_i(\varepsilon(\boldsymbol{v}(x))) = \max_{\|\boldsymbol{n}\|=1} \varepsilon_n(\boldsymbol{v}(x)) \qquad \min_{1\leq i\leq n} \lambda_i(\varepsilon(\boldsymbol{v}(x))) = \min_{\|\boldsymbol{n}\|=1} \varepsilon_n(\boldsymbol{v}(x)) \tag{3.83}$$

The above equalities are known extremal characteristics of the smallest and the greatest eigenvalues of tensor $\varepsilon = (\varepsilon_{ij})$. Let $B(\sigma_C, \sigma_T)$ be the set of tensors $\varepsilon \in \mathbb{E}^2_s$ satisfying

$$-\frac{\sigma_0}{\sigma_C} \leq \lambda_i(\varepsilon) \leq \frac{\sigma_0}{\sigma_T}, \quad i = 1, \ldots, n \tag{3.84}$$

The problem (3.82) may be rearranged to the form

$$\begin{aligned}\tilde{\mathcal{Z}}_1 = \sup\{&\mathcal{L}(\boldsymbol{v}) \mid \boldsymbol{v} \text{ kinematically admissible in } \Omega \text{ such that}\\ &\varepsilon(\boldsymbol{v}(x)) \in B(\sigma_C, \sigma_T) \quad \text{a.e. in } \Omega\}\end{aligned} \tag{3.85}$$

The set $B(\sigma_C, \sigma_T)$ is called a locking locus for virtual strains.

The volume of the optimal structure is given by (3.21), where now $\mathcal{Z} = \tilde{\mathcal{Z}}_1$.

Remark 3.4.1 The formulation (3.85) does not encompass the conditions imposed on strains on the surface of the body; the body may be reinforced by membrane shells S_r on the boundary or inside the body, or reinforced by curvilinear bars along the spatial lines Γ_r. The conditions bounding the virtual strains in (3.85) should be augmented with

$$\varepsilon(\boldsymbol{v}(x)) \in B(\sigma_C, \sigma_T) \quad \text{in} \quad S_r \tag{3.86}$$

where $\varepsilon(\cdot)$ is defined by (1.136) ; it refers to the membrane theory of shells and

$$-\frac{\sigma_0}{\sigma_C} \le \epsilon_t(\boldsymbol{v}(x)) \le \frac{\sigma_0}{\sigma_T} \tag{3.87}$$

The deformation measure ϵ_t along the curve Γ_r is determined as in the theory of spatial curved bars, which will be commented in Remark 16.1. The formulae for ε on S and ϵ_t on Γ involve geometric characteristics of the surface S and the curve Γ. The formula (3.79) was concerned only with strains within the body.

3.5 Static Formulation

The aim of the present section is the extension of the problem (3.33) to the case of $\sigma_C \neq \sigma_T$. Similarly as in Sect. 3.2, we assume that the load $\boldsymbol{p}$ is applied on Γ_1 and the part Γ_2 of the boundary of Ω is supported and there the kinematically admissible displacements vanish. Let us define the function

$$\eta_i(\boldsymbol{\tau}) = \begin{cases} \dfrac{\sigma_0}{\sigma_T} & \text{if } \operatorname{sgn}(\lambda_i(\boldsymbol{\tau})) > 0 \\ \dfrac{\sigma_0}{\sigma_C} & \text{if } \operatorname{sgn}(\lambda_i(\boldsymbol{\tau})) < 0 \\ \text{arbitrary value} & \text{if } \lambda_i(\boldsymbol{\tau}) = 0 \end{cases} \tag{3.88}$$

Consider now the functional

$$\boldsymbol{\tau} \to \int_\Omega \rho(\boldsymbol{\tau}) dx, \qquad \boldsymbol{\tau} \in \Sigma_p(\Omega) \tag{3.89a}$$

where

$$\rho(\boldsymbol{\tau}) = \sum_{i=1}^{n} \eta_i(\boldsymbol{\tau}) |\lambda_i(\boldsymbol{\tau})| \tag{3.89b}$$

The form of the functional (3.89a) is motivated by the form of Eq. (2.46) concerning the areas of cross sections of a truss. Let us note that if $\lambda_i(\boldsymbol{\tau} \neq 0)$ then Eq. (3.88) may be rewritten as below

$$\eta_i(\boldsymbol{\tau}) = \frac{1}{2|\lambda_i(\boldsymbol{\tau})|} \left[\left(\frac{\sigma_0}{\sigma_T} - \frac{\sigma_0}{\sigma_C} \right) \lambda_i(\boldsymbol{\tau}) + \left(\frac{\sigma_0}{\sigma_T} + \frac{\sigma_0}{\sigma_C} \right) |\lambda_i(\boldsymbol{\tau})| \right] \tag{3.90}$$

Thus, the integrand in (3.89a) can be written in the form

$$\rho(\boldsymbol{\tau}) = \frac{1}{2} \sum_{i=1}^{n} \left[\left(\frac{\sigma_0}{\sigma_T} - \frac{\sigma_0}{\sigma_C} \right) \lambda_i(\boldsymbol{\tau}) + \left(\frac{\sigma_0}{\sigma_T} + \frac{\sigma_0}{\sigma_C} \right) |\lambda_i(\boldsymbol{\tau})| \right] \tag{3.91}$$

Note that if $\sigma_C = \sigma_T = \sigma_0$ then this expression reduces to the form

$$\sum_{i=1}^{n} |\lambda_i(\boldsymbol{\tau})| = ||\boldsymbol{\tau}||^{\star}_{\text{spec}} \tag{3.92}$$

or it coincides with the integrand of problem (3.33). We note that

$$\rho(\boldsymbol{\tau}) = \frac{1}{2}\left(\frac{\sigma_0}{\sigma_T} - \frac{\sigma_0}{\sigma_C}\right) \operatorname{tr} \boldsymbol{\tau} + \frac{1}{2}\left(\frac{\sigma_0}{\sigma_T} + \frac{\sigma_0}{\sigma_C}\right) ||\boldsymbol{\tau}||^{\star}_{\text{spec}} \tag{3.93}$$

where $\operatorname{tr} \boldsymbol{\tau} = \sum_{i=1}^{n} \lambda_i(\boldsymbol{\tau})$ is the trace of $\boldsymbol{\tau}$.

Moreover

$$\begin{aligned} \rho(\alpha\boldsymbol{\tau}) &= \alpha\rho(\boldsymbol{\tau}) \quad \text{if} \quad \alpha \geq 0 \\ \rho(\boldsymbol{\tau}^1 + \boldsymbol{\tau}^2) &\leq \rho(\boldsymbol{\tau}^1) + \rho(\boldsymbol{\tau}^2) \qquad \forall \ \boldsymbol{\tau}^1, \boldsymbol{\tau}^2 \in \mathbb{E}^2_s \\ \rho(\boldsymbol{\tau}) &= 0 \quad \Rightarrow \quad \boldsymbol{\tau} = \mathbf{0} \in \mathbb{E}^2_s \\ \rho(\boldsymbol{\tau}) &\geq 0 \ \ \forall \ \ \boldsymbol{\tau} \in \mathbb{E}^2_s \end{aligned} \tag{3.94}$$

We conclude that $\rho(\boldsymbol{\tau})$ is not a norm in $\mathbb{E}^2_s$, since

$$\rho(\alpha\boldsymbol{\tau}) \neq |\alpha|\rho(\boldsymbol{\tau}) \quad \text{if} \quad \alpha < 0$$

yet, the function $\rho(\boldsymbol{\tau})$ is positively homogeneous of degree 1; we note that ρ is a gauge, see Rockafellar (1970, sec. 15).

Consider the problem

$$\tilde{\mathcal{Z}}_2 = \inf\left\{ \int_{\Omega} \rho(\boldsymbol{\tau}) dx \mid \boldsymbol{\tau} \in \Sigma_p(\Omega) \right\} \tag{3.95}$$

Our aim is to construct the problem dual to (3.95). This construction reduces to Fenchel's conjugate for the function (3.93)

$$\rho^{\star}(\boldsymbol{\varepsilon}) = \sup\left\{ \boldsymbol{\tau} \cdot \boldsymbol{\varepsilon} - \rho(\boldsymbol{\tau}) \mid \boldsymbol{\tau} \in \mathbb{E}^2_s \right\} \tag{3.96}$$

Since $\rho(\cdot)$ is a homogeneous function of degree 1, the function dual to it must have the form

$$\rho^{\star}(\boldsymbol{\varepsilon}) = \begin{cases} 0 & \text{if} \ \ \boldsymbol{\varepsilon} \in B \\ \infty & \text{otherwise} \end{cases} \tag{3.97}$$

and B is a set in $\mathbb{E}^2_s$. Thus, the construction of the function dual to $\rho(\cdot)$ reduces to the construction of the set B. This set is usually called a locking locus.

Let $n = 2$. Let φ be the angle between the principal directions of tensors $\boldsymbol{\tau}$ and $\boldsymbol{\varepsilon}$. We shall make use of the identity

$$\boldsymbol{\tau} \cdot \boldsymbol{\varepsilon} = (\tau_I \varepsilon_I + \tau_{II} \varepsilon_{II}) - \sin^2 \varphi (\tau_I - \tau_{II})(\varepsilon_I - \varepsilon_{II}) \tag{3.98}$$

where $\varepsilon_I \geq \varepsilon_{II}$, $\tau_I \geq \tau_{II}$. The supremum operation chooses $\varphi = 0$ or $\varphi = \pi$. Therefore, the eigendirections of $\boldsymbol{\tau}$ and $\boldsymbol{\varepsilon}$ coincide. Thus, the condition $\boldsymbol{\varepsilon} \in B$ will refer only to the principal strains. One should perform the operation

$$\rho^\star(\varepsilon) = \sup \{\tau_I \varepsilon_I + \tau_{II} \varepsilon_{II} - \alpha(\tau_I + \tau_I) - \beta(|\tau_I| + |\tau_{II}|) \mid \tau_i \in \mathbb{R}\} \tag{3.99}$$

where

$$\alpha = \frac{1}{2}\left(\frac{\sigma_0}{\sigma_T} - \frac{\sigma_0}{\sigma_C}\right), \qquad \beta = \frac{1}{2}\left(\frac{\sigma_0}{\sigma_T} + \frac{\sigma_0}{\sigma_C}\right) \tag{3.100}$$

hence

$$\alpha - \beta = -\frac{\sigma_0}{\sigma_C} < 0, \qquad \alpha + \beta = \frac{\sigma_0}{\sigma_T} > 0$$

Therefore

$$\rho^\star(\boldsymbol{\varepsilon}) = \sup_{\tau_I = x \in \mathbb{R}} \{x\varepsilon_I - \alpha x - \beta|x|\} + \sup_{\tau_{II} = x \in \mathbb{R}} \{x\varepsilon_{II} - \alpha x - \beta|x|\} \tag{3.101}$$

Let γ be an arbitrary number; β given by (3.100) is positive. We note that

$$\sup_{x \in \mathbb{R}} \{\gamma x - \beta|x|\} = \sup_{x \in \mathbb{R}} \{|\gamma| x - \beta|x|\}$$

We compute

$$\sup_{x \in \mathbb{R}} \{|\gamma| x - \beta|x|\} = \begin{cases} 0 & \text{if } \ |\gamma| \leq \beta \\ \infty & \text{otherwise} \end{cases}$$

We put $\gamma = \varepsilon_I - \alpha$ and find

$$\sup_{x \in \mathbb{R}} \{(\varepsilon_i - \alpha)x - \beta|x|\} = \begin{cases} 0 & \text{if } \ |\varepsilon_i - \alpha| \leq \beta \\ \infty & \text{otherwise} \end{cases}$$

where $\varepsilon_1 = \varepsilon_I$ and $\varepsilon_2 = \varepsilon_{II}$. The function $\rho^\star(\boldsymbol{\varepsilon})$ assumes the form

$$\rho^\star(\boldsymbol{\varepsilon}) = \begin{cases} 0 & \text{if } \alpha - \beta \leq \varepsilon_I \leq \alpha + \beta \text{ and } \alpha - \beta \leq \varepsilon_{II} \leq \alpha + \beta \\ \infty & \text{otherwise} \end{cases} \tag{3.102}$$

Thus, the domain B is determined by the conditions

$$-\frac{\sigma_0}{\sigma_C} \leq \lambda_i(\boldsymbol{\varepsilon}) \leq \frac{\sigma_0}{\sigma_T} \qquad i = 1, 2 \tag{3.103}$$

and this domain coincides with the domain $B(\sigma_C, \sigma_T)$ defined by (3.84).

Now we are prepared to the construction of the problem dual to (3.95). We introduce the condition $\boldsymbol{\tau} \in \Sigma_p(\Omega)$ to this formulation; the virtual fields v_i in (3.31) play the role of the Lagrangian multipliers

$$\tilde{\mathcal{Z}}_2 = \inf_{\boldsymbol{\tau}\in\mathcal{S}(\Omega,\mathbb{E}_s^2)} \sup_{\boldsymbol{v}\in\mathcal{V}(\Omega)} \left\{ \mathcal{L}(\boldsymbol{v}) + \int_\Omega [\rho(\boldsymbol{\tau}) - \boldsymbol{\tau}\cdot\boldsymbol{\varepsilon}(\boldsymbol{v})]dx \right\} \tag{3.104}$$

Like in the theory of linear programming, we interchange the inf and sup operations to obtain

$$\tilde{\mathcal{Z}}_2 = \sup_{\boldsymbol{v}\in\mathcal{V}(\Omega)} \{\mathcal{L}(\boldsymbol{v}) + \mathcal{R}(\boldsymbol{\varepsilon}(\boldsymbol{v}))\} \tag{3.105}$$

where

$$\mathcal{R}(\boldsymbol{\varepsilon}) = - \sup_{\boldsymbol{\tau}\in\mathcal{S}(\Omega,\mathbb{E}_s^2)} \int_\Omega (\boldsymbol{\tau}\cdot\boldsymbol{\varepsilon} - \rho(\boldsymbol{\tau}))dx \tag{3.106}$$

Since the condition $\boldsymbol{\tau} \in S(\Omega, \mathbb{E}_s^2)$ does not involve the differential constraints, one can shift the operation sup below the integral

$$\mathcal{R}(\boldsymbol{\varepsilon}) = - \int_\Omega \rho^\star(\boldsymbol{\varepsilon})dx \tag{3.107}$$

where $\rho^\star(\boldsymbol{\varepsilon})$ is given by (3.97). Thus

$$\mathcal{R}(\boldsymbol{\varepsilon}) = \begin{cases} 0 & \text{if } \ \boldsymbol{\varepsilon} \in B(\sigma_C, \sigma_T) \\ -\infty & \text{otherwise} \end{cases} \tag{3.108}$$

with $B(\sigma_C, \sigma_T)$ is the set of fields $\boldsymbol{\varepsilon} \in \mathbb{E}_s^2$ satisfying (3.103), see Fig. 3.3.

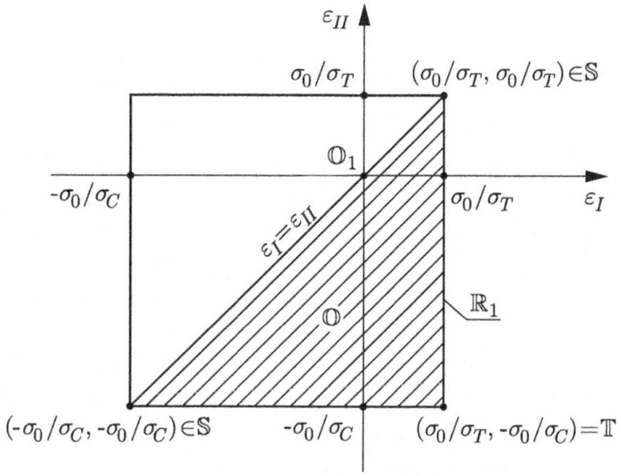

Fig. 3.3 The locking locus $B(\sigma_C, \sigma_T)$ in the 2D setting. Case of $\sigma_T \geq \sigma_C$

We come back to (3.105)

$$\tilde{\mathcal{Z}}_2 = \sup\{\mathcal{L}(\boldsymbol{v}) \mid \boldsymbol{v} \in \mathcal{V}(\Omega), \quad \varepsilon(\boldsymbol{v}(x)) \in B(\sigma_C, \sigma_T) \quad \text{a.e. in } \Omega\} \tag{3.109}$$

The above result is also true for $n = 3$. Thus, (3.95) has been rearranged to (3.85), which proves the equality

$$\tilde{\mathcal{Z}}_1 = \tilde{\mathcal{Z}}_2 \tag{3.110}$$

The volume of the optimal truss is given by (3.53) where now the volume fraction of the material equals

$$\upsilon = \frac{1}{\sigma_0}\rho(\breve{\boldsymbol{\tau}}) \tag{3.111}$$

and $\breve{\boldsymbol{\tau}}$ is the minimizer of (3.95). The physical meaning of υ will be cleared up in Sect. 3 in the context of planar problems.

Concluding, the Michell problem consists of two problems:

- the primal problem (3.85), where the maximizer $\breve{\boldsymbol{v}}$ is the unknown;
- the dual problem (3.95), where the statically admissible minimizer $\breve{\boldsymbol{\tau}}$ is the unknown.

The Michell problem is solved if both the above problems are solved and the equality (3.110) holds. The fields $\breve{\boldsymbol{v}}$ and $\breve{\boldsymbol{\tau}}$ are not linked by the constitutive equations, but by certain relations which assure that the equality (3.110) holds true. The 2D and 3D cases will be discussed independently.

Case of $n = 2$.
In the 2D setting, the domain $B(\sigma_C, \sigma_T)$ is a half of the square on the plane: $\varepsilon_I, \varepsilon_{II}$, see Fig. 3.3, concerning the case of $\sigma_T \geq \sigma_C$.

We expect that the supremum operation in (3.109) will choose the pairs $(\varepsilon_I, \varepsilon_{II})$ lying on the edge of the domain $B(\sigma_C, \sigma_T)$ in the greatest distance from the origin; these pairs correspond to the vertices of the locking locus. Let us introduce the notation of the subsequent segments of the locking locus (see Fig. 3.3): the vertices are denoted by $\mathbb{S}$ and $\mathbb{T}$, the edges are: $\mathbb{R}_1$, $\mathbb{R}_2$, the interior is denoted by $\mathbb{O}$ and the origin by $\mathbb{O}_1$. Consequently, the design domain Ω will be divided into subdomains (called: *regimes*, see Hu and Shield (1961) and Cherkaev (2000) corresponding to these segments of the locking locus. It turns out that the subdomains of the design domain Ω working in the regime $\mathbb{T}$ are filled up with two families of fibres: in tension and in compression. The mutual relations between the principal virtual strains $\varepsilon(\breve{\boldsymbol{v}})_I$, $\varepsilon(\breve{\boldsymbol{v}})_{II}$ which maximize the functional in (3.109) and the principal virtual stresses $\lambda_1(\breve{\boldsymbol{\tau}})$, $\lambda_2(\breve{\boldsymbol{\tau}})$ which minimize the functional in (3.95) are set up in Table 3.1. The quantities $\lambda_1(\breve{\boldsymbol{\tau}})$, $\lambda_2(\breve{\boldsymbol{\tau}})$ are equal to $\breve{\tau}_I$, $\breve{\tau}_{II}$, if $\breve{\tau}_I \geq \breve{\tau}_{II}$. The relations between $\varepsilon(\breve{\boldsymbol{v}})$ and $\breve{\boldsymbol{\tau}}$, set up in Table 3.1 follow from the condition (3.110), hence from the analysis of the sup problem in (3.101). In the regime $\mathbb{S}$, the layout of fibres is not determined. Table 3.1 shows that the load is transmitted to the support along the fibres forming fans (of straight or curved lines) or orthogonal nets.

Along the boundaries of the regimes, the equilibrium equations should be fulfilled. Thus, the solutions are composed of planar substructures of different underlying microstructures. Due to the possible presence of bars of finite cross sections, the theory encompasses the solutions to the spectacular problems of optimal transmission of the point loads to the supports, or concerning optimal girders carrying self-equilibrated systems of point loads. The simplest class of this kind: the three-force problem in the plane will be discussed in Sect. 4.15.

Case of $n = 3$.
Consider the vertices of the locking locus $B(\sigma_C, \sigma_T)$ in the frame: $\varepsilon_I, \varepsilon_{II}, \varepsilon_{III}$:

$$A\left(\frac{\sigma_0}{\sigma_T}, -\frac{\sigma_0}{\sigma_C}, -\frac{\sigma_0}{\sigma_C}\right) \quad B\left(-\frac{\sigma_0}{\sigma_C}, -\frac{\sigma_0}{\sigma_C}, -\frac{\sigma_0}{\sigma_C}\right)$$

$$C\left(\frac{\sigma_0}{\sigma_T}, \frac{\sigma_0}{\sigma_T}, -\frac{\sigma_0}{\sigma_C}\right) \quad D\left(\frac{\sigma_0}{\sigma_T}, \frac{\sigma_0}{\sigma_T}, \frac{\sigma_0}{\sigma_T}\right).$$

In the 3D case, instead of one Table 3.1, one can build three such tables referring to the planes: $(\varepsilon_I, \varepsilon_{II})$, $(\varepsilon_I, \varepsilon_{III})$, $(\varepsilon_{II}, \varepsilon_{III})$. The design domain Ω will be divided into regimes with fibres going in one, two or three directions. We expect that in many cases, the final solution will contain shells with fibrous microstructure. Moreover, due to the presence of point loads, there should appear bars of finite cross section transmitting them to the support or transmitting them to the vertices of the fans. These bars may also reinforce the edges of the membranes. Thus, the three-dimensional solutions may have very rich structure; this topic will be addressed in Chap. 5.

Remark 3.5.1 Let us refer to Sect. 3.3, where we have reconstructed an optimal (least compliant) body whose properties are governed by the Michell problem. Let us introduce the energy density by analogy with (3.62)

$$j^{\star}(\sigma) = \frac{1}{2E}[\rho(\sigma)]^2 \tag{3.112}$$

where ρ is given by (3.93) and consider the problem (3.65) with a new density of energy involved in $\wp(E)$ given by (3.63). Theorem (3.66) holds, where now $Z = \mathcal{Z}_1$ or $Z = \mathcal{Z}_2$. By solving problem (3.95), we find the minimizer $\breve{\tau}$ which determines the optimal layout of the modulus E according to (3.77).

3.6 Bibliographical Comments

The history of the theory of optimum design of trusses is described in the books by Cox (1965) and Hemp (1973). The lightest fully stressed trusses with all bars in tension (or compression) have been discovered by Maxwell (1870), see Michell (1904) and Hemp (1973). Maxwell is also the author of the theorem outlined in Sect. 2.3. James Clerk Maxwell's results were the inspiration for A.G.M. Michell (1904). The biographical note on Michell written by Cherry (1962) shows that this work of 1904 was only an episode of Michell's professional career. His main papers were on lubrication and crankless engines. The paper of 1904 was preceded by the paper: Michell (1899) on lateral buckling and the research on the topic of the main paper of 1904 was not continued. It is thought appropriate to repeat after Cherry (1962) that the surname Michell should be 'pronounced with stress on the first syllable and with that syllable rhyming with <rich> and <which>'.

In Michell's paper (1904), one can find the formulations of the problems (3.18) and (3.85), i.e. the kinematic formulations of the optimum design of trusses of minimal volume in which stresses assume the values between the bounds $(-\sigma_C, \sigma_T)$. The derivation of the formulations (3.18) and (3.85) has been performed by Hemp (1958, 1973), Chan (1960), discussed in Lagache (1981) and put in the mathematically strict form by Bouchitté et al. (2008).

The problem (3.33) dual to (3.18) has been given by Rozvany (1976) in the context of flexural problems and by Strang and Kohn (1983). The formulation (3.33) can be found neither in Cox (1965) nor in Hemp (1973). The problem (3.95) or the problem dual to (3.85) involving the norm $\rho(\cdot)$ given by (89_{b}) can be found in Stavroulakis and Tzaferopoulos (1994) and can be viewed as a special case of the general approach by Prager and Shield (1967). Its reformulation (3.93) is probably published here for the first time. The locking locus in Fig. 3.3 is a correction of Fig. 5 of Lewiński and Sokół (2014).

A first mathematically strict proof of equivalence of problems (3.18) and (3.33), or the proof of the equality $\mathcal{Z}_1 = \mathcal{Z}_2$, has been delivered in Theorem 2.1 in Bouchitté et al. (2008) under the condition of the design domain being convex. Till now, this result has been extended neither to the case of non-convex design domains, nor to the case of $\sigma_T \neq \sigma_C$. The crucial result of equivalence of the conditions (i) and (ii) in (3.14) has been noted in Bouchitté and Buttazzo (2001). The condition (i) introduces the bounds on the virtual deformations of potential bars linking arbitrary points of the design domain. On the other hand, the condition (ii) refers to the 2D continuum deformation and is expressed by the strain tensor. Thus, equivalence of (i) and (ii) is crucial, since it links the theory of plane trusses with the theory of in-plane loaded continuum plates.

Theorem 2.1 in Bouchitté et al. (2008) expressing the equality $\mathcal{Z}_1 = \mathcal{Z}_2$ is not constructive, since it does not specify the stress field which makes the minimum of the static problem attained. The known exact solutions suggest that the stress field has its support on the families of curves, but Theorem 3.1 (the most important result of Bouchitté et al. 2008) does not confirm that the Michell structures are composed of fibres along curves. Nevertheless, neither of the known Michell structures extend beyond this class.

In case $\sigma_T = \sigma_C$, the equality $\mathcal{Z}_1 = \mathcal{Z}_2$ implies the relations (3.48) between the maximizer $\breve{\mathbf{v}}$ of the kinematic problem and the minimizer $\breve{\boldsymbol{\tau}}$ of the static problem, see (3.51a), (3.52). The extension towards $\sigma_T \neq \sigma_C$ is given in Table 3.1. The idea of dividing the design domain into regimes $\mathbb{S}$, $\mathbb{T}$, $\mathbb{R}_1$, $\mathbb{R}_2$, $\mathbb{O}$, $\mathbb{O}_1$ can be found in many papers by Prager and Rozvany, see Rozvany (1976, 1989), see also Lewiński and Sokół (2014).

Table 3.1 Properties of the optimal solutions. Case of $\sigma_T \geq \sigma_C$

Regime	Characterization of the maximizer $\check{\boldsymbol{v}}$	Characterization of the maximizer $\check{\boldsymbol{\tau}}$	Description of the microstructure
$\mathbb{S}$	$\varepsilon_I(\check{\boldsymbol{v}}) = \varepsilon_{II}(\check{\boldsymbol{v}}) = \frac{\sigma_0}{\sigma_T}$ or $\varepsilon_I(\check{\boldsymbol{v}}) = \varepsilon_{II}(\check{\boldsymbol{v}}) = \frac{-\sigma_0}{\sigma_C}$	$\lambda_1(\check{\boldsymbol{\tau}}) \geq 0 \quad \lambda_2(\check{\boldsymbol{\tau}}) \geq 0$ $\lambda_1(\check{\boldsymbol{\tau}}) \leq 0 \quad \lambda_2(\check{\boldsymbol{\tau}}) \leq 0$	Undetermined directions of fibres
$\mathbb{T}$	$\varepsilon_I(\check{\boldsymbol{v}}) = \frac{\sigma_0}{\sigma_T}$ $\varepsilon_{II}(\check{\boldsymbol{v}}) = \frac{-\sigma_0}{\sigma_C}$	$\lambda_1(\check{\boldsymbol{\tau}}) \geq 0 \quad \lambda_2(\check{\boldsymbol{\tau}}) \leq 0$	Two families of fibres in tension and in compression
$\mathbb{R}_1$	$\varepsilon_I(\check{\boldsymbol{v}}) = \frac{\sigma_0}{\sigma_T}$ $-\frac{\sigma_0}{\sigma_C} \leq \varepsilon_{II}(\check{\boldsymbol{v}}) \leq \frac{\sigma_0}{\sigma_T}$	$\lambda_1(\check{\boldsymbol{\tau}}) \geq 0 \quad \lambda_2(\check{\boldsymbol{\tau}}) = 0$	One family of fibres in tension
$\mathbb{R}_2$	$-\frac{\sigma_0}{\sigma_C} \leq \varepsilon_I(\check{\boldsymbol{v}}) \leq \frac{\sigma_0}{\sigma_T}$ $\varepsilon_{II}(\check{\boldsymbol{v}}) = -\frac{\sigma_0}{\sigma_C}$	$\lambda_1(\check{\boldsymbol{\tau}}) = 0 \quad \lambda_2(\check{\boldsymbol{\tau}}) \leq 0$	One family of fibres in compression
$\mathbb{O}$	$-\frac{\sigma_0}{\sigma_C} < \varepsilon_I(\check{\boldsymbol{v}}) < \frac{\sigma_0}{\sigma_T}$ $\varepsilon_I(\check{\boldsymbol{v}}) \geq \varepsilon_{II}(\check{\boldsymbol{v}})$ $-\frac{\sigma_0}{\sigma_C} < \varepsilon_{II}(\check{\boldsymbol{v}}) < \frac{\sigma_0}{\sigma_T}$	$\lambda_1(\check{\boldsymbol{\tau}}) = 0 \quad \lambda_2(\check{\boldsymbol{\tau}}) = 0$	Empty domain
$\mathbb{O}_1$	$\varepsilon_I(\check{\boldsymbol{v}}) = \varepsilon_{II}(\check{\boldsymbol{v}}) = 0$	$\lambda_1(\check{\boldsymbol{\tau}}) = 0 \quad \lambda_2(\check{\boldsymbol{\tau}}) = 0$	Empty domain undergoing a rigid motion

The kinematic problems (3.18), (3.85) are similar to the static problems of bodies with locking, see Čyras (1972), Demengel and Suquet (1986), Telega and Jemioło (1998).

The permissible stresses may be viewed as random, thus leading to a biomaterial approach, see Ostoja-Starzewski (2001).

References

Allaire, G. (2002). *Shape optimization by the homogenization method*. New York: Springer.

Bouchitté, G., & Buttazzo, G. (2001). Characterization of optimal shapes and masses through Monge-Kantorovich equation. *Journal of the European Mathematical Society, 3*, 139–168.

Bouchitté, G., Gangbo, W., & Seppecher, P. (2008). Michell trusses and lines of principal action. *Mathematical Models and Methods in Applied Sciences, 18*, 1571–1603.

Chan, A. S. L. (1960). *The design of Michell optimum structures*. Cranfield: The College of Aeronautics (Report 142).

Cherkaev, A. V. (2000). *Variational methods for structural optimization*. New York: Springer.

Cherry, T. M. (1962). Anthony George Maldon Michell: 1870–1959. Biographical Memoirs of Fellows of the Royal Society (pp. 90–103). www.jstor.org.

Cox, H. L. (1965). *The design of structures of least weight*. Oxford: Pergamon Press.

Čyras, A. (1972). Optimization theory of perfectly locking bodies. *Archives of Mechanics*, *24*, 203–210.
Demengel, F., & Suquet, P. (1986). On locking materials. *Acta Applicandae Mathematicae*, *6*, 185–211.
Desmorat, B., & Duvaut, G. (2003). Compliance optimization with nonlinear elastic materials. Application to constitutive laws dissymmetric in tension-compression. *European Journal of Mechanics A/Solids*, *22*, 179–192.
Hemp, W. S. (1958). Theory of structural design. The College of Aeronautics, Cranfield. Report no 115.
Hemp, W. S. (1973). *Optimum structures*. Oxford: Clarendon Press.
Hu, T.-C., & Shield, T. T. (1961). Minimum weight design of discs. *Zeitschrift für Angewandte Mathematik und Physik*, *12*, 414–433.
Lagache, J.-M. (1981). Developments in Michell theory, pp. 4–9 to 4–16. In A. Atrek & R. H. Gallagher (Eds.), *Proceedings of the International Symposium on Optimal Structural design. The 11th ONR Naval Structural Mechanics Symposium, 19–22 October 1981*.
Lewiński, T., & Sokół, T. (2014). On basic properties of Michell's structures. In G. I. N. Rozvany & T. Lewiński (Eds.), *Topology optimization in structural and continuum mechanics* (Vol. 549). CISM international centre for mechanical sciences. Courses and lectures. Wien: Springer.
Lewiński, T., & Telega, J. J. (2000). *Plates, laminates and shells. Asymptotic analysis and homogenization*. Singapore, New Jersey, London, Hong Kong: World Scientific.
Maxwell, C. (1870). On reciprocal figures, frames and diagrams of forces. *Scientific Papers II*, *26*, 161–207.
Michell, A. G. M. (1899). Elastic stability of long beams under transverse forces. *Philosophical Magazine. Series 5*, *48*(292), 298–309.
Michell, A. G. M. (1904). The limits of economy of material in frame structures. *Philosophical Magazine. Series 6*, *8*(47), 589–597.
Naghdi, P. M. (1963). Foundation of elastic shell theory. In I. N. Sneddon & R. Hill (Eds.), *Progress in solid mechanics* (pp. 2–90). Amsterdam: North Holland.
Ostoja-Starzewski, M. (2001). Michell trusses in the presence of microscale material randomness: limitation of optimality. *Proceedings of the Royal Society of London*, *457*, 1787–1797.
Prager, W., & Shield, R. T. (1967). A general theory of optimal plastic design. *Journal of Applied Mechanics, Transactions ASME*, *34*, 184–186.
Rockafellar, R. T. (1970). *Convex analysis*. New Jersey: Princeton University Press.
Rozvany, G. I. N. (1976). *Optimal design of flexural systems*. London: Pergamon Press.
Rozvany, G. I. N. (1989). *Structural design via optimality criteria*. Dordrecht: Kluwer Academic Publishers.
Sigmund, O., Aage, N., & Andreassen, E. (2016). On the (non-) optimality of Michell's structures. *Structural and Multidisciplinary Optimization*, *54*, 361–373.
Stavroulakis, G. E., & Tzaferopoulos, M. Ap. (1994). A variational inequality approach to optimal plastic design of structures via the Prager-Rozvany theory. *Structural Optimization*, *7*, 160–169.
Strang, G., & Kohn, R. V. (1983). Hencky-Prandtl nets and constrained Michell trusses. *Computer Methods in Applied Mechanics and Engineering*, *36*, 207–222.
Telega, J. J., & Jemioło, S. (1998). Macroscopic behaviour of locking materials with microstructure. Part I. Primal and dual locking potential. Relaxation. *Bulletin of the Polish Academy of Sciences*, *46*, 265–276.

Chapter 4
Constructing Michell Structures in Plane. Single Load Case

This chapter introduces the reader into the methods of construction of the planar frameworks being exact solutions to the Michell problems of optimum design. The layout of bars of these structures follows the trajectories of specific strain fields. The methods of their construction are given in Sects. 4.1–4.4. The simplest Michell structures are composed of straight and circular bars; they are described in Sect. 4.5. The next sections outline the construction of all available nowadays exact solutions of the Michell's theory; some constructions are checked by the static method. The analytical results are compared with their numerical predictions found by the ground structure method.

4.1 Orthogonal Curvilinear Parameterization

Constructions of the exact solutions to Michell's problems (3.85) and (3.95), the solutions of which: $\breve{\boldsymbol{v}}$ and $\breve{\boldsymbol{\tau}}$ are linked by the formulae of Table 3.1, require introducing a curvilinear orthogonal parameterization which follows the trajectories of the strain field $\boldsymbol{\varepsilon}(\breve{\boldsymbol{v}})$ and of the stress field $\breve{\boldsymbol{\tau}}$. The aim of the present section is to recall the main terms and properties of curvilinear orthogonal parameterizations.

4.1.1 Geometric and Kinematic Description

Let $\boldsymbol{e}_1, \boldsymbol{e}_2$ be basis vectors of a Cartesian coordinate system; $\boldsymbol{e}_i \cdot \boldsymbol{e}_j = \delta_{ij}$, δ_{ij} being Kronecker's delta. Each point $\boldsymbol{x}$ of the plane is determined by the position vector $\boldsymbol{r} = x^1 \boldsymbol{e}_1 + x^2 \boldsymbol{e}_2$, where x^i are Cartesian coordinates of a point $\boldsymbol{x}$, identified with the pair (x^1, x^2).

T. Lewiński et al., *Michell Structures*,
https://doi.org/10.1007/978-3-319-95180-5_4

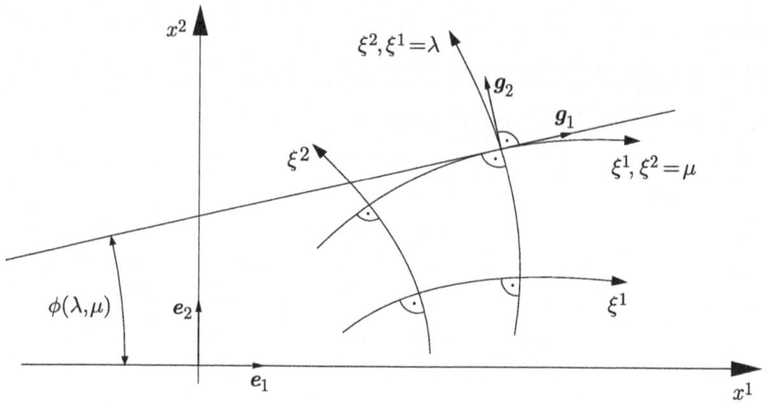

Fig. 4.1 An orthogonal curvilinear parametrization in the plane

Assume that the mapping

$$(\xi^1, \xi^2) \to \boldsymbol{r}\left[x^1(\xi^1, \xi^2), x^2(\xi^1, \xi^2)\right] \tag{4.1}$$

is given. It determines two families of curves on the plane, see Fig. 4.1. The quantities ξ^1, ξ^2 parameterize this family of curves. We define the vectors

$$\boldsymbol{g}_i = \frac{\partial \boldsymbol{r}}{\partial \xi^i}, \qquad i = 1, 2 \tag{4.2}$$

tangent to the parametric lines, see (1.127) in the context of shells. The metric tensor is defined by (see (1.130))

$$g_{ij} = \boldsymbol{g}_i \cdot \boldsymbol{g}_j \tag{4.3}$$

The condition of correctness of the parameterization (4.1) is the linear independence of the vectors $\boldsymbol{g}_1, \boldsymbol{g}_2$ in the whole domain parameterized by ξ^1, ξ^2.

In the sequel, only the orthogonal parameterizations will be used; hence, we assume $g_{12} = 0$. We introduce the Lamé coefficients

$$A(\xi^1, \xi^2) = \sqrt{g_{11}(\xi^1, \xi^2)}, \qquad B(\xi^1, \xi^2) = \sqrt{g_{22}(\xi^1, \xi^2)} \tag{4.4}$$

which determine the natural parameterization of the parametric lines. The elementary lengths of the arcs of the line ξ^1 (or $\xi^2 = \text{const}$) and the line ξ^2 (or $\xi^1 = \text{const}$) are expressed by $ds_\alpha = A(\xi_1, \xi_2)d\xi^1$, $ds_\beta = B(\xi_1, \xi_2)d\xi^2$, respectively.

Let us define the vectors $\boldsymbol{g}^1, \boldsymbol{g}^2$ according to the rule: $\boldsymbol{g}^i \cdot \boldsymbol{g}_j = \delta^i_j$. Thus, the vectors $\boldsymbol{g}^1, \boldsymbol{g}^2$ can be expressed by $\boldsymbol{g}_1, \boldsymbol{g}_2$ by the rules

$$\boldsymbol{g}^1 = \frac{1}{A^2}\boldsymbol{g}_1, \qquad \boldsymbol{g}^2 = \frac{1}{B^2}\boldsymbol{g}_2 \tag{4.5}$$

Let the field $\boldsymbol{v}$ be given. At each point $(x^1(\xi^1, \xi^2), x^2(\xi^1, \xi^2))$ the vector $\boldsymbol{v}$ can be decomposed in both the bases: $\boldsymbol{g}_i$ and $\boldsymbol{g}^i$, i.e.

$$\boldsymbol{v} = v^1\boldsymbol{g}_1 + v^2\boldsymbol{g}_2 \quad \text{or} \quad \boldsymbol{v} = v_1\boldsymbol{g}^1 + v_2\boldsymbol{g}^2 \tag{4.6}$$

It is helpful to introduce the basis of unit vectors along the basis vectors $\boldsymbol{g}_i$, or

$$\boldsymbol{a}_1 = \frac{1}{A}\boldsymbol{g}_1, \qquad \boldsymbol{a}_2 = \frac{1}{B}\boldsymbol{g}_2 \tag{4.7}$$

Thus, $\boldsymbol{a}_i \boldsymbol{\cdot} \boldsymbol{a}_j = \delta_{ij}$. The vector $\boldsymbol{v}$ is decomposed in this basis

$$\boldsymbol{v} = u\boldsymbol{a}_1 + v\boldsymbol{a}_2 \tag{4.8}$$

Thus, u is the displacement along the ξ^1 line and v- along the ξ^2 line. By comparing with (4.6), we find

$$u = \frac{1}{A}v_1, \qquad v = \frac{1}{B}v_2 \tag{4.9}$$

Let us introduce the Christoffel symbols, cf. (1.133)

$$\Gamma_{ij}^k = \boldsymbol{g}^k \cdot \frac{\partial \boldsymbol{g}_i}{\partial \xi^j} \tag{4.10}$$

In the orthogonal parameterization ($g_{12} = 0$) these symbols are expressed by

$$\begin{aligned}
\Gamma_{11}^1 &= \frac{1}{A}\frac{\partial A}{\partial \xi^1}, \qquad \Gamma_{11}^2 = -\frac{A}{B^2}\frac{\partial A}{\partial \xi^2} \\
\Gamma_{12}^1 &= \Gamma_{21}^1 = \frac{1}{A}\frac{\partial A}{\partial \xi^2} \\
\Gamma_{12}^2 &= \Gamma_{21}^2 = \frac{1}{B}\frac{\partial B}{\partial \xi^1} \\
\Gamma_{22}^1 &= -\frac{B}{A^2}\frac{\partial B}{\partial \xi^1} \\
\Gamma_{22}^2 &= \frac{1}{B}\frac{\partial B}{\partial \xi^2}
\end{aligned} \tag{4.11}$$

Let us define the covariant derivative of the vector field $\boldsymbol{v}$, see (1.134)

$$v_{i|j} = \frac{\partial v_i}{\partial \xi^j} - \sum_{k=1}^{2} \Gamma_{ij}^k v_k \tag{4.12}$$

The summation convention being not adopted here. We define now the physical components $(v_{i|j})^*$ of the covariant derivatives (4.12) by

$$v_{i|j}\boldsymbol{g}^i \otimes \boldsymbol{g}^j = (v_{i|j})^* \boldsymbol{a}^i \otimes \boldsymbol{a}^j \tag{4.13}$$

or by referring the components to the basis of unit vectors. Using (4.5), (4.7), we find

$$\begin{aligned}(v_{1|1})^* &= \frac{1}{A^2}(v_{1|1}), \quad (v_{2|2})^* = \frac{1}{B^2}(v_{2|2}) \\ (v_{1|2})^* &= \frac{1}{AB}(v_{1|2}), \quad (v_{2|1})^* = \frac{1}{AB}(v_{2|1})\end{aligned} \tag{4.14}$$

Let us define the tensor of strain $\varepsilon(\boldsymbol{v})$ by (1.136) with $b_{ij} = 0$ and the tensor of rigid rotation by

$$\omega_{ij}(\boldsymbol{v}) = \frac{1}{2}(v_{i|j} - v_{j|i}) \tag{4.15}$$

Now, we introduce a more convenient notation of the physical components of strains and rigid rotation

$$\begin{aligned}\epsilon_{11} &= (v_{1|1})^*, \qquad \epsilon_{22} = (v_{2|2})^* \\ \epsilon_{12} &= \epsilon_{21} = (\varepsilon_{12}(v))^* \\ 2\epsilon_{12} &= \omega_1 + \omega_2, \\ \omega_1 &= (v_{2|1})^*, \qquad \omega_2 = (v_{1|2})^* \\ \omega &= \frac{1}{2}(\omega_1 - \omega_2)\end{aligned} \tag{4.16}$$

In order to refer to the classical literature (cf. Hemp (1973)), we change the notation of the parameters: $\xi^1 = \alpha$, $\xi^2 = \beta$ and simplify the notation of the Cartesian axes: $x^1 = x$, $x^2 = y$. The substitution $v_1 = Au$ and $v_2 = Bv$ into (4.12), with using the formulae (4.11), gives the formulae for the physical components of strains and rigid rotation (4.15) in terms of u, v, A, B, being functions of arguments α, β

$$\begin{aligned}\epsilon_{11} &= \frac{1}{A}\left(\frac{\partial u}{\partial \alpha} + \frac{1}{B}\frac{\partial A}{\partial \beta}v\right) \\ \epsilon_{22} &= \frac{1}{B}\left(\frac{\partial v}{\partial \beta} + \frac{1}{A}\frac{\partial B}{\partial \alpha}u\right) \\ \omega_1 &= \frac{1}{A}\left(\frac{\partial v}{\partial \alpha} - \frac{1}{B}\frac{\partial A}{\partial \beta}u\right) \\ \omega_2 &= \frac{1}{B}\left(\frac{\partial u}{\partial \beta} - \frac{1}{A}\frac{\partial B}{\partial \alpha}v\right) \\ 2\epsilon_{12} &= \frac{B}{A}\frac{\partial}{\partial \alpha}\left(\frac{v}{B}\right) + \frac{A}{B}\frac{\partial}{\partial \beta}\left(\frac{u}{A}\right)\end{aligned} \tag{4.17}$$

The Lamé coefficients $A(\alpha, \beta)$, $B(\alpha, \beta)$ are not independent. They satisfy the condition

$$K_G = 0, \qquad K_G = \frac{1}{AB}\left[\frac{\partial}{\partial \alpha}\left(\frac{1}{A}\frac{\partial B}{\partial \alpha}\right) + \frac{\partial}{\partial \beta}\left(\frac{1}{B}\frac{\partial A}{\partial \beta}\right)\right] \tag{4.18}$$

Here, K_G represents Gauss curvature of the plane; hence, it vanishes identically.

The essential characteristic of the parametric lines is the angle ϕ of inclination of the tangent to the line $\xi = \alpha$, measured with respect to the x axis

$$\phi(\alpha, \beta) = \sphericalangle(\boldsymbol{g}_1(\alpha, \beta), \boldsymbol{e}_1) \tag{4.19}$$

We shall prove that

$$\begin{aligned} \frac{1}{A}\frac{\partial x}{\partial \alpha} &= \cos\phi, & \frac{1}{B}\frac{\partial x}{\partial \beta} &= -\sin\phi \\ \frac{1}{A}\frac{\partial y}{\partial \alpha} &= \sin\phi, & \frac{1}{B}\frac{\partial y}{\partial \beta} &= \cos\phi \end{aligned} \tag{4.20}$$

The point of departure are definitions of vectors $\boldsymbol{g}_i$, Eqs. (4.2). The components of vectors $\boldsymbol{g}_1, \boldsymbol{g}_2$ in the basis $(\boldsymbol{e}_1, \boldsymbol{e}_2)$ are

$$\boldsymbol{g}_1 = \left[\frac{\partial x}{\partial \alpha}, \frac{\partial y}{\partial \alpha}\right], \qquad \boldsymbol{g}_2 = \left[\frac{\partial x}{\partial \beta}, \frac{\partial y}{\partial \beta}\right] \tag{4.21}$$

where

$$\boldsymbol{r} = [x(\alpha, \beta), y(\alpha, \beta)]$$

The conditions (4.3) and Eqs. (4.4) are rewritten as follows:

$$(z_1)^2 + (z_2)^2 = 1, \qquad (z_3)^2 + (z_4)^2 = 1, \qquad z_1 z_3 + z_2 z_4 = 0 \tag{4.22}$$

with

$$\begin{aligned} z_1 &= \frac{1}{A}\frac{\partial x}{\partial \alpha}, & z_2 &= \frac{1}{A}\frac{\partial y}{\partial \alpha}, \\ z_3 &= \frac{1}{B}\frac{\partial x}{\partial \beta}, & z_4 &= \frac{1}{B}\frac{\partial y}{\partial \beta} \end{aligned} \tag{4.23}$$

From Eqs. (4.22), one can infer existence of angles ψ, φ such that

$$z_1 = \cos\psi, \quad z_2 = \sin\psi, \quad z_3 = \cos\varphi, \quad z_4 = \sin\varphi, \quad \cos(\varphi - \psi) = 0 \tag{4.24}$$

Let $\psi = \phi$ and $\varphi = \phi + \pi/2$. Having this notation, one can rearrange (4.24) to the form (4.20). The latter equations may be expressed as follows:

$$\begin{aligned} dx &= \cos\phi(\alpha, \beta) ds_\alpha, & dx &= -\sin\phi(\alpha, \beta) ds_\beta \\ dy &= \sin\phi(\alpha, \beta) ds_\alpha, & dy &= \cos\phi(\alpha, \beta) ds_\beta \end{aligned} \tag{4.25}$$

with

$$ds_\alpha = A d\alpha, \qquad ds_\beta = B d\beta$$

Let us write now the formulae for the differentials

$$dx = \frac{\partial x}{\partial \alpha} d\alpha + \frac{\partial x}{\partial \beta} d\beta, \qquad dy = \frac{\partial y}{\partial \alpha} d\alpha + \frac{\partial y}{\partial \beta} d\beta \tag{4.26}$$

and insert (4.20) to obtain

$$dx = A \cos\phi d\alpha - B \sin\phi d\beta, \qquad dy = A \sin\phi d\alpha + B \cos\phi d\beta \tag{4.27}$$

These equations allow for finding $x(\alpha, \beta)$, $y(\alpha, \beta)$ upon finding first the Lamé fields $A(\alpha, \beta)$, $B(\alpha, \beta)$. We shall prove that the conditions

$$\frac{\partial A}{\partial \beta} = -B \frac{\partial \phi}{\partial \alpha}, \qquad \frac{\partial B}{\partial \alpha} = A \frac{\partial \phi}{\partial \beta} \tag{4.28}$$

are the integrability conditions of Eqs. (4.27). Using (4.20), we find

$$\begin{aligned} \frac{\partial^2 x}{\partial \alpha \partial \beta} &= \frac{\partial}{\partial \beta}(A \cos\phi) = -\frac{\partial}{\partial \alpha}(B \sin\phi) \\ \frac{\partial^2 y}{\partial \alpha \partial \beta} &= \frac{\partial}{\partial \beta}(A \sin\phi) = \frac{\partial}{\partial \alpha}(B \cos\phi) \end{aligned} \tag{4.29}$$

The above formulae lead to the set of equations

$$\begin{bmatrix} \cos\phi & \sin\phi \\ \sin\phi & -\cos\phi \end{bmatrix} \begin{bmatrix} \dfrac{\partial A}{\partial \beta} + B \dfrac{\partial \phi}{\partial \alpha} \\ \dfrac{\partial B}{\partial \alpha} - A \dfrac{\partial \phi}{\partial \beta} \end{bmatrix} = \begin{bmatrix} 0 \\ 0 \end{bmatrix} \tag{4.30}$$

the solution of which is expressed by (4.28). Let us integrate (4.27) from the point x_0, y_0, where $(\alpha = 0, \beta = 0)$ to the point (x, y) where $(\alpha = \alpha_P, \beta = \beta_P)$

$$\begin{aligned} x &= x_0 + \int_0^{\alpha_P} A(\alpha, 0) \cos\phi(\alpha, 0) d\alpha - \int_0^{\beta_P} B(\alpha_P, \beta) \sin\phi(\alpha_P, \beta) d\beta \\ y &= y_0 + \int_0^{\alpha_P} A(\alpha, 0) \sin\phi(\alpha, 0) d\alpha + \int_0^{\beta_P} B(\alpha_P, \beta) \cos\phi(\alpha_P, \beta) d\beta \end{aligned} \tag{4.31}$$

It is expedient to combine these formulae as follows:

$$x + iy = x_0 + iy_0 + \int_0^{\alpha_P} e^{\phi(\alpha,0)i} A(\alpha, 0) d\alpha + i \int_0^{\beta_P} e^{\phi(\alpha_P,\beta)i} B(\alpha_P, \beta) d\beta \tag{4.32}$$

The above expression is independent of the integration path if

$$\frac{\partial}{\partial\beta}\left[e^{i\phi(\alpha,\beta)}A(\alpha,\beta)\right]=\frac{\partial}{\partial\alpha}\left[ie^{i\phi(\alpha,\beta)}B(\alpha,\beta)\right] \tag{4.33}$$

hence Eqs. (4.28) are derived. Thus, the latter are integrability conditions of Eqs. (4.27). Let us now integrate (4.28) with respect to ϕ from the point ($\alpha = 0$, $\beta = 0$) to the point ($\alpha = \alpha_P$, $\beta = \beta_P$)

$$\begin{aligned}\phi(\alpha_P,\beta_P)=\phi(0,0)+&\int_0^{\alpha_P}\left(-\frac{1}{B(\alpha,0)}\frac{\partial A}{\partial\beta}(\alpha,0)\right)d\alpha+\\&\int_0^{\beta_P}\frac{1}{A(\alpha_P,\beta)}\frac{\partial B}{\partial\alpha}(\alpha_P,\beta)d\beta\end{aligned} \tag{4.34}$$

The first term will be denoted by $\phi(0,0)=\phi_0$ in the sequel. The formula (4.34) determines the value $\phi(\alpha_P,\beta_P)$ uniquely, if the integral does not depend on the integration path. This is satisfied if

$$\frac{\partial}{\partial\beta}\left[-\frac{1}{B(\alpha,\beta)}\frac{\partial A}{\partial\beta}(\alpha,\beta)\right]=\frac{\partial}{\partial\alpha}\left[\frac{1}{A(\alpha,\beta)}\frac{\partial B}{\partial\alpha}(\alpha,\beta)\right] \tag{4.35}$$

Let us note that this equality is equivalent to the condition (4.18) of vanishing of the Gauss curvature of the plane.

Let us rewrite now the formulae for the components of deformations (4.14) using (4.28); they assume the form

$$\begin{aligned}\epsilon_{11}&=\frac{1}{A}\left(\frac{\partial u}{\partial\alpha}-\frac{\partial\phi}{\partial\alpha}v\right) &\quad \epsilon_{22}&=\frac{1}{B}\left(\frac{\partial v}{\partial\beta}+\frac{\partial\phi}{\partial\beta}u\right)\\ \omega_1&=\frac{1}{A}\left(\frac{\partial v}{\partial\alpha}+\frac{\partial\phi}{\partial\alpha}u\right) &\quad \omega_2&=\frac{1}{B}\left(\frac{\partial u}{\partial\beta}-\frac{\partial\phi}{\partial\beta}v\right)\end{aligned} \tag{4.36}$$

Note that according to (4.36)

$$\begin{aligned}L_1=&\;A(\epsilon_{11}+i\omega_1)d\alpha+B(\omega_2+i\epsilon_{22})d\beta=\\&\left(\frac{\partial u}{\partial\alpha}-\frac{\partial\phi}{\partial\alpha}v\right)d\alpha+i\left(\frac{\partial v}{\partial\alpha}+\frac{\partial\phi}{\partial\alpha}u\right)d\alpha+\\&\left(\frac{\partial u}{\partial\beta}-\frac{\partial\phi}{\partial\beta}v\right)d\beta+i\left(\frac{\partial v}{\partial\beta}+\frac{\partial\phi}{\partial\beta}u\right)d\beta\end{aligned} \tag{4.37}$$

hence

$$L_1=du+idv+iud\phi-vd\phi \tag{4.38}$$

or

$$d(u+iv)+i(u+iv)d\phi = A(\epsilon_{11}+i\omega_i)d\alpha + B(\omega_2 + i\epsilon_{22})d\beta \tag{4.39}$$

Using the identity

$$df + ifd\phi = e^{-i\phi}d(e^{i\phi}f) \tag{4.40}$$

we rearrange (4.39) to the form

$$d[e^{i\phi}(u+iv)] = e^{i\phi}\left[A(\epsilon_{11}+i\omega_1)d\alpha + B(\omega_2+i\epsilon_{22})d\beta\right] \tag{4.41}$$

This formula makes it possible to recover displacements having the deformation measures $\epsilon_{11}, \epsilon_{22}, \omega_1, \omega_2$.

4.1.2 Equilibrium Equations in Lines of Trajectories of Principal Stresses

Let us consider the problem of equilibrium of a segment of a plane structure parameterized by the lines $\xi^1 = \alpha$, $\xi^2 = \beta$ coinciding with trajectories of principal stresses $\lambda_i(\boldsymbol{\tau})$. Thus, the basis $\boldsymbol{g}_1, \boldsymbol{g}_2$ is chosen such that $\boldsymbol{\tau} = \tau^{ij}\boldsymbol{g}_i \otimes \boldsymbol{g}_j$, $\tau^{12} = 0$. Let us assume that the edges of the body coincide with the lines ξ^i: $\alpha = \alpha_0, \alpha = \alpha_1, \beta = \beta_0, \beta = \beta_1$, see Fig. 4.2.

Let the edges A_1A_2, A_2A_3 of the body be subject to the tractions of intensity $\boldsymbol{p}$ of units [N/m]. Let the edge A_0A_1 be clamped and there the displacement field $\boldsymbol{v}$ vanishes. The virtual work of stresses

$$\int_{\Omega} \boldsymbol{\tau} \cdot \boldsymbol{\varepsilon}(\boldsymbol{v})dx \tag{4.42}$$

has now the form

$$\int_{\alpha_0}^{\alpha_1}\int_{\beta_0}^{\beta_1} \left[\lambda_1(\boldsymbol{\tau})\epsilon_{11}(\boldsymbol{v}) + \lambda_2(\boldsymbol{\tau})\epsilon_{22}(\boldsymbol{v})\right] AB d\alpha d\beta \tag{4.43}$$

since the shear stresses τ^{12} vanish. The virtual work of the load reads

$$\mathcal{L}(\boldsymbol{v}) = \int_{\Gamma} \boldsymbol{p}\cdot \boldsymbol{v} ds \tag{4.44}$$

Let us introduce Hemp's (1973) stress resultants (or internal forces) per unit angles

$$T_1 = B\lambda_1(\boldsymbol{\tau}) \qquad T_2 = A\lambda_2(\boldsymbol{\tau}) \tag{4.45}$$

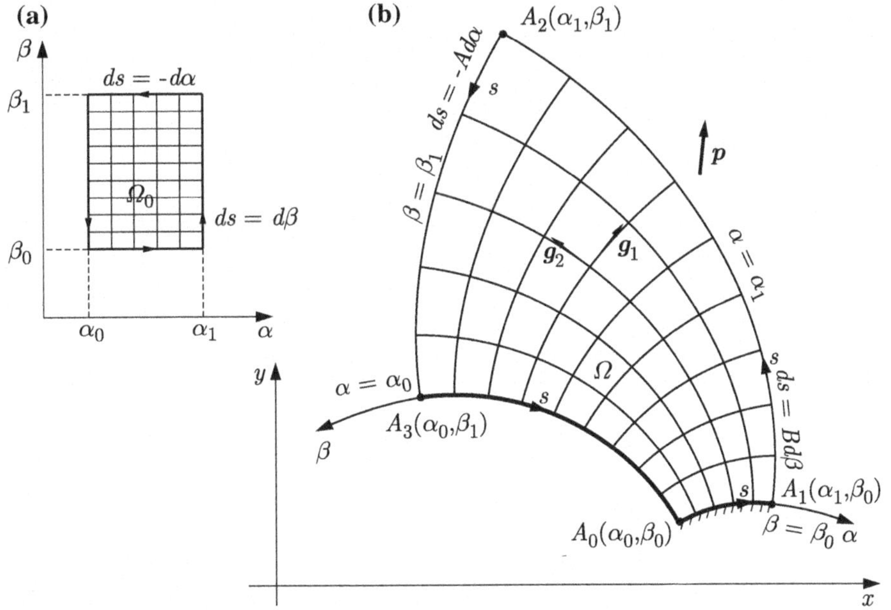

Fig. 4.2 The design domain and its parameterization

The units of Hemp's stress resultants are [N], hence the same as the units of axial forces in the bars of trusses. Now, we express the virtual work of stresses in terms of Hemp's internal forces

$$\int_{\Omega} \boldsymbol{\tau} \cdot \boldsymbol{\varepsilon}(\boldsymbol{u}) dx = \int_{\alpha_0}^{\alpha_1} \int_{\beta_0}^{\beta_1} \mathcal{W} d\alpha d\beta \tag{4.46a}$$

where, see (4.45),

$$\mathcal{W} = T_1 A \epsilon_{11}(\boldsymbol{v}) + T_2 B \epsilon_{22}(\boldsymbol{v}) \tag{4.46b}$$

Upon substitution of the first two equations of (4.17), we write

$$\mathcal{W} = T_1 \left(\frac{\partial u}{\partial \alpha} + \frac{1}{B} \frac{\partial A}{\partial \beta} v \right) + T_2 \left(\frac{\partial v}{\partial \beta} + \frac{1}{A} \frac{\partial B}{\partial \alpha} u \right) \tag{4.47}$$

or

$$\mathcal{W} = \frac{\partial}{\partial \alpha}(T_1 u) + \frac{\partial}{\partial \beta}(T_2 v) + \left[-\frac{\partial T_1}{\partial \alpha} + \frac{1}{A} \frac{\partial B}{\partial \alpha} T_2 \right] u + \left[-\frac{\partial T_2}{\partial \beta} + \frac{1}{B} \frac{\partial A}{\partial \beta} T_1 \right] v \tag{4.48}$$

Let p_1, p_2 be the components of $\boldsymbol{p}$ along the edges of the domain Ω, referred to the directions (α, β). The virtual work of this load has the form

$$\mathcal{L}(\boldsymbol{v}) = \int_{\beta_0}^{\beta_1} (p_1 u + p_2 v)|_{\alpha=\alpha_1} B|_{\alpha=\alpha_1} d\beta + \int_{\alpha_1}^{\alpha_0} (p_1 u + p_2 v)|_{\beta=\beta_1} \left(-A|_{\beta=\beta_1} d\alpha\right) \tag{4.49}$$

According to Green's theorem applied to the domain: $\alpha_0 < \alpha < \alpha_1$, $\beta_0 < \beta < \beta_1$, we find

$$\int_{\Omega} \tau \cdot \varepsilon(\boldsymbol{v}) dx = \int_{\alpha_0}^{\alpha_1} (T_2 v)|_{\beta=\beta_1} d\alpha + \int_{\beta_0}^{\beta_1} (T_1 u)|_{\alpha=\alpha_1} d\beta +$$
$$\int_{\alpha_0}^{\alpha_1} \int_{\beta_0}^{\beta_1} \left\{ \left[-\frac{\partial T_1}{\partial \alpha} + \frac{1}{A} \frac{\partial B}{\partial \alpha} T_2 \right] u + \left[-\frac{\partial T_2}{\partial \beta} + \frac{1}{B} \frac{\partial A}{\partial \beta} T_1 \right] v \right\} d\alpha d\beta \tag{4.50}$$

The variational equation (3.31) is fulfilled for each choice of u, v such that

$$u(\alpha, \beta_0) = 0, \qquad v(\alpha, \beta_0) = 0, \qquad \alpha_0 \le \alpha \le \alpha_1$$

By equating (4.50) with (4.49), one gets

$$-\frac{\partial T_1}{\partial \alpha} + \frac{1}{A} \frac{\partial B}{\partial \alpha} T_2 = 0, \qquad -\frac{\partial T_2}{\partial \beta} + \frac{1}{B} \frac{\partial A}{\partial \beta} T_1 = 0 \tag{4.51}$$

and

$$\begin{aligned} T_1 &= B p_1 \quad \text{on the line } \alpha = \alpha_1 \text{ or } A_1 A_2 \\ T_2 &= A p_2 \quad \text{on the line } \beta = \beta_1 \text{ or } A_2 A_3 \\ T_1 &= 0 \quad \text{on the line } \alpha = \alpha_0 \text{ or } A_0 A_3 \end{aligned} \tag{4.52}$$

Moreover

$$p_2 = 0 \text{ along } \alpha = \alpha_1 \qquad p_1 = 0 \text{ along } \beta = \beta_1 \tag{4.53}$$

We see that Michell's structure composed of two families of orthogonal fibres cannot be loaded by forces tangent to the edges.

Remark 4.1.1 Equation (4.51) can be inferred from the invariant version of equilibrium equations

$$\sum_{j=1}^{2} \tau^{ij}|_j = 0 \qquad i = 1, 2 \tag{4.54}$$

where

$$\tau^{ij}|_r = \frac{\partial \tau^{ij}}{\partial \xi^r} + \sum_{k=1}^{2} \Gamma^i_{rk} \tau^{kj} + \sum_{k=1}^{2} \Gamma^j_{rk} \tau^{ik} \tag{4.55}$$

and Christoffel's symbols are given by (4.11). In Eqs. (4.54), one should insert $\tau^{12} = 0$ and express the tensorial components τ^{11}, τ^{22} by the physical components

τ^{*11}, τ^{*22} by the rules

$$\tau^{*11} = A^2\tau^{11}, \qquad \tau^{*22} = B^2\tau^{22} \tag{4.56}$$

Let us expand (4.54) according to the rules (4.55) using (4.56) to obtain

$$\begin{aligned} -\frac{\partial}{\partial\alpha}B\tau^{*11} + \frac{\partial B}{\partial\alpha}\tau^{*22} &= 0 \\ -\frac{\partial}{\partial\beta}A\tau^{*22} + \frac{\partial A}{\partial\beta}\tau^{*11} &= 0 \end{aligned} \tag{4.57}$$

Recall that $\tau^{*11} = \lambda_1(\boldsymbol{\tau})$, $\tau^{*22} = \lambda_2(\boldsymbol{\tau})$. Due to (4.45), it is seen that (4.57) imply (4.51). Let us substitute (4.28) into (4.51)

$$-\frac{\partial T_1}{\partial\alpha} + \frac{\partial\phi}{\partial\beta}T_2 = 0, \qquad -\frac{\partial T_2}{\partial\beta} - \frac{\partial\phi}{\partial\alpha}T_1 = 0 \tag{4.58}$$

The coefficients $\frac{\partial\phi}{\partial\beta}, \frac{\partial\phi}{\partial\alpha}$ affect the form of Eqs. (4.58). The following cases should be considered separately:

$$\text{(i)} \qquad \frac{\partial\phi}{\partial\alpha} \neq 0, \qquad \frac{\partial\phi}{\partial\beta} \neq 0 \tag{4.59}$$

Then, the lines (α, β) form an orthogonal curvilinear system the lines of which have no inflection points

$$\text{(ii)} \qquad \frac{\partial\phi}{\partial\alpha} = 0, \qquad \frac{\partial\phi}{\partial\beta} \neq 0 \tag{4.60}$$

Then, the lines α are straight, while the lines β are orthogonal to these lines, see Fig. 4.3.

The family of straight lines α arises as the lines are tangent to a given curve, called an evolute. Assume that α measures the distance from the tangent point to the given point on the straight line. The assumption $\phi = \beta$ means that this parameter measures the angle between the line α and the axis x. Since $ds_1 = A d\alpha$ and α is a distance, then $A = 1$. Let us insert $\phi = \beta$ and $A = 1$ into (4.28). The first equation is satisfied, while the second reads

$$\frac{\partial B}{\partial\alpha} = 1 \tag{4.61}$$

Upon integration, one obtains

$$B = \alpha + g(\beta) \tag{4.62}$$

Along the evolute the field B vanishes. Thus, the evolute is given by

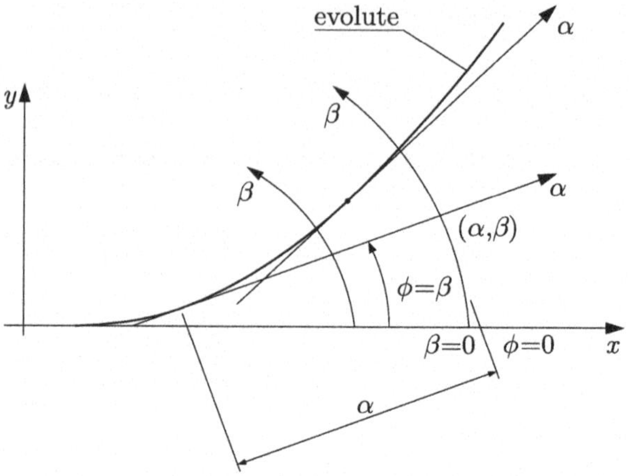

Fig. 4.3 Parameterization with one family of lines being straight

$$\alpha + g(\beta) = 0 \tag{4.63}$$

The last case:

$$\text{(iii)} \qquad \frac{\partial \phi}{\partial \alpha} = 0 \quad \text{and} \quad \frac{\partial \phi}{\partial \beta} = 0 \tag{4.64}$$

Then, (α, β) form the Cartesian frame. One can assume $A = 1$, $B = 1$ and then α, β measure distances in both the directions.

4.2 Hencky and Michell Nets Composed of Two Families of Curves. Kinematic Consideration

4.2.1 Geometric Properties of Hencky Nets

According to Table 3.1 within the regimes $\mathbb{S}$ and $\mathbb{T}$, the virtual principal strains $\varepsilon_I(\breve{\boldsymbol{v}})$, $\varepsilon_{II}(\breve{\boldsymbol{v}})$ have constant values. We assume that the lines (α, β) are trajectories of the principal strains $\varepsilon_I(\breve{\boldsymbol{v}})$, $\varepsilon_{II}(\breve{\boldsymbol{v}})$ and they assume constant values at each point of the considered subdomain. We assume now

$$\begin{gathered} \epsilon_{11} = \text{const}\,, \qquad \epsilon_{22} = \text{const}\,, \qquad \epsilon_{12} = 0, \\ \omega_1 = \omega(\alpha, \beta), \qquad \omega_2 = -\omega(\alpha, \beta) \end{gathered} \tag{4.65}$$

where ω stands for angle of the rigid rotation. The conditions (4.65) determine the geometry of the parametric lines (α, β). This narrow class of mutually orthogonal lines, being trajectories of strains of constant distribution within a certain subdomain, are called Hencky nets. They have been the subject of deep studies in the past, let us remind only Geiringer (1937), Carathéodory and Schmidt (1923), Chakrabarty (2006). In the sequel, we shall assume the property (4.59) corresponding to the case of both the families of parametric lines being curved. We come back to the formula (4.41) assuming now (4.65). The integration along the parametric line α from $\alpha = 0$ to $\alpha = \alpha_P$, and then, along the line β, from $\beta = 0$ to $\beta = \beta_P$ gives the formula

$$\begin{aligned} u(\alpha_P, \beta_P) + iv(\alpha_P, \beta_P) = e^{i(\phi_0 - \phi_P)}(u_0 + iv_0) + \\ e^{-i\phi_P} \int_0^{\alpha_P} e^{i\phi(\alpha,0)}(\epsilon_{11} + i\omega(\alpha, 0))A(\alpha, 0)d\alpha + \\ e^{-i\phi_P} i \int_0^{\beta_P} e^{i\phi(\alpha_P,\beta)}(\epsilon_{22} + i\omega(\alpha_P, \beta))B(\alpha_P, \beta)d\beta \end{aligned} \tag{4.66}$$

where

$$\phi_P = \phi(\alpha_P, \beta_P), \qquad u_0 = u(0,0), \qquad v_0 = v(0,0)$$

The condition of making the differential (4.41) complete has the form

$$\frac{\partial}{\partial \beta}\left[e^{i\phi} A(\epsilon_{11} + i\omega)\right] = \frac{\partial}{\partial \alpha}\left[e^{i\phi} B(-\omega + i\epsilon_{22})\right] \tag{4.67}$$

where

$$\omega = \omega(\alpha, \beta), \qquad \phi = \phi(\alpha, \beta), \qquad \epsilon_{11} = \text{const}\,, \qquad \epsilon_{22} = \text{const}$$

The condition (4.67) implies

$$\begin{aligned} -\frac{\partial \phi}{\partial \beta} A\omega + \frac{\partial A}{\partial \beta}\epsilon_{11} &= -\frac{\partial \phi}{\partial \alpha} B\epsilon_{22} - \frac{\partial B}{\partial \alpha}\omega - B\frac{\partial \omega}{\partial \alpha} \\ -\frac{\partial \phi}{\partial \alpha} B\omega + \frac{\partial B}{\partial \alpha}\epsilon_{22} &= \frac{\partial \phi}{\partial \beta} A\epsilon_{11} + \frac{\partial A}{\partial \beta}\omega + A\frac{\partial \omega}{\partial \beta} \end{aligned} \tag{4.68}$$

Let us make use of the relationships (4.28) to find

$$\frac{\partial \omega}{\partial \alpha} = (\epsilon_{11} - \epsilon_{22})\frac{\partial \phi}{\partial \alpha}, \qquad \frac{\partial \omega}{\partial \beta} = -(\epsilon_{11} - \epsilon_{22})\frac{\partial \phi}{\partial \beta} \tag{4.69}$$

since $A > 0$, $B > 0$ and due to the strains ϵ_{11}, ϵ_{22} being constant. Let us note that if $\epsilon_{11} = \epsilon_{22}$, which is the case within the regimes $\mathbb{S}$ and $\mathbb{O}_1$, see Table 3.1, then $\omega = \omega_0 = \text{const}$, $\omega_0 = \omega(0, 0)$. In this case, we have no additional condition to determine the angle ϕ, since Eq. (4.69) is satisfied identically for each choice of the function $\phi(\alpha, \beta)$.

On the other hand, if $\epsilon_{11} \neq \epsilon_{22}$, then Eq. (4.69) determines the distribution of the field $\omega(\alpha, \beta)$ having the field $\phi(\alpha, \beta)$, provided that Eqs. (4.69) are not mutually contradictory, or if

$$\frac{\partial^2 \phi}{\partial \alpha \partial \beta} = 0 \tag{4.70}$$

This equation, along with (4.28) gives the equation

$$\frac{\partial}{\partial \alpha}\left(\frac{1}{A}\frac{\partial B}{\partial \alpha}\right) = \frac{\partial}{\partial \beta}\left(\frac{1}{B}\frac{\partial A}{\partial \beta}\right) \tag{4.71}$$

yet before we have got (4.35). Combining these equations, we find the two differential equations

$$\frac{\partial}{\partial \alpha}\left(\frac{1}{A}\frac{\partial B}{\partial \alpha}\right) = 0, \qquad \frac{\partial}{\partial \beta}\left(\frac{1}{B}\frac{\partial A}{\partial \beta}\right) = 0 \tag{4.72}$$

Let us come back to (4.69). If the function ϕ satisfies (4.70), then Eq. (4.69) determines the field $\omega(\alpha, \beta)$ independently of the integration path. In particular, one can find the value of this field at (α_P, β_P), namely

$$\omega(\alpha_P, \beta_P) = \omega(0,0) + (\epsilon_{11} - \epsilon_{22})\left[\int_0^{\alpha_P} \frac{\partial \phi(\alpha, 0)}{\partial \alpha} d\alpha - \int_0^{\beta_P} \frac{\partial \phi(\alpha_P, \beta)}{\partial \beta} d\beta\right] \tag{4.73}$$

The solution to Eq. (4.70) has the general form $\phi = f_1(\alpha) + f_2(\beta)$, but the assumption

$$\phi(\alpha, \beta) = \phi_0 - a\alpha + b\beta \tag{4.74}$$

with $a, b \in \{-1, 1\}$ does not introduce restrictions to the analysis. In majority of problems to be discussed, we shall assume $a = 1$, $b = 1$ or

$$\phi(\alpha, \beta) = \phi_0 - \alpha + \beta \tag{4.75}$$

The parameter α measures the angle between the line tangent to the α line (there $\beta = \text{const}$) and the axis x; the parameter β represents an angle between the tangent to the β lines (or $\alpha = \text{const}$) and the vertical axis y, see Fig. 4.4.

Let us substitute (4.74) into (4.28) to find

$$\frac{\partial A}{\partial \beta} = aB, \qquad \frac{\partial B}{\partial \alpha} = bA \tag{4.76}$$

hence

$$\frac{\partial^2 A}{\partial \alpha \partial \beta} - cA = 0, \qquad \frac{\partial^2 B}{\partial \alpha \partial \beta} - cB = 0 \tag{4.77}$$

where $c = ab$. If ϕ is given by (4.75), then the integration of (4.69) leads to

$$\omega(\alpha_P, \beta_P) = \omega(0,0) - (\epsilon_{11} - \epsilon_{22})(\alpha + \beta) \tag{4.78}$$

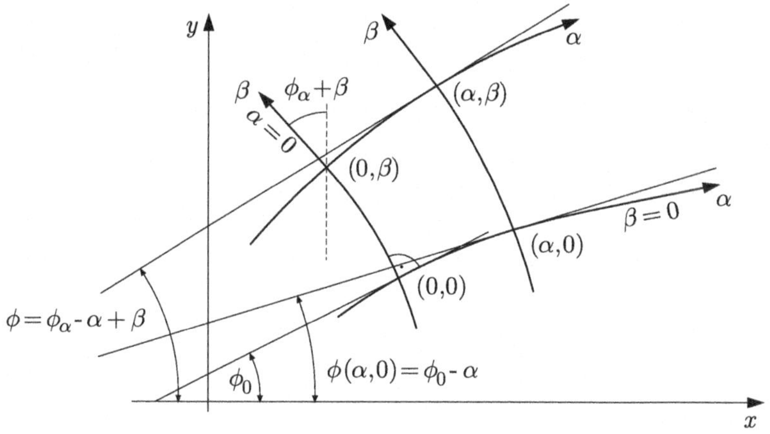

Fig. 4.4 The Hencky net and the geometrical meaning of its parameterization

The condition (4.70) is crucial for the whole book. Thus, it is thought appropriate to provide an alternative derivation of this equation, without referring to the idea of completeness of a differential. To this end, let us rewrite the first of Eqs. (4.36)

$$\frac{\partial u}{\partial \alpha} = A\epsilon_{11} + \frac{\partial \phi}{\partial \alpha} v \tag{4.79}$$

and perform differentiation;

$$\frac{\partial^2 u}{\partial \beta \partial \alpha} = \epsilon_{11}\frac{\partial A}{\partial \beta} + \frac{\partial^2 \phi}{\partial \alpha \partial \beta} v + \frac{\partial \phi}{\partial \alpha}\frac{\partial v}{\partial \beta} \tag{4.80}$$

But, see (4.28),

$$\frac{\partial A}{\partial \beta} = -B\frac{\partial \phi}{\partial \alpha} \tag{4.81}$$

The second of Eqs. (4.36) reads

$$\frac{\partial v}{\partial \beta} = B\epsilon_{22} - \frac{\partial \phi}{\partial \beta} u \tag{4.82}$$

or

$$\frac{\partial^2 u}{\partial \alpha \partial \beta} = \epsilon_{11}\left(-B\frac{\partial \phi}{\partial \alpha}\right) + \frac{\partial^2 \phi}{\partial \alpha \partial \beta} v - u\frac{\partial \phi}{\partial \beta}\frac{\partial \phi}{\partial \alpha} + B\epsilon_{22}\frac{\partial \phi}{\partial \alpha} \tag{4.83}$$

The fourth of Eqs. (4.36) can be put in the form

$$\frac{\partial u}{\partial \beta} = -B\omega + \frac{\partial \phi}{\partial \beta} v \tag{4.84}$$

since $\omega_2 = -\omega$. Differentiation of (4.84) gives

$$\frac{\partial^2 u}{\partial \alpha \partial \beta} = -\omega \frac{\partial B}{\partial \alpha} - B \frac{\partial \omega}{\partial \alpha} + \frac{\partial v}{\partial \alpha} \frac{\partial \phi}{\partial \beta} + v \frac{\partial^2 \phi}{\partial \alpha \partial \beta} \tag{4.85}$$

Now

$$\frac{\partial B}{\partial \alpha} = A \frac{\partial \phi}{\partial \beta} \tag{4.86}$$

and we take the third of Eqs. (4.36)

$$\frac{\partial v}{\partial \alpha} = A\omega - u \frac{\partial \phi}{\partial \alpha} \tag{4.87}$$

since $\omega_1 = \omega$. Let us insert (4.86), (4.87) into (4.85) to arrive at

$$\frac{\partial^2 u}{\partial \alpha \partial \beta} = -A \frac{\partial \phi}{\partial \beta} \omega - B \frac{\partial \omega}{\partial \alpha} + \left(A\omega - u \frac{\partial \phi}{\partial \alpha} \right) \frac{\partial \phi}{\partial \beta} + v \frac{\partial^2 \phi}{\partial \alpha \partial \beta} \tag{4.88}$$

Let us subtract (4.83) and (4.88) by sides to get

$$B \left[\frac{\partial \omega}{\partial \alpha} - (\epsilon_{11} - \epsilon_{22}) \frac{\partial \phi}{\partial \alpha} \right] = 0 \tag{4.89}$$

In a similar manner, we find

$$\frac{\partial \omega}{\partial \beta} + (\epsilon_{11} - \epsilon_{22}) \frac{\partial \phi}{\partial \beta} = 0 \tag{4.90}$$

Thus, for the second time we have arrived at (4.69) which imply (4.70). This ends the derivation.

Summing up, the Hencky nets are given by (4.31), where Lamé fields satisfy (4.76) with $a, b \in \{-1, 1\}$.

In domains characterized by the regime $\mathbb{T}$, upon assuming $\sigma = \sigma_T$ and $a = b = 1$, the Hencky net is determined by the formulae

$$\epsilon_{11} = 1, \qquad \epsilon_{22} = -\kappa, \qquad \epsilon_{12} = 0, \qquad \kappa = \frac{\sigma_T}{\sigma_C} \tag{4.91}$$

$$\omega(\alpha, \beta) = \omega(0, 0) - (1 + \kappa)(\alpha + \beta), \qquad \phi = \beta - \alpha \tag{4.92}$$

The Lamé fields satisfy

$$\frac{\partial A}{\partial \beta} = B, \qquad \frac{\partial B}{\partial \alpha} = A \tag{4.93}$$

The coordinates $x(\alpha, \beta)$, $y(\alpha, \beta)$ of the parametric lines are given by the integrals (4.31).

Let us introduce the auxiliary unknowns $\bar{x}(\alpha, \beta)$, $\bar{y}(\alpha, \beta)$ by

$$\bar{x} + i\bar{y} = e^{-i\phi}(x + iy) \tag{4.94}$$

where ϕ is given by (4.92). Let us write down (4.95) in the form

$$\bar{x} = x\cos\phi + y\sin\phi, \qquad \bar{y} = -x\sin\phi + y\cos\phi \tag{4.95}$$

or

$$x = \bar{x}\cos\phi - \bar{y}\sin\phi, \qquad y = \bar{x}\sin\phi + \bar{y}\cos\phi \tag{4.96}$$

One can easily prove that the fields $\bar{x}(\alpha, \beta)$, $\bar{y}(\alpha, \beta)$ satisfy

$$\frac{\partial \bar{x}}{\partial \beta} = \bar{y}, \qquad \frac{\partial \bar{y}}{\partial \alpha} = \bar{x} \tag{4.97}$$

The fields A, B are determined by the fields $\bar{x}$, $\bar{y}$ according to

$$A(\alpha, \beta) = \frac{\partial \bar{x}(\alpha, \beta)}{\partial \alpha} + \bar{y}(\alpha, \beta), \qquad B(\alpha, \beta) = \frac{\partial \bar{y}(\alpha, \beta)}{\partial \beta} + \bar{x}(\alpha, \beta) \tag{4.98}$$

To prove the above equations, one should make use of (4.20).

To construct the fields $x(\alpha, \beta)$, $y(\alpha, \beta)$ having the fields $A(\alpha, \beta)$, $B(\alpha, \beta)$, one should make use of (4.27). In Sect. 4.1, the integral equations (4.31) have been derived. Now, we rearrange them to make them dependent on the parameter $\theta < \lambda$:

$$x(\lambda, \mu) = x(\theta, \mu) + \int_\theta^\lambda \cos(\mu - \alpha)A(\alpha, \mu)d\alpha \tag{4.99}$$

$$y(\lambda, \mu) = y(\theta, \mu) + \int_\theta^\lambda \sin(\mu - \alpha)A(\alpha, \mu)d\alpha$$

Substitution of these formulae into (4.95) leads to the formulae for $\bar{x}$, $\bar{y}$

$$\bar{x}(\lambda, \mu) = \tilde{x}(\lambda, \mu) + \int_\theta^\lambda \cos(\lambda - \alpha)A(\alpha, \mu)d\alpha \tag{4.100}$$

$$\bar{y}(\lambda, \mu) = \tilde{y}(\lambda, \mu) + \int_\theta^\lambda \sin(\lambda - \alpha)A(\alpha, \mu)d\alpha$$

where

$$\tilde{x}(\lambda, \mu) = x(\theta, \mu)\cos(\mu - \lambda) + y(\theta, \mu)\sin(\mu - \lambda) \tag{4.101}$$
$$\tilde{y}(\lambda, \mu) = -x(\theta, \mu)\sin(\mu - \lambda) + y(\theta, \mu)\cos(\mu - \lambda)$$

In the case of $\theta = 0$, the integrals (4.100) are of convolution form and hence made suitable for the integration method based on the Laplace transform. The integrals in (4.99) are not of convolution type, hence less practical. This justifies introduction of the auxiliary fields $\bar{x}$, $\bar{y}$.

Remark 4.2.1 If the plane body is reinforced by a rib along a curve Γ, this curve must lie along one of the parametric lines. Let the curve Γ coincide with the α line. Along this curve there should be $\epsilon_\tau = \text{const}$ and then it is expressed by the same equation as ϵ_{11}, see the first of Eqs. (4.17) or the first of Eqs. (4.36). This formula coincides with the formula known from the theory of curved bars:

$$\epsilon_\tau = \frac{\partial u}{\partial s_1} + \frac{v}{R_1} \tag{4.102}$$

where R_1 is the radius of curvature of the line α, since $ds_1 = A d\alpha$, $R_1 = A$, $d\phi = -ds_1/R_1$; $\boldsymbol{\tau}$ is the unit vector tangent to the curve. We see that if the ribs lie along the parametric lines, then the condition (3.87) is an extension of one of the conditions (3.84) towards the edge of the design domain. Thus, there is no necessity to add the condition (3.87) to the formulation (3.85). It is sufficient to introduce the closure of the domain Ω, or $\bar{\Omega}$ in the formulation (3.85).

4.2.2 Construction of the Displacement Fields Within Hencky's Parameterization. Michell's Nets

We come back to the rule (4.41) for the displacements (u, v). Along the line α, where $d\beta = 0$ we have

$$d\left(e^{i\phi}(u + iv)\right) = e^{i\phi}\left[A(\epsilon_{11} + i\omega_1)d\alpha\right] \tag{4.103}$$

and, according to (4.65): $\epsilon_{11} = \text{const}$, $\omega_1 = \omega$. We perform the integration from $\alpha = \theta$ to $\alpha = \lambda$; thus, Eq. (4.103) is rewritten as follows:

$$d\left[e^{i(\mu-\lambda)}(u(\lambda, \mu) + iv(\lambda, \mu))\right] = e^{i(\mu-\alpha)}[\epsilon_{11} + i\omega(\alpha, \mu)]A(\alpha, \mu)d\alpha \tag{4.104}$$

Hence,

$$e^{i(\mu-\lambda)}\left[u(\lambda,\mu)+iv(\lambda,\mu)\right]=e^{i(\mu-\theta)}\left[u(\theta,\mu)+iv(\theta,\mu)\right]+\int_{\theta}^{\lambda}e^{i(\mu-\alpha)}[\epsilon_{11}+i\omega(\alpha,\mu)]A(\alpha,\mu)d\alpha \tag{4.105}$$

and

$$u(\lambda,\mu)+iv(\lambda,\mu)=e^{i(\lambda-\theta)}\left[u(\theta,\mu)+iv(\theta,\mu)\right]+\int_{\theta}^{\lambda}e^{i(\lambda-\alpha)}[\epsilon_{11}+i\omega(\alpha,\mu)]A(\alpha,\mu)d\alpha \tag{4.106}$$

Consider the regime $\mathbb{T}$, where (4.91)–(4.92) hold. Let us substitute $\epsilon_{11}=1$

$$\begin{aligned}u(\lambda,\mu)&=\hat{u}_\theta(\lambda,\mu)+\int_{\theta}^{\lambda}\left[\cos(\lambda-\alpha)-\omega(\alpha,\mu)\sin(\lambda-\alpha)\right]A(\lambda,\mu)d\alpha\\ v(\lambda,\mu)&=\hat{v}_\theta(\lambda,\mu)+\int_{\theta}^{\lambda}\left[\sin(\lambda-\alpha)+\omega(\alpha,\mu)\cos(\lambda-\alpha)\right]A(\lambda,\mu)d\alpha\end{aligned} \tag{4.107}$$

where

$$\begin{aligned}\hat{u}_\theta(\lambda,\mu)&=\cos(\lambda-\theta)u(\theta,\mu)-\sin(\lambda-\theta)v(\theta,\mu)\\ \hat{v}_\theta(\lambda,\mu)&=\sin(\lambda-\theta)u(\theta,\mu)+\cos(\lambda-\theta)v(\theta,\mu)\end{aligned} \tag{4.108}$$

Now, we perform integration along the line β. Along this line $d\alpha=0$, and Eq. (4.41) results in

$$d\left(e^{i\phi}(u+iv)\right)=e^{i\phi}[B(\omega_2+i\epsilon_{22})]d\beta \tag{4.109}$$

and, according to (4.74) we have $\epsilon_{22}=\text{const}$, $\omega_2=-\omega$. We perform integration from $\beta=\theta$ to $\beta=\mu$. We report the final formulae, upon substitution of $\epsilon_{22}=-\kappa$:

$$\begin{aligned}u(\lambda,\mu)&=\breve{u}_\theta(\lambda,\mu)+\int_{\theta}^{\mu}\left[-\cos(\beta-\mu)\omega(\lambda,\beta)+\kappa\sin(\beta-\mu)\right]B(\lambda,\beta)d\beta\\ v(\lambda,\mu)&=\breve{v}_\theta(\lambda,\mu)+\int_{\theta}^{\mu}\left[-\kappa\cos(\beta-\mu)-\sin(\beta-\mu)\omega(\lambda,\beta)\right]B(\lambda,\beta)d\beta\end{aligned} \tag{4.110}$$

where

$$\begin{aligned}\breve{u}_\theta(\lambda,\mu)&=\cos(\theta-\mu)u(\lambda,\theta)-\sin(\theta-\mu)v(\lambda,\theta)\\ \breve{v}_\theta(\lambda,\mu)&=\sin(\theta-\mu)u(\lambda,\theta)+\cos(\theta-\mu)v(\lambda,\theta)\end{aligned} \tag{4.111}$$

Let us introduce the operator

$$\mathscr{L} = \frac{\partial^2}{\partial\alpha\partial\beta} - 1 \tag{4.112}$$

We see that (4.93), (4.97) imply

$$\mathscr{L}w = 0, \qquad w = A, B, \bar{x}, \bar{y} \tag{4.113}$$

Moreover, by assuming ϕ in the form: $\phi(\alpha, \beta) = \beta - \alpha$ we simplify (4.36) to the form

$$\epsilon_{11} = \frac{1}{A}\left(\frac{\partial u}{\partial\alpha} + v\right), \qquad \epsilon_{22} = \frac{1}{B}\left(\frac{\partial v}{\partial\beta} + u\right) \tag{4.114}$$

$$\omega_1 = \frac{1}{A}\left(\frac{\partial v}{\partial\alpha} - u\right), \qquad \omega_2 = \frac{1}{B}\left(\frac{\partial u}{\partial\alpha} - v\right) \tag{4.115}$$

In the regime $\mathbb{T}$, see Table 3.1, upon assuming (4.91)–(4.92), we have

$$\begin{aligned} &\frac{1}{A}\left(\frac{\partial u}{\partial\alpha} + v\right) = 1, \qquad \frac{1}{B}\left(\frac{\partial v}{\partial\beta} + u\right) = -\kappa, \\ &\frac{1}{A}\left(\frac{\partial v}{\partial\alpha} - u\right) = \omega, \qquad \frac{1}{B}\left(\frac{\partial u}{\partial\beta} - v\right) = -\omega \end{aligned} \tag{4.116}$$

where ω is given by (4.92).

Let us note that the fields (u, v) determined by (4.107)–(4.108) satisfy the first of equations (4.116), but do not have to satisfy the second of equations (4.116).

Let us note that the fields (u, v) satisfying (4.116) satisfy also

$$\mathscr{L}u = (1 + \kappa)B, \qquad \mathscr{L}v = -(1 + \kappa)A \tag{4.117}$$

The derivation of the above equations makes use of (4.93).

Let us introduce new fields

$$\begin{aligned} u^0(\alpha, \beta) = &\ u(\alpha, \beta) - (1 + \kappa)\alpha A(\alpha, \beta) \\ v^0(\alpha, \beta) = &\ v(\alpha, \beta) + (1 + \kappa)\beta B(\alpha, \beta) \end{aligned} \tag{4.118a}$$

and note that

$$\frac{\partial u^0}{\partial\beta} - v^0 = -\omega(0, 0)B, \qquad \frac{\partial v^0}{\partial\alpha} - u^0 = \omega(0, 0)A \tag{4.118b}$$

where ω is given by (4.92). Thus,

$$\mathscr{L}u^0 = 0, \qquad \mathscr{L}v^0 = 0 \tag{4.119}$$

or the fields u^0, v^0 satisfy the same Eq. (4.113), as the fields A, B, $\bar{x}$, $\bar{y}$ do. This equation is of hyperbolic type and can be solved by Riemann's method, outlined in the sequel.

The fields (u, v) given along the lines α, β of the Hencky net satisfy the differential equations (4.116) where ω is given by (4.92). The fields (u, v) satisfy Eqs. (4.117), where A, B are linked by (4.93). If the fields (u, v) satisfy these equations as well as the required boundary conditions, then the net of lines α, β will be called Michell's net. Thus, Michell's nets are a narrow subclass of Hencky's nets.

4.2.3 *The Riemann Method of Solving the Problems Governed by the Equation $\mathcal{L}w = f$*

The fields A, B, u^0, v^0, $\bar{x}$, $\bar{y}$ are governed by the same hyperbolic equation of the form $\mathscr{L}w = 0$. In the present section we shall outline the Riemann method of solving this equation, if the boundary conditions are prescribed along the parametric lines $\alpha = \theta_1, \beta = \theta_2$. It is helpful to consider a problem governed by the non-homogeneous equation: $\mathscr{L}w = f$. Let the function $f = f(\alpha, \beta)$ be given in the domain Ω_0: $\theta_1 \le \alpha \le \lambda$, $\theta_2 \le \beta \le \mu$. Let us define the functions

$$G_n(\alpha, \beta) = \alpha^n \sum_{k=0}^{\infty} \frac{(\alpha\beta)^k}{k!(n+k)!}, \qquad n = 0, 1, 2, \ldots \tag{4.120}$$

Alternatively, these functions may be defined by

$$G_n(\alpha, \beta) = \left(\frac{\alpha}{\beta}\right)^{\frac{n}{2}} I_n(2\sqrt{\alpha\beta}) \tag{4.121}$$

where I_n is modified Bessel function. Just this definition justifies the properties of the functions G_n and confirms the correctness of its definition (4.120). Since

$$I_n(x) = \frac{\left(\frac{x}{2}\right)^n}{n!} + \frac{\left(\frac{x}{2}\right)^{n+2}}{1!(n+1)!} + \frac{\left(\frac{x}{2}\right)^{n+4}}{2!(n+2)!} + \cdots \tag{4.122}$$

hence

$$I_n(2\sqrt{\alpha\beta}) = (\alpha\beta)^{\frac{n}{2}} \sum_{k=0}^{\infty} \frac{(\alpha\beta)^k}{k!(n+k)!} \tag{4.123}$$

which confirms (4.121). A particular role will be played by the function

$$G_0(\alpha\beta) = I_0(2\sqrt{\alpha\beta}), \qquad I_0(2\sqrt{\alpha\beta}) = \sum_{k=0}^{\infty} \frac{(\alpha\beta)^k}{(k!)^2} \tag{4.124}$$

The functions $G_n(\alpha, \beta)$ satisfy the equation

$$\mathscr{L} G_n = 0 \tag{4.125}$$

and the recurrence-differential equations

$$\frac{\partial G_n(\beta, \alpha)}{\partial \alpha} = G_{n+1}(\beta, \alpha), \qquad \frac{\partial G_n(\beta, \alpha)}{\partial \beta} = G_{n-1}(\beta, \alpha) \tag{4.126}$$

$$G_{-n}(\alpha, \beta) = G_n(\beta, \alpha) \tag{4.127}$$

$$\alpha G_{n-1}(\alpha, \beta) - \beta G_{n+1}(\alpha, \beta) = n G_n(\alpha, \beta) \tag{4.128}$$

In particular, there hold

$$\alpha G_1(\beta, \alpha) = \beta G_1(\alpha, \beta), \qquad \alpha G_0(\alpha, \beta) - \beta G_2(\alpha, \beta) = G_1(\alpha, \beta) \tag{4.129}$$

and

$$G_n(\alpha, 0) = \frac{\alpha^n}{n!}, \qquad G_n(0, \beta) = \delta_{0n}, \qquad n \geq 0 \tag{4.130}$$

Let the function $w(\alpha, \beta)$ satisfy within Ω_0 the non-homogeneous equation

$$\mathscr{L} w = f \tag{4.131}$$

The value of w at a point (λ, μ) is expressed by Riemann's formula

$$\begin{aligned} w(\lambda, \mu) = {} & w(\theta_1, \theta_2) G_0(\lambda - \theta_1, \mu - \theta_2) + \\ & \int_{\theta_1}^{\lambda} G_0(\lambda - \alpha, \mu - \theta_2) \frac{\partial w}{\partial \alpha}(\alpha, \theta_2) d\alpha + \\ & \int_{\theta_2}^{\mu} G_0(\lambda - \theta_1, \mu - \beta) \frac{\partial w}{\partial \beta}(\theta_1, \beta) d\beta + \\ & \int_{\theta_1}^{\lambda} \int_{\theta_2}^{\mu} G_0(\lambda - \alpha, \mu - \beta) f(\alpha, \beta) d\alpha d\beta \end{aligned} \tag{4.132}$$

Thus, the value $w(\lambda, \mu)$ is given by $w(\theta_1, \theta_2)$ and by the values $w(\alpha, \theta_2)$ on the line α and by the values $w(\theta_1, \beta)$ on the line β, as well as by the distribution of f inside the domainΩ_0, see Fig. 4.5.

The formula (4.132) will be frequently used; hence, it is helpful to give here its derivation. Let

$$G(\alpha, \beta) = G_0(\lambda - \alpha, \mu - \beta) \tag{4.133}$$

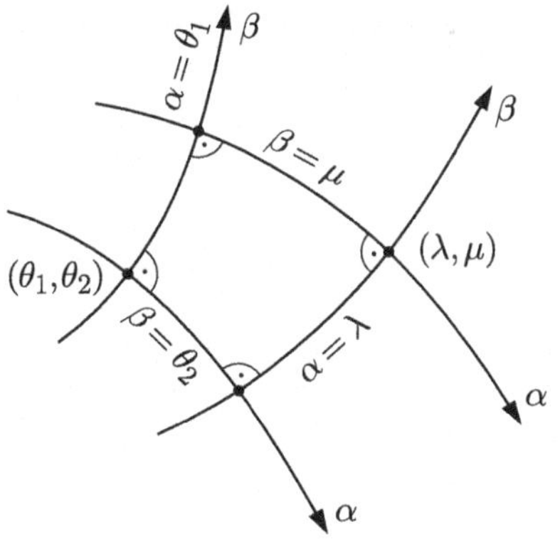

Fig. 4.5 The setting of the Riemann problem

According to (4.125), the function G satisfies $\mathscr{L}G = 0$. Since w satisfies (4.98), the following equality holds:

$$\int_{\Omega_0} 2[G(\mathscr{L}w) - w(\mathscr{L}G)]d\alpha d\beta = \int_{\Omega_0} 2Gf d\alpha d\beta \tag{4.134}$$

Let us recall Green's formula

$$\int_{\Omega_0} \left(\frac{\partial Q}{\partial \alpha} - \frac{\partial P}{\partial \beta}\right) d\alpha d\beta = \int_{\partial\Omega_0} (P d\alpha + Q d\beta) \tag{4.135}$$

The integrand of the left-hand side (4.134) can be written in the form

$$\frac{\partial}{\partial \alpha}\left(G\frac{\partial w}{\partial \beta} - w\frac{\partial G}{\partial \beta}\right) - \frac{\partial}{\partial \beta}\left(-G\frac{\partial w}{\partial \alpha} + w\frac{\partial G}{\partial \alpha}\right) \tag{4.136}$$

Integration of this expression using (4.135) and by taking into account: $G_0(0,\beta) = 1$, $G_0(\alpha, 0) = 1$, leads to (4.132).

Let us define a new family of functions:

$$F_n(\beta, \alpha) = \int_0^\beta \cos(\beta - t)G_{n-1}(t, \alpha)dt, \qquad n \geq 1 \tag{4.137}$$

or, equivalently

$$F_n(\beta, \alpha) = \int_0^\beta \sin(\beta - t)G_{n-2}(t, \alpha)dt, \qquad n \geq 2 \tag{4.138}$$

and

$$F_0(\beta, \alpha) = F_2(\alpha, \beta) + \cos(\beta - \alpha) \tag{4.139}$$

The functions $F_n(\cdot,\cdot)$ thus defined can also be defined by

$$F_n(\alpha,\beta)=\sum_{m=0}^{\infty}(-1)^m G_{2m+n}(\alpha,\beta) \tag{4.140}$$

or by

$$F_n(\beta,\alpha)=U_n(2\beta,2i\sqrt{\alpha\beta}) \tag{4.141}$$

with $U_n(\cdot,\cdot)$ being Lommel's function belonging to the family of Bessel functions, see Watson (1966). These functions satisfy the following recurrence-differential equations:

$$\frac{\partial F_n(\beta,\alpha)}{\partial\alpha}=F_{n+1}(\beta,\alpha),\qquad \frac{\partial F_n(\beta,\alpha)}{\partial\beta}=F_{n-1}(\beta,\alpha) \tag{4.142}$$

Thus, they satisfy the same equations, as given before, see (4.126). Moreover, $\mathscr{L}F_n=0$ and

$$\begin{gathered} F_n(\alpha,\beta)+F_{n+2}(\alpha,\beta)=G_n(\alpha,\beta)\\ F_n(\beta,\alpha)-F_{2-n}(\alpha,\beta)=(-1)^n\cos\left(\frac{n\pi}{2}+\beta-\alpha\right) \end{gathered} \tag{4.143}$$

In particular, the following equations hold:

$$\begin{gathered} F_1(\beta,\alpha)-F_1(\alpha,\beta)=\sin(\beta-\alpha)\\ F_1(\beta,0)=\sin\beta\\ F_2(\beta,0)=1-\cos\beta\\ F_n(0,\alpha)=\delta_{n0},\qquad n=0,1,2,\ldots \end{gathered} \tag{4.144}$$

Let Z_n represent either G_n or F_n. The functions Z_n satisfy (4.131) for $f=0$ and hence can be represented by means of Eq. (4.132) as follows:

$$\begin{aligned} Z_n(\lambda,\mu)=&\,Z_n(\theta_1,\theta_2)G_0(\lambda-\theta_1,\mu-\theta_2)+\\ &\int_{\theta_1}^{\lambda}G_0(\lambda-\alpha,\mu-\theta_2)Z_{n-1}(\alpha,\theta_2)d\alpha+\\ &\int_{\theta_2}^{\mu}G_0(\lambda-\theta_1,\mu-\beta)Z_{n+1}(\theta_1,\beta)d\beta \end{aligned} \tag{4.145}$$

Consider now the case of $\theta_1=\theta_2=\theta$. Changing the variables (λ,μ) into (μ,λ) one can rewrite Eq. (4.145) in the form

$$\begin{aligned} Z_n(\mu,\lambda)=&\,Z_n(\theta,\theta)G_0(\lambda-\theta,\mu-\theta)+\\ &\int_{\theta}^{\lambda}G_0(\lambda-\alpha,\mu-\theta)Z_{n+1}(\theta,\alpha)d\alpha+\\ &\int_{\theta}^{\mu}G_0(\lambda-\theta,\mu-\beta)Z_{n-1}(\beta,\theta)d\beta \end{aligned} \tag{4.146}$$

In the sequel, the following auxiliary functions will be used:

$$\begin{aligned} T_n(\lambda, \mu) &= G_n(\lambda, \mu) + G_{n+1}(\lambda, \mu) \\ H_n(\lambda, \mu) &= F_n(\lambda, \mu) + F_{n+1}(\lambda, \mu) \qquad n = 0, 1, \ldots \end{aligned} \tag{4.147}$$

Let us collect now all the further used integration rules involving the special functions defined above. Let us define first:

$$\begin{aligned} f_n^{\theta}(\lambda, \mu) &= \int_{\theta}^{\lambda} G_0(\lambda - \alpha, \mu - \theta) F_n(\alpha - \theta, 2\theta) d\alpha \\ G_n^{\theta}(\lambda, \mu) &= \int_{\theta}^{\lambda} G_0(\lambda - \alpha, \mu - \theta) \frac{\partial}{\partial \alpha} G_n(\alpha - \theta, 2\theta) d\alpha \\ F_n^{\theta}(\lambda, \mu) &= \int_{\theta}^{\lambda} G_0(\lambda - \alpha, \mu - \theta) \frac{\partial}{\partial \alpha} F_n(\alpha - \theta, 2\theta) d\alpha \\ t_n(\lambda, \mu) &= \int_{0}^{\lambda} \alpha \sin(\lambda - \alpha) G_n(\alpha, \mu) d\alpha \\ h_n(\lambda, \mu) &= \int_{0}^{\lambda} \alpha \cos(\lambda - \alpha) G_n(\alpha, \mu) d\alpha \end{aligned} \tag{4.148}$$

These integrals can be computed analytically; the final results read

$$\begin{aligned} f_n^{\theta}(\lambda, \mu) &= F_{n+1}(\lambda - \theta, \mu + \theta) \\ G_n^{\theta}(\lambda, \mu) &= G_n(\lambda - \theta, \mu + \theta) - \delta_{n0} G_0(\lambda - \theta, \mu - \theta) \\ F_n^{\theta}(\lambda, \mu) &= F_n(\lambda - \theta, \mu + \theta) - \delta_{n0} G_0(\lambda - \theta, \mu - \theta) \\ t_n(\lambda, \mu) &= (n + 1) F_{n+3}(\lambda, \mu) + \mu F_{n+4}(\lambda, \mu) \\ h_n(\lambda, \mu) &= (n + 1) F_{n+2}(\lambda, \mu) + \mu F_{n+3}(\lambda, \mu) \end{aligned} \tag{4.149}$$

Let us define

$$\begin{aligned} g_1(\lambda, \mu) &= \int_{0}^{\lambda} \cos(\lambda - \alpha) G_1(\mu, \alpha) d\alpha \\ g_2(\lambda, \mu) &= \int_{0}^{\lambda} \sin(\lambda - \alpha) G_1(\mu, \alpha) d\alpha \\ g_3(\lambda, \mu) &= \int_{0}^{\lambda} \alpha \sin(\lambda - \alpha) G_0(\alpha, \mu) d\alpha \\ g_4(\lambda, \mu) &= \int_{0}^{\lambda} \alpha \sin(\lambda - \alpha) G_1(\mu, \alpha) d\alpha \\ g_5(\theta; \lambda, \mu) &= \int_{\theta}^{\lambda} \alpha \sin(\lambda - \alpha) G_0(\alpha - \theta, \mu + \theta) d\alpha \\ g_6(\theta; \lambda, \mu) &= \int_{\theta}^{\lambda} \alpha \sin(\lambda - \alpha) G_1(\alpha - \theta, \mu + \theta) d\alpha \end{aligned} \tag{4.150}$$

Let $\theta < \lambda$; then

$$\begin{aligned}
g_1(\lambda, \mu) &= -\cos\lambda + G_0(\lambda, \mu) - F_2(\lambda, \mu) \\
g_2(\lambda, \mu) &= -\sin\lambda + F_1(\lambda, \mu) \\
g_3(\lambda, \mu) &= \lambda G_0(\lambda, \mu) - \mu F_2(\lambda, \mu) - F_1(\lambda, \mu) \\
g_4(\lambda, \mu) &= \mu F_3(\lambda, \mu) \\
g_5(\theta; \lambda, \mu) &= (\lambda - \theta) G_0(\lambda - \theta, \mu + \theta) - \mu F_2(\lambda - \theta, \mu + \theta) - F_1(\alpha - \theta, \mu + \theta) \\
g_6(\theta; \lambda, \mu) &= -\mu F_3(\lambda - \theta, \mu + \theta) - 2F_2(\alpha - \theta, \mu + \theta) + (\lambda - \theta) G_1(\lambda - \theta, \mu + \theta)
\end{aligned} \tag{4.151}$$

Moreover

$$\int_\theta^\lambda \alpha\cos(\lambda - \alpha) G_n(\alpha - \theta, \mu + \theta) d\alpha = (n+1) F_{n+2}(\lambda - \theta, \mu + \theta) + (\mu + \theta) F_{n+3}(\lambda - \theta, \mu + \theta) + \theta F_{n+1}(\lambda - \theta, \mu + \theta) \tag{4.152}$$

$$\int_\theta^\lambda \alpha\sin(\lambda - \alpha) G_n(\alpha - \theta, \mu + \theta) d\alpha = (n+1) F_{n+3}(\lambda - \theta, \mu + \theta) + (\mu + \theta) F_{n+4}(\lambda - \theta, \mu + \theta) + \theta F_{n+2}(\lambda - \theta, \mu + 0) \tag{4.153}$$

$$\int_{\theta_1}^\lambda G_0(\lambda - \alpha, \mu) G_{n-1}(\alpha - \theta_1, \theta_2) d\alpha = G_n(\lambda - \theta_1, \mu + \theta_2) - \delta_{n0} G_0(\lambda - \theta_1, \mu), \qquad n = 0, 1, \ldots \tag{4.154}$$

$$\int_0^\lambda G_0(\lambda - \alpha, \mu) G_1(\theta, \alpha) d\alpha = G_0(\lambda, \mu + \theta) - G_0(\lambda, \mu), \tag{4.155}$$

$$\int_0^\lambda G_0(\lambda - \alpha, \mu) F_n(\alpha, \theta) d\alpha = F_{n+1}(\lambda, \mu + \theta), \qquad n \geq 0$$

Let

$$\begin{aligned}
N_1 &= \int_0^\lambda G_0(\lambda - \alpha, \mu) Z_{n-1}(\alpha + \theta, \theta) d\alpha \\
N_2 &= \int_0^\mu G_0(\mu - \beta, \lambda) Z_{n+1}(\theta, \beta + \theta) d\beta
\end{aligned} \tag{4.156}$$

where Z_n stands for G_n or F_n. One can derive

$$N_1 + N_2 = Z_n(\lambda + \theta, \mu + \theta) - Z_n(\theta, \theta) G_0(\lambda, \mu) \tag{4.157}$$

The explicit forms of N_1, N_2 separately, are unknown.

In the sequel the functions G_n, F_n, T_n, H_n, $n = 0, 1, \ldots$ will appear in numerous analytical results. They belong to a wide family of Bessel functions and are linked with Lommel's functions. In the present book, these functions will be called Chan's functions, to stress a great role of Henry Chan in developing new methods of solving Michell's problems.

4.3 Statics of a Continuum Body Parameterized by Curvilinear Michell Nets

4.3.1 A Body Without Reinforcing Ribs

We come back to the problems of Sect. 4.1.2 under the assumption that the parameterization (α, β), see Fig. 4.2, is of Michell class of curved lines, hence has the properties outlined in Sect. 4.2. Let us recall that in Sect. 4.1.2 we have discussed the parameterization formed by the trajectories of principal stresses. This assumption holds good here; nevertheless, we impose additional geometric conditions discussed in Sect. 4.2, being the consequence of the assumption that the parametric lines follow also the trajectories of virtual principal strains of constant distribution within the domain, see (4.65). The latter condition has its strong consequences in the form of (4.74).

Substitution of (4.74) into (4.58) gives a specific form of equations linking the forces in the fibres

$$-\frac{\partial T_1}{\partial \alpha} + bT_2 = 0, \qquad -\frac{\partial T_2}{\partial \beta} + aT_1 = 0 \tag{4.158}$$

where $a, b \in \{-1, 1\}$, depending on the method of the net's construction.

According to (4.36) and (4.74), the relative elongation of fibres along α, β is given by

$$A\epsilon_{11}(\boldsymbol{v}) = \frac{\partial u}{\partial \alpha} + av, \qquad B\epsilon_{22}(\boldsymbol{v}) = \frac{\partial v}{\partial \beta} + bu \tag{4.159}$$

Thus, according to (46_b)

$$\mathcal{W} = T_1\left(\frac{\partial u}{\partial \alpha} + av\right) + T_2\left(\frac{\partial v}{\partial \beta} + bu\right) \tag{4.160}$$

and the equation of the virtual work (3.3) assumes the form

$$\int_{\alpha_0}^{\alpha_1}\int_{\beta_0}^{\beta_1}\left(T_1\frac{\partial u}{\partial \alpha} + T_2\frac{\partial v}{\partial \beta} + aT_1 v + bT_2 u\right)d\alpha d\beta = \mathcal{L}(\boldsymbol{v}) \tag{4.161}$$

where $\mathcal{L}(\cdot)$ has the form (4.49); the fields (u, v) are kinematically admissible. The localization of (4.161) leads to Eqs. (4.158) satisfied in Ω and to the natural boundary conditions (4.52).

Remark 4.3.1 The variational equation of equilibrium (4.161) is the continuum counterpart of the variational equation of equilibrium of trusses (2.2). Indeed, recall now the formula for the relative elongation of a curved bar

$$\epsilon_\tau = \frac{\partial u_\tau(s)}{\partial s} + \frac{1}{R(s)} u_n(s) \tag{4.162}$$

where $R(s)$ is the radius of curvature of the neutral axis of the bar; u_τ, u_n are tangential and normal components of the displacement vector, see Fig. 4.6.

Let us apply the formula (4.162) to compute the deformation along the lines $\xi^1 = \alpha$, $\xi^2 = \beta$ of Hencky's nets. We find

$$\epsilon_1(u, v) = \frac{\partial u}{\partial s_1} + \frac{v}{R_1}, \qquad \epsilon_2(u, v) = \frac{\partial v}{\partial s_2} + \frac{u}{R_2} \tag{4.163}$$

where R_1, R_2 are radii of curvature of the line $\alpha(\beta = \text{const})$ and of the line $\beta(\alpha = \text{const})$. Moreover, we know that $R_1 = A(\alpha, \beta)$, $R_2 = B(\alpha, \beta)$, and $ds_1 = Ad\alpha$, $ds_2 = Bd\beta$. Thus

$$\epsilon_1(u, v) = \frac{1}{A}\left(\frac{\partial u}{\partial \alpha} + v\right), \qquad \epsilon_2(u, v) = \frac{1}{B}\left(\frac{\partial v}{\partial \beta} + u\right) \tag{4.164}$$

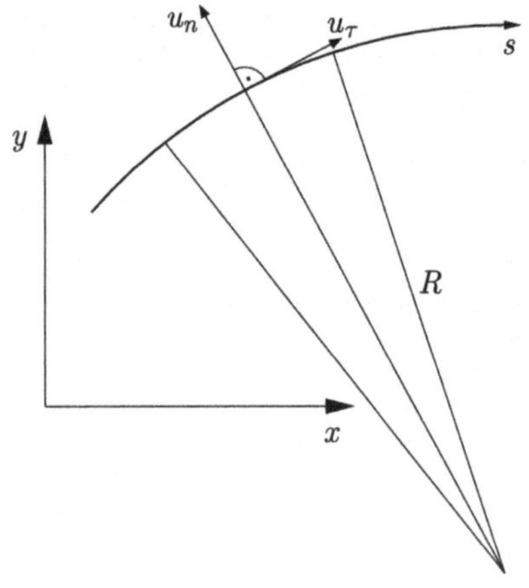

Fig. 4.6 Displacements along a curved line

and these equations coincide with (4.159), if $a = b = 1$; just these values of a, b will be used in the sequel. The virtual work equation (4.161) can be written in the form

$$\iint [T_1 d\beta \epsilon_1 ds_1 + T_2 d\alpha \epsilon_2 ds_2] = \mathcal{L}(\boldsymbol{v}) \tag{4.165}$$

where $T_1 d\beta$ represents the force in the strip of width $Bd\beta = ds_2$ and $T_2 d\alpha$ represents the force in the strip of width $Ad\alpha = ds_1$, see Fig. 4.7 and $\epsilon_1 ds_1 = A\epsilon_1 d\alpha = d\Delta_1$ is the elongation of the fibre of length $Ad\alpha = ds_1$, $\epsilon_2 ds_2 = B\epsilon_2 d\beta = d\Delta_2$ is the elongation of the fibre of length $Bd\beta = ds_2$. Assume that the fields $T_1, T_2, \epsilon_1, \epsilon_2$ are constant within the rectangles of dimensions $\Delta s_1 \times \Delta s_2$. We compute the integral at the left-hand side of (4.165) over this rectangle

$$\int_{s_1}^{s_1+\Delta s_1} \int_{\beta_1}^{\beta_1+\Delta\beta_1} T_1 d\beta \epsilon_1 ds_1 + \int_{\alpha_1}^{\alpha_1+\Delta\alpha_1} \int_{s_2}^{s_2+\Delta s_2} T_2 d\alpha \epsilon_2 ds_2 =$$
$$T_1 \Delta\beta_1 \epsilon_1 \Delta s_1 + T_2 \Delta\alpha_1 \epsilon_2 \Delta s_2 = \tilde{T}_1 \tilde{\Delta}_1 + \tilde{T}_2 \tilde{\Delta}_2 \tag{4.166}$$

where $\tilde{T}_1 = T_1 \Delta\beta_1$ is the resultant of the force T_1 referred to the width $\Delta\beta_1$, and $\tilde{T}_2 = T_2 \Delta\alpha_1$ is the resultant of the force T_2 referred to the width $\Delta\alpha_1$. Moreover, $\tilde{\Delta}_i = \epsilon_i \Delta s_i$ represent elongations of fibres of length Δs_i. Thus, the sum of the quantities (4.166) assumes the form

$$\sum_k \tilde{T}_k \Delta_k = \tilde{\boldsymbol{T}} \cdot \tilde{\boldsymbol{\Delta}} \tag{4.167}$$

identical to that involved in the virtual work equation for trusses, see (2.1). The above truss analogy means that the body is replaced with the system of bars; $\tilde{T}_k$ assume the meaning of axial forces in the bars of the truss. The bars are linked by hinge nodes to prevent from appearing of the shear. The right-hand side of (4.165) can be interpreted as the virtual work of a certain system of loads on the virtual displacements $\bar{\boldsymbol{u}}$ of the nodes, see Sect. 1.2.

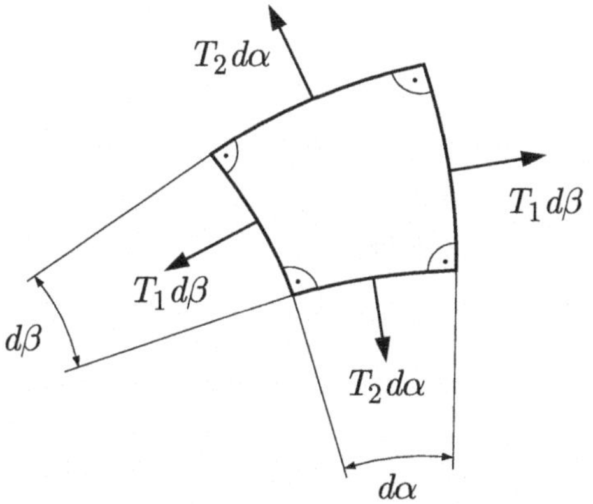

Fig. 4.7 The problem setting

Remark 4.3.2 The volume of Michell trusses composed of two families of fibres is given by (3.53), where the volume fraction of the constituent material is given by (3.111), (3.89$_b$). In the considered planar problem this volume fraction can be expressed by Hemp's forces (4.45), or

$$v = \frac{1}{\sigma_0}\rho(\boldsymbol{\tau}), \qquad \rho(\boldsymbol{\tau}) = \eta_1(\boldsymbol{\tau})\frac{|T_1|}{B} + \eta_2(\boldsymbol{\tau})\frac{|T_2|}{A} \tag{4.168}$$

where $\eta_i(\boldsymbol{\tau})$ are given by (3.88). If the fibres α are in tension (then $T_1 > 0$) and the fibres β are in compression (then $T_2 < 0$), we have $\eta_1 = \dfrac{\sigma_0}{\sigma_T}$, $\eta_2 = \dfrac{\sigma_0}{\sigma_C}$ and then

$$v(\alpha, \beta) = \frac{1}{\sigma_T}\frac{T_1(\alpha, \beta)}{B(\alpha, \beta)} - \frac{1}{\sigma_C}\frac{T_2(\alpha, \beta)}{A(\alpha, \beta)} \tag{4.169}$$

The first component above measures the volume fraction of the fibres in tension:

$$v_1(\alpha, \beta) = \frac{1}{\sigma_T}\frac{T_1(\alpha, \beta)}{B(\alpha, \beta)} \tag{4.170}$$

while the second one represents the volume fraction of the fibres in compression

$$v_2(\alpha, \beta) = \frac{1}{\sigma_C}\frac{-T_2(\alpha, \beta)}{A(\alpha, \beta)} \tag{4.171}$$

The integral of $v = v_1 + v_2$ over the design domain determines the volume of the material to be used. We shall see in the sequel that the fields T_1, T_2 suffer jumps along certain curves, while the fields A, B suffer jumps along a different set of curves. The cross section of these two divisions make a new and much more complicated division of the design domain; consequently, the field v suffers jumps along many curves within the design domain.

The volume of the whole structure is expressed by the integral of the volume fraction v over the design domain plus the sum of volumes of the reinforcing ribs which appear always, if the point loads are applied, or if the optimal structure is pin-supported.

4.3.2 *The Equilibrium of a Fibrous Structure Reinforced by Ribs*

Assume that a Michell truss transmits a given point load P applied at point P and the edge load $\boldsymbol{p}$ applied along the sides $\alpha = \alpha_P$, $\beta = \beta_{P^-}$ to the support along the curve Γ_0, cf. Fig. 4.8. Assume that the Michell truss consists of

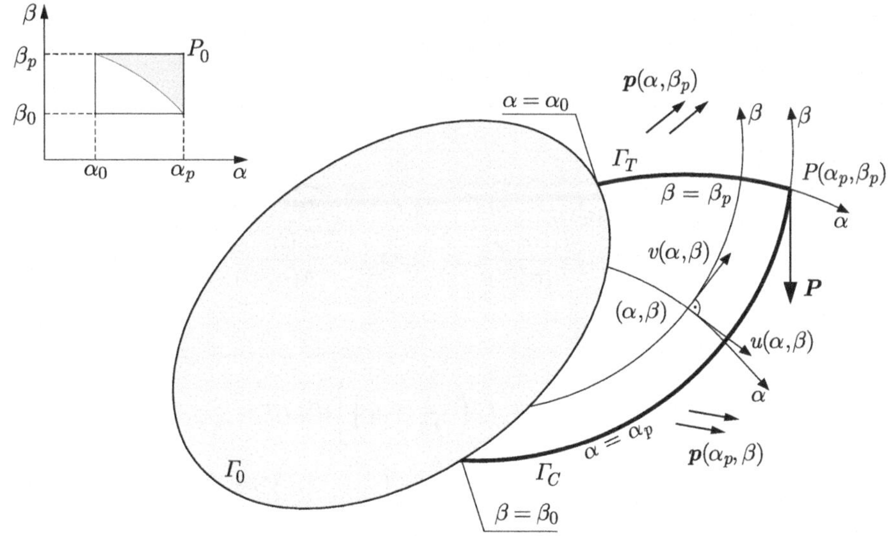

Fig. 4.8 The problem setting

- the fibrous domain Ω composed of two families of fibres: in tension (the lines $\xi^1 = \alpha$), and in compression (the lines $\xi^2 = \beta$);
- the rib in tension along $\beta = \beta_P$;
- the rib in compression along $\alpha = \alpha_P$.

The load causes the internal forces:

- Hemp's forces T_1, T_2 in fibres;
- the axial force F_T in the rib in tension;
- the axial force F_C in the rib in compression.

The units of all these forces are the same: [N]. We assume that $a = b = 1$ in (4.74). Let $\boldsymbol{p} = (p_1, p_2)$, where p_1 is directed along the $\xi^1 = \alpha$ lines and p_2 is applied along the line $\xi^2 = \beta$. Moreover, the point load P is decomposed into: P_1 tangent to the $\xi^1 = \alpha$ line and P_2- tangent to the $\xi^2 = \beta$ line.

The virtual work equation encompasses the work of the internal forces in fibres and the virtual work of the force F_T in the rib in tension and the virtual work of the force F_C in the rib in compression. Thus, the virtual work equation assumes the form

$$\int_\Omega [\lambda_1(\boldsymbol{\tau})\epsilon_{11}(u, v) + \lambda_2(\boldsymbol{\tau})\epsilon_{22}(u, v)]\, d\Omega + \\ \int_{\Gamma_T} F_T \epsilon_1(u, v) ds_1 + \int_{\Gamma_C} F_C \epsilon_2(u, v) ds_2 = \mathcal{L}(\boldsymbol{v}) \qquad (4.172)$$

where $\epsilon_i(u, v)$ is defined by (4.164); the virtual work of the load is

$$\mathcal{L}(\boldsymbol{v}) = \int_{\Gamma_T} (p_1 u + p_2 v)|_{\beta=\beta_P} ds_1 + \int_{\Gamma_C} (p_1 u + p_2 v)|_{\alpha=\alpha_P} ds_2 + P_1 u(\alpha_P, \beta_P) + P_2 v(\alpha_P, \beta_P) \tag{4.173}$$

Taking into account that

$$d\Omega = ds_1 ds_2, \quad ds_1 = A d\alpha, \quad ds_2 = B d\beta$$
$$\lambda_1(\boldsymbol{\tau}) = \frac{T_1(\alpha, \beta)}{B(\alpha, \beta)}, \quad \lambda_2(\boldsymbol{\tau}) = \frac{T_2(\alpha, \beta)}{A(\alpha, \beta)} \tag{4.174}$$

we rewrite (4.172) in the form

$$\int_{\Omega_0} \left[T_1 \left(\frac{\partial u}{\partial \alpha} + v \right) + T_2 \left(\frac{\partial v}{\partial \beta} + u \right) \right] d\alpha d\beta +$$
$$\int_{\Gamma_T} F_T \left(\frac{\partial u}{\partial \alpha} + v \right) d\alpha + \int_{\Gamma_C} F_C \left(\frac{\partial v}{\partial \beta} + u \right) d\beta = \mathcal{L}(\boldsymbol{v}) \tag{4.175}$$

where $\boldsymbol{v} = (u, v)$ and Ω_0 is the domain of variation of the parameters (α, β). Localization of (4.175) leads to the equilibrium equations (4.158) which, for $a = b = 1$, assume the form, see. Fig. 4.9

$$-\frac{\partial T_1}{\partial \alpha} + T_2 = 0, \quad -\frac{\partial T_2}{\partial \beta} + T_1 = 0 \tag{4.176}$$

Moreover, we arrive at the equilibrium equations of the domains adjacent to the ribs. A standard procedure leads to

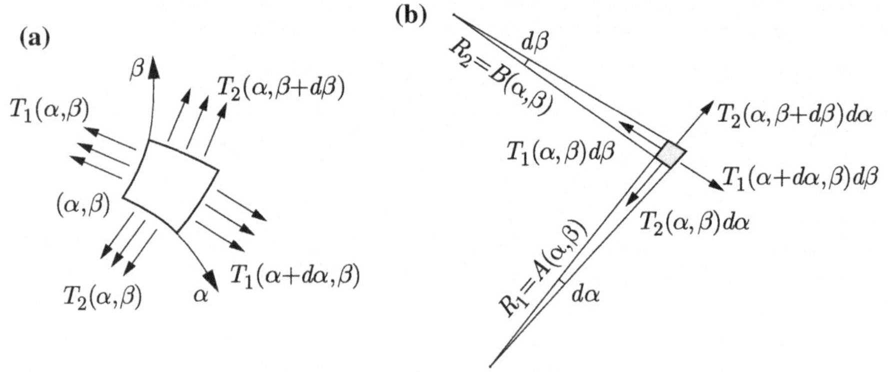

Fig. 4.9 Stress resultants within the fibrous domain: **a** Hemp's stress resultants and **b** forces acting on the elementary segment of the body

$$\int_{\Omega_0} T_1 \frac{\partial u}{\partial \alpha} d\alpha d\beta = \int_{\Omega_0} \left[\frac{\partial}{\partial \alpha}(T_1 u) - \frac{\partial T_1}{\partial \alpha} u \right] d\alpha d\beta = \int_{\Omega_0} -\frac{\partial T_1}{\partial \alpha} u d\alpha d\beta + \int_{\beta_0}^{\beta_P} (T_1 u)|_{\alpha=\alpha_P} d\beta \quad (4.177)$$

since $u = 0$ on Γ_0. By analogy, we find

$$\int_{\Omega_0} T_2 \frac{\partial v}{\partial \beta} d\alpha d\beta = \int_{\Omega_0} -\frac{\partial T_2}{\partial \beta} v d\alpha d\beta + \int_{\alpha_0}^{\alpha_P} (T_2 v)|_{\beta=\beta_P} d\alpha \quad (4.178)$$

since $v = 0$ on Γ_0. Now, we proceed further

$$\int_{\Gamma_T} F_T \frac{\partial u}{\partial \alpha} = \int_{\alpha_0}^{\alpha_P} \left(-\frac{\partial F_T}{\partial \alpha} u \right) \Big|_{\beta=\beta_P} d\alpha + F_T(\alpha_P, \beta_P) u(\alpha_P, \beta_P) \quad (4.179)$$

then

$$\int_{\Gamma_C} F_C \frac{\partial v}{\partial \beta} = \int_{\beta_0}^{\beta_P} \left(-\frac{\partial F_C}{\partial \beta} v \right) \Big|_{\alpha=\alpha_P} d\beta + F_C(\alpha_P, \beta_P) v(\alpha_P, \beta_P) \quad (4.180)$$

Equation (4.175) implies (4.176) and

$$\begin{aligned}
&\int_{\alpha_0}^{\alpha_P} \left[-\frac{dF_T(\alpha)}{d\alpha} - p_1(\alpha, \beta_P) A(\alpha, \beta_P) \right] u(\alpha, \beta_P) d\alpha + \\
&\int_{\beta_0}^{\beta_P} \left[-\frac{dF_C(\beta)}{d\beta} - p_2(\alpha_P, \beta) B(\alpha_P, \beta) \right] v(\alpha_P, \beta) d\beta + \\
&\int_{\beta_0}^{\beta_P} \left[F_C(\beta) + T_1(\alpha_P, \beta) - p_1(\alpha_P, \beta) B(\alpha_P, \beta) \right] u(\alpha_P, \beta) d\beta + \\
&\int_{\alpha_0}^{\alpha_P} \left[F_T(\alpha) + T_2(\alpha, \beta_P) - p_2(\alpha, \beta_P) A(\alpha, \beta_P) \right] v(\alpha, \beta_P) d\alpha + \\
&\quad [F_T(\alpha_P) - P_1] u(\alpha_P, \beta_P) + [F_C(\beta_P) - P_2] v(\alpha_P, \beta_P) = 0
\end{aligned} \quad (4.181)$$

Hence, we obtain the equilibrium equations:

- linking the forces in the fibrous plate with the forces in the ribs

$$\begin{aligned}
T_1(\alpha_P, \beta) + F_C(\beta) &= B(\alpha_P, \beta) p_1(\alpha_P, \beta), \\
T_2(\alpha, \beta_P) + F_T(\alpha) &= A(\alpha, \beta_P) p_2(\alpha, \beta_P)
\end{aligned} \quad (4.182)$$

- linking the forces in the ribs with the boundary load

$$-\frac{dF_T(\alpha)}{d\alpha} = A(\alpha, \beta_P)p_1(\alpha, \beta_P),$$
$$-\frac{dF_C(\beta)}{d\beta} = B(\alpha_P, \beta)p_2(\alpha_P, \beta) \tag{4.183}$$

- of the node P

$$F_T(\alpha_P) = P_1, \qquad F_C(\beta_P) = P_2 \tag{4.184}$$

see Figs. 4.10–4.11.

Let us provide the illustration of the second of the conditions (4.182). The equilibrium equation can be written elementary as, see Fig. 4.10:

$$F_T(\alpha)\sin\frac{d\alpha}{2} + F_T(\alpha + d\alpha)\sin\frac{d\alpha}{2} + T_2(\alpha, \beta_P)d\alpha - p_2(\alpha, \beta_P)A(\alpha, \beta_P)d\alpha = 0 \tag{4.185}$$

If we let $d\alpha$ tend to zero, then the above equation assumes the form of the second equation of (4.182).

Let us now consider the axial equilibrium of an elementary segment of the rib in tension lying along the α line. There is $ds_1 = Ad\alpha$ and the equilibrium condition assumes the form

$$\frac{dF_T}{ds_1} + p_1 = 0 \quad \text{along the line } \beta = \beta_P \tag{4.186}$$

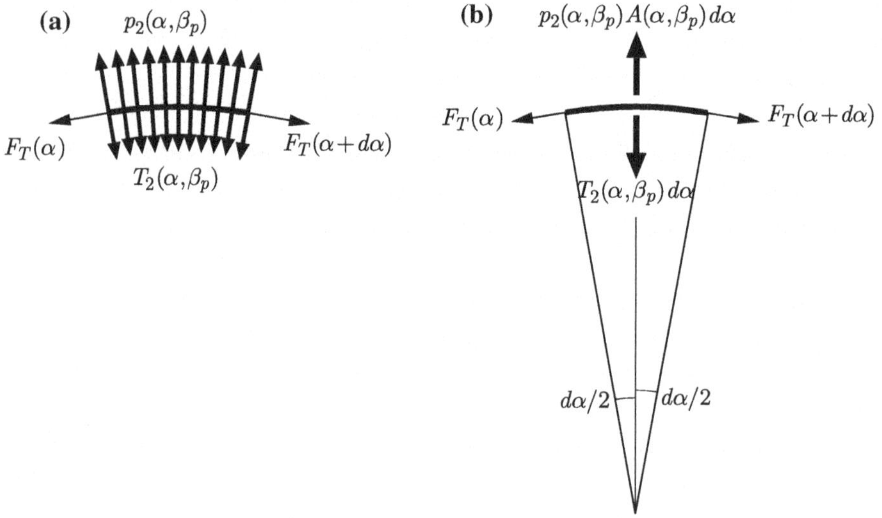

Fig. 4.10 The load normal to the edge: **a** the forces acting on an elementary segment of the rib in tension, and **b** equilibrium of the stress resultants acting on the segment. The tangential load p_1 is omitted for simplicity

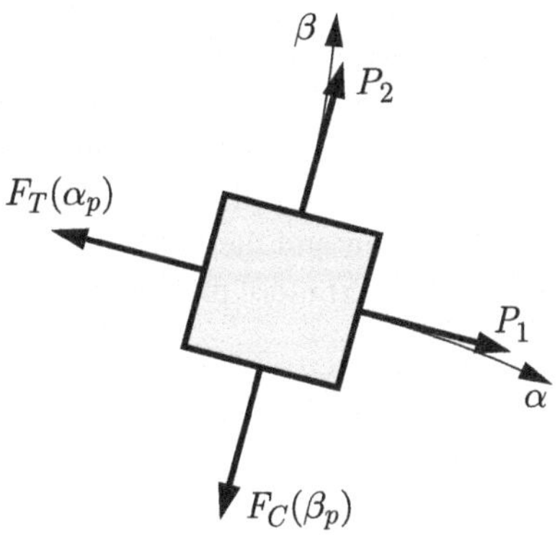

Fig. 4.11 The forces acted on the node at point P

This complies with the first of Eqs. (4.183). Moreover, Eqs. (4.184) may be easily interpreted as equilibrium equations of the node P, see Fig. 4.11.

The equilibrium equations (4.176), (4.182), (4.183), (4.184) link: the Hemp's forces T_1, T_2 and the axial forces F_T, F_C in the ribs with the load: $\boldsymbol{p}$ and P. This link will be shortly denoted by $(\boldsymbol{T}, \boldsymbol{F}) \in \Sigma$, Σ being the set of fields T_1, T_2, F_T, F_C which satisfy the mentioned equilibrium conditions. This set will be a one element set in those problems in which the Michell net will be uniquely determined by the load and the support. These problems will be viewed as statically determinate.

Remark 4.3.3 Due to the presence of ribs in tension and in compression along the lines Γ_T, Γ_C and due to the specific form of the stress tensor field $\boldsymbol{\tau}$ in Michell's structures, the formulation (3.95) with the integrand given by (3.91) is not put in a convenient and explicit form. Let Σ be the set of fields T_1, T_2 such that $T_1 > 0$, $T_2 < 0$ and the fields F_T, F_C in ribs such that $F_T > 0$, $F_C < 0$ which satisfy all the equilibrium conditions, thus making these fields compatible with the load. The problem (3.95) can be written in the following explicit form:

$$\tilde{Z}_2 = \inf \left\{ \int_\Omega \left[\frac{\sigma_0}{\sigma_T} \frac{|T_1|}{B} + \frac{\sigma_0}{\sigma_C} \frac{|T_2|}{A} \right] dx + \right.$$
$$\left. \int_{\Gamma_T} \frac{\sigma_0}{\sigma_T} |F_T| ds + \int_{\Gamma_C} \frac{\sigma_0}{\sigma_C} |F_C| ds \;\middle|\; (\boldsymbol{T}, \boldsymbol{F}) \in \Sigma \right\} \tag{4.187}$$

The above formulation is equivalent to (3.95) under the assumption that the stress field may concentrate along the curves. The formula (3.81) and the result $\tilde{Z}_1 = \tilde{Z}_2$ determines the volume of the optimum Michell truss by $V = \tilde{Z}_2/\sigma_0$.

Let us define the volume fraction of the constituent material in the ribs in compression and in tension by

$$v_C = \frac{1}{\sigma_C}|F_C|, \qquad v_T = \frac{1}{\sigma_T}|F_T| \tag{4.188}$$

Using the notation (4.170), (4.171), we may write the volume in the form

$$V = \int_{\Omega} (v_1 + v_2) dx + \int_{\Gamma_T} v_T ds + \int_{\Gamma_C} v_C ds \tag{4.189}$$

Remark 4.3.4 Let us come back to the problem (3.85) with the modification of Remark 3.4.1, assuming the 2D setting. The field $\boldsymbol{v}$ determines the strain field $(\varepsilon_{ij}(\boldsymbol{v}))$ in the domain and the fields $\epsilon_\tau(\boldsymbol{v})$ along the curve Γ, see (4.162). Let us write the problem (3.85) more precisely, as follows:

$$\begin{aligned} \mathcal{Z}_1 = \sup \Big\{ & \mathcal{L}(\boldsymbol{v}) \mid \boldsymbol{v} \text{ kinematically admissible in } \Omega \text{ such that} \\ & -\frac{\sigma_0}{\sigma_C} \le \lambda_i(\varepsilon(\boldsymbol{v})) \le \frac{\sigma_0}{\sigma_T} \quad \text{in } \Omega \\ & -\frac{\sigma_0}{\sigma_C} \le \epsilon_\tau(\boldsymbol{v}) \le \frac{\sigma_0}{\sigma_T} \quad \text{on } \Gamma \Big\} \end{aligned} \tag{4.190}$$

If Γ coincides with the parametric lines, then $\epsilon_\tau(\boldsymbol{v})$ is equal either to $\epsilon_1(u, v)$ or to $\epsilon_2(u, v)$, see (4.163), (4.164). Equations (4.190), (3.85) for $n = 2$ are in fact equivalent. But Eq. (4.190) delivers more precisely the conditions imposed on the virtual field $\boldsymbol{v}$.

In the sequel the semi-inverse methods are used: we construct the field $\boldsymbol{v}$ which in Ω and on its boundary satisfies (4.190). We then compute $\mathcal{L}(\boldsymbol{v})$ and form the optimum structure according to the rules of Table 3.1, to make it statically admissible. Then, we compute the Hemp's forces, the axial forces in the ribs and compute the integral in (4.187). If this integral is equal to $\mathcal{L}(\boldsymbol{v})$, then the duality gap vanishes, and the equality $\tilde{\mathcal{Z}}_1 = \tilde{\mathcal{Z}}_2$ holds. The optimum structure is then correctly constructed.

4.4 Parameterization of the Subdomains in the Regimes $\mathbb{R}_1$, $\mathbb{R}_2$: The Nets Including Families of Straight Lines. Kinematics and Equilibrium Conditions

In Sect. 4.1, the family of orthogonal nets has been divided into three families: (i), (ii) and (iii), see (4.59), (4.60), (4.64). The case (i) has been discussed in Sects. 4.2 and 4.3.

4.4.1 The Nets Including One Family of Straight Lines

Now, the case (ii) will be discussed, where the family of the α lines is composed of straight lines while the β lines form the family of curves crossing the straight lines at right angles, see Fig. 4.3.

Thus, we shall outline the consequences of the equality: $\phi(\alpha, \beta) = \beta$, $A(\alpha, \beta) = 1$, $B(\alpha, \beta) = \alpha + g(\beta)$, see (4.62). The geometry of the net is determined by (4.27). By integrating these formulae, we arrive at a specific form of Eq. (4.32) or

$$x(\alpha_p, \beta_p) + iy(\alpha_p, \beta_p) = x_0 + iy_0 + \alpha_p e^{i\beta_p} + i\int_0^{\beta_p} e^{i\beta} g(\beta) d\beta \tag{4.191}$$

Assume now that the virtual strain refers to the regime $\mathbb{R}_1$ in which

$$\varepsilon_I(\boldsymbol{v}) = \frac{\sigma_0}{\sigma_T}, \qquad -\frac{\sigma_0}{\sigma_C} \leq \varepsilon_{II}(\boldsymbol{v}) \leq \frac{\sigma_0}{\sigma_T}$$

Equation (4.36) leads to

$$\epsilon_I(\boldsymbol{v}) = \frac{\partial u}{\partial \alpha} = \frac{\sigma_0}{\sigma_T} \qquad \varepsilon_{II}(\boldsymbol{u}) = \frac{1}{\alpha + g(\beta)}\left(\frac{\partial v}{\partial \beta} + u\right) \tag{4.192}$$

and

$$\omega_1 = \frac{\partial v}{\partial \alpha} \qquad \omega_2(\boldsymbol{u}) = \frac{1}{\alpha + g(\beta)}\left(\frac{\partial u}{\partial \beta} - v\right) \tag{4.193}$$

The condition $\omega_1 + \omega_2 = 0$ gives

$$\frac{\partial v}{\partial \alpha} + \frac{1}{\alpha + g(\beta)}\left(\frac{\partial u}{\partial \beta} - v\right) = 0 \tag{4.194}$$

We integrate the first of Eqs. (4.192)

$$u(\alpha, \beta) = \frac{\sigma_0}{\sigma_T}\alpha + G(\beta) \tag{4.195}$$

and insert this result into (4.194) to find

$$\frac{\partial v}{\partial \alpha} + \frac{1}{\alpha + g(\beta)}\left(\frac{\partial G}{\partial \beta} - v\right) = 0 \tag{4.196}$$

Let

$$v_1(\alpha, \beta) = v(\alpha, \beta) - \frac{dG}{d\beta} \tag{4.197}$$

Then

$$\frac{\partial v_1}{\partial \alpha} = \frac{1}{\alpha + g(\beta)} v_1(\alpha, \beta) \tag{4.198}$$

The solution can be written as follows:

$$v_1(\alpha, \beta) = (\alpha + g(\beta))H(\beta) \tag{4.199}$$

where $H(\beta)$ is an arbitrary function. Thus,

$$v(\alpha, \beta) = \frac{dG}{d\beta} + (\alpha + g(\beta))H(\beta) \tag{4.200}$$

Let us compute ω_1, ω_2 using (4.193)

$$\omega_1(\alpha, \beta) = H(\beta), \qquad \omega_2(\alpha, \beta) = -H(\beta) \tag{4.201}$$

which confirms $\omega_1 + \omega_2 = 0$; now we compute the angle of rigid rotation

$$\omega = H(\beta) \tag{4.202}$$

Moreover, we compute the strain in the β direction

$$\epsilon_{22} = \frac{1}{\alpha + g(\beta)} \left[\frac{d^2G}{d\beta^2} + \frac{dg}{d\beta} H(\beta) + \frac{\sigma_0}{\sigma_T}\alpha + G(\beta) \right] + \frac{dH}{d\beta} \tag{4.203}$$

In the regime $\mathbb{R}_1$, there should be: $-\frac{\sigma_0}{\sigma_C} \leq \epsilon_{22} \leq \frac{\sigma_0}{\sigma_T}$.

Let us examine the equilibrium equations (4.158). In the regime $\mathbb{R}_1$, one of the principal stresses vanishes; namely, this stress component which is transverse to the straight fibres, or $T_2 = 0$. The second of Eqs. (4.158) is satisfied identically, since $a = 0$, $\phi = \beta$, $T_2 = 0$. The first of these equations leads to

$$T_1 = T_1(\beta) \tag{4.204}$$

which means that the force T_1 is constant along the straight fibres. Within the regime $\mathbb{R}_1$, the volume fraction of the material is given by

$$v(\alpha, \beta) = \frac{1}{\sigma_T} \frac{T_1(\beta)}{B(\alpha, \beta)} \tag{4.205}$$

where now $B(\alpha, \beta) = \alpha + g(\beta)$. This result follows from (4.169).

4.4.2 *The Michell Nets Parameterized by the Polar Coordinate System*

The specific case of the nets discussed here are the simplest nets formed by the parametric lines of the polar coordinate system. One family of the lines are straight lines of a given origin, called the pole, see Fig. 4.12.

Let us recall the formulae for the strain components referred to the polar coordinate system

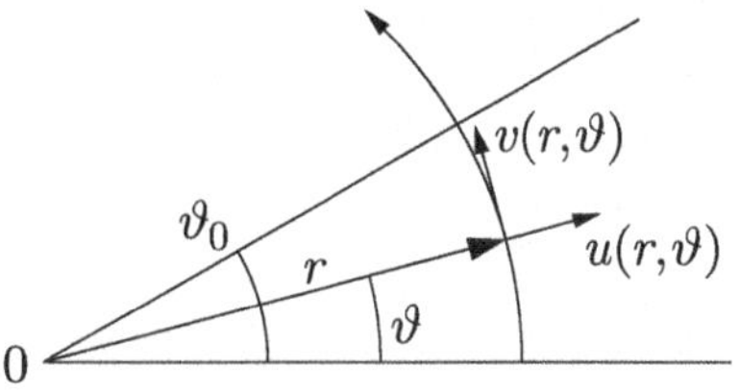

Fig. 4.12 The polar coordinate system

$$
\begin{aligned}
\varepsilon_r &= \frac{\partial u}{\partial r} \\
\varepsilon_\vartheta &= \frac{1}{r}\frac{\partial v}{\partial \vartheta} + \frac{u}{r} \\
2\varepsilon_{r\vartheta} &= \frac{1}{r}\frac{\partial u}{\partial \vartheta} + \frac{\partial v}{\partial r} - \frac{v}{r}
\end{aligned} \tag{4.206}
$$

In the regimes $\mathbb{R}_1$, $\mathbb{R}_2$, the radial strains may saturate the bound, i.e. $\varepsilon_r = \dfrac{\sigma_0}{\sigma_T}$, while the circumferential strain ε_ϑ lies between the bounds $\left(-\dfrac{\sigma_0}{\sigma_C}, \dfrac{\sigma_0}{\sigma_T}\right)$. In the examples to be discussed, stronger conditions will be satisfied, like in the regime $\mathbb{T}$; then we shall have

$$
\varepsilon_r = 1, \qquad \varepsilon_\vartheta = -\kappa, \qquad \varepsilon_{r\vartheta} = 0, \qquad \kappa = \frac{\sigma_T}{\sigma_C} \tag{4.207}
$$

where $\sigma_0 = \sigma_T$.

Let us consider a more general case:

$$
\varepsilon_r = -\epsilon_1 \qquad \varepsilon_\vartheta = \epsilon_2 \qquad \varepsilon_{r\vartheta} = 0 \tag{4.208}
$$

where ϵ_1, ϵ_2 are constants. We shall find the displacement fields (u, v) corresponding to the strain state (4.208). The conditions (4.208) lead to the solution of the form

$$
\begin{aligned}
u &= -\epsilon_1 r + C_1 \cos(\vartheta - \varphi) \\
v &= (\epsilon_1 + \epsilon_2) r\vartheta + d_1 r - C_1 \sin(\vartheta - \varphi)
\end{aligned} \tag{4.209}
$$

where d_1, C_1, φ are arbitrary constants. Since the angles ω_1, ω_2 are given by

$$
\omega_1 = \omega_r = \frac{\partial v}{\partial r}, \qquad \omega_2 = \omega_\vartheta = \frac{1}{r}\frac{\partial u}{\partial \vartheta} - \frac{v}{r} \tag{4.210}
$$

hence

$$
\omega_1 = (\epsilon_1 + \epsilon_2)\vartheta + d_1, \qquad \omega_2 = -\omega_1 \tag{4.211}
$$

Thus, the angle of rigid rotation equals

$$\omega = (\epsilon_1 + \epsilon_2)\vartheta + d_1 \tag{4.212}$$

The formulae above will be used to construct the virtual displacement fields within circular fans, frequently appearing in planar Michell structures.

4.5 The Simplest Michell Structures

This is an introductory section preceding more advanced solutions to be discussed later. Just these simplest Michell structures are referred to in many textbooks on structural mechanics, frequently without any assumptions and details. This is an additional argument not to omit them here. The layouts of these simplest Michell structures will be expressed by elementary functions. In the each example, the aim is to find a truss of the least volume which transmits a given load to the given supports or to find a truss which carries a given system of self-equilibrated forces. The stresses in the trusses are subject to the conditions: $-\sigma_C \leq \sigma \leq \sigma_T$, where σ is the axial stress in a bar.

4.5.1 *The Two-Force Problem in the Plane*

If self-equilibrated, the two forces should act along the same line and be of equal magnitudes, see Fig. 4.13a. The forces form the vectors in the directions AB and BA. The optimal structure, designed in the whole plane, carrying these two forces will be found by guessing the solutions to the problems (3.85) and (3.95). The equality $\tilde{\mathcal{Z}}_1 = \tilde{\mathcal{Z}}_2$ will guarantee the correctness of both the solutions and of the truss layout.

We introduce the Cartesian parameterization $\alpha = x, \beta = y, A = 1, B = 1$. We guess the maximizer in (3.85):

$$\boldsymbol{v} = \left(\frac{\sigma_0}{\sigma_T} x, 0\right) \tag{4.213}$$

Then, $\varepsilon_{11} = \frac{\sigma_0}{\sigma_T}$, $\varepsilon_{12} = 0$, $\varepsilon_{22} = 0$, hence

$$\lambda_1(\varepsilon) = \frac{\sigma_0}{\sigma_T}, \qquad \lambda_2(\varepsilon) = 0 \tag{4.214}$$

or $\varepsilon(\boldsymbol{v}(\boldsymbol{x})) \in B(\sigma_C, \sigma_T)$ at each point $\boldsymbol{x} = (x, y)$ of the plane. We compute $\boldsymbol{v}^B = \left(\frac{\sigma_0}{\sigma_T} l, 0\right)$, hence $\mathcal{L}(\boldsymbol{v}) = \frac{\sigma_0}{\sigma_T} P l$. We predict that $\tilde{\mathcal{Z}}_1 = \mathcal{L}(\boldsymbol{v})$.

Now, we guess that the optimal structure is a straight bar AB, see Fig. 4.13b. Then, the set Σ is a single element set: $T_1 = P = \text{const}$ along AB. Thus, the minimization operation in (3.95) is trivial, and the result reads

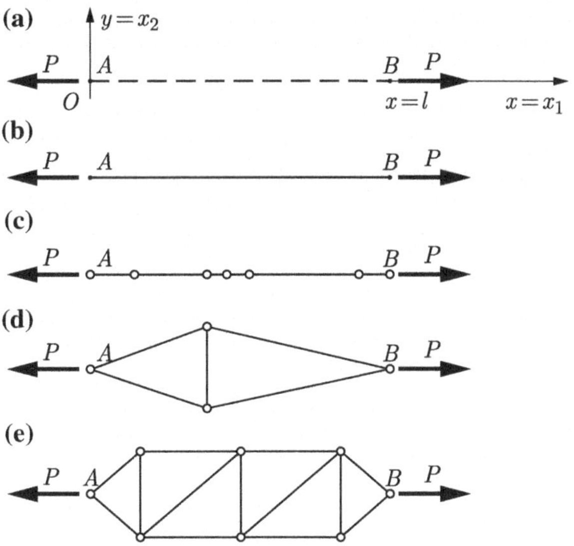

Fig. 4.13 The two-force problem (**a**); the one-bar solution (**b**); the chain-like solution (**c**); and suboptimal trusses (**d**, **e**)

$$\tilde{\mathcal{Z}}_2 = \int_0^l \frac{\sigma_0}{\sigma_T}\frac{T_1}{B}dx = \frac{\sigma_0}{\sigma_T}Pl \tag{4.215}$$

It is seen that $\tilde{\mathcal{Z}}_1 = \tilde{\mathcal{Z}}_2$ and according to (3.21) $V = \frac{Pl}{\sigma_T}$. Other structures carrying these two forces and satisfying the same stress conditions are chains of collinear bars, see Fig. 4.13c. The number of hinges in these structures is arbitrary. These structures are the simplest funiculars.

Let us note that the problem discussed satisfies the assumptions of Maxwell's theorem: the forces are known and the structure can be subject to uniform dilatation, cf. Sect. 2.2.3. Thus, if one finds a structure of all members in tension and equilibrating the given system of loads, no other structure can be lighter. For instance, by the Maxwell theorem (Sect. 2.2.3), we know that the trusses of Fig. 4.13d, e in which some members are in compression must be heavier than the structures of Fig. 4.13b, c in which all members are in tension.

Let us note that the whole design domain corresponds to the regime $\mathbb{R}_1$, see (4.214).

4.5.2 *The Lightest Cantilever Transmitting a Point Load to a Straight Support*

We consider the problem of the paper by Hu and Shield (1961): find the least volume truss which transmits the given force at point $P(d, y_P)$ to the support along the vertical axes $x_0 = 0$; the stresses are subject to the bounds: $-\sigma_C \le \sigma \le \sigma_T$; the design domain is the half-plane, right to the axes $x_0 = 0$.

Thus, $x_0 = 0$ is the edge of the design domain $x_0 \ge 0$. The force P forms the angle φ with the axis y_0, see Fig. 4.14. In this problem, not only the structure is unknown, but also the position of supporting nodes along the axis y_0 and the number of these supporting nodes are unknown. We shall show that the optimal structure is the two-bar truss the bars of which form a right angle, see Fig. 4.15.

We guess the virtual displacement field $\boldsymbol{v}$ which satisfies the optimality condition $\varepsilon(\boldsymbol{v}(\boldsymbol{x})) \in B(\sigma_C, \sigma_T)$ at each point $\boldsymbol{x} = (x_1, x_2)$ of the half-plane $x_1 \ge 0$:

$$\boldsymbol{v} = (\tilde{a}x_1, \tilde{d}x_1) \tag{4.216}$$

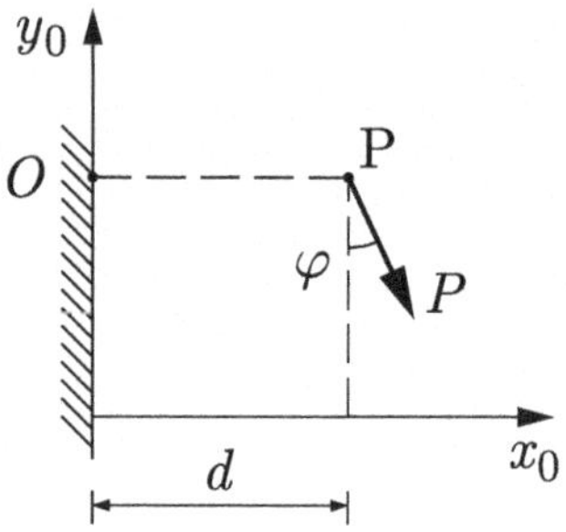

Fig. 4.14 Problem formulation: transmit the given force to the vertical support

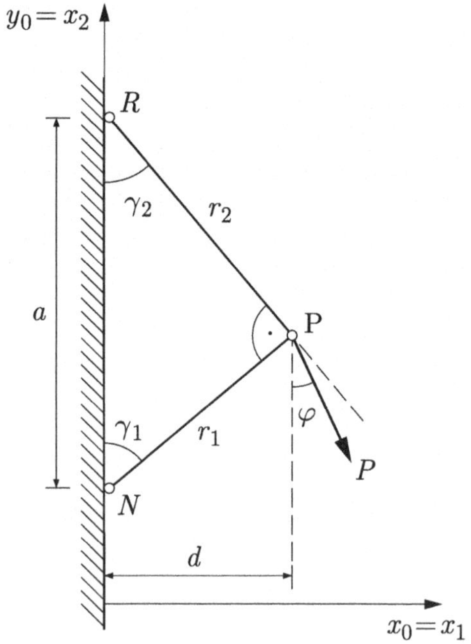

Fig. 4.15 Geometry of the two-bar truss; $\gamma_1 + \gamma_2 = \dfrac{\pi}{2}$

It is seen that $\boldsymbol{v} = \mathbf{0}$ along $x_1 = 0$. The constants $\tilde{a}, \tilde{d}$ will be chosen such that the virtual strain will satisfy the desired optimality condition. The strains associated with (4.216) are

$$\varepsilon_{11}(\boldsymbol{v}) = \tilde{a}, \qquad \varepsilon_{12}(\boldsymbol{v}) = \frac{1}{2}\tilde{d}, \qquad \varepsilon_{22}(\boldsymbol{v}) = 0$$

The principal virtual strains or the eigenvalues of the strain tensor are the roots of the characteristic equation

$$\lambda^2 - \tilde{a}\lambda - \frac{1}{4}\tilde{d}^2 = 0 \tag{4.217}$$

We require that

$$\lambda_2 = -\frac{\sigma_0}{\sigma_C}, \qquad \lambda_1 = \frac{\sigma_0}{\sigma_T} \tag{4.218}$$

We take $\sigma_0 = \sigma_T$ and write $\dfrac{\sigma_T}{\sigma_C} = \kappa$. At least for metal structures, it is natural to assume that $\kappa > 1$, since the permissible tension stress should be greater than the permissible stress at compression, as compression leads to buckling. That is why in Fig. 4.15, we have assumed that $r_2 > r_1$. Nevertheless, we assume only that $\kappa > 0$. The conditions (4.218) give the values of the constants

$$\tilde{a} = 1 - \kappa, \qquad \tilde{d} = \pm 2\sqrt{\kappa} \tag{4.219}$$

We compute the virtual work done by the force P

$$\mathcal{L}(\boldsymbol{v}) = P(v_{x_0} \sin\varphi - v_{y_0} \cos\varphi) \tag{4.220}$$

where $v_{x_0} = (1-\kappa)d$, $v_{y_0} = \pm 2\sqrt{\kappa}d$. To make the value $\mathcal{L}(\boldsymbol{v})$ maximal, we choose $\tilde{d} = -2\sqrt{\kappa}$. Thus

$$\mathcal{L}(\boldsymbol{v}) = Pd[(1-\kappa)\sin\varphi + 2\sqrt{\kappa}\cos\varphi] \tag{4.221}$$

We guess that $\tilde{\mathcal{Z}} = \mathcal{L}(\boldsymbol{v})$.

Let us find the principal directions of the strain tensor of components

$$\varepsilon_{11} = 1 - \kappa, \qquad \varepsilon_{12} = -\sqrt{\kappa}, \qquad \varepsilon_{22} = 0 \tag{4.222}$$

The first direction is given by the angle γ_I for which

$$\tan(2\gamma_I) = \frac{2\varepsilon_{12}}{\varepsilon_{11} - \varepsilon_{22}} \tag{4.223}$$

or

$$\tan(2\gamma_I) = \frac{-2\sqrt{\kappa}}{1-\kappa} \tag{4.224}$$

Using the identity

$$\tan(2\gamma_I) = \frac{2\tan\gamma_I}{1-\tan^2\gamma_I} \tag{4.225}$$

we obtain

$$\tan\gamma_I = -\sqrt{\kappa}, \qquad \gamma_I = -\arctan(\sqrt{\kappa}) \tag{4.226}$$

The optimal eigendirections are shown in Fig. 4.16. The angles γ_1, γ_2 are given by

$$\gamma_1 = \arctan(\sqrt{\kappa}), \qquad \gamma_2 = \arctan\left(\frac{1}{\sqrt{\kappa}}\right) \tag{4.227}$$

or $\gamma_1 = |\gamma_I|$. The bars of the optimal truss will be directed along the principal directions of the virtual strain field (4.222). The lengths of the bars: $r_1 = |PN|$, $r_2 = |PR|$ are such that

$$\frac{r_2}{r_1} = \sqrt{\kappa} \tag{4.228}$$

The solution is a two-bar truss, as in Fig. 4.15, if the bar RP is in tension and the bar NP-in compression. This is the case, if

$$-\gamma_1 < \varphi < \gamma_2 \tag{4.229}$$

The axial forces in bars are

$$T_{RP} = \sin(\gamma_1 + \varphi)P, \qquad T_{PN} = -\sin(\gamma_2 - \varphi)P \tag{4.230}$$

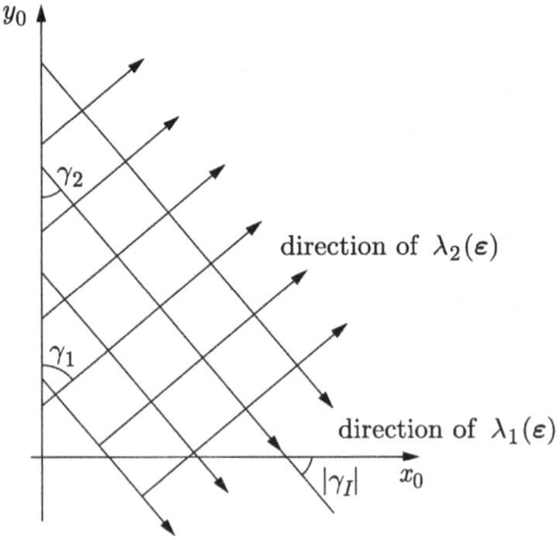

Fig. 4.16 The eigendirections of the optimal virtual strain

The lengths of the bars are given by

$$l_{RP} = \frac{d}{\sin\gamma_2}, \qquad l_{PN} = \frac{d}{\sin\gamma_1} \tag{4.231}$$

We can now compute

$$\tilde{\mathcal{Z}}_2 = T_{RP} l_{RP} - \kappa T_{PN} l_{PN} \tag{4.232}$$

and check whether the equality $\tilde{\mathcal{Z}}_1 = \tilde{\mathcal{Z}}_2$ holds. Since this equality holds true we conclude that the volume of the optimal truss equals $V = \frac{1}{\sigma_T}\mathcal{Z}_1$.

The equality $\tilde{\mathcal{Z}}_1 = \tilde{\mathcal{Z}}_2$ means that if the condition (4.229) concerning the direction of the force is satisfied, then the structure of Fig. 4.15 has the volume smaller than any other structures which transmit this force to the vertical support. If $\varphi = \gamma_2$, then the structure is composed of a single-bar RP. If $\varphi = -\gamma_1$, then the structure is composed of a single-bar PN. If φ lies outside the range (4.229), the solutions are one-bar trusses; the bar is inclined with the same angle as the force.

Let us stress once again: the inclinations of the bars in Fig. 4.15 are independent of the value of the angle φ provided that this angle is included in the range (4.229). This property is the consequence of the net of Fig. 4.16 being independent of the angle φ; this net depends only on the ratio $\kappa = \frac{\sigma_T}{\sigma_C}$. We shall see in the sequel that such a stability of the optimal trajectories of the virtual strain takes place only in specific problems with a deciding role of the kinematic boundary conditions.

If $\sigma_T = \sigma_C$ and if $-\frac{\pi}{4} < \varphi < \frac{\pi}{4}$, the two-bar solution has a specific form shown in Fig. 4.17; the bars have equal lengths and are inclined by angles $\frac{\pi}{4}$ to the support.

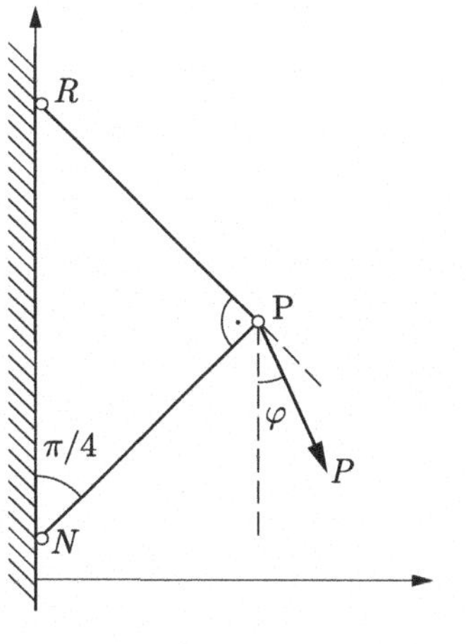

Fig. 4.17 The optimal solution referring to the case of $\sigma_T = \sigma_C$ and $-\frac{\pi}{4} < \varphi < \frac{\pi}{4}$

Let us note that the whole design domain is characterized by the regime $\mathbb{T}$, see Table 3.1: the operation *maximum* in the kinematic problem chooses the same point of the locking locus—its lower vertex.

Remark 4.5.1 Let us assume that the optimal solution to the problem discussed is a two-bar truss. Under this assumption, one can reconstruct the optimal solution found above by an elementary analysis. This problem amounts to finding the positions of points R and N along the supporting line to make the truss as light as possible, keeping the assumed stress levels for tension and compression. The problem thus formulated involves four design variables: $\gamma_1, \gamma_2, A_1, A_2$, A_i being areas of the cross sections of bars. This problem can be solved using classical tools of calculus and the final solution coincides with that reported above; in particular, the orthogonality property can be confirmed. Despite the drastic assumption on the number of bars, the problem is not an easy task. It is worth performing all the computations; this is recommended for younger readers.

4.5.3 The Simplest Three-Force Problem: Case of Three Parallel Forces

Consider the problem of transmitting a vertical force $2P$ to two, symmetrically located vertical supports, one of them being sliding, see Fig. 4.18a. The stresses in bars are subject to the constraints $-\sigma_C \le \sigma \le \sigma_T$. The design domain is the half-plane over the line linking the supports.

Instead of solving the initial problem, we shall solve the equivalent problem of the three self-equilibrated forces, see Fig. 4.18b, to get rid of kinematic conditions. We guess the Hencky net in which the optimal solution will be inscribed, see Fig. 4.19.

We assume that the whole design refers to the regime $\mathbb{R}_1$, or $\epsilon_{11} = \dfrac{\sigma_0}{\sigma_T}$, $-\dfrac{\sigma_0}{\sigma_C} \le \epsilon_{22} \le \dfrac{\sigma_0}{\sigma_T}$. Nonetheless, we assume here: $\epsilon_{22} = -\dfrac{\sigma_0}{\sigma_C}$ to assure continuity of the strain ϵ_{22} in the circular fan and the strain in the circular rib.

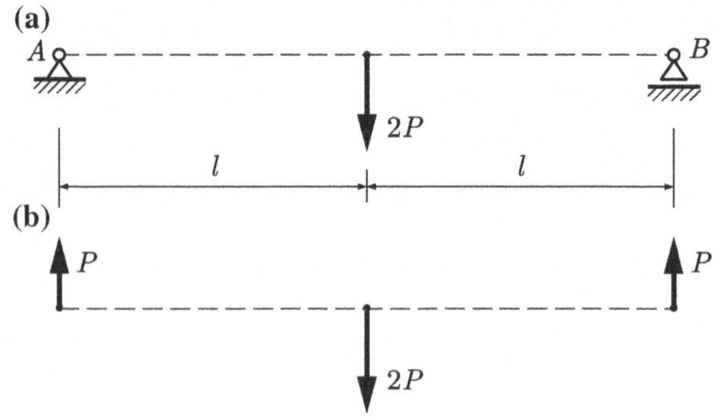

Fig. 4.18 **a** Problem formulation and **b** the equivalent three-force problem

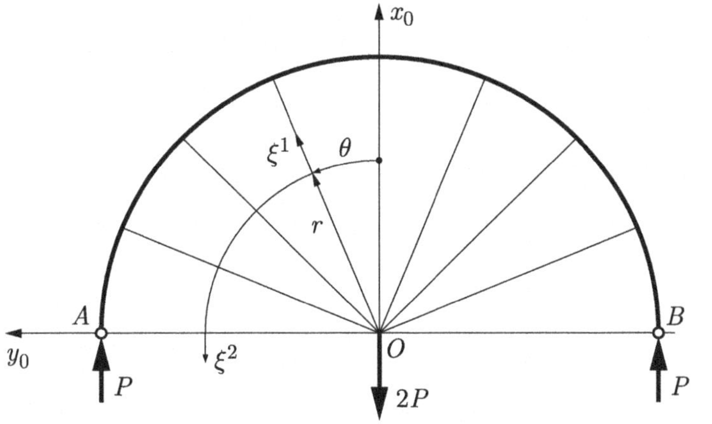

Fig. 4.19 The Hencky net for the Michell problem of Fig. 4.18b

The lines $\xi^1 = \alpha$ go in radial directions and $\alpha = r$ measures the distance r from the origin O to a given point. Thus, $A = 1, ds_1 = dr$. The lines $\xi^2 = \beta$ are circumferences and $\beta = \theta$ measures the angle from the vertical axis, see Fig. 4.19; $ds_2 = Bd\theta$. It is seen that $\phi(\alpha, \beta) = \beta$; hence $\phi(r, \theta) = \theta$. We assume $\omega(0, 0) = 0$, $u_0 = 0$, $v_0 = 0$, $\epsilon_{11} = \frac{\sigma_0}{\sigma_T}$, $\epsilon_{22} = -\frac{\sigma_0}{\sigma_C}$ and set $\omega(\alpha, \beta) = \omega(r, \theta)$ according to Eq. (4.73), or

$$\omega(r, \theta) = -(\epsilon_{11} - \epsilon_{22}) \int_0^\theta d\beta = -\left(\frac{\sigma_0}{\sigma_T} + \frac{\sigma_0}{\sigma_C}\right)\theta \tag{4.233}$$

We shall make use of the formulae (4.191)–(4.203). Assume that $u(\alpha, \beta) = u_r(r, \theta)$ does not depend on $\beta = \theta$. According to (4.195), assuming $G(\beta) = 0$, we have

$$u_r(r, \theta) = \frac{\sigma_0}{\sigma_T} r \tag{4.234}$$

Following (4.200), we take $v(\alpha, \beta)$ in the form

$$v(\alpha, \beta) = \alpha H(\beta) \tag{4.235}$$

where, according to (4.202), (4.233), assuming $g(\beta) = 0$, we get

$$H(\beta) = -\left(\frac{\sigma_0}{\sigma_T} + \frac{\sigma_0}{\sigma_C}\right)\beta \tag{4.236}$$

Let us introduce the conventional notation: $u_\theta = v$, $\alpha = r$, $\beta = \theta$. Hence,

$$u_\theta(r, \theta) = -\left(\frac{\sigma_0}{\sigma_T} + \frac{\sigma_0}{\sigma_C}\right) r\theta \tag{4.237}$$

Let us check the value of ϵ_{22}. By (4.192), we compute

$$\epsilon_{22} = \frac{1}{r}\left(\frac{\partial u_\theta}{\partial \theta} + u_r\right) = -\frac{\sigma_0}{\sigma_C} \tag{4.238}$$

Let us compute the circumferential strain along the rib for $r = l$. By (4.163), we have $\epsilon_2 = \epsilon_{22}$, since $ds_2 = r d\theta$, $R_2 = r = l$.

Let us note that the displacement field (u_r, u_θ) is symmetric with respect to the axis $y_0 = 0$, similarly as the set of the forces in Fig. 4.18. This field satisfies the conditions of regime $\mathbb{T}$, hence also of the regime $\mathbb{R}$ within the whole half-plane $x_0 \geq 0$, and in the regime $r \geq 0$, $\theta \in [-\pi + \gamma, \pi - \gamma]$ if $\gamma > 0$. If $\gamma = 0$ the field u_θ ceases to be continuous along the line $\theta = \pi$ and $\theta = -\pi$. Thus, our formulae cannot apply to the problem of Fig. 4.18 if the design domain is the whole plane.

We shall show that the field (u_r, u_θ) constructed is the maximizer in problem (3.82); the proof will follow from the equality $\tilde{\mathcal{Z}}_1 = \tilde{\mathcal{Z}}_2$. We start from computing $\tilde{\mathcal{Z}}_1$:

$$\tilde{\mathcal{Z}}_1 = \mathcal{L}(\boldsymbol{u}) = -P u_\theta\left(l, \frac{\pi}{2}\right) + P u_\theta\left(l, -\frac{\pi}{2}\right) + 2P(-u_r(0, \theta)) \tag{4.239}$$

$$\tilde{\mathcal{Z}}_1 = \pi\left(\frac{\sigma_0}{\sigma_T} + \frac{\sigma_0}{\sigma_C}\right) P l$$

Now, we shall compute $\tilde{\mathcal{Z}}_2$; we start from computing the internal forces within the Michell truss of Fig. 4.19. Within the circular fan, we construct the state of internal forces T_1, T_2 according to the rules of regime $\mathbb{R}$, see Sect. 4.4. Thus, we write $T_1(r, \theta) = T_1(l, \theta)$, $T_2 = 0$, see Eq. (4.204). Along the edge $r = l$ the load is absent, hence $p_1(\theta) = 0$. By the first of Eqs. (4.182), we find

$$T_1(l, \theta) + F_C(\theta) = 0 \tag{4.240}$$

where $F_C(\theta)$ is the axial force in the rib along the contour $r = l$.

The equation of equilibrium along the rib has the form of the first of Eqs. (4.183). Since $p_2 = 0$ along the rib we conclude that $F_C = \text{const}$. From the equilibrium of the node A, we have $F_C(\pi/2) = -P$. Thus, $T_1(l, \theta) = P$, and since T_1 does not depend on $\alpha = r$, hence $T_1(r, \theta) = P$ within the fan: $0 < r < l$, $-\pi/2 < \theta < \pi/2$. The condition of equilibrium of the node at the centre of the structure

$$\int_{-\frac{\pi}{2}}^{\frac{\pi}{2}} T_1(0, \theta) \cos\theta d\theta = 2P \tag{4.241}$$

is satisfied identically, which is the important necessary condition of correctness of the whole analysis.

Let us compute the volume of the structure of Fig. 4.19. The area of the cross section of the bar in compression equals $v = \dfrac{|F_C|}{\sigma_C}$, hence is equal $\dfrac{P}{\sigma_C}$. Its length equals πl; the volume of this bar equals

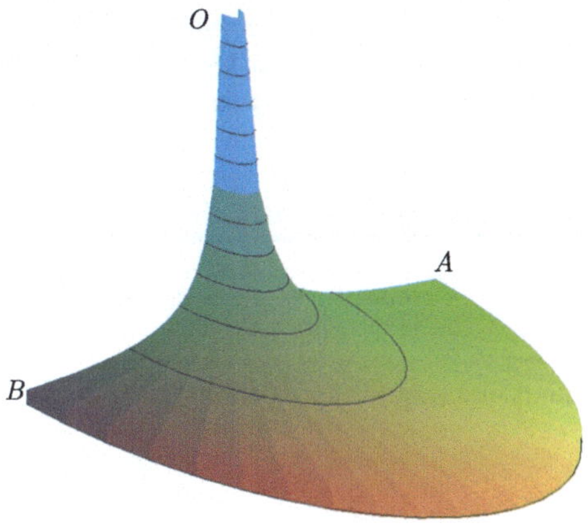

Fig. 4.20 The optimal distribution of the volume fraction $\tilde{v}_1 = \frac{v_1 \sigma_T}{P}$ of fibres within the fan

$$V^F = \frac{\pi}{\sigma_C} Pl \tag{4.242}$$

The volume of the fan is expressed by the integral of the volume fraction (4.170), or

$$v_1(r, \theta) = \frac{1}{\sigma_T} \frac{P}{r} \tag{4.243}$$

see Fig. 4.20.

This function has a singularity at the central point where the force $2P$ is applied. The volume of the fan is the integral of v_1, see (4.189); hence

$$V^T = \int_{-\frac{\pi}{2}}^{\frac{\pi}{2}} \int_0^l v_1(r, \theta) r d\theta \tag{4.244}$$

and the result is $V^T = \frac{\pi}{\sigma_T} Pl$. The volume of the structure equals

$$V = V^F + V^T = \pi \left(\frac{1}{\sigma_T} + \frac{1}{\sigma_C} \right) Pl \tag{4.245}$$

and the same result is found by $V = \frac{1}{\sigma_0} \tilde{\mathcal{Z}}_1$, where $\tilde{\mathcal{Z}}_1$ is given by (4.239).

If $\sigma_C = \sigma_T$, then the volume of the rib V^F is equal to the volume V^T of the fibres of the fan, which confirms Maxwell's theorem, see Sect. 2.2.3. Its assumptions are here satisfied, since a uniform dilatation is kinematically admissible.

Remark 4.5.2 Equations (4.234), (4.237) can be obtained by putting $\phi = \beta$, $\epsilon_{11} = \frac{\sigma_0}{\sigma_T}$, $\epsilon_{22} = -\frac{\sigma_0}{\sigma_C}$ and ω by (4.233) in the formula (4.66).

Remark 4.5.3 The displacement fields u_r, u_θ do not satisfy the kinematic conditions at A and B, see Fig. 4.18a. One can add a rigid body motion to make these conditions fulfilled. In fact, it is easy to check that the fields

$$\tilde{u}_r(r,\theta) = \frac{\sigma_0}{\sigma_T}(r - l\sin\theta) - \frac{\pi}{2}\left(\frac{\sigma_0}{\sigma_T} + \frac{\sigma_0}{\sigma_C}\right) l\cos\theta \tag{4.246}$$

$$\tilde{u}_\theta(r,\theta) = -\frac{\sigma_0}{\sigma_T}\cos\theta l + \left(\frac{\sigma_0}{\sigma_T} + \frac{\sigma_0}{\sigma_C}\right)\left(-r\theta + \frac{\pi}{2} l\sin\theta\right)$$

satisfy the required conditions:

$$\tilde{u}_r\left(l, \frac{\pi}{2}\right) = 0, \qquad \tilde{u}_\theta\left(l, \frac{\pi}{2}\right) = 0, \qquad \tilde{u}_\theta\left(l, -\frac{\pi}{2}\right) = 0 \tag{4.247}$$

The component of the displacement along the axis x_0 is expressed by

$$u_x(r,\theta) = \tilde{u}_r(r,\theta)\cos\theta - \tilde{u}_\theta\sin\theta \tag{4.248}$$

hence

$$u_x(0,\theta) = -\frac{\pi}{2}\left(\frac{\sigma_0}{\sigma_T} + \frac{\sigma_0}{\sigma_C}\right) l \tag{4.249}$$

We see that this result is θ-independent. The virtual work is now done only by the force $2P$; it is equal to

$$\mathcal{L}(\boldsymbol{v}) = -2Pu_x(0,\theta) = \pi\left(\frac{\sigma_0}{\sigma_T} + \frac{\sigma_0}{\sigma_C}\right) l \tag{4.250}$$

and we come back to the result (4.239) obtained using the fields (4.234), (4.237).

4.5.4 Hemp's Problem of Transmitting a Vertical Force to Two Fixed Supports. The Full-Plane Design

This section is aimed at delivering a complete solution to the problem posed and partly solved by Hemp (1973, Sect. 4.5): transmit a vertical force of magnitude $2P$ to the two pin supports at A and B, see Fig. 4.21. The design domain is the full plane, while the line of the force is the mid-perpendicular of the interval AB of length $2l$. The force is applied at point O of the ordinate y_0, which will be replaced by the non-dimensional parameter $\eta = y_0/l$. To simplify some formulae, we shall also make use of the angle $\chi = \arctan(\eta)$, see Fig. 4.21. Without loss of generality, the force can be assumed as directed downwards. We shall examine how the layout of the optimal solution depends on the ordinate of the point of the force application. The data of the problem are as follows: P, l, σ_T, σ_C, while η will be treated as a parameter; η

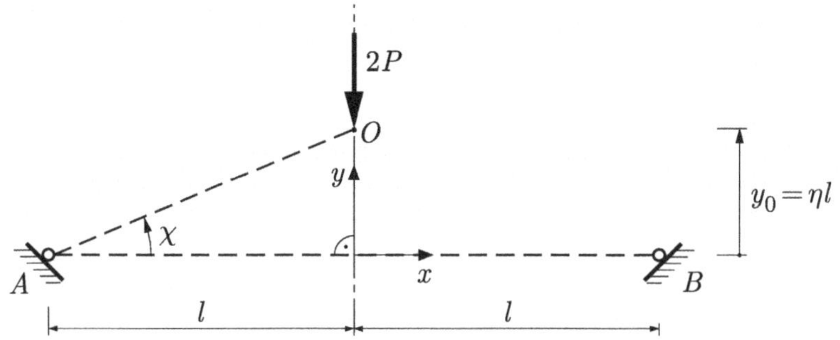

Fig. 4.21 Setting of Hemp's problem for the full plane

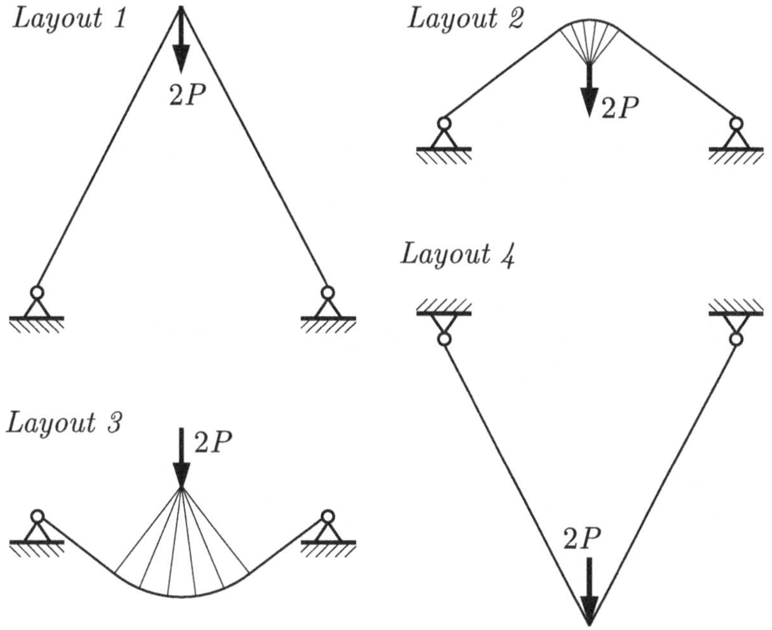

Fig. 4.22 The four basic layouts of Hemp's problem. The *Layouts 1* and *4* correspond to two-bar trusses. The *Layouts 2* and *3* represent trusses with upper and lower fans, respectively

may be negative, zero, or positive, since the position of the point O may be over or below the line AB. The volume of the optimal structure will be treated as a function $V = V(\eta)$.

By a numerical search with the help of the ground structure method, the four basic layouts of the solutions have been detected, cf. Fig. 4.22.

In contrast to the previous sections, we shall first deliver the solutions by the static method and only then these solutions will be checked by the kinematic method by covering the whole design domain (the full plane) by the virtual displacement field satisfying all the desired optimality conditions thus proving the correctness of the layouts.

4.5.4.1 Static Method—Preliminary Investigations for Different Layouts

In the static method, the equations of equilibrium are directly used to determine the areas of the cross sections of the bars such that the stresses saturate the bounds. Let F be the cross-sectional member force (or the axial force) in a bar. The areas of the cross sections of the bars in tension and compression are assumed as $A = F/\sigma_T$ and $A = -F/\sigma_C$, respectively. We put below the formulae for the volumes of the trusses of layouts shown in Fig. 4.22.

Layouts 1 and 4: Two-Bar Trusses

In the case of the *Layout 1*, if $\eta > 0$ both the bars are in compression. It is easy to check that the volume of the fully stressed two-bar truss equals

$$V_1 = \frac{2Pl}{\sigma_C}\left(\eta + \frac{1}{\eta}\right) = \frac{4Pl}{\sigma_C \sin(2\chi)}, \quad \text{here } \chi > 0 \tag{4.251}$$

By analogy, in the case of the *Layout 4*, if $\eta < 0$ both the bars are in tension, hence the volume is expressed by

$$V_4 = \frac{-2Pl}{\sigma_T}\left(\eta + \frac{1}{\eta}\right) = \frac{-4Pl}{\sigma_T \sin(2\chi)}, \quad \text{here } \chi < 0 \tag{4.252}$$

We note that the formulae (4.251)–(4.252) do not encompass the case of $\eta = 0$, since in this case the equilibrium of the two-bar truss cannot be satisfied. This suggests formation of more complex layouts, capable of carrying the given vertical force. These layouts are considered below.

Layout 2: A Michell Truss with an Upper Fan

Geometry of the truss of Fig. 4.22b should be characterized by an additional, still unknown, parameter θ_2 representing the half of the angle of the circular fan OCD, see Fig. 4.23a. Let us note that the circular arch CD must be continuously extended into the straight members AC and BD, which means that $AC \perp CO$ and $BD \perp DO$. Thus, the axial force in CD must be equal to the member forces in AC and BD.

The following geometric relations hold:

$$\begin{aligned} l_{AO} &= \frac{l}{\cos\chi}, \qquad l_{AC} = l_{AO}\cos(\theta_2 - \chi), \\ l_{CO} &= l_{AO}\sin(\theta_2 - \chi), \qquad l_{CD} = 2l_{CO}\theta_2 \end{aligned} \tag{4.253}$$

These formulae are true provided that the lengths l_{AC}, l_{CO} of the bars AC and CO are non-negative, which holds if $0 \le \theta_2 - \chi \le \pi/2$, or if $\chi \le \theta_2 \le \pi/2 + \chi$. On the

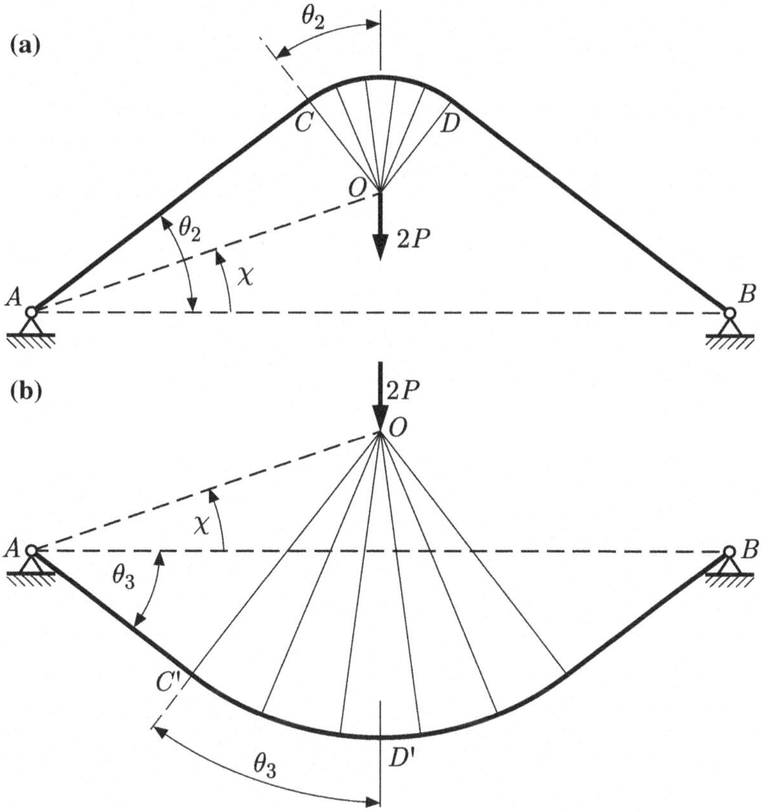

Fig. 4.23 Geometry of Michell trusses with: **a** upper fan and **b** lower fan

other hand, the angle θ_2 must be non-negative and not greater than $\pi/4$. This latter condition will be cleared up later. Thus, the range of applicability of the formulae considered here implies the following bounds for the angle θ_2:

$$\max(0, \chi) \leq \theta_2 \leq \min(\pi/4, \pi/2 + \chi) \tag{4.254}$$

By equilibrium equations of the node A, we find the axial force in the bar AC

$$F_{AC} = \frac{-P}{\sin\theta_2} \tag{4.255}$$

The same axial forces appear in the bars BD and CD: $F_{CD} = F_{BD} = F_{AC}$. Thus, the volumes of the bars in compression are expressed as follows:

$$V_{AC} = V_{BD} = \frac{-F_{AC} l_{AC}}{\sigma_C} = \frac{Pl}{\sigma_C}(\eta + \cot\theta_2) \tag{4.256}$$

$$V_{CD} = \frac{-F_{CD} l_{CD}}{\sigma_C} = \frac{2Pl}{\sigma_C} \theta_2 (1 - \eta \cot \theta_2) \tag{4.257}$$

To compute the volume of the fibres in the fan OCD, we shall make use of Maxwell's theorem (see Sect. 2.3) which applies not only to trusses but also to Michell structures. If applied to the fan OCD including the reinforcing ribs, it gives directly the formula

$$V_{OCD} = \frac{\sigma_C}{\sigma_T} V_{CD} \tag{4.258}$$

linking the volumes of the bars (or fibres) in tension and in compression.

Combining the formulae (4.256)–(4.257), the volume of the Michell structure with the upper fan may be written as below:

$$\begin{aligned} V_2 &= 2V_{AC} + V_{CD} + V_{OCD} = \\ &2Pl \left(\frac{\eta + \theta_2 + (1 - \eta\theta_2) \cot \theta_2}{\sigma_C} + \frac{\theta_2 (1 - \eta \cot \theta_2)}{\sigma_T} \right) \end{aligned} \tag{4.259}$$

The above formula involves the angle θ_2, still unknown. The formula generates a family of statically admissible structures for given η and P, l, σ_T, σ_C. The optimal angle θ_2 corresponding to the Michell structure (i.e. of the smallest volume) satisfies the equation $\partial V_2 / \partial \theta_2 = 0$. This equation can be reduced to the form

$$\frac{\sigma_T \cos^2 \theta_2 - \sigma_C \sin^2 \theta_2}{(\sigma_T + \sigma_C)\,(\theta_2 - \sin(2\theta_2)/2)} = \eta \tag{4.260}$$

Note that for $\theta_2 > 0$ the denominator of the left-hand side is positive while the numerator may assume both the negative and positive values depending on the values of the permissible stresses σ_T, σ_C. Thus, the *Layout 2* may be optimal also for negative values of η.

For given σ_T, σ_C and for fixed η, the process of computation of the angle θ_2 reduces to solving the transcendental equation (4.260). This can only be performed numerically. In general, the function $\eta(\theta_2)$ given by (4.260) is not a monotone function, which makes it difficult to find θ_2 for a given η.

The shapes of the plots of this function depend on the parameter $\kappa = \sigma_T / \sigma_C$, see Fig. 4.24. The grey regions correspond to non-feasible solutions which violate the constraints (4.254). Remarkably, within the admissible region the function $\eta(\theta_2)$ is a decreasing function (this can be proved analytically; the proof will be omitted), which makes it possible to solve the equation (4.260) with respect to θ_2 uniquely: this is simply the first positive root of this equation.

Upon substituting (4.260) into (4.259), we eventually arrive at the formula for the least volume Michell structure composed of two bars and an upper fan

$$V_2 = \frac{2Pl}{\sigma_T + \sigma_C} \cdot \frac{(\sigma_T + \sigma_C)^2 \theta_2^2 - \sigma_T \sigma_C}{(\theta_2 - \sin(2\theta_2)/2)\,\sigma_T \sigma_C} \tag{4.261}$$

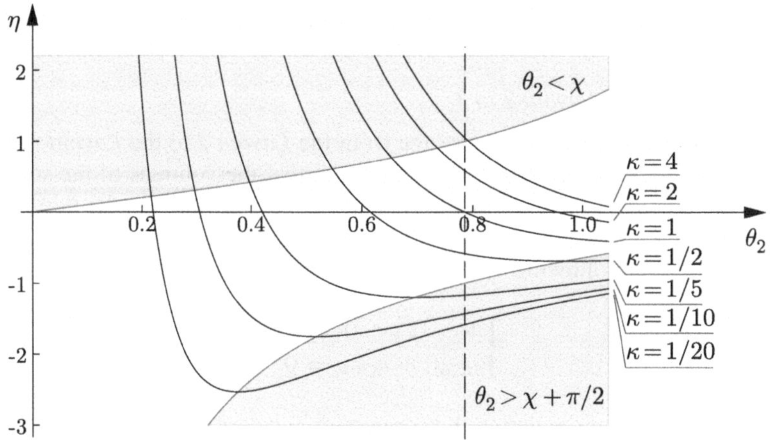

Fig. 4.24 Plots of functions $\eta(\theta_2)$ given by (4.260) for selected values of the parameter $\kappa = \sigma_T/\sigma_C$

with the angle θ_2 satisfying (4.260) and the bounds (4.254). Due to the mentioned conditions, the applicability range of this formula is limited to the appropriately chosen range for the parameter η which depends on the parameter κ.

Layout 3: A Michell Truss with a Lower Fan

The geometry of the structure is shown in Fig. 4.23b. By the arguments similar to those used in the previous case, by interchanging the roles of bars in tension and in compression, one can derive the following formula for the volume of the fully stressed structure with the lower fan:

$$V_3 = \frac{2Pl}{\sigma_T + \sigma_C} \cdot \frac{(\sigma_T + \sigma_C)^2\theta_3^2 - \sigma_T\sigma_C}{(\theta_3 - \sin(2\theta_3)/2)\,\sigma_T\sigma_C} \tag{4.262}$$

where the angle θ_3 satisfies the equation

$$\frac{\sigma_T \sin^2\theta_3 - \sigma_C \cos^2\theta_3}{(\sigma_T + \sigma_C)\,(\theta_3 - \sin(2\theta_3)/2)} = \eta \tag{4.263}$$

and its values lie in the range:

$$\max(0, -\chi) \le \theta_3 \le \min(\pi/4, \pi/2 - \chi) \tag{4.264}$$

4.5.4.2 Static Method—Comparison of Volumes of Selected Layouts

In the previous section, we have found the formulae for the volumes of four types of Michell structures. Seemingly, to choose the structure of the least volume it is sufficient to compare these values for given η and χ. However, this cannot be done directly since the formulae for the volumes involve parameters which are linked by non-linear equations.

A Snap-Through from Layout 2 to 3

Let us start with considering the passage from the *Layout* 2 to the *Layout 3*, which can be viewed as a snap-through. It is surprising that the volumes of the structures given by (4.261) and (4.262) coincide if $\theta_2 = \theta_3$. Yet one should take into account that the parameters η expressed by (4.260) and (4.263), respectively, are different. We come across the following system of equations with θ_2 and θ_3 as unknowns:

$$\begin{cases} V_2 - V_3 = 0 \\ \eta_{(261)} - \eta_{(264)} = 0 \end{cases} \tag{4.265}$$

The last equation means equating the parameters η determined by Eqs. (4.260) and (4.263), respectively.

In case of $\theta_2 = \theta_3$, the first of these equations is satisfied identically, while the second one gives

$$\theta_2 = \theta_3 = \frac{\pi}{4} \tag{4.266}$$

which corresponds to the parameter η given by

$$\eta_{23} = \frac{2(\sigma_T - \sigma_C)}{(\pi - 2)(\sigma_T + \sigma_C)} = \frac{2(\kappa - 1)}{(\pi - 2)(\kappa + 1)} \tag{4.267}$$

The notation η_{23} indicates that this parameter corresponds to the passage from the *Layout* 2 to the *Layout 3*.

A more accurate analysis shows that for certain values of κ one can find other solutions of the system (4.265), yet these solutions lie outside the range of applicability of Eqs. (4.260)–(4.263). We conclude that a snap-through between the layouts discussed takes place for $\theta_2 = \theta_3 = \pi/4$. Moreover, note that for $\eta > \eta_{23}$ the *Layout* 2 is better than the *Layout 3*, since $V_2 < V_3$ and then $\theta_2 < \pi/4$ and $\theta_3 > \pi/4$. And vice versa, for $\eta < \eta_{23}$, the *Layout 3* is better than the *Layout* 2, since then $V_3 < V_2$ and $\theta_2 > \pi/4$ and $\theta_3 < \pi/4$.

This means that the angles of the circular fans appearing at the point of application of the force should not exceed $\pi/2$, since the force (or its part) can be better transmitted using a fan of inverse orientation in which the role of the bars in tension and in compression is inverted. This remark is true for a general problem of designing Michell structures in the full plane, as noted already by Cox (1965), Chan (1966) and Hemp (1973). We shall come back to this question later.

Summing up the discussion, let us note that the bounds: $\theta_2 \leq \pi/4$ and $\theta_3 \leq \pi/4$, taken into account in (4.254), (4.264), are consequence of the fans with obtuse angles being non-optimal. That is why the limit angle $\pi/4$ has been indicated in Fig. 4.24 by a dashed vertical axis; the solutions from the domain right to this line are formally feasible but are non-optimal.

On the other hand, in the problems referring to the bounded design domains, like in the problems of designing structures within a half-plane, the angular measure of fans may exceed $\pi/2$, which is well seen already in the solution to the original Michell's problem discussed in Sect. 4.5.3.

Passing from the Layout 1 to the Layout 2

In this case, we should equate the volumes given by (4.251) and (4.261). Upon elimination of the parameter η, or upon substituting (4.260) into (4.251) and equating the volumes V_1, V_2, we come across the equation of one unknown

$$\theta_2 \tan\theta_2 = \frac{\sigma_T}{\sigma_T + \sigma_C} \tag{4.268}$$

The function at the left-hand side is monotonically increasing for $0 < \theta_2 < \pi/2$, which makes it possible to compute the angle θ_2. Upon using (4.268) in (4.261) and making some rearrangements, we obtain the equation $\tan\theta_2 = \tan\chi$ from which we infer that $\theta_2 = \chi$. This means that the passage from the one layout to the next one is accompanied by forming the upper fan; originally its radius is zero, i.e. $l_{OC} = 0$ in (4.253).

The passage form the *Layout 1* to the *Layout 2* is governed by the equation

$$\chi_{12} \tan\chi_{12} = \frac{\kappa}{1+\kappa} \quad \text{for} \quad 0 < \chi_{12} < \frac{\pi}{2} \tag{4.269}$$

The angle χ_{12} determines the ordinate of the point O of application of the force: $\eta_{12} = \tan(\chi_{12})$. In particular, for $\kappa = 1$ we have $\chi_{12} = 0.653271 = 37.4297°$.

The limit value κ_{max} for which this passage is possible follows from the above analysis bounding the value of the angle θ_2 (hence bounding χ) by the angle $\pi/4$. From this condition, we find

$$\kappa_{max} = \frac{\pi}{4-\pi} \approx 3.659792366 \tag{4.270}$$

Let us note that if $\kappa > \kappa_{max}$ then *Layout 2* cannot occur, since then the angle θ_2 would be greater than $\pi/4$, which would lead, in the light of the remarks above, to a non-optimal solution. In this case, the *Layout 1* passes directly to the *Layout 3*.

Passing from Layout 1 to 3

Upon using Eqs. (4.251) and (4.262)–(4.263), by arguments similar to those invoked previously, we find

$$\theta_3 \cot\theta_3 = \frac{\sigma_T}{\sigma_T + \sigma_C} \quad \text{and} \quad \theta_3 = \frac{\pi}{2} - \chi \tag{4.271}$$

hence

$$\left(\frac{\pi}{2} - \chi_{13}\right)\tan\chi_{13} = \frac{\kappa}{1+\kappa} \quad \text{for} \quad 0 < \chi_{13} < \frac{\pi}{2} \tag{4.272}$$

and $\eta_{13} = \tan(\chi_{13})$.

Passing from Layout 3 to 4

This passage is analysed as before when the passage from the *Layout 1* to the *Layout 2* was considered, only now the role of the parameters σ_T, σ_C is interchanged. Now, we analyse Eqs. (4.252) and (4.262)–(4.263) remembering that the first of these equations applies only for negative values of the parameters η and χ. We obtain

$$\theta_3 \tan\theta_3 = \frac{\sigma_C}{\sigma_T + \sigma_C} \quad \text{and} \quad \theta_3 = -\chi \tag{4.273}$$

These equations lead to the formula for the limit value of χ_{34}

$$\chi_{34} \tan\chi_{34} = \frac{1}{1+\kappa} \quad \text{for} \quad -\frac{\pi}{2} < \chi_{34} < 0 \tag{4.274}$$

hence $\eta_{34} = \tan(\chi_{34})$.

One can note that for sufficiently small κ the angle χ_{34} is smaller than $-\pi/4$, which means that the angle θ_3 would be greater than $\pi/4$, which would lead to a non-optimal solution. This takes place for $\kappa = \kappa_{\min}$, where

$$\kappa_{\min} = \frac{4-\pi}{\pi} = \frac{1}{\kappa_{\max}} \approx 0.2732395447 \tag{4.275}$$

and then (for $\kappa < \kappa_{\min}$) the *Layout 3* does not occur; for the decreasing η the *Layout 2* passes directly to the *Layout 4*.

Passing from Layout 2 to Layout 4

In this case, one should equate the volumes given by (4.252) and (4.260)–(4.261), remembering that η and χ are negative. We find

$$\theta_2 \cot\theta_2 = \frac{\sigma_C}{\sigma_T + \sigma_C} \quad \text{and} \quad \theta_2 = \frac{\pi}{2} + \chi \tag{4.276}$$

hence

$$-\left(\frac{\pi}{2} + \chi_{24}\right) \tan\chi_{24} = \frac{1}{1+\kappa} \quad \text{for} \quad -\frac{\pi}{2} < \chi_{24} < 0 \tag{4.277}$$

and finally $\eta_{24} = \tan(\chi_{24})$.

4.5.4.3 Static Method—Synthesis of Results for Various σ_T/σ_C

A complete analysis of Hemp's problem turned out to be complicated, the main reason being non-linearity of the conditions (4.251)–(4.277). The formulae for the volumes of Michell trusses of the layouts of Fig. 4.22 have been given in Sect. 4.5.4.1. In the previous section, the formulae for the limit values of the angles χ_{ij} (and hence for $\eta_{ij} = \tan(\chi_{ij})$) have been derived. These angles determine the passages between the possible layouts from the layout of number 'i' to the layout of number 'j'.

Let us sum up the results of the analysis by delivering the decision algorithm of constructing the optimal solution for the given parameters κ and η.

1. If $\kappa \geq \kappa_{max}$, then the *Layout 2* does not occur and then
   for $\eta \geq \eta_{13}$, the *Layout 1* is optimal;
   for $\eta_{13} \geq \eta > \eta_{34}$, the *Layout 3* is optimal;
   for $\eta \leq \eta_{34}$, the *Layout 4* is optimal.
2. If $\kappa \leq \kappa_{max}$, then the *Layout 3* does not occur, and then
   for $\eta \geq \eta_{12}$, the *Layout 1* is optimal;
   for $\eta_{12} > \eta \geq \eta_{24}$, the *Layout 2* is optimal;
   for $\eta \leq \eta_{24}$, the *Layout 4* is optimal.
3. If $\kappa_{min} < \kappa < \kappa_{max}$, then
   for $\eta \geq \eta_{12}$, the *Layout 1* is optimal;
   for $\eta_{12} > \eta \geq \eta_{23}$, the *Layout 2* is optimal;
   for $\eta_{23} \geq \eta > \eta_{34}$, the *Layout 3* is optimal;
   for $\eta \leq \eta_{34}$, the *Layout 4* is optimal.

It is worth noting that some inequalities are strict to eliminate the layouts with circular fans of zero radii, since then they represent two-bar trusses. In other cases, we introduce the non-strict inequalities; this is justified since for the limit values of η_{ij} the layouts of numbers i and j are essentially different, but correspond to the same volume. Then, the solution ceases to be unique, since any arbitrary convex linear combination of the solutions i and j forms an optimal solution, too.

The admissible limit values of the parameters η_{ij} for various values of the parameter κ are set up in Table4.1. For the limit values κ_{max} and κ_{min}, the values $+1$ or -1 of the parameter η are put in grey colour; they refer to the *Layouts 2* and *3* with circular fans of radii equal 0. These values are shown only to indicate the agreements between the formulae for these limit cases.

The volumes are shown as plots of continuous functions with respect to the argument η; the four illustrative graphs are given in Fig. 4.25 for selected values of κ. The passages between the layouts are marked by circles. Note that the passages from the *Layout 1* to the *Layout 2* and from the *Layout 3* to the *Layout 4* are smooth, which can be checked analytically. Moreover, according to (4.251)–(4.252), the functions $V_1(\eta)$, $V_4(\eta)$ have local minima: $V_1(1) = 4Pl/\sigma_C$ and $V_4(-1) = 4Pl/\sigma_T$. One of these values (or two, in the case of $\kappa = 1$) refers to the minimal value $V(\eta)$ for a fixed κ. This topic will be discussed in Chap. VI in the context of the transmissible loads.

The graphical results in Fig. 4.25 suggest that $\bar{V}(0)$ equals 5.0 for each value of κ, but this is not the case. The selected results are:

(a) $\kappa = 4$; $\bar{V}(0) = \bar{V}_3(0) = 5.3978$;
(b) $\kappa = 2$; $\bar{V}(0) = \bar{V}_3(0) = 4.8909$;
(c) $\kappa = 1$; $\bar{V}(0) = \bar{V}_2(0) = \bar{V}_3(0) = 2 + \pi = 5.14159$;
(d) $\kappa = 1/4$; $\bar{V}(0) = \bar{V}_2(0) = 5.3978$.

Table 4.1 The limit values of the parameters η_{ij} for selected values of the parameter κ

κ	η_{12}	η_{13}	η_{23}	η_{24}	η_{34}
∞	–	∞	–	–	−0
100	–	5.750540	–	–	−0.099668
10	–	1.756652	–	–	−0.306072
5	–	1.198265	–	–	−0.419553
4	–	1.053591	–	–	−0.462064
$\kappa_{\max}$	1.000000	1.000000	1.000000	–	−0.479752
3	0.972309	–	0.875969	–	−0.520731
2	0.905791	–	0.583979	–	−0.609205
1	0.765379	–	0.000000	–	−0.765379
1/2	0.609205	–	−0.583979	–	−0.905791
1/3	0.520731	–	−0.875969	–	−0.972309
$\kappa_{\min}$	0.479752	–	1.000000	−1.000000	1.000000
1/4	0.462064	–	–	−1.053591	–
1/5	0.419553	–	–	−1.198265	–
1/10	0.306072	–	–	−1.756652	–
1/100	0.099668	–	–	−5.750540	–
+0	+0	–	–	$-\infty$	–

A natural question arises: for which κ the value of $\bar{V}(0)$ is minimal? To solve the problem, we fix $\eta = 0$ and consider the problem of minimizing $\bar{V}(0)$ over κ. It turns out that

- among the Michell trusses with the upper fan the minimal value $\bar{V}(0) = 4.87598$ is attained for $\kappa = 0.573916$;
- among the Michell trusses with the lower fan the same minimal value $\bar{V}(0) = 4.87598$ is attained for $\kappa = 1.74241 = 1/0.573916$.

4.5.4.4 Kinematic Method—Confirmation and Justification of Results

The kinematic method is based on the construction of a virtual displacement field defined within the whole design domain (here the full plane) such that it is kinematically admissible, is continuous together with the field of rigid rotation ω and satisfies Michell's optimality conditions, cf. Table 3.1. The construction of such a field along with the divisions of the whole plane into subdomains referring to the regimes $\mathbb{T}$, $\mathbb{S}$ has been proposed by Hemp (1973, Sect. 4.5). This division is shown in Fig. 4.26; it is composed of two circular fans (upper fan UF and the lower fan LF) of the same angles $2\theta_0$, two quadrants (left LQ and right RQ) and two angular regions of hydrostatic compression LC and RC.

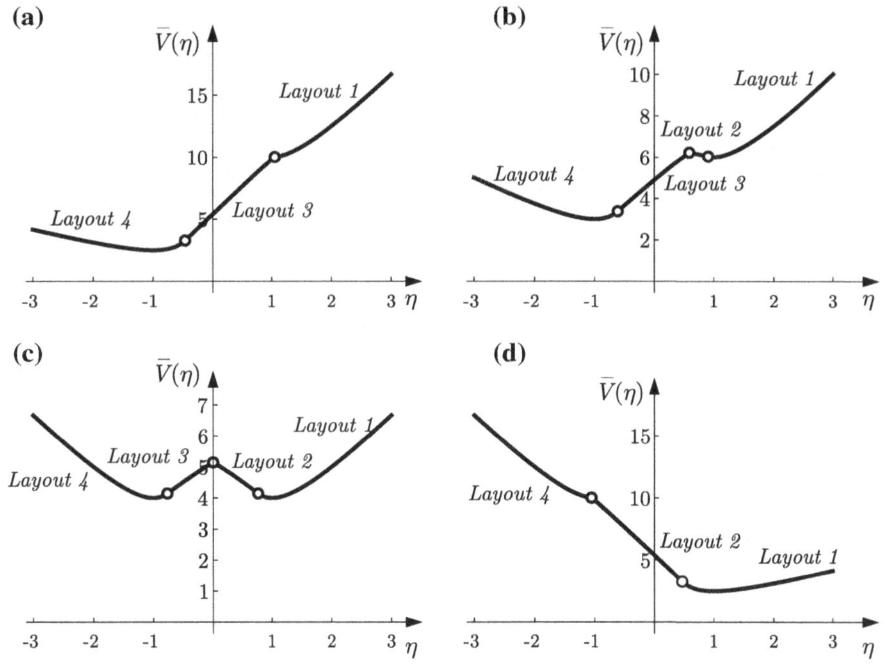

Fig. 4.25 Plots of the normalized volume $\bar{V}(\eta)$ for selected values of κ: **a** $\kappa = 4$, **b** $\kappa = 2$, **c** $\kappa = 1$ and **d** $\kappa = 1/4$, where $\bar{V}(\eta) = V(\eta)/V_0$, in which $V_0 = 2Pl/(\sigma_T + \sigma_C)$

The division of the plane into subdomains shown in Fig. 4.26 is the basis for the construction of the virtual displacement fields corresponding to the *Layouts 1* and *2*, while the angle θ_0 corresponds to the angle θ_2 used before as the measure of the angle of the upper fan. It is helpful to introduce now a new notation for this angle, since the angle θ_0 plays two roles: it not only represents the angle of the upper fan but also the angle of the lower fan to make the displacement field within the lower fan compatible with the displacement field in the adjacent domains. For a correct solution of *Layouts 3* and *4*, one should apply the division in which the regions LC and RC are replaced by the subdomains of the hydrostatic tension adjacent to the upper fan. The construction is then similar and leads to similar results to those shown below; hence, the details of this case are omitted.

It should be stressed that the division applied is based on the assumption: $\pi/2 - 2\theta_0 \geq 0$ or refers to the case of $\theta_0 \leq \pi/4$. This assumption appeared already in the static analysis—it imposed the angles of fans to be acute angles (not greater than $\pi/2$), but it was not obvious. Here, this assumption appears at the very beginning and is a natural condition of the kinematic construction. If $\theta_0 = \pi/4$, then the regions LC and RC disappear, and the whole plane refers to the regime $\mathbb{T}$. It is a composition of displacement fields referring to the snap-through from the *Layout 2* to the *Layout 3*.

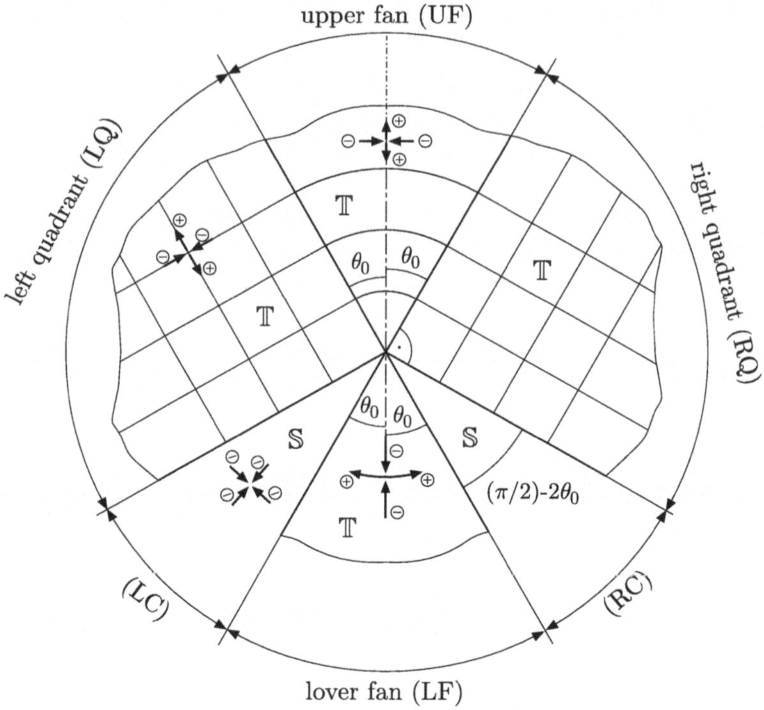

Fig. 4.26 Division of the full plane into subdomains corresponding to the regimes $\mathbb{T}$ and $\mathbb{S}$

The parameterization of the subsequent subdomains is shown in Fig. 4.27. The circular fans UF and LF are parameterized by the polar systems: (r, θ), where $r \geq 0$ and $-\theta_0 \leq \theta \leq \theta_0$. The domains LQ and RQ are parameterized by (x, y), $x \geq 0$, $y \geq 0$. Moreover, the domains LC and RC are parameterized by polar coordinates ρ, ψ, where $\rho \geq 0$, $0 \leq \psi \leq \psi_0$, where $\psi_0 = \pi/2 - 2\theta_0$.

The problem considered is statically equivalent to the problem posed in Fig. 4.28 where the point O is supported such that its displacements vanish.

This setting is more convenient for the kinematic approach to be used now; it is based on adopting Hemp's (1973) virtual displacements which satisfy just the modified kinematic conditions assumed in Fig. 4.28 as well as the conditions of continuity (along with continuity of the field ω) and optimality conditions. The components of the virtual displacement fields are assumed as follows.
The upper fan:

$$u_r^{UF}(r,\theta) = \frac{\sigma_0}{\sigma_T}, \qquad u_\theta^{UF}(r,\theta) = -\left(\frac{\sigma_0}{\sigma_T} + \frac{\sigma_0}{\sigma_C}\right) r\theta \tag{4.278}$$

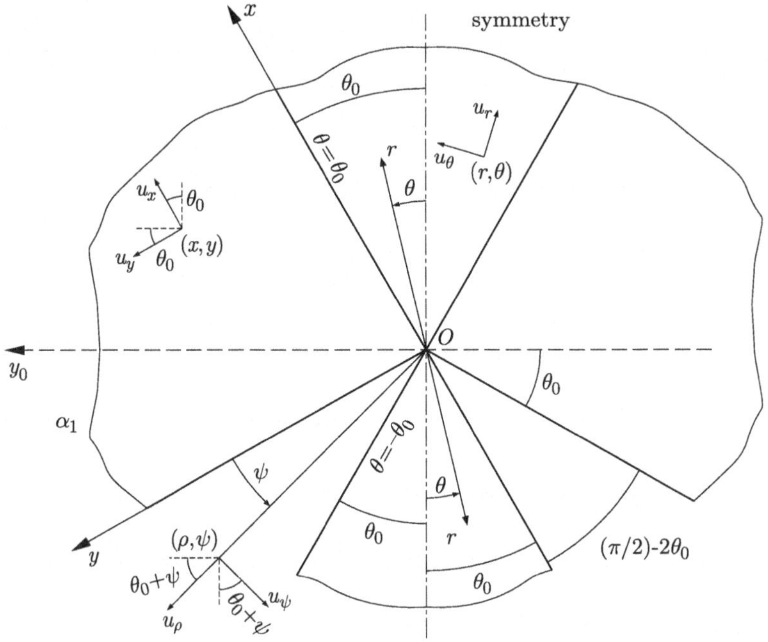

Fig. 4.27 Parameterization of the subdomains

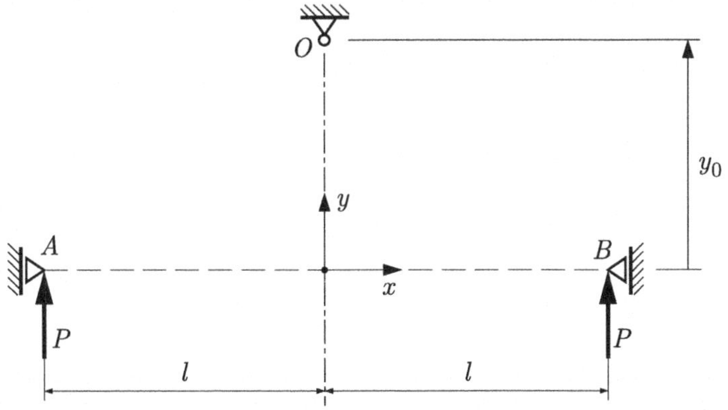

Fig. 4.28 A statically equivalent setting of Hemp's problem

The left quadrant:

$$u_x^{LQ}(x, y) = \frac{\sigma_0}{\sigma_T} x + \left(\frac{\sigma_0}{\sigma_T} + \frac{\sigma_0}{\sigma_C}\right)\theta_0 y,$$
$$u_y^{LQ}(x, y) = -\left(\frac{\sigma_0}{\sigma_T} + \frac{\sigma_0}{\sigma_C}\right)\theta_0 x - \frac{\sigma_0}{\sigma_C} y \tag{4.279}$$

The left domain of hydrostatic compression:

$$u_\rho^{LC}(\rho, \psi) = -\frac{\sigma_0}{\sigma_C}\rho,$$
$$u_\psi^{LC}(\rho, \psi) = -\left(\frac{\sigma_0}{\sigma_T} + \frac{\sigma_0}{\sigma_C}\right)\theta_0 \rho \tag{4.280}$$

The lower fan:

$$u_r^{LF}(r, \theta) = -\frac{\sigma_0}{\sigma_C} r, \qquad u_\theta^{LF}(r, \theta) = \left(\frac{\sigma_0}{\sigma_T} + \frac{\sigma_0}{\sigma_C}\right) r\theta \tag{4.281}$$

Let us note that the following continuity conditions along the boundary lines of the subdomains are fulfilled:

$$u_x^{LQ}(x, 0) = u_r^{UF}(r, \theta_0) \qquad u_y^{LQ}(x, 0) = u_\theta^{UF}(r, \theta_0), \qquad \text{for } x = r$$

$$u_\rho^{LC}(\rho, 0) = u_y^{LQ}(0, y) \qquad u_\psi^{LC}(\rho, 0) = u_x^{LQ}(0, y), \qquad \text{for } \rho = y$$

$$u_r^{LF}(r, -\theta_0) = u_\rho^{LC}(\rho, \psi_0), \qquad u_\theta^{LF}(r, -\theta_0) = u_\psi^{LC}(\rho, \psi_0), \qquad \text{for } \rho = r$$

Consequently, the tangent component ε_τ, see (4.162), is made compatible. However, the normal components ε_n may suffer jumps along these boundaries.

For instance, along the boundary of the domains LQ and LC the normal component of strain changes from: σ_0/σ_T to $-\sigma_0/\sigma_C$.

The angle of rigid rotation ω, see the last of Eqs. (4.16), must be continuous and this condition is satisfied in the whole design domain, which can be checked as follows. Let us find

$$\omega^{UF}(r, \theta) = -\left(\frac{\sigma_0}{\sigma_T} + \frac{\sigma_0}{\sigma_C}\right)\theta$$

Thus, along the boundary line of the domain UF, i.e. for $\theta = \theta_0$ we have

$$\omega^{UF}(r, \theta_0) = -\left(\frac{\sigma_0}{\sigma_T} + \frac{\sigma_0}{\sigma_C}\right)\theta_0 = \omega_0$$

while within the domains LQ and LC the field ω is constant:

$$\omega^{LQ} = \omega^{LC} = \text{const} = \omega_0$$

which proves continuity. Within the domain LF, we have

$$\omega^{LF}(r, \theta) = \left(\frac{\sigma_0}{\sigma_T} + \frac{\sigma_0}{\sigma_C}\right)\theta$$

and for $\theta = -\theta_0$ we compute $\omega^{LF}(r, -\theta_0) = \omega_0$, which confirms continuity between the domains LF and LC. It should be stressed that this continuity holds only if the angles of the fans UF and LF are identical.

The displacement field (4.278)–(4.281) satisfies the kinematic conditions shown in Fig. 4.28. The horizontal displacements of points of the domain LQ are expressed by

$$u^{LQ} = u_x \sin\theta_0 + u_y \cos\theta_0 \tag{4.282}$$

while the horizontal displacements of points in the domain LC are given by

$$u^{LC} = u_\rho \cos(\psi + \theta_0) - u_\psi \sin(\psi + \theta_0) \tag{4.283}$$

The points of the domain LQ satisfying the condition $u_{LQ} = 0$ form a straight line $x = a_0 y$ where

$$a_0 = \frac{(\sigma_T + \sigma_C)\theta_0 \sin\theta_0 - \sigma_T \cos\theta_0}{(\sigma_T + \sigma_C)\theta_0 \cos\theta_0 - \sigma_C \sin\theta_0} \tag{4.284}$$

but our attention is focused on the ray $\mathcal{L}_1$ within the domain LQ, i.e. for $x \geq 0$, $y \geq 0$, hence $a_0 \geq 0$, see Fig. 4.29a.

The points of the domain LC satisfying $u^{LC} = 0$ lie on the ray $\mathcal{L}_2$, see Fig. 4.29b, defined for $\rho \geq 0$ and $\psi = \psi_0$ with ψ_0 given by

$$\tan(\psi_0 + \theta_0) = \frac{\sigma_T}{\sigma_T + \sigma_C}\frac{1}{\theta_0} \tag{4.285}$$

Let χ^{LQ} and χ^{LC} represent the angles between the rays $\mathcal{L}_1$, $\mathcal{L}_2$ and the horizontal line

$$\chi^{LQ} = \theta_0 - \arctan a_0, \qquad \chi^{LC} = \psi_0 + \theta_0 \tag{4.286}$$

The first of these angles may be expressed as follows:

$$\tan\chi^{LQ} = \frac{\sigma_T \cos^2\theta_0 - \sigma_C \sin^2\theta_0}{(\sigma_T + \sigma_C)(\theta_0 - \sin\theta_0 \cos\theta_0)} \tag{4.287}$$

giving the formula being equivalent to (4.260) derived by different arguments.

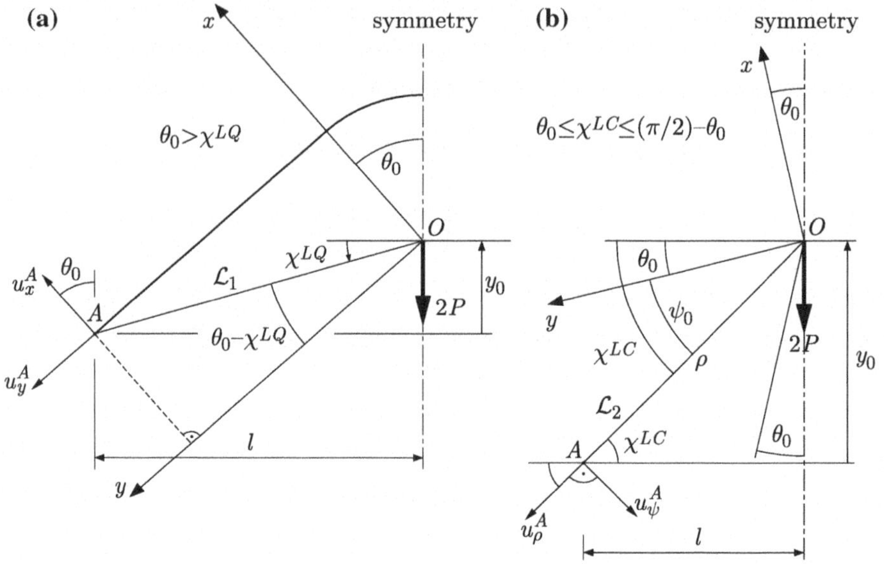

Fig. 4.29 Parameterization of layouts in case of the support A lying in **a** domain LQ and **b** domain LC

One should take into account that the equation above applies for $\theta_0 \geq \chi^{LQ}$, since the equation refers to the domain LQ. This condition implies

$$\tan\theta_0 \geq \frac{\sigma_T \cos^2\theta_0 - \sigma_C \sin^2\theta_0}{(\sigma_T + \sigma_C)(\theta_0 - \sin(2\theta_0)/2)} \tag{4.288}$$

Note that for $\theta_0 > 0$ the denominator of the right-hand side of the above inequality is positive. Thus

$$(\sigma_T + \sigma_C)\tan\theta_0(\theta_0 - \sin\theta_0\cos\theta_0) \geq \sigma_T\cos^2\theta_0 - \sigma_C\sin^2\theta_0 \tag{4.289}$$

hence, by simple rearrangements, we find

$$\theta_0 \tan\theta_0 \geq \frac{\sigma_T}{\sigma_T + \sigma_C}. \tag{4.290}$$

This result is compatible with the formula (4.269) derived before; it means that the solution considered applies to the case of $\chi_{12} \geq \theta_0 \geq \pi/4$ and corresponds to the *Layout 2* analysed in Sect. 4.5.4.1. In this case, the point of support A (see Fig. 4.28) lies in the domain LQ. The volume of such a structure (i.e. a Michell truss with an upper fan) has been expressed by (4.261). Let us check whether the kinematic method leads to the same formula. To this end, we have to compute vertical displacements of point A. We know that this point lies on the line given by (4.287) at the distance l to the line of symmetry. Hence,

$$x^A = l\frac{\sin(\theta_0 - \chi)}{\cos\chi}, \qquad y^A = l\frac{\cos(\theta_0 - \chi)}{\cos\chi} \tag{4.291}$$

where $\chi = \chi^{LQ}$.

The vertical displacement of this point (viewed as positive upwards) is given by

$$v^A = u_x^A \cos\theta_0 - u_y^A \sin\theta_0 \tag{4.292}$$

Taking into account the symmetry of the problem, i.e. $v^B = v^A$ and $v^O = 0$, see Fig. 4.28, we compute the virtual work done by the load

$$\mathcal{L} = 2Pv^A \tag{4.293}$$

The volume of the structure equals $V = \mathcal{L}/\sigma_0$ or

$$V = \frac{2Pl}{\sigma_T + \sigma_C} \cdot \frac{(\sigma_T + \sigma_C)^2\theta_0^2 - \sigma_T\sigma_C}{(\theta_0 - \sin(2\theta_0)/2)\,\sigma_T\sigma_C} \tag{4.294}$$

Upon substitution: $\theta_0 = \theta_2$, this equation coincides with (4.261).

The second of Eqs. (4.286) refers to the case of supports lying in the domains of the hydrostatic compression LC and RC and leads to the solution in the form of the *Layout 1*, i.e. the two-bar truss, both bars in compression. To find the volume of the optimal truss, we make use of the formula $V = \mathcal{L}/\sigma_0$. The formula (4.291) holds now with $\chi = \chi^{LC}$. The vertical displacement of the supporting point A is expressed by

$$v^A = -u_\rho^A \sin\chi^{LC} - u_\phi^A \cos\chi^{LC} \tag{4.295}$$

Using (4.285) and the second of Eqs. (4.286) and taking into account the formula $\tan(\chi^{LC}) = y_0/l$ we reduce (4.295) to the form

$$v^A = \frac{\sigma_0}{\sigma_C} l\left(\frac{y_0}{l} + \frac{l}{y_0}\right) \tag{4.296}$$

Thus, the volume of the two-bar truss equals

$$V = \frac{2Pl}{\sigma_C}\left(\frac{y_0}{l} + \frac{l}{y_0}\right) \tag{4.297}$$

which coincides with (4.251).

In the case of *Layout 1*, the angle θ_0 cannot be longer interpreted as the angle of the upper fan, since the fan disappears. Nevertheless, this angle is necessary to construct the kinematically admissible virtual displacement field in the whole plane; in the case considered $0 \le \theta_0 \le \chi$ and any value of θ_0 from this range is suitable, including a special case of vanishing fans with $\theta_0 = 0$.

Remarks

1. A similar analysis can be performed for the case of the division of the plane into subdomains characterized by the regimes $\mathbb{T}$ and $\mathbb{S}^+$ (i.e. hydrostatic tension). This leads to the formulae compatible with (4.252), (4.262)–(4.263) which is left to the reader.
2. Let us stress that the scheme of displacement and strain fields proposed in Figs. 4.26 and 4.27 is not stable with respect to the position of the load, since the angle θ_0 varies along with the variation of the parameter η, see (4.263). Thus, the Hemp's problems for the subsequent load positions cannot be solved by cutting out the structures from the fixed net of trajectories of the principal strains being load independent.
3. The virtual displacement field put forward by Hemp covers the whole design domain (i.e. the plane) and satisfies all the optimality conditions referring to the regimes $\mathbb{T}$ and $\mathbb{S}$. Hence, the Michell optimality conditions are fulfilled in the whole plane, which proves the optimality of the structures constructed. Note that only the kinematic method assures the global optimality of the solutions. Yet this method requires covering the whole design domain by the appropriate displacement field; this requirement is difficult to satisfy in many problems.
4. The solution found by Hemp (1973) has been here completed by the cases omitted by Hemp. In particular, we have added solutions referring to η being both positive and negative and $\kappa > \kappa_{\max}$ or $\kappa < \kappa_{\min}$. It turned out that some passages between the optimal layouts are more complex than those studied by Hemp (1973, Fig. 4.7).

4.5.5 *Symmetric Problem of Three Parallel Forces in the Plane*

Consider the problem of the three vertical forces, see Fig. 4.30.

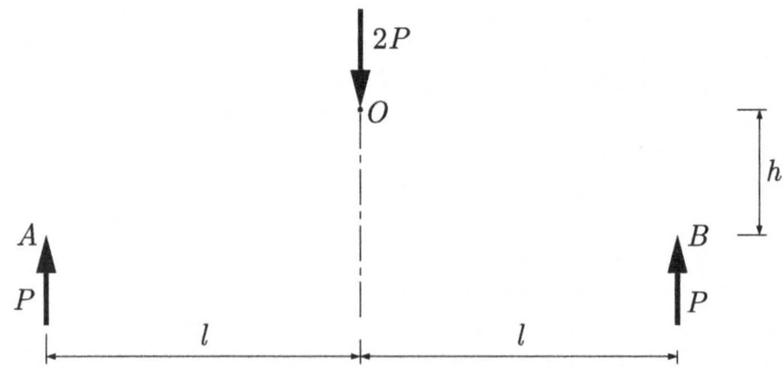

Fig. 4.30 Three parallel forces problem. The design domain is the full plane

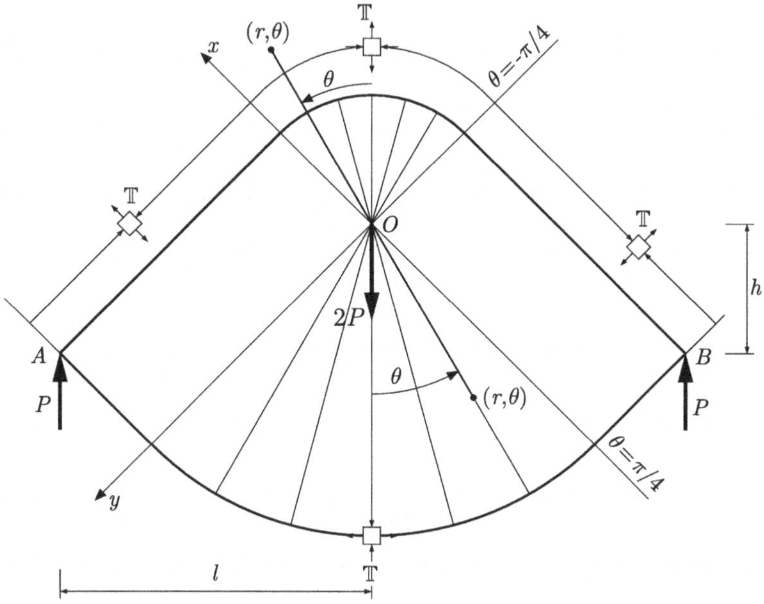

Fig. 4.31 The construction of the virtual displacement field and the division into regimes

The solution will be constructed by the kinematic method using Hemp's virtual displacement field proposed in the previous section. It is sufficient to assume $\theta_0 = \pi/4$. Then, the regime $\mathbb{S}$ from Fig. 4.26 vanishes; this figure simplifies to Fig. 4.31.

The components (u, v) of the displacement field are given by

- Equations (4.278)—the upper fan;
- Equations (4.279)—the left quadrant;
- Equations (4.281)—the lower fan.

This field is continuous in the whole plane; the displacements u, v vanish at the point O. We have no additional conditions on the values of the displacement field at points of application of the forces.

Let us inscribe the triangle AOB into the net of Fig. 4.31. The pair (A, B) lies in the quadrants or in the fan regions. Thus, we analyse two cases in the sequel.

A. The Pair (A,B) Lies in the Quadrants

Let (x^A, y^A) are coordinates of point A, see Fig. 4.31. They are determined by the data: h, l:

$$x^A = \frac{l-h}{\sqrt{2}}, \qquad y^A = \frac{l+h}{\sqrt{2}} \tag{4.298}$$

The displacements of point A are given by (4.279) where $\theta_0 = \pi/4$. Let us compute the vertical displacement of the point A; using (4.292) we find

$$v^A = \frac{\sqrt{2}}{2}(u_x^A - u_y^A) \tag{4.299}$$

By (4.279), we have

$$\begin{aligned} u_x^A &= \frac{\sigma_0}{\sigma_T} x^A + \left(\frac{\sigma_0}{\sigma_T} + \frac{\sigma_0}{\sigma_C}\right)\frac{\pi}{4} y^A \\ u_y^A &= -\left(\frac{\sigma_0}{\sigma_T} + \frac{\sigma_0}{\sigma_C}\right)\frac{\pi}{4} x - \frac{\sigma_0}{\sigma_C} y^A \end{aligned} \tag{4.300}$$

Substitution of (4.298) gives v^A. The virtual work of the forces P, $-2P$, P equals $\mathcal{L}(\boldsymbol{v}) = 2Pv^A$, since at the point O the displacement field vanishes. The volume $V = \dfrac{1}{\sigma_0}\mathcal{L}(\boldsymbol{v})$ is expressed by

$$V = \frac{Pl}{\sigma_T \sigma_C}\left[(\sigma_C + \sigma_T)\left(1 + \frac{\pi}{2}\right) + (\sigma_T - \sigma_C)\frac{h}{l}\right] \tag{4.301}$$

see Fig. 4.32a.

In the particular case of $h = 0$, the points A, O, B lie on one line, see Fig. 4.32b. Then, the volume of the optimal structure equals

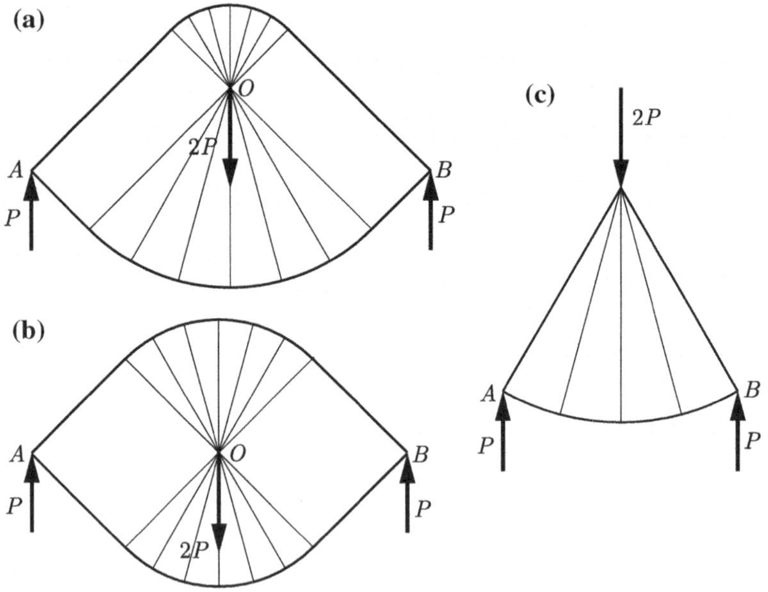

Fig. 4.32 Three optimal layouts for $0 < h < l$ (**a**); $h = 0$ (**b**); and $h \geq l$ (**c**)

$$V = \left(\frac{1}{\sigma_C} + \frac{1}{\sigma_T}\right)\left(1 + \frac{\pi}{2}\right) Pl \tag{4.302}$$

Let us check the formula (4.301) by adding the volumes of fragments of the structures of Fig. 4.32.

The Straight Bars

The bars in compression have the lengths $\sqrt{2}(h + l)/2$; the axial forces are $-P\sqrt{2}/2$. The volume of these bars equals

$$V^{(1)} = \frac{2}{\sigma_C} \cdot \frac{P\sqrt{2}}{2} \cdot \frac{\sqrt{2}}{2}(h + l) = \frac{P(h + l)}{\sigma_C} \tag{4.303}$$

In a similar manner, we compute the volume of the bars in tension; we get

$$V^{(2)} = \frac{P(l - h)}{\sigma_T} \tag{4.304}$$

The Curved Bar in Compression

The axial force in this bar equals $F_C = -P\sqrt{2}/2$. In case of $h \le l$, its length equals $\sqrt{2}\pi(l - h)/4$ and its volume is expressed by

$$V^{(3)} = \frac{P\sqrt{2}}{2\sigma_C}\frac{\sqrt{2}}{4}\pi(l - h) = \frac{\pi}{4}(l - h)\frac{P}{\sigma_C} \tag{4.305}$$

The Curved Bar in Tension

In a similar way, we find

$$V^{(4)} = \frac{\pi}{4}(l + h)\frac{P}{\sigma_T} \tag{4.306}$$

The Upper Fan

According to (4.182), the Hemp force T_1 equals $T_1 = -F_C = P\sqrt{2}/2$. The volume fraction of the fibres is given by (4.205); hence, we find

$$\upsilon(r, \theta) = \frac{1}{\sigma_T}\frac{T_1}{r}$$

The volume of the material in the upper fan equals

$$V^{(5)} = \int_0^{\frac{\pi}{2}} \int_0^{\frac{\sqrt{2}}{2}(l-h)} \frac{1}{\sigma_T}\frac{1}{r}\frac{P\sqrt{2}}{2} r\, d\theta \qquad V^{(5)} = \frac{\pi}{4}\frac{P}{\sigma_T}(l - h) \tag{4.307}$$

The Lower Fan

Proceeding in a similar way we compute the volume of the material in the lower fan

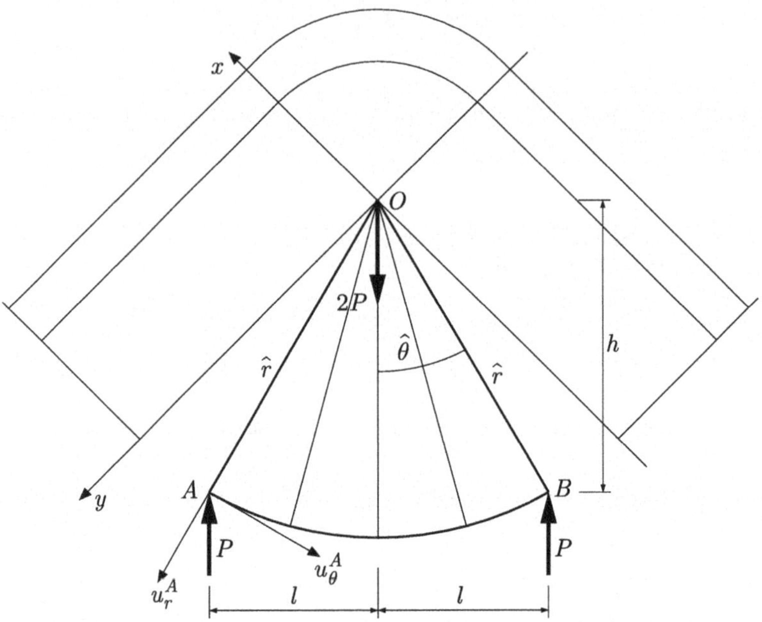

Fig. 4.33 Case of the pair (A, B) lying in the lower fan

$$V^{(6)} = \frac{\pi P}{4\sigma_C}(l+h) \tag{4.308}$$

The total volume of the structure equals $V = V^{(1)} + \cdots + V^{(6)}$; this leads to the formula (4.301). Since both the methods: static and kinematic gave the same results, we conclude that the solution given in Fig. 4.32 and the result (4.301) are correct.

B. The Pair (A, B) Lies in the Lower Fan

We consider the case of $h \geq l$ shown in Figs. 4.32c and 4.33.

Let us compute the lengths: $|OA| = |OB| = \hat{r} = \sqrt{h^2 + l^2}$, $\tan\hat{\theta} = l/h$. By (4.281), we have

$$u_r^A = -\frac{\sigma_0}{\sigma_C}\hat{r}, \qquad u_\theta^A = \left(\frac{\sigma_0}{\sigma_T} + \frac{\sigma_0}{\sigma_C}\right)\hat{r}(-\hat{\theta}) \tag{4.309}$$

The vertical displacement of the point A is expressed by

$$v^A = -u_r^A \cos\hat{\theta} - u_\theta^A \sin\hat{\theta} \tag{4.310}$$

Thus

$$v^A = \frac{\sigma_0}{\sigma_C}h + \left(\frac{\sigma_0}{\sigma_T} + \frac{\sigma_0}{\sigma_C}\right)l\hat{\theta} \tag{4.311}$$

The virtual work equals $2Pv^A$ and the volume of the optimal structure is expressed by

$$V = 2P\left(\frac{h}{\sigma_C} + \left(\frac{1}{\sigma_T} + \frac{1}{\sigma_C}\right) l\hat{\theta}\right) \tag{4.312}$$

Let us check this result by computing the volumes of the segments of the structure.

The axial force in the curved bar equals $F_T = P \sin\hat{\theta}$. Its length equals $l_1 = 2\hat{\theta}\dfrac{l}{\sin\hat{\theta}}$. The volume is expressed by

$$V^{(1)} = \frac{F_T}{\sigma_T} l_1 = \frac{2Pl\hat{\theta}}{\sigma_T} \tag{4.313}$$

The axial force in the bars OA and OB equals $F_C = -P\cos\hat{\theta}$ and their lengths are $\hat{r} = \dfrac{h}{\cos\hat{\theta}}$. Thus, the volume of the material of the bars is

$$V^{(2)} = 2\frac{|F_C|}{\sigma_C}\hat{r} = \frac{2P}{\sigma_C}h \tag{4.314}$$

According to (4.182), the Hemp force T_1 equals $T_1 = -F_T$, and according to (4.205), the volume fraction of the material in the fan equals

$$v = \frac{1}{\sigma_C}\frac{|F_T|}{r} \tag{4.315}$$

Thus, the volume of the material in the fan is expressed by

$$V^{(3)} = \int_0^{2\hat{\theta}}\int_0^{\hat{r}} v r d\theta = \frac{2P}{\sigma_C} l\hat{\theta} \tag{4.316}$$

By summing up: $V^{(1)} + V^{(2)} + V^{(3)}$, we obtain V found already in (4.312), which ends the derivation.

Remark 4.5.4 The formulations of the problems of Figs. 4.19 and 4.32b differ in the design domains. In case of the design domain being a half-plane, the solution is a fibrous half-circle; if the design domain is the full plane, two empty square domains appear along with two fans of angles $\pi/2$.

Let us draw once again the solution of Fig. 4.19 as a junction of two solutions referring to the half-planes, see Fig. 4.34.

The volume of this structure equals

$$V = \left(\frac{1}{\sigma_T} + \frac{1}{\sigma_C}\right)\pi Pl \tag{4.317}$$

and is greater than the volume of the structure of Fig. 4.32 given by Eq. (4.302).

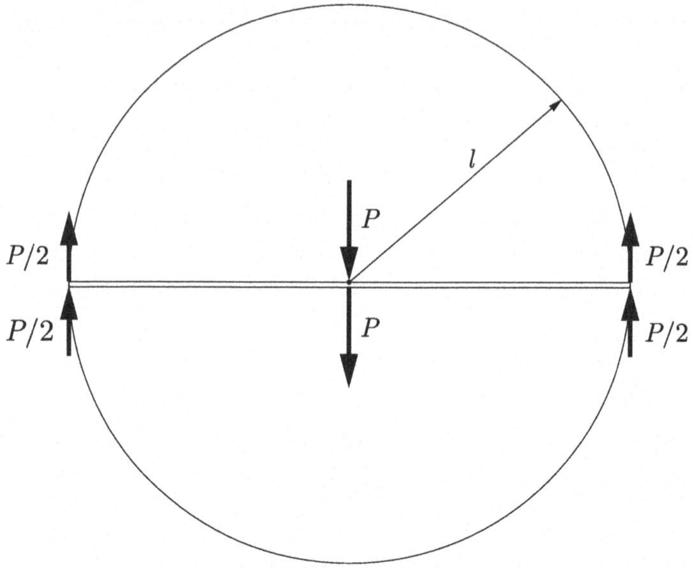

Fig. 4.34 Junction of two solutions for the half-planes

Remark 4.5.5 The problem of Fig. 4.30 for the full plane was solved for arbitrary ratios h/l. If the design domain is the half-plane over the line AB, then, up to this place of the book, we know the solution only for $h = 0$, see Fig. 4.19. The solutions to the problem of Fig. 4.30 referring to the half-plane, for $h \neq 0$ are much more complicated, as will be shown in the next sections.

4.6 The Problem of the Exterior of a Circle

We consider the least volume frameworks transmitting a point load P applied at point P—lying in the exterior of a circle—to the boundary of the circle. Let the circle have its centre at O and be of radius R; let the distance OP be equal l. The construction will be restricted to the case of $\sigma_T = \sigma_C = \sigma_0$. It turns out that this assumption is essential. Depending on the force inclination the solution assumes three possible shapes: (a) Michell's cantilever, (b) Michell's cantilever with an additional straight bar linking the tip of the cantilever with the point P and (c) a single straight bar.

Section 4.6.1 is devoted to the mathematical construction of the Hencky net, proposed already in 1923 by Prandtl. This net determines the Michell cantilever (a). In Sects. 4.6.2–4.6.4, a complete characterization of the Michell cantilever is delivered, thus proving that the Hencky net is a Michell net. This construction enables solving the torsion problem: constructing the least volume structure—an optimal wheel—transmitting a tangent circumferential load to the boundary of a circle, cf. Sect. 4.6.5.

The truss approximations of Michell's cantilevers are discussed in Sect. 4.6.6. The layouts of the kind (b) and (c) have been found in Hu and Shield (1961), Prager (1985).

If $\sigma_T \neq \sigma_C$, the optimal solutions are not known; in particular, a counterpart of Michell's cantilever is unknown.

4.6.1 The Hencky Net Composed of Logarithmic Spirals

According to (4.75) and (4.76), we assume a Hencky net given by

$$\begin{gathered} \phi = \phi_0 + \beta - \alpha \\ \frac{\partial A}{\partial \beta} = B, \qquad \frac{\partial B}{\partial \alpha} = A \end{gathered} \tag{4.318}$$

and the parametric lines (4.31) are given by

$$\begin{aligned} x(\alpha_P, \beta_P) &= x_0 + \int_0^{\alpha_P} A(\alpha, 0)\cos(\phi_0 - \alpha)d\alpha - \int_0^{\beta_P} B(\alpha_P, \beta)\sin(\phi_0 - \alpha_P + \beta)d\beta \\ y(\alpha_P, \beta_P) &= y_0 + \int_0^{\alpha_P} A(\alpha, 0)\sin(\phi_0 - \alpha)d\alpha + \int_0^{\beta_P} B(\alpha_P, \beta)\cos(\phi_0 - \alpha_P + \beta)d\beta \end{aligned} \tag{4.319}$$

Let us follow the arguments of Carathéodory and Schmidt (1923) and consider a particular form of the field A

$$A(\alpha, \beta) = f(p\alpha + q\beta) \tag{4.320}$$

where f is an unknown function of one argument and p, q are parameters. Let us assume that the function f is a positive and increasing function. Substitution of (4.320) into (4.318) leads to the differential equation

$$\frac{d^2 f(t)}{dt^2} - f(t) = 0 \tag{4.321}$$

and to the identity $pq = 1$. We assume the solution of (4.321) in the form $f(t) = C_1 \exp(t)$ or

$$A(\alpha, \beta) = C_1 \exp\left(\frac{\alpha}{q} + q\beta\right), \qquad B(\alpha, \beta) = qC_1 \exp\left(\frac{\alpha}{q} + q\beta\right) \tag{4.322}$$

We shall show later that the following choice of the constants C_1, q is helpful:

$$C_1 = R\frac{\sqrt{1+\kappa}}{\sqrt{\kappa}}, \qquad q = \sqrt{\kappa} \tag{4.323}$$

where R will represent a radius of a circle and κ will be equal to $\dfrac{\sigma_T}{\sigma_C}$. Thus, we assume

$$\begin{aligned} \frac{A(\alpha,\beta)}{R} &= \frac{\sqrt{1+\kappa}}{\sqrt{\kappa}}\exp\left(\frac{\alpha}{\sqrt{\kappa}}+\sqrt{\kappa}\beta\right), \\ \frac{B(\alpha,\beta)}{R} &= \sqrt{1+\kappa}\exp\left(\frac{\alpha}{\sqrt{\kappa}}+\sqrt{\kappa}\beta\right) \end{aligned} \tag{4.324}$$

and substitute these equations into (4.319), putting $x_0 = R$ and $y_0 = 0$, $\phi_0 = -\arctan(\sqrt{\kappa})$. The formulae (4.319) result in

$$\begin{aligned} \frac{x(\alpha,\beta)}{R} &= \exp\left(\frac{\alpha}{\sqrt{\kappa}}+\sqrt{\kappa}\beta\right)\cos(\beta-\alpha), \\ \frac{y(\alpha,\beta)}{R} &= \exp\left(\frac{\alpha}{\sqrt{\kappa}}+\sqrt{\kappa}\beta\right)\sin(\beta-\alpha) \end{aligned} \tag{4.325}$$

These are just the equations of the logarithmic spirals of Prandtl, see Carathéodory and Schmidt (1923). Let us note that the points of coordinates $(-\kappa\beta, \beta)$ lie on the circle of origin at (0, 0) and of radius R.

Let us consider an infinitesimal triangle of sides $Ad\alpha$, $Bd\beta$, see Fig. 4.35. On the circle, we have $\alpha = -\kappa\beta$ or $d\alpha = -\kappa d\beta$. We compute the tangent of the angle γ_1 of inclination of the β line to the circle:

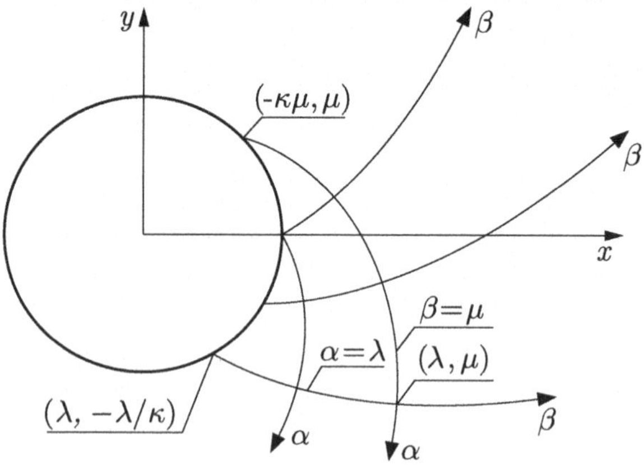

Fig. 4.35 The Hency net in the exterior of a circle

$$|\tan\gamma_1| = \left|\frac{Ad\alpha}{Bd\beta}\right| = \frac{1}{\sqrt{\kappa}}\kappa = \sqrt{\kappa} \tag{4.326}$$

We see that this angle is the same as in problem of Sect. 4.5.2, where the support was a straight line.

4.6.2 *Non-existence of Displacement Fields* (u, v) *Associated with the Hencky Net for* $\kappa \neq 1$

If the circle of radius R is the line of the support, where displacements u, v vanish, then, according to (4.41), the following differential condition

$$A(\epsilon_{11} + i\omega_1)d\alpha + B(\omega_2 + i\epsilon_{22})d\beta = 0 \tag{4.327}$$

should hold along the boundary of the circle. Let us substitute

$$\epsilon_{11} = 1, \qquad \epsilon_{22} = -\kappa, \qquad \omega_1 = \omega, \qquad \omega_2 = -\omega \tag{4.328}$$

into (4.327), guessing that the solution of the problem (3.85) will lie in the regime $\mathbb{T}$, see Table 3.1, where one should put $\sigma_0 = \sigma_T$. Equations (4.327), (4.328) lead to the equations

$$\begin{bmatrix} A & -B\omega \\ A\omega & -B\kappa \end{bmatrix} \begin{bmatrix} d\alpha \\ d\beta \end{bmatrix} = \begin{bmatrix} 0 \\ 0 \end{bmatrix} \tag{4.329}$$

The condition of solvability of the above system has the form $\omega^2 = \kappa$ or

$$\omega(-\kappa\beta, \beta) = -\sqrt{\kappa} \tag{4.330}$$

Hence, $\omega(0, 0) = -\sqrt{\kappa}$. The first of Eqs. (4.329) is compatible with the equality (4.326). According to (4.78), (4.328), we have

$$\omega(\alpha, \beta) = -\sqrt{\kappa} - (1 + \kappa)(\alpha + \beta) \tag{4.331}$$

We note that on the circle, we have $\omega = \text{const}$ only if $\kappa = 1$. This is the first reason why the net considered becomes Michell's net only if $\sigma_T = \sigma_C$.

The displacement fields can be constructed by the rules (4.107), (4.108). Along the circle the fields u, v vanish, provided that

$$u(-\kappa\mu, \mu) = 0, \qquad v(-\kappa\mu, \mu) = 0 \tag{4.332}$$

We apply the formulae (4.107) by putting $\theta = -\kappa\mu$:

$$u(\lambda, \mu) = \int_{-\mu\kappa}^{\mu} [\cos(\lambda - \alpha) - \omega(\alpha, \mu)\sin(\lambda - \alpha)]\, A(\alpha, \mu) d\alpha \quad (4.333)$$

$$v(\lambda, \mu) = \int_{-\mu\kappa}^{\mu} [\sin(\lambda - \alpha) + \omega(\alpha, \mu)\cos(\lambda - \alpha)]\, A(\alpha, \mu) d\alpha$$

The displacement fields thus constructed satisfy the first of Eqs. (4.116), but do not satisfy the second one, if $\kappa \neq 1$.

If one constructs the displacement fields u, v by integrating by the rule (4.110) from $\beta = -\dfrac{\lambda}{\kappa}$ to $\beta = \mu$, then we obtain other fields, non-compatible with (4.333), if $\kappa \neq 1$.

We conclude that if $\kappa \neq 1$ then the Hencky net given by (4.325), discovered by Prandtl in 1923 is not the Michell net, since this net does not allow to fulfil the conditions (4.116).

The assumption of $\kappa = 1$ (or $\sigma_C = \sigma_T$) simplifies the construction of the solution: the fields (u, v) given by (4.107)–(4.111) satisfy all the conditions (4.116).

4.6.3 The Case of $\sigma_C = \sigma_T$ and $|\varphi| < \frac{\pi}{4}$. The Kinematic Solution

Let us consider now the classical problem of Michell (1904): find the least volume structure transmitting the given force to the boundary of the circle; the stress is bounded from both sides $-\sigma_0 \leq \sigma \leq \sigma_0$, see Fig. 4.36.

Substitution of $\kappa = 1$ into (4.325) gives the formulae for the parametric lines, see Fig. 4.37

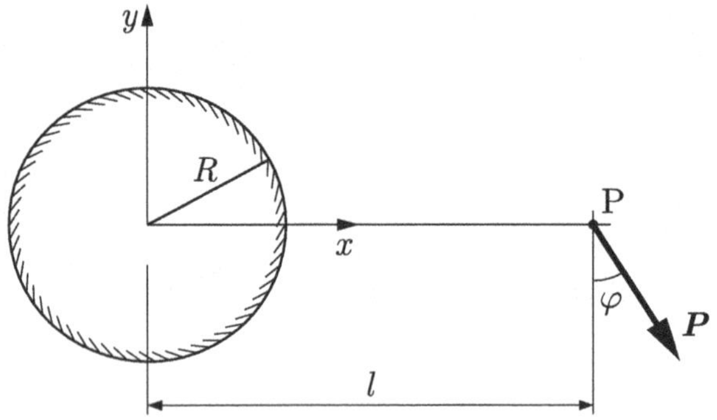

Fig. 4.36 The problem setting: transmit the force P to the boundary of the circle

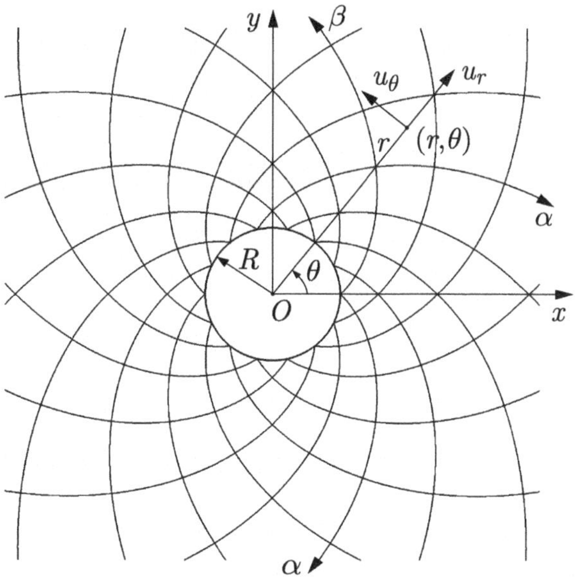

Fig. 4.37 Parametric lines of the Michell net

$$\frac{x(\alpha,\beta)}{R} = \exp(\alpha+\beta)\cos(\beta-\alpha) \tag{4.334}$$
$$\frac{y(\alpha,\beta)}{R} = \exp(\alpha+\beta)\sin(\beta-\alpha)$$

These fields determine the Lamé coefficients

$$\frac{A(\alpha,\beta)}{R} = \sqrt{2}\exp(\alpha+\beta), \qquad B(\alpha,\beta) = A(\alpha,\beta) \tag{4.335}$$

Using (4.333), we find

$$\frac{u(\alpha,\beta)}{R} = \sqrt{2}(\alpha+\beta)\exp(\alpha+\beta), \qquad \frac{v(\alpha,\beta)}{R} = -\frac{u(\alpha,\beta)}{R} \tag{4.336}$$

It is seen that u, v vanish on the boundary of the circle, i.e. for $\alpha = -\beta$. Thus, the net of the logarithmic spirals (4.334) is a Michell net. The radial displacement u_r and the circumferential displacement u_θ are expressed by

$$u_r(\alpha,\beta) = 0, \qquad u_\theta(\alpha,\beta) = -2R(\alpha+\beta)\exp(\alpha+\beta) \tag{4.337}$$

or

$$u_\theta(r,\theta) = -2r\ln\left(\frac{r}{R}\right) \tag{4.338}$$

The virtual work of the force P applied at the point $\theta = 0$, $r = l$ equals

$$\mathcal{L}(\boldsymbol{u}) = -Pu_\theta(l,0)\cos\varphi \tag{4.339}$$

We guess that this value is the maximal value of the virtual work attained in the problem (3.85); thus, $\mathcal{L}(\boldsymbol{u}) = \tilde{Z}_1$ and the volume of the optimal structure equals $V_{\min} = \dfrac{\tilde{Z}_1}{\sigma_0}$. Thus

$$V_{\min} = \frac{2P}{\sigma_0} l\ln\left(\frac{l}{R}\right)\cos\varphi \tag{4.340}$$

The result thus obtained does not deliver the structure itself. The form of the optimal structure is determined by the stress analysis discussed in the next section.

4.6.4 Case of $\sigma_C = \sigma_T$ and $|\varphi| < \frac{\pi}{4}$. The Static Solution

Assume that the point P lies on the axis x, or $\alpha_P = \beta_P$. This argument is computed from the condition $l = R\exp(2\alpha_P)$.

The optimal structure shown in Fig. 4.38 is composed of

- the bar in tension: AP;
- the bar in compression: BP;

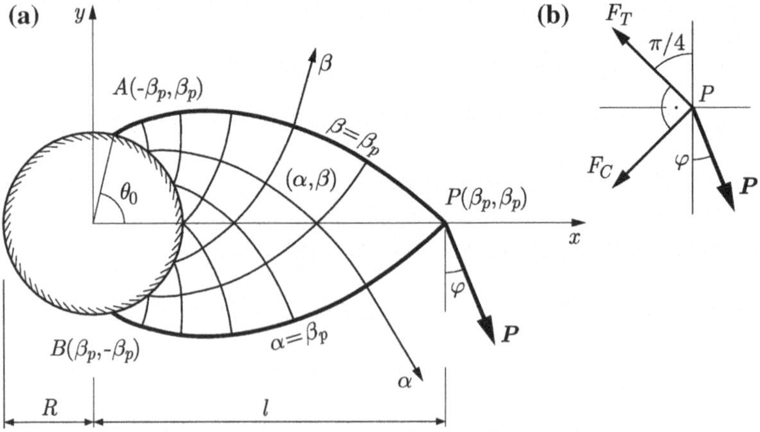

Fig. 4.38 The optimal layout: **a** general view and **b** equilibrium of the node P

- the fibrous domain APB of varying distribution of the volume fraction of the constituent material.

The equilibrium conditions of the node P deliver the values of the forces F_T, F_C at the adjacent cross sections of the reinforcing bars: AP and BP:

$$F_T(\alpha) = P_1, \qquad P_1 = P \sin\left(\frac{\pi}{4} + \varphi\right), \qquad (4.341)$$
$$F_C(\beta) = P_2, \qquad P_2 = -P \cos\left(\frac{\pi}{4} + \varphi\right)$$

The conditions $F_T \geq 0$, $F_C \leq 0$ are satisfied if $-\frac{\pi}{4} \leq \varphi \leq \frac{\pi}{4}$ and only then the whole construction is correct. The Hemp's forces $T_1(\alpha, \beta)$, $T_2(\alpha, \beta)$ in the fibres satisfy the boundary conditions (4.182). Substitution of $p_1 = 0$, $p_2 = 0$ gives

$$T_1(\alpha_P, \beta) + F_C(\beta) = 0, \qquad T_2(\alpha, \beta_P) + F_T(\alpha) = 0 \qquad (4.342)$$

where $-\alpha_P < \alpha < \alpha_P$, $-\beta_P < \beta < \beta_P = \alpha_P$ and the functions F_C, F_T are constant. Let us note that the angle $\theta_0 = \angle(\text{axis}\, x, OA)$ equals $2\beta_P$, since the coordinates of A are as follows: $\alpha_A = -\beta_P, \beta_A = \beta_P$ or $x_A = R\cos(2\beta_P)$ but $\cos\theta_0 = x_A/R$ or $\theta_0 = 2\beta_P$. For the sake of convenience, we introduce now the new parameterization

$$\bar{\alpha} = \alpha_P - \alpha, \qquad \bar{\beta} = \beta_P - \beta \qquad (4.343)$$

such that the point P from which the static analysis starts has the coordinates: $\bar{\alpha}_P = 0$, $\bar{\beta}_P = 0$. The unknowns are the fields: $\bar{T}_i(\bar{\alpha}, \bar{\beta}) = T_i(\alpha_P - \bar{\alpha}, \beta_P - \bar{\beta})$, $i = 1, 2$, satisfying the boundary conditions

$$\bar{T}_1(0, \bar{\beta}) = -P_2 \qquad \bar{\beta} \in [0, 2\beta_P] \qquad (4.344)$$
$$\bar{T}_2(\bar{\alpha}, 0) = -P_1 \qquad \bar{\alpha} \in [0, 2\beta_P]$$

The equilibrium equations (4.176) assume the form

$$\frac{\partial \bar{T}_1}{\partial \bar{\alpha}} + \bar{T}_2 = 0, \qquad \frac{\partial \bar{T}_2}{\partial \bar{\beta}} + \bar{T}_1 = 0 \qquad (4.345)$$

The field $\bar{T}_1(\bar{\alpha}, \bar{\beta})$ satisfies:

- the hyperbolic equation

$$\bar{\mathscr{L}}\bar{T}_1 = 0, \qquad \bar{\mathscr{L}} = \frac{\partial^2}{\partial \bar{\alpha} \partial \bar{\beta}} - 1 \qquad (4.346)$$

- the boundary conditions:

$$
\begin{aligned}
&\bar{T}_1(0,0) = -P_2 \\
&\frac{\partial \bar{T}_1}{\partial \bar{\beta}}(0,\bar{\beta}) = 0, \qquad \bar{\beta} \in [0, 2\beta_P] \\
&\frac{\partial \bar{T}_1}{\partial \bar{\alpha}}(\bar{\alpha},0) = P_1, \qquad \bar{\alpha} \in [0, 2\beta_P]
\end{aligned} \tag{4.347}
$$

which follow from (4.344), (4.345). The above problem can be solved by Riemann's method. Using the formula (4.132) for $w = \bar{T}_1$, $\theta_1 = \theta_2 = 0$, we get

$$
\bar{T}_1(\bar{\lambda},\bar{\mu}) = -P_2 G_0(\bar{\lambda},\bar{\mu}) + \int_0^{\bar{\lambda}} G_0(\bar{\lambda}-\bar{\alpha},\bar{\mu}) P_1 d\bar{\alpha} \tag{4.348}
$$

According to (4.126), we have

$$
G_0(\bar{\lambda}-\bar{\alpha},\bar{\mu}) = -\frac{\partial G_1(\bar{\lambda}-\bar{\alpha},\bar{\mu})}{\partial \bar{\alpha}} \tag{4.349}
$$

hence, taking into account $G_1(0,\bar{\mu}) = 0$, we find

$$
\bar{T}_1(\bar{\lambda},\bar{\mu}) = -P_2 G_0(\bar{\lambda},\bar{\mu}) + P_1 G_1(\bar{\lambda},\bar{\mu}) \tag{4.350}
$$

Thus, the value of the force T_1 at a point (α, β) equals

$$
\begin{aligned}
&T_1(\alpha,\beta) = T_1^*(\alpha,\beta) \\
&T_1^*(\alpha,\beta) = -P_2 G_0(\alpha_P-\alpha, \beta_P-\beta) + P_1 G_1(\alpha_P-\alpha, \beta_P-\beta)
\end{aligned} \tag{4.351}
$$

Proceeding in a similar manner we find

$$
\begin{aligned}
&T_2(\alpha,\beta) = T_2^*(\alpha,\beta) \\
&T_2^*(\alpha,\beta) = -P_1 G_0(\alpha_P-\alpha, \beta_P-\beta) + P_2 G_1(\beta_P-\beta, \alpha_P-\alpha)
\end{aligned} \tag{4.352}
$$

Thus, the problem of computing the internal forces and the reactions at the circular boundary turned out to be statically determinate. The forces F_T, F_C in bars AP, PB as well as the internal forces T_1, T_2 in the fibrous interior were found by equilibrium equations. We did not have to minimize the functional (4.187).

Having solved the static problem we compute the volume by summing up the volumes of the bars and of the interior

$$
\tilde{\mathcal{Z}}_2 = Z^{(F)} + Z^{(T)} + Z^{(C)} \tag{4.353}
$$

where

$$Z^{(F)} = \int_{\Omega_0} \left(\frac{|T_1|}{B} + \frac{|T_2|}{A} \right) d\Omega_0 \tag{4.354}$$

$$Z^{(T)} = \int_{AP} |F_T| ds_T, \qquad Z^{(C)} = \int_{PB} |F_C| ds_C$$

and $d\Omega_0 = ABd\alpha d\beta$, $ds_T = Ad\alpha$, $ds_C = Bd\beta$. Upon taking into account (4.341), (4.335), (4.351), (4.352), one obtains

$$Z^{(F)} = \sqrt{2}R(P_1 - P_2)\mathcal{J}(\alpha_P, \beta_P) \tag{4.355}$$

$$\mathcal{J}(\alpha_P, \beta_P) = \int_{-\alpha_P}^{\beta_P} \int_{-\beta}^{\alpha_P} e^{\alpha+\beta}[G_0(\alpha_P - \alpha, \beta_P - \beta) + G_1(\alpha_P - \alpha, \beta_P - \beta)] d\alpha d\beta \tag{4.356}$$

and

$$\begin{aligned} Z^{(T)} &= \sqrt{2}RP_1[\exp(\alpha_P + \beta_P) - 1] \\ Z^{(C)} &= -\sqrt{2}RP_2[\exp(\alpha_P + \beta_P) - 1] \end{aligned} \tag{4.357}$$

Now, we shall show that $\mathcal{J}$ given by (4.356) can be reduced to the form

$$\mathcal{J}(\alpha_P, \beta_P) = [(\alpha_P + \beta_P) - 1] \exp(\alpha_P + \beta_P) + 1 \tag{4.358}$$

To this end, we define

$$j(s) = e^s \int_0^s e^{-y} F(s, y) dy \tag{4.359}$$

with

$$F(s, y) = \int_0^{s-y} e^{-x}[G_0(x, y) + G_1(x, y)] dx \tag{4.360}$$

Let us note that the expression (4.356) is equal to $j(\alpha_P + \beta_P)$. It is sufficient to prove that

$$j(s) = (s - 1)e^s + 1 \tag{4.361}$$

which would confirm the thesis (4.358).

Let us differentiate $F(s, \eta)$ with respect to s

$$\frac{\partial F(s, y)}{\partial s} = \exp(y - s)[G_0(s - y, y) + G_1(s - y, y)] \tag{4.362}$$

and then differentiate both sides of (4.359) over s. Upon using (4.362), one finds

$$e^s \frac{d}{ds}[e^{-s} j(s)] = f_1(s) \tag{4.363}$$

where

$$f_1(s) = \int_0^s [G_0(s-y, y) + G_1(s-y, y)]dy \tag{4.364}$$

Let us now differentiate (4.363) over s and make use of (4.126). We get

$$\frac{d}{ds}\left\{e^s \frac{d}{ds}(e^{-s} j(s))\right\} = f_1(s) + 1 \tag{4.365}$$

By combining this result with (4.363), we have

$$\frac{d}{ds} f_1(s) = f_1(s) + 1 \tag{4.366}$$

with $f_1(0) = 0$. Hence, $f_1(s) = \exp(s) - 1$. We put this result into (4.363)

$$e^s \frac{d}{ds}(e^{-s} j(s)) = e^s - 1 \tag{4.367}$$

hence

$$j(s) = se^s + 1 + C_1 e^s \tag{4.368}$$

with C_1 being a constant. Since $j(0) = 0$ we find $C_1 = -1$ which leads to (4.361). This ends the derivation.

Now, we can compute $\tilde{Z}_2$ given by (4.353), (4.357), (4.358):

$$\tilde{Z}_2 = \sqrt{2}R(P_1 - P_2)(\alpha_P + \beta_P)\exp(\alpha_P + \beta_P) \tag{4.369}$$

The volume given by $\dfrac{\tilde{Z}_2}{\sigma_0}$ coincides with the result (4.340) for $\theta = 0$.

Thus, the equality $\tilde{Z}_1 = \tilde{Z}_2$ has been proved. We see that the solutions to the problems (4.187) and (4.190) are identical, which proves that the solution is correct. However, the uniqueness is an open question; theoretically, there could be other solutions delivering the same volume.

Let us examine the distribution of fibres in the interior of the structure. This distribution is given by (4.169) or

$$v = \frac{1}{\sigma_0} \frac{T_1(\alpha, \beta) - T_2(\alpha, \beta)}{A(\alpha, \beta)} \tag{4.370}$$

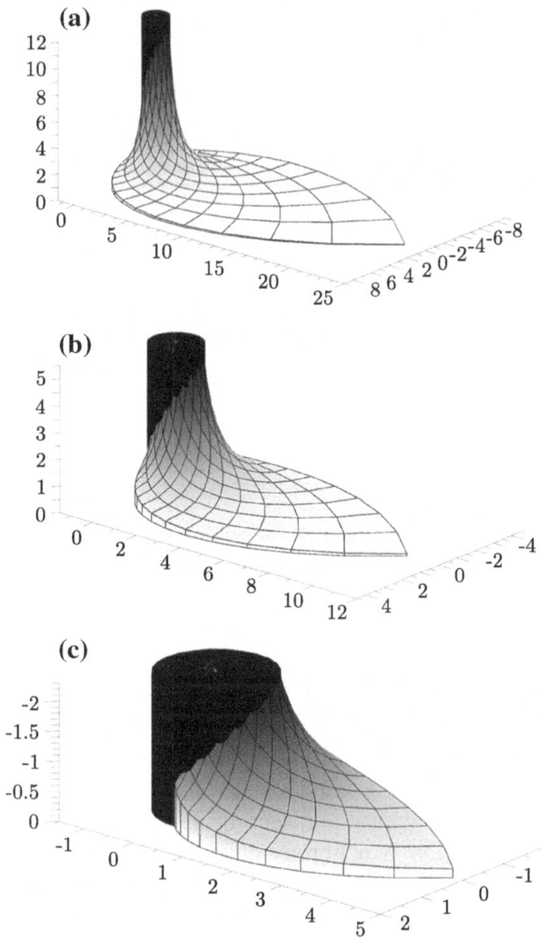

Fig. 4.39 Density of fibres υ: **a** the case of $l/R = 23.14$, $\varphi = 0$; **b** the case of $l/R = 10.55$, $\varphi = \pi/6$; and **c** the case of $l/R = 4.81$, $\varphi = \pi/4$

where $T_2 < 0$. If $\varphi = 0$, then the distribution of υ is symmetric with respect to the line x. The distribution of υ depends strongly on the ratio l/R, see Fig. 4.39.

Let us examine the ratio: $\dfrac{V_R}{V_{\min}}$, where V_R is the volume of the reinforcing bars and $V_{\min}$ is the total volume of the structure. According to (4.357), we have

$$V_R = \frac{2PR}{\sigma_0}[\exp(\alpha_P + \beta_P) - 1]\cos\varphi \tag{4.371}$$

hence

$$\frac{V_R}{V_{\min}} = \frac{\exp(\alpha_P + \beta_P) - 1}{(\alpha_P + \beta_P)\exp(\alpha_P + \beta_P)} = \frac{\dfrac{l}{R} - 1}{\dfrac{l}{R}\ln\left(\dfrac{l}{R}\right)} \tag{4.372}$$

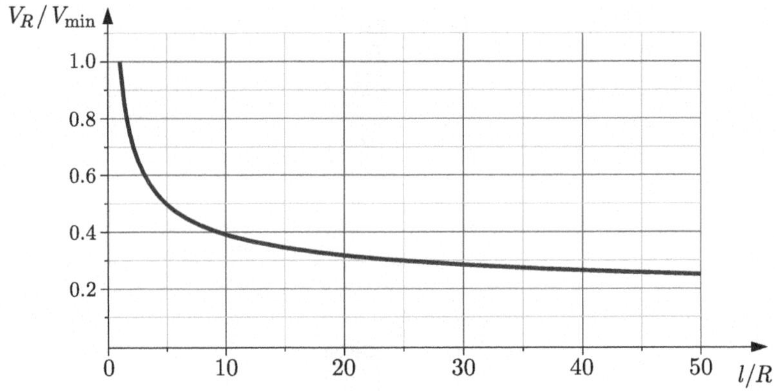

Fig. 4.40 The variation of $\dfrac{V_R}{V_{\min}}$ as a function of the ratio l/R

It is seen that if the point P of application of the force lies close to the support, the majority of the material concentrates in the reinforcing bars, see Fig. 4.40.

4.6.5 *The Optimal Wheel*

We consider the problem of optimum design of an annular plate clamped along the internal contour $r = l$ and subjected to a circumferential traction $q = \text{const}$ along the external contour $r = l$, see Fig. 4.41.

The solution is a fibrous plate whose external contour is not reinforced; the fibres run along the spirals (4.334). The final design turns out to be a superposition of cantilevers corresponding to the elementary forces $dP = qld\theta$ acting orthogonally to the radii, hence tangentially to the external contour. The optimal volume is computed as a sum of elementary volumes of the optimal Michell structures constructed in the previous section, corresponding to the elementary forces $dP = qld\theta$. The virtual work of this load done on the virtual displacement field (4.337) equals $d\mathcal{L} = -dPu_\theta(l, \theta) = -qld\theta(-2l) \ln\left(\dfrac{l}{R}\right) = 2ql^2 \ln\left(\dfrac{l}{R}\right) d\theta$. The total virtual work is

$$\mathcal{L}(\boldsymbol{u}) = \int_0^{2\pi} 2ql^2 \ln\left(\frac{l}{R}\right) d\theta = 4\pi ql^2 \ln\left(\frac{l}{R}\right) \tag{4.373}$$

while the volume equals

$$V_{\min} = \frac{4\pi ql^2}{\sigma_0} \ln\left(\frac{l}{R}\right) \tag{4.374}$$

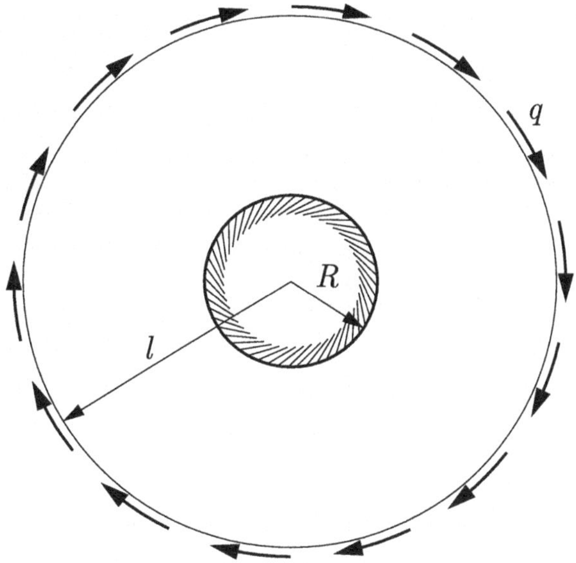

Fig. 4.41 The problem setting: design a wheel subject to a uniform torsion

Now, we shall find the distribution of the volume fraction of fibres by the static analysis. The stress state of the annular plate subject to the uniform torsion is $\tau_r = 0$, $\tau_\theta = 0$, $\tau_{r\theta} = \tau_{r\theta}(r)$, and the shear stress satisfies the equilibrium equation

$$\frac{d\tau_{r\theta}}{dr} + \frac{2}{r}\tau_{r\theta} = 0 \tag{4.375}$$

hence $\tau_{r\theta} = Cr^{-2}$. The boundary condition $\tau_{r\theta}(r = l) = -q$ gives

$$\tau_{r\theta} = -q\left(\frac{l}{r}\right)^2 \tag{4.376}$$

The eigenvalues of the stress tensor are as follows: $\lambda_1(\boldsymbol{\tau}) = |\tau_{r\theta}|$, $\lambda_2(\boldsymbol{\tau}) = -|\tau_{r\theta}|$ while the density of fibres is

$$\upsilon(r, \theta) = \frac{2}{\sigma_0}|\tau_{r\theta}| = \frac{2q}{\sigma_0}\left(\frac{l}{r}\right)^2 \tag{4.377}$$

The volume of the fibres equals

$$V = \int_0^{2\pi}\int_R^l \upsilon(r, \theta) r d\theta dr \tag{4.378}$$

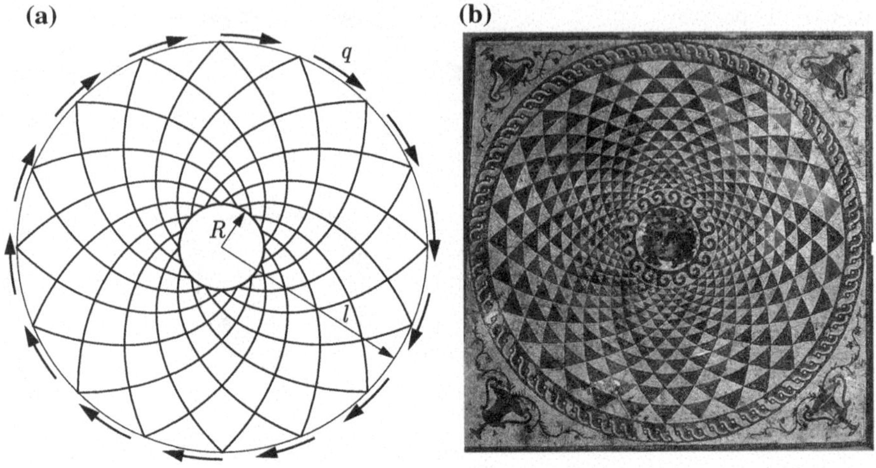

Fig. 4.42 **a** The optimal layout of fibres and **b** central panel from a tesselated floor of a Roman villa, second half second century AD, Archaeological Museum of Ancient Corinth, Greece. The black and white version of the photo by Carole Raddato. *The source* https://commons.wikimedia.org/wiki/File:Dionysos_with_fruit_and_ivy_in_his_hair,_Archaeologi-cal_Museum_of_Ancient_Corinth,_Greece.jpg; https://www.flickr.com/photos/carolemage//14099190391/

and the integration gives the result known already, see (4.374). This proves that the whole construction is for sure correct. The fibres run along the logarithmic spirals which are trajectories of the principal stresses, see Fig. 4.42.

The optimal wheel solution forms the figure used in the antic art, see the Roman floor shown in Fig. 4.42b. Similar patterns encrust the floor of Saint Mark's Basilica in Venice.

4.6.6 A Family of Suboptimal Solutions—Illustrative Examples of Trusses of Finite Number of Bars

We make use of the Michell net found in this section to construct trusses approximating the Michell structure corresponding to the data: $l = 4.81R$, $\varphi = 0$. The trusses are be based on the nets composed of n parametric lines, cf. Fig. 4.43, where $n = 2, 3, 4, 6, 8, 12, 100$. This figure shows both the nets and the corresponding trusses, except the case of $n = 100$. All the trusses are statically determinate.

The simplest considered truss is composed of six members, and it is heavier by 11.1% than Michell's cantilever. The weight of the external members constitutes 79.2% of the total weight of the truss. Let s be the number of parametric lines of each family. The number of bars in the truss equals $p = s(s+1)$, while the number of nodes is given by $w = 1 + s(s+3)/2$.

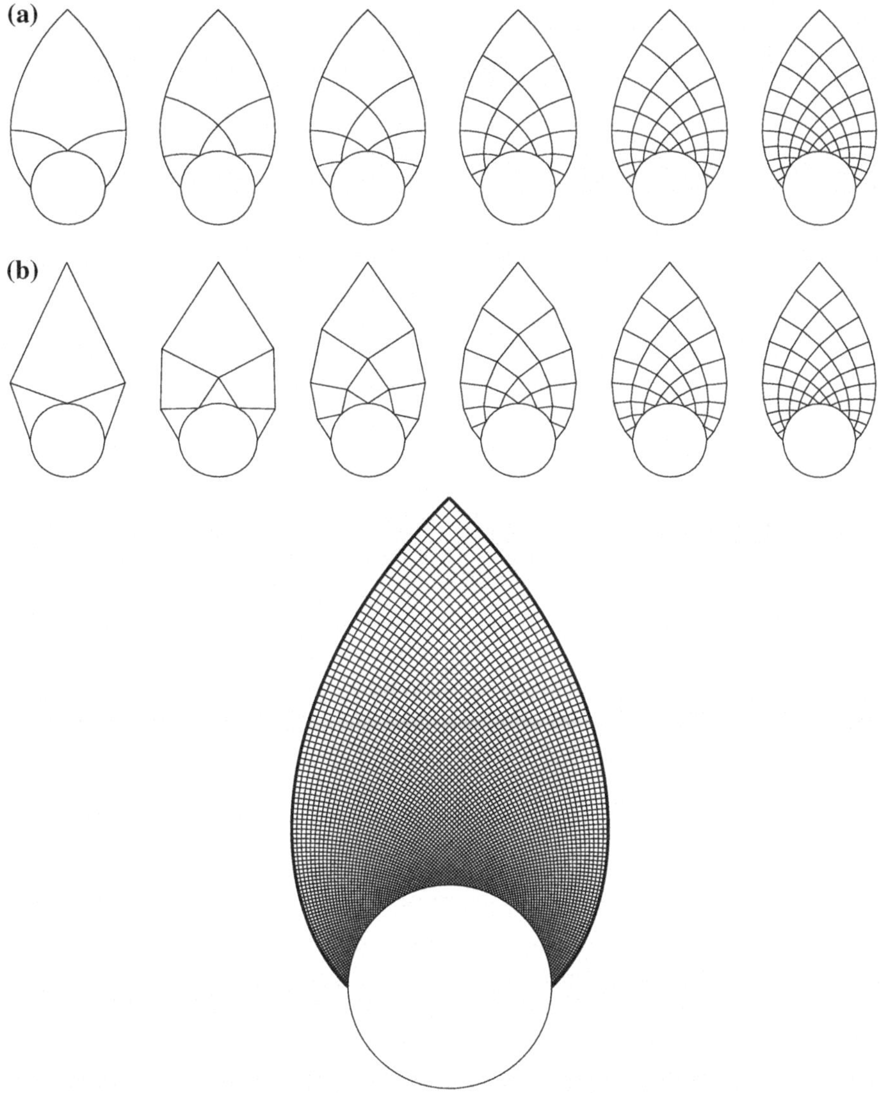

Fig. 4.43 A sequence of nets (**a**) and a sequence of trusses (**b**) of increasing number of members and decreasing weight; case of $l = 4.81R$, $\varphi = 0$. The numerical results are set up in Table 4.2

Table 4.2 The non-dimensional volumes $\bar{V} = \dfrac{V}{PR/\sigma_0}$ of external members, internal members and the non-dimensional volumes of the trusses as functions of total number of members

Number of parametric lines	Number of members	Number of nodes	Volume of external members	Volume of internal members	Total volume of the truss
2	6	6	13.31266	3.49094	16.80360
3	12	10	11.13618	4.69297	15.82916
4	20	15	10.17524	5.33388	15.50911
5	30	21	9.62977	5.73469	15.36446
6	42	28	9.27726	6.00952	15.28678
7	56	36	9.03041	6.20984	15.24025
8	72	45	8.84779	6.36238	15.21017
9	90	55	8.70716	6.48244	15.18960
10	110	66	8.59552	6.57940	15.17491
12	156	91	8.42942	6.72640	15.15582
14	210	120	8.31174	6.83258	15.14432
16	272	153	8.22399	6.91287	15.13687
18	342	190	8.15604	6.97572	15.13176
20	420	231	8.10186	7.02626	15.12811
25	650	351	8.00472	7.11779	15.12251
30	930	496	7.94024	7.17923	15.11947
40	1640	861	7.85993	7.25651	15.11645
50	2550	1326	7.81191	7.30314	15.11505
60	3660	1891	7.77996	7.33433	15.11429
100	10100	5151	7.71621	7.39697	15.11318

The increase in the total number of members results in the decrease of the total weight and leads to the transfer of the mass from external to internal bars. The truss of 10100 bars has the volume $15.11318V_0$, $V_0 = PR/\sigma_0$, while the volume of Michell's cantilever equals $15.11256V_0$. In such a truss, the weight of the external bars constitutes 51.056% of total weight, while in Michell's cantilever this ratio equals 50.428%. All results are set up in Table 4.2.

The trusses of finite number of bars constructed on the basis on the Michell net, as shown in Fig. 4.43, are not optimal among the trusses of the same number of bars. The best possible trusses of finite number of bars should satisfy the optimality conditions discovered by Prager (1978a, b) and improved by Mazurek (2012), see Sect. 4.4.

4.6.7 Case of $\sigma_T = \sigma_C$ and Arbitrary φ: Hu and Shield Optimal Structures

The analytical solution given in Sects. 4.6.1–4.6.4 concerns the case of $\sigma_T = \sigma_C = \sigma_0$ and $|\varphi| \le \pi/4$. Then, from the equilibrium equations of the node P one concludes that the bar PA is in tension and the bar PB is in compression. In the case of $\frac{\pi}{4} < \varphi < \frac{\pi}{2}$, this solution ceases to hold and the optimal structure must change its shape.

New layouts corresponding to the case of $\varphi > \frac{\pi}{4}$ have been discovered by Hu and Shield (1961). These layouts will be repeated, their proofs are left to the reader.

Let us show that if $\frac{\pi}{4} < \varphi < \beta_0$ and $\cos(\beta_0) = \frac{R}{l\sqrt{2}}$, the solution has the shape of Fig. 4.44a. The structure is composed of a bar P_1P which is linked tangentially with the rib AP_1. From the equilibrium of the node P_1, one can note that the structure is not reinforced along the boundary BP_1; this line is only a boundary of the fibrous domain AP_1B.

Consider the triangle OP_1P. Let us note that

$$\angle OP_1P = \frac{3\pi}{4}, \quad \angle OPP_1 = \frac{\pi}{2} - \varphi, \quad |OP| = l, \quad |OP_1| = R + |P_2P_1|$$

By the sinus rule, we have

$$\frac{|OP_1|}{\sin \angle OPP_1} = \frac{|OP|}{\sin \angle OP_1P}$$

hence

$$|P_2P_1| = l\sqrt{2}\cos\varphi - R$$

The solution of Fig. 4.44a refers to the case of $|P_2P_1| \ge 0$, which results in the condition $\varphi \le \beta_0$ where

$$\beta_0 = \arccos\left(\frac{R}{l\sqrt{2}}\right).$$

If $\varphi > \beta_0$ and $\varphi < \frac{\pi}{2}$, then the optimum structure is composed of a single bar linking the point P with the point P_0 at the boundary of the circle, see Fig. 4.44b.

Concluding, the solution (only for the case of $\sigma_T = \sigma_C$) is given by:

(i) Figure 4.38, for the case of $0 < \varphi < \frac{\pi}{4}$;
(ii) Figure 4.44a if $\frac{\pi}{4} < \varphi < \beta_0$;
(iii) Figure 4.44b if $\beta_0 < \varphi < \frac{\pi}{2}$.

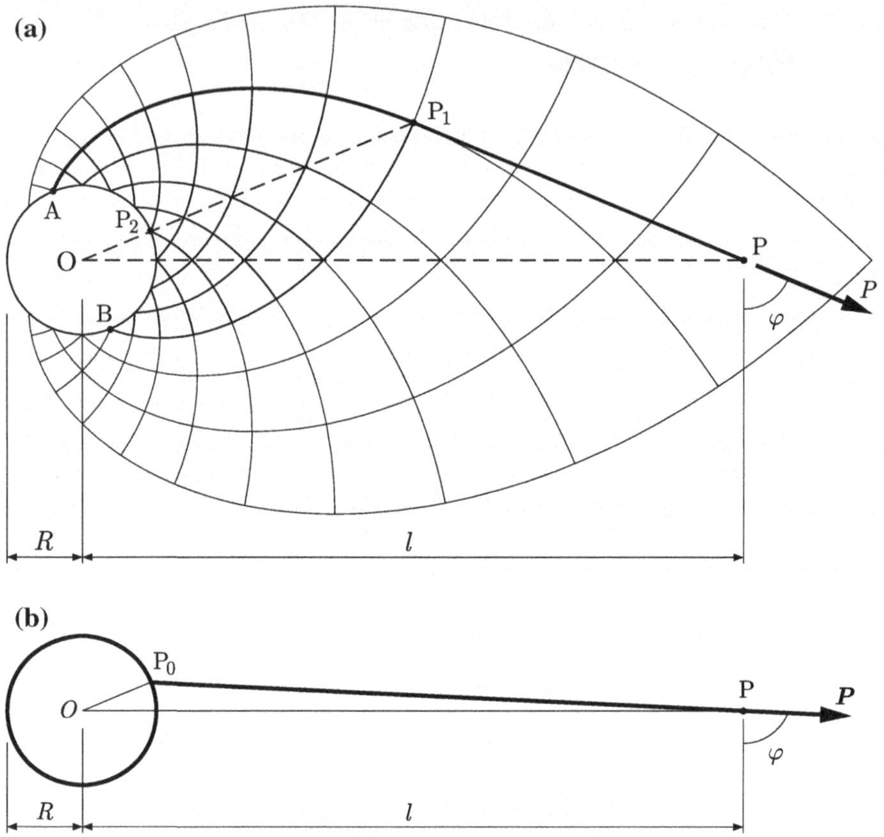

Fig. 4.44 Optimum designs for the cases of $\pi/4 < \varphi < \beta_0$ (**a**) and $\beta_0 < \varphi < \pi/2$ (**b**)

The results above can be confirmed by showing the virtual field $\boldsymbol{v}$ within the exterior of the circle satisfying all the conditions set up in Table 3.1. This construction can be found in Hu and Shield (1961) and in Prager (1985).

4.7 The Problem of the Interior of a Circle

We consider the problems similar to those of Sect. 4.6, but now the loads are applied in the interior of a circle. We still assume that $\sigma_T = \sigma_C = \sigma_0$.

4.7.1 *The Optimal Wheel*

Consider the problem of optimal transmitting a circumferential load of intensity $p = \text{const}$ around the circular contour $r = R$ within the circular domain to be fixed around the external contour $r = l$, see Fig. 4.45. We predict that the optimal structure will have an opening of radius $r = R$, while the load p becomes the traction load.

The loads intensities q of Fig. 4.41 and p of Fig. 4.45 are linked by $ql^2 = -pR^2$. According to (4.376), the shear stress equals

$$\tau_{r\theta} = p\left(\frac{R}{r}\right)^2 \tag{4.379}$$

and this state of stress determines the volume fraction of the fibres and the volume of the optimal structure:

$$V_{\min} = \frac{4\pi pR^2}{\sigma_0}\ln\left(\frac{l}{R}\right) \tag{4.380}$$

The fibres of the optimal structure form the same net as that constructed in Sect. 4.6.5, see Fig. 4.42a.

4.7.2 *The Point Load Within the Domain*

The problem of optimal transmitting a force $\boldsymbol{P}$ at point P to the boundary of the circle involves two parameters: an angle α between the line of the force and the radius $|OP|$ and the parameter $\xi = |OP|/l$ which measures the distance $|OP|$ to the middle of

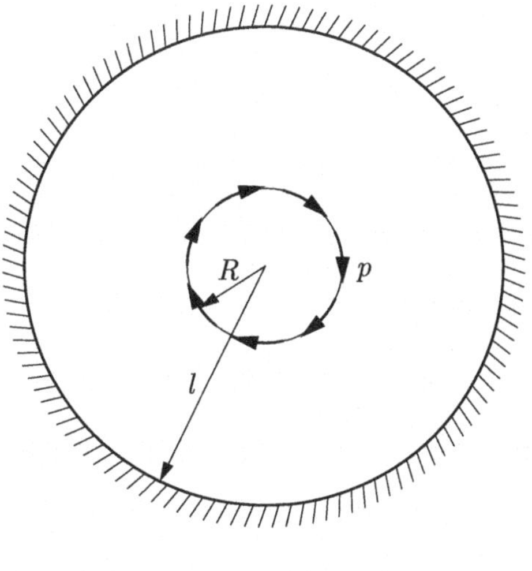

Fig. 4.45 The problem setting: design a wheel subject to a uniform torsion load applied along the contour $r = R$ within the circular domain

the domain. The numerical experiments performed by the present authors teach that the solution to this problem is either

- a single bar (in tension or in compression)
  or
- a two-bar truss (with axial forces of different signs) whose bars do not (in general) form a right angle.

Thus, the solutions observed are not inscribed into the Prandtl net shown in Fig. 4.42a. It is worth emphasizing that the Prandtl net determines the solution to the rotationally symmetric problem of Fig. 4.45 but does not appear in the solutions of problems concerning point loads.

4.8 Transmitting a Point Load to a Segment of a Straight Support. Full Plane Designs

In Sect. 4.5, the fundamental problem of optimal transmitting of a force to an infinitely long straight support is solved, leading to a two-bar solution. Usually, however, the support has a finite length, which changes the optimal solution drastically. This class of problems will be solved in the present section.

4.8.1 Formulation of the Problem and the Simplest Forms of the Solutions

Given is the vertical support NR (i.e. their points at the right-hand side); the design domain is either the whole plane (see Fig. 4.46a) or the domain bounded by RN and by the rays RR_0, NN_0, see Fig. 4.46b, this latter case will be analysed later.

The force P is inclined by the angle φ with respect to the vertical direction. The force should be transmitted to the right-hand side of the interval RN. We do not specify the supporting points; their position is unknown of the optimum design problem considered. The data of the problem are position of points: N, R, P; the permissible stresses: σ_T, σ_C; angle φ and the magnitude of the force P.

Let the angles γ_1, γ_2 be defined by (4.227). According to Sect. 4.5.2, we form a triangle RAN such that $\angle NRA = \gamma_2$ and the angle $\angle RAN$ is a right angle, see Fig. 4.47a.

If point P lies within the triangle RAN, the optimum structure is (a) a two-bar truss of the bars parallel to RA and AN, if $-\gamma_1 < \varphi < \gamma_2$, see Fig. 4.47b; (b) a single-bar structure if the direction of the force lies within the angle $\angle RAN$, see Fig. 4.47c. Thus, the case of P lying within RAN has been already solved in Sect. 4.5.2.

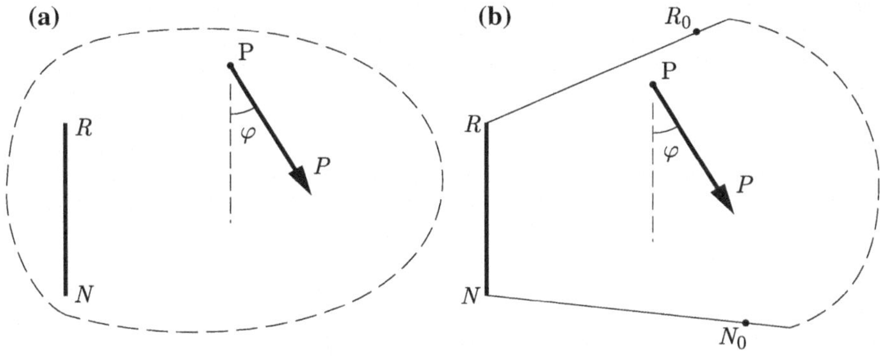

Fig. 4.46 The problem setting. The design domain is: **a** the whole plane and **b** restricted by RN and two rays

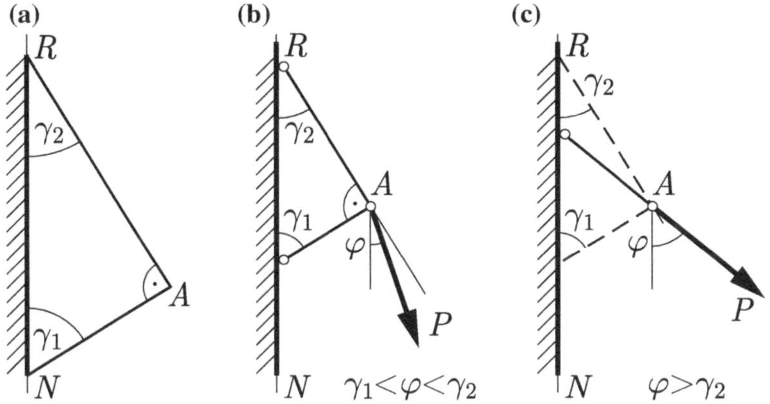

Fig. 4.47 Optimum designs in case of P lying within RAN: **a** the triangle RAN, **b** a two-bar solution and **c** one-bar solution

Let us draw two sectors of circles: of origin at R, of radius $|RA| = r_2$ and of angle $\theta_2 \le \frac{\pi}{2}$, and of origin at N, of radius $|NA| = r_1$ and of angle $\theta_1 \le \frac{\pi}{2}$. The bounds: $\theta_i \le \frac{\pi}{2}$ will be cleared up later.

The domains BRA and CAN will be called: the upper and lower fans. If point P_2 lies in the upper fan and the force P_2 is directed within the shaded region, the optimal structure looks like in Fig. 4.48c.

If point P_1 lies in the lower fan and if the force P_1 is directed within the shaded regions, then the optimal structure looks like in Fig. 4.48b. The bars R_1A_1, NP_1 are straight and prismatic (they have constant areas of the cross sections). The bar A_1P_1 is curved and prismatic, of the cross section the same as in bar R_1A_1. The domain A_1NP_1 is filled up with fibres of centre at N. The triangle R_1A_1N is empty.

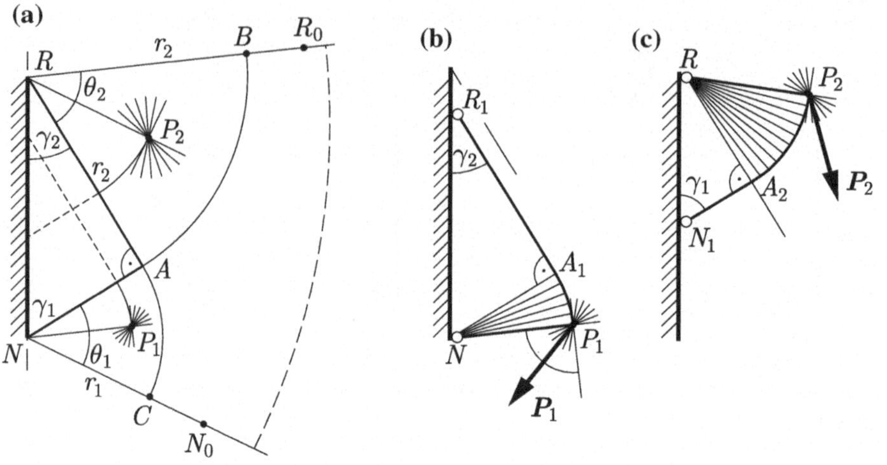

Fig. 4.48 Construction of the optimum design if the point P lies within the domain: $RBACNR$: **a** geometry of the domain $RBACNR$, **b** case of P lying within RBA and **c** case of P lying within NAC

Fig. 4.49 Parameterization of the fan domains

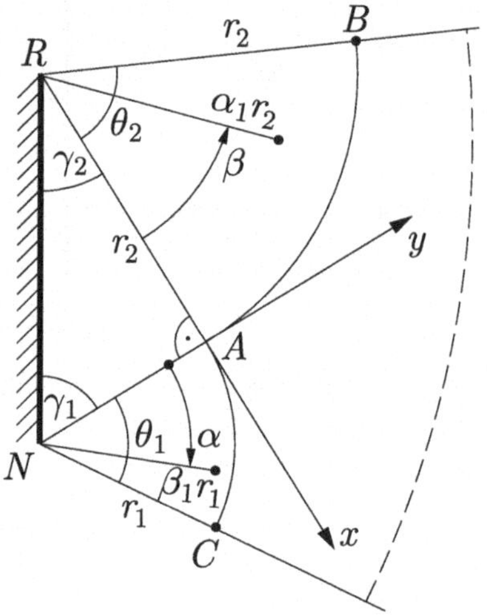

The structure of Fig. 4.48c is similar: it is composed of a circular fan, two straight bars and one curved bar of circular shape.

Both the structures are inscribed into the net shown in Fig. 4.48, generated from the triangle RAN. We shall show now that the structures of Fig. 4.48b, c are optimal.

Let us introduce the Cartesian parameterization (x, y) determined by the triangle RAN and the polar parameterizations of the fans, see Fig. 4.49.

The polar system within the lower fan NAC is (α, β_1) such that

$$\begin{aligned} ds_1 &= A d\alpha, \qquad A = \beta_1 r_1 \\ ds_2 &= B d\beta_1, \qquad B = r_1 \end{aligned} \tag{4.381}$$

In the upper fan RAB the polar system is (α_1, β) such that

$$\begin{aligned} ds_1 &= A d\alpha_1, \qquad A = r_2 \\ ds_2 &= B d\beta, \qquad B = \alpha_1 r_2 \end{aligned} \tag{4.382}$$

Let us introduce the virtual displacement field:
• within NAC

$$u^{NAC} = (1+\kappa)\alpha\beta_1 r_1 + \beta_1 r_2, \qquad v^{NAC} = -\kappa\beta_1 r_1 \tag{4.383}$$

Here, u, v are measured along α, β_1
• within RAN

$$u^{RAN} = x + \frac{r_2}{r_1} y + r_2, \qquad v^{RAN} = -\kappa y - \frac{r_2}{r_1} x - \kappa r_1 \tag{4.384}$$

Here, u, v are measured along x, y
• within RBA

$$u^{RBA} = \alpha_1 r_2, \qquad v^{RBA} = -(1+\kappa)\alpha_1\beta r_2 - \kappa\alpha_1 r_1 \tag{4.385}$$

Here, u, v are measured along α_1, β.

The field (u, v) given above satisfies:

- kinematic conditions: $u = 0$, $v = 0$ along the support RN;
- continuity conditions along RA and NA;
- optimality condition of type $\mathbb{T}$: $\epsilon_I = 1$, $\epsilon_{II} = -\kappa$ within RAB, RAN, NAC; we have assumed: $\sigma_0 = \sigma_T$, as in Sect. 4.5.2.

Let us examine the optimal structure corresponding to the load P at point P within NAC, see Fig. 4.50.

Let us denote: $\psi = \angle NP$, vertical direction $= \gamma_1 + \alpha_P$.

From the equilibrium equation of the node P, we find the adjacent forces in the bars

$$F_T = P\sin(\psi + \varphi), \qquad F_C = -P\cos(\psi + \varphi) \tag{4.386}$$

where $\psi + \varphi = \psi - |\varphi|$, since $\varphi < 0$, see Fig. 4.50. Equation (4.386) refers to the bar $R_1 A_1 P$ being in tension and to the bar NP being in compression, which is the case if

$$0 \le \gamma_1 + \alpha_P + \varphi \le \frac{\pi}{2} \tag{4.387}$$

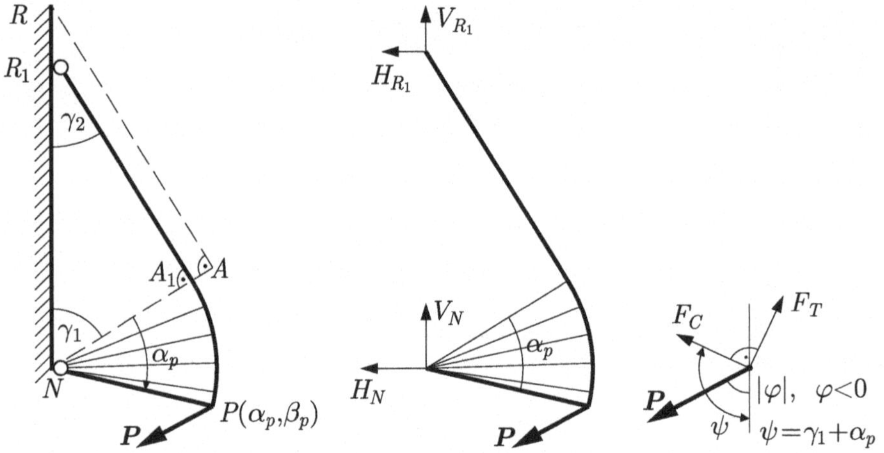

Fig. 4.50 The optimal structure for the case of P lying within NAC

According to (3.81), the volume of the optimal structure (under the convention: $\sigma_0 = \sigma_T$) is equal to $V = \dfrac{1}{\sigma_T}\tilde{\mathcal{Z}}_1$, where $\tilde{\mathcal{Z}}_1$ is the value of the virtual work of the force P done on the computed displacement, or

$$\begin{aligned}\tilde{\mathcal{Z}}_1 &= P u_p \\ u_p &= u^{NAC}(\alpha_P, \beta_P)\sin(\psi+\varphi) - v^{NAC}(\alpha_P, \beta_P)\cos(\psi+\varphi)\end{aligned} \tag{4.388}$$

Thus

$$V = \frac{P}{\sigma_T}\left[\sin(\psi+\varphi)[r_1\alpha_P(1+\kappa)\beta_P + r_2\beta_P] + \cos(\psi+\varphi)\kappa r_1\beta_P\right] \tag{4.389}$$

The same result can be confirmed by computing the volumes of the bars and of the fans:

$$\begin{aligned}&V^{R_1A_1} = \frac{F_T}{\sigma_T} r_2\beta_P, \qquad V^{A_1P} = \frac{F_T}{\sigma_T} r_1\beta_P\alpha_P, \qquad V^{NP} = \frac{F_C}{\sigma_C} r_1\beta_P \\ &V^{NA_1P} = \int_0^{\alpha_P}\int_0^{\beta_P} v A B d\alpha d\beta_1\end{aligned} \tag{4.390}$$

where

$$v = -\frac{T_2}{\sigma_C A} \tag{4.391}$$

and according to (4.182), we have $T_2 = -F_T$, since $p_2 = 0$. By summing up the above volumes, we arrive at the formula (4.389), which confirms that the construction of Fig. 4.50 is correct.

Let us compute the reactions at R_1 and N, see Fig. 4.50.

$$\begin{aligned} H_{R_1} &= F_T \sin\gamma_2, \qquad V_{R_1} = F_T \cos\gamma_2 \\ H_N &= F_C \cos\left(\alpha+\gamma_1-\frac{\pi}{2}\right) - \int_0^{\alpha_P} F_T \cos\left(\alpha+\gamma_1-\frac{\pi}{2}\right) d\alpha \\ V_N &= F_C \sin\left(\alpha+\gamma_1-\frac{\pi}{2}\right) - \int_0^{\alpha_P} F_T \sin\left(\alpha+\gamma_1-\frac{\pi}{2}\right) d\alpha \end{aligned} \tag{4.392}$$

The reactions thus computed satisfy the equilibrium equations

$$\begin{aligned} V_{R_1} + V_N &= P\cos\varphi \\ H_{R_1} + H_N &= P\sin\varphi \\ H_{R_1}|R_1N| &= |NN_1|P\sin(\gamma_1+\alpha_P+\varphi) \end{aligned} \tag{4.393}$$

Now, let us examine the optimal structure referring to the load at point P within the fan RAB.

From the equilibrium equations of the node P, we compute the axial forces at the adjacent cross sections:

$$\begin{aligned} F_T &= P\sin(\psi+\varphi), \\ F_C &= -P\cos(\psi+\varphi) \end{aligned} \tag{4.394}$$

with $\psi = \gamma_1 - \beta_P$. Now, see Fig. 4.51, the angle φ is positive, see Fig. 4.46. To keep: $F_T > 0$ and $F_C < 0$, one should assume that $\psi + \varphi \in \left(0, \frac{\pi}{2}\right)$. The virtual work of the force P is given by (4.388) where now u, v refer to the domain RAB and are

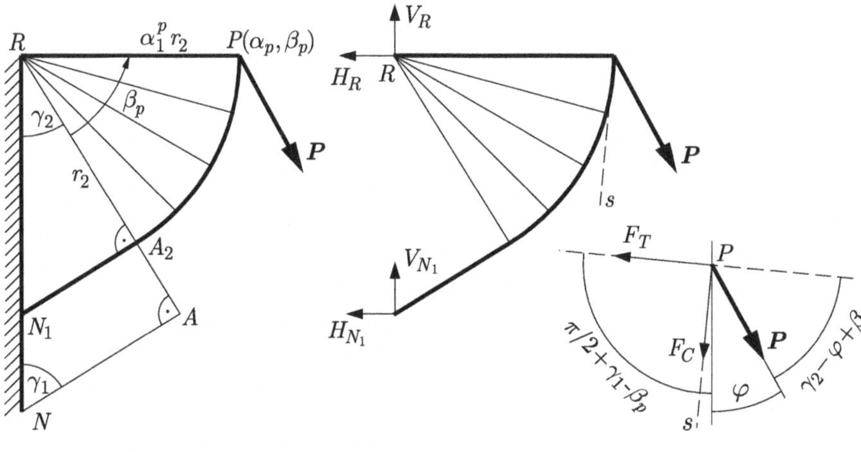

Fig. 4.51 The optimal structure for the case of P lying within RAB

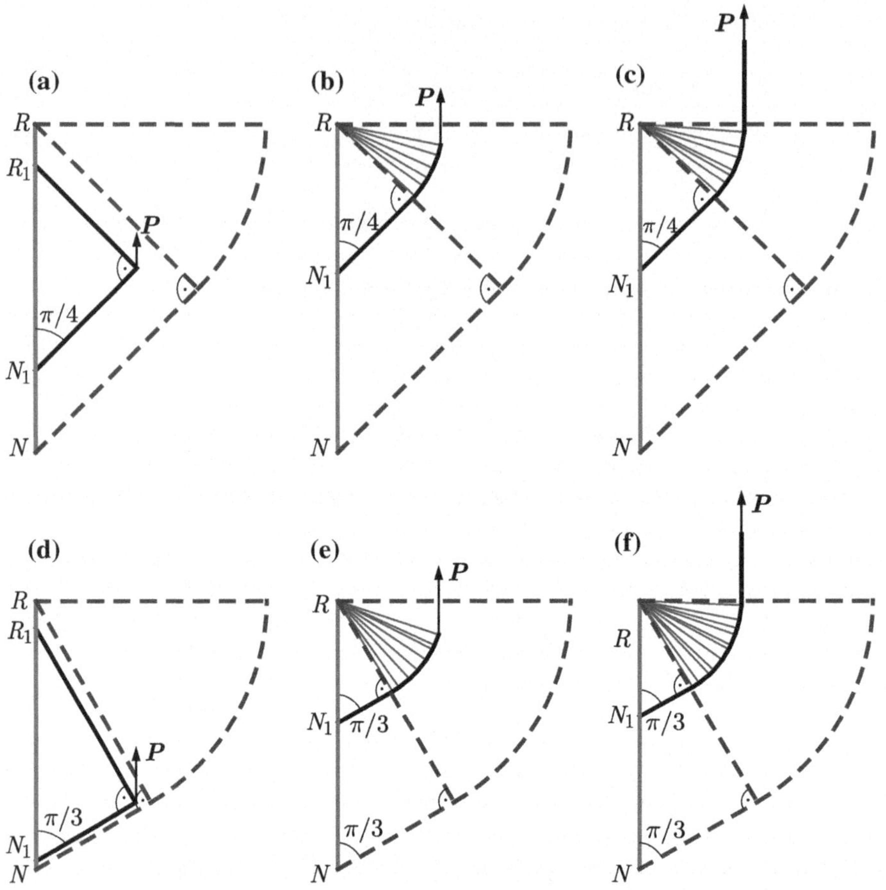

Fig. 4.52 Selected numerical solutions to the problem discussed in the present section. The results: **a**, **b**, **c** three solutions for the case of $\sigma_C = \sigma_T$ corresponding to the three different positions of the point load. The results: **d**, **e**, **f** three solutions for the case of $\sigma_T/\sigma_C = 1/3$ corresponding to the three different positions of the point load

given by (4.385) with $\alpha_1 = \alpha_P$, $\beta = \beta_P$. The volume of the structure of Fig. 4.51 equals

$$V = \frac{P}{\sigma_P}[\sin(\psi + \varphi)r_2\alpha_P + \cos(\psi + \varphi)[(1 + \kappa)r_2\alpha_P\beta_P + \kappa r_1\alpha_P]] \quad (4.395)$$

The same result can be confirmed by summing up the volumes of the bars N_1A_2, A_2P and the fan RPA_2; the details of this computation are omitted.

Some selected numerical solutions are given in Fig. 4.52.

If $\sigma_T \neq \sigma_C$, then the optimal structure depends on the direction of the load. For illustration, two solutions corresponding to two cases of the load: $\boldsymbol{P}$ and $-\boldsymbol{P}$ and for the same position of the load are shown in Fig. 4.53.

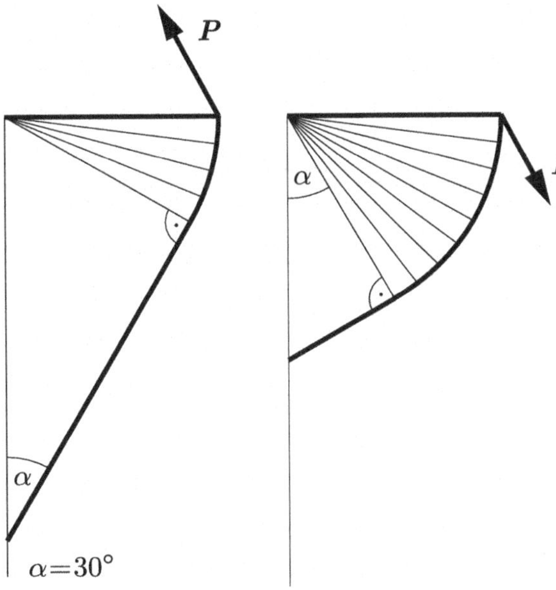

Fig. 4.53 Two different optimal structures corresponding to the loads: $\boldsymbol{P}$ and $-\boldsymbol{P}$, respectively. Here, $\kappa = 3$

4.8.2 *The Cantilevers Supported on Two Circles*

4.8.2.1 Construction of the Michell Net

Let us sum up the hitherto found results of this section. If the point P lies in the triangle RAN, then the solution is a single-bar truss or a two-bar truss. If the point P lies in the sectors of the circles: BRA or ANC, then the solutions have the form like in Fig. 4.48b, c, provided that the force is appropriately directed. In the sequel, it will turn out that these solutions are only valid if $\theta_i \leq \frac{\pi}{2}$. Thus, assume that the angles θ_i are fixed and such that $\theta_i \leq \frac{\pi}{2}$. The lines RB and NC encompass a certain domain right to the line RN. Below we shall find the boundaries BD, CD of the domain $BDCAB$ which is a subsequent singled out domain determined by the position of point P. The optimal structures transmitting the force P (applied at point P lying within $BDCAB$) to the support RN are generated by the sectors of circles, see Fig. 4.54.

The aim of this section is an exact construction of the solution outlined in Fig. 4.54.

We shall construct the Hencky net generated on two circles, see Figs. 4.55, 4.56. We assume that the line $\beta = 0$ being a boundary line of a circle of origin at N of radius r_1 is the first parametric line α, while the line $\alpha = 0$—a boundary of the circle of origin R and of radius r_2 is the first parametric line β. These conditions determine

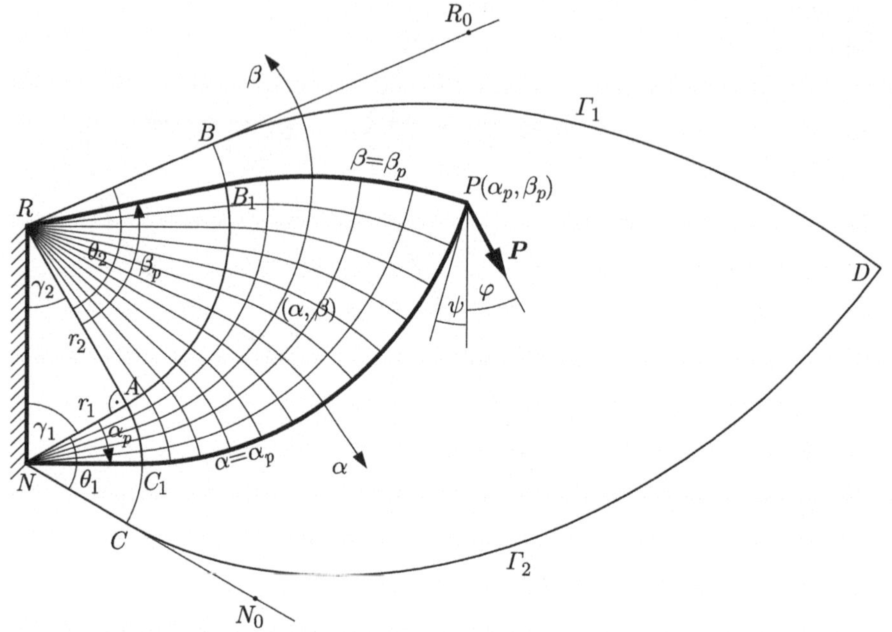

Fig. 4.54 An illustrative optimum design of the class of problems discussed in the present section. Case of $\kappa = 3$

the Lamé coefficients along the contours of the circular fans

$$A(\alpha, 0) = r_1, \qquad B(0, \beta) = r_2 \tag{4.396}$$

We postulate that $\phi = \beta - \alpha$ and that the fields A, B are linked by (4.93). Thus,

$$\frac{\partial A}{\partial \beta}(0, \beta) = B(0, \beta) = r_2 \tag{4.397}$$

The field $A(\alpha, \beta)$ satisfies (4.131) with $f = 0$; hence, the value $A(\lambda, \mu)$ for $0 \le \lambda \le \theta_1$, $0 \le \mu \le \theta_2$ is expressed by (4.132)

$$A(\lambda, \mu) = A(0, 0)G_0(\lambda, \mu) + \int_0^{\lambda} G_0(\lambda - \alpha, \mu)\frac{\partial A}{\partial \alpha}(\alpha, 0)d\alpha + \int_0^{\mu} G_0(\lambda, \mu - \beta)\frac{\partial A}{\partial \beta}(0, \beta)d\beta \tag{4.398}$$

Substitution of (4.396), (4.397) gives

$$A(\lambda, \mu) = r_1 G_0(\lambda, \mu) + r_2 \int_0^{\mu} G_0(\lambda, \mu - \beta)d\beta \tag{4.399}$$

According to (4.127), (4.126), we have

$$G_0(\lambda, \mu - \beta) = G_0(\mu - \beta, \lambda) = -\frac{\partial}{\partial \beta} G_1(\mu - \beta, \lambda) \tag{4.400}$$

Substitution of this equation into (4.399) gives

$$A(\lambda, \mu) = r_1 G_0(\lambda, \mu) + r_2 G_1(\mu, \lambda) \tag{4.401}$$

Similar arguments lead to

$$B(\lambda, \mu) = r_2 G_0(\lambda, \mu) + r_1 G_1(\lambda, \mu) \tag{4.402}$$

Using (4.126)–(4.130), it is easy to check that the conditions (4.396) and the equations (4.93) are fulfilled (Fig. 4.55).

Let us note that if $\sigma_T = \sigma_C$, then $\kappa = 1$, $r_1 = r_2$ and then the net is symmetric with respect to the axis $x = y$ and $A(\alpha, \beta) = B(\beta, \alpha)$.

We shall find equations which determine the form of the parametric lines (α, β). The direct substitution of (4.401), (4.382) into (4.99) is not helpful; this leads to the integral expressions that cannot be integrated analytically. It turns out to be helpful to make use of Eqs. (4.100) which lead to integrable expressions. The starting point are the boundary conditions

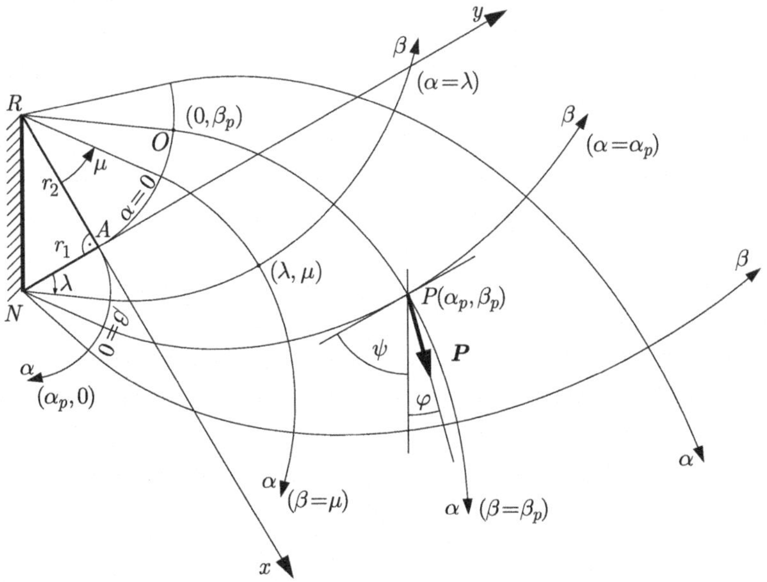

Fig. 4.55 The optimum design supported on two circles: geometry of the net

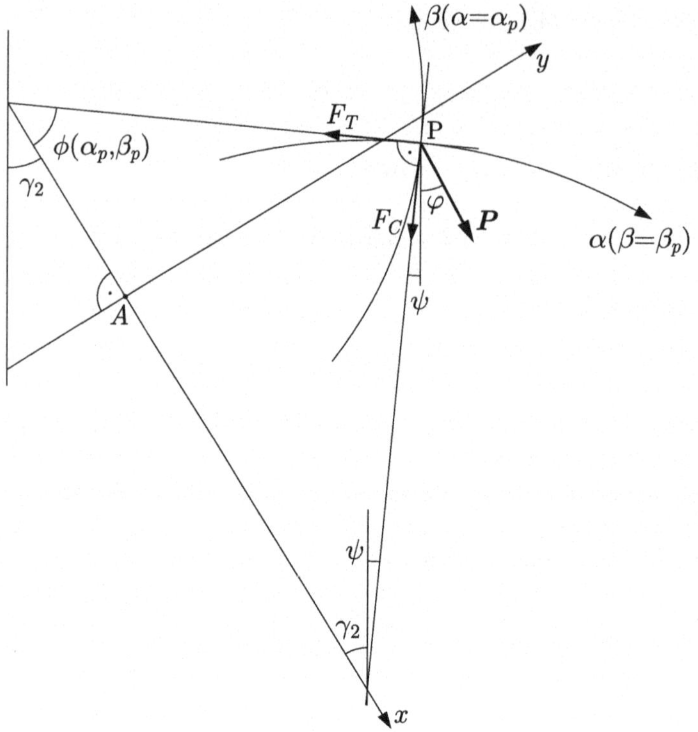

Fig. 4.56 Equilibrium of the node *P*

$$x(0, \mu) = -r_2 + r_2 \cos \mu, \qquad y(0, \mu) = r_2 \sin \mu \tag{4.403}$$

For the coordinates of points on the circle $\alpha = 0$, see Fig. 4.56. Let us substitute these into (4.101) and put $\theta = 0$; one obtains

$$\begin{aligned} \tilde{x}(\lambda, \mu) &= r_2[-\cos(\lambda - \mu) + \cos \lambda] \\ \tilde{y}(\lambda, \mu) &= r_2[\sin(\mu - \lambda) + \sin \lambda] \end{aligned} \tag{4.404}$$

Upon substituting (4.401), (4.404) into (4.100), we arrive at

$$\begin{aligned} \bar{x}(\lambda, \mu) &= \tilde{x}(\lambda, \mu) + r_1 f_1(\lambda, \mu) + r_2 f_2(\lambda, \mu) \\ \bar{y}(\lambda, \mu) &= \tilde{y}(\lambda, \mu) + r_1 g_1(\lambda, \mu) + r_2 g_2(\lambda, \mu) \end{aligned} \tag{4.405}$$

where

$$
\begin{aligned}
f_1(\lambda, \mu) &= \int_0^{\lambda} \cos(\lambda - \alpha) G_0(\alpha, \mu) d\alpha \\
f_2(\lambda, \mu) &= \int_0^{\lambda} \cos(\lambda - \alpha) G_1(\mu, \alpha) d\alpha \\
g_1(\lambda, \mu) &= \int_0^{\lambda} \sin(\lambda - \alpha) G_0(\alpha, \mu) d\alpha \\
g_2(\lambda, \mu) &= \int_0^{\lambda} \sin(\lambda - \alpha) G_1(\mu, \alpha) d\alpha
\end{aligned}
\tag{4.406}
$$

According to (4.137), (4.138), we get

$$
f_1(\lambda, \mu) = F_1(\lambda, \mu), \qquad g_1(\lambda, \mu) = F_2(\lambda, \mu) \tag{4.407}
$$

Moreover, one can derive

$$
\begin{aligned}
f_2(\lambda, \mu) &= \cos(\mu - \lambda) - \cos\lambda + F_2(\mu, \lambda) \\
g_2(\lambda, \mu) &= -\sin\lambda - \sin(\mu - \lambda) + F_1(\mu, \lambda)
\end{aligned}
\tag{4.408}
$$

Substitution of (4.404), (4.406)–(4.408) into (4.405) leads to

$$
\begin{aligned}
\bar{x}(\lambda, \mu) &= r_1 F_1(\lambda, \mu) + r_2 F_2(\mu, \lambda) \\
\bar{y}(\lambda, \mu) &= r_1 F_2(\lambda, \mu) + r_2 F_1(\mu, \lambda)
\end{aligned}
\tag{4.409}
$$

The coordinates x, y are given by (4.96), hence

$$
\begin{aligned}
x(\lambda, \mu) &= \bar{x}(\lambda, \mu)\cos(\mu - \lambda) - \bar{y}(\lambda, \mu)\sin(\mu - \lambda) \\
y(\lambda, \mu) &= \bar{x}(\lambda, \mu)\sin(\mu - \lambda) + \bar{y}(\lambda, \mu)\cos(\mu - \lambda)
\end{aligned}
\tag{4.410}
$$

The fields thus constructed satisfy the conditions (4.403) along the line $\lambda = 0$ and satisfy

$$
x(\lambda, 0) = r_1 \sin\lambda, \qquad y(\lambda, 0) = -r_1 + r_1\cos\lambda \tag{4.411}
$$

along the line $\mu = 0$. Moreover, the functions (4.409) satisfy (4.97). One can prove that the fields $\bar{x}$, $\bar{y}$, A, B are interrelated by the conditions (4.98).

It turns out that the nets given by (4.410) apply to problems corresponding to a rich class of point loads. The optimal structures are cut out from the domain RDN by the lines: $\alpha = \alpha_P$ and $\beta = \beta_P$, where (α_P, β_P) are coordinates of the point of application of the point load, see Fig. 4.57.

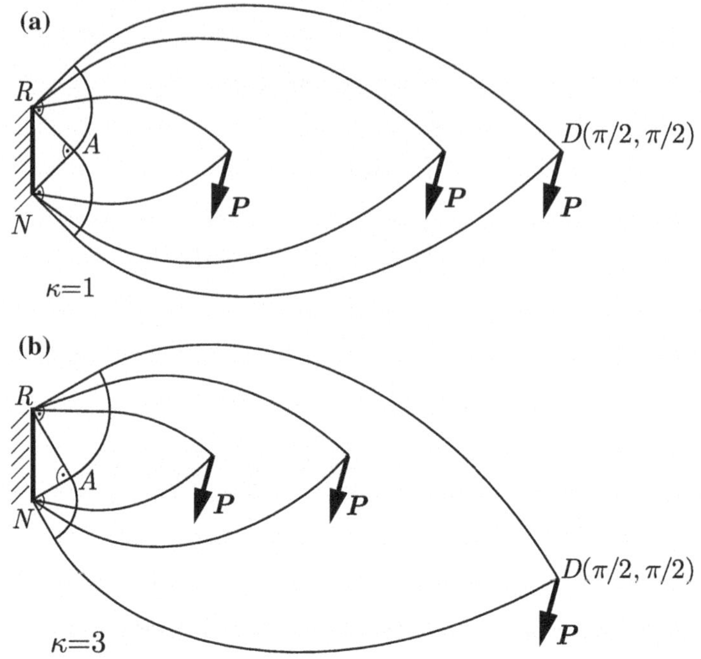

Fig. 4.57 The largest optimal solutions RDN for the problem of Fig. 4.54: **a** case of $\kappa = 1$; **b** $\kappa = 3$

The numerical experiments teach us that the angles θ_1, θ_2 cannot exceed $\frac{\pi}{2}$. Thus, if the point P lies outside the domain RDN in Fig. 4.57, the solutions will assume new shapes, see Fig. 4.58.

At this stage, we do not know whether the Hencky nets constructed above are Michell's nets as well. In other words, we do not know whether there exist the associated virtual fields u, v corresponding to the regime $\mathbb{T}$, where $\varepsilon_I = 1, \varepsilon_{II} = -\kappa$ along the α, β lines.

4.8.2.2 Construction of the Virtual Displacement Fields (u, v) Within the Domain $ABDC$

We shall find the fields (u, v) within $ABDC$ satisfying the optimality conditions (4.116). These fields are continuous extension of the fields (u, v) defined previously within the fans RBA, NAC and within the triangular domain RAN.

Along the line AB the fields (u, v) are given by

$$u(0, \beta) = r_2, \qquad v(0, \beta) = -\kappa r_1 - (1 + \kappa) r_2 \beta \tag{4.412}$$

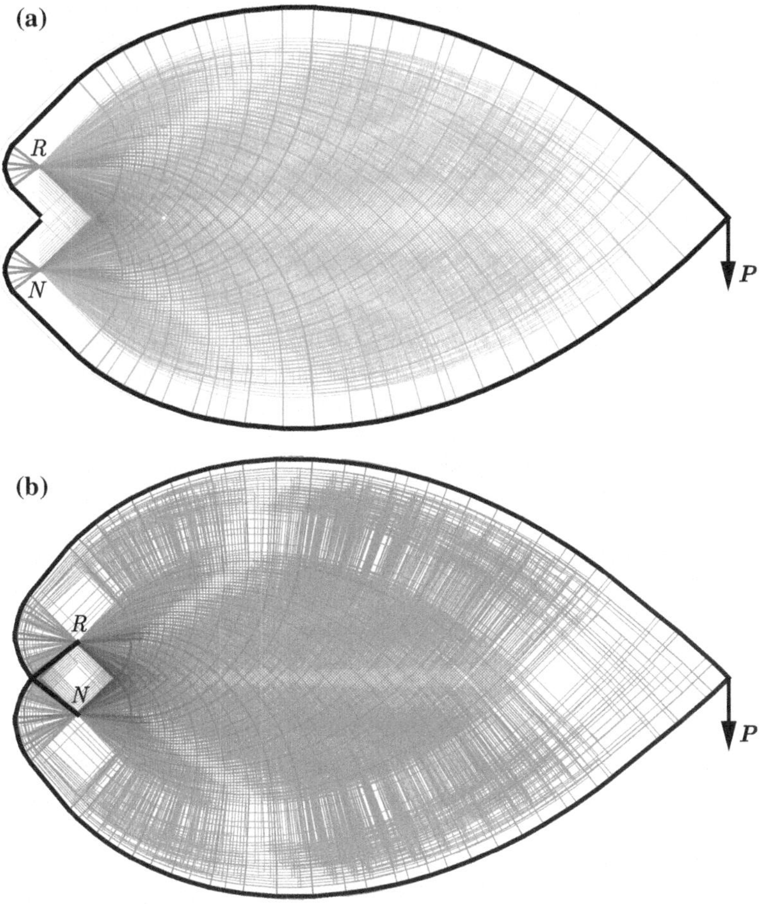

Fig. 4.58 The numerical solutions corresponding to the point load being applied outside the domain RDN. The force of magnitude P is parallel to the line RN. Case **a**: the distance of the force to the supporting line equals $6.6a$, $a = |RN|$. The optimal volume equals $V = 40.2986761\,Pa/\sigma_0$. Case **b**: the distance of the force to the supporting line equals $9a$. The optimal volume equals $V = 60.6259859\,Pa/\sigma_0$

while the Lamé coefficients are

$$A(0, \beta) = r_1\beta, \qquad B(0, \beta) = r_2 \tag{4.413}$$

The boundary values of the fields u^0, v^0, see (4.118a), are constant and equal

$$u^0(0, \beta) = r_2, \qquad v^0(0, \beta) = -\kappa r_1 \tag{4.414}$$

Similarly, along the line CA the fields u, v are given by

$$u(\alpha, 0) = (1 + \kappa)\alpha r_1 + r_2, \qquad v(\alpha, 0) = -\kappa r_1 \tag{4.415}$$

while the Lamé coefficients on CA are

$$A(\alpha, 0) = r_1, \qquad B(\alpha, 0) = r_2 + r_1\alpha \tag{4.416}$$

The boundary values of the fields u^0, v^0 are constant

$$u^0(\alpha, 0) = r_2, \qquad v^0(\alpha, 0) = -\kappa r_1 \tag{4.417}$$

The fields u^0, v^0 satisfy (4.119); hence, their values at an arbitrary point λ, μ within the domain $ABDC$ are given by the Riemann formula (4.132). Substitution of (4.414), (4.417) gives

$$\begin{aligned} u^0(\lambda, \mu) &= u^0(0, 0)G_0(\lambda, \mu) = r_2 G_0(\lambda, \mu) \\ v^0(\lambda, \mu) &= v^0(0, 0)G_0(\lambda, \mu) = -\kappa r_1 G_0(\lambda, \mu) \end{aligned} \tag{4.418}$$

We insert now (4.401), (4.402), (4.418) into (4.118a) to find

$$\begin{aligned} u(\lambda, \mu) &= [r_2 + (1 + \kappa)r_1\lambda]G_0(\lambda, \mu) + (1 + \kappa)r_2\lambda G_1(\mu, \lambda) \\ v(\lambda, \mu) &= -[\kappa r_1 + (1 + \kappa)r_2\mu]G_0(\lambda, \mu) - (1 + \kappa)r_1\mu G_1(\lambda, \mu) \end{aligned} \tag{4.419}$$

One can check that these fields satisfy (4.116) identically and they are continuous extensions of the fields u, v found previously. The same formulae can be found using the integral formulae (4.107), (4.108) or (4.110), (4.111) by taking the field ω in the form

$$\omega(\alpha, \beta) = -\sqrt{\kappa} - (1 + \kappa)(\alpha + \beta) \tag{4.420}$$

Since the fields u, v satisfy all the conditions of the regime $\mathbb{T}$, the Hencky net α, β is also a Michell net. This net forms directions of the fibres within the domain $B_1PC_1AB_1$ cut out along the lines α, β of the domain $BDCAB$, see Fig. 4.54. The boundaries of this domain are determined by the position of point P of coordinates (α_P, β_P); that is why these lines are parametric lines: $(\alpha = \alpha_P)$ and $(\beta = \beta_P)$.

4.8.2.3 The Force Fields T_1, T_2 Within the Domains $B_1PC_1AB_1$

Assume that the force P is directed like in Figs. 4.54, 4.56: it forms an angle φ with respect to the vertical line set by RN. The conditions of equilibrium of the node P determine the values of the forces: within the bar B_1P an axial force $F_T(\alpha) > 0$ appears and within the bar C_1P there appears the axial force $F_C(\beta) < 0$. Their values are constant and equal

$$
\begin{aligned}
F_T(\alpha) &= P_1, \qquad P_1 = P\sin(\psi+\varphi)\\
F_C(\beta) &= P_2, \qquad P_2 = -P\cos(\psi+\varphi)
\end{aligned}
\tag{4.421}
$$

where ψ is an angle between the tangent to the line ($\alpha = \alpha_P$) and the vertical direction; from Fig. 4.56, one can read off that $\gamma_2 + \psi = \frac{\pi}{2} - \phi$, where $\phi = \beta_P - \alpha_P$. Thus

$$
\psi = \gamma_1 + \alpha_P - \beta_P \tag{4.422}
$$

The conditions $F_T(\alpha) > 0$, $F_C(\beta) < 0$ are fulfilled, if

$$
0 \le \gamma_1 + \alpha_P - \beta_P + \varphi \le \frac{\pi}{2} \tag{4.423}
$$

In the fibrous domainAB_1PC_1A, there appear the force fields T_1, T_2. They satisfy the homogeneous equations (4.176) and the boundary conditions

$$
T_1(\alpha_P, \beta) + F_C(\beta) = 0, \qquad T_2(\alpha, \beta_P) + F_T(\alpha) = 0 \tag{4.424}
$$

The problem of computing the forces T_1, T_2 have the same mathematical structure as the problem of computing the forces in the cantilever supported on a circle (Sect. 4.6.4), although in the latter problem $\kappa = 1$. However, the parameter κ does not appear directly in the problem of constructing the forces T_1, T_2. Thus, the solution to the problem considered is given by (4.351), (4.352), where now P_1, P_2 are given by Eqs. (4.421).

4.8.2.4 Computing the Forces T_1, T_2 Within the Circular Fans

Having the fields T_1, T_2 in the domain AB_1PC_1A, one can find these fields in the domains of the fans: RB_1A, NAC_1.

The Fan NAC_1

The parameterization in the fan is assumed as in Fig. 4.49. In Sect. 4.4, we have noted that the force T transverse to the fibres is zero and the force T in the radial direction is constant along the radius. Thus

$$
T_1^{NAC_1}(\alpha, \beta_1) = 0, \qquad T_2^{NAC_1}(\alpha, \beta_1) = T_2^{AB_1PC_1A}(\alpha, 0) \tag{4.425}
$$

or, in the domain NAC_1, we have

$$
T_1 = 0, \qquad T_2 = -P_1G_0(\alpha_P - \alpha, \beta_P) + P_2G_1(\beta_P, \alpha_P - \alpha) \tag{4.426}
$$

The Fan RAB_1

In a similar way we find

$$T_1(\alpha_1, \beta) = -P_2 G_0(\alpha_P, \beta_P - \beta) + P_1 G_1(\alpha_P, \beta_P - \beta), \qquad T_2(\alpha_1, \beta) = 0 \tag{4.427}$$

Note that the force T_1 suffers a jump along the line AC_1 and the force T_2 suffers a jump along AB_1.

4.8.2.5 Computing the Volume of the Structural Segments and the Total Volume

According to (4.169), the volume fraction v of fibres in the domain AB_1PC_1A is expressed by the fields T_1, T_2, A, B found above. The volume of the fibres in this domain is equal to the integral of v in the limits: $0 \le \alpha \le \alpha_P$, $0 \le \beta \le \beta_P$. This integral can be computed analytically. The final result reads

$$\begin{aligned} V^{AB_1PC_1} &= \\ &\frac{P_1}{\sigma_T}\{\beta_P(1+\kappa)[r_1 G_2(\alpha_P, \beta_P) + r_2 G_2(\alpha_P, \beta_P)] + r_2[1 - G_0(\alpha_P, \beta_P)]\} - \\ &-\frac{P_2}{\sigma_T}\{\alpha_P(1+\kappa)[r_2 G_2(\beta_P, \alpha_P) + r_1 G_1(\beta_P, \alpha_P)] + r_1\kappa[1 - G_0(\alpha_P, \beta_P)]\} \end{aligned} \tag{4.428}$$

The volume of the fibres in the domain NAC_1 is expressed by

$$V^{NAC_1} = \int_0^{\alpha_P}\int_0^1 \left(-\frac{T_2}{\sigma_C A}\right) AB d\alpha d\beta_1 \tag{4.429}$$

where T_2 is given by (4.426), A, B being given by (4.381). Upon integration one finds

$$V^{NAC_1} = \frac{r_1}{\sigma_T}\{P_1\kappa G_1(\alpha_P, \beta_P) + P_2\kappa[1 - G_0(\alpha_P, \beta_P)]\} \tag{4.430}$$

The volume of the fibres in the domain RB_1A is expressed by

$$V^{RB_1A} = \int_0^{\beta_P}\int_0^1 \frac{T_1}{\sigma_T B} AB d\alpha_1 d\beta \tag{4.431}$$

where T_1 is given by (4.427); A, B are given by (4.382). Upon integration we have

$$V^{RB_1A} = \frac{r_2}{\sigma_T}\{P_1[-1 + G_0(\alpha_P, \beta_P)] - P_2 G_1(\beta_P, \alpha_P)\} \tag{4.432}$$

The volume of the bar in tension RB_1P equals

$$V^{RB_1P} = \frac{P_1}{\sigma_T}\left(r_2 + \int_0^{\alpha_P} A(\alpha, \beta_P)d\alpha\right) \tag{4.433}$$

where A is given by (4.401). Integration gives

$$V^{RB_1P} = \frac{P_1}{\sigma_T}[r_1 G_1(\alpha_P, \beta_P) + r_2 G_0(\alpha_P, \beta_P)] \tag{4.434}$$

The volume of the bar in compression NC_1P is expressed by

$$V^{NC_1P} = -\frac{P_2}{\sigma_C}\left[r_1 + \int_0^{\beta_P} B(\alpha_P, \beta)d\beta\right] \tag{4.435}$$

where B is given by (4.402). Integration leads to

$$V^{NC_1P} = -\frac{\kappa P_2}{\sigma_C}[r_2 G_1(\beta_P, \alpha_P) + r_1 G_0(\alpha_P, \beta_P)] \tag{4.436}$$

By summing up these volumes, we compute the total volume of the structure RB_1PC_1N

$$V_{\min} = \frac{P_1}{\sigma_T}a_1 + \frac{P_2}{\sigma_C}a_2 \tag{4.437}$$

where

$$\begin{aligned} a_1 &= [r_2 + (1+\kappa)r_1\alpha_P]G_0(\alpha_P, \beta_P) + \alpha_P(1+\kappa)r_2 G_1(\beta_P, \alpha_P) \\ a_2 &= -[\kappa r_1 + (1+\kappa)\beta_P r_2]G_0(\alpha_P, \beta_P) - (1+\kappa)\beta_P r_1 G_1(\beta_P, \alpha_P) \end{aligned} \tag{4.438}$$

Let us note that

$$a_1 = u(\alpha_P, \beta_P), \qquad a_2 = v(\alpha_P, \beta_P) \tag{4.439}$$

where u, v are given by (4.419). Thus, let us note that according (3.81), we have $V_{\min} = \dfrac{1}{\sigma_0}\tilde{\mathcal{Z}}_1$, where $\sigma_0 = \sigma_T$ and

$$\tilde{\mathcal{Z}}_1 = P[\sin(\psi + \varphi)u(\alpha_P, \beta_P) - \cos(\psi + \varphi)v(\alpha_P, \beta_P)] \tag{4.440a}$$

Hence, $\tilde{\mathcal{Z}}_1$ is the value of the virtual work done by the force P on the virtual displacement of the point P. The results (4.437) is compatible with the formula $V_{\min} = \dfrac{1}{\sigma_0}\tilde{\mathcal{Z}}_2$, $\sigma_0 = \sigma_T$, the quantity $\tilde{\mathcal{Z}}_2$ is given by (4.187). Thus, we have proved that the solution found above delivers:

- the maximizer of the problem (3.85) or the field (u, v) satisfying the desired kinematic conditions;

- the minimizer of problem (3.95) or the fields T_1, T_2, F_T, F_C satisfying the desired static conditions.

From the mathematical point of view, we have proved that the duality gap between the problems (3.85) and (3.95) vanishes.

The optimum design is characterized by the Michell net, the fields of internal forces and the volume fractions of the fibres within each subdomain. The most important component of the solution is the shape of the boundary of the structure, which is defined as a sharp contour cutting out the solution from the design domain.

4.8.2.6 Case Studies

a. Symmetric Cantilevers. A Study of the Layouts and the Volume

Let us examine the family of solutions of problems in Fig. 4.57 for the case of $\kappa = 1$ and for the case of the force P being vertical and applied at a point lying between A and D. The volume of the optimal structure equals $V = \mathcal{L}/\sigma_0$, where $\mathcal{L} = P u_P$, u_P being

$$u_P = \frac{\sqrt{2}}{2}[u(P) - v(P)] = \sqrt{2}u(P) \tag{4.440b}$$

since along the line $\alpha = \beta$ we have $v(P) = -u(P)$. If $\kappa = 1$, then $r_1 = r_2 = r_0$, $r_0 = \sqrt{2}a/2$, $a = |RN|$. If the coordinates of point P are (α, α), then $u(P) = u(\alpha, \alpha)$, where

$$u(\alpha, \alpha) = r_0\left[(1 + 2\alpha)G_0(\alpha, \alpha) + 2\alpha G_1(\alpha, \alpha)\right] \tag{4.441}$$

Thus, the volume equals $V = \bar{V}(\alpha)V_0$, $V_0 = Pa/\sigma_0$ where

$$\bar{V}(\alpha) = (1 + 2\alpha)G_0(\alpha, \alpha) + 2\alpha G_1(\alpha, \alpha) \tag{4.442}$$

The distance from point P of coordinates (α, α) to the point A equals

$$|AP| = a[F_1(\alpha, \alpha) + F_2(\alpha, \alpha)] \tag{4.443}$$

with $|RN| = a$. The volume increases if the distance PA_0 increases, see Fig. 4.59, where $\bar{V} = V/V_0$.

We shall perform the complete analysis of the solution for the data: $\alpha_P = \beta_P = \frac{7}{20}\pi$, $\sigma_C = \sigma_T = \sigma_0$; the force is vertical, hence $\varphi = 0$, see Fig. 4.60.

Denote $|RA| = |AN| = r_0$, $|RN| = a = r_0\sqrt{2}$. Let $V_0 = \dfrac{Pr_0}{\sigma_0}$. The coordinates (x_P, y_P) of point P (in the coordinate system x, y) are given by (4.410). We find $x_P/r_0 = y_P/r_0 = 2.444960002$.

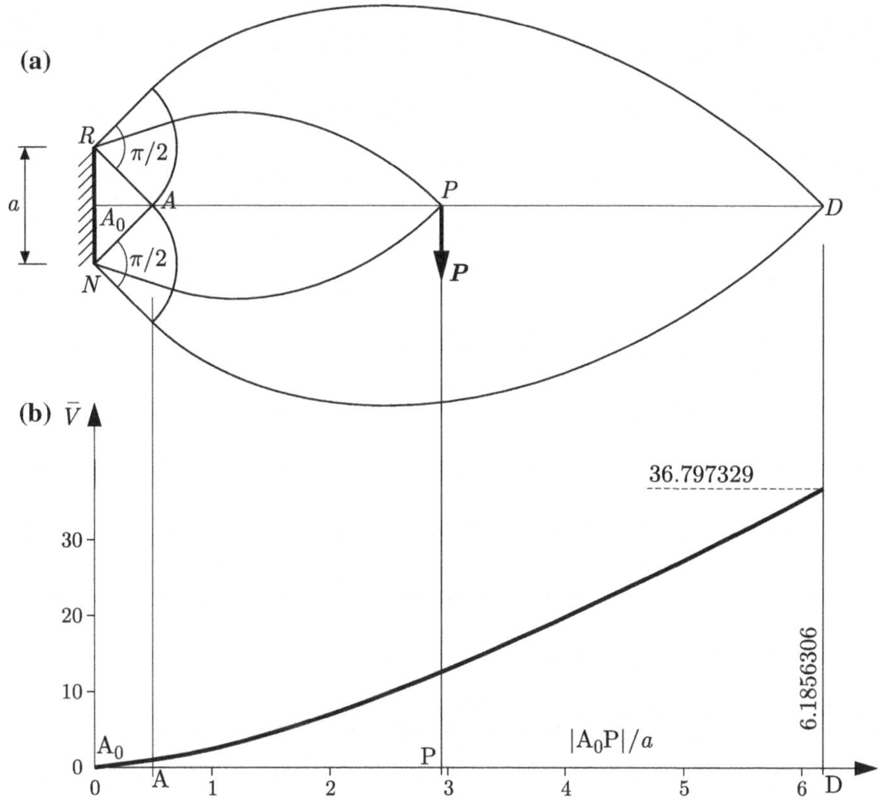

Fig. 4.59 An illustrative problem for the case of $\kappa = 1$ (**a**); the plot of the volumes (**b**)

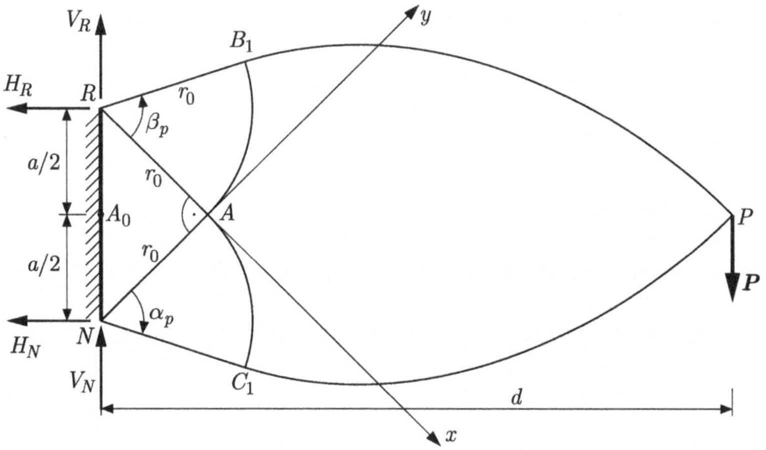

Fig. 4.60 An illustrative example. Case of $\alpha_P = \beta_P = \dfrac{7}{20}\pi$ and $\kappa = 1$

The distance of point P to the line RN is $d = 4.164802374 r_0$. The forces in the ribs are as follows:

the bar RB_1P: $F_T = P_1$, $P_1 = \frac{\sqrt{2}}{2}P$

the bar NC_1P: $F_C = P_2$, $P_2 = -\frac{\sqrt{2}}{2}P$.

The reactions at point R are expressed by

$$\begin{aligned} H_R &= P_1 \sin\left(\tfrac{\pi}{4} + \beta_P\right) + \int_0^{\beta_P} T_1^{RAB_1}(\alpha_1, \beta) \sin\left(\tfrac{\pi}{4} + \beta\right) d\beta \\ V_R &= P_1 \cos\left(\tfrac{\pi}{4} + \beta_P\right) + \int_0^{\beta_P} T_1^{RAB_1}(\alpha_1, \beta) \cos\left(\tfrac{\pi}{4} + \beta\right) d\beta \end{aligned} \tag{4.444}$$

where $T_1^{RAB_1}$ is given by (4.427). Hence, we compute $H_R = 2.944960P$, $V_R = 0.49999P$. Similarly, one can compute the reactions at N

$$\begin{aligned} H_N &= P_2 \sin\left(\tfrac{\pi}{4} + \alpha_P\right) + \int_0^{\alpha_P} T_2^{NAC_1}(\alpha, \beta_1) \sin\left(\tfrac{\pi}{4} + \alpha\right) d\alpha \\ V_N &= -P_2 \cos\left(\tfrac{\pi}{4} + \alpha_P\right) - \int_0^{\alpha_P} T_2^{NAC_1}(\alpha, \beta_1) \cos\left(\tfrac{\pi}{4} + \alpha\right) d\alpha \end{aligned} \tag{4.445}$$

where $T_2^{NAC_1}$ is given by (4.426). We find $H_N = -2.944960P$, $V_N = 0.499999P$ The reactions found above satisfy the equilibrium equations; in particular, the equation $H_R a - Pd = 0$ holds. Using skew-symmetry, one can find the reactions from the conditions of the global equilibrium. In general, if $\alpha_P \neq \beta_P$ the analytical method of computing the reactions used here is the only one which is effective.

Using the formulae of Sect. 4.8.2.5, one can compute the volumes of subsequent segments of the structure.

The volumes of the fans

$V^{NAC_1}/V_0 = 2.503149$, $V^{RB_1A}/V_0 = 2.503149$

The volume of the domain AB_1PC_1

$V^{AB_1PC_1}/V_0 = 6.408407$

The volume of the rib in tension: $V^{RB_1P}/V_0 = 3.210255$

The volume of the rib in compression NC_1P is the same. The volume of the whole structure is $V/V_0 = 17.83521753$.

Let us compute now this volume by the kinematic method. According to (4.442), we obtain the same result.

The field T_1 suffers a jump along AC_1 and the field T_2 has a jump along AB_1, see Fig. 4.61. Both these lines are the lines of discontinuity of the volume fraction υ given by (4.169), where $\sigma_T = \sigma_C = \sigma_0$ and the fields A, B are given by (4.401), (4.402) in the domain AB_1PC_1, and—by (4.381), (4.382)—in the fans. The volume fraction υ is infinite at points R, N, see Fig. 4.62.

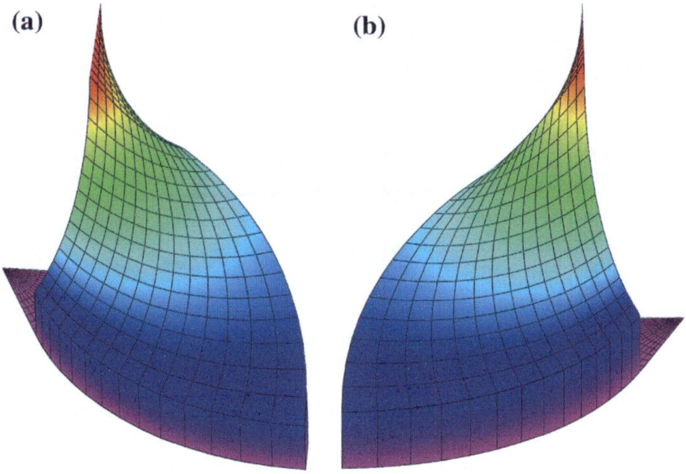

Fig. 4.61 The plot of Hemp's forces T_1 and T_2

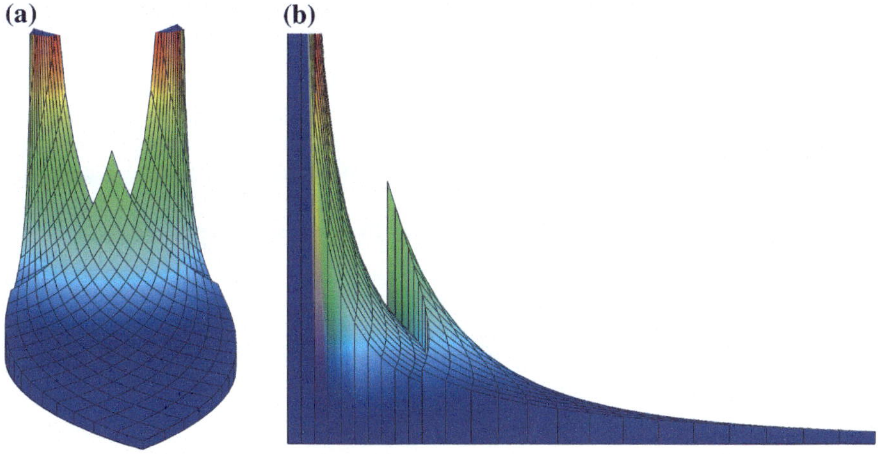

Fig. 4.62 The plot of the volume fraction v

b. Selected Solutions in the Case of $\sigma_T = 3\sigma_C$

Let us examine a family of problems corresponding to the case of $\kappa = 3$, $\kappa = \dfrac{\sigma_T}{\sigma_C}$. Using the formulae of the previous section, we find the optimal shapes and volumes of the Michell structures for subsequent values of the distance of the point P to the line RN. The volume is an increasing function of the distance of point P to RN, see Fig. 4.63b.

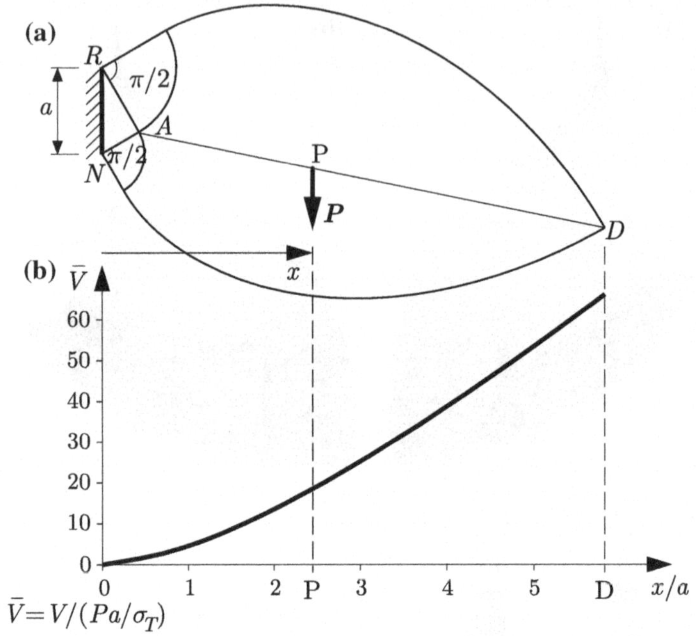

Fig. 4.63 Case of $\kappa = 3$. The largest admissible domain for points P (**a**); the plot of the volumes corresponding to the position P of the force between A and D (**b**)

4.9 Cantilevers with a Three-Sided Polygonal Support. Case of $\sigma_T = \sigma_C$

4.9.1 Construction of the Michell Net and the Adjoint Displacement Fields. Symmetric Case

Consider the Michell problem of forming the least volume structure transmitting a point load to a support in the shape of a straight segment RN and two rays RH_3, NH_3'. The permissible stresses in tension and compression are equal: $\sigma_T = \sigma_C$. Assume that the angles $\angle H_3RN$ and $\angle H_3'NR$ are the same and equal γ, see Fig. 4.64. Let $|RN| = a$ and $r_0 = a\sqrt{2}/2$.

If the point P lies within the triangle RAN, then the solution is a two-bar truss, discussed in Sect. 4.5.2, see also Sect. 4.8.1. If the point P lies within the sector A_1RA of a circle and sufficiently close to the line RA, then the solution will be of the form of Fig. 4.51. Similarly, if the point P lies within the sector ANA_2, sufficiently close to the line AN, then the solution will have the form shown in Fig. 4.50. If, however, the angle $\angle PRA$ (or $\angle ANP$) will be sufficiently big, then the optimum structure will make use of the support along the ray RH_3 (or the ray NH_3'). That is why the solutions of Sect. 4.8 have a limited application here.

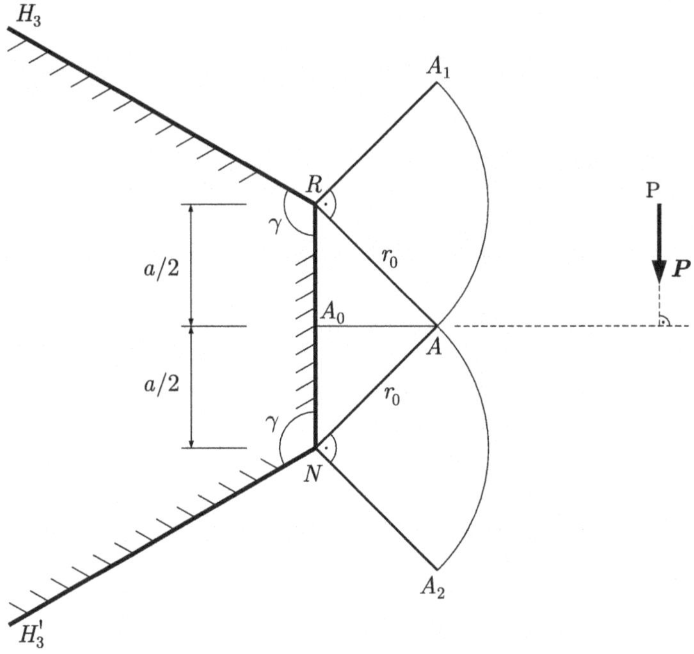

Fig. 4.64 Problem setting: transmit the force P to the boundary H_3RNH_3'

The Hencky net will be shown to be the one in Fig. 4.65. To find the optimal structures, we shall construct the fields u, v vanishing on the support H_3RNH_3' and satisfying the optimality conditions (4.116) (here $\kappa = 1$); thus we predict that the solutions will correspond to the regimes $\mathbb{T}$ or $\mathbb{R}$; the fibres along the α lines will be in tension and the fibres along the β lines will be in compression.

We shall prove that $\theta_1 = \theta$, which results in $\theta = \dfrac{1}{2}(\pi - \gamma)$ and which fixes geometry of the whole Hencky net. We use the parameterization as in Fig. 4.49 but now $\kappa = 1$, $r_1 = r_2 = r_0$, $\gamma_1 = \gamma_2 = \pi/4$.

The field (u, v) is known in RAN and is given by (see (4.384) for $\kappa = 1$)

$$u = x + y + r_0, \qquad v = -u \tag{4.446}$$

where $r_0 = |AR|$; the displacements u, v are directed along the axes x, y of the system of origin at A. Thus, $\omega = -1$ within RAN. Previously, in Fig. 4.49 we have introduced the polar system $(r = \alpha_1 r_2, \beta)$. Now, we introduce the polar system (r, β); the displacements u, v are

$$u = r, \qquad v = -2r\beta - r \tag{4.447}$$

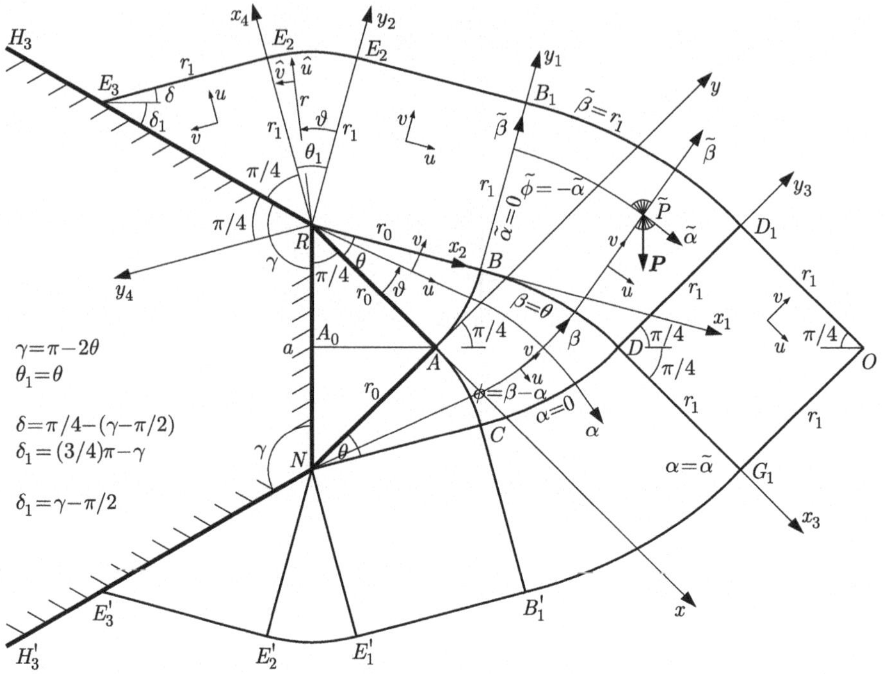

Fig. 4.65 The Hencky net. The optimality conditions imply $\theta_1 = \theta$

see (4.385). The field (4.447) is a continuous extension of the field (4.446), since along AR we have $y = 0$, $\beta = 0$, $x = r - r_0$. Therefore, along RB

$$u^{RB} = r, \qquad v^{RB} = -2r\theta - r \tag{4.448}$$

or

$$u^{RB} = x_2, \qquad v^{RB} = -(1 + 2\theta)x_2 \tag{4.449}$$

where the axis x_2 has its origin at R.

The Domain ABDC

We apply the formulae of Sect. 4.8, e.g. from Figs. 4.54 and 4.55; now we have $\theta_1 = \theta_2 = \theta$. According to (4.401), (4.402), we have

$$\frac{A(\alpha, \beta)}{r_0} = G_0(\alpha, \beta) + G_1(\beta, \alpha) \tag{4.450}$$
$$\frac{B(\alpha, \beta)}{r_0} = G_0(\alpha, \beta) + G_1(\alpha, \beta)$$

The angle of a rigid rotation equals

$$\omega(\alpha, \beta) = -1 - 2(\alpha + \beta) \tag{4.451}$$

The displacements are, see (4.419),

$$\begin{aligned} \frac{u(\alpha, \beta)}{r_0} &= (1 + 2\alpha)G_0(\alpha, \beta) + 2\alpha G_1(\beta, \alpha) \\ \frac{v(\alpha, \beta)}{r_0} &= -(1 + 2\beta)G_0(\alpha, \beta) - 2\beta G_1(\alpha, \beta) \end{aligned} \tag{4.452}$$

The Domain BDD_1B_1

This domain corresponds to the regimes $\mathbb{R}_1$, $\mathbb{R}_2$, see (4.60) and Sect. 4.4. We introduce the parameterization $(\tilde{\alpha}, \tilde{\beta})$ within the domain parameterized by the Cartesian coordinate system (x_1, y_1) of origin at B. The lines $\tilde{\beta}$ are straight; they are extensions of the lines β arising along the line BD of the domain $ABDC$. The coordinate $\tilde{\beta}$ measures the distance to the line BD. Along the line B_1D_1 we have $\tilde{\beta} = r_1$. Along BD there is $\tilde{\alpha} = \alpha$, where α is the coordinate determined within the domain $ABDC$. The lines $\tilde{\alpha}$ are orthogonal to $\tilde{\beta}$. The angle ϕ of inclination of tangents to the lines $\tilde{\alpha}$ to the axis x_1 does not depend on $\tilde{\beta}$: $\tilde{\phi} = -\tilde{\alpha}$. According to Sect. 4.4, the Lamé fields read

$$\tilde{A}(\tilde{\alpha}, \tilde{\beta}) = \tilde{\beta} + f(\tilde{\alpha}), \qquad \tilde{B} = 1 \tag{4.453}$$

The field $\tilde{A}$ is a continuous extension of the field A along BD: $\tilde{A}(\tilde{\alpha}, 0) = A(\tilde{\alpha}, \theta)$, where A is given by (4.450). Thus

$$\tilde{A}(\tilde{\alpha}, \tilde{\beta}) = \tilde{\beta} + r_0[G_0(\tilde{\alpha}, \theta) + G_1(\theta, \tilde{\alpha})], \qquad \tilde{B} = 1 \tag{4.454}$$

along B_1D_1, where $\tilde{\beta} = r_1$, we have

$$\tilde{A}^{B_1D_1} = r_1 + r_0[G_0(\tilde{\alpha}, \theta) + G_1(\theta, \tilde{\alpha})] \tag{4.455}$$

Similarly as in (1.36), the rigid rotation is given by

$$\tilde{\omega} = -1 - 2(\theta + \tilde{\alpha}) \tag{4.456}$$

Proceeding like in Sect. 4.4 we integrate the optimality conditions

$$\frac{\partial \tilde{v}}{\partial \tilde{\beta}} = -1, \qquad \frac{\partial \tilde{u}}{\partial \tilde{\beta}} = -\tilde{\omega} \tag{4.457}$$

The integration of these equations with the use of continuity conditions along BD gives

$$\begin{aligned} \tilde{u}(\tilde{\alpha}, \tilde{\beta}) &= (1 + 2\theta + 2\tilde{\alpha})\tilde{\beta} + r_0\left[(1 + 2\tilde{\alpha})G_0(\tilde{\alpha}, \theta) + 2\tilde{\alpha}G_1(\theta, \tilde{\alpha})\right] \\ \tilde{v}(\tilde{\alpha}, \tilde{\beta}) &= -\tilde{\beta} - r_0\left[(1 + 2\theta)G_0(\tilde{\alpha}, \theta) + 2\theta G_1(\tilde{\alpha}, \theta)\right] \end{aligned} \tag{4.458}$$

Thus, along the line BB_1 we have

$$\tilde{u}^{BB_1} = (1+2\theta)\tilde{\beta} + r_0, \qquad \tilde{v}^{BB_1} = -\tilde{\beta} - r_0(1+2\theta) \tag{4.459}$$

and, on the line DD_1 we find

$$\begin{aligned}\tilde{u}^{DD_1} &= (1+4\theta)\tilde{\beta} + r_0\left[(1+2\theta)G_0(\theta,\theta) + 2\theta G_1(\theta,\theta)\right]\\ \tilde{v}^{DD_1} &= -\tilde{\beta} - r_0\left[(1+2\theta)G_0(\theta,\theta) + 2\theta G_1(\theta,\theta)\right]\end{aligned} \tag{4.460}$$

Domain DD_1OG_1

We introduce the Cartesian coordinate system (x_3, y_3) of origin at D. According to the optimality conditions, we have $\epsilon_{11} = 1$, $\epsilon_{22} = -1$, which is assured by

$$u(x_3, y_3) = x_3 + C_1 y_3 + D_1, \qquad v(x_3, y_3) = -y_3 - C_1 x_3 + D_2 \tag{4.461}$$

The continuity conditions of u, v along DD_1 define the constants:

$$\begin{aligned}C_1 &= 1+4\theta, \qquad D_1 = -D_1\\ D_1 &= r_0\left[(1+2\theta)G_0(\theta,\theta) + 2\theta G_1(\theta,\theta)\right]\end{aligned} \tag{4.462}$$

Domain RBB_1E_1

We introduce the Cartesian parameterization x_2, y_2 of origin at R. The fields u, v are given by

$$u(x_2, y_2) = x_2 + C_1 y_2 + D_1, \qquad v(x_2, y_2) = -y_2 - C_1 x_2 + D_2 \tag{4.463}$$

The constants involved above are determined by the continuity requirements along RB and BB_1:

$$C_1 = 1+2\theta, \qquad D_1 = 0, \qquad D_2 = 0 \tag{4.464}$$

Domain RE_1E_2

We introduce the polar coordinate system (r, ϑ); $0 \le \vartheta \le \theta_1$; $0 \le r \le r_1$. The displacement fields

$$\hat{u}(r,\vartheta) = -r, \qquad \hat{v}(r,\vartheta) = 2r\vartheta - (1+2\vartheta)r \tag{4.465}$$

are continuous extensions of the fields u, v given in RBB_1E_1. Along RE_2 we have

$$\hat{u}^{RE_2} = -x_4, \qquad \hat{v}^{RE_2} = (2\theta_1 - 1 - 2\vartheta)x_4 \tag{4.466}$$

Domain $E_3 E_2 R$

Le us introduce the Cartesian system x_4, y_4 of origin at R. The fields

$$
\begin{aligned}
u(x_4, y_4) &= -x_4 + (-2\theta_1 + 1 + 2\theta) y_4 \\
v(x_4, y_4) &= y_4 + (2\theta_1 - 1 - 2\theta) x_4
\end{aligned} \tag{4.467}
$$

are continuous extensions of the fields u, v from the domain RE_1E_2. The condition of fixing the edge E_3R reads

$$
u(x_4, -x_4) = 0, \qquad v(x_4, -x_4) = 0 \tag{4.468}
$$

and implies $\theta_1 = \theta$. Thus

$$
u(x_4, x_4) = -x_4 + y_4, \qquad v(x_4, x_4) = u(x_4, y_4) \tag{4.469}
$$

The condition $\theta_1 = \theta$ implies

$$
\theta = \frac{1}{2}(\pi - \gamma) \tag{4.470}
$$

and fixes the geometry of Michell's net around the support E_3RNE_3'.

Let us proceed to the construction of the parametric lines in the subsequent domains.

Domain ABDC

The parametric lines are given by (4.410), where, according to (4.409),

$$
\begin{aligned}
\frac{\bar{x}(\alpha, \beta)}{r_0} &= F_1(\alpha, \beta) + F_2(\beta, \alpha) \\
\frac{\bar{y}(\alpha, \beta)}{r_0} &= F_2(\alpha, \beta) + F_1(\beta, \alpha)
\end{aligned} \tag{4.471}
$$

Domain BB_1D_1D

The parametric lines are determined by the integral formulae (4.31). Thus

$$
\begin{aligned}
x_1(\tilde{\lambda}, \tilde{\mu}) &= x_1(0, 0) + \int_0^{\tilde{\lambda}} \tilde{A}(\tilde{\alpha}, 0) \cos(\tilde{\phi}(\tilde{\alpha}, 0)) d\tilde{\alpha} - \\
&\int_0^{\tilde{\mu}} \tilde{B}(\tilde{\lambda}, \tilde{\beta}) \sin(\tilde{\phi}(\tilde{\lambda}, \tilde{\beta})) d\tilde{\beta}, \\
y_1(\tilde{\lambda}, \tilde{\mu}) &= y_1(0, 0) + \int_0^{\tilde{\lambda}} \tilde{A}(\tilde{\alpha}, 0) \sin(\tilde{\phi}(\tilde{\alpha}, 0)) d\tilde{\alpha} + \\
&\int_0^{\tilde{\mu}} \tilde{B}(\tilde{\lambda}, \tilde{\beta}) \cos(\tilde{\phi}(\tilde{\lambda}, \tilde{\beta})) d\tilde{\beta},
\end{aligned} \tag{4.472}
$$

where $\tilde{\phi}(\tilde{\alpha}, \tilde{\beta}) = -\tilde{\alpha}$ and $\tilde{A}$, $\tilde{B}$ are given by (4.454); moreover, $x_1(0,0) = 0$, $y_1(0,0) = 0$. Therefore

$$x_1(\tilde{\lambda}, \tilde{\mu}) = \tilde{\mu} \sin \tilde{\lambda} + r_0 \int_0^{\tilde{\lambda}} [G_0(\tilde{\alpha}, \theta) + G_1(\theta, \tilde{\alpha})] \cos \tilde{\alpha} d\tilde{\alpha} \tag{4.473}$$

$$y_1(\tilde{\lambda}, \tilde{\mu}) = \tilde{\mu} \cos \tilde{\lambda} - r_0 \int_0^{\tilde{\lambda}} [G_0(\tilde{\alpha}, \theta) + G_1(\theta, \tilde{\alpha})] \sin \tilde{\alpha} d\tilde{\alpha}$$

Using the integral formulae of Sect. 4.2, we perform the integration and obtain

$$\begin{aligned} x_1(\tilde{\lambda}, \tilde{\mu}) &= \tilde{\mu} \sin \tilde{\lambda} + r_0[-1 + \cos \tilde{\lambda} H_0(\tilde{\lambda}, \theta) + \sin \tilde{\lambda} H_1(\tilde{\lambda}, \theta)] \\ y_1(\tilde{\lambda}, \tilde{\mu}) &= \tilde{\mu} \cos \tilde{\lambda} - r_0[\sin \tilde{\lambda} H_0(\tilde{\lambda}, \theta) - \cos \tilde{\lambda} H_1(\tilde{\lambda}, \theta)] \end{aligned} \tag{4.474}$$

where functions H_n are defined by (4.147). One can check that along the line BD the coordinates x_1, y_1 coincide with the coordinates x, y given by (4.471), (4.410) in the domain $ABDC$.

The Michell net found satisfies all the required conditions. The field (u, v) is continuous and has been determined uniquely. Moreover, this field can be extended to the whole exterior of the domain E_3RNE_3'.

Let us compute the distance $|AD|$. According to (4.410), we have $x(\theta, \theta) = \bar{x}(\theta, \theta)$, $y(\theta, \theta) = \bar{y}(\theta, \theta)$, where $\bar{x}$, $\bar{y}$ are given by (4.471). Since $x_D = x(\theta, \theta) = \bar{x}(\theta, \theta)$, hence $|AD| = \sqrt{2} x_D$, but $\sqrt{2} r_0 = a = |RN|$, which gives

$$|AD| = a[F_1(\theta, \theta) + F_2(\theta, \theta)] \tag{4.475}$$

with $\theta = \dfrac{1}{2}(\pi - \gamma)$.

4.9.2 *Case Studies*

The Michell net found outlines the optimal structures generated by the load which causes the axial forces in the fibres, with the prescribed signs, see Fig. 4.65. For instance, the force applied at point $\tilde{P}$ of coordinates $(\tilde{\alpha}, \tilde{\beta})$ within the domain BDD_1B_1 causes the axial forces introducing tension in one family and compression in the second family of fibres, if the force is directed within the fans shown in Fig. 4.65. Then, the optimal solution is cut out along the parametric lines. In the problem discussed here, the optimal structure corresponding to a system of point loads is a superposition of structures corresponding to the subsequent loads. This property will be called stability of the Michell net with respect to the load applied. This property is a characteristic feature of cantilevers; this feature was noted previously in Sect. 4.5.2, and in Sects. 4.6–4.8. But this property did not characterize the solution in Sect. 4.5.4, where the net was sensitive to the position of the point load $2P$.

We shall consider selected optimal structures corresponding to the Michell net of Fig. 4.65. Their volumes will be computed by the formula $V = \dfrac{1}{\sigma_0}\mathcal{L}$, where $\mathcal{L}$ is the virtual work of the force P, or $\mathcal{L} = Pu_P$, where u_P stands for the adjoint displacement along the direction of the force P.

The Problem of the Exterior of the Equilateral Triangle

Assume that the support has the shape of an equilateral triangle of side a. Since $\gamma = \pi/3$, hence $\theta = \pi/3$. Consider the family of optimal structures generated by the forces P directed vertically at points along the axis A_0O, see Fig. 4.66a. According to (4.475) we have $|AD| = 2.217004a$; $|DO| = a$.

If the point P lies left to point D, then the optimal structure is supported only on the part RN of the support, see Fig. 4.59. If the point P lies between D and O, then the optimal structure makes use of the support along the sides RE and EN.

If the point P lies between A_0 and O, then the volume of the optimal structure equals $V = \mathcal{L}/\sigma_0$, where the virtual work of the force P is equal to

$$\mathcal{L} = Pu_p, \qquad u_p = \frac{\sqrt{2}}{2}[u(P) - v(P)] = \sqrt{2}u(P) \tag{4.476}$$

since along the line A_0O we have $v(P) = -u(P)$.

If the point P lies between A and D, then $u(P)$ is given by (4.441), and the volume is expressed by (4.442). If the point P lies between D and O, then $u(P)$ is given by (4.461), or

$$u(P) = \left(2 + \frac{4}{3}\pi\right)x_3 + \frac{a\sqrt{2}}{2}\left[\left(1 + \frac{2}{3}\pi\right)G_0\left(\frac{\pi}{3}, \frac{\pi}{3}\right) + \frac{2}{3}\pi G_1\left(\frac{\pi}{3}, \frac{\pi}{3}\right)\right] \tag{4.477}$$

Thus, on the interval DO the plot of the volume is linear, see Fig. 4.66b. If the point P lies right to point O, then the optimal structure is not described by the net of Fig. 4.65.

Cantilevers Supported on a Rectangle

Consider the Michell problem of optimal transmitting a force P to the boundary of a long rectangle, see Fig. 4.67.

In the case discussed $\gamma = \dfrac{\pi}{2}$, $\theta = \dfrac{\pi}{4}$. Thus, the net of Fig. 4.65 is such that the lines RB and NC are horizontal. If point P lies on the line A_0D right to D, then the optimum solution has the form of Fig. 4.68. The volume of this structure is expressed by

$$V = \frac{Pa}{\sigma_0}\left\{\left[\left(1 + \frac{\pi}{2}\right)G_0\left(\frac{\pi}{4}, \frac{\pi}{4}\right) + \frac{\pi}{2}G_1\left(\frac{\pi}{4}, \frac{\pi}{4}\right)\right] + (2+\pi)\sqrt{2}\frac{r_1}{a}\right\} \tag{4.478}$$

For instance, if $r_1 = r_0$ then $V = 9.7096\dfrac{Pa}{\sigma_0}$.

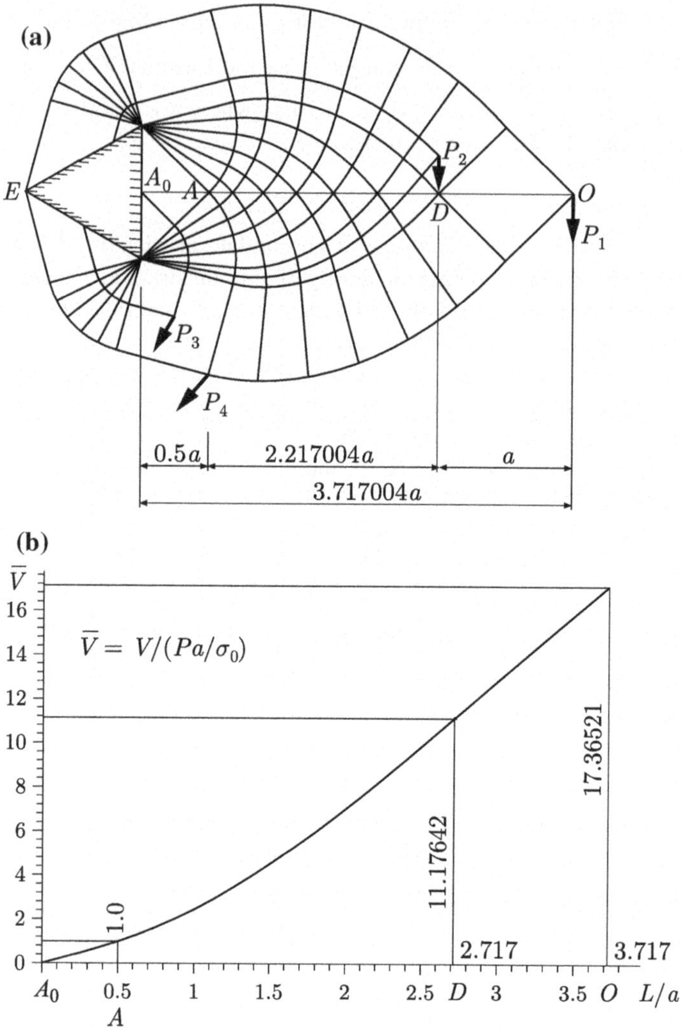

Fig. 4.66 Illustrative solutions corresponding to selected positions of the point load (**a**); the non-dimensional volume $\bar{V} = \dfrac{V\sigma_0}{Pa}$ for the subsequent position of the point load A_0O (**b**)

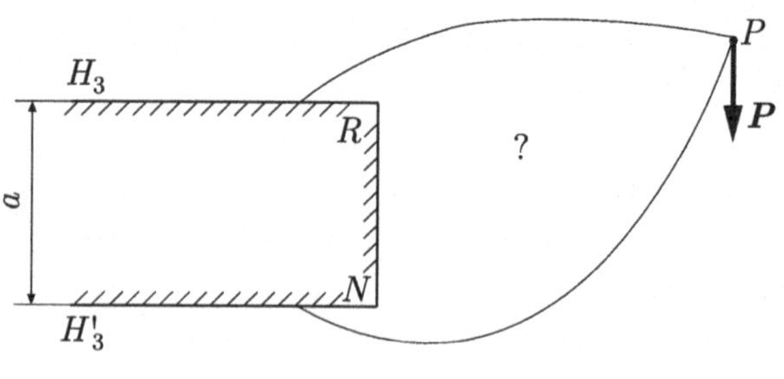

Fig. 4.67 The problem setting: optimal transmitting of a point load P to the boundary of a rectangle

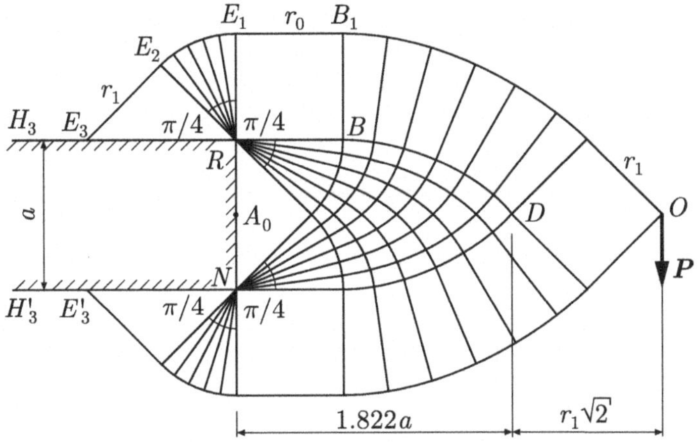

Fig. 4.68 Michell's structure supported on a rectangle

4.9.3 The Case of Asymmetry

The construction of the net of Fig. 4.65 can be extended to the case of asymmetric problems, for which $\angle E_3RN = \gamma_1$, $\angle RNE'_3 = \gamma_2$, $\gamma_1 \neq \gamma_2$. The net has the form as depicted in Fig. 4.69.

The parameters r_1, r_2 are arbitrary and independent. The angles α and δ of fan's apertures are $\alpha = \frac{1}{2}(\pi - \gamma_1)$, $\delta = \frac{1}{2}(\pi - \gamma_2)$. Thus, the data of the problem: $|RN| = a$, γ_1, γ_2 determine the position of point D.

4.10 Cantilevers Supported on a Square. Case of $\sigma_C = \sigma_T$

The aim of this section is the construction of Michell structures transmitting a point load to the boundary of a square domain, for the case of $\sigma_C = \sigma_T$. We predict that the solution will correspond to the regimes $\mathbb{T}$ and $\mathbb{R}_1$, $\mathbb{R}_2$.

4.10.1 The Michell Net

The Michell net is predicted as natural extension of the net of Fig. 4.69 in case of $r_1 = r_0 = a\sqrt{2}/2$, cf. Fig. 4.70.

The net of Fig. 4.70 fills up the whole exterior of the square RNC_3E_3. The nets in the domain $RE_1B_1D_1OG_1K_1C_1NR$ has been constructed in Sect. 4.9. By reflection the known net by symmetry with respect to the vertical line and by rotation by angle

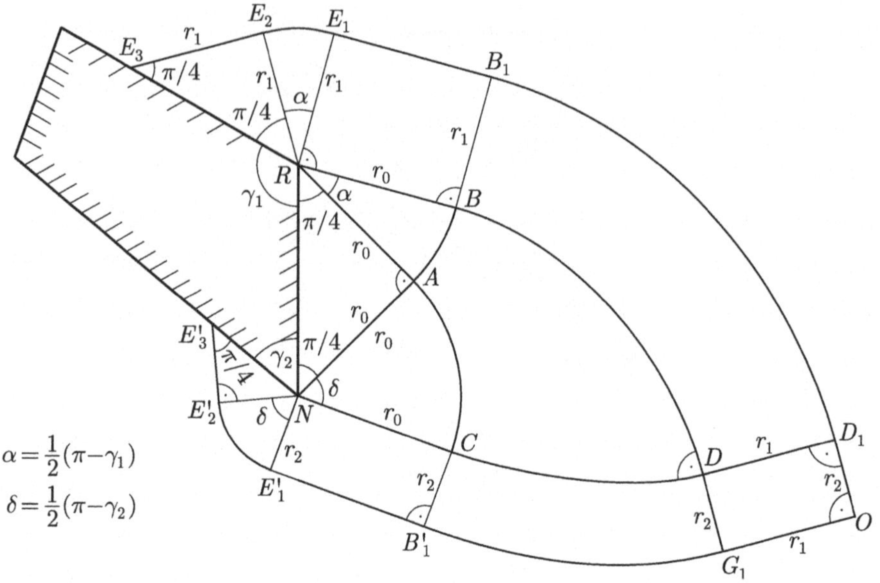

Fig. 4.69 An outline of an optimum layout in case of asymmetry

$\pi/2$, one obtains the net covering the whole completion of the square. The net is invariant under rotation by the angle $\pi/2$ with respect to the centre T of the square.

We shall proceed as in Sect. 4.9. First, we construct the fields A, B, then the fields u^0, v^0 and finally the fields u, v. The parametric lines α, β will be found using the auxiliary fields $\bar{x}$, $\bar{y}$. Thus, we start from the construction of the Lamé fields A, B.

The Domain $B_1 D_1 H D_1'$

The Hencky net is built on two arcs: $B_1 D_1$ and $B_1 D_1'$. We introduce the coordinate system (α, β)such that B_1 has coordinates $(0, 0)$, $B_1 D_1$ is an α line where $\beta = 0$, $D_1 D_1'$ is the β line where $\alpha = 0$. According to (4.454) along the α line, we have

$$A(\alpha, 0) = r_0 \left[1 + T_0\left(\frac{\pi}{4}, \alpha\right)\right] \tag{4.479}$$

Similarly, along the β line we have

$$B(0, \beta) = r_0 \left[1 + T_0\left(\frac{\pi}{4}, \beta\right)\right] \tag{4.480}$$

where T_0 is a special function given by (4.147), for $n = 0$. Let us note that $A(0, 0) = \left(2 + \frac{\pi}{4}\right) r_0$. We substitute the boundary values (4.479), (4.480) into (4.398) and perform the integration using the integration results collected in Sect. 4.2.3. We arrive at

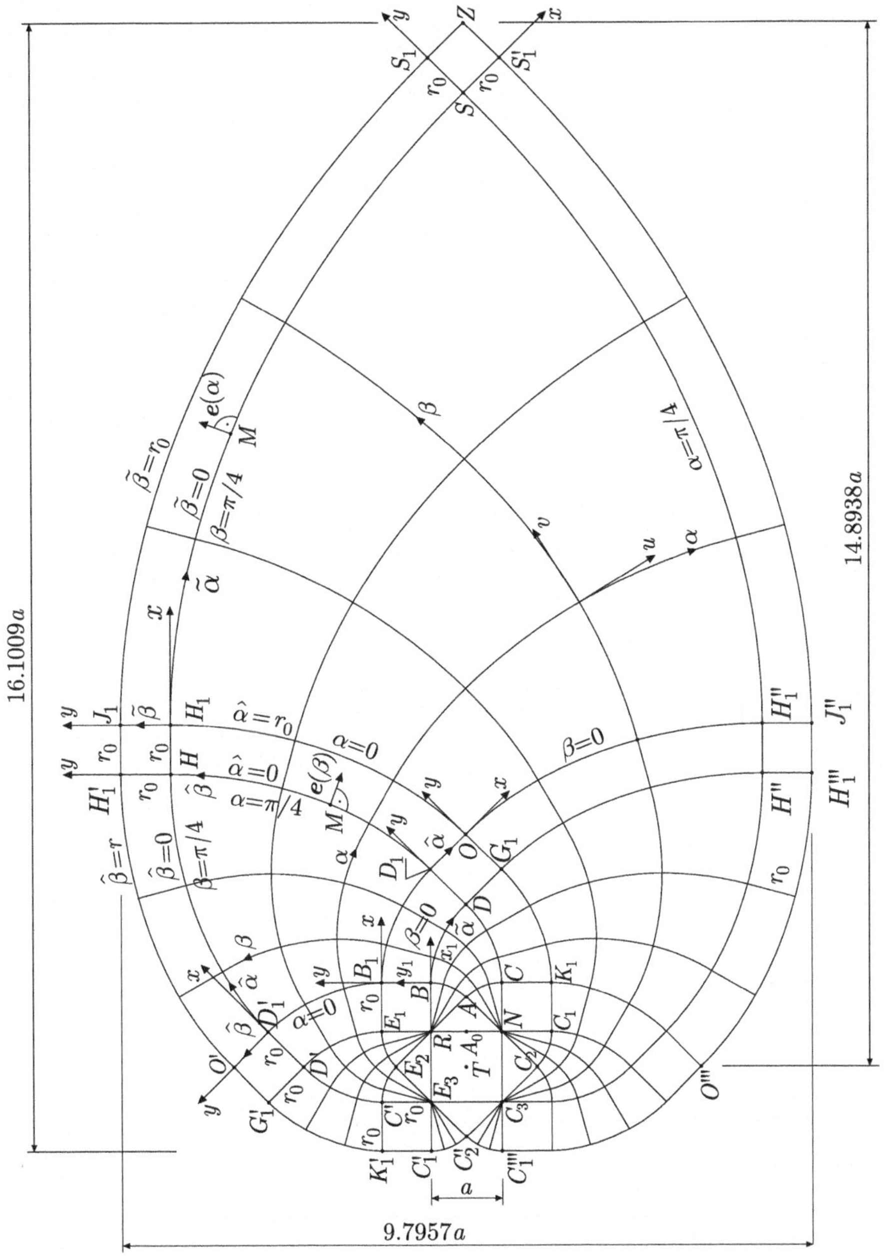

Fig. 4.70 The Michell net for the problem of the exterior of the square

$$\frac{A(\lambda,\mu)}{r_0} = T_0\left(\mu+\frac{\pi}{4},\lambda\right) + T_0\left(\mu,\lambda+\frac{\pi}{4}\right), \qquad B(\lambda,\mu) = A(\mu,\lambda) \tag{4.481}$$

It is not difficult to check that these fields satisfy (4.93) and the boundary conditions (4.479), (4.480).

The Domain $D_1'HH_1'O'$

The fields (4.481) will be extended into the domain $D_1'HH_1'O'$ through the line $D_1'H$, where $\beta = \pi/4$. We introduce the coordinate system $(\hat{\alpha}, \hat{\beta})$ such that $\hat{\beta} = 0$ line coincides with $\beta = \pi/4$ and there $\hat{\alpha} = \alpha$. The $\hat{\beta}$ lines (or $\hat{\alpha} = \text{const}$ lines) are straight.

We shall construct the fields $A(\hat{\alpha}, \hat{\beta})$, $B(\hat{\alpha}, \hat{\beta})$ in the domain considered. According to (4.62)

$$A(\hat{\alpha}, \hat{\beta}) = \hat{\beta} + f_1(\hat{\alpha}) \tag{4.482}$$

and $f_1(\hat{\alpha}) = A(\hat{\alpha}, \hat{\beta} = 0)$. Since A is continuous along $D_1'H$, we have $f_1(\hat{\alpha}) = A\left(\hat{\alpha}, \frac{\pi}{4}\right)$, where A is given by (4.481). Thus, we find

$$A(\hat{\alpha}, \hat{\beta}) = \hat{\beta} + r_0\left[T_0\left(\frac{\pi}{2},\hat{\alpha}\right) + T_0\left(\frac{\pi}{4},\hat{\alpha}+\frac{\pi}{4}\right)\right] \qquad B(\hat{\alpha}, \hat{\beta}) = 1 \tag{4.483}$$

The Domain D_1HH_1O

The Hencky net is symmetric with respect to the line B_1H. Thus, the Lamé fields are given by

$$\begin{aligned} A(\hat{\alpha}, \hat{\beta}) &= 1 \\ B(\hat{\alpha}, \hat{\beta}) &= \hat{\alpha} + r_0\left[T_0\left(\frac{\pi}{2},\hat{\beta}\right) + T_0\left(\frac{\pi}{4},\hat{\beta}+\frac{\pi}{4}\right)\right] \end{aligned} \tag{4.484}$$

cf. Eq. (4.483).

The Domain OH_1SH_1''

We introduce the parametric lines (α, β) such that $\phi = \beta - \alpha$ with respect to the Cartesian frame (x, y) of origin at O. We note that $\beta = 0$ along OH_1'' and $\alpha = 0$ along OH_1. According to (4.483), we find

$$\begin{aligned} A(\alpha, 0) &= r_0\left[1 + T_0\left(\frac{\pi}{2},\alpha\right) + T_0\left(\frac{\pi}{4},\alpha+\frac{\pi}{4}\right)\right] \\ B(0, \beta) &= A(\beta, 0) \end{aligned} \tag{4.485}$$

The field A is determined according to (4.398). Upon performing the integration using the rules of Sect. 4.2.3, we get

$$\frac{A(\lambda,\mu)}{r_0} = T_0\left(\mu+\frac{\pi}{4},\lambda+\frac{\pi}{4}\right) + T_0\left(\mu,\lambda+\frac{\pi}{2}\right) + T_0\left(\mu+\frac{\pi}{2},\lambda\right)$$
$$B(\lambda,\mu) = A(\mu,\lambda) \tag{4.486}$$

The Domain $H_1SS_1J_1$

We introduce the parameterization $(\tilde{\alpha},\tilde{\beta})$ such that $\tilde{\beta}=0$ determines the line: $\beta = \pi/4$, β defined within OH_1SH_1''. The $\tilde{\beta}$ lines are straight such that $\phi = -\tilde{\alpha}$ with respect to the (x, y) frame with origin at H_1. The boundary line J_1S is given by $\tilde{\beta} = r_0$. Thus, $B(\tilde{\alpha},\tilde{\beta}) = 1$. We use the boundary condition $A(\tilde{\alpha}, 0) = A\left(\tilde{\alpha}, \beta = \frac{\pi}{4}\right)$, where the latter function $A(\cdot,\cdot)$ is defined by (4.486). This gives

$$A(\tilde{\alpha},\tilde{\beta}) = \tilde{\beta} + r_0\left[T_0\left(\frac{\pi}{2},\tilde{\alpha}+\frac{\pi}{4}\right) + T_0\left(\frac{\pi}{4},\tilde{\alpha}+\frac{\pi}{2}\right) + T_0\left(\frac{3\pi}{4},\tilde{\alpha}\right)\right]$$
$$B(\tilde{\alpha},\tilde{\beta}) = 1 \tag{4.487}$$

The Lamé fields within other domains of Fig. 4.70 can be constructed by symmetry.

4.10.2 The Adjoint Displacement Fields

Having constructed the Lamé fields we are ready to find the adjoint displacement fields in the subsequent domains.

The adjoint displacement fields (u, v) in the domains: RAN, RBA, NAC, ABDC, BB_1D_1D, BB_1E_1R, RE_1E_2, E_2E_3R have been already constructed in Sect. 4.9. One should put there $\gamma = \pi/2$ and $\theta = \pi/4$. The adjoint displacements (u, v) within $B_1D_1'D'E_1$ can be constructed by antisymmetric reflection with respect to the axis B_1H.

The Domain $B_1D_1HD_1'$

The fields (u, v) will be constructed by means of the auxiliary fields (u^0, v^0) defined by (4.118a). Here, for $\kappa = 1$, they read

$$\begin{aligned} u^0(\alpha,\beta) &= u(\alpha,\beta) - 2\alpha A(\alpha,\beta) \\ v^0(\alpha,\beta) &= v(\alpha,\beta) + 2\beta B(\alpha,\beta) \end{aligned} \tag{4.488}$$

As they fulfil (4.119), they can be directly found using Riemann's formula (4.132) with $f = 0$. We need to have the boundary values

$$v^0(\alpha, 0) = v(\alpha, 0), \qquad v^0(0, \beta) = v(0, \beta) + 2\beta B(0, \beta) \tag{4.489}$$

They can be deduced from the results in Sect. 4.8:

$$\begin{aligned} v^0(\alpha, 0) &= r_0\left[-1 - \left(1 + \frac{\pi}{2}\right) G_0\left(\alpha, \frac{\pi}{4}\right) - \frac{\pi}{2} G_1\left(\alpha, \frac{\pi}{4}\right)\right] \\ v^0(0, \beta) &= r_0\left[-1 - \frac{\pi}{2} - G_0\left(\beta, \frac{\pi}{4}\right)\right] \end{aligned} \tag{4.490}$$

Note that $v^0(0, 0) = -2 - \pi/2$. Upon inserting (4.490) into the formula (4.132) and performing integration, one finds

$$\frac{v^0(\lambda, \mu)}{r_0} = -\left(1 + \frac{\pi}{2}\right) G_0\left(\lambda, \mu + \frac{\pi}{4}\right) - G_0\left(\mu, \lambda + \frac{\pi}{4}\right) - \frac{\pi}{2} G_1\left(\lambda, \mu + \frac{\pi}{4}\right) \tag{4.491}$$

The field $v(\lambda, \mu)$ is given by (4.488), (4.491) and (4.481). The field u is given by $u(\lambda, \mu) = -v(\mu, \lambda)$. Moreover

$$\omega(\lambda, \mu) = -1 - \frac{\pi}{2} - 2(\lambda + \mu) \tag{4.492}$$

and $\omega(0, 0) = \omega_{B_1} = -1 - \dfrac{\pi}{2}$.

The Domain $D_1' H H_1' O'$

The angle of the rigid rotation equals

$$\omega(\hat{\alpha}, \hat{\beta}) = -1 - \pi - 2\hat{\alpha} \tag{4.493}$$

Now, we make use of (4.457) to find u and v:

$$\frac{\partial u}{\partial \hat{\beta}} = -\omega(\hat{\alpha}, \hat{\beta}), \qquad \frac{\partial v}{\partial \hat{\beta}} = -1 \tag{4.494}$$

Integration gives

$$u(\hat{\alpha}, \hat{\beta}) = (1 + \pi + 2\hat{\alpha})\hat{\beta} + f_2(\hat{\alpha}), \qquad v(\hat{\alpha}, \hat{\beta}) = -\hat{\beta} + f_1(\hat{\alpha}) \tag{4.495}$$

with

$$f_1(\hat{\alpha}) = v(\hat{\alpha}, 0), \qquad f_2(\hat{\alpha}) = u(\hat{\alpha}, 0) \tag{4.496}$$

The latter functions are defined along $D_1'H$ and can be found by the continuity condition, using (4.491), (4.488). This gives

$$u(\hat{\alpha}, \hat{\beta}) = f_1(\hat{\alpha}, \hat{\beta}), \qquad v(\hat{\alpha}, \hat{\beta}) = f_2(\hat{\alpha}, \hat{\beta}) \tag{4.497}$$

where

$$\begin{aligned} \frac{f_1(\hat{\alpha}, \hat{\beta})}{r_0} = \frac{(1+\pi+2\hat{\alpha})\hat{\beta}}{r_0} + \left(1+\frac{\pi}{2}+2\hat{\alpha}\right) G_0\left(\hat{\alpha}+\frac{\pi}{4}, \frac{\pi}{4}\right) + \\ (1+2\hat{\alpha})G_0\left(\hat{\alpha}, \frac{\pi}{2}\right) + 2\hat{\alpha}G_1\left(\frac{\pi}{2}, \hat{\alpha}\right) + \left(2\hat{\alpha}+\frac{\pi}{2}\right) G_1\left(\frac{\pi}{4}, \hat{\alpha}+\frac{\pi}{4}\right) \\ \frac{f_2(\hat{\alpha}, \hat{\beta})}{r_0} = -\frac{\hat{\beta}}{r_0} - (1+\pi)G_0\left(\hat{\alpha}, \frac{\pi}{2}\right) - \left(1+\frac{\pi}{2}\right) G_0\left(\frac{\pi}{4}, \hat{\alpha}+\frac{\pi}{4}\right) - \\ \pi G_1\left(\hat{\alpha}, \frac{\pi}{2}\right) - \frac{\pi}{2} G_1\left(\hat{\alpha}+\frac{\pi}{4}, \frac{\pi}{4}\right) \end{aligned} \tag{4.498}$$

The Domain $D_1 H H_1 O$

Due to skew-symmetry, we find

$$u(\hat{\alpha}, \hat{\beta}) = -f_2(\hat{\beta}, \hat{\alpha}), \qquad v(\hat{\alpha}, \hat{\beta}) = -f_1(\hat{\beta}, \hat{\alpha}), \qquad \omega(\hat{\alpha}, \hat{\beta}) = -1-\pi-2\hat{\beta} \tag{4.499}$$

where f_1, f_2 are given by (4.498). Note that ω is symmetric with respect the axis TZ due to skew-symmetry of the problem setting.

The Domain $H H_1 J_1 H_1'$

We use the Cartesian frame (x, y) of origin at H, such that $y = 0$ along HH_1 and $x = 0$ along HH_1', see Fig. 4.70. By analogy with (4.461), we can put the displacement fields as

$$u = x + C_1 y + D_1, \qquad v = -y - C_1 x - D_1 \tag{4.500}$$

where now

$$\begin{aligned} C_1 = 1 + \frac{3\pi}{2} \\ \frac{D_1}{r_0} = \left(2+\frac{3\pi}{2}\right) G_0\left(\frac{\pi}{4}, \frac{\pi}{2}\right) + \frac{\pi}{2} G_1\left(\frac{\pi}{2}, \frac{\pi}{4}\right) + \pi G_1\left(\frac{\pi}{4}, \frac{\pi}{2}\right) \end{aligned} \tag{4.501}$$

The angle of rigid rotation equals $\omega = -1 - \dfrac{3}{2}\pi$ and is constant. The fields (4.500) coincide with the fields (4.497), (4.499) along HH_1', HH_1.

The Domain OH_1SH_1''

Note that $\alpha = 0$ along OH_1; $\alpha = \frac{\pi}{4}$ along $H_1''S$; $\beta = 0$ along OH_1''; $\beta = \frac{\pi}{4}$ along H_1S.

Note that

$$\omega(\alpha, \beta) = -1 - \pi - 2(\alpha + \beta) \tag{4.502}$$

We know u, v and A, B along OH_1'' and OH_1. Thus, we can find the values of v^0 along these lines:

$$\begin{aligned} \frac{v^0(\alpha, 0)}{r_0} &= -1 - (1+\pi)G_0\left(\alpha, \frac{\pi}{2}\right) - \left(1 + \frac{\pi}{2}\right)G_0\left(\frac{\pi}{4}, \alpha + \frac{\pi}{4}\right) - \\ &\quad \pi G_1\left(\alpha, \frac{\pi}{4}\right) - \frac{\pi}{2}G_1\left(\alpha + \frac{\pi}{4}, \frac{\pi}{4}\right) \\ \frac{v^0(0, \beta)}{r_0} &= -(1+\pi) - \left(1 + \frac{\pi}{2}\right)G_0\left(\beta + \frac{\pi}{4}, \frac{\pi}{4}\right) - G_0\left(\beta, \frac{\pi}{2}\right) - \\ &\quad \frac{\pi}{2}G_1\left(\frac{\pi}{4}, \beta + \frac{\pi}{4}\right) \end{aligned} \tag{4.503}$$

hence

$$\frac{v^0(0, 0)}{r_0} = -2 - \pi - \left(1 + \frac{\pi}{2}\right)G_0\left(\frac{\pi}{4}, \frac{\pi}{4}\right) - \frac{\pi}{2}G_1\left(\frac{\pi}{4}, \frac{\pi}{4}\right) \tag{4.504}$$

Now, we are ready to construct $v^0(\alpha, \beta)$ at arbitrary point in this domain using Riemann's formula (4.132). Upon performing the integrations, one finds

$$\begin{aligned} v^0(\alpha, \beta) &= -u^0(\beta, \alpha) \\ \frac{u^0(\alpha, \beta)}{r_0} &= (1+\pi)G_0\left(\beta, \alpha + \frac{\pi}{2}\right) + G_0\left(\alpha, \beta + \frac{\pi}{4}\right) + \\ &\quad \left(1 + \frac{\pi}{2}\right)G_0\left(\alpha + \frac{\pi}{4}, \beta + \frac{\pi}{4}\right) + \pi G_1\left(\beta, \alpha + \frac{\pi}{2}\right) + \\ &\quad \frac{\pi}{2}G_1\left(\beta + \frac{\pi}{4}, \alpha + \frac{\pi}{4}\right) \end{aligned} \tag{4.505}$$

The fields A, B are given by (4.486). Having these fields and the fields (u^0, v^0), one can find the fields (u, v) by (4.488). Note that $u(\alpha, \beta) = -v(\beta, \alpha)$.

The Domain $H_1SS_1J_1$

The domain is parameterized by $(\tilde{\alpha}, \tilde{\beta})$. Here

$$\omega(\tilde{\alpha}, \tilde{\beta}) = -1 - \frac{3\pi}{2} - 2\tilde{\alpha} \tag{4.506}$$

which is compatible with (4.502). The displacement fields are given by

$$
\begin{aligned}
u(\tilde{\alpha}, \tilde{\beta}) &= \left(1 + \frac{3\pi}{2} + 2\tilde{\alpha}\right)\tilde{\beta} + u^0\left(\tilde{\alpha}, \frac{\pi}{4}\right) + 2\tilde{\alpha}A\left(\tilde{\alpha}, \frac{\pi}{4}\right) \\
v(\tilde{\alpha}, \tilde{\beta}) &= -\tilde{\beta} + v^0\left(\tilde{\alpha}, \frac{\pi}{4}\right) - \frac{\pi}{2}B\left(\tilde{\alpha}, \frac{\pi}{4}\right)
\end{aligned}
\tag{4.507}
$$

where the fields above: u^0, v^0, A, B correspond to the domain OH_1SH_1''. Along the line SS_1 we find

$$
\begin{aligned}
u\left(\frac{\pi}{4}, \tilde{\beta}\right) &= (1 + 2\pi)\tilde{\beta} + u^0\left(\frac{\pi}{4}, \frac{\pi}{4}\right) + \frac{\pi}{2}A\left(\frac{\pi}{4}, \frac{\pi}{4}\right) \\
v\left(\frac{\pi}{4}, \tilde{\beta}\right) &= -\tilde{\beta} + v^0\left(\frac{\pi}{4}, \frac{\pi}{4}\right) - \frac{\pi}{2}B\left(\frac{\pi}{4}, \frac{\pi}{4}\right)
\end{aligned}
\tag{4.508}
$$

The Domain SS_1ZS_1''

We introduce the Cartesian frame (x, y) of origin at S such that Z has coordinates (r_0, r_0). The displacements (u, v) along (x, y) are given by

$$
u = x + C_1 y + D_1, \qquad v = -y - C_1 x - D_1 \tag{4.509}
$$

where here

$$
C_1 = 1 + 2\pi, \qquad D_1 = u^0\left(\frac{\pi}{4}, \frac{\pi}{4}\right) + \frac{\pi}{2}A\left(\frac{\pi}{4}, \frac{\pi}{4}\right) \tag{4.510}
$$

and u^0, A refer to the domain OH_1SH_1''. The rigid rotation equals $\omega = -1 - 2\pi$. Note that $u(x, y) = -v(y, x)$.

4.10.3 Geometry of the Michell Net

The net in Fig. 4.70 is an extension of the net given in Fig. 4.65 and elaborated in Sect. 4.9. The geometry of the Michell net in the domain $E_3E_2E_1B_1D_1OG_1B_1'E_1'E_2'C_3$ is already known. We shall derive analytical expressions defining the fields $x(\alpha, \beta)$, $y(\alpha, \beta)$ outside the mentioned domain.

The Domain $B_1D_1HD_1'$

The origin of the Cartesian coordinate system is at point B_1, see Fig. 4.70. The Michell net is denoted by (α, β). The fields $(x(\alpha, \beta), y(\alpha, \beta))$ will be constructed by the auxiliary fields $(\bar{x}(\alpha, \beta), \bar{y}(\alpha, \beta))$, cf. (4.95). Using the results (4.474), one can find the boundary values of these fields on the lines B_1D_1', B_1D_1

$$\frac{\bar{x}(\alpha,0)}{r_0} = \sin\alpha - \cos\alpha + H_0\left(\alpha, \frac{\pi}{4}\right),$$
$$\frac{\bar{y}(0,\beta)}{r_0} = \sin\beta - \cos\beta + H_0\left(\beta, \frac{\pi}{4}\right) \tag{4.511}$$

where $H_0 = F_0 + F_1$, see (4.147). By (4.97): $\dfrac{\partial \bar{x}(0,\beta)}{\partial\beta} = \bar{y}(0,\beta)$.

Thus, the Riemann formula (4.132) gives

$$\bar{x}(\lambda,\mu) = \bar{x}(0,0)G_0(\lambda,\mu) + \int_0^\lambda G_0(\lambda-\alpha,\mu)\frac{\partial\bar{x}}{\partial\alpha}(\alpha,0)d\alpha +$$
$$+ \int_0^\mu G_0(\mu-\beta,\lambda)\bar{y}(0,\beta)d\beta \tag{4.512}$$

By substituting (4.511) into (4.512) and performing the integrations along α and β lines, using the integral rules given in Sect. 4.3, one finds

$$\frac{\bar{x}(\lambda,\mu)}{r_0} = \sin(\lambda-\mu) - \cos(\lambda-\mu) + H_0\left(\lambda, \mu+\frac{\pi}{4}\right) + H_1\left(\mu, \lambda+\frac{\pi}{4}\right) \tag{4.513}$$

and $\bar{y}(\lambda,\mu) = \bar{x}(\mu,\lambda)$. To find x, y, we make use of (4.96). Having found $x(\alpha,\beta)$, $y(\alpha,\beta)$, we can draw the boundary line D_1H (where $\alpha = \pi/4$) and $D_1'H$ (where $\beta = \pi/4$) and obtain the position of point H.

The Michell net found in $B_1D_1HD_1'$ can be extended into the adjacent domains: $D_1'HH_1'O'$ and D_1OH_1H; the coordinate lines there are denoted by $(\hat{\alpha}, \hat{\beta})$. Along D_1H we have $\hat{\alpha} = 0$.

Within D_1OH_1H, the $\hat{\beta}$ lines are parallel to the line $\hat{\alpha} = 0$ and $\hat{\alpha}$ lines are straight. The line $\hat{\alpha} = r_0$ determines OH_1. In a similar manner, we determine the line OH_1''. Within OH_1SH_1'', we construct the (α,β) lines such that $\alpha = 0$ coincides with OH_1 and $\beta = 0$ with OH_1''. The parametric lines $x(\alpha,\beta)$, $y(\alpha,\beta)$ are referred to the system (x,y) of origin at O. The functions $x(\alpha,\beta)$, $y(\alpha,\beta)$ are constructed within OH_1SH_1'' via the functions $\bar{x}(\alpha,\beta)$, $\bar{y}(\alpha,\beta)$ given by (4.95). Knowing the lines OH_1, OH_1'' one can find the boundary values of $\bar{x}(\alpha,\beta)$, $\bar{y}(\alpha,\beta)$ along these lines. Along OH_1'':

$$\frac{\bar{x}(\alpha,0)}{r_0} = -(1+\xi)\cos\alpha + (2+\xi)\sin\alpha + H_0\left(\alpha, \frac{\pi}{2}\right) + H_1\left(\frac{\pi}{4}, \alpha+\frac{\pi}{4}\right) \tag{4.514}$$
$$\frac{\bar{y}(\alpha,0)}{r_0} = 1 - (2+\xi)\cos\alpha - (1+\xi)\sin\alpha + H_0\left(\frac{\pi}{4}, \alpha+\frac{\pi}{4}\right) + H_1\left(\alpha, \frac{\pi}{2}\right)$$

where $\xi = H_1\left(\dfrac{\pi}{4}, \dfrac{\pi}{4}\right)$.

Along OH_1:

$$\begin{aligned}\frac{\bar{x}(0,\beta)}{r_0} &= 1-(2+\xi)\cos\beta-(1+\xi)\sin\beta+H_0\left(\frac{\pi}{4},\beta+\frac{\pi}{4}\right)+H_1\left(\beta,\frac{\pi}{2}\right)\\ \frac{\bar{y}(0,\beta)}{r_0} &= -(1+\xi)\cos\beta+(2+\xi)\sin\beta+H_0\left(\beta,\frac{\pi}{2}\right)+H_1\left(\frac{\pi}{4},\beta+\frac{\pi}{4}\right)\end{aligned} \tag{4.515}$$

To find $\bar{x}$, $\bar{y}$, we make use of (4.512). We put (4.513), (4.515) into (4.512) and perform integrations using the rules collected in Sect. 4.2.3. Finally, we find

$$\begin{aligned}\frac{\bar{x}(\lambda,\mu)}{r_0} &= (2+\xi)\sin(\lambda-\mu)-(1+\xi)\cos(\lambda-\mu)+H_0\left(\lambda,\mu+\frac{\pi}{2}\right)+\\ &\quad H_1\left(\mu,\lambda+\frac{\pi}{2}\right)+H_1\left(\mu+\frac{\pi}{4},\lambda+\frac{\pi}{4}\right)\\ \bar{y}(\lambda,\mu) &= \bar{x}(\mu,\lambda)\end{aligned} \tag{4.516}$$

To find $x(\alpha,\beta)$, $y(\alpha,\beta)$, we come back to (4.96). The results (4.516) determine the shapes of the boundary lines H_1S and $H_1''S$ and the position of point S.

The above results determine the distances

$$\begin{aligned}|AO| &= \sqrt{2}(1+\xi)r_0\\ \frac{|AS|}{r_0} &= \sqrt{2}\left[H_0\left(\frac{\pi}{4},\frac{3\pi}{4}\right)+H_1\left(\frac{\pi}{4},\frac{3\pi}{4}\right)+H_1\left(\frac{\pi}{2},\frac{\pi}{2}\right)\right]\end{aligned} \tag{4.517}$$

or $|AS| = 12.8938a$. Let us compute

$$\frac{|H_1H_1''|}{r_0} = \sqrt{2}+2H_0\left(\frac{\pi}{4},\frac{\pi}{2}\right)+2H_1\left(\frac{\pi}{4},\frac{3\pi}{4}\right)+2H_1\left(\frac{\pi}{4},\frac{\pi}{2}\right) \tag{4.518}$$

or $|H_1H_1''|/r_0 = 8.381482a$. Hence, $|JJ_1''| = 9.7957a$ and

$$\frac{|A_0Z|}{a} = \frac{3}{2}+H_0\left(\frac{\pi}{4},\frac{3\pi}{4}\right)+H_1\left(\frac{\pi}{4},\frac{3\pi}{4}\right)+H_1\left(\frac{\pi}{2},\frac{\pi}{2}\right) \tag{4.519}$$

or $|A_0Z| = 14.39382a$.

Thus, we have found the main dimensions of the Michell structure transmitting a force at point Z to the boundary of the square, see Fig. 4.70.

4.10.4 Optimal Structures for Selected Positions of the Point Load

4.10.4.1 Vertical Point Load Between B_1 and H

Assume that point P of coordinates (α_P, β_P) lies within the domain $B_1 D_1 H D_1'$ between B_1 and H, see Fig. 4.71.

Assume that the force P is orthogonal to the line $B_1 H$. Thus, the virtual (or adjoint) displacement along the force direction equals

$$u_P = \frac{\sqrt{2}}{2}[u(\alpha_P, \alpha_P) - v(\alpha_P, \alpha_P)] \tag{4.520}$$

or $u_P = -\sqrt{2}v(\alpha_P, \alpha_P)$. According to (4.488), the value $v(\alpha_P, \alpha_P)$ is determined by $v^0(\alpha_P, \alpha_P)$, see Eq. (4.491), and by $B(\alpha_P, \alpha_P)$, see Eq. (4.481). We find

$$\frac{v(\alpha, \alpha)}{r_0} = -\left[\left(2 + \frac{\pi}{2} + 4\alpha\right) G_0\left(\alpha, \alpha + \frac{\pi}{4}\right) + \left(\frac{\pi}{2} + 2\alpha\right) G_1\left(\alpha, \alpha + \frac{\pi}{4}\right) + 2\alpha G_1\left(\alpha + \frac{\pi}{4}, \alpha\right)\right] \tag{4.521}$$

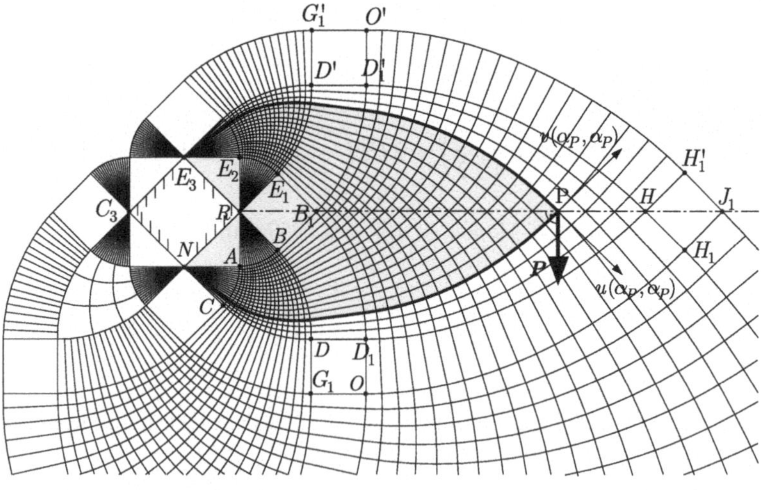

Fig. 4.71 Case of point P between B_1 and H

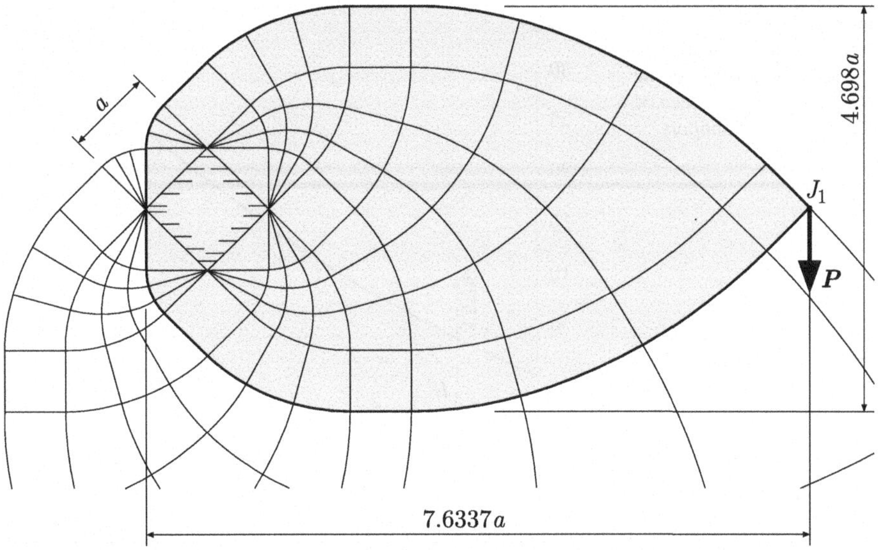

Fig. 4.72 Case of $P = J_1$

The volume of the optimal structure transmitting the load P to the boundary of the square equals $V = Pu_P/\sigma_0$. An illustrative optimal structure is shown in Fig. 4.71. In particular

$$\text{If } P = B_1, \quad V = \frac{\left(2 + \frac{\pi}{2}\right) Pa}{\sigma_0} \tag{4.522}$$

$$\text{If } P = H, \quad V = \left[\left(2 + \frac{3\pi}{2}\right) G_0\left(\frac{\pi}{4}, \frac{\pi}{2}\right) + \pi G_1\left(\frac{\pi}{4}, \frac{\pi}{2}\right) + \frac{\pi}{2} G_1\left(\frac{\pi}{2}, \frac{\pi}{4}\right)\right] \frac{Pa}{\sigma_0}$$

or $V = 26.59927 Pa/\sigma_0$.

4.10.4.2 Vertical Force at J_1

Assume $P = J_1$ and the force P is still orthogonal to RJ_1. The volume of the optimal structure shown in Fig. 4.72 equals

$$V = \left[\left(2 + \frac{3\pi}{2}\right)\left(1 + G_0\left(\frac{\pi}{4}, \frac{\pi}{2}\right)\right) + \frac{\pi}{2} G_1\left(\frac{\pi}{2}, \frac{\pi}{4}\right) + \pi G_1\left(\frac{\pi}{4}, \frac{\pi}{2}\right)\right] \frac{Pa}{\sigma_0} \tag{4.523}$$

or $V = 33.31166 Pa/\sigma_0$.

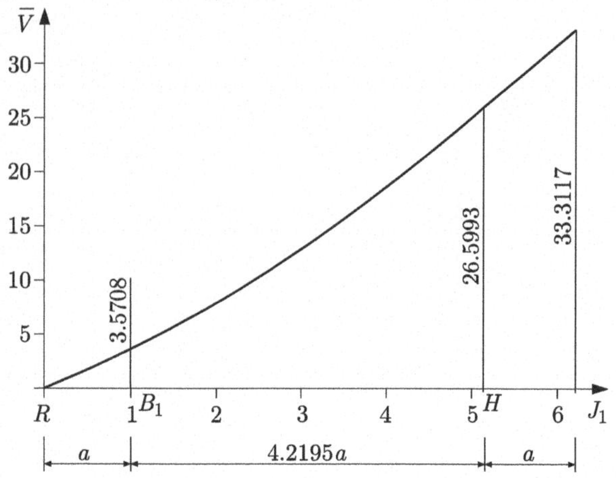

Fig. 4.73 The plot of the volumes $\bar{V} = V/(Pa/\sigma_0)$ of the structures corresponding to the increasing distance of point P from the support

In this case, the whole boundary of the square is used as the support. Thus, the volume is an increasing function of the distance of point P to the boundary, see Fig. 4.73.

4.10.4.3 Vertical Point Load Between A_0 and Z

Let the point P of coordinates (α_P, α_P) lie between O and S. The volume of the optimal structure is computed as before, using the displacement field $u(\alpha, \beta)$. We find $V = \left(\frac{Pa}{\sigma_0}\right)\bar{V}$, $\bar{V} = f(\alpha_P)$, where

$$f(\alpha) = G_0\left(\alpha, \alpha + \frac{\pi}{4}\right) + (1 + \pi + 4\alpha)G_0\left(\alpha, \alpha + \frac{\pi}{2}\right) + \left(1 + \frac{\pi}{2} + 2\alpha\right)G_0\left(\alpha + \frac{\pi}{4}, \alpha + \frac{\pi}{4}\right) + (\pi + 2\alpha)G_1\left(\alpha, \alpha + \frac{\pi}{2}\right) + \left(\frac{\pi}{2} + 2\alpha\right)G_1\left(\alpha + \frac{\pi}{4}, \alpha + \frac{\pi}{4}\right) + 2\alpha G_1\left(\alpha + \frac{\pi}{2}, \alpha\right) \quad (4.524)$$

If $P = S$, then $\alpha = \pi/4$ and $\bar{V} = 84.6279$. The optimal structure for $P = S$ is shown in Fig. 4.74a. The layout for $P = Z$ is given in Fig. 4.74b, where the geometry of the fibres is shown.

The volume $\bar{V}$ is an increasing function of the distance of P to the support, see Fig. 4.75.

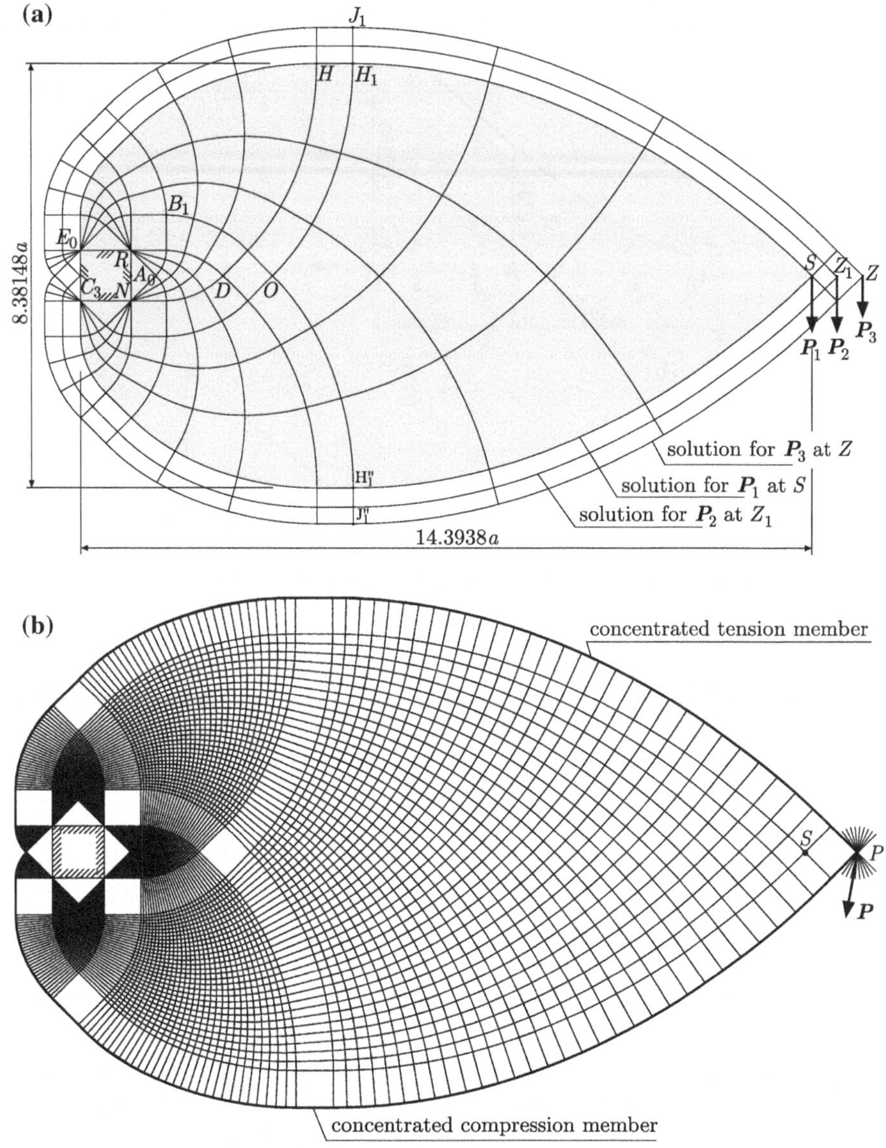

Fig. 4.74 Cases of $P = S$ (**a**); $P = Z$ (**b**)

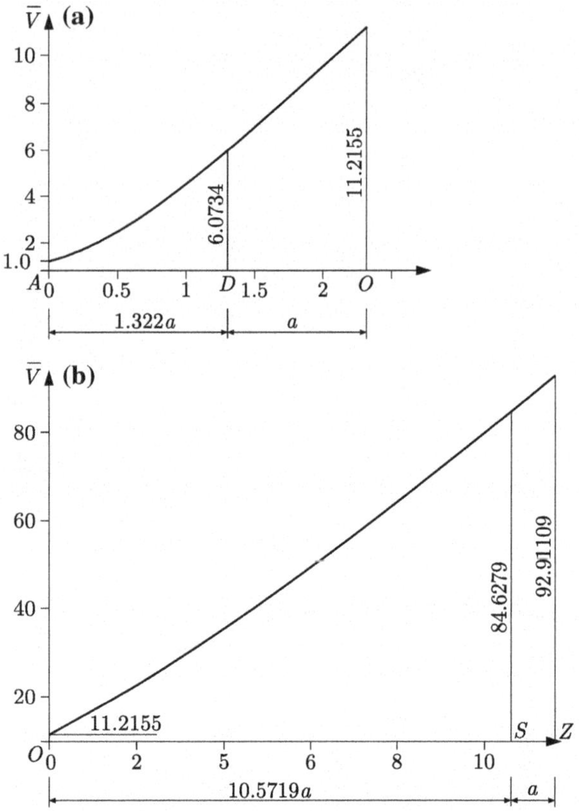

Fig. 4.75 The plot of the volumes

If $P = Z_1$, see Fig. 4.74, the whole boundary of the square is used. If $P = Z$, the left vertical support E_3C_3 is fully used, see Fig. 4.74. For $P = Z$, the volume equals $(Pa/\sigma_0)\bar{V}$, with

$$\bar{V} = 2(1+\pi) + G_0\left(\frac{\pi}{4}, \frac{\pi}{2}\right) + (1+2\pi)G_0\left(\frac{\pi}{4}, \frac{3\pi}{4}\right) + (1+\pi)G_0\left(\frac{\pi}{2}, \frac{\pi}{2}\right) + \frac{3\pi}{2}G_1\left(\frac{\pi}{4}, \frac{3\pi}{4}\right) + \pi G_1\left(\frac{\pi}{2}, \frac{\pi}{2}\right) + \frac{\pi}{2}G_1\left(\frac{3\pi}{4}, \frac{\pi}{4}\right) \tag{4.525}$$

hence $\bar{V} = 92.911089$, see Fig. 4.75. This result can be confirmed by the ground structure method, see Fig. 4.76.

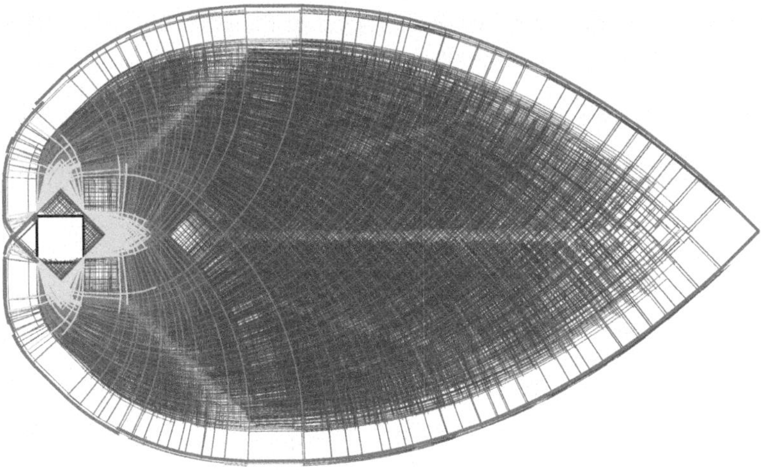

Fig. 4.76 The numerical solution referring to the case of the force at point Z. The ground structure method involved ca 100 million of potential bars. The volume found by the numerical method is greater by 1.4% than the analytical value. The CPU time was 9574.95 s

4.11 Towards Finding Michell Nets in the Bounded Domains: Additional Mathematical Tools

The aim of this section is to prepare new tools for solving Michell problems for bounded design domains, in particular, for the domains with straight line boundaries. One of these tools is the Laplace transform. Let us note that some of the integration rules collected in Sect. 4.2 have been derived using the Laplace transform method.

4.11.1 Selected Laplace Transforms of Chan's Functions

The Laplace transform of a function $t \to f(t)$ is defined by

$$f^*(p) = L_t[f(t)] = \int_0^\infty e^{-pt} f(t)dt \tag{4.526}$$

where p is a complex number. Let us recall the rule for the Laplace transform of the first derivative:

$$L_t[f'(t)] = pf^*(p) - f(0+) \tag{4.527}$$

If a function is defined in a convolution form as below:

$$F(t) = \int_0^t f(t-x)g(x)dx \tag{4.528}$$

its Laplace transform is expressed as a product of Laplace transforms of both the functions involved in the convolution

$$F^*(p) = f^*(p)g^*(p) \tag{4.529}$$

Let us list the Laplace transforms used in this book, starting from the well-known rules

$$L_t[\cos(t)] = \frac{p}{p^2+1}, \qquad L_t[\sin(t)] = \frac{1}{p^2+1} \tag{4.530}$$

The Laplace transforms of Chan's functions introduced in Sect. 4.2.3 can be expressed by elementary functions:

$$L_t[G_n(t,\beta)] = \frac{1}{p^{n+1}} \exp\left(\frac{\beta}{p}\right) \tag{4.531}$$

$$L_t[F_n(t,\beta)] = \frac{1}{p^{n-1}(p^2+1)} \exp\left(\frac{\beta}{p}\right) \tag{4.532}$$

for $n \geq 0$. Moreover,

$$L_t[G_n(t, t-\xi)] = \frac{2^n \exp\left[-\frac{1}{2}\left(p - \sqrt{p^2-4}\right)\xi\right]}{\sqrt{p^2-4}\left(p+\sqrt{p^2-4}\right)^n} \tag{4.533}$$

$$L_t[F_n(t, t-\xi)] = \frac{2^{n-1} \exp\left[-\frac{1}{2}\left(p - \sqrt{p^2-4}\right)\xi\right]}{p\sqrt{p^2-4}\left(p+\sqrt{p^2-4}\right)^{n-1}} \tag{4.534}$$

for $n \geq 0$. Let us introduce

$$K_n(\xi,\eta;t) = \int_0^t G_0(\xi-x, t-x)F_n(x, x-\eta)dx \tag{4.535}$$

$$M_n(\xi,\eta;t) = \int_0^t G_0(\xi-x, t-x)G_n(x, x-\eta)dx \tag{4.536}$$

$$w_n(\xi,\eta,t) = 2K_n(\xi,\eta;t) - M_n(\xi,\eta;t) \tag{4.537a}$$

or

$$w_n(\xi,\eta,t) = M_n(\xi,\eta;t) - 2K_{n+2}(\xi,\eta;t) \tag{4.537b}$$

One can then derive (for $n \geq 0$)

$$L_t[K_n(\xi,\eta;t)] = \frac{\exp\left(\frac{\xi-\eta}{p}\right)}{p^{n-2}(p^2-1)(p^2+1)} \tag{4.538}$$

$$L_t[M_n(\xi,\eta;t)] = \frac{\exp\left(\frac{\xi-\eta}{p}\right)}{p^{n}(p^2-1)} \tag{4.539}$$

and

$$w_n(\xi,\eta;t) = F_{n+1}(t,\xi-\eta) \tag{4.540}$$

Moreover, for f appropriately regular:

$$L_t\left[\int_0^t G_0(\xi-x,t-x)f(x)dx\right] = \frac{1}{p}f^*\left(p+\frac{1}{p}\right)\exp\left(\frac{\xi}{p}\right) \tag{4.541}$$

Let

$$D_n(\alpha,\beta) = \left(\frac{\beta}{\alpha}\right)^{\frac{n}{2}} J_n(2\sqrt{\alpha\beta}) \tag{4.542}$$

where $J_n(\cdot)$ is Bessel's function of order n. Note that for $n=0$ we have $D_0(\alpha,\beta) = J_0(2\sqrt{\alpha\beta})$ and

$$D_0(\alpha,\beta) = G_0(-\alpha,\beta) \tag{4.543}$$

where $G_0(\cdot)$ is defined by (4.124). For the special case of $\xi=0$, the rule (4.541) reduces to

$$L_t\left[\int_0^t D_0(x,t-x)f(x)dx\right] = \frac{1}{p}f^*\left(p+\frac{1}{p}\right) \tag{4.544}$$

The following integration results will be used in the sequel:

$$\int_0^t D_0(x,t-x)dx = \sin t \tag{4.545}$$

$$\int_0^t \frac{t}{2\sqrt{x}\sqrt{t-x}} J_1(2\sqrt{x}\sqrt{t-x})dx = 1-\cos t \tag{4.546}$$

$$\int_0^{t_\lambda} D_0(t_\lambda-\alpha,\alpha)[G_n(t-\alpha,t-\alpha+\theta) - G_{n+2}(t-\alpha,t-\alpha+\theta]d\alpha = G_{n+1}(t,t-t_\lambda+\theta) - G_{n+1}(t-t_\lambda,t+\theta) \tag{4.547}$$

for $n \geq -1$. By putting $t_\lambda = t$, we get

$$\int_0^t D_0(t-\alpha,\alpha)[G_n(t-\alpha,t-\alpha+\theta)-G_{n+2}(t-\alpha,t-\alpha+\theta)]d\alpha =$$

$$\begin{cases} G_0(t,\theta)-1 & \text{for } n=-1 \\ G_{n+1}(t,\theta) & \text{for } n\geq 0 \end{cases} \tag{4.548}$$

4.11.2 Riemann's Formula: An Alternative Version

Let $w = w(\alpha,\beta)$ satisfy the differential equation: $\mathscr{L}w = 0$ within the domain CDG or for $\theta_1 < \alpha < \theta_1+\theta_2$, $\alpha-\theta_1 < \beta < \theta_2$, see Fig. 4.77. Let

$$G(\alpha,\beta) = D_0(\alpha-\lambda,\mu-\beta) \tag{4.549}$$

$$\varphi(\alpha) = \left.\left[\frac{\partial w(\alpha,\beta)}{\partial\alpha}-\frac{\partial w(\alpha,\beta)}{\partial\beta}\right]\right|_{\beta=\alpha-\theta_1} \tag{4.550}$$

and

$$w_F = w(\mu+\theta_1,\mu), \qquad w_E = w(\lambda,\lambda-\theta_1), \qquad w_Q = w(\lambda,\mu) \tag{4.551}$$

One can express w_Q by

$$w_Q = \frac{1}{2}(w_F+w_E) - \frac{1}{2}\int_\lambda^{\mu+\theta_1} w(\alpha,\alpha-\theta_1)\left.\left(\frac{\partial G}{\partial\beta}-\frac{\partial G}{\partial\alpha}\right)\right|_{\beta=\alpha-\theta_1} d\alpha -$$
$$\frac{1}{2}\int_\lambda^{\mu+\theta_1} G(\alpha,\alpha-\theta_1)\varphi(\alpha)d\alpha \tag{4.552}$$

Let us define

$$g(\lambda,\mu) = \int_\lambda^{\mu+\theta_1}\left.\left(\frac{\partial G}{\partial\beta}-\frac{\partial G}{\partial\alpha}\right)\right|_{\beta=\alpha-\theta_1} d\alpha \tag{4.553}$$

Using the rule: $J_1(z) = -\dfrac{dJ_0(z)}{dz}$ known from the theory of Bessel functions one can rearrange g to the form

$$g(\lambda,\mu) = \int_\lambda^{\mu+\theta_1} \frac{\mu+\theta_1-\lambda}{\sqrt{(\alpha-\lambda)(\mu+\theta_1-\lambda)}} J_1\left(2\sqrt{(\alpha-\lambda)(\mu+\theta_1-\lambda)}\right) d\alpha \tag{4.554}$$

and by applying (4.546) one finds

$$g(\lambda,\mu) = 2[1-\cos(\mu+\theta_1-\lambda)] \tag{4.555}$$

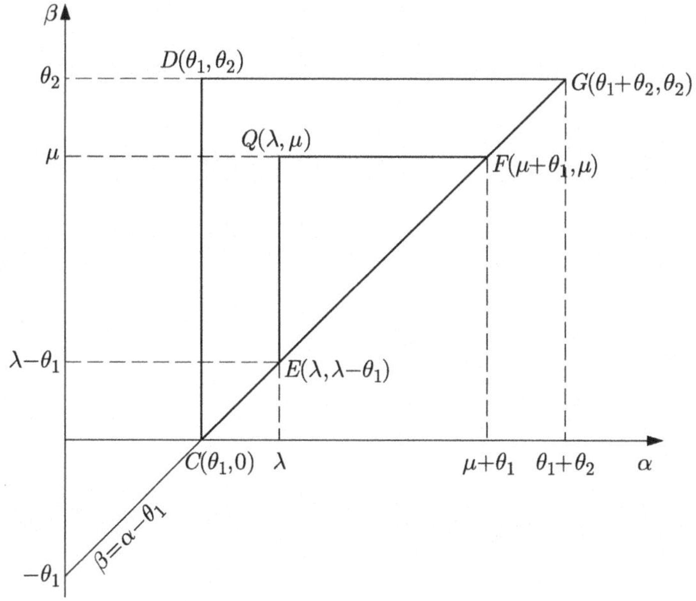

Fig. 4.77 The parameterization around the point Q. The first case

Let us proceed to derive a counterpart of the formula (4.552) for the domain QFE, see Fig. 4.78. Let T satisfy the equation $\mathscr{L}T = 0$ within the domain BHD, where $0 < \alpha < \theta_1$, $\theta_2 < \beta < \theta_1 + \theta_2$, see Fig. 4.78. Let

$$G(\alpha, \beta) = D_0(\lambda - \alpha, \beta - \mu) \tag{4.556}$$

For $(\alpha, \beta) \in EQF$, we have $\lambda - \alpha > 0, \beta - \mu > 0$; hence, the arguments of $D_0(\lambda - \alpha, \beta - \mu)$ are positive. Proceeding as in Sect. 4.2.3, we obtain

$$T_Q = \frac{1}{2}(T_F + T_E) + \frac{1}{2}\int_{\mu-\theta_2}^{\lambda}\left[T(\alpha, \alpha+\theta_2)\left(\frac{\partial G}{\partial \beta} - \frac{\partial G}{\partial \alpha}\right)\bigg|_{\beta=\alpha+\theta_2} + G(\alpha, \alpha+\theta_2)\left(\frac{\partial T}{\partial \alpha} - \frac{\partial T}{\partial \beta}\right)\bigg|_{\beta=\alpha+\theta_2}\right] d\alpha \tag{4.557}$$

with

$$T_Q = T(\lambda, \mu), \qquad T_F = T(\lambda, \lambda+\theta_2), \qquad T_E = T(\mu - \theta_2, \mu)$$

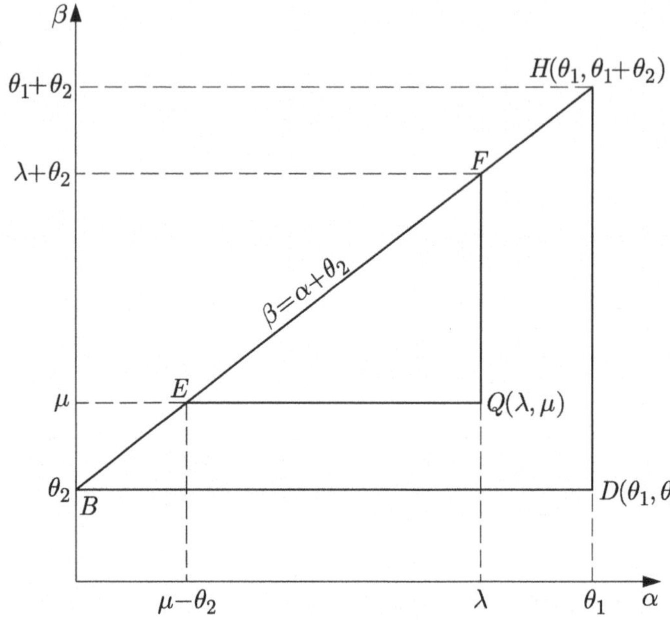

Fig. 4.78 The parameterization around the point Q. The second case

4.12 Cantilevers Constructed Within Trapezoidal Domains. Case of $\sigma_T \neq \sigma_C$

In this section, Michell's cantilevers transmitting a point load at point P to the support RN will be constructed within the domain bounded by the half-lines RR_0, NN_0, cf. Fig. 4.46b.

4.12.1 Geometry of Michell's Nets Including H.S.Y. Chan's Domains

Consider the problem of Fig. 4.46b: the feasible domain is bounded by the half-lines RR_0, NN_0. The cantilevers constructed in Sect. 4.8 did not touch the half-lines BR_0, CN_0, see Fig. 4.54; now this condition will be active. The solutions will depend on the values of the angles θ_1, θ_2 of the fans NAC, RBA, see Figs. 4.54 and 4.79.

We shall show first that if point P lies in the domain $DHJG$ and the force P is directed such that one family of fibres is in compression and the second one-in tension, then the optimal cantilever is cut out by the lines $\alpha = \alpha_P$, $\beta = \beta_P$ or by the lines linking the points $(\alpha_P, \alpha_P - \theta_1)$ and (α_P, β_P) and the points $(\beta_P - \theta_2, \beta_P)$

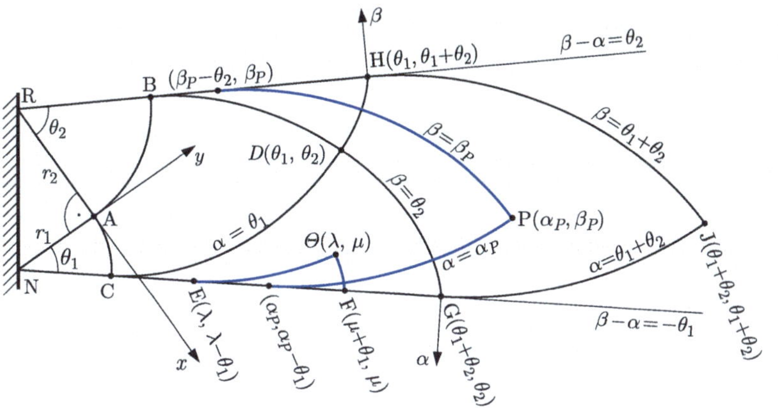

Fig. 4.79 Kinematic division of the cantilever bounded by two rays

and (α_P, β_P). The case of the point P lying outside the domain $RHJGN$ will be discussed later.

The solution to the problem of Fig. 4.79 reduces to extending the net from the $ABDC$ domain to the domains: BDH, CDG, called further H.S.Y. Chan's (or simply: Chan's) domains. The construction of the nets in these domains is fairly complex, since the Riemann formula (4.132) cannot be directly applied, as the given fields are not known along two initial arcs.

Domain CDG

As the main unknown, the field $\bar{x}(\alpha, \beta)$ is chosen. This field is continuous along CD, where $\alpha = \theta_1$. The result (4.409) determines $\bar{x}$ on CD, or

$$\bar{x}(\theta_1, \beta) = r_1 F_1(\theta_1, \beta) + r_2 F_2(\beta, \theta_1) \tag{4.558}$$

The equation of the line CG, referred to the system (x, y) of the origin at A, has the form

$$x \cos\theta_1 - y \sin\theta_1 = r_1 \sin\theta_1 \tag{4.559}$$

Let us compute $\bar{x}(\alpha, \alpha - \theta_1)$ using (4.95)

$$\begin{aligned} \bar{x}(\alpha, \beta)|_{\beta=\alpha-\theta_1} &= [x(\alpha, \beta)\cos(\beta - \alpha) + y(\alpha, \beta)\sin(\beta - \alpha)]|_{\beta=\alpha-\theta_1} \\ &= [x(\alpha, \beta)\cos\theta_1 - y(\alpha, \beta)\sin\theta_1]|_{\beta=\alpha-\theta_1} \end{aligned} \tag{4.560}$$

According to (4.559), one finds

$$\bar{x}(\alpha, \alpha - \theta_1) = r_1 \sin\theta_1 \tag{4.561}$$

which is just the boundary condition of the problem. Now, we shall apply the formula (4.552) to the field $w = \bar{x}$. We insert (4.561), make use of (4.555) and get

$$\bar{x}(\lambda, \mu) = r_1 \sin\theta_1 \cos(\eta + \theta_1 - \lambda) - \frac{1}{2}\int_{\lambda}^{\mu+\theta_1} D_0(\alpha - \lambda, \mu + \theta_1 - \alpha)\varphi(\alpha)d\alpha \tag{4.562}$$

where φ is defined by (4.550) and $w = \bar{x}$. The function φ is the main unknown of the problem. It will be determined from the compatibility of the condition (4.562) with (4.558). Making use of the identity $(4.144)_1$ we write down this condition in the form

$$\int_{\theta_1}^{\mu+\theta_1} D_0(\alpha - \lambda, \mu + \theta_1 - \alpha)\varphi(\alpha)d\alpha = -2f_1(\mu)$$

$$f_1(\mu) = r_1 F_1(\mu, \theta_1) + r_2 F_2(\mu, \theta_1) - r_1 \sin(\mu - \theta_1) - r_1 \cos\mu \sin(\theta_1) \tag{4.563}$$

Let $\bar{\varphi}(x) = \varphi(x + \theta_1)$. We rewrite (4.563) in the form

$$\int_0^{\mu} D_0(x, \mu - x)\varphi(x)dx = -2[r_1 F_1(\mu, \theta_1) + r_2 F_2(\mu, \theta_1) - r_1 \cos\theta_1 \sin\mu] \tag{4.564}$$

The above equation is of convolution type. This suggests to apply the Laplace transform. By performing the Laplace transform using the rules (4.544), (4.534) and (4.530), one arrives at

$$\frac{1}{p}\bar{\varphi}^*\left(p + \frac{1}{p}\right) = -2\left[r_1\frac{\exp\left(\frac{\theta_1}{p}\right)}{p^2+1} + r_2\frac{\exp\left(\frac{\theta_1}{p}\right)}{p(p^2+1)} - r_1\frac{\cos\theta_1}{p^2+1}\right] \tag{4.565}$$

Now, we change the variables as below:

$$\begin{aligned} &p + \frac{1}{p} = s, \qquad p = \frac{1}{2}(s + R), \qquad R = \sqrt{s^2 - 4},\\ &\frac{1}{p} = \frac{2}{s+R}, \qquad \frac{1}{p} = \frac{1}{2}(s - R), \qquad \frac{p}{p^2+1} = \frac{1}{s} \end{aligned} \tag{4.566}$$

and rearrange (4.565) to the form

$$\bar{\varphi}^*(s) = -2\frac{\exp\left[\frac{\theta_1}{2}(s - R)\right]}{R}\left[\frac{2r_2}{s} - \frac{4r_1}{s(s+R)} + r_1 - \frac{2r_2}{s+R}\right] + 2r_1\frac{\cos\theta_1}{s} \tag{4.567}$$

By taking advantage of the Laplace transforms (4.533), (4.534), one finds

$$\begin{aligned} \bar{\varphi}(x) = &-2[2r_2 F_1(x, x+\theta_1) - 2r_1 F_2(x, x+\theta_1) + r_1 G_0(x, x+\theta_1)\\ &- r_2 G_1(x, x+\theta_1) - r_1\cos\theta_1] \end{aligned} \tag{4.568}$$

and come back to the unknown φ

$$\varphi(\alpha) = -2[2r_2F_1(\alpha-\theta_1,\alpha) - 2r_1F_2(\alpha-\theta_1,\alpha) + r_1G_0(\alpha-\theta_1,\alpha) - r_2G_1(\alpha-\theta_1,\alpha) - r_1\cos\theta_1] \quad (4.569)$$

Thus, the problem (4.563) is solved. To construct the unknown field $\bar{x}$, we substitute (4.569) into (4.562). Using definitions (4.535)–4.537a and the result (4.540), one finds $\bar{x}$ and then

$$\bar{y}(\lambda,\mu) = \frac{\partial\bar{x}(\lambda,\mu)}{\partial\mu}$$

Thus, we find

$$\begin{aligned}\bar{x}(\lambda,\mu) &= \bar{x}^*(\lambda,\mu) + \bar{x}.(\lambda-\theta_1,\mu+\theta_1)\\ \bar{y}(\lambda,\mu) &= \bar{y}^*(\lambda,\mu) + \bar{y}.(\lambda-\theta_1,\mu+\theta_1)\end{aligned} \quad (4.570)$$

where

$$\begin{aligned}\bar{x}^*(\lambda,\mu) &= r_1F_1(\lambda,\mu) + r_2F_2(\mu,\lambda)\\ \bar{y}^*(\lambda,\mu) &= r_1F_2(\lambda,\mu) + r_2F_1(\mu,\lambda)\end{aligned} \quad (4.571)$$

and

$$\begin{aligned}\bar{x}.(\lambda,\mu) &= -r_1F_1(\lambda,\mu) - r_2F_2(\lambda,\mu)\\ \bar{y}.(\lambda,\mu) &= -r_1F_2(\lambda,\mu) - r_2F_3(\lambda,\mu)\end{aligned} \quad (4.572)$$

The coordinates x, y are given by (4.96). To find Lamé fields, we make use of (4.98). The results are

$$\begin{aligned}A(\alpha,\beta) &= A^*(\alpha,\beta) + A.(\alpha-\theta_1,\beta+\theta_1)\\ B(\alpha,\beta) &= B^*(\alpha,\beta) + B.(\alpha-\theta_1,\beta+\theta_1)\end{aligned} \quad (4.573)$$

where

$$\begin{aligned}A^*(\alpha,\beta) &= r_1G_0(\alpha,\beta) + r_2G_1(\beta,\alpha)\\ B^*(\alpha,\beta) &= r_1G_1(\alpha,\beta) + r_2G_0(\alpha,\beta)\end{aligned} \quad (4.574)$$

and

$$\begin{aligned}A.(\alpha,\beta) &= -r_1G_0(\alpha,\beta) - r_2G_1(\alpha,\beta)\\ B.(\alpha,\beta) &= -r_1G_1(\alpha,\beta) - r_2G_2(\alpha,\beta)\end{aligned} \quad (4.575)$$

Selected illustrative nets are shown in Fig. 4.80.

Let us note that B is continuous along CD, while A suffers a jump

$$[A]_{CD} = -r_1(\theta_1+\beta) \quad (4.576)$$

Let us note that β lines are tangent to CG, hence $A(\alpha,\alpha-\theta_1) = 0$ or A vanishes along CG.

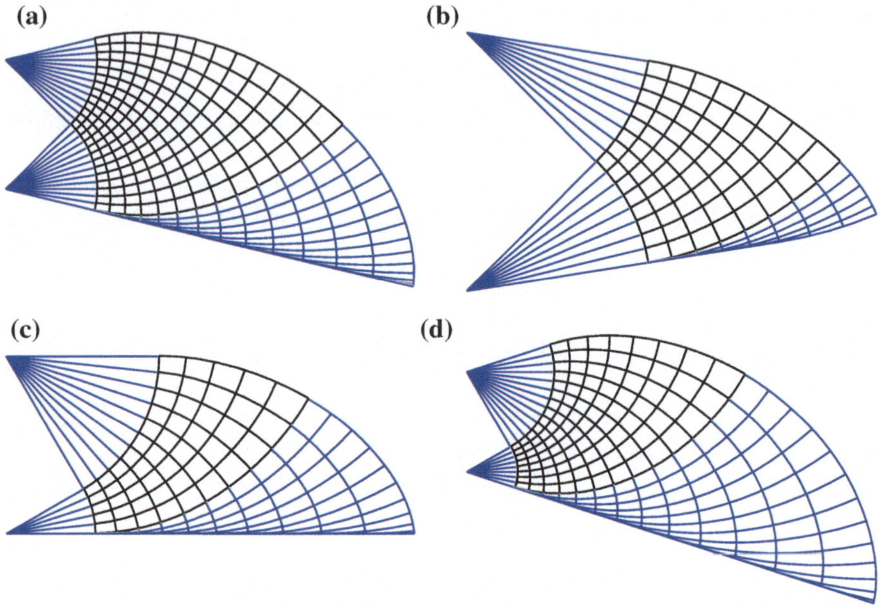

Fig. 4.80 Michell cantilevers: **a** $\kappa = 1$, $\theta_1 = \frac{13\pi}{40}$, $\alpha_P = \frac{13\pi}{20}$, $\beta_P = \frac{13\pi}{40}$; **b** $\kappa = 1$, $\theta_1 = \frac{\pi}{5}$, $\alpha_P = \frac{13\pi}{40}$, $\beta_P = \frac{\pi}{5}$; **c** $\kappa = 3$, $\theta_1 = \frac{\pi}{6}$, $\alpha_P = \frac{\pi}{2}$, $\beta_P = \frac{\pi}{3}$; and **d** $\kappa = 3$, $\theta_1 = \frac{4\pi}{15}$, $\alpha_P = \frac{7\pi}{10}$, $\beta_P = \frac{13\pi}{30}$

Domain BDH

The results are similar. Substitution: $\beta \mapsto \alpha, \theta_1 \mapsto \theta_2, r_1 \mapsto r_2, \bar{x} \mapsto \bar{y}$, changes the formulae for CDG into the formulae for BDH. Thus, one obtains

$$\begin{aligned} \bar{x}(\alpha, \beta) &= \bar{x}^*(\alpha, \beta) + \bar{x}^{\cdot}(\alpha + \theta_2, \beta - \theta_2) \\ \bar{y}(\alpha, \beta) &= \bar{y}^*(\alpha, \beta) + \bar{y}^{\cdot}(\alpha + \theta_2, \beta - \theta_2) \end{aligned} \tag{4.577}$$

where

$$\begin{aligned} \bar{x}^{\cdot}(\alpha, \beta) &= -r_1 F_3(\beta, \alpha) - r_2 F_2(\beta, \alpha) \\ \bar{y}^{\cdot}(\alpha, \beta) &= -r_1 F_2(\beta, \alpha) - r_2 F_1(\beta, \alpha) \end{aligned} \tag{4.578}$$

Then, we can construct the Lamé fields

$$\begin{aligned} A(\alpha, \beta) &= A^*(\alpha, \beta) + A^{\cdot}(\alpha + \theta_2, \beta - \theta_2) \\ B(\alpha, \beta) &= B^*(\alpha, \beta) + B^{\cdot}(\alpha + \theta_2, \beta - \theta_2) \end{aligned} \tag{4.579}$$

Here, A^*, B^* are given by (4.574) and

$$\begin{aligned} A^{\cdot}(\alpha, \beta) &= -r_1 G_2(\alpha, \beta) - r_2 G_1(\beta, \alpha) \\ B^{\cdot}(\alpha, \beta) &= -r_1 G_1(\beta, \alpha) - r_2 G_0(\alpha, \beta) \end{aligned} \tag{4.580}$$

The field A is continuous along BD, while B suffers a jump

$$[B]_{BD} = r_2(\theta_2 + \alpha) \tag{4.581}$$

Moreover, $B = 0$ along BH. The α lines are tangent to BH. In case of $\sigma_T = \sigma_C$, we have $r_1 = r_2$. If $\theta_1 = \theta_2$, the domains CDG and BDH are mutually symmetric with respect to the AD line.

Domain DHJG

The fields can be guessed. It is sufficient to check that the fields

$$\begin{aligned} \bar{x}(\alpha, \beta) &= \bar{x}^*(\alpha, \beta) + \bar{x}_\cdot(\alpha - \theta_1, \beta + \theta_1) + \bar{x}^\cdot(\alpha + \theta_2, \beta - \theta_2) \\ \bar{y}(\alpha, \beta) &= \bar{y}^*(\alpha, \beta) + \bar{y}_\cdot(\alpha - \theta_1, \beta + \theta_1) + \bar{y}^\cdot(\alpha + \theta_2, \beta - \theta_2) \end{aligned} \tag{4.582}$$

satisfy all the desired conditions. Consequently, the Lamé fields are given by

$$\begin{aligned} A(\alpha, \beta) &= A^*(\alpha, \beta) + A_\cdot(\alpha - \theta_1, \beta + \theta_1) + A^\cdot(\alpha + \theta_2, \beta - \theta_2) \\ B(\alpha, \beta) &= B^*(\alpha, \beta) + B_\cdot(\alpha - \theta_1, \beta + \theta_1) + B^\cdot(\alpha + \theta_2, \beta - \theta_2) \end{aligned} \tag{4.583}$$

Selected nets are shown in Fig. 4.81. In the case of $\kappa = 1, \theta_1 = \theta_2 = \pi/4$; the feasible domain is a strip; the BH and CG lines are parallel. The lines BA, AC, BDG, CDH will be called lines of kinematic division. They are discontinuity lines of A and B fields, cf. Fig. 4.82.

Domain GJG_2

This domain will be called H.S.Y. Chan's domain of 2nd rank, see Fig. 4.83. The net already constructed is extended such that the coordinates of G_2 are $(2\theta_1 + \theta_2, \theta_1 + \theta_2)$, see Fig. 4.83.

The $\bar{x}$, $\bar{y}$ fields are given by the rule

$$\begin{aligned} f(\alpha, \beta) = {} & f^*(\alpha, \beta) + f_\cdot(\alpha - \theta_1, \beta + \theta_1) + f^\cdot(\alpha + \theta_2, \beta - \theta_2) \\ & + f_{\cdot\cdot}(\alpha - \theta_1 - \theta_2, \beta + \theta_1 + \theta_2) \end{aligned} \tag{4.584}$$

where f represents $\bar{x}$ or $\bar{y}$ and

$$\begin{aligned} \bar{x}_{\cdot\cdot}(\alpha, \beta) &= r_1 F_3(\alpha, \beta) + r_2 F_2(\alpha, \beta) \\ \bar{y}_{\cdot\cdot}(\alpha, \beta) &= r_1 F_4(\alpha, \beta) + r_2 F_3(\alpha, \beta) \end{aligned} \tag{4.585}$$

The Lamé fields are expressed in a similar manner, by the rule (4.584) and

$$\begin{aligned} A_{\cdot\cdot}(\alpha, \beta) &= r_1 G_2(\alpha, \beta) + r_2 G_1(\alpha, \beta) \\ B_{\cdot\cdot}(\alpha, \beta) &= r_1 G_3(\alpha, \beta) + r_2 G_2(\alpha, \beta) \end{aligned} \tag{4.586}$$

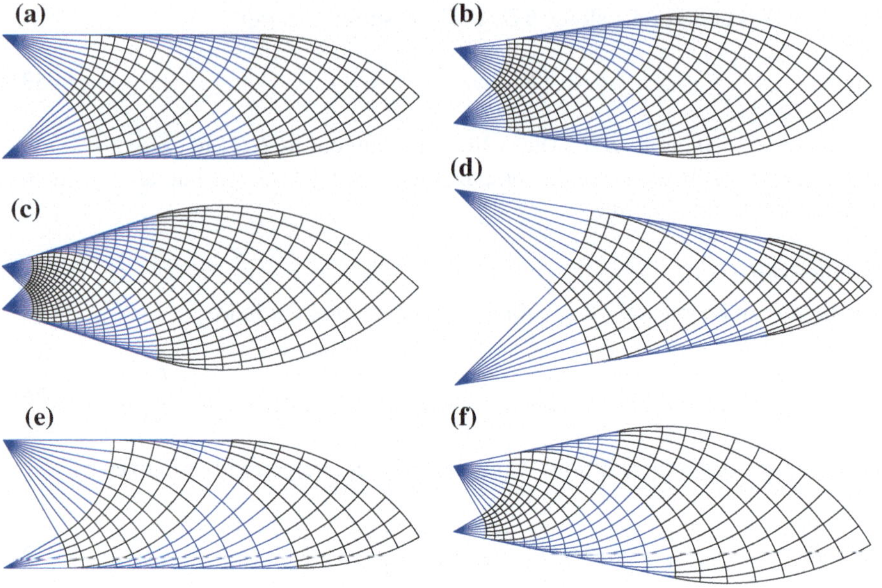

Fig. 4.81 Michell cantilever: **a** $\kappa = 1$, $\theta_1 = \dfrac{\pi}{4}$, $\theta_2 = \dfrac{\pi}{4}$, $\alpha_P = \dfrac{\pi}{2}$, $\beta_P = \dfrac{\pi}{2}$; **b** $\kappa = 1$, $\theta_1 = \dfrac{3\pi}{10}$, $\theta_2 = \dfrac{3\pi}{10}$, $\alpha_P = \dfrac{3\pi}{5}$, $\beta_P = \dfrac{3\pi}{5}$; **c** $\kappa = 1$, $\theta_1 = \dfrac{7\pi}{20}$, $\theta_2 = \dfrac{7\pi}{20}$, $\alpha_P = \dfrac{7\pi}{10}$, $\beta_P = \dfrac{7\pi}{10}$; **d** $\kappa = 1$, $\theta_1 = \dfrac{\pi}{5}$, $\theta_2 = \dfrac{\pi}{5}$, $\alpha_P = \dfrac{2\pi}{5}$, $\beta_P = \dfrac{2\pi}{5}$; **e** $\kappa = 3$, $\theta_1 = \dfrac{\pi}{6}$, $\theta_2 = \dfrac{2\pi}{6}$, $\alpha_P = \dfrac{\pi}{2}$, $\beta_P = \dfrac{\pi}{2}$; and **f** $\kappa = 3$, $\theta_1 = \dfrac{7\pi}{30}$, $\theta_2 = \dfrac{2\pi}{5}$, $\alpha_P = \dfrac{19\pi}{30}$, $\beta_P = \dfrac{19\pi}{30}$

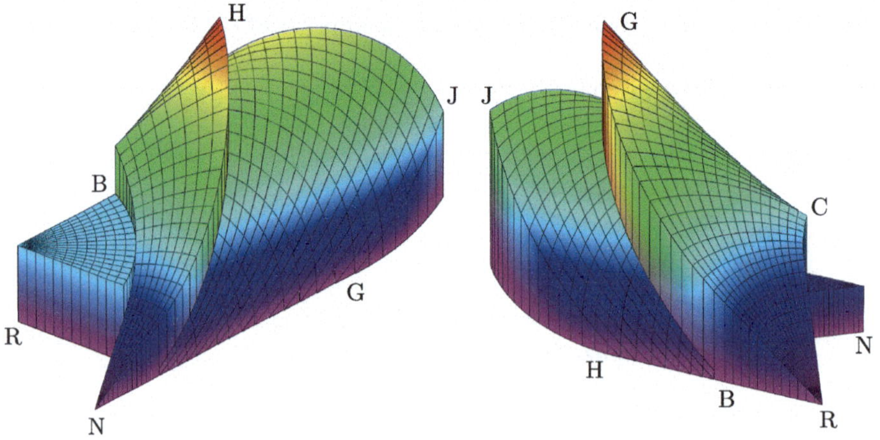

Fig. 4.82 **a** Lamé field $A(\alpha, \beta)$ for $\kappa = 3$, $\theta_1 = \pi/6$, $\theta_2 = \pi/3$; **b** Lamé field $B(\alpha, \beta)$; κ, θ_1, θ_2 as before

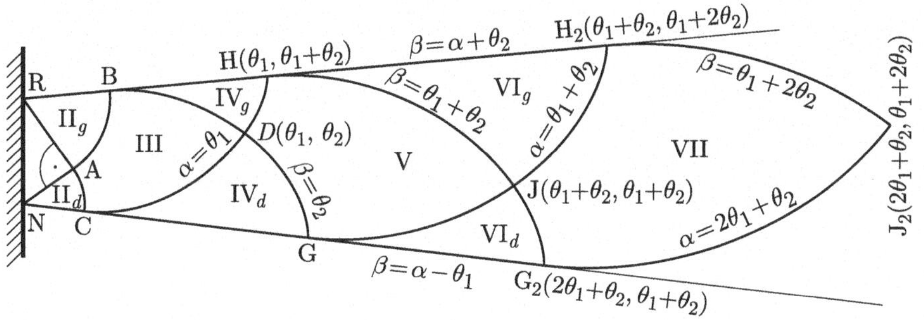

Fig. 4.83 Kinematic division of the cantilever extended to the region $JH_2J_2G_2$

Domain HJH_2

The parametric lines are determined by the rule

$$f(\alpha,\beta) = f^*(\alpha,\beta) + f_.(\alpha-\theta_1,\beta+\theta_1) + f^{\cdot}(\alpha+\theta_2,\beta-\theta_2) + f^{\cdot\cdot}(\alpha+\theta_1+\theta_2,\beta-\theta_1-\theta_2) \quad (4.587)$$

where f represents $\bar{x}$ or $\bar{y}$ and

$$\begin{aligned} \bar{x}^{\cdot\cdot}(\alpha,\beta) &= r_1F_3(\beta,\alpha) + r_2F_4(\beta,\alpha) \\ \bar{y}^{\cdot\cdot}(\alpha,\beta) &= r_1F_2(\beta,\alpha) + r_2F_3(\beta,\alpha) \end{aligned} \quad (4.588)$$

The Lamé fields are decomposed by the same rule (4.587) where f represents A or B and

$$\begin{aligned} A^{\cdot\cdot}(\alpha,\beta) &= r_1G_2(\beta,\alpha) + r_2G_3(\beta,\alpha) \\ B^{\cdot\cdot}(\alpha,\beta) &= r_1G_1(\beta,\alpha) + r_2G_2(\beta,\alpha) \end{aligned} \quad (4.589)$$

The field B vanishes along $\beta = \alpha + \theta_2$.

Domain $JH_2J_2G_2$

The parametric lines (via $\bar{x}$ or $\bar{y}$) as well as Lamé fields A, B are determined by the rule

$$\begin{aligned} f(\alpha,\beta) = {} & f^*(\alpha,\beta) + f_.(\alpha-\theta_1,\beta+\theta_1) + f^{\cdot}(\alpha+\theta_2,\beta-\theta_2) \\ & + f_{..}(\alpha-\theta_1-\theta_2,\beta+\theta_1+\theta_2) \\ & + f^{\cdot\cdot}(\alpha+\theta_1+\theta_2,\beta-\theta_1-\theta_2) \end{aligned} \quad (4.590)$$

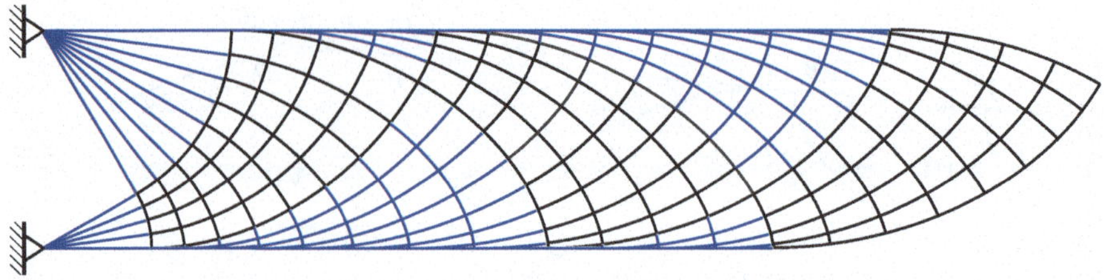

Fig. 4.84 Michell cantilever: $\kappa = 3, \theta_1 = \frac{\pi}{6}, \theta_2 = \frac{\pi}{3}, \alpha_P = \frac{2\pi}{3}, \beta_P = \frac{5\pi}{6}$

where f represents $\bar{x}$, $\bar{y}$, A or B. The illustrative example of a cantilever is shown in Fig. 4.84.

4.12.2 Adjoint Displacement Fields

When P is located in $ABCD$, then adjoint displacement fields are given by (4.419). Using the formulae (4.107), we will expand this field in a continuous way through the lines CD and BD.

Domain CDG

We will use the formula (4.107) where ω is given by (4.420). We assume $\theta = 0$ and we take into account that then

$$A(\alpha, \mu) = \begin{cases} A^*(\alpha, \mu) & \text{if } 0 \leq \alpha \leq \theta_1 \\ A^*(\alpha, \mu) + A.(\alpha - \theta_1, \beta + \theta_1) & \text{if } \alpha \geq \theta_1 \end{cases} \tag{4.591}$$

$\hat{u}_\theta(\lambda, \mu) = \hat{u}_0(\lambda, \mu)$, where according to (4.108)

$$\hat{u}_0(\lambda, \mu) = \cos \lambda u(0, \mu) - \sin \lambda v(0, \mu) \tag{4.592}$$

Substitution of (4.412) gives

$$\hat{u}_0(\lambda, \mu) = r_2 \cos \lambda + [\kappa r_1 + (1 + \kappa) r_2 \mu] \sin \lambda \tag{4.593}$$

We introduce (4.591), (4.593) to (4.107) and we obtain

$$u(\lambda, \mu) = \hat{u}_0(\lambda, \mu) + \int_0^{\theta_1} U.(\lambda, \mu; \alpha) A^*(\alpha, \mu) d\alpha + \int_{\theta_1}^{\lambda} U.(\lambda, \mu; \alpha)[A^*(\alpha, \mu) + A.(\alpha - \theta_1, \mu + \theta_1)] d\alpha \tag{4.594}$$

where $A^*(\cdot,\cdot)$, $A.(\cdot,\cdot)$ we defined before and

$$U.(\lambda,\mu;\alpha) = \cos(\lambda-\alpha) + [\sqrt{\kappa} + (1+\kappa)(\alpha+\mu)]\sin(\lambda-\alpha) \tag{4.595}$$

The field $u(\lambda,\mu)$ is defined within the limits $\theta_1 \le \lambda \le \mu+\theta_1$, $0 \le \lambda \le \theta_2$.

We rewrite the formula (4.594) as follows:

$$u(\lambda,\mu) = u^*(\lambda,\mu) + u_1(\lambda,\mu) \tag{4.596}$$

where

$$u^*(\lambda,\mu) = \hat{u}_0(\lambda,\mu) + \int_0^\lambda U.(\lambda,\mu;\alpha)A^*(\alpha,\mu)d\alpha \tag{4.597}$$

$$u_1(\lambda,\mu) = \int_{\theta_1}^\lambda U.(\lambda,\mu;\alpha)A.(\alpha-\theta_1,\mu+\theta_1)d\alpha \tag{4.598}$$

Let us note that if $\alpha \leqslant \theta_1$ then $u^*(\lambda,\mu)$ represents the displacement $u(\lambda,\mu)$ in the region $ABCD$. Thus, $u^*(\lambda,\mu)$ is expressed by the function $(4.419)_1$ also when $\lambda \geqslant \theta_1$. Instead, the integration in (4.598) is conducted with the use of formulae collected in Sect. 4.2. We obtain

$$u_1(\lambda,\mu) = u.(\lambda-\theta_1,\mu+\theta_1) \tag{4.599}$$

where

$$\begin{aligned}\frac{u.(\lambda,\mu)}{r_1} &= \kappa F_1(\lambda,\mu) - \kappa F_3(\lambda,\mu) - (1+\kappa)\lambda G_0(\lambda,\mu) + \\ &\sqrt{\kappa}[2\kappa F_2(\lambda,\mu) - (1+\kappa)\lambda G_1(\lambda,\mu)]\end{aligned} \tag{4.600}$$

Thus

$$u(\lambda,\mu) = u^*(\lambda,\mu) + u.(\lambda-\theta_1,\mu+\theta_1) \tag{4.601}$$

where $u^*(\cdot,\cdot)$ is given by r.h.s. of the first Eq. (4.419) Similarly, using the formula $(4.107)_2$ and the result (4.591), we find

$$v(\lambda,\mu) = v^*(\lambda,\mu) + v.(\lambda-\theta_1,\mu+\theta_1) \tag{4.602}$$

where $v^*(\lambda,\mu)$ is given by the r.h.s. of $(4.419)_2$ and

$$\begin{aligned}\frac{v.(\lambda,\mu)}{r_1} &= 2\kappa F_2(\lambda,\mu) + (1+\kappa)\mu G_1(\lambda,\mu) + \\ &\sqrt{\kappa}[(1+\kappa)\mu G_2(\lambda,\mu) + (1+2\kappa)F_3(\lambda,\mu) + F_1(\lambda,\mu)]\end{aligned} \tag{4.603}$$

To verify this result, it is sufficient to check

$$\frac{\partial u_{\cdot}(\alpha,\beta)}{\partial\alpha} + v_{\cdot}(\alpha,\beta) = A_{\cdot}(\alpha,\beta) \qquad (4.604)$$
$$\frac{\partial v_{\cdot}(\alpha,\beta)}{\partial\beta} + u_{\cdot}(\alpha,\beta) = -\kappa B_{\cdot}(\alpha,\beta)$$

cf. (4.116), where $A_{\cdot}$, $B_{\cdot}$ are defined by (4.575).

Domain BDM

In order to determine (u, v), it is necessary to use the formulae (4.110), where the integration is performed along the line β. For the sake of brevity, we present only the final results:

$$u(\lambda,\mu) = u^{*}(\lambda,\mu) + u^{\cdot}(\lambda+\theta_2, \mu-\theta_2) \qquad (4.605)$$
$$v(\lambda,\mu) = v^{*}(\lambda,\mu) + v^{\cdot}(\lambda+\theta_2, \mu-\theta_2)$$

where u^*, v^* are defined by the formulae (4.419) and

$$\frac{u^{\cdot}(\lambda,\mu)}{r_1} = -(\kappa+2)F_3(\mu,\lambda) - \kappa F_1(\mu,\lambda) - (1+\kappa)\lambda G_2(\mu,\lambda) +$$
$$\sqrt{\kappa}[-2F_2(\mu,\lambda) - (1+\kappa)\lambda G_1(\mu,\lambda)], \qquad (4.606)$$
$$\frac{v^{\cdot}(\lambda,\mu)}{r_1} = -2F_2(\mu,\lambda) + (1+\kappa)\mu G_1(\mu,\lambda) +$$
$$\sqrt{\kappa}[F_3(\mu,\lambda) - F_1(\mu,\lambda) + (1+\kappa)\mu G_0(\mu,\lambda)]$$

The fields $A^{\cdot}$, $B^{\cdot}$, $u^{\cdot}$, $v^{\cdot}$ satisfy Eqs. (4.116), which confirms their correctness.

When $\sigma_T = \sigma_C$, thus $\kappa = 1, r_1 = r_2 = r_0, \theta_1 = \theta_2 = \theta$ then the determined above fields u, v satisfy the relations

$$u(\alpha,\beta)^{ABDC} = -v(\beta,\alpha)^{ABDC}$$
$$u(\alpha,\beta)^{BDH} = -v(\beta,\alpha)^{CDG} \qquad (4.607)$$
$$v(\alpha,\beta)^{BDH} = -u(\beta,\alpha)^{CDG}$$

Domain DHJG

One can check that the adjoint displacement fields are expressed by the rule

$$f(\alpha,\beta) = f^{*}(\alpha,\beta) + f_{\cdot}(\alpha-\theta_1, \beta+\theta_1) + f^{\cdot}(\alpha+\theta_2, \beta-\theta_2) \qquad (4.608)$$

where $f = u$ or $f = v$ and u^*, v^*, $u_{\cdot}$, $v_{\cdot}$, $u^{\cdot}$, $v^{\cdot}$ are derived by (4.419), (4.600), (4.603), (4.606).

In case of $\sigma_T = \sigma_C$, for $r_1 = r_2 = r_0$, $\theta_1 = \theta_2 = \theta$, we find $u(\alpha,\beta) = u^{**}(\alpha,\beta)$, $v(\alpha,\beta) = -u^{**}(\alpha,\beta)$ and

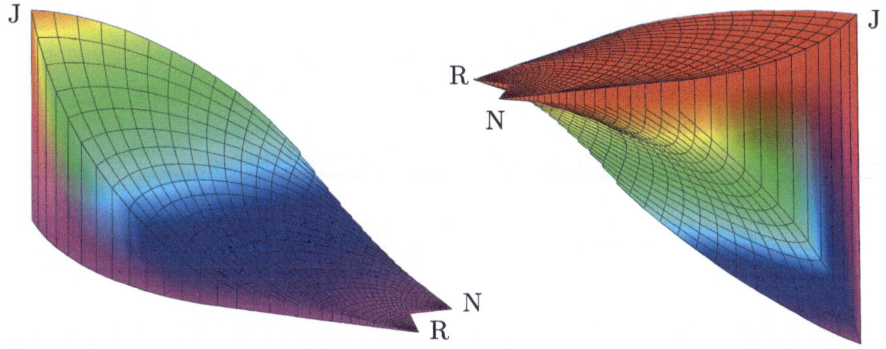

Fig. 4.85 Virtual displacement fields $u(\alpha, \beta)$ and $v(\alpha, \beta)$, $\kappa = 1, \theta_1 = \theta_2 = \dfrac{13}{40}\pi$

$$\begin{aligned}\frac{u^{**}(\alpha,\beta)}{r_0} = {} & (1+2\alpha)G_0(\alpha,\beta) + 2\alpha G_1(\beta,\alpha) - \\ & 2(\beta-\theta)T_0(\beta-\theta,\alpha+\theta) - 2(\alpha-\theta)T_0(\alpha-\theta,\beta+\theta) + \\ & F^{\cdot}(\beta-\theta,\alpha+\theta) + F_{.}(\alpha-\theta,\beta+\theta)\end{aligned} \tag{4.609a}$$

where

$$F^{\cdot} = F_1 - 2F_2 - F_3, \qquad F_{.} = F_1 + 2F_2 - F_3 \tag{4.609b}$$

and $T_0 = G_0 + G_1$, see Eq. (4.147).

An illustrative diagram of $u(\alpha, \beta)$ is shown in Fig. 4.85.

Domain GJG_2

For the sake of simplicity, we will assume $\sigma_T = \sigma_C$ and thus $\kappa = 1$, $r_1 = r_2 = r_0$, $\theta_1 = \theta_2 = \theta$. The counterpart of the formula (4.596) assumes the form

$$u(\lambda,\mu) = u^{**}(\lambda,\mu) + u_1(\lambda,\mu) \tag{4.610}$$

where

$$u_1(\lambda,\mu) = \int_{2\theta}^{\lambda} U_{.}(\lambda,\mu;\alpha)A_{..}(\alpha-2\theta,\mu+2\theta)d\alpha \tag{4.611}$$

and $U_{.}(\lambda, \mu; \alpha)$ takes now the form:

$$U_{.}(\lambda,\mu;\alpha) = \cos(\lambda-\alpha) + (1+2\alpha+2\mu)\sin(\lambda-\alpha) \tag{4.612}$$

The integral (4.611) can be expressed with the use of Chan's functions:

$$u_1(\lambda,\mu) = u_{..}(\lambda-2\theta,\mu+2\theta) \tag{4.613}$$

where

$$\frac{1}{r_0} u_{..}(\lambda, \mu) = 2\lambda G_1(\lambda, \mu) - (1 + 2\lambda) G_2(\lambda, \mu) - 4F_2(\lambda, \mu) - 4F_3(\lambda, \mu) \tag{4.614}$$

Therefore, u is given explicitly by (4.610), (4.609a), (4.614). Similarly, we find the field v

$$v(\lambda, \mu) = v^{**}(\lambda, \mu) + v_{..}(\lambda - 2\theta, \mu + 2\theta) \tag{4.615}$$

$$\frac{v_{..}(\lambda, \mu)}{r_0} = -2\lambda G_0(\lambda, \mu) - G_2(\lambda, \mu) - 2(1 + \lambda) G_1(\lambda, \mu) + 4F_1(\lambda, \mu) + 4F_2(\lambda, \mu) \tag{4.616}$$

The condition of correctness is

$$\frac{\partial u_{..}(\alpha, \beta)}{\partial \alpha} + v_{..}(\alpha, \beta) = A_{..}(\alpha, \beta) \tag{4.617}$$

where $A_{..}(\alpha, \beta)$ is given by (4.589).

Domain HJH_2

We still assume $\sigma_T = \sigma_C$. In a similar way, we find

$$\begin{aligned} u(\lambda, \mu) &= u^{**}(\lambda, \mu) - v_{..}(\mu - 2\theta, \lambda + 2\theta) \\ v(\lambda, \mu) &= v^{**}(\lambda, \mu) - u_{..}(\mu - 2\theta, \lambda + 2\theta) \end{aligned} \tag{4.618}$$

Domain $JH_2J_2G_2$

We further assume $\sigma_T = \sigma_C$. The fields u, v are given by

$$\begin{aligned} u(\alpha, \beta) &= u^{**}(\alpha, \beta) + u_{..}(\lambda - 2\theta, \mu + 2\theta) - v_{..}(\mu - 2\theta, \lambda + 2\theta) \\ v(\alpha, \beta) &= -u(\beta, \alpha) \end{aligned} \tag{4.619}$$

4.12.3 Force Fields

As long as the boundary RB_1PC_1N does not touch the rays RR_0, NN_0 (cf. Fig. 4.54), the static solution found in Sect. 4.8 holds also when the design domain is confined by the above rays. The fields T_1, T_2 in the domain $B_1PC_1AB_1$ (Fig. 4.54) are given by (4.351), (4.352).

Let us look at Fig. 4.79. If the point P is located outside the domain $RBDCN$, then the solution found in Sect. 4.8 ceases to be valid. We will further, respectively, assume that the point P is located inside the regions marked in Fig. 4.79.

4.12.3.1 The Force in Domain BDH

Let P be located in the domain BDH—the upper Chan's domain. Two cases should be considered separately. First, we consider the case of point P lying on the boundary RR_0. Then, we deal with the case of P within the BDH domain. These two cases result in different divisions of the cantilever called static divisions.

A. Consider now the case of P of coordinates (α_P, β_P) lying on RR_0 which corresponds to $\alpha_P \leqslant \theta_1$, $\beta_P = \alpha_P + \theta_2$, see Fig. 4.86.

The force P is directed in such a way that it compresses the bar PD_PC_PN and extends the bar RBP. The force in the bar PD_PC_P is constant since all connected bars are perpendicular. The force in the bar BP changes since α-lines of fibres are connected tangentially and they alter the value of longitudinal force along the BP.

We will solve the problem of statics in the domain BD_PP. We consider $T_2 = T_2(\alpha, \beta)$ as a main unknown. In the domain BDH, the field T_2 satisfies the equation $\mathscr{L}T_2 = 0$. Hence, the value of T_2 in an arbitrary point (λ, μ) of this domain is expressed by the formula (4.557). The field T_2 vanishes on the line BH, thus $T_2(\alpha, \alpha + \theta_2) = 0$, $\mu - \theta_2 \leq \alpha \leq \lambda$ which simplifies (4.557) to the form

$$T_2(\lambda, \mu) = \frac{1}{2}\int_{\mu-\theta_2}^{\lambda} D_0(\lambda - \alpha, \alpha + \theta_2 - \mu)\chi(\alpha)d\alpha \tag{4.620}$$

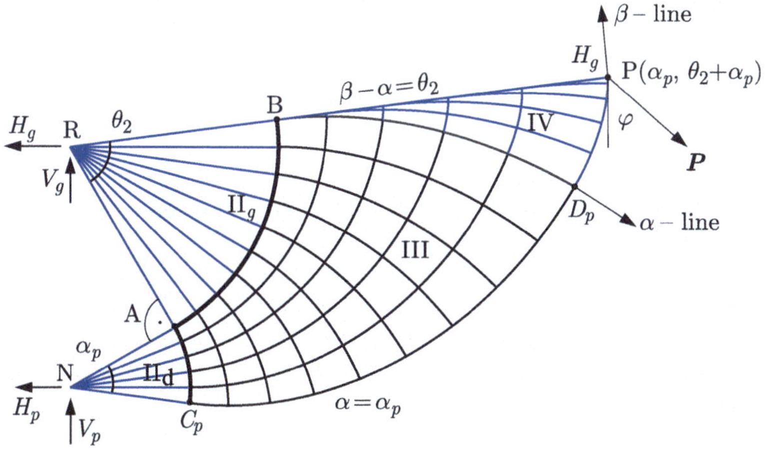

Fig. 4.86 Michell's cantilever with the force applied on the straight boundary

where

$$\chi(\alpha) = \left(\frac{\partial T_2}{\partial \alpha} - \frac{\partial T_2}{\partial \beta} \right) \bigg|_{\beta=\alpha+\theta_2} \tag{4.621}$$

The equilibrium condition of the node P yields formulae (4.421). According to boundary condition (4.182) along the line β holds

$$T_1(\alpha_P, \beta) + F_C(\beta) = 0 \tag{4.622}$$

where $F_C(\beta) = P_2 = \text{const}$, cf. (4.421), $P_2 = -P\cos(\psi + \varphi)$. From the equilibrium equation $(4.176)_2$

$$T_2 = -P_2\beta + C \tag{4.623}$$

and the constant C is obtained from the condition $T_2 = 0$ at P or $T_2(\alpha_P, \theta_2 + \alpha_P) = 0$. Thus, on the line $\alpha = \alpha_P$, we have

$$T_2(\alpha_P, \beta) = -P_2(\beta - \alpha_P - \theta_2) \tag{4.624}$$

The above boundary condition leads to the integral equation

$$-P_2(\mu - \alpha_P - \theta_2) = \frac{1}{2}\int_{\mu-\theta_2}^{\alpha_P} D_0(\alpha_P - \alpha, \alpha + \theta_2 - \mu)\chi(\alpha)d\alpha \tag{4.625}$$

We will prove that the solution has the form:

$$\chi(\alpha) = 2P_2[G_0(\alpha_P - \alpha, \alpha_P - \alpha) - G_2(\alpha_P - \alpha, \alpha_P - \alpha)] \tag{4.626}$$

while the substitution of the above result to (4.620) renders the final form of the field T_2

$$T_2(\lambda, \mu) = P_2[G_1(\beta_P - \mu, \alpha_P - \lambda) - G_1(\alpha_P - \lambda, \beta_P - \mu)] \tag{4.627}$$

and Eqs. $(4.176)_2$ render the field

$$T_1(\lambda, \mu) = P_2[-G_0(\beta_P - \mu, \alpha_P - \lambda) + G_2(\alpha_P - \lambda, \beta_P - \mu)] \tag{4.628}$$

The above results define T_1, T_2 in the domain BPD_PC_PAB. Division of this region by the line BD_P does not change the form of the above formulae. Thus, the crucial issue is to find the solution of the equation (4.625). This problem is similar to the problem (4.563), whose solution was the condition of the construction of the lines of Michell's net in the domain CDG.

The change of variables

$$\tilde{\alpha} = \alpha - \mu + \theta_2, \qquad t = \theta_2 + \alpha_P - \mu, \qquad \chi(\alpha) = \tilde{\chi}(\alpha - \mu + \theta_2) \tag{4.629}$$

transfers (4.625) to the form

$$P_2 t = \frac{1}{2}\int_0^t D_0(t-\tilde{\alpha},\tilde{\alpha})\tilde{\chi}(\tilde{\alpha})d\tilde{\alpha} \tag{4.630}$$

We perform the Laplace transform of both sides to obtain

$$\frac{P_2}{p^2} = L_t\left\{\frac{1}{2}\int_0^t D_0(t-\tilde{\alpha},\tilde{\alpha})\tilde{\chi}(\tilde{\alpha})d\tilde{\alpha}\right\} \tag{4.631}$$

Let us decompose the l.h.s. by

$$\frac{P_2}{p^2} = \frac{P_2}{p^2-1} - \frac{P_2}{p^2(p^2-1)} \tag{4.632}$$

The rules (4.531)–(4.539) imply

$$L_t\left\{\frac{1}{2}\int_0^t D_0(x,t-x)G_n(t-x,t-x)dx\right\} = \frac{1}{p^n(p^2-1)} \tag{4.633}$$

and the formulae (4.632) can be rewritten as

$$\frac{P_2}{p^2} = P_2 L_t\left\{\int_0^t D_0(\tilde{\alpha},t-\tilde{\alpha})[G_0(t-\tilde{\alpha},t-\tilde{\alpha}) - G_2(t-\tilde{\alpha},t-\tilde{\alpha})]d\tilde{\alpha}\right\} \tag{4.634}$$

By equating the above with (4.631), we find two possible solutions
(a) $\tilde{\chi}(\tilde{\alpha}) = 2P_2[G_0(t-\tilde{\alpha},t-\tilde{\alpha}) - G_2(t-\tilde{\alpha},t-\tilde{\alpha})]$
or
(b) $\tilde{\chi}(\tilde{\alpha}) = 2P_2[G_0(\tilde{\alpha},\tilde{\alpha}) - G_2(\tilde{\alpha},\tilde{\alpha})]$

The solution (a) leads to the result (4.626). It can be proved that the result (b) is not correct since it leads to contradiction in the equations of equilibrium.

In order to obtain (4.627), we substitute (4.626) to (4.620) and we perform integration using the formula (4.547) for $\theta = 0$, $n = 0$. This concludes derivation (4.627), (4.628).

Let us repeat that the fields T_1, T_2 given by (4.627), (4.628) are defined by the same formulae in two domain BPD_P and ABD_PC_PA.

The line AB is the line of continuity of the field T_1 and the line of jump of the field T_2. The line AC_P is the line of continuity of the field T_2 and the line of jump of the field T_1.

In the domain NAC_P, the fields T_1, T_2 are given by

$$\begin{aligned} T_1(\alpha,\beta_1) &= 0 \\ T_2(\alpha,\beta_1) &= P_2[G_1(\beta_P,\alpha_P-\alpha) - G_1(\alpha_P-\alpha,\beta_P)] \end{aligned} \tag{4.635}$$

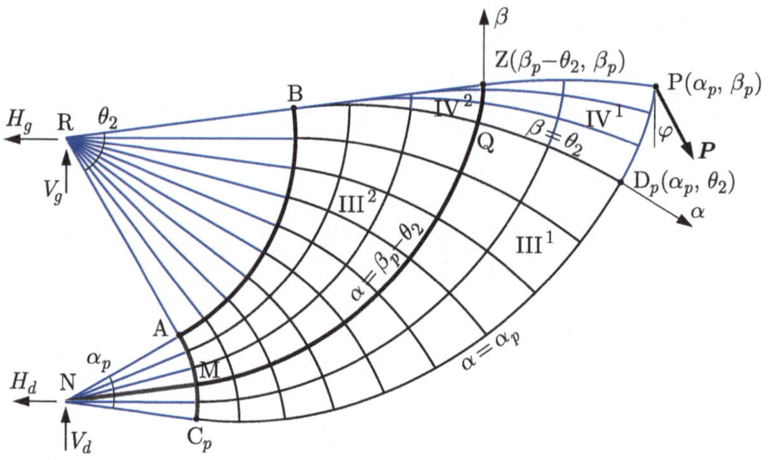

Fig. 4.87 Michell's cantilever with the force applied in H.S.Y. Chan's domain

In the domain RBA, the fields T_1 and T_2 are given by

$$T_1(\alpha_1, \beta) = P_2[-G_0(\beta_P - \beta, \alpha_P) + G_2(\alpha_P, \beta_P - \beta)] \qquad (4.636)$$
$$T_2(\alpha_1, \beta) = 0$$

the parametric lines in these domains being shown in Fig. 4.49.

B. Let us further consider the case when P is located inside domain BDH. Now $0 < \alpha_P < \theta_1$, $\theta_2 < \beta_P < \alpha_P + \theta_2$, cf. Fig. 4.87.

The Michell's structure is now confined by the lines $RBZP$, NC_PD_PP; point Z lies at the line α tangent to RB and it has coordinates $(\beta_P - \theta_2, \beta_P)$. The point Z determines the position of points $M(\beta_P - \theta_2, 0)$ and $Q(\beta_P - \theta_2, \theta_2)$ on the line $\alpha = \beta_P - \theta_2$.

The structure is composed of:

- circular fans RBA, NAM, NMC_P;
- Chan's domain $BQZ = IV^2$;
- non-singular domains: $ABQM = III^2$, $MQDC_P = III^1$, $ZPD_PQ = IV^1$.

The condition of the equilibrium of the node P defines the values of the forces F_C, F_P in the reinforcing ribs PZ, PD_P in the vicinity of the point P; these values are given by the formulae (4.421). The axial force F_C remains constant along the rib PD_PC_PN. The force $F_T = P_1 = \text{const}$ on PZ. We see that the kinematic division (Roman numerals in Fig. 4.83) is additionally divided by the lines (bold lines in Fig. 4.86) of the static division. Some of these lines are the lines of discontinuity of the fields T_1 or T_2. Indices of the static division are the top indices, e.g. IV^2 is the second static domain of the domain IV resulting from the kinematic analyses. We can see that static domain no 1 is composed of domains III^1 and IV^1.

On the other hand, the force F_T is varying along the section BZ and then it is constant on the section RB. The fields of forces T_1, T_2 in the domain ZPD_PC_PMQZ are defined by the values at the boundaries ZP and PD_PC_P and they are given by the formulae (4.351) and (4.352), which were derived for the first time while solving the problem of the exterior of the circle.

The line $ZQMN$ is the line of the jump of the field T_2 and the line of the continuity of the field T_1. According to (4.351), we write T_1 on the line $ZQMN$:

$$T_1(\beta_P - \theta_2, \beta) = -P_2G_0(\alpha_P - \beta_P + \theta_2, \beta_P - \beta) + P_1G_1(\alpha_P - \beta_P + \theta_2, \beta_P - \beta) \tag{4.637}$$

and according to (4.176)

$$\frac{\partial T_2(\beta_P - \theta_2, \beta)}{\partial \beta} = T_1(\beta_P - \theta_2, \beta) \tag{4.638}$$

and thus

$$T_2(\beta_P - \theta_2, \beta) = P_2G_1(\beta_P - \beta, \alpha_P - \beta_P + \theta_2) - P_1G_0(\alpha_P - \beta_P + \theta_2, \beta_P - \beta) + C \tag{4.639}$$

But $T_2 = 0$ at the cross section located left from Z, e.g. for $\beta = \beta_P$, we have $T_2(\beta_P - \theta_2, \beta_P) = 0$ hence $C = P_1$. Therefore, T_2 on the line ZQ is equal to

$$T_2(\beta_P - \theta_2, \beta) = P_1[1 - G_0(\alpha_P - \beta_P + \theta_2, \beta_P - \beta)] + P_2G_1(\beta_P - \beta, \alpha_P - \beta_P + \theta_2) \tag{4.640}$$

On the other hand, the values $T_2(\beta_P - \theta_2, \mu)$ are represented by (4.620) for $\lambda = \beta_P - \theta_2$. Therefore, we obtain an integral equation for $\chi(\alpha)$:

$$\begin{aligned}\frac{1}{2}\int_{\mu-\theta_2}^{\beta_P-\theta_2} & D_0(\beta_P - \theta_2 - \alpha, \alpha + \theta_2 - \mu)\chi(\alpha)d\alpha = \\ & P_1[1 - G_0(\alpha_P - \beta_P + \theta_2, \beta_P - \mu)] + \\ & P_2G_1(\beta_P - \mu, \alpha_P - \beta_P + \theta_2)\end{aligned} \tag{4.641}$$

The solution of this equation has the form:

$$\begin{aligned}\chi(\alpha) = {} & 2P_1[G_1(\beta_P - \theta_2 - \alpha, \alpha_P - \alpha) - G_{-1}(\beta_P - \theta_2 - \alpha, \alpha_P - \alpha)] + \\ & 2P_2[G_0(\beta_P - \theta_2 - \alpha, \alpha_P - \alpha) - G_2(\beta_P - \theta_2 - \alpha, \alpha_P - \alpha)]\end{aligned} \tag{4.642}$$

Substitution of the result (4.642) into the representation (4.620) and performing integration with the use of (4.547) yield the result in the form:

$$T_2(\lambda, \mu) = T_2^*(\lambda, \mu) + T_2^{(2)}(\lambda, \mu) \tag{4.643a}$$

where $T_2^*(\lambda, \mu)$ is given by (4.352) and

$$T_2^{(2)}(\alpha,\beta) = P_1 G_0(\beta_P - \theta_2 - \alpha, \alpha_P + \theta_2 - \beta) - P_2 G_1(\beta_P - \theta_2 - \alpha, \alpha_P + \theta_2 - \beta) \tag{4.643b}$$

and Eq. (4.187) gives the second force field

$$T_1(\lambda,\mu) = T_1^*(\lambda,\mu) + T_1^{(2)}(\lambda,\mu) \tag{4.644a}$$

where $T_1^*(\lambda,\mu)$ is given by (4.351) and

$$T_1^{(2)}(\alpha,\beta) = -P_1 G_1(\beta_P - \theta_2 - \alpha, \alpha_P + \theta_2 - \beta) + P_2 G_2(\beta_P - \theta_2 - \alpha, \alpha_P + \theta_2 - \beta) \tag{4.644b}$$

The above formulae defining T_1, T_2 hold in the domain $AMQZB$. These results define distribution of forces in the domains of circular fans. Using the conditions of the continuity of the field T_1 on the line AB, we find the forces inside the domain RBA:

$$\begin{aligned} T_1(\alpha_1,\beta) = {} & P_1[G_1(\alpha_P, \beta_P - \beta) - G_1(\beta_P - \theta_2, \alpha_P + \theta_2 - \beta)] + \\ & P_2[G_2(\beta_P - \theta_2, \alpha_P + \theta_2 - \beta) - G_0(\beta_P - \beta, \alpha_P)] \\ T_2(\alpha_1,\beta) = {} & 0 \end{aligned} \tag{4.645}$$

the parameterization (α_1, β) being shown in Fig. 4.49. The fields T_1, T_2 in $MZPD_PC_PM$ are given by (4.351), (4.352). The conditions of the continuity of T_2 along MC_P determine the fields T_1, T_2 in the fan NMC_P:

$$\begin{aligned} T_1(\alpha,\beta_1) &= 0 \\ T_2(\alpha,\beta_1) &= P_2 G_1(\beta_P, \alpha_P - \alpha) - P_1 G_0(\beta_P, \alpha_P - \alpha) \end{aligned} \tag{4.646}$$

where (α, β_1) are the parametric lines shown in Fig. 4.49.

The fields T_1, T_2 in the fan NAM are defined by the fields T_1, T_2 in $AMQZB$. Due to conditions of continuity of the field T_2 on the line AM, we obtain

$$\begin{aligned} T_1(\alpha,\beta_1) = {} & 0, \\ T_2(\alpha,\beta_1) = {} & P_1[G_0(\beta_P - \theta_2 - \alpha, \alpha_P + \theta_2) - G_0(\beta_P, \alpha_P - \alpha)] + \\ & P_2[G_1(\beta_P, \alpha_P - \alpha) - G_1(\beta_P - \theta_2 - \alpha, \alpha_P + \theta_2)] \end{aligned} \tag{4.647}$$

where α, β_1 parameterize the fan, see Fig. 4.49. Thus, the line NM is the line of a jump of the field T_2 in the region NAC_P.

Derivation of the formula (4.642)

For the sake of completeness, we will present derivation of the solution of the integral equation (4.641). We change the variables

$$\tilde{\alpha} = \alpha - \mu + \theta_2, \qquad t = \beta_P - \mu \tag{4.648}$$

and thus

$$t - \tilde{\alpha} = \beta_P - \alpha - \theta_2, \qquad t - \tilde{\alpha} + \theta = \alpha_P - \alpha \tag{4.649}$$

where

$$\theta = \alpha_P - \beta_P + \theta_2$$

We rewrite (4.641) in the form:

$$\frac{1}{2}\int_0^t D_0(t-\tilde{\alpha}, \tilde{\alpha})\tilde{\chi}(t-\tilde{\alpha})d\tilde{\alpha} = P_1[1 - G_0(\theta, t)] + P_2 G_1(t, \theta) \tag{4.650}$$

where $\chi(\alpha) = \tilde{\chi}(t - \tilde{\alpha})$.

We perform the Laplace transform with respect to t. Using (4.544) and (4.531), (4.532), one finds

$$\frac{1}{2}\frac{1}{p}\tilde{\chi}^*\left(p + \frac{1}{p}\right) = \frac{P_1}{p} - \frac{P_1}{p}\exp\left(\frac{\theta}{p}\right) + \frac{P_2}{p^2}\exp\left(\frac{\theta}{p}\right) \tag{4.651}$$

We change the variable p into s by (4.566). Hence,

$$\frac{1}{2}\tilde{\chi}^*(s) = P_1[1 - \exp(2\theta h)] + 2P_2 h \exp(2\theta h) \tag{4.652a}$$

where $h = 1/(s + R)$. Let us differentiate both sides of (652) and rewrite the result as below

$$\frac{1}{2}\frac{d\tilde{\chi}^*(s)}{ds} = P_1\theta\frac{2^1}{R(s+R)}\exp\left[\frac{\theta}{2}(s-R)\right] - P_2\frac{2^1}{R(s+R)}\exp\left[\frac{\theta}{2}(s-R)\right] - \\ - P_2\theta\frac{2^2}{R(s+R)}\exp\left[\frac{\theta}{2}(s-R)\right] \tag{4.652b}$$

Now, we perform the Laplace transform with respect to s and make use of the formulae (4.533), (4.534). We get

$$-\frac{1}{2}L_t[t\tilde{\chi}(t)] = P_1\theta L_t[G_1(t, t+\theta)] - P_2 L_t[G_1(t, t+\theta)] - P_2\theta L_t[G_2(t, t+\theta)] \tag{4.653}$$

hence

$$-\frac{1}{2}t\tilde{\chi}(t) = P_1\theta G_1(t, t+\theta) - P_2[G_1(t, t+\theta) + \theta G_2(t, t+\theta)] \tag{4.654}$$

Now, one can come back to previous variables using the identities (4.127), (4.128); we arrive at (4.642).

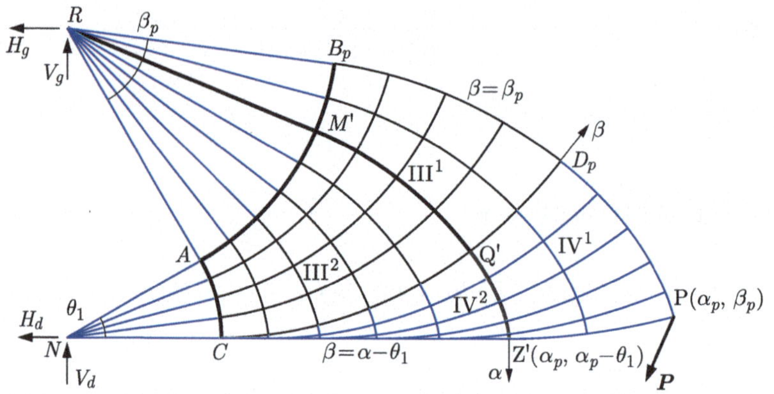

Fig. 4.88 Optimal cantilever with the force applied within lower Chan domain. Here, $\varphi < 0$

The remaining part is to determine forces in the surrounding cables. The force in the bar NC_PD_PP is constant and equal $F_C = P_2$, $P_2 = -P\cos(\psi + \phi)$. This result was known and applied before.

The force in the bar ZP is constant and equal $F_T = P_1$, $P_1 = P\sin(\psi + \phi)$. The force in bar BZ is varying and equals

$$F_T^{BZ}(\alpha) = P_1 + \int_{\alpha}^{\beta_P - \theta_2} T_1(\bar{\alpha}, \bar{\alpha} + \theta_2)d\bar{\alpha} \tag{4.655}$$

where $T_1(\alpha, \beta)$ concerns the domain BQZ and it is given by (4.644a). The force in the bar RB is constant and equals $F_T^{BZ}(\alpha = 0)$.

4.12.3.2 Force in the Domain CDG

We consider the case when point P is located inside domain CDG, see Fig. 4.88.

The location of the point P determines the cutting from the net of the domain $RB_PPZ'CNAR$ and distinguishes the points: $B_P(0, \beta_P)$, $M'(0, \alpha_P - \theta_1)$, $D(\theta_1, \beta_P)$, $Q'(\theta_1, \alpha_P - \theta_1)$, $Z'(\alpha_P, \alpha_P - \theta_1)$, $C(\theta_1, 0)$ which define the subdomains: III^1, III^2, IV^1, IV^2. The upper index enumerates static division, while the Roman indices refer to the kinematic division from Fig. 4.83.

The construction of the internal forces T_1, T_2 is similar and will not be discussed herein. We will present exclusively the final formulae.

In the domain $AM'Q'Z'CA$, the force fields T_1, T_2 are expressed by

$$T_i(\lambda, \mu) = T_i^*(\lambda, \mu) + T_i^{(3)}(\lambda, \mu), \qquad i = 1, 2 \tag{4.656a}$$

where $T_i^*(\alpha, \beta)$ are given by (4.351), (4.352) and

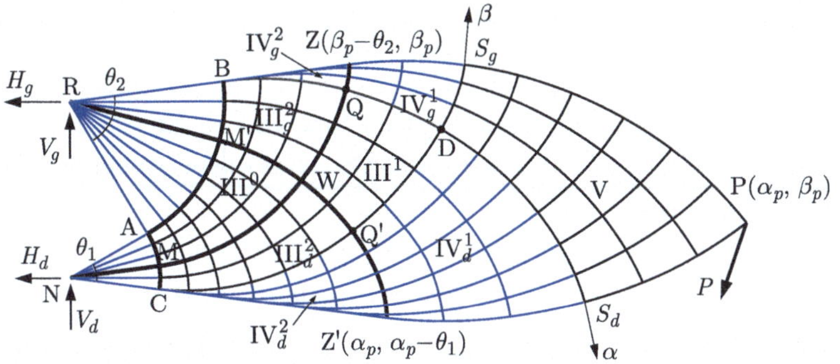

Fig. 4.89 The static division when the load is applied within $DHJG$ domain

$$
\begin{aligned}
T_1^{(3)}(\alpha,\beta) &= -P_1 G_1(\alpha_P-\theta_1-\beta,\beta_P+\theta_1-\alpha)+\\
&\quad P_2 G_0(\alpha_P-\theta_1-\beta,\beta_P+\theta_1-\alpha),\\
T_2^{(3)}(\alpha,\beta) &= P_1 G_2(\alpha_P-\beta-\theta_1,\beta_P+\theta_1-\alpha)-\\
&\quad P_2 G_1(\alpha_P-\theta_1-\beta,\beta_P+\theta_1-\alpha)
\end{aligned}
\tag{4.656b}
$$

The axial force in the bar CZ' is non-uniformly distributed, according to

$$
F_C(\beta) = P_2 + \int_\beta^{\alpha_P-\theta_1} T_2(\bar\beta+\theta_1,\bar\beta)d\bar\beta \tag{4.657}
$$

where $T_2(\cdot,\cdot)$ is given by (4.656a).

The compression force in NC is constant and equal to $F_C(0)$.

4.12.3.3 Force in the Domain $DHJG$

When the force is applied at point P located in domain $DHJG$, then the Michell structure is confined as presented in Fig. 4.89 and covers the area $RBZPZ'CNAR$. The point $P(\alpha_P,\beta_P)$ determines the location of the points:

$Z(\beta_P-\theta_2,\beta_P)$, $Z'(\alpha_P,\alpha_P-\theta_1)$ on the lines RR_0 and NN_0 (see Fig. 4.46b) and points $Q'(\theta_1,\alpha_P-\theta_1)$, $S_g(\theta_1,\beta_P)$, $M(\beta_P-\theta_2,0)$, $W(\beta_P-\theta_2,\alpha_P-\theta_1)$, $Q(\beta_P-\theta_2,\theta_2)$, $S_d(\alpha_P,\theta_2)$ which define the subdomains: $III^0=AM'WM$, $III^1=WQDQ'$, $III_d^2=CMWQ'$, $III_g^2=BM'WQ$, $IV_g^2=BQZ$, $IV_d^2=CQ'Z'$, $IV_g^1=QZS_gD$, $IV_d^1=Q'DS_dZ'$, $V=DS_gPS_d$.

The upper indices refer to the static division.

The coordinates (α_P,β_P) of the point P in the domain $DHJG$ are in the range: $\theta_1<\alpha_P\le\theta_1+\theta_2$, $\theta_2<\beta_P\le\theta_1+\theta_2$.

Let us compare the kinematic division from Fig. 4.83 and the kinematic division in Fig. 4.89. The division in Fig. 4.83 results from the kinematic analysis. This kinematic division overlaps with the lines of static division (Fig. 4.89), which results in generation of many smaller subdomains.

Michell's structure from Fig. 4.89 is reinforced by the bar in tension $RBZS_GP$ of variable thickness at the section BZ and by the compressed bar $NCZ'S_dP$ of variable thickness at the section CZ'. The force fields (T_1, T_2) are defined as follows:

- domain $WZPZ'W$: $T_i = T_i^*(\alpha, \beta)$;
- domain $M'WQZBM'$: $T_i = T_i^*(\alpha, \beta) + T_i^{(2)}(\alpha, \beta)$.

These formulae determine T_1, T_2 within the fan domain $RM'B$.

- domain $MWQ'Z'CM$: $T_i = T_i^*(\alpha, \beta) + T_i^{(3)}(\alpha, \beta)$. These formulae determine T_1, T_2 within the fan domain NCM.
- domain $MWM'A$:

$$T_i(\alpha, \beta) = T_i^*(\alpha, \beta) + T_i^{(2)}(\alpha, \beta) + T_i^{(3)}(\alpha, \beta) \tag{4.658}$$

The formulae above determine the force fields within the fan domain RAM' and NMA; these formulae will not be displayed.

4.12.3.4 Force in the Domain HJH_2

The coordinates (α_P, β_P) of the point P satisfy the conditions: $\theta_1 < \alpha_P < \theta_1 + \theta_2$, $\theta_1 + \theta_2 < \beta_P < \alpha_P + \theta_2$ cf. Fig. 4.90.

Imposition of the kinematic division from Fig. 4.83 on the static division decomposes the entire structure $RBHZPD_5B_1Z'Z_1CNAR$ into the set of subdomains: $III^5 = AA_1M_1C$, $III^4 = A_1M'D_3M_1$, $III^2 = M'BDD_3$, $IV_g^2 = BDH$,

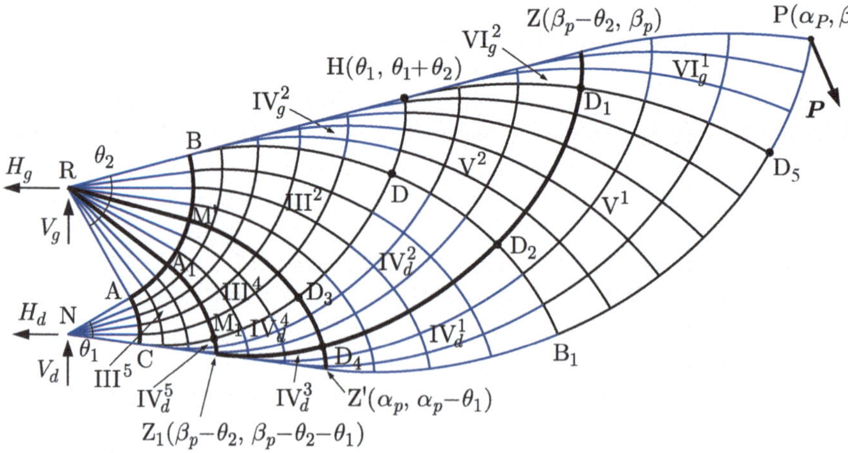

Fig. 4.90 The static division when the force is applied within the upper Chan's domain of the second rank

$IV_d^2 = D_3DD_2D_4, IV_d^4 = M_1D_3D_4Z_1, IV_d^3 = Z_1D_4Z', IV_d^1 = D_4D_2B_1Z', IV_d^5 = CM_1Z_1$, $V^1 = D_2D_1D_5B_1$, $V^2 = DHD_1D_2$, $VI_g^2 = HD_1Z$, $VI_g^1 = ZD_1D_5P$.

The coordinates of the vertices are given by $A(0,0)$, $B(0,\theta_2)$, $M'(0,\alpha_P-\theta_1)$, $D(\theta_1,\theta_2)$, $M_1(\theta_1,\beta_P-\theta_1-\theta_2)$, $C(\theta_1,0)$, $Z(\beta_P-\theta_2,\beta_P$ $A_1(0,\beta_P-\theta_1-\theta_2)$, $Z_1(\beta_P-\theta_2\beta_P-\theta_1-\theta_2)$, $D_3(\theta_1,\alpha-\theta_1)$, $D_4(\beta_P-\theta_2,\alpha_P-\theta_1)$, $H(\theta_1,\theta_1+\theta_2)$ $D_2(\beta_P-\theta_2,\theta_2)$, $B_1(\alpha_P,\theta_2)$, $D_5(\alpha_P,\theta_1+\theta_2)$, $D_1(\beta_P-\theta_2,\theta_1+\theta_2)$.

It can be seen that static domain no. 1 is composed of subdomains: IV_d^1, V^1, VI_g^1. The boundaries of the static domains are marked in bold.

In the domain of static division no. 1, the force fields T_1, T_2 are given by $T_1^*(\alpha,\beta)$, $T_2^*(\alpha,\beta)$. In the domain of static division no. 2, the force fields are defined by the formulae (4.643a), (4.644a). In the domain of static division no. 3, the force fields are given by (4.656a). In the domain of static division no. 4, the force fields are defined by (4.658).

Construction of the fields T_1, T_2 in subdomains IV_d^5, III^5 requires solution of the integral equation on the line CZ', where $\beta = \beta_P - \theta_1 - \theta_2$. We omit this derivation and we present the final results. The fields T_1, T_2 in the domains IV_d^5, III^5 take the form:

$$T_i^{(2)}(\alpha,\beta) = T_i^*(\alpha,\beta) + T_i^{(2)}(\alpha,\beta) + T_i^{(3)}(\alpha,\beta) + T_i^{(5)}(\alpha,\beta) \tag{4.659}$$

where

$$\begin{aligned} T_1^{(5)}(\alpha,\beta) &= P_1G_1(\beta_P-\theta_1-\theta_2-\beta,\alpha_P+\theta_1+\theta_2-\alpha) - \\ &\quad P_2G_2(\beta_P-\theta_1-\theta_2-\beta,\alpha_P+\theta_1+\theta_2-\alpha) \\ T_2^{(5)}(\alpha,\beta) &= -P_1G_2(\beta_P-\theta_1-\theta_2-\beta,\alpha_P+\theta_1+\theta_2-\alpha) + \\ &\quad P_2G_3(\beta_P-\theta_1-\theta_2-\beta,\alpha_P+\theta_1+\theta_2-\alpha) \end{aligned} \tag{4.660}$$

Now, we can determine internal fields T_i in the lower fan, cf. Fig. 4.49, where we had introduced the parameterization:

$$T_1(\alpha,\beta_1) = 0, \qquad T_2(\alpha,\beta_1) = T_2^{[5]}(\alpha,0) \tag{4.661}$$

where $T_2^{[5]}$ is given by (4.659).

The upper fan in Fig. 4.90 is divided into three static subdomains

- upper subdomain: $\alpha_P - \theta_1 < \beta < \theta_2$:

$$T_1(\alpha_1,\beta) = T_1^{[2]}(0,\beta), \qquad T_2^{[2]}(\alpha_1,\beta) = 0 \tag{4.662}$$

  where $T_i^{[2]}$ is given by (4.644a);
- middle subdomain: $\beta_P - \theta_2 - \theta_1 < \beta < \alpha_P - \theta_1$:

$$T_1(\alpha_1,\beta) = T_1^{[4]}(0,\beta), \qquad T_2^{[4]}(\alpha_1,\beta) = 0 \tag{4.663}$$

where $T_i^{[4]}$ is given by (4.658);

- lower subdomain: $0 < \beta < \alpha_P - \theta_1 - \theta_2$:

$$T_1(\alpha_1, \beta) = T_1^{[5]}(0, \beta), \quad T_2^{[5]}(\alpha_1, \beta) = 0 \tag{4.664}$$

where $T_i^{[5]}$ is given by (4.659).

Let us calculate the longitudinal force in the compressed reinforcing bar: at the segment $Z'P$ we have $F_C = P_2$, at the segment Z_1Z' this force varies:

$$F_C(\beta) = P_2 \int_{\beta}^{\alpha_P - \theta_1} T_2^{[3]}(\bar{\beta} + \theta_1, \bar{\beta})d\bar{\beta} \tag{4.665}$$

where $T_2^{[3]}$ is given by (4.656a).

At the segment CZ_1, the longitudinal force changes according to the formula:

$$F_C(\beta) = P_2 + \int_{\beta_P - \theta_1 - \theta_2}^{\alpha_P - \theta_1} T_2^{[3]}(\bar{\beta} + \theta_1, \bar{\beta})d\bar{\beta} + \int_{\beta}^{\beta_P - \theta_1 - \theta_2} T_2^{[5]}(\bar{\beta} + \theta_1, \bar{\beta})d\bar{\beta} \tag{4.666}$$

where $T_2^{[5]}$ is given by (4.659).

4.12.3.5 Force in the Domain GJG_2

Results from Sect. 4.12.3.4 define the fields T_1, T_2 corresponding to the case when point P is located inside the domain GJG_2 through the exchange

$$T_1(\alpha, \beta) \leftrightarrow T_2(\alpha, \beta); \qquad \alpha \leftrightarrow \beta, \qquad \alpha_P \leftrightarrow \beta_P, \qquad \theta_1 \leftrightarrow \theta_2 \tag{4.667}$$

4.12.3.6 Force in the Domain $JH_2J_2G_2$

The force applied in the domain $JH_2J_2G_2$ introduces the static division of the cantilever domain into seven subdomains (cf. Fig. 4.91) and the static division of the fans.

The coordinates α_P, β_P of the point P lie in the range: $\theta_1 + \theta_2 < \alpha_P < 2\theta_1 + \theta_2$, $\theta_1 + \theta_2 < \beta_P < \theta_1 + 2\theta_2$.

The formulae defining fields T_1, T_2 in the domains of static division indexed from 1 to 5 are identical to the ones found in previous sections. The internal fields T_1, T_2 in the domain no. 6 of the static division are defined by

$$T_i(\alpha, \beta) = T_i^*(\alpha, \beta) + T_i^{(2)}(\alpha, \beta) + T_i^{(3)}(\alpha, \beta) + T_i^{(6)}(\alpha, \beta) \tag{4.668}$$

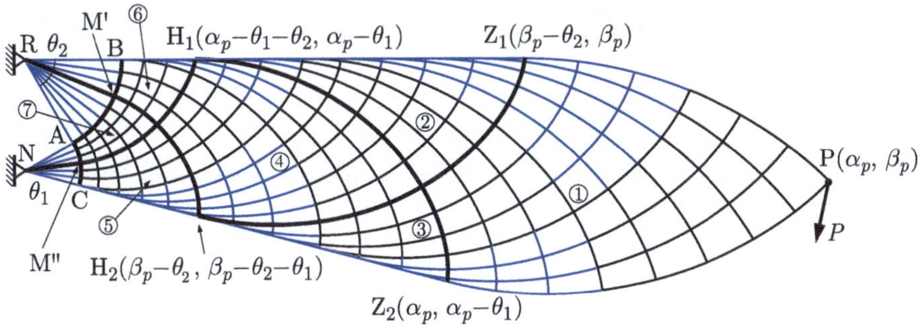

Fig. 4.91 Static division of the cantilever when the load is applied within $JH_2J_2G_2$ domain

where

$$
\begin{aligned}
T_1^{(6)}(\alpha, \beta) &= P_1 G_3(\beta_P - \theta_1 - \theta_2 - \alpha, \beta_P + \theta_1 + \theta_2 - \beta) - \\
&\quad P_2 G_2(\alpha_P - \theta_1 - \theta_2 - \alpha, \beta_P + \theta_1 + \theta_2 - \beta) \\
T_2^{(6)}(\alpha, \beta) &= -P_1 G_2(\beta_P - \theta_1 - \theta_2 - \alpha, \beta_P + \theta_1 + \theta_2 - \beta) + \\
&\quad P_2 G_1(\alpha_P - \theta_1 - \theta_2 - \alpha, \beta_P + \theta_1 + \theta_2 - \beta)
\end{aligned}
\tag{4.669}
$$

Finally, the internal fields T_1, T_2 in the domain no. 7 can be expressed as

$$T_i = T_i^* + T_i^{(2)} + T_i^{(3)} + T_i^{(5)} + T_i^{(6)} \tag{4.670}$$

Forces in fan domains.

The parameterization of the fans is presented in Fig. 4.49.

Fan NAM'', $0 < \alpha < \alpha_P - \theta_1 - \theta_2$,

$$T_1(\alpha, \beta_1) = 0, \qquad T_2(\alpha, \beta_1) = T_2^{[7]}(\alpha, 0) \tag{4.671}$$

where $T_2^{[7]}$ is given by (4.670).

Fan $NM''C$, $\alpha_P - \theta_2 - \theta_1 < \alpha < \theta_1$,

$$T_1(\alpha, \beta_1) = 0, \qquad T_2(\alpha, \beta_1) = T_2^{[5]}(\alpha, 0) \tag{4.672}$$

where $T_2^{[5]}$ is given by (4.659)

Fan RAM', $0 < \beta < \beta_P - \theta_1 - \theta_2$,

$$T_1(\alpha_1, \beta) = T_1^{[7]}(0, \beta), \qquad T_2(\alpha_1, \beta) = 0 \tag{4.673}$$

where $T_1^{[7]}$ is given by (4.670)

Fan $RM'B$, $\beta_P - \theta_1 - \theta_2 < \beta < \theta_2$,

$$T_1(\alpha_1, \beta) = T_1^{[6]}(0, \beta), \qquad T_2(\alpha_1, \beta) = 0 \tag{4.674}$$

where $T_1^{[6]}$ is given by (4.661)

The Longitudinal Forces in the Surrounding Reinforcing Bars

The Bar RBH_1Z_1P in Tension
At the segment Z_1P, we have $F_T(\alpha) = P_1 = \text{const}$, P_1 given by (4.421), $P_1 > 0$.
At the segment H_1Z_1, we have $F_T(\alpha) = F_T^{[2]}(\alpha)$

$$F_T^{[2]}(\alpha) = P_1 + \int_{\alpha}^{\beta_P - \theta_2} T_1^{[2]}(\bar{\alpha}, \bar{\alpha} + \theta_2) d\bar{\alpha} \tag{4.675}$$

where $T_1^{[2]}$ is given by (4.644a).
At the segment BH_1, we have $F_T(\alpha) = F_T^{[6]}(\alpha)$, where

$$F_T^{[6]}(\alpha) = F_T^{[2]}(\alpha_P - \theta_1 - \theta_2) + \int_{\alpha}^{\alpha_P - \theta_1 - \theta_2} T_1^{[6]}(\bar{\alpha}, \bar{\alpha} + \theta_2) d\bar{\alpha} \tag{4.676}$$

where $T_1^{[6]}$ is given by (4.670)
At the segment RB, we have $F_T = F_T^{[6]}(0) = \text{const}$.

Compressed Bar NCH_2Z_2P
At the segment Z_2P, we have $F_C(\alpha) = P_2 = \text{const}$, P_2 given by (4.421), $P_2 < 0$.
At the segment H_2Z_2, we have $F_C(\beta) = F_C^{[3]}(\beta)$, where

$$F_C^{[3]}(\beta) = P_2 + \int_{\beta}^{\alpha_P + \theta_1} T_2^{[3]}(\bar{\beta} + \theta_1, \bar{\beta}) d\bar{\beta} \tag{4.677}$$

where $T_2^{[3]}$ is given by (4.656a).
At the segment CH_2, we have $F_C(\beta) = F_C^{[5]}(\beta)$, where

$$F_C^{[5]}(\beta) = F_C^{[3]}(\beta_P - \theta_1 - \theta_2) + \int_{\beta}^{\beta_P - \theta_1 - \theta_2} T_2^{[5]}(\bar{\beta} + \theta_1, \bar{\beta}) d\bar{\beta} \tag{4.678}$$

where $T_2^{[5]}$ is given by (4.659).
At the segment NC, we have $F_C = F_C^{[5]}(0)$.

Knowing the forces F_T, F_C at the bars RB, NC and the forces T_1, T_2 in the fans surrounding R and N one can compute the reactions at nodes R and N.

4.12.4 Density of Fibres and Computation of the Volumes of Michell's Cantilevers

In the basis of distribution of forces $T_1 > 0$, $T_2 < 0$ and Lamé coefficients A, B one can determine the fibre volume content $\upsilon(\alpha, \beta)$ according to the formula (4.169).

Forces T_1, T_2 are defined by various formulae in particular domains of static division, cf. Fig. 4.91, where the thick lines indicate the boundaries of the static division. Fields A, B are defined by various formulae in particular domains of the kinematic division, cf. Fig. 4.83. Superimposing static and kinematic division results in generation of 23 subdomains. In each of these subdomains, different formula defining fibre volume content υ holds.

The volume of the fibrous domains is defined as

$$V_{\text{fibrous}} = \int_{\Omega} \upsilon(\alpha, \beta) A(\alpha, \beta) B(\alpha, \beta) d\alpha d\beta \tag{4.679}$$

where Ω is the domain composed of two families of fibres.

The volume of the curved cable in tension equals

$$V_{T,C} = \int_{\beta_P - \theta_2}^{\alpha_P} \frac{1}{\sigma_T} P_1 A(\alpha, \beta_P) d\alpha \tag{4.680}$$

where A refers to the domain 1.

The volume of the straight cable in tension $H_1 Z_1$ equals

$$V_{T,S} = \int_{\alpha_P - \theta_1 - \theta_2}^{\beta_P - \theta_2} \frac{1}{\sigma_T} F_T^{[2]}(\alpha) A(\alpha, \alpha + \theta_2) d\alpha \tag{4.681}$$

where A refers to the domain 2.

The volumes of the remaining segments of the surrounding cables can be calculated in a similar way.

4.12.5 Case Studies

4.12.5.1 Symmetric Cantilevers Within a Half Strip. Case of $\sigma_C = \sigma_T$. Study of the Total Volume

Let us consider the case $\theta_1 = \theta_2 = \theta = \dfrac{\pi}{4}$ and $\sigma_C = \sigma_T = \sigma_0$. The design domain is a half strip of thickness $a = |RN|$. Point P lies on mid-perpendicular of RN.

The force P is parallel to RN, see Fig. 4.92.

The problem: construct Michell's structure inside half strip, which transfers the force P located on the symmetry axis to the support RN. Investigate the increase

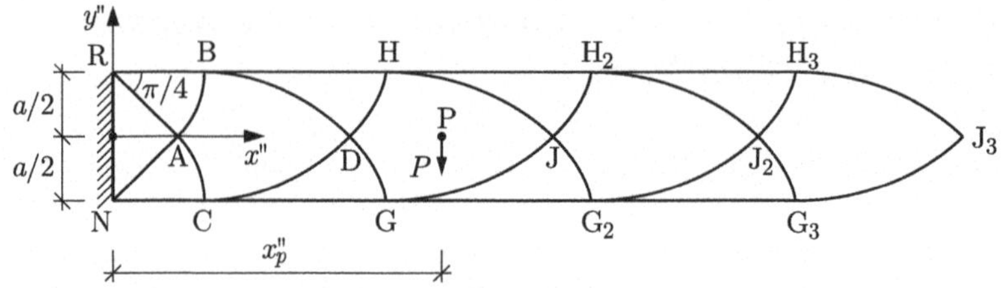

Fig. 4.92 Symmetric cantilever within a half strip

in volume when the distance between the force P and the support increases. The kinematic division of the design domain is presented in Fig. 4.92.

The location of the point P introduces static division, which will not be presented herein. When P is located in points A, D, J, J_2, J_3 then both divisions overlap.

The solution from this section precisely defines geometry of the Michell's net. It can be proved that horizontal ordinate $\xi = \frac{x''}{a}$ of the points marked in Fig. 4.92 are equal (with precision up to four signs in decimal)

$$\begin{aligned} \xi_A = 0.5, \quad \xi_B = 0.7071, \quad \xi_D = 1.822, \\ \xi_H = 2.0971, \quad \xi_J = 3.3589, \quad \xi_{J_2} = 4.9270 \end{aligned} \tag{4.682}$$

The volume of the Michell's structure raises along with increase in the distance of force P from the support RN, see Fig. 4.93. When P is vertical then adjoint displacement of the points on the axis $y'' = 0$ (then $\alpha_P = \beta_P$) measured in the direction of the force is equal: $u_p = \frac{\sqrt{2}}{2}(u(\alpha_P, \alpha_P) - v(\alpha_P, \alpha_P))$ but $u(\alpha_P, \alpha_P) = -v(\alpha_P, \alpha_P)$, thus $u_p = \sqrt{2}u(\alpha_P, \alpha_P)$. If P lies between A and D, then the volume of the optimal structure equals $V = \bar{V} V_0$, $V_0 = \frac{Pa}{\sigma_0}$, $\bar{V} = \frac{u_p}{a}$ and $\bar{V} = \bar{V}(\alpha)$ is given by the formula (4.442) which, however, concerns: $\alpha \leq \frac{\pi}{4}$. One can compute: $\bar{V}(0) = 1$, $\bar{V}\left(\frac{\pi}{4}\right) = 6.07386$, which is confirmed by the ordinate of the volume plot at points A and D, cf. Fig. 4.93.

The ordinate of the function $\bar{V}$ at point J, where $\alpha_P = \beta_P = \frac{\pi}{2}$ can be found according to the formula $\bar{V}\left(\frac{\pi}{2}\right) = u^{**}\left(\frac{\pi}{2}, \frac{\pi}{2}\right)/r_0$ where $u^{**}(\alpha, \beta)$ is given by the formula (4.609a) in which one has to substitute $\theta = \frac{\pi}{4}$.

The ordinate of the function $\bar{V}$ at point J_2, where $\alpha_P = \beta_P = \frac{3\pi}{4}$ can be found according to the formula $\bar{V}\left(\frac{3\pi}{4}\right) = u\left(\frac{3\pi}{4}, \frac{3\pi}{4}\right)/r_0$ where $u(\cdot, \cdot)$ is given by the formula (4.619).

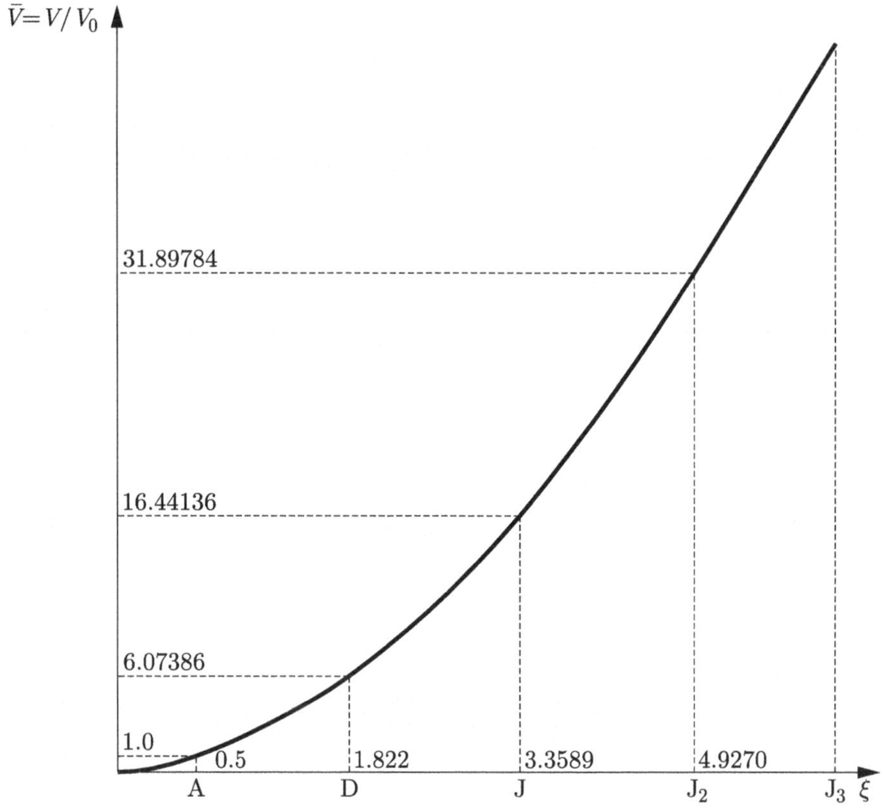

Fig. 4.93 The plot of volume in terms of position of the force P; $V_0 = \dfrac{Pa}{\sigma_0}$

The ordinate of the function $\bar{V}$ at point J_3, where $\alpha_P = \beta_P = \pi$ can be found according to the formula $\bar{V}(\pi) = u^{IV}(\pi, \pi)/r_0$ where $u^{IV}(\pi, \pi)$ is the virtual displacement field within the $J_2H_3J_3G_3$ domain.

The entire plot of structure's volume is presented in Fig. 4.93. The optimal designs are depicted in Fig. 4.94.

4.12.5.2 Asymmetric Cantilevers Within a Half Strip. Selected Cases of $\sigma_T \geq \sigma_C$

Let us solve the problem from the Sect. 4.12.5.1, but now assuming that $\sigma_T \geq \sigma_C$. Let point P have the coordinates (x''_p, y''_p) and lie within a given half strip: $x'' \geq 0$, $-\dfrac{a}{2} \leq y'' \leq \dfrac{a}{2}$.

Table 4.3 Coordinates of the points A, B, C, D, H, G, J

κ	$x''(A)$	$y''(A)$	$x''(B)$	$x''(C)$	$x''(D)$	$y''(D)$	$x''(H)$	$x''(G)$	$x''(J)$	$y''(J)$
1	0.5000	0.0000	0.7071	0.7071	1.8220	0.0000	2.0971	2.0971	3.3589	0.0000
2	0.4714	−0.1667	0.8165	0.5774	1.7728	0.1749	1.9096	2.2395	3.3025	−0.1704
3	0.4330	−0.2500	0.8660	0.5000	1.7072	0.2608	1.7920	2.2966	3.2272	−0.2548
5	0.3727	−0.3333	0.9129	0.4082	1.6051	0.3448	1.6489	2.3442	3.1096	−0.3383
7	0.3307	−0.3750	0.9354	0.3536	1.5348	0.3857	1.5624	2.3641	3.0283	−0.3795
9	0.3000	−0.4000	0.9487	0.3162	1.4837	0.4097	1.5031	2.3747	2.9691	−0.4040

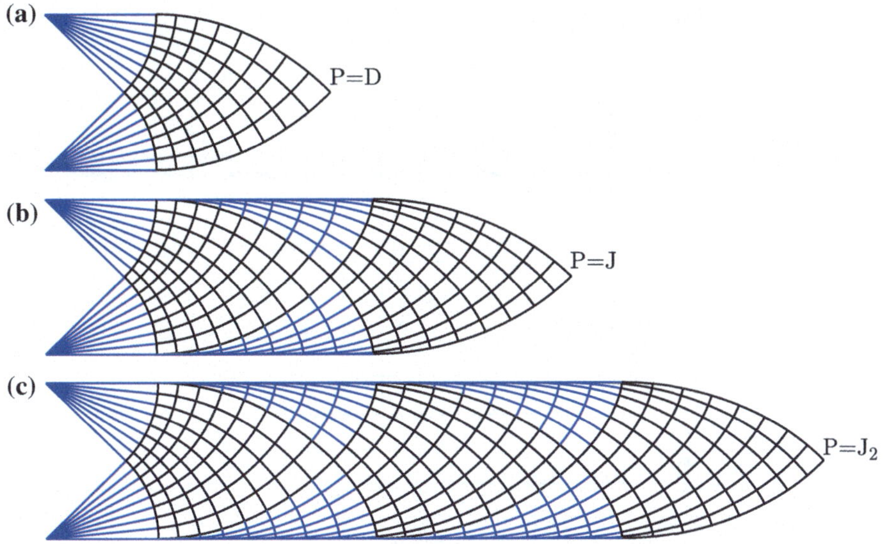

Fig. 4.94 Cantilevers constructed within a half strip: **a**, **b**, **c** analytical solutions

The kinematic division depends on the parameter $\kappa = \frac{\sigma_T}{\sigma_C}$; in particular the coordinates (x'', y'') of the nodes A, B, C, D, H, G, J depend on the value of κ, cf. Fig. 4.94 and Table 4.3, where we assumed: $\kappa = 1, 2, 3, 4, 7, 9$.

If P coincides with one of the points: A, B, C, D, H, G, then optimal cantilevers are confined by the lines from Fig. 4.95. In general, the optimal structure is composed of many subdomains resulting from imposition of the static and kinematic divisions. The illustrative solution given by the data: $\kappa = 5$, $x''_P = 2.75a$, $y''_P = -0.25a$ is presented in Fig. 4.96.

Using the formulae from Sect. 4.12.1, we can find the curvilinear coordinates of the point P: $\alpha_P = 1.315713099$, $\beta_P = 1.408664644$, $\theta_2 = 0.120534$ and plot the net of the lines (α, β) inside the structure. In particular, we can find the nodes of the subdomains to which the optimal structure is divided, i.e. the location of the points:

$$\begin{array}{llll} A(0,0), & C(\theta_1,0), & M(\beta_P-\theta_2,0), & M'(0,\alpha_P-\theta_1), \\ Z(\beta_P-\theta_2,\beta_P), & Z'(\alpha_P,\alpha_P-\theta_1), & & \\ Q(\beta_P-\theta_2,\theta_2), & W(\beta_P-\theta_2,\alpha_P-\theta_1), & & \\ Q'(\theta_1,\alpha_P-\theta_1), & S_g(\theta_1,\beta_P), & S_d(\alpha_P,\theta_2) & \end{array} \tag{4.683}$$

Further, we can calculate volume of the Michell's structure: $V = 34.66641774V_0$, $V_0 = \dfrac{Pa}{\sigma_T}$. The fields T_1, T_2 are discontinuous within the domain of the structure, see Fig. 4.97.

The plot of fibres' density υ given by (4.169) is even more complex since entangled fields A, B have different lines of discontinuity then the fields T_1, T_2, see Fig. 4.98.

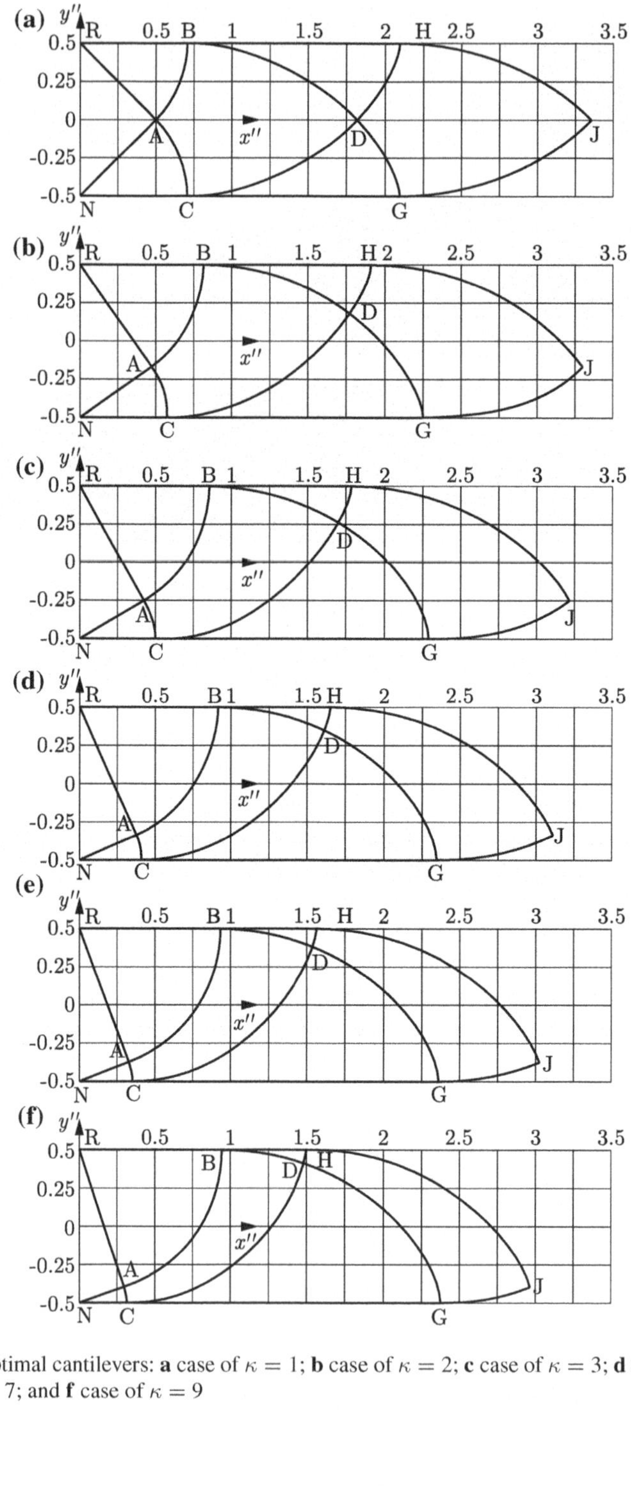

Fig. 4.95 Optimal cantilevers: **a** case of $\kappa = 1$; **b** case of $\kappa = 2$; **c** case of $\kappa = 3$; **d** case of $\kappa = 5$; **e** case of $\kappa = 7$; and **f** case of $\kappa = 9$

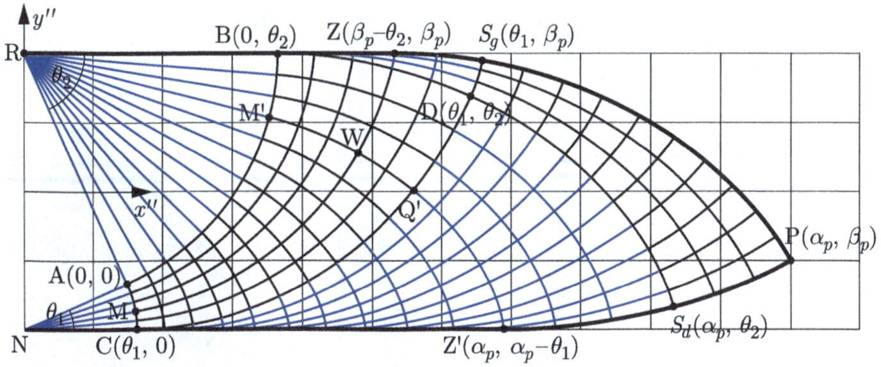

Fig. 4.96 Optimal cantilever, the case of $\kappa = 5$, $\theta_2 = 0.420534$ rad

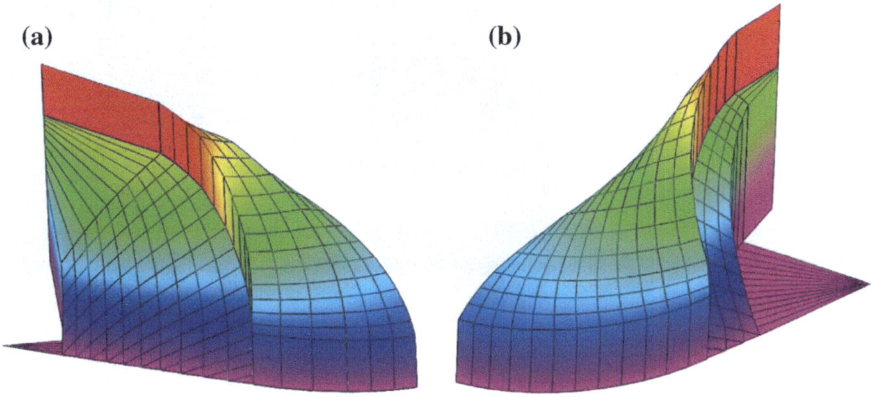

Fig. 4.97 The fields T_1, T_2 concerning Fig. 4.96

Similarly, we can find the volumes of the Michell structures generated by location of the vertical force P in grid points from Fig. 4.95, i.e. in the coordinate points: $x''_P = n\frac{a}{4}$, $n = 1, \ldots, 14$, $y''_P = m\frac{a}{4}$, $m = -2, -1, 0, 1, 2$. The results are tabulated in Graczykowski and Lewiński (2010).

Let us note that when the ratio of admissible stresses in tension and compression tends to infinity ($\sigma_T/\sigma_C \to \infty$) the lower circular fan and the following central and Chan's regions disappear (cf. Fig. 4.95). The plot of parametric lines within a half strip becomes identical as in the case of deep beam transmitting point load to a straight boundary with lower sliding support (see Fig. 4.131a). Indeed, since the admissible stresses in tension are infinite the horizontal support of the structure is not required.

4.12.5.3 Cantilevers Within Trapezoidal Domains. Case Studies

In this section, we will look more closely at selected examples of optimal cantilevers. Their analyses will include partition into regions of kinematic and static divisions as

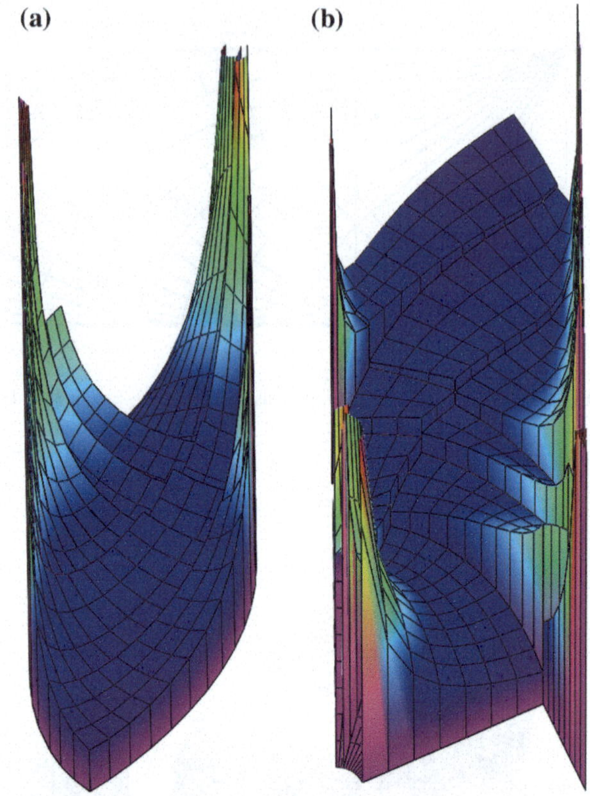

Fig. 4.98 The field of fibres' density corresponding to Fig. 4.96: **a** front view and **b** back view

well as graphs of the internal forces and fibres density function. In each case, we will calculate the total volume of the optimal cantilever by two distinct methods. At first, we will use the stress-based formulation (3.95) and compute the total volume of the cantilever by integration of the fibre density function υ and summation of volumes of all fibrous domains and reinforcing bars. Second, we will use the dual formulation (3.85) and calculate the total volume from the algebraic formula (3.81) involving virtual displacement field. The equivalence of both approaches will prove that the *duality gap* vanishes and that all the results concerning geometry of fibres, Lamé coefficients, virtual displacements, internal forces in reinforcing bars and fibrous domains as well as fibres density functions are determined correctly. Moreover, selected cantilevers will be approximated by trusses of finite number of members and it will be shown that volumes of these trusses tend to the volumes of the optimal Michell cantilevers.

A. Prager–Hill Cantilevers

A1. The Point Load in the Domain $ABDC$

The Prager–Hill cantilevers arise when the point load is applied within the region of kinematic division $ABDC$ (Fig. 4.99). Then, the optimal cantilever is composed of

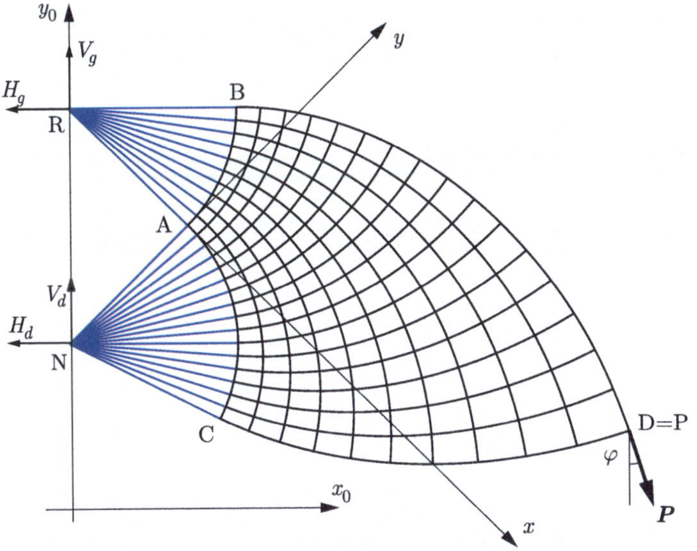

Fig. 4.99 Prager–Hill cantilever; the force ***P*** is tangent at $D = P$ to the BD bar

two fans, Prager–Hill region and of two reinforcing bars. We will show the distribution of internal forces and fibre density function within the design. We will also find the volume of such cantilevers by summing up the volumes of their parts and then verify this result by computing the virtual work.

We consider the problem of Fig. 4.99 for the case of $\sigma_T = \sigma_C = \sigma_0$, which implies $\kappa = 1$; $|RA| = |NA| = r_0$. The force ***P*** is directed in a specific way, perpendicularly to the lower reinforcing bar DC, which corresponds to the case: $\varphi = \pi/10$. Thus, the whole force is transmitted by the tension bar RBD: $F_T = P$, while in the compression bar NCD we have $F_C = 0$. Position of point P is given by Cartesian coordinates with respect to the (x, y) frame: $x_P = 2.73874732 r_0$; $y_P = 0.9680382174 r_0$, or by the values: $\alpha_P = \frac{2\pi}{5}$; $\beta_P = \frac{\pi}{4}$ characterizing also the angles of the lower and upper fans. Distance of the point P to the supporting line equals $d = |x_0(P)| = 3.32819983 r_0$. We compute the reactions at the upper supporting node R:

$$H_g = \int_0^{\beta_P} T_1^{RBA}(\alpha_1, \beta) \sin\left(\frac{\pi}{4} + \beta\right) d\beta + F_T \sin\left(\frac{\pi}{4} + \beta_P\right)$$
$$= 2.119128363P$$

$$V_g = \int_0^{\beta_P} T_1^{RBA}(\alpha_1, \beta) \cos\left(\frac{\pi}{4} + \beta\right) d\beta + F_T \cos\left(\frac{\pi}{4} + \beta_P\right)$$
$$= 0.5035531419P$$

and at the lower supporting node N

$$H_d = \int_0^{\alpha_P} T_2^{NAC}(\alpha, \beta_1) \sin\left(\frac{\pi}{4} + \alpha\right) d\alpha + F_C \sin\left(\frac{\pi}{4} + \alpha_P\right)$$
$$= -1.810111369P$$

$$V_d = \int_0^{\alpha_P} \left(-T_2^{NAC}(\alpha, \beta_1)\right) \cos\left(\frac{\pi}{4} + \alpha\right) d\alpha - F_C \cos\left(\frac{\pi}{4} + \alpha_P\right)$$
$$= 0.4475033745P$$

The reactions H_g, H_d are directed opposite to the axis x_0 and V_g, V_d are directed along the axis y_0. Further, we compute the non-dimensional volumes of all components of the cantilever defined by $\bar{V} = V/V_0$, $V_0 = Pr_0/\sigma_0$. Primarily, we compute the volumes of the fibrous domains. The volumes of the fans NAC, RBA are $\bar{V}_{NAC} = 1.987586823$, $\bar{V}_{RBA} = 1.258902591$, respectively. The volume of the Prager–Hill domain $ABDC$ equals $\bar{V}_{ABDC} = 3.565259273$. The total volume of fibres in the fibrous domains is $\bar{V}_S = 6.811748687$. Second, we compute the volumes of the reinforcing bars: the tension bar: $\bar{V}_T = 4.246489413$ and the compression bar: $\bar{V}_C = 0$. The total volume of the reinforcing bars equals $\bar{V}_K = 4.246489413$. The ratio of the volume of the fibrous domain to the total volume of the cantilever equals 61.599%. The total volume of the material of the structure is $\bar{V} = \bar{V}_S + \bar{V}_K = 11.05823810$. The volume of the same structure by Michell's formulae (3.81) and (3.85) turns out to be: $\bar{V} = 11.05823810$, hence is given by the same number.

The example considered here is specific since the force is applied such that the force in the tension bar is equal to the point force applied, while the compression bar does not work. This makes it possible to figure out how the internal force fields T_1 and T_2 satisfy the boundary conditions: the boundary condition along the compression bar reads: $T_1 = -F_C$, but here $F_C = 0$, see Fig. 4.100. The boundary condition for T_2 along the tension bar assumes the form: $T_2 = -F_T$, see Fig. 4.101 showing that the values of T_2 are constant along the tension bar. Furthermore, the force field T_1 suffers a jump along AC, while the force field T_2 has a jump along the line AB.

Both the lines AB and AC are the lines of discontinuity of the field v, since along these lines at least one of the fields A, B, T_1, T_2 suffers a jump. The big values of the function v at the neighbourhood of point A is caused by small values of Lamé coefficients in the vicinity of this point. Function v grows to infinity at points N and R where one of the Lamé coefficients vanishes (Fig. 4.102).

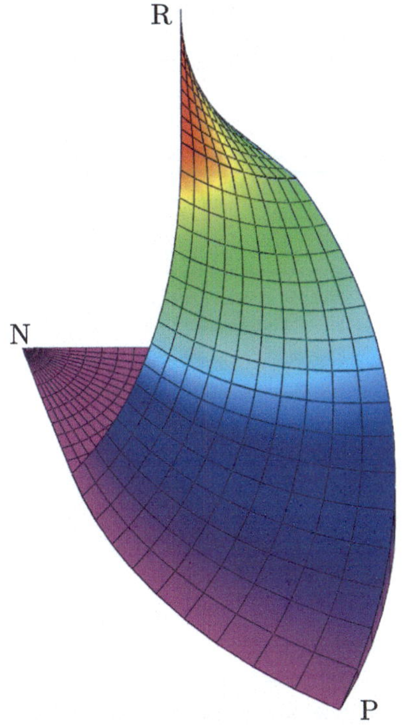

Fig. 4.100 The plot of the Hemp's field T_1 within Prager–Hill cantilever

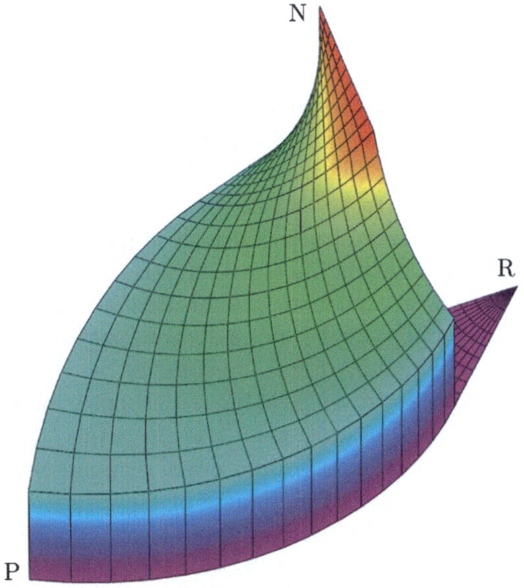

Fig. 4.101 The plot of the Hemp's field T_2 within Prager–Hill cantilever

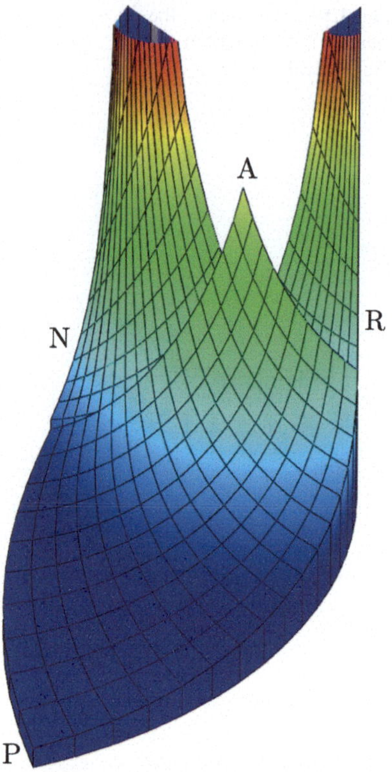

Fig. 4.102 The plot of the fibre density function v within Prager–Hill cantilever

The optimal cantilever found becomes a discrete-continuous structure which can be viewed as a limit of a sequence of trusses of finite number of members. We show below that the optimal cantilever can indeed be constructed by forming an appropriate sequence of trusses in which the number of members tends to infinity. To this end, we consider the sequence of trusses of joints lying at the selected points of intersection of the parametric lines. All the trusses constructed are statically determinate—just this is the crucial feature of this design sequence, enabling a direct determination of cross sections by the values of the forces in the members and, consequently, enabling for computation of the total volume.

The truss presented in Fig. 4.103 approximating the Michell cantilever of Fig. 4.99 possesses 108 members and 56 joints. By assuming that the members are fully stressed, we fix the cross sections and compute the volume of the truss: $11.08633976V_0$. The volume of the external bars is $4.780978146V_0$. The volume of the internal bars equals $6.305361614V_0$. The volume of the lower (compression) external members is $0.2826225365V_0$. The volume of the upper (tension) external members equals $4.498355609V_0$.

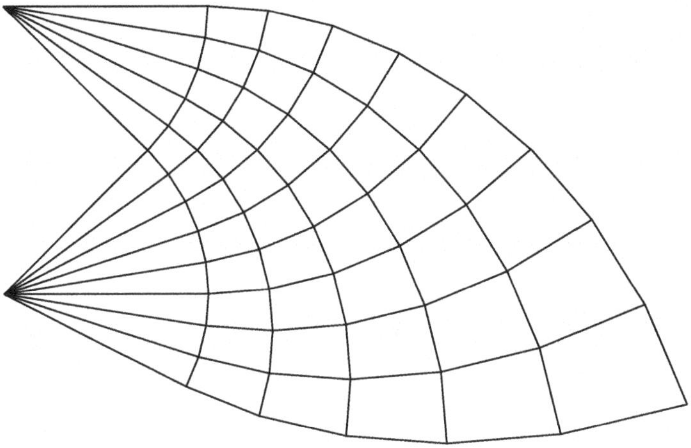

Fig. 4.103 A truss of 108 members approximating Prager–Hill cantilever

The volume of the truss turns out to be greater than the volume of the Michell structure by about 0.25%. A greater difference is visible in the division of the volume into the volumes of the external and internal parts. The ratio of volume of the internal members and total volume of the cantilever equals 56.875%, and it is considerably smaller than in optimal cantilever. Moreover, the volume of the external compressed members of the truss is not exactly zero, as in the optimal solution. The values of forces in the external members are greater than the exact constant value in the optimal solution. The biggest is the force in the member at the support; it is greater by 10% than the force in the member at node P. The variation of the values of the forces in the external tension members is caused by interaction with internal members which do not form exactly the right angles with external members.

Consider now a truss of larger number of members, loaded as before, see Fig. 4.104. It has 374 members and 189 joints. The volume of the truss: $11.06524488V_0$. The volume of external members: $4.508915026V_0$. The volume of internal members: $6.556329854V_0$. The volume of external (lower) compression members: $0.1392986978V_0$. The volume of external (upper) tension members: $4.369616328V_0$.

Now, the total volume of the truss differs from the volume of the Michell structure by only 0.063%. The ratio between the volume of the internal members and the total volume of the cantilever increases, equals 59.252% and is slightly smaller to the corresponding ratio in the Michell structure. Moreover, the volume of the external compression members becomes closer to zero. The variation of the values of longitudinal forces in the external members is now smaller; the deviation from the exact constant value of the force is below 5%, see Fig. 4.105.

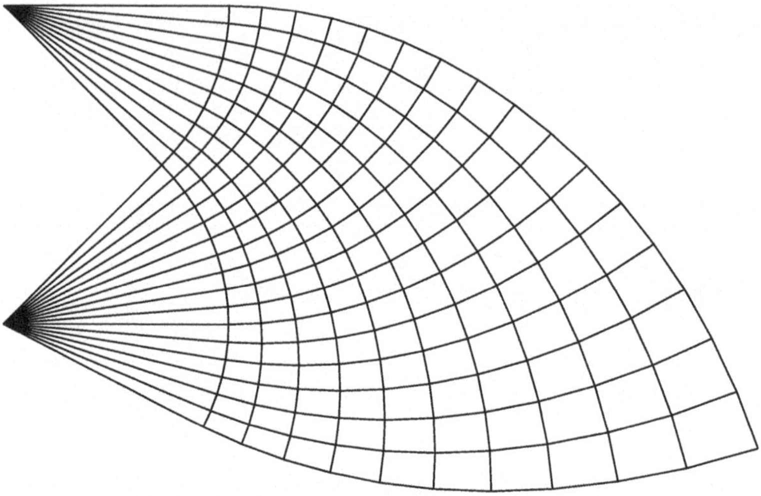

Fig. 4.104 A truss of 374 members approximating Prager–Hill cantilever

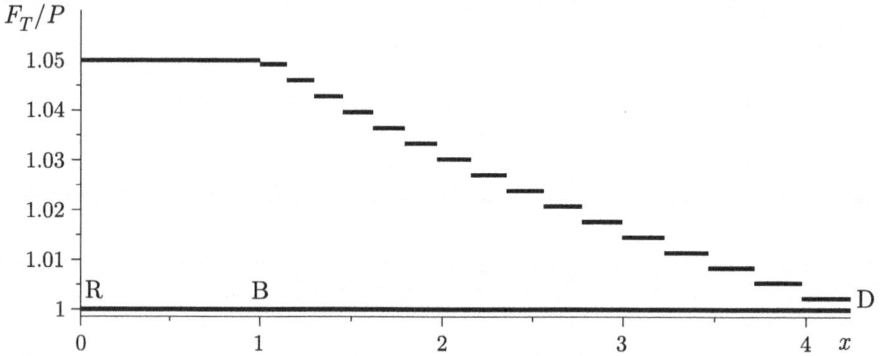

Fig. 4.105 Change of longitudinal forces (divided by P) in the upper external members of the truss of Fig. 4.104

B. Chan-Like Cantilevers

The Chan-like cantilevers emerge when the point load is applied within the regions of kinematic division BDH or CDG. The main difference in comparison to Prager–Hill cantilever is the occurrence of the Chan's regions whose boundaries are partially straight and adjacent to the boundary of the feasible domain. The force in the reinforcing bars located along the straight boundaries varies due to tangential connection between the fibres and the boundary. In these regions, the volume of the reinforcing bars is calculated by proper line integration, while the volume of fibrous domain is computed by an iterated integral.

B1. The Point Load at the Boundary BH

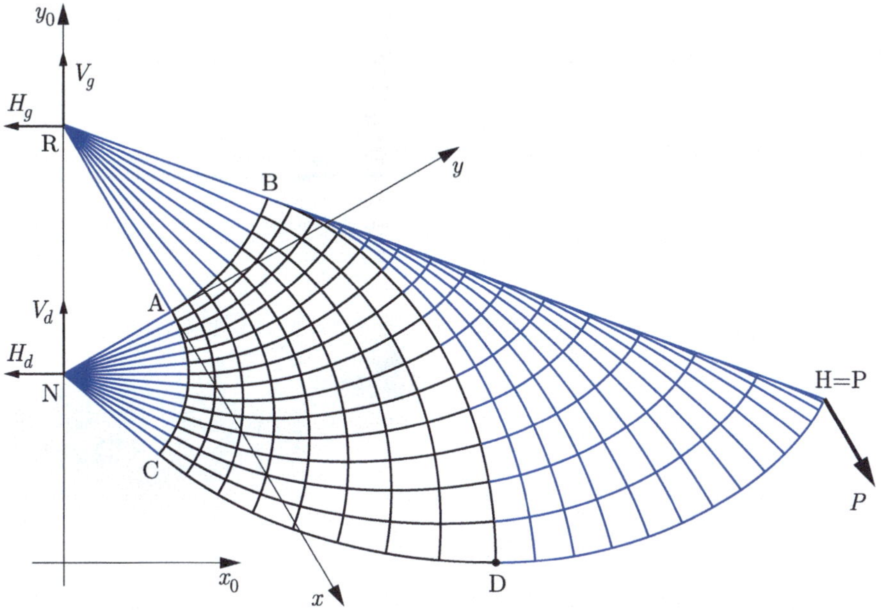

Fig. 4.106 The Chan-like cantilever with point force located at the line BH. Here, $P = H$

The analysis of the properties of such an optimal cantilever will be shown by a concrete example, see Fig. 4.106. In the considered case, the entire boundary of the Chan's domain is straight.

In the analysed example, we have $r_1 = |AN| = a/2$, $r_2 = |RA| = a\sqrt{3}/2$, $a = |RN|$, $\kappa = 3$. The angle of the upper circular fan: $\theta_2 = 2\pi/9$, while the angle of the lower circular fan θ_1 is smaller than α_P and smaller than $\pi/2$. The curvilinear coordinates of the point P are $\alpha_P = 7\pi/18$; $\beta_P = 11\pi/18$, which corresponds to its Cartesian coordinates $x_P = 1.593994062a$; $y_P = 2.064201430a$. The distance of point P to the supporting line RN: $d = 3.017660610a$. Direction of the force is defined by the angle $\varphi = \pi/5$. We can compute the forces in the reinforcing bars at node $P = H$:

$$F_T(P) = 0.8290375726P; \qquad F_C(P) = -0.5591929034P$$

Reactions at the upper supporting node R:

$$H_g = 2.383536711P, \qquad V_g = 1.284829035P$$

Reactions at the lower supporting node N:

$$H_d = -1.795751459P, \qquad V_d = -0.4758120425P$$

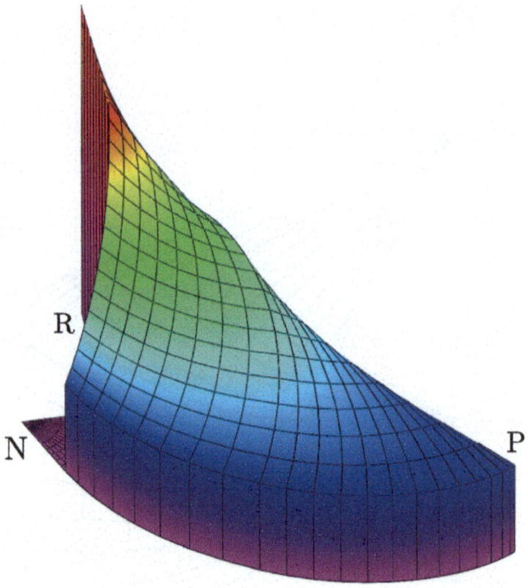

Fig. 4.107 The Hemp's field T_1 within Chan-like cantilever

The non-dimensional volume is defined by $\bar{V} = V/\tilde{V}_0$ with $\tilde{V}_0 = Pa/\sigma_T$. The volumes of fibres in the fibrous domains:

$$\bar{V}_{NAC} = 2.186540186, \qquad \bar{V}_{RBA} = 0.9237368385,$$

$$\bar{V}_{ABDC} = 3.750368950, \qquad \bar{V}_{BDH} = 1.633409910.$$

The total volume of fibres in the fibrous domains: $\bar{V}_S = 8.494055884$.

We compute the volume of reinforcing bars. Volumes of the tension bar and of the compression bars are $\bar{V}_T = 4.146566250$, $\bar{V}_C = 5.796540063$, respectively. The total volume of the reinforcing bars: $\bar{V}_K = 9.943106313$. The total volume of the cantilever equals $\bar{V} = \bar{V}_S + \bar{V}_K = 18.43716220$. The same total volume, but computed by (3.81), (3.85) equals $\bar{V} = 18.43716218$.

The force field T_1 suffers a jump along the line AC or along the boundary between Hill's domain (III) and the lower circular domain II_d, see Fig. 4.107. This force is continuous along BD, where the Hill domain touches Chan's domain IV. Moreover, the force T_1 is constant along the compression bar. The force field T_2 suffers a jump along AB or along the boundary between Hill's domain (III) and the upper circular domain. The force T_2 is continuous along BD and vanishes along the straight boundary BH, see Fig. 4.108.

Every line of the geometric division is a line of discontinuity of the effective density of fibres v, since this quantity depends both on Lamé coefficients and force fields. The density of fibres v tends to infinity at points where one of the Lamé coefficients vanishes. This holds at points N and R and along the whole line BH. Indeed, the lines α approach the boundary BH tangentially, which means that the density of these lines around BH must be infinite, see Fig. 4.109.

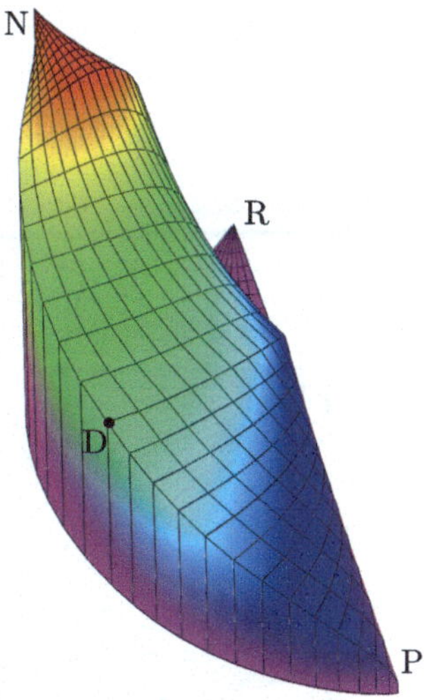

Fig. 4.108 The Hemp's field T_2 within Chan-like cantilever

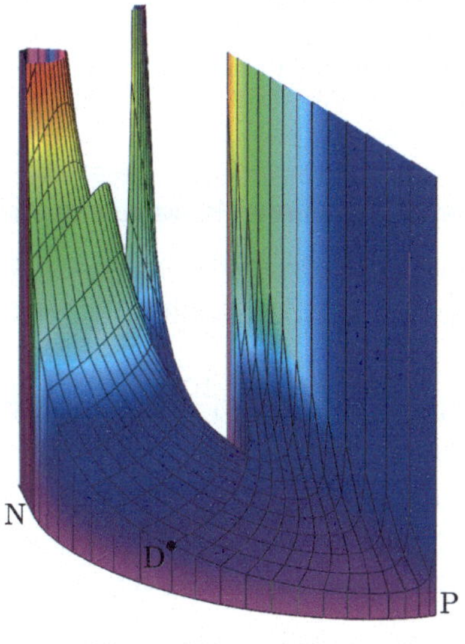

Fig. 4.109 Density of fibres v within Chan-like cantilever

It is worth emphasizing here that this infinitely dense package of fibres along BH is not sufficient to strengthen the boundary. This boundary must be additionally reinforced by a bar of finite cross section.

B2. The Point Load Applied Within Domain BDH

To be specific, we consider a concrete example presented in Fig. 4.110.

In considered case, the part of the boundary of the Chan's domain is straight and adjacent to boundary of the feasible domain, while the other part of the boundary is curved and located along the parametric line.

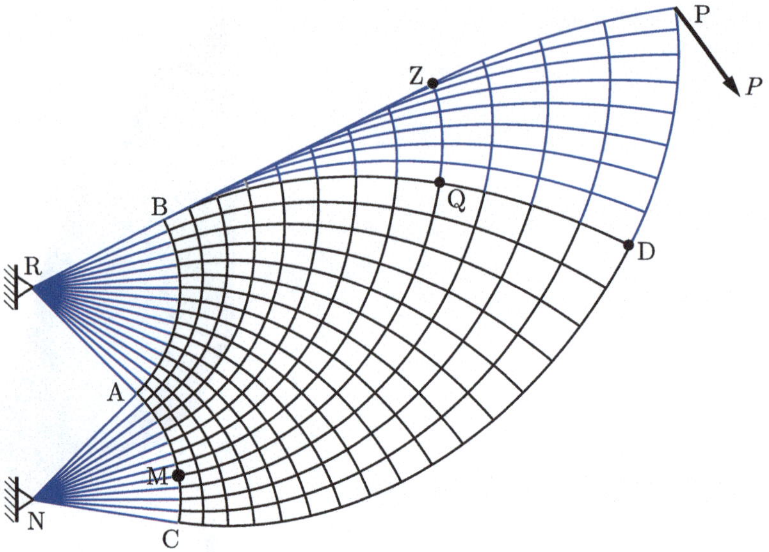

Fig. 4.110 The Chan-like cantilever considered. The segment BZ is straight, as lying along the boundary of the feasible domain Ω_0

We assume the allowable stresses $\sigma_T = \sigma_C = \sigma_0$ and thus $\kappa = 1$; $|RA| = |NA| = r_0$. The angle of the upper fan: $\theta_2 = \frac{2\pi}{5}$, while the angle of the lower circular fan θ_1 is smaller than α_P and smaller than $\pi/2$. Curvilinear coordinates of point P are $\alpha_P = \frac{3\pi}{10}$; $\beta_P = \frac{3\pi}{5}$, which corresponds to its Cartesian coordinates $x_P = 0.706548421 r_0$; $y_P = 4.347572630 r_0$. The distance between the point P and the supporting line is $d = 4.2809100497 r_0$. Finally, the angle between the force direction and the supporting line RN is $\varphi = \frac{\pi}{5}$.

The optimal cantilever is characterized by the following quantities.
The forces in the reinforcing bars:

$$F_T = 0.4539904998P; \qquad F_C = -0.8910065242P$$

Reaction forces at the upper supporting node R:

$$H_g = 3.812906196P, \qquad V_g = 0.1353238651P$$

Reaction forces at the lower supporting node:

$$H_d = -3.225120945P, \qquad V_d = 0.6736931261P$$

Volumes of material within the fibrous regions:

$$\bar{V}_{NMC} = 0.8003150410, \qquad \bar{V}_{NAM} = 1.774085275,$$

$$\bar{V}_{RBA} = 3.001533208, \qquad \bar{V}_{MQDC} = 2.835794214, \qquad \bar{V}_{ABQM} = 4.567136096,$$

$$\bar{V}_{ZPDQ} = 1.261828008, \qquad \bar{V}_{BZQ} = 0.8789743173$$

The total volume of the material within the fibrous domain: $\bar{V}_S = 15.11966616$.

Volumes of the reinforcing tension and compression bars are $\bar{V}_T = 3.518088666$; $\bar{V}_C = 5.697963659$, respectively. The total volume of the reinforcing bars: $\bar{V}_K = 9.216052325$.

The ratio of the volume of the fibrous domain to total volume: 62.129%.

The total volume of the material within the whole structure: $\bar{V} = \bar{V}_S + \bar{V}_K = 24.33571848$.

The total volume of the material within the whole structure computed by (3.81), (3.85) equals $\bar{V} = 24.33571848$.

The graph of T_1 is similar to the case of the point load being applied at the boundary of Chan's domain IV_g (Fig. 4.111). Along PC, the Hemp force T_1 is constant and equal (up to the sign) to the longitudinal force in the compression bar. The force T_1 achieves a maximal constant value on the straight segment RA. On the lower circular fan, the force T_1 vanishes. Thus, the arc AC is the only discontinuity line for this force field.

The graph of the force T_2 has a specific shape, not occurring till now. Along the segment PZ of the boundary of the region IV, see Fig. 4.112, this force assumes the absolute value equal to the longitudinal force in the tension bar. On the other hand, this force vanishes along the straight segment ZB of the boundary of the Chan-like domain. A discontinuity of this force occurs at point Z, where the natural boundary condition changes its character. This discontinuity extends towards the domains IV, III and II along the line $ZQMN$. The value of the jump is constant along this line and equals $F_T(P)$; this is just the value of T_2 along ZP. The values of T_2 increase from the boundary of Chan's domain IV_g towards the beginning of the parametric line along which this force is acting. The maximum is achieved on the straight line AN.

Thus, the discontinuity lines of the density function v are as follows:
BD—the interface between Hill's domain III and Chan's domain IV_g, see Fig. 4.113 (there A jumps);
ZN—the interface between the domains of type 1 and 2 (there the force T_2 jumps);
AB—there both A and T_2 suffer jumps;
AC—there B and T_1 suffer jumps.

We note that the subdomains of continuity of function v appear as a result of the overlapping of the static and kinematic subdivision. This leads to a complicated form of the graph of the function v. This function tends to infinity along the straight segment BZ and at the supporting nodes N and R. A highly irregular behaviour of function v is observed around point A.

The above analytical results are checked by analysis of trusses of finite number of joints. We consider first a truss of 162 members and 83 nodes, see Fig. 4.114. We

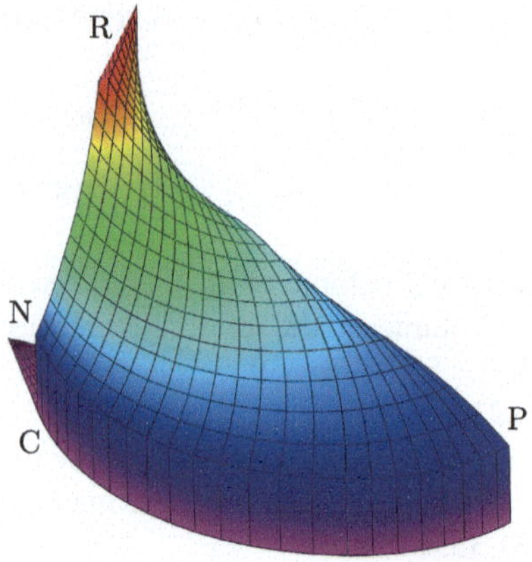

Fig. 4.111 The Hemp's field T_1 within Chan-like cantilever

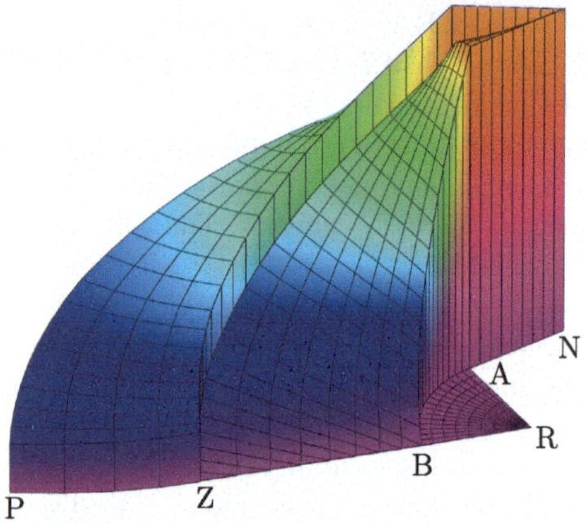

Fig. 4.112 The Hemp's field T_2 within Chan-like cantilever

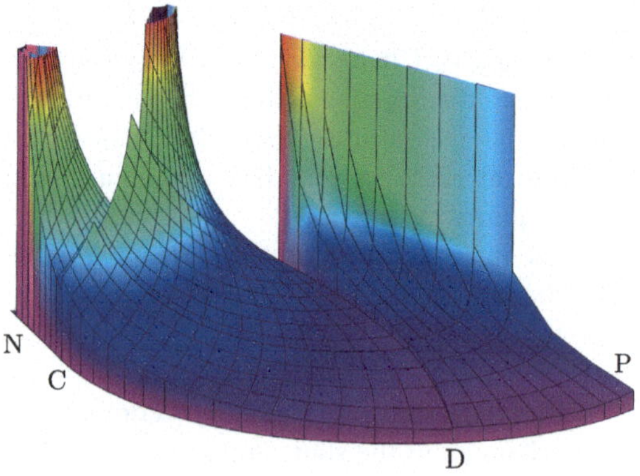

Fig. 4.113 Effective density of the fibres within Chan-like cantilever

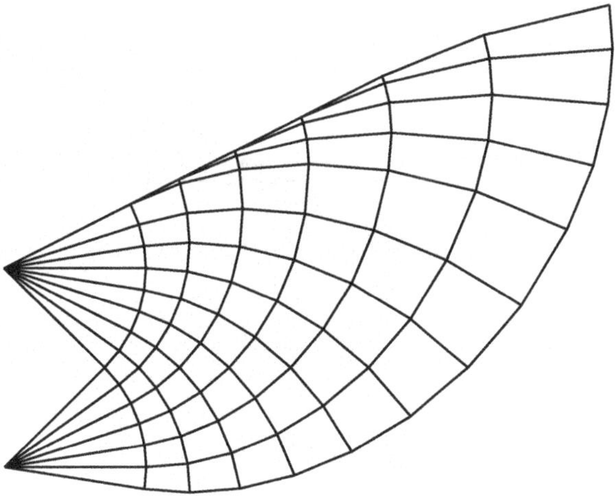

Fig. 4.114 A truss of 83 nodes approximating Chan-like cantilever

assume that the members are fully stressed and compute the volume of the material of the truss which equals $24.39800388\,V_0$, while the volume of the external members is $10.24421848\,V_0$ and the volume of the internal members is $14.15378540\,V_0$. The volume of the upper external tension members is $3.790527422\,V_0$ and the volume of the lower external compression members equals $6.453691055\,V_0$.

The volume of the truss is greater than the volume of the material used for the Michell structure by 0.256%. The internal members contribute to 58.012% of the total volume and this ratio is smaller than in the optimal cantilever.

The force in the external members subject to tension located near the point P is approximately constant since in this region the internal members are linked with the external ones almost orthogonally. In turn, the force in the external members more remote from point P increases 2.5 times, which is caused by the appearance of members approaching nearly tangentially. On the other hand, the change of the force in the external compression members is relatively small since there no tangent members are present.

We consider now the truss of 1091 joints and of 2178 members, see Fig. 4.115. Now, the volume of the material of the truss equals $24.33960822\,V_0$, while the volume of external members is $9.458673821\,V_0$ and the volume of internal members is $14.88093440\,V_0$. The volume of the upper external tension members is $3.580848790\,V_0$ and the volume of the lower external tension members equals $5.877825031\,V_0$.

The volume of the material used for constructing the truss of Fig. 4.115 is almost the same as the volume of the material of the Michell structure; the difference is 0.016%. The contribution of the internal members is now greater than in the previous approximation and equals 61.139% of the total volume. Moreover, the change of internal force in external members in tension resembles the change of tension of external cable of optimal cantilever. In turn, the force in external compressed members remains almost constant.

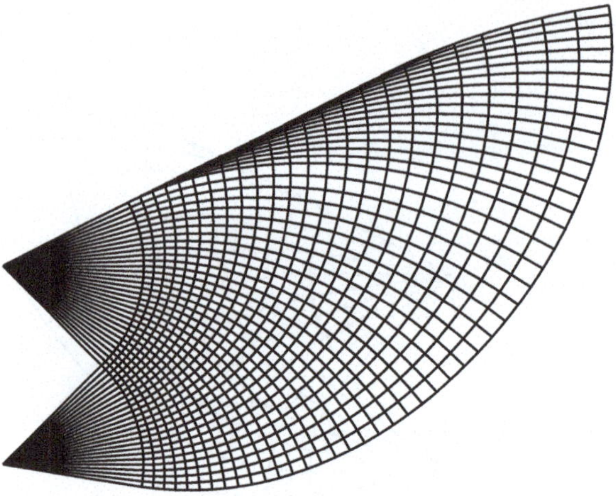

Fig. 4.115 The truss of 1091 joints approximating Chan-like cantilever

C. The Cantilevers Associated with the Force Applied Within Domain $DHJG$

If the point load P is applied within $DHJG$, then the optimal cantilever will be included within $RBHJGCN$. Position of the point load introduces the external boundaries: $RBZS_gP$ and $NCZ'S_dP$, see Fig. 4.116 where the intervals BZ and CZ' are straight. Consequently, the cantilever contains Hill's region, two Chan's region divided into parts with straight boundaries and parts confined by parametric lines as well as one additional central region confined by two parametric lines whose vertex is the point of application of the point force.

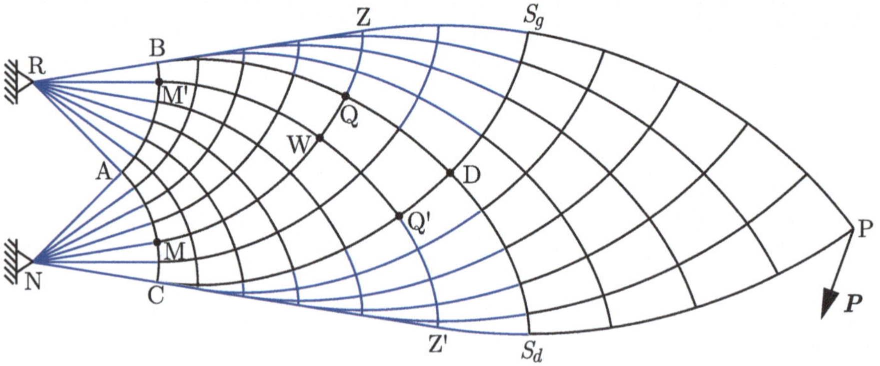

Fig. 4.116 The cantilever considered. Segments ZR, $Z'N$ lie along the boundaries of the feasible domain Ω_0

C1. Case of $\kappa = 1$

The aim of this section is to analyse the static work of the optimal cantilever by using the results concerning geometry of Michell nets, virtual displacement fields and force fields. To be specific, we assume a concrete data to fix geometry of the feasible domain, the position and direction of the concentrated force of magnitude P:

$$\theta_1 = \theta_2 = \frac{3\pi}{10}, \quad x_P = 4.346573703r_0; \quad y_P = 3.737498915r_0, \quad \varphi = -\frac{\pi}{10}$$

where $|RN| = r_0\sqrt{2}$; coordinates x_P, y_P are referred to the (x, y) system, as in Fig. 4.99. Then

$$d = 6.423409349r_0, \quad \alpha_P = \frac{11\pi}{20}, \quad \beta_P = \frac{\pi}{2}$$

We compute the forces in the reinforcing bars:

$$F_T = 0.5877852524P, \quad F_C = -0.8090169943P,$$

the reactions at R:

$$H_g = 4.259331967P, \quad V_g = 0.4880608891P$$

and the reaction forces at N:

$$H_d = -4.568348964P, \quad V_d = 0.4629956217P.$$

These reactions are directed along $(-x_0)$ and y_0. Having the Lamé coefficients and the force fields we may compute the distribution of the density function υ. Appropriate integration gives the non-dimensional volumes $\bar{V} = V/V_0$ of the material within the subdomains shown in Fig. 4.116; here $V_0 = Pr_0/\sigma_0$; $\sigma_0 = \sigma_T = \sigma_C$.

$$\bar{V}_{NAM} = 1.77494717, \quad \bar{V}_{NMC} = 0.9002697526,$$

$$\bar{V}_{RM'A} = 2.163813476, \quad \bar{V}_{RBM'} = 0.4821436945,$$

$$\bar{V}_{AM'WM} = 3.126098073, \quad \bar{V}_{MWQ'C} = 1.898471189$$

$$\bar{V}_{BQWM'} = 0.7987535187, \quad \bar{V}_{CQ'Z'} = 1.785009056,$$

$$\bar{V}_{BZQ} = 1.221821872, \quad \bar{V}_{QDQ'W} = 0.5861877908,$$

$$\bar{V}_{Q'DS_dZ'} = 0.979949195473, \quad \bar{V}_{ZS_gDQ} = 1.583623832,$$

$$\bar{V}_{DS_gPS_d} = 3.180358924$$

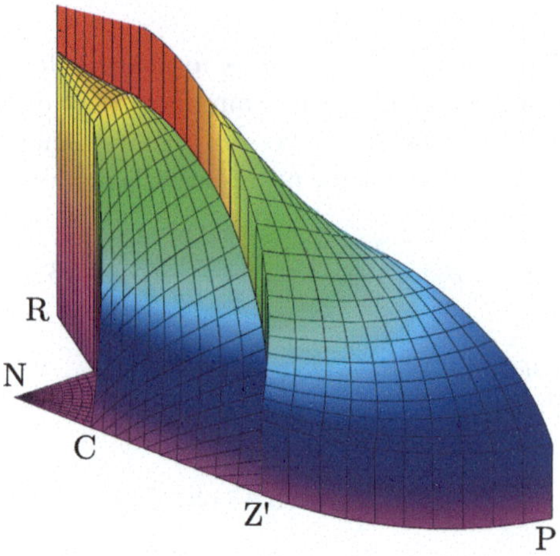

Fig. 4.117 Graph of the Hemp's force field T_1

Thus, the total volume of the material within the fibrous domains equals $\bar{V}_S = 20.48144755$. Then, we compute the volumes of the tension (T) and compression (C) bars

$$\bar{V}_T = 6.180644310, \qquad \bar{V}_C = 7.831983019, \qquad \bar{V}_K = \bar{V}_T + \bar{V}_C = 14.01262733$$

The volume of the material used for the whole structure:
$\bar{V} = \bar{V}_S + \bar{V}_K = 34.49407488$.

The same volume can be computed by the alternative formula based on virtual displacement field. Using the results of Sect. 4.12.5.1, one finds: $\bar{V} = 34.49407452$, which confirms the previous result based on the force field analysis.

The division into static subdomains is now more complicated since each internal force has its line of discontinuity starting at the points where boundary conditions change. The main part of the cantilever (without circular fans) is divided into four subdomains separated by the lines MZ and $M'Z'$. The graph of T_1 has a jump along the line $Z'Q'WM'R$ equal to $F_C(P)$ and it is continuous across the line $ZQWMN$, see Fig. 4.117. The graph of T_2 is similar to that of previous example, where the point load was applied within upper Chan's domain. In contrast to force T_1, the force T_2 has a jump of value $F_T(P)$ along the line $ZQWMN$, and it is continuous across the line $Z'Q'WM'R$, see Fig. 4.118.

The graph of the density of fibres function υ is constructed by overlapping the graphs of Lamé coefficients and the internal forces T_1 and T_2. Six discontinuity lines appear: two discontinuity lines of the force fields ZN and $Z'R$; the lines of geometric division CS_g and BS_d; and the circular arcs AC and AB. The optimal cantilever considered is composed of 13 subdomains in which density υ is determined by different formulae. Four subdomains of the final division appear on the two circular domains, four on the Hill's domain, two of them lie on Chan's upper domain and

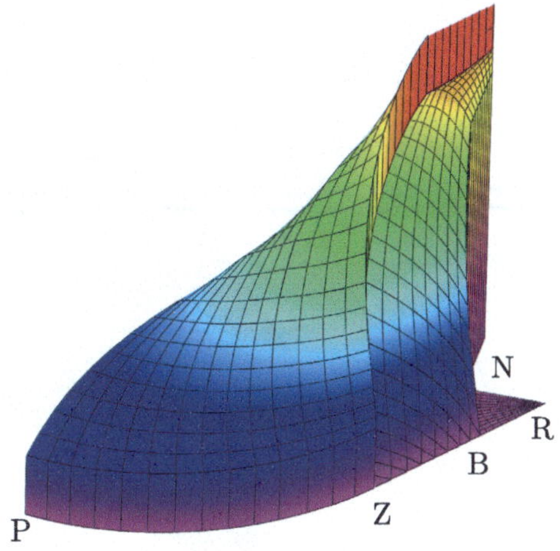

Fig. 4.118 Graph of the Hemp's force field T_2

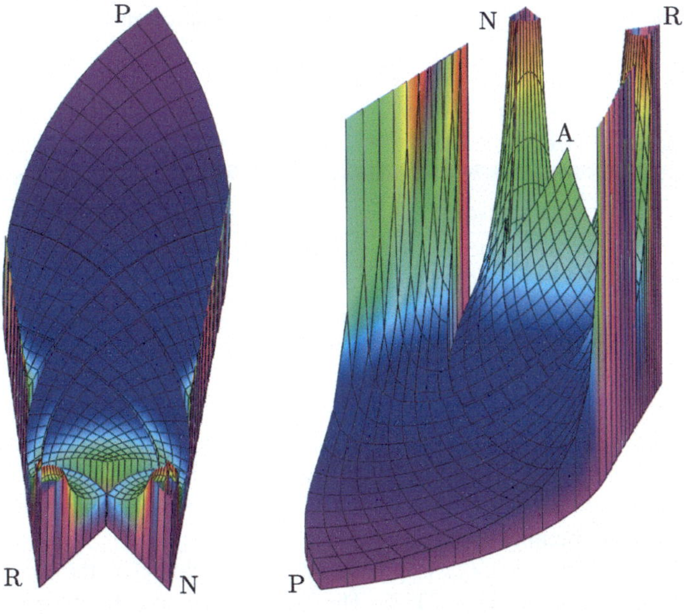

Fig. 4.119 Graph of the density of fibres

two on Chan's lower domain; one subdomain appears in domain V. Thus, the density function graph is fairly complicated, see Fig. 4.119.

The results presented above will be verified by considering two trusses (1,2) of finite number of joints fixed at the selected intersection points of the Hencky net of our problem. The truss (1) of 214 and truss (2) of 2998 members are shown in Fig. 4.120. One can prove that both these trusses are statically determinate. This

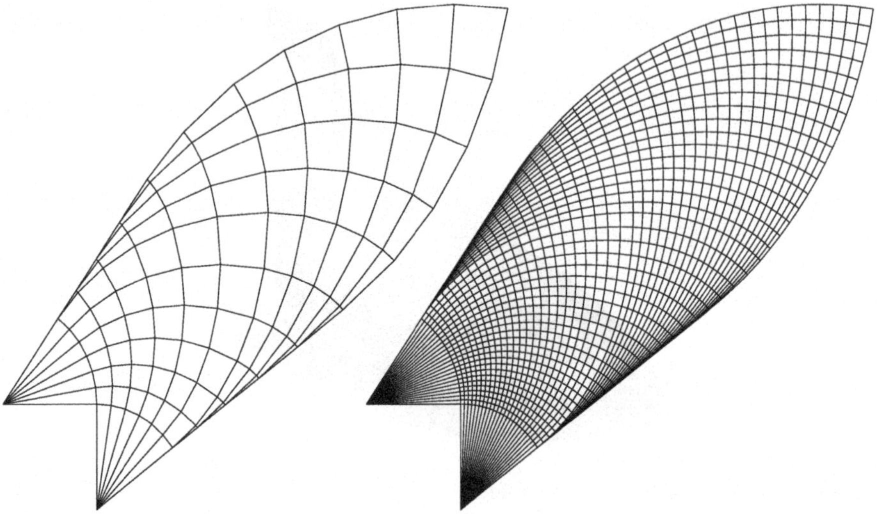

Fig. 4.120 The trusses (1) and (2) of 214 and 2998 members, respectively

Table 4.4 The characteristics of the trusses of Fig. 4.120

	Truss 1	Truss 2	Michell structure
Number of members	214	2998	Infinite
Number of joints	109	1501	Infinite
$\bar{V}$	34.57834674	34.49933802	34.49407
$\bar{V}_T$	6.851381387	6.336352716	6.1806443
$\bar{V}_C$	8.319861742	7.94368072	7.831983
$\bar{V}_K$	15.17124313	14.28003344	14.012627
$\bar{V}_S$	19.40710361	20.21930458	20.481447

property makes it easy to fix the cross sections to saturate the stress level. The results of computations are set up in Table 4.4, where $\bar{V}_S$ for trusses represents the total volume of internal members.

The volume of the truss (1) is greater than the volume of the material used for Michell-like cantilever by 0.244%. The volume of internal members occupies 56.125% of the total volume in comparison to 59.376% in the optimal cantilever. Thus, we note a discrepancy much larger than that concerning approximation of the total volume.

The truss (2) is heavier than the optimal cantilever only by 0.015%. The internal members contribute to 58.61% of the total volume, which is close to the corresponding ratio in the optimal cantilever.

C.2. Case of $\kappa = 3$
To be specific, we assume concrete data:

$$\theta_1 = \frac{\pi}{6}; \qquad \theta_2 = \frac{5\pi}{12}, \qquad x_P = 1.557879337a;$$

$$y_P = 2.994292715a, \qquad \varphi = \frac{\pi}{8}$$

where $a = |RN|$; coordinates x_P, y_P are referred to the (x, y) system, as in Fig. 4.99. Then

$$d = 3.805085930a, \qquad \alpha_P = \frac{\pi}{2}, \qquad \beta_P = \frac{\pi}{2}$$

We compute the forces in the reinforcing bars:

$$F_T = 0.6087614290P; \qquad F_C = -0.7933533403P$$

Then, we compute the reactions at R:

$$H_g = 3.363139403P, \qquad V_g = 0.5613526265P$$

and the reaction forces at N:

$$H_d = -3.74582283P, \qquad V_d = 0.3625269006P.$$

Having the Lamé coefficients and the force fields, we may find the distribution of the fibre density function v. Appropriate integration gives the non-dimensional volumes $\bar{V} = V/\tilde{V}_0$ of the material within the subdomains shown in Fig. 4.121; here $\tilde{V}_0 = Pa/\sigma_T$;

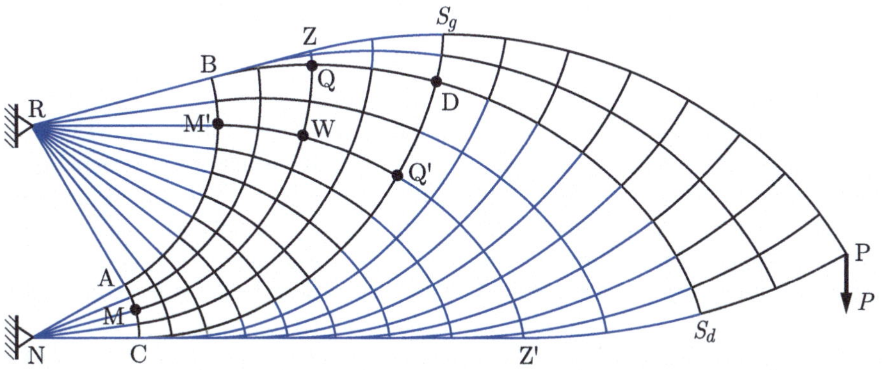

Fig. 4.121 The cantilever considered with the force located within the second central domain

$$\bar{V}_{NAM} = 1.03291971, \quad \bar{V}_{NMC} = 1.095218421,$$

$$\bar{V}_{RM'A} = 1.738224285, \quad \bar{V}_{RBM'} = 0.5492020058,$$

$$\bar{V}_{AM'WM} = 1.957767017, \quad \bar{V}_{MWQ'C} = 2.749061275,$$

$$\bar{V}_{BQWM'} = 0.3951255746, \quad \bar{V}_{CQ'Z'} = 4.825841085,$$

$$\bar{V}_{BZQ} = 0.1280375442, \quad \bar{V}_{QDQ'W} = 0.7147046372,$$

$$\bar{V}_{Q'DS_dZ'} = 2.807816353, \quad \bar{V}_{ZS_gDQ} = 0.3647799660,$$

$$\bar{V}_{DS_gPS_d} = 1.823721477$$

Thus, the total volume of the material within the fibrous domains equals $\bar{V}_S = 20.18241935$. Then, we compute the volumes of the tension and compression bars

$$\bar{V}_T = 3.086166189, \quad \bar{V}_C = 14.67228125, \quad \bar{V}_K = \bar{V}_T + \bar{V}_C = 17.75844744$$

The volume of the material used for the whole structure: $\bar{V} = \bar{V}_S + \bar{V}_K = 37.94086679$.

The same volume computed by the alternative formula based on virtual displacement fields allows to find $\bar{V} = 37.94086663$, which confirms the previous result based on the force field analysis.

D. The Case of the Concentrated Force Applied at a Point Within $JH_2J_2G_2$

Now, we consider a longer cantilever, transmitting the point load applied within $JH_2J_2G_2$ to the supporting line RN. From the kinematic point of view, the cantilever is composed of Hill's domain, four Chan's domains and two central domains, see Fig. 4.121.

We assume that

$$\sigma_0 = \sigma_T = \sigma_C \quad (\kappa = 1) \text{ and } \theta_1 = \theta_2 = \theta = 3\pi/10; \quad |RA| = |NA| = r_0$$

The key to the solution of the equilibrium problem of Figs. 4.122 and 4.123 is to figure out the division of the cantilever domain which is a result of an overlapping of the kinematic division denoted by Roman numerals (I, II, ..., VII) and of the static division denoted by Arabic numbers $(1, 2, \ldots, 7)$ as superscripts. The indices g and d mean the upper or lower domain, respectively. This example is fairly complicated, but since it could serve as a benchmark, the indispensable details will be reported.

Thus, the domain of the cantilever is divided into the following 23 subdomains: $RAM' = II_g^7$, $NAM = II_d^7$, $AM'WM = III^7$, $RBM' = II_g^6$, $BM'WQ = III^6$, $BQZ = IV_g^6$, $NMC = II_d^5$, $CMWQ' = III^5$, $CQ'Z' = IV_d^5$, $WQDQ' = III^4$, $Q'DZ_1'Z' = IV_d^4$, $QZZ_1D = IV_g^4$, $DZ_1W_1Z' = V^4$, $Z'Z_1'G = IV_d^3$, $GZ_1'W_1Z_d =$

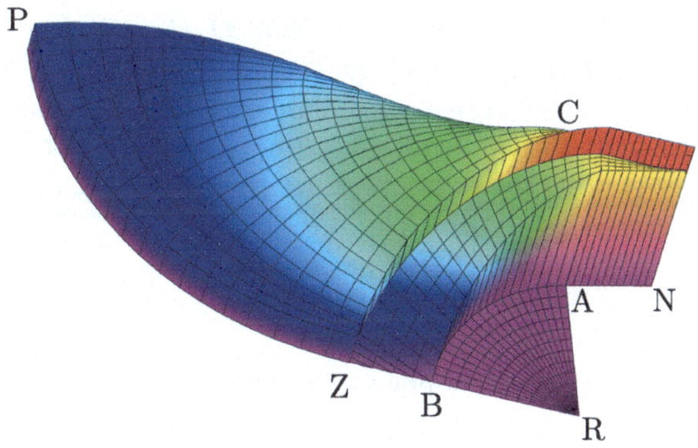

Fig. 4.122 Graph of the Hemp force field T_2

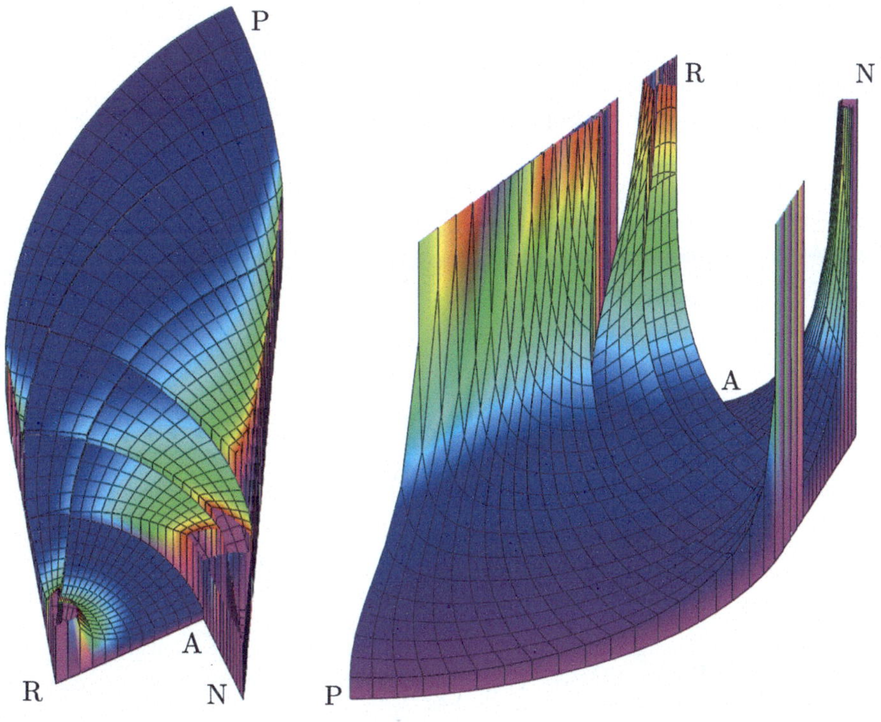

Fig. 4.123 Graph of the density of fibres function υ

V^3, $GQ_1'Z_d = VI_d^3$, $ZZ_1H = IV_g^2$, $Z_1W_1Q_1H = V^2$, $HQ_1Z_g = VI_g^2$, $W_1Q_1D_1Q_1' = V^1$, $Q_1'D_1H_1Z_d = VI_d^1$, $Q_1D_1G_1Z_g = VI_g^1$, $D_1G_1PH_1 = VII^1$.

The vertices of these domains have the following curvilinear coordinates: (α, β) $P = (\alpha_P, \beta_P)$, $B = (0, \theta)$, $M' = (0, \beta_P - 2\theta)$, $A = (0, 0)$, $M = (\alpha_P - 2\theta, 0)$, $C = (\theta, 0)$, $W = (\alpha_P - 2\theta, \beta_P - 2\theta)$, $Q' = (\theta, \beta_P - 2\theta)$, $Q = (\alpha_P - 2\theta, \theta)$, $D = (\theta, \theta)$, $Z = (\alpha_P - 2\theta, \alpha_P - \theta)$, $Z' = (\beta_P - \theta, \beta_P - 2\theta)$, $G = (2\theta, \theta)$, $Z_1' = (\beta_P - \theta, \theta)$, $Z_1 = (\theta, \alpha_P - \theta)$, $H = (\theta, 2\theta)$, $W_1 = (\beta_P - \theta, \alpha_P - \theta)$, $Q_1 = (\beta_P - \theta, 2\theta)$, $Q_1' = (2\theta, \alpha_P - \theta)$, $D_1 = (2\theta, 2\theta)$, $Z_g = (\beta_P - \theta, \beta_P)$, $Z_d = (\alpha_P, \alpha_P - \theta)$, $G_1 = (2\theta, \beta_P)$, $H_1 = (\alpha_P, 2\theta)$.

We shall report the results corresponding to the following data:

$$\alpha_P = \frac{17\pi}{20}; \qquad \beta_P = \frac{4\pi}{5}; \qquad \varphi = \frac{\pi}{10}; \qquad d = 13.01824557r_0$$

The Cartesian coordinates of point P read $x_P = 9.188744886r_0$; $y_P = 8.221834561r_0$. We compute the forces in the reinforcing bars at node P: $F_T(P) = 0.9510565165P$; $F_C(P) = -0.3090169938P$.

The reaction forces at the upper supporting node are computed as follows:

$$H_g = \int_0^{\beta_P-2\theta} T_1^{IIg7}(1, \beta) \sin\left(\frac{\pi}{4} + \beta\right) d\beta + \int_{\beta_P-2\theta}^{\theta} T_1^{IIg6}(1, \beta) \sin\left(\frac{\pi}{4} + \beta\right) d\beta + F_T(B) \sin\left(\frac{\pi}{4} + \theta\right),$$

$$V_g = \int_0^{\beta_P-2\theta} T_1^{III7}(1, \beta) \cos\left(\frac{\pi}{4} + \beta\right) d\beta + \int_{\beta_P-2\theta}^{\theta} T_1^{III5}(1, \beta) \cos\left(\frac{\pi}{4} + \beta\right) d\beta + F_T(B) \cos\left(\frac{\pi}{4} + \theta\right),$$

which gives: $H_g = 8.759863521P$, $V_g = 0.4486582529P$. The reaction forces at the lower supporting node are computed as follows:

$$H_d = \int_0^{\alpha_P-2\theta} T_2^{IId7}(\alpha, 1) \sin\left(\frac{\pi}{4} + \alpha\right) d\alpha + \int_{\alpha_P-2\theta}^{\theta} T_2^{IId5}(\alpha, 1) \sin\left(\frac{\pi}{4} + \alpha\right) d\alpha + F_C(C) \sin\left(\frac{\pi}{4} + \theta\right)$$

$$V_d = -\int_0^{\alpha_P-2\theta} T_2^{IId7}(\alpha, 1) \cos\left(\frac{\pi}{4} + \alpha\right) d\alpha - \int_{\alpha_P-2\theta}^{\theta} T_2^{IId5}(\alpha, 1) \cos\left(\frac{\pi}{4} + \alpha\right) d\alpha - F_C(C) \cos\left(\frac{\pi}{4} + \theta\right),$$

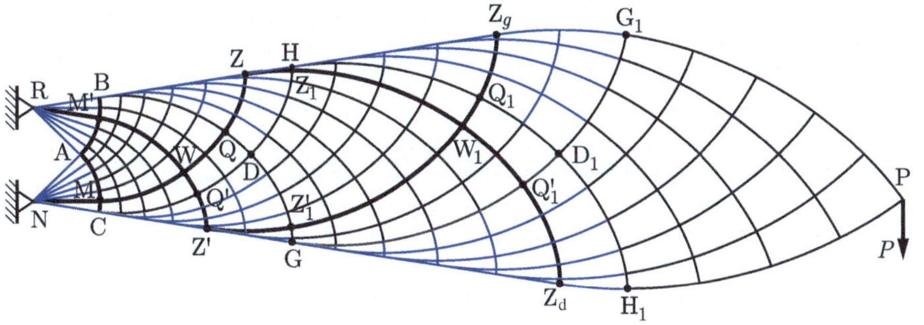

Fig. 4.124 Subdivision of the optimal cantilever into the domains of static and geometric divisions. Case of point P lying within the domain VII

which gives: $H_d = -8.450846479P$, $V_d = 0.5023981407P$.

The volumes of the material within the fibrous domains are denoted according to the notation of Fig. 4.124. To make the analysis complete, we shall put forward all the formulae for the volumes of the subdomains into which the structure is divided.

The lower circular internal fan NAM, see Fig. 4.124:

$$V^{IId7} = \int_0^1 \int_0^{\alpha_P - 2\theta} A^{IId7}(\alpha, \beta_1) B^{IId7}(\alpha, \beta_1) v^{IId7}(\alpha, \beta_1) d\alpha d\beta_1;$$

$$V^{IId7} = 3.475700031 V_0$$

where $V_0 = \dfrac{P r_0}{\sigma_0}$. The lower circular external fan NMC:

$$V^{IId5} = \int_0^1 \int_{\alpha_P - 2\theta}^{\theta} A^{IId5}(\alpha, \beta_1) B^{IId5}(\alpha, \beta_1) h^{IId5}(\alpha, \beta_1) d\alpha d\beta_1;$$

$$V^{IId5} = 0.6455328237 V_0$$

The upper circular internal fan RAM':

$$V^{IIg7} = \int_0^{\beta_P - 2\theta} \int_0^1 A^{IIg7}(\alpha_1, \beta) B^{IIg7}(\alpha_1, \beta) v^{IIg7}(\alpha_1, \beta) d\alpha_1 d\beta;$$

$$V^{IIg7} = 2.780971414 V_0$$

The upper circular external fan RBM':

$$V^{IIg6} = \int_{\beta_P-2\theta}^{\theta} \int_0^1 A^{IIg6}(\alpha_1, \beta) B^{IIg6}(\alpha_1, \beta) v^{IIg6}(\alpha_1, \beta) d\alpha_1 d\beta;$$

$$V^{IIg6} = 1.345820099 V_0$$

The Hill's region $AM'WM$ of the static division 7:

$$V^{III7} = \int_0^{\beta_P-2\theta} \int_0^{\alpha_P-2\theta} A^{III7}(\alpha, \beta) B^{III7}(\alpha, \beta) v^{III7}(\alpha, \beta) d\alpha d\beta;$$

$$V^{III7} = 4.997932218 V_0$$

The Hill's region $BM'WQ$ of the static division 6:

$$V^{III6} = \int_{\beta_P-2\theta}^{\theta} \int_0^{\alpha_P-2\theta} A^{III6}(\alpha, \beta) B^{III6}(\alpha, \beta) v^{III6}(\alpha, \beta) d\alpha d\beta;$$

$$V^{III6} = 2.944450720 V_0$$

The Hill's region $CMWQ'$ of the static division 5:

$$V^{III5} = \int_0^{\beta_P-2\theta} \int_{\alpha_P-2\theta}^{\theta} A^{III5}(\alpha, \beta) B^{III5}(\alpha, \beta) v^{III5}(\alpha, \beta) d\alpha d\beta;$$

$$V^{III5} = 1.094712407 V_0$$

The Hill's region $WQDQ'$ of the static division 4:

$$V^{III4} = \int_{\beta_P-2\theta}^{\theta} \int_{\alpha_P-2\theta}^{\theta} A^{III4}(\alpha, \beta) B^{III4}(\alpha, \beta) v^{III4}(\alpha, \beta) d\alpha d\beta;$$

$$V^{III4} = 0.7912442817 V_0$$

Chan's lower domain $CQ'Z'$ of the static division 5:

$$V^{IVd5} = \int_0^{\beta_P-2\theta} \int_{\theta}^{\theta+\beta} A^{IVd5}(\alpha, \beta) B^{IVd5}(\alpha, \beta) v^{IVd5}(\alpha, \beta) d\alpha d\beta;$$

$$V^{IVd5} = 1.949485842 V_0$$

Chan's lower domain $Q'DZ_1'Z'$ of the static division 4:

$$V^{IVd4} = \int_{\beta_P-2\theta}^{\theta} \int_{\theta}^{\beta_P+\theta} A^{IVd4}(\alpha,\beta)B^{IVd4}(\alpha,\beta)\upsilon^{IVd4}(\alpha,\beta)d\alpha d\beta;$$

$$V^{IVd4} = 2.613805605V_0$$

Chan's lower domain $Z'Z_1'G$ of the static division 3:

$$V^{IVd3} = \int_{\beta_P-2\theta}^{\theta} \int_{\beta_P-\theta}^{\theta+\beta} A^{IVd3}(\alpha,\beta)B^{IVd3}(\alpha,\beta)\upsilon^{IVd3}(\alpha,\beta)d\alpha d\beta;$$

$$V^{IVd3} = 0.7153223805V_0$$

Chan's upper domain BQZ of the static division 6:

$$V^{IVg6} = \int_{0}^{\alpha_P-2\theta} \int_{\theta}^{\theta+\alpha} A^{IVg6}(\alpha,\beta)B^{IVg6}(\alpha,\beta)\upsilon^{IVg6}(\alpha,\beta)d\alpha d\beta;$$

$$V^{IVg6} = 3.345627065V_0$$

Chan's upper domain QZZ_1D of the static division 4:

$$V^{IVg4} = \int_{\alpha_P-2\theta}^{\theta} \int_{\theta}^{\alpha_P-\theta} A^{IVg4}(\alpha,\beta)B^{IVg4}(\alpha,\beta)\upsilon^{IVg4}(\alpha,\beta)d\alpha d\beta;$$

$$V^{IVg4} = 1.715672751V_0$$

Chan's upper domain ZZ_1H of the static division 2:

$$V^{IVg2} = \int_{\alpha_P-2\theta}^{\theta} \int_{\alpha_P-\theta}^{\theta+\alpha} A^{IVg2}(\alpha,\beta)B^{IVg2}(\alpha,\beta)\upsilon^{IVg2}(\alpha,\beta)d\alpha d\beta;$$

$$V^{IVg2} = 0.1564714358V_0$$

Domain V of the static division 4, or DZ_1W_1Z':

$$V^{V4} = \int_{\theta}^{\alpha_P-\theta} \int_{\theta}^{\beta-\theta} A^{V4}(\alpha,\beta)B^{V4}(\alpha,\beta)\upsilon^{V4}(\alpha,\beta)d\alpha d\beta;$$

$$dV^{V4} = 6.338378990V_0$$

Domain V of the static division 3, or $GZ_1'W_1Z_d$:

$$V^{V3} = \int_{\theta}^{\alpha_P - \theta} \int_{\beta_P - \theta}^{2\theta} A^{V3}(\alpha, \beta) B^{V3}(\alpha, \beta) \upsilon^{V3}(\alpha, \beta) d\alpha d\beta;$$

$$V^{V3} = 3.543679041 V_0$$

Domain V of the static division 2, or $Z_1 W_1 Q_1 H$:

$$V^{V2} = \int_{\alpha_P - \theta}^{2\theta} \int_{\theta}^{\beta_P - \theta} A^{V2}(\alpha, \beta) B^{V2}(\alpha, \beta) \upsilon^{V2}(\alpha, \beta) d\alpha d\beta;$$

$$V^{V2} = 1.213272591 V_0$$

Domain V of the static division 1, or $W_1 Q_1 D_1 Q_1'$:

$$V^{V1} = \int_{\alpha_P - \theta}^{2\theta} \int_{\beta_P - \theta}^{2\theta} A^{V1}(\alpha, \beta) B^{V1}(\alpha, \beta) \upsilon^{V1}(\alpha, \beta) d\alpha d\beta;$$

$$V^{V1} = 0.7663860998 V_0$$

The second lower Chan's domain $G Q_1' Z_d$ of the static division 3:

$$V^{VId3} = \int_{\theta}^{\alpha_P - \theta} \int_{2\theta}^{\theta + \beta} A^{VId3}(\alpha, \beta) B^{VId3}(\alpha, \beta) \upsilon^{VId3}(\alpha, \beta) d\alpha d\beta;$$

$$V^{VId3} = 3.211119411 V_0$$

The second lower Chan's domain $Q_1' D_1 H_1 Z_d$ of the static division 1:

$$V^{VId1} = \int_{\alpha_P - \theta}^{2\theta} \int_{2\theta}^{\alpha_P} A^{VId1}(\alpha, \beta) B^{VId1}(\alpha, \beta) \upsilon^{VId1}(\alpha, \beta) d\alpha d\beta;$$

$$dV^{VId1} = 1.350077647 V_0$$

The second upper Chan's domain $H Q_1 Z_g$ of the static division 2:

$$V^{VIg2} = \int_{\theta}^{\beta_P - \theta} \int_{2\theta}^{\theta + \alpha} A^{VIg2}(\alpha, \beta) B^{VIg2}(\alpha, \beta) \upsilon^{VIg2}(\alpha, \beta) d\alpha d\beta;$$

$$V^{VIg2} = 1.969535422 V_0$$

The second upper Chan's domain $Q_1 D_1 G_1 Z_g$ of the static division 1:

$$V^{VIg1} = \int_{\beta_P-\theta}^{2\theta} \int_{2\theta}^{\beta_P} A^{VIg1}(\alpha,\beta)B^{VIg1}(\alpha,\beta)v^{VIg1}(\alpha,\beta)d\alpha d\beta;$$

$$V^{VIg1} = 2.286771699V_0$$

The domain VII of the static division 1, or $D_1G_1PH_1$:

$$V^{VII1} = \int_{2\theta}^{\beta_P} \int_{2\theta}^{\alpha_P} A^{VII1}(\alpha,\beta)B^{VII1}(\alpha,\beta)v^{VII1}(\alpha,\beta)d\alpha d\beta;$$

$$V^{VII1} = 4.911550402V_0$$

The total volume of the material used for the fibrous domains equals $V_S = 54.16352037V_0$.

We compute the volumes of the reinforcing bars. The tension reinforcing bar along the domain RBM':

$$V_k^{IIg6} = \sigma_0^{-1}F_T^6(B)r_0; \qquad V_k^{IIg6} = 5.039527441V_0$$

The tension reinforcing bar along the domain BQZ:

$$V_k^{IVg6} = \sigma_0^{-1}\int_0^{\alpha_P-2\theta} A^{IVg6}(\alpha,\alpha+\theta)F_T^{(6)}d\alpha;$$

$$V_k^{IVg6} = 7.772438081V_0$$

The tension reinforcing bar along the domain ZZ_1H:

$$V_k^{IVg2} = \sigma_0^{-1}\int_{\alpha_P-2\theta}^{\theta} A^{IVg2}(\alpha,\alpha+\theta)F_T^{(2)}d\alpha;$$

$$V_k^{IVg2} = 1.629394719V_0$$

The tension reinforcing bar along the domain HQ_1Z_g:

$$V_k^{VIg2} = \sigma_0^{-1}\int_{\theta}^{\beta_P-\theta} A^{VIg2}(\alpha,\alpha+\theta)F_T^{(2)}d\alpha;$$

$$V_k^{VIg2} = 4.37933693V_0$$

The tension reinforcing bar along the domain $Q_1D_1G_1Z_g$:

$$V_k^{VIg1} = \sigma_0^{-1} \int_{\beta_P-\theta}^{2\theta} A^{VIg1}(\alpha, \beta_P) F_T d\alpha;$$

$$V_k^{VIg1} = 1.853498369 V_0$$

The tension reinforcing bar along the domain $D_1 G_1 P H_1$:

$$V_k^{VII1} = \sigma_0^{-1} \int_{2\theta}^{\alpha_P} A^{VII1}(\alpha, \beta_P) F_T d\alpha;$$

$$V_k^{VII1} = 4.714622429 V_0$$

The total volume of the tension reinforcing bar: $V_T = 5.38881797 V_0$.

The reinforcing compression bar along the domain NMC:

$$V_k^{IId5} = -\sigma_0^{-1} F_C^{(5)}(C) r; \qquad V_k^{IId5} = 4.733490212 V_0$$

The reinforcing compression bar along the domain $CQ'Z'$:

$$V_k^{IVd5} = -\sigma_0^{-1} \int_0^{\beta_P-2\theta} B^{IVd5}(\beta+\theta, \beta) F_C^{(5)} d\beta;$$

$$V_k^{IVd5} = 5.711466545 V_0$$

The reinforcing compression bar along the domain $Z'Z'_1G$:

$$V_k^{IVd3} = -\sigma_0^{-1} \int_{\beta_P-2\theta}^{\theta} B^{IVd3}(\beta+\theta, \beta) F_C^{(3)} d\beta;$$

$$V_k^{IVd3} = 2.753625735 V_0$$

The reinforcing compression bar along the domain GQ'_1Z_d:

$$V_k^{VId3} = -\sigma_0^{-1} \int_{\theta}^{\alpha_P-2\theta} B^{VId3}(\beta+\theta, \beta) F_C^3 d\beta;$$

$$V_k^{VId3} = 3.43389751 V_0$$

The reinforcing compression bar along the domain $Q'_1 D_1 H_1 Z_d$:

$$V_k^{VId1} = -\sigma_0^{-1} \int_{\alpha_P-\theta}^{2\theta} B^{VId1}(\alpha_P, \beta) F_C d\beta;$$

$$V_k^{VId1} = 0.3160621721 V_0$$

The reinforcing compression bar along the domain $D_1G_1PH_1$:

$$V_k^{VII1} = -\sigma_0^{-1} \int_{2\theta}^{\beta_P} B^{VII1}(\alpha_P, \beta) F_C d\beta;$$

$$V_k^{VII1} = 1.363841418 V_0$$

The total volume of the compression reinforcing bar: $V_C = 18.31238360 V_0$.

The total volume of the reinforcing bars: $V_K = 43.70120157 V_0$.

The total volume of the material used for the structure:
$V = V_S + V_K = 97.86472194 V_0$.

The same volume computed by the alternative formula (3.81), (3.85):
$V = 97.86471582 V_0$.

The analysis of the concrete examples of Michell's cantilevers reveals great complexity of the emerging internal structure of optimal designs. In particular, it shows the influence of kinematic and static fields and their discontinuities on the final, surprisingly complicated distribution of the fibre densities. Moreover, the direct computation of the volume with the use of stress-based formulation and using the dual kinematic formulation eventually proves the correctness of all the optimal structures constructed.

4.13 Deep Beam Problem. Case of $\sigma_C = \sigma_T$

In this section, we will confine our consideration to the case of $\sigma_T = \sigma_C = \sigma_0$. The subject of the research are Michell's structures transmitting vertical loads to two supports transferring vertical loads, see Fig. 4.125.

Thus, the horizontal reactions are equal to zero. All the problems to be analysed have a vertical symmetry axis, and thus, the solutions can be constructed by analysing a half of the structure.

4.13.1 Solution of the Auxiliary Problem

We consider the right-hand part of the problem of Fig. 4.125a by adopting the proper symmetry conditions, see Fig. 4.126a.

The solution of the problem of Fig. 4.126a differs from the solution of the problem in Fig. 4.126b exclusively in the kinematic conditions which, however, do not change the form of the Michell structure. It will turn out that the optimal structure does not make use of the horizontal intermediate supports, and thus, it is sufficient to solve

Fig. 4.125 Design domains and loadings

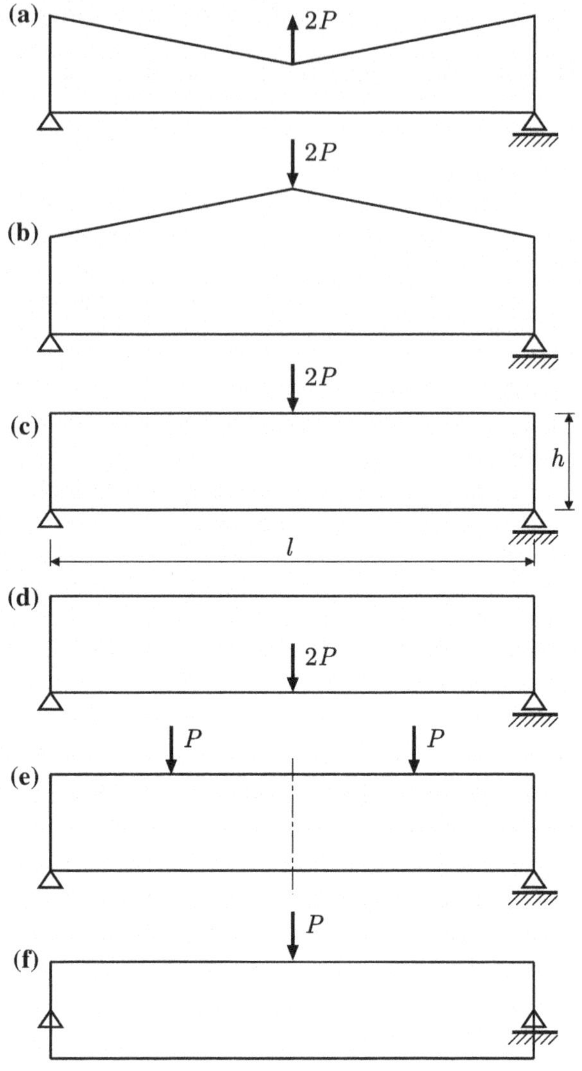

the problem from Fig. 4.126c: transmit the force P at point P to supports B and C in a given design domain. The solution will depend on the ratio $\dfrac{|PC|}{r_0}$, $r_0 = |BC|$. The solution is elementary only if $P = C$. Then, the optimal structure is composed of a single bar in tension along BC. The volume of this structure is $V_0 = \dfrac{Pr_0}{\sigma_0}$. We predict the solutions as inscribed into the net depicted in Fig. 4.127.

The Michell net is constructed on the circle of the centre at point B and the radius r_0. In the sector of the circle BDC, we introduce the polar parameterization (α_1, β),

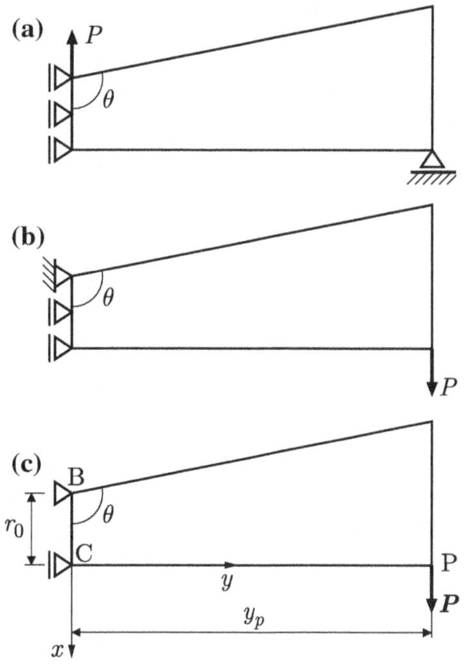

Fig. 4.126 Problem **a** is equivalent to **b** and the latter one simplifies to the setting (**c**)

where $r = \alpha_1 r_0$, $0 \leq \alpha_1 \leq 1$, $0 \leq \beta \leq \theta$, where $\theta = \measuredangle(CB, DB)$ see Fig. 4.127b. In the following steps, we will build the net in the domain CDM, further in DMN and then in MNM_1.

4.13.1.1 The Adjoint Displacement Field Within BDC

The construction of fields (u, v) is similar as in Sect. 4.8. In the framework of the parameterization (α_1, β), the Lamé coefficients are given by

$$A(\alpha_1, \beta) = r_0, \qquad B(\alpha_1, \beta) = \alpha_1 r_0 \tag{4.684}$$

Let us notice that the field of components

$$u(\alpha_1, \beta) = r_0 \alpha_1, \qquad v(\alpha_1, \beta) = -2 r_0 \alpha_1 \beta \tag{4.685}$$

satisfies the given conditions of the optimality (4.116) of the regime $\mathbb{T}$, cf. Table 3.1, here $\alpha = 1$.

The result (4.685) can be obtained from (4.209) assuming $C_1 = 0$, $d_1 = 0$, $\epsilon_1 = -1$, $\epsilon_2 = -1$, $r = r_0 \alpha_1$, $\beta = \vartheta$.

The field ω given by (4.212) is equal

$$\omega = -2\beta \tag{4.686}$$

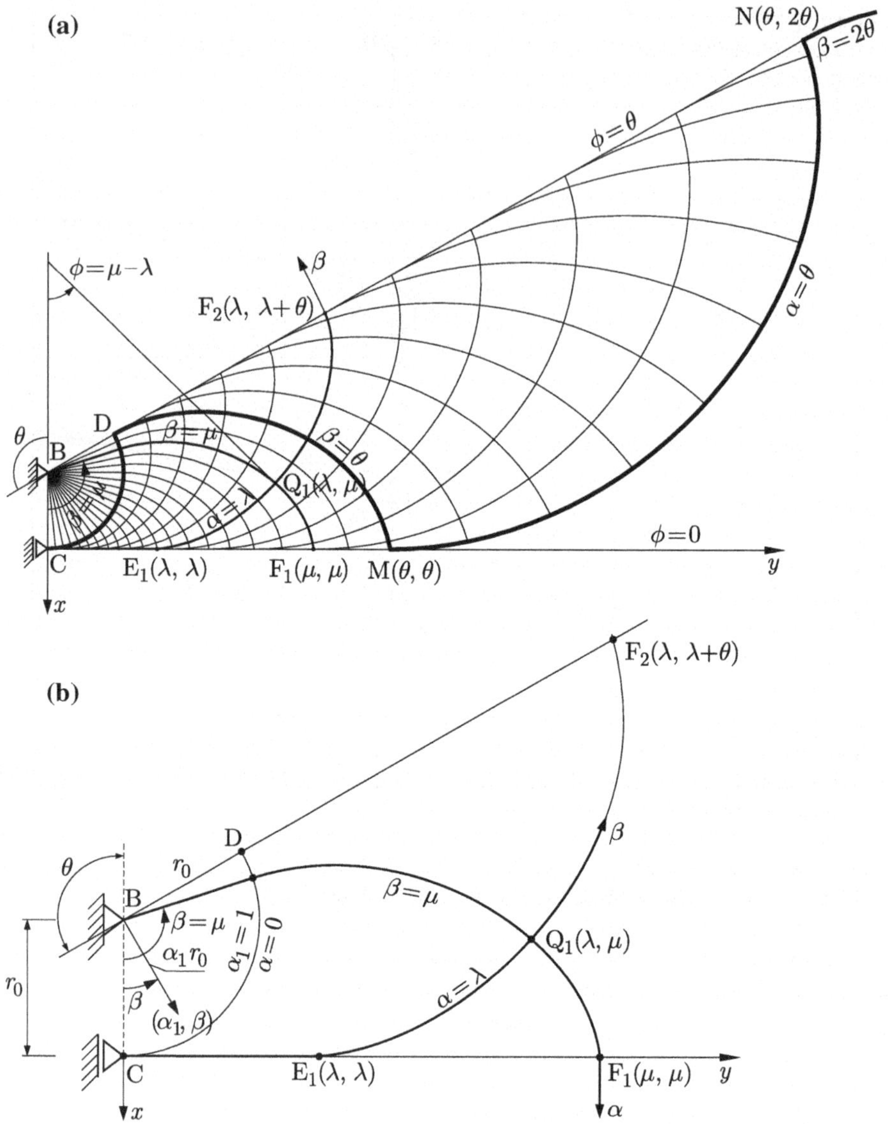

Fig. 4.127 The Michell net for problem of Fig. 4.126c (**a**); more details of parameterization in BDC (**b**)

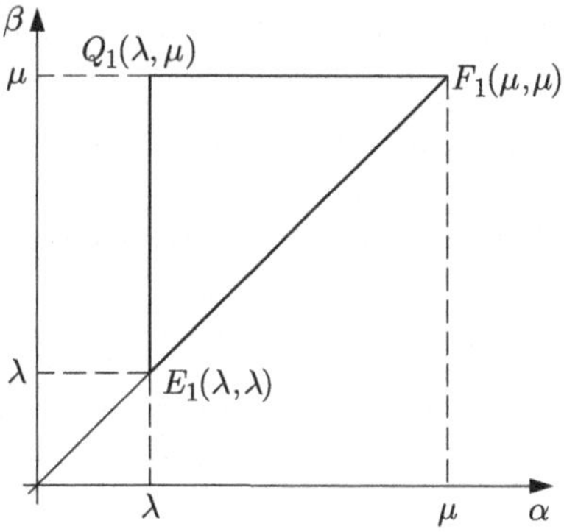

Fig. 4.128 Juxtaposition of points Q_1, F_1, E_1 in the (α, β)-plane

and now $\omega(0, 0) = 0$.

4.13.1.2 The Hencky Net in CDM

We will apply the methods described in Sect. 4.12. The net (α_1, β) defined in CBD will be extended to the domain CDM in such a way that $\alpha = 0$ is the line $\alpha_1 = 1$ of the circle CD, while the lines β are tangent to the axis x.

If $\phi(\alpha, \beta) = \beta - \alpha$ is the angle defined by the tangent to the line α given at the point (α, β) with respect to the axis x, then $\phi = 0$ along the axis x, thus there $\alpha = \beta$.

Let the point $Q_1(\lambda, \mu)$ lie in the domain CDM. The point Q_1 defines the location of the points $E_1(\lambda, \lambda)$ and $F_1(\mu, \mu)$ at the line $x = 0$. We consider the domain $(E_1 Q_1 F_1)$. Let (α, β) lie inside this domain. Then: $\lambda \le \alpha \le \mu$, $\lambda \le \beta \le \mu$.

The domain $E_1 Q_1 F_1$ is the image of a triangle, see Fig. 4.128.

Let $G(\alpha, \beta)$ be given by (4.549). We notice that both the arguments $(\alpha - \lambda, \mu - \beta)$ of the function D_0 are non-negative. The main unknown is the field $\bar{x}(\alpha, \beta)$ introduced by the formula (4.95), which satisfies the equation $\mathscr{L}\bar{x} = 0$, where $\mathscr{L}$ is given by (4.112). The value $\bar{x}(\alpha, \beta)$ can be expressed by the formula (4.552) assuming $w = \bar{x}$, $\theta_1 = 0$. Thus

$$\bar{x}(\lambda, \mu) = \frac{1}{2}\left(\bar{x}(\lambda, \lambda) + \bar{x}(\mu, \mu)\right) - \frac{1}{2}\int_{\lambda}^{\mu}\left[\bar{x}(\alpha, \alpha)\left(\frac{\partial G}{\partial \beta} - \frac{\partial G}{\partial \alpha}\right)\bigg|_{\beta=\alpha} + G(\alpha, \alpha)\hat{\varphi}(\alpha)\right] d\alpha \quad (4.687)$$

where, according to (4.550)

$$\hat{\varphi}(\alpha) = \left(\frac{\partial \bar{x}}{\partial \alpha} - \frac{\partial \bar{x}}{\partial \beta}\right)\bigg|_{\beta=\alpha} \quad (4.688)$$

We note that $\bar{x}(\lambda, \lambda) = 0$, $\bar{x}(\mu, \mu) = 0$ and this simplifies (4.687):

$$\bar{x}(\lambda, \mu) = -\frac{1}{2}\int_{\lambda}^{\mu} D_0(\alpha - \lambda, \mu - \alpha)\hat{\varphi}(\alpha)d\alpha \tag{4.689}$$

The function $\bar{x}(\lambda, \mu)$ is given for $\lambda = 0$.

$$\bar{x}(0, \mu) = r_0(1 - \cos\mu) \tag{4.690}$$

which leads to integral equation of the convolution type:

$$-\frac{1}{2}\int_{0}^{\mu} D_0(\alpha, \mu - \alpha)\varphi(\alpha)d\alpha = 1 - \cos\mu \tag{4.691}$$

where

$$\varphi(\alpha) = \frac{\hat{\varphi}(\alpha)}{r_0}$$

Similarly as in Sect. 4.12, we solve Eq. (4.691) with the method of Laplace transform. By applying the Laplace transform L_μ to both sides of (4.691) and using (4.544), (4.530), we have

$$\varphi^*\left(p + \frac{1}{p}\right) = -\frac{2}{p^2 + 1} \tag{4.692}$$

We change the variables by (4.566), rearrange (4.692) to the form:

$$\varphi^*(s) = -\frac{4}{s(s + \sqrt{s^2 - 4})} \tag{4.693}$$

and rewrite this result as follows:

$$-\frac{1}{4}\varphi^*(s) = \frac{1}{s\sqrt{s^2 - 4}} - \frac{1}{\sqrt{s^2 - 4}(s + \sqrt{s^2 - 4})} \tag{4.694}$$

The components at the r.h.s. can be recognized in the formulae (4.533), (4.534) for $\xi = 0$. Thus, we note that:

$$-\frac{1}{4}\varphi(\alpha) = F_1(\alpha, \alpha) - \frac{1}{2}G_1(\alpha, \alpha) \tag{4.695}$$

In this manner we had found the solution of the equation (4.691). We go back to the formula (4.689) taking $\hat{\varphi}(\alpha) = r_0\varphi(\alpha)$:

$$\frac{\bar{x}(\lambda, \mu)}{r_0} = \int_{\lambda}^{\mu} D_0(\alpha - \lambda, \mu - \alpha)[2F_1(\alpha, \alpha) - G_1(\alpha, \alpha)]d\alpha \tag{4.696}$$

By taking into account (4.543) and using the notation (4.537a), we write the result (4.696) in the form:

$$\bar{x}(\lambda, \mu) = r_0[w_1(\lambda, 0; \mu) - w_1(\mu, 0; \lambda)] \tag{4.697}$$

Using the result (4.540), we find

$$\frac{\bar{x}(\lambda, \mu)}{r_0} = F_2(\mu, \lambda) - F_2(\lambda, \mu) \tag{4.698}$$

We define the field $\bar{y}$ according to (4.97). We differentiate both sides of (4.698) with respect to μ using the rules (4.142). We obtain

$$\frac{\bar{y}(\lambda, \mu)}{r_0} = F_1(\mu, \lambda) - F_3(\lambda, \mu) \tag{4.699}$$

The coordinates $x(\lambda, \mu)$, $y(\lambda, \mu)$ are defined by rules (4.96), where $\phi(\lambda, \mu) = \mu - \lambda$. In particular, it can be checked that

$$\frac{x(0, \mu)}{r_0} = -(1 - \cos\mu), \qquad \frac{y(0, \mu)}{r_0} = \sin\mu \tag{4.700}$$

and thus the parametric line $\lambda = 0$ coincides with the circle CD.

The Lamé coefficient can be constructed by (4.98), using the differential rules (4.142) and the identities (4.143):

$$\frac{A(\lambda, \mu)}{r_0} = G_1(\mu, \lambda) - G_1(\lambda, \mu), \qquad \frac{B(\lambda, \mu)}{r_0} = G_0(\lambda, \mu) - G_2(\lambda, \mu) \tag{4.701}$$

Note that along the $x = 0$ line

$$A(\lambda, \lambda) = 0, \qquad B(\lambda, \lambda) > 0 \tag{4.702}$$

4.13.1.3 The Adjoint Displacements Within CDM

The field (u, v) in the domain CDM is the continuous extension of the field (u, v) in the domain BDC. Therefore:

$$u^{CDM}(0, \mu) = u^{BDC}(1, \mu), \qquad v^{CDM}(0, \mu) = v^{BDC}(1, \mu) \tag{4.703}$$

and according to (4.685)

$$u(0, \mu) = r_0, \qquad v(0, \mu) = -2r_0\mu \tag{4.704}$$

where (u, v) concern the domain CDM.

The fields (u, v) depend on the form of field ω. According to (4.78), we have

$$\omega(\lambda, \mu) = \omega(0, 0) - 2(\lambda + \mu) \tag{4.705}$$

and the field has to be continuous extension of the ω field in BDC. The condition of continuity of ω at point C simplifies (4.705) to the form:

$$\omega(\lambda, \mu) = -2(\lambda + \mu) \tag{4.706}$$

Now, we can define the fields (u, v) in CDM using the representation (4.107). First, we find

$$u(\lambda, \mu) = \hat{u}_{\theta=0}(\lambda, \mu) + u_3(\lambda, \mu) \tag{4.707}$$

where according to (4.108), (4.704), we have

$$\frac{\hat{u}_{\theta=0}(\lambda, \mu)}{r_0} = \cos\lambda + 2\mu \sin\lambda \tag{4.708}$$

$$u_3(\lambda, \mu) = \int_0^{\lambda} [\cos(\lambda - \alpha) + 2(\alpha + \mu)\sin(\lambda - \alpha)] A(\alpha, \mu) d\alpha$$

By substitution (4.701) and integration with the use of rules (4.150), (4.151), we obtain

$$\frac{u(\lambda, \mu)}{r_0} = G_0(\lambda, \mu) + 2(\mu - \lambda)G_1(\lambda, \mu) + 2F_2(\lambda, \mu) \tag{4.709}$$

or, according to (4.118a),

$$u(\lambda, \mu) = u^0(\lambda, \mu) + 2\lambda A(\lambda, \mu) \tag{4.710}$$

where A is given by (4.701) and

$$\frac{u^0(\lambda, \mu)}{r_0} = G_0(\lambda, \mu) + 2F_2(\lambda, \mu) \tag{4.711}$$

The field v can be found according to (4.116). We find

$$\frac{v(\lambda, \mu)}{r_0} = G_1(\lambda, \mu) + 2(\lambda - \mu)G_0(\lambda, \mu) - 2F_1(\lambda, \mu) \tag{4.712}$$

or

$$v(\lambda, \mu) = v^0(\lambda, \mu) - 2\mu B(\lambda, \mu) \tag{4.713}$$

where B is given by (4.701) and

$$\frac{1}{r_0} v^0(\lambda, \mu) = 3G_1(\lambda, \mu) - 2F_1(\lambda, \mu) \tag{4.714}$$

4.13.1.4 Hencky Net in DMN

The Hencky net in CDM will be extended to DMN through the line $DM(\beta = \theta)$.

As the main unknown, we choose

$$z(\alpha, \beta) = \bar{y}(\alpha, \beta) - r_0 \sin(\beta - \alpha) \tag{4.715}$$

where $\bar{y}$ is defined by (4.95).

The function z satisfies $\mathscr{L}\bar{z} = 0$, $\mathscr{L}$ being defined by (4.112) and the boundary conditions

$$z(\eta - \theta, \eta) = 0, \qquad \frac{z(\alpha, \theta)}{r_0} = F_1(\alpha, \theta) - F_3(\alpha, \theta) \tag{4.716}$$

In order to find $z(\lambda, \eta)$, i.e. at Q_2, see Fig. 4.129, we apply the representation (4.557), which now reduces to the form

$$z(\lambda, \eta) = \frac{1}{2} \int_{\eta-\theta}^{\lambda} D_0(\lambda - \alpha, \alpha + \theta - \eta)\chi(\alpha)d\alpha \tag{4.717}$$

where

$$\chi(\alpha) = \left(\frac{\partial z}{\partial \alpha} - \frac{\partial z}{\partial \beta}\right)\Bigg|_{\beta=\alpha+\theta} \tag{4.718}$$

The function χ we be determined from the boundary condition $(4.716)_2$, i.e.

$$\int_0^{\lambda} D_0(\lambda - \alpha, \alpha)\bar{\chi}(\alpha)d\alpha = F_1(\alpha, \theta) - F_3(\alpha, \theta) \tag{4.719}$$

where

$$\bar{\chi}(\alpha) = \frac{\chi(\alpha)}{2r_0}$$

We perform the Laplace operation on Eq. (4.719) using the rules (4.544), (4.534). We obtain

$$\bar{\chi}^*\left(p + \frac{1}{p}\right) = \left(\frac{p}{p^2+1} - \frac{1}{p(p^2+1)}\right)\exp\left(\frac{\theta}{p}\right) \tag{4.720}$$

We introduce the variable s, cf. (4.566)

$$\bar{\chi}^*(s) = 2\left(\frac{1}{s} - \frac{1}{s + \sqrt{s^2-4}}\right)\exp\left(\frac{\theta}{2}(s - \sqrt{s^2-4})\right) \tag{4.721}$$

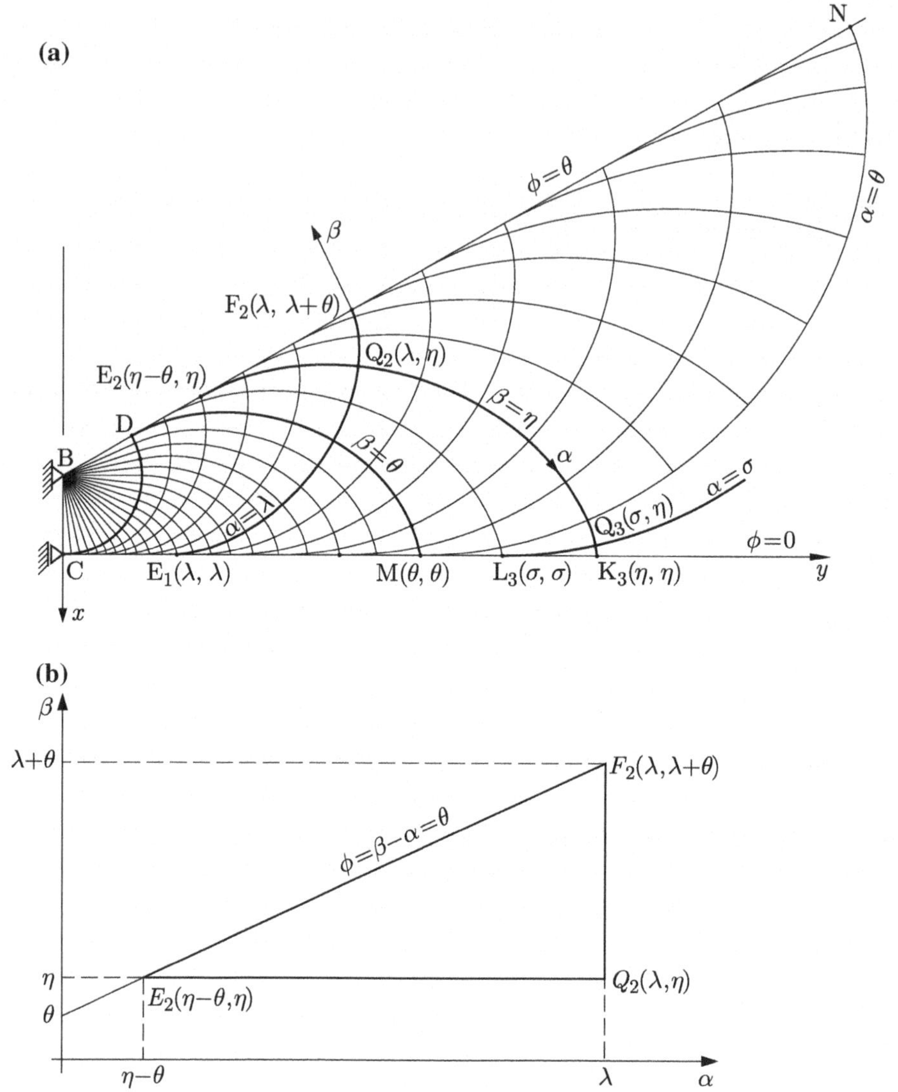

Fig. 4.129 Analysis of the domain DMN. The Michell net (**a**); juxtaposition of points E_2, F_2, Q_2 in the $(\alpha,\,\beta)$-plane

We will apply decomposition

$$\frac{1}{s} - \frac{1}{s+R} = \frac{1}{2R} - \frac{4}{sR(s+R)} + \frac{2}{R(s+R)^2} \tag{4.722}$$

where $R = \sqrt{s^2 - 4}$. We substitute the above formula to (4.721) and we perform the inverse Laplace transform with the use of formulae (4.533), (4.534), we find

$$\bar{\chi}(\alpha) = G_0(\alpha, \alpha + \theta) - 4F_2(\alpha, \alpha + \theta) + G_2(\alpha, \alpha + \theta) \tag{4.723}$$

Substitution to (4.717) and application of the notation (4.535), (4.536) gives

$$\begin{aligned}\frac{z(\lambda, \eta)}{r_0} &= M_0(\eta - \theta, -\theta; \lambda) - 4K_2(\eta - \theta, -\theta; \lambda) + M_2(\eta - \theta, -\theta; \lambda) - \\ &\quad M_0(\lambda, -\theta; \eta - \theta) + 4K_2(\lambda, -\theta; \eta - \theta) - M_2(\lambda, -\theta; \eta - \theta)\end{aligned} \tag{4.724}$$

and now we apply the notation (4.537a)

$$\begin{aligned}\frac{z(\lambda, \eta)}{r_0} &= w_0(\eta - \theta, -\theta; \lambda) - w_2(\eta - \theta, -\theta; \lambda) - \\ &\quad w_0(\lambda, -\theta; \eta - \theta) + w_2(\lambda, -\theta; \eta - \theta)\end{aligned} \tag{4.725}$$

The result (4.540) leads (4.725) to the final form

$$\frac{z(\lambda, \eta)}{r_0} = F_1(\lambda, \eta) - F_1(\eta - \theta, \lambda + \theta) + F_3(\eta - \theta, \lambda + \theta) - F_3(\lambda, \eta) \tag{4.726}$$

and according to (4.715) we find

$$\frac{\bar{y}(\alpha, \beta)}{r_0} = F_1(\beta, \alpha) - F_3(\alpha, \beta) + F_3(\beta - \theta, \alpha + \theta) - F_1(\beta - \theta, \alpha + \theta) \tag{4.727}$$

while from the differential formula (4.97) we obtain

$$\frac{\bar{x}(\alpha, \beta)}{r_0} = F_2(\beta, \alpha) - F_2(\alpha, \beta) + F_4(\beta - \theta, \alpha + \theta) - F_2(\beta - \theta, \alpha + \theta) \tag{4.728}$$

It can be find out that $(\bar{x}, \bar{y})$ defined above are the continuous extensions (4.698), (4.699) to the domain DMN. Moreover, $(\bar{x}, \bar{y})$ define (x, y) in such a way that $x + r_0 = y \cot\theta$ along $\beta = \alpha + \theta$.

In order to find the fields A, B, we can use the formulae (4.98). We find

$$\frac{A(\alpha,\beta)}{r_0} = G_1(\beta,\alpha) - G_1(\alpha,\beta) + G_3(\beta-\theta,\alpha+\theta) - G_1(\beta-\theta,\alpha+\theta),$$
$$\frac{B(\alpha,\beta)}{r_0} = G_0(\beta,\alpha) - G_2(\alpha,\beta) + G_2(\beta-\theta,\alpha+\theta) - G_0(\beta-\theta,\alpha+\theta) \tag{4.729}$$

Let us note that A is a continuous extension of the field given by (4.701) while B suffers a jump along DM: $[\![B]\!]_{DM} = -r_0$. Moreover, B vanishes along DN.

4.13.1.5 The Adjoint Displacement Field in DMN

The fields (u, v) in DMN can be found using the formulae (4.110), (4.111) through the integration along β lines. Taking into account the formula (4.706) defining the field ω, we find u at the point $Q_2(\lambda,\eta)$:

$$u(\lambda,\eta) = \breve{u}_\theta(\lambda,\eta) + \int_\theta^\eta [2(\lambda+\beta)\cos(\beta-\eta) + \sin(\beta-\eta)]B(\lambda,\beta)d\beta \tag{4.730}$$

where $\breve{u}_\theta(\lambda,\mu)$ is given by (4.111) and $B(\lambda,\beta)$ is defined by (4.729). Let us write B in the form

$$B(\lambda,\beta) = B_1(\lambda,\beta) - B_1(\beta-\theta,\lambda+\theta) \tag{4.731}$$

where

$$B_1(\lambda,\beta) = r_0[G_0(\lambda,\beta) - G_2(\lambda,\beta)] \tag{4.732}$$

i.e. $B_1 = B$ in the domain CDM, see (4.701). Substitution $B_1 = B$ to (4.730) gives the formula, which specifies $u(\lambda,\eta)$ according to (4.709). Hence,

$$u(\lambda,\eta) = r_0[G_0(\lambda,\eta) + 2(\eta-\lambda)G_1(\lambda,\eta) + 2F_2(\lambda,\eta)] - u_2^\theta(\lambda,\eta) \tag{4.733}$$

where

$$\frac{u_2^\theta(\lambda,\eta)}{r_0} = \int_\theta^\eta [2(\lambda+\beta)\cos(\eta-\beta) - \sin(\eta-\beta)]$$
$$[G_0(\beta-\theta,\lambda+\theta) - G_2(\beta-\theta,\lambda+\theta)]d\beta \tag{4.734}$$

or

$$\frac{u_2^\theta(\lambda,\eta)}{r_0} = \int_0^t [2(\lambda+\theta+\delta)\cos(t-\delta) - \sin(t-\delta)]$$
$$[G_0(\delta,\lambda+\theta) - G_2(\delta,\lambda+\theta)]d\delta \tag{4.735}$$

where $t = \eta - \theta$, $\delta = \beta - \theta$.

The integral (27.61) can be calculated analytically with the use of formulae (4.137), (4.138), $(4.148)_5$. We obtain

$$\frac{u_2^\theta(\lambda,\eta)}{r_0} = 2(\lambda+\theta)[F_1(\eta-\theta,\lambda+\theta) - F_3(\eta-\theta,\lambda+\theta)] - [F_2(\eta-\theta,\lambda+\theta) - F_4(\eta-\theta,\lambda+\theta)] + 2[h_0(\eta-\theta,\lambda+\theta) - h_2(\eta-\theta,\lambda+\theta)] \quad (4.736)$$

where h_0, h_2 is given by $(4.148)_5$. Substitution of the results yields the final form of the formula (4.736). After substitution of this result to (4.733), we obtain cf. $(4.118\text{a})_1$.

$$u(\lambda,\eta) = u^0(\lambda,\eta) + 2\lambda A(\lambda,\eta) \quad (4.737)$$

where $A(\lambda,\eta)$ is defined by $(4.729)_1$ and

$$\frac{u_0(\lambda,\eta)}{r_0} = G_0(\lambda,\eta) + 2F_2(\lambda,\eta) + 2\theta[F_5(\eta-\theta,\lambda+\theta) - F_1(\eta-\theta,\lambda+\theta)] + 5F_4(\eta-\theta,\lambda+\theta) - F_2(\eta-\theta,\lambda+\theta) \quad (4.738)$$

Let us note that u^0 satisfies (4.119). Moreover, $u^0(\lambda,\theta)$ is in agreement with (4.711), i.e. u^0 determined above is a continuous extension of the field u^0 in the domain CDM. Moreover, A specified by (4.729) is a continuous extension of the field A in the domain CDM. Thus, u defined by (4.737) is a continuous extension of the field u in the domain CDM.

Having the field u, the field v can be calculated with the use of (4.116). This field can be expressed according to (4.118a), i.e.

$$v(\lambda,\eta) = v^0(\lambda,\eta) - 2\eta B(\lambda,\eta) \quad (4.739)$$

where B is given by (4.729) and

$$\frac{v^0(\lambda,\eta)}{r_0} = G_1(\lambda,\eta) + 2F_3(\lambda,\eta) + 2\theta[F_4(\eta-\theta,\lambda+\theta) - F_0(\eta-\theta,\lambda+\theta)] + 5F_3(\eta-\theta,\lambda+\theta) - F_1(\eta-\theta,\lambda+\theta) \quad (4.740)$$

It is observed that v^0 satisfies (4.119). The field v is a continuous extension of the field v given in the domain CDM by (4.712)—along the line DM.

4.13.1.6 Hencky Net in the Domain MNS

The Hencky net will be extended to the MNS domain, along the boundary line MN along which $\alpha = \theta$, see Figs. 4.127, 4.129. The field $\bar{x}(\alpha,\beta)$ is chosen as the main unknown. Proceeding as in Sect. 4.13.1.2 devoted to CDM domain, we obtain the

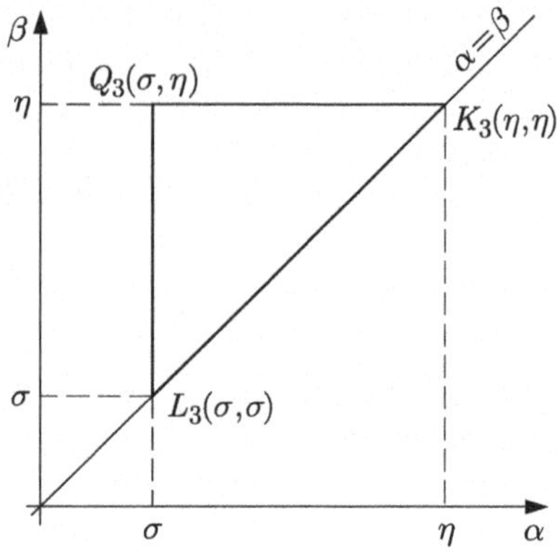

Fig. 4.130 Juxtaposition of points Q_3 K_3 L_3 in the (α, β)-plane

following representation (Fig. 4.130):

$$\bar{x}(\sigma, \eta) = -\frac{1}{2}\int_{\sigma}^{\eta} D_0(\alpha - \sigma, \eta - \alpha)T(\alpha)d\alpha \tag{4.741}$$

see (4.689), or

$$\frac{\bar{x}(\sigma, \eta)}{r_0} = \int_{\sigma}^{\eta} D_0(\alpha - \sigma, \eta - \alpha)\bar{T}(\alpha)d\alpha \tag{4.742}$$

with

$$\bar{T}(\alpha) = \frac{-T(\alpha)}{2r_0}$$

According to (4.728), $\bar{x}$ along MN is given by

$$\frac{\bar{x}(\theta, \eta)}{r_0} = F_2(\eta, \theta) - F_2(\theta, \eta) + F_4(\eta - \theta, 2\theta) - F_2(\eta - \theta, 2\theta) \tag{4.743}$$

or

$$\frac{\bar{x}(\theta, \eta)}{r_0} = \xi_1^\theta(\eta - \theta) + \xi_2^\theta(\eta) \tag{4.744}$$

where

$$\xi_1^\theta(t) = F_4(t, 2\theta) - F_2(t, 2\theta), \qquad \xi_2^\theta(\eta) = F_2(\eta, \theta) - F_2(\theta, \eta) \tag{4.745}$$

Let us consider two integral equations

$$\xi_1^\theta(\eta - \theta) = \int_{\theta}^{\eta} D_0(\alpha - \theta, \eta - \alpha)\bar{T}_1(\alpha)d\alpha \tag{4.746}$$

$$\xi_2^\theta(\eta) = \int_{\theta}^{\eta} D_0(\alpha - \theta, \eta - \alpha)\bar{T}_2(\alpha)d\alpha$$

The solution $\bar{T}(\alpha) = \bar{T}_1(\alpha) + \bar{T}_2(\alpha)$ satisfies the condition (4.742) when $\sigma = \theta$ and l.h.s. of (4.742) is given by (4.743). After calculation of $\bar{T}_1(\alpha)$ and $\bar{T}_2(\alpha)$, we construct the functions

$$\xi_i(\sigma, \eta) = \int_{\sigma}^{\eta} D_0(\alpha - \sigma, \eta - \alpha)\bar{T}_i(\alpha)d\alpha \tag{4.747}$$

which yield

$$\frac{\bar{x}(\sigma, \eta)}{r_0} = \xi_1(\sigma, \eta) + \xi_2(\sigma, \eta) \tag{4.748}$$

The function $\xi_2(\sigma, \eta)$ can be easily guessed without the construction of $\bar{T}_2(\alpha)$. Indeed, let us note that function

$$\xi_2(\sigma, \eta) = F_2(\eta, \sigma) - F_2(\sigma, \eta) \tag{4.749}$$

satisfies $\mathscr{L}\xi_2 = 0$ and satisfies the conditions

$$\xi_2(\theta, \eta) = \xi_2^{\theta}(\eta), \qquad \xi_2(\alpha, \alpha) = 0 \tag{4.750}$$

Construction of the function $\xi_1(\sigma, \eta)$ requires solution of the equation $(4.755)_1$. Let us rewrite this equation in the form

$$\int_0^t D_0(\gamma, t - \gamma)\tilde{T}_1(\gamma)d\gamma = F_4(t, 2\theta) - F_2(t, 2\theta) \tag{4.751}$$

where

$$\tilde{T}_1(\gamma) = \bar{T}_1(\gamma + \theta)$$

We perform the Laplace operation on both sides of (4.760) with the use of rules (4.544), (4.532). We obtain

$$\tilde{T}_1{}^*\left(p + \frac{1}{p}\right) = \left[\frac{1}{p^2(p^2+1)} - \frac{1}{p^2+1}\right]\exp\left(\frac{2\theta}{p}\right) \tag{4.752}$$

We introduce the variable s, cf. (4.752)

$$\tilde{T}_1{}^*(s) = -\frac{4}{s+R}\left(\frac{1}{s} - \frac{1}{s+R}\right)\exp[\theta(s-R)] \tag{4.753}$$

where $R = \sqrt{s^2 - 4}$. Once again we apply the decomposition (4.722) and we perform the inverse Laplace transform using the rules (4.533), (4.534) for $\xi = -2\theta$. We obtain

$$\tilde{T}_1(t) = -G_1(t, t + 2\theta) + 4F_3(t, t + 2\theta) - G_3(t, t + 2\theta) \tag{4.754}$$

and we return to function $\bar{T}_1$

$$\bar{T}_1(\alpha) = -G_1(\alpha-\theta,\alpha+\theta) + 4F_3(\alpha-\theta,\alpha+\theta) - G_3(\alpha-\theta,\alpha+\theta) \tag{4.755}$$

We introduce the above result into (4.747). We perform the integration according to the rules (4.535)–(4.537a). After application of (4.540), we obtain

$$\begin{aligned}\xi_1(\sigma,\eta) = {} & F_2(\sigma-\theta,\eta+\theta) - F_2(\eta-\theta,\sigma+\theta) - \\ & F_4(\sigma-\theta,\eta+\theta) + F_4(\eta-\theta,\sigma+\theta)\end{aligned} \tag{4.756}$$

Therefore, according to (4.748), (4.749), we eventually find

$$\begin{aligned}\frac{\bar{x}(\alpha,\beta)}{r_0} = {} & F_2(\beta,\alpha) - F_2(\alpha,\beta) + F_2(\alpha-\theta,\beta+\theta) - F_2(\beta-\theta,\alpha+\theta) + \\ & F_4(\beta-\theta,\alpha+\theta) - F_4(\alpha-\theta,\beta+\theta)\end{aligned} \tag{4.757}$$

By (4.97) and the differentiation rules (4.142), the function $\bar{y}$ is given by

$$\begin{aligned}\frac{\bar{y}(\alpha,\beta)}{r_0} = {} & F_1(\beta,\alpha) - F_3(\alpha,\beta) + F_3(\alpha-\theta,\beta+\theta) - F_1(\beta-\theta,\alpha+\theta) + \\ & F_3(\beta-\theta,\alpha+\theta) - F_5(\alpha-\theta,\beta+\theta)\end{aligned} \tag{4.758}$$

The Lamé fields are constructed by (4.98)

$$\begin{aligned}\frac{A(\alpha,\beta)}{r_0} = {} & G_1(\beta,\alpha) - G_1(\alpha,\beta) + G_3(\beta-\theta,\alpha+\theta) - G_1(\beta-\theta,\alpha+\theta) + \\ & G_1(\alpha-\theta,\beta+\theta) - G_3(\alpha-\theta,\beta+\theta) \qquad (4.759)\\ \frac{B(\alpha,\beta)}{r_0} = {} & G_0(\alpha,\beta) - G_2(\alpha,\beta) + G_2(\beta-\theta,\alpha+\theta) - G_0(\beta-\theta,\alpha+\theta) + \\ & G_2(\alpha-\theta,\beta+\theta) - G_4(\alpha-\theta,\beta+\theta)\end{aligned}$$

4.13.1.7 Adjoint Displacements in the Domain *MNS*

To compute $u(\sigma,\eta)$ at Q_3, we make use of the integral formula (4.107). Taking into account the expression (4.706) for ω, we obtain

$$u(\sigma,\eta) = \hat{u}_\theta(\sigma,\eta) + \int_\theta^\sigma [\cos(\sigma-\alpha) + 2(\alpha+\eta)\sin(\sigma-\alpha)]A(\alpha,\eta)d\alpha \tag{4.760}$$

with $\hat{u}_\theta(\sigma,\eta)$ given by (4.108), where $u(\theta,\eta)$, $v(\theta,\eta)$ are defined in Sect. 4.13.1.5. Let us decompose

$$A(\alpha,\eta) = A^{DMN}(\alpha,\eta) + r_0\Delta A(\alpha,\eta) \tag{4.761}$$

where $A^{DMN}(\alpha,\eta)$ are given by first four terms in (4.759) for $\beta=\eta$

$$\Delta A(\alpha,\eta) = G_1(\alpha-\theta,\eta+\theta) - G_3(\alpha-\theta,\eta+\theta) \tag{4.762}$$

One can express $u(\sigma,\eta)$ as follows:

$$u(\sigma,\eta) = u^{DMN}(\sigma,\eta) + r_0 \Delta u(\sigma,\eta) \tag{4.763}$$

$u^{DMN}(\sigma,\eta)$ is given by (4.737), as if (σ,η) belonged to the domain DMN. This is not the case, but the formula itself holds here. Moreover

$$\Delta u(\sigma,\eta) = \int_\theta^\sigma [\cos(\sigma-\alpha) + 2(\alpha+\eta)\sin(\sigma-\alpha)]\Delta A(\alpha,\eta)d\alpha \tag{4.764}$$

This integral can be expressed by known functions. By inserting (4.762) into (4.764) and using the integration rules (4.137), (4.148), (4.149), one finds

$$\begin{aligned}\Delta u(\sigma,\eta) = {} & F_2(\sigma-\theta,\eta+\theta) + 3F_4(\sigma-\theta,\eta+\theta) - 8F_6(\sigma-\theta,\eta+\theta) + \\ & 2(\eta+\theta)[F_3(\sigma-\theta,\eta+\theta) - F_7(\sigma-\theta,\eta+\theta)]\end{aligned} \tag{4.765}$$

coming back to (4.763) to find $u(\sigma,\eta)$. The function $u(\alpha,\beta)$ may be expressed by

$$u(\alpha,\beta) = u^0(\alpha,\beta) + 2\alpha A(\alpha,\beta) \tag{4.766}$$

with $A(\alpha,\beta)$ given by (4.759) and

$$\begin{aligned}\frac{u^0(\alpha,\beta)}{r_0} = {} & G_0(\alpha,\beta) + 2F_2(\alpha,\beta) + 2\theta[F_5(\beta-\theta,\alpha+\theta) - F_1(\beta-\theta,\alpha+\theta) + \\ & F_5(\alpha-\theta,\beta+\theta) - F_1(\alpha-\theta,\beta+\theta)] + 5F_4(\beta-\theta,\alpha+\theta) - \\ & F_2(\beta-\theta,\alpha+\theta) + 7F_4(\alpha-\theta,\beta+\theta) - 3F_2(\alpha-\theta,\beta+\theta)\end{aligned} \tag{4.767}$$

The field v^0 is constructed by (54_b), using $\omega(0,0) = 0$. The final result reads

$$\begin{aligned}\frac{v^0(\alpha,\beta)}{r_0} = {} & G_1(\alpha,\beta) + 2F_3(\alpha,\beta) + 2\theta[F_4(\beta-\theta,\alpha+\theta) - F_0(\beta-\theta,\alpha+\theta) + \\ & F_6(\alpha-\theta,\beta+\theta) - F_2(\alpha-\theta,\beta+\theta)] + \\ & 5F_3(\beta-\theta,\alpha+\theta) - F_1(\beta-\theta,\alpha+\theta) + 7F_5(\alpha-\theta,\beta+\theta) - \\ & 3F_3(\alpha-\theta,\beta+\theta)\end{aligned} \tag{4.768}$$

4.13.1.8 Analysis of the Volume of the Optimal Cantilever of Fig. 4.126c

Let us consider the family of optimal design problems of the cantilever loaded along the bottom edge with boundary conditions shown in Fig. 4.126c, when the design domain is a trapezoidal semi-strip. The location of the force P is defined by the ordinate y_P, which corresponds to coordinates α_P, $\beta_P = \alpha_P$ on the line CS, see Fig. 4.127.

The field $\boldsymbol{v} = (u, v)$ determined by the formulae in previous sections is the maximizer of the problem (3.18). We will calculate the virtual work done on the optimal field $\boldsymbol{v}$

$$\mathcal{Z}_1 = Pu(\alpha_P, \alpha_P) \tag{4.769}$$

where according to (4.709) and (4.766)

$$\frac{u(\alpha, \alpha)}{r_0} = \begin{cases} G_0(\alpha, \alpha) + 2F_2(\alpha, \alpha) & \text{if } \alpha \in [0, \theta] \\ G_0(\alpha, \alpha) + 2F_2(\alpha, \alpha) + 4f(\alpha - \theta, \alpha + \theta) & \text{if } \alpha \in [\theta, 2\theta] \end{cases} \tag{4.770}$$

where

$$f(\alpha, \beta) = \theta[G_3(\alpha, \beta) - G_1(\alpha, \beta)] + 3G_2(\alpha, \beta) - 4F_2(\alpha, \beta) \tag{4.771}$$

The volume of the optimal cantilever structure is defined by the formula (3.21), i.e. $V_{\min} = \dfrac{\mathcal{Z}_1}{\sigma_0}$.

Let us consider the family of optimal trusses in the half strip $\left(\theta = \dfrac{\pi}{2}\right)$ of height r_0, generated by changing location y_P of the force P. The volume $V_{\min}$ (referred to $V_0 = \dfrac{Pr_0}{\sigma_0}$), is the smooth function of the variable y_P, see Fig. 4.131. If $y_P = 0$, then $\dfrac{V_{\min}}{V_0} = 1$ since the optimal structure is a single bar in tension BC of the volume V_0. Let us note that point M is the location of smooth transition between two formulae (4.770).

4.13.2 Optimal Designs of Deep Beams Subjected to Point Loads at the Upper Edge

We consider the problem from Fig. 4.125c with various ratios l/h. Thus, the design domain is a rectangle, and the beam is simply supported. Due to the vertical load, we can assume that both supports are sliding. The force $2P$ is applied to the centre of the upper edge. The problem thus posed is equivalent to the problem of Fig. 4.126c, which was already solved in Sect. 4.13.1. Taking advantage of the results from Sect. 4.13.1 it is possible to construct optimal solution of the family of problems from Fig. 4.125c.

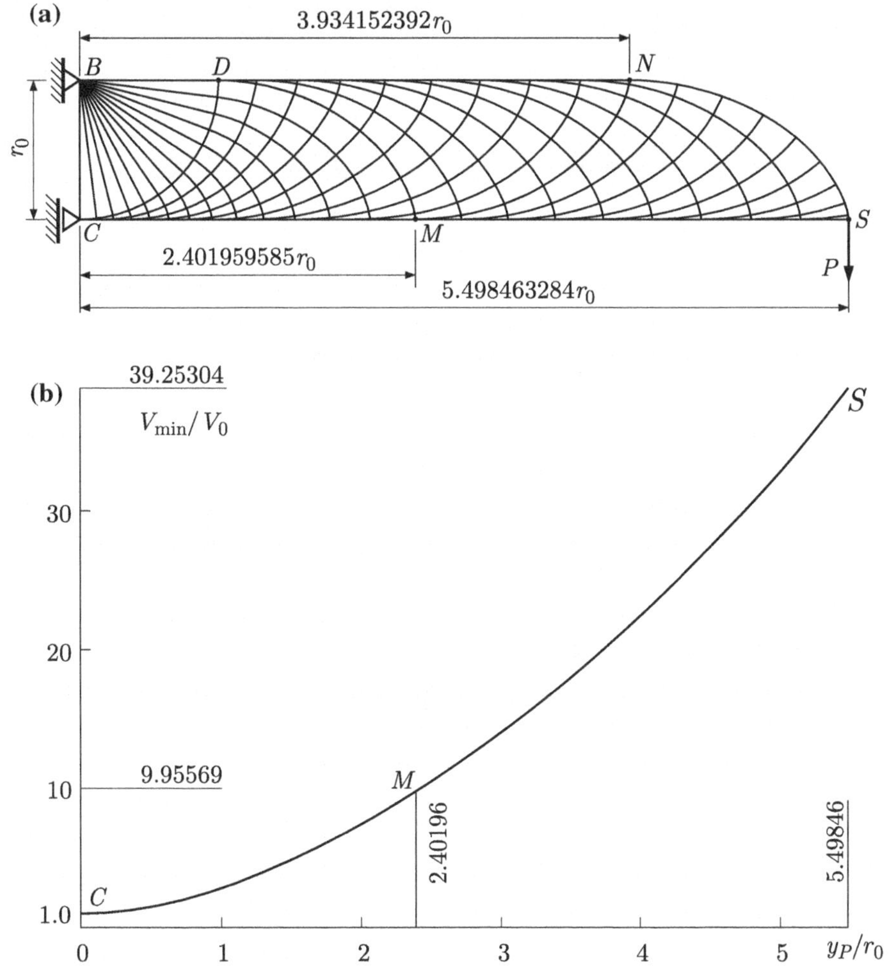

Fig. 4.131 Solution to the problem of Fig. 4.126c for the case of $\theta = \pi/2$. The optimal layout (**a**). The volume of the structure for subsequent values of y_P (**b**)

We will confine our considerations to the cases: $l/h = 1, 2, 3, 5, 7$, see Fig. 4.132a, b, Fig. 4.133a, b, Fig. 4.134a, b, Fig. 4.135a, b and 4.136a, b respectively. The numerical solutions have been constructed by the ground structure method using appropriate number of nodes, see Table 4.5.

The case of $l/h = 0$ refers to the structure composed of a single bar subjected to the compression force $2P$.

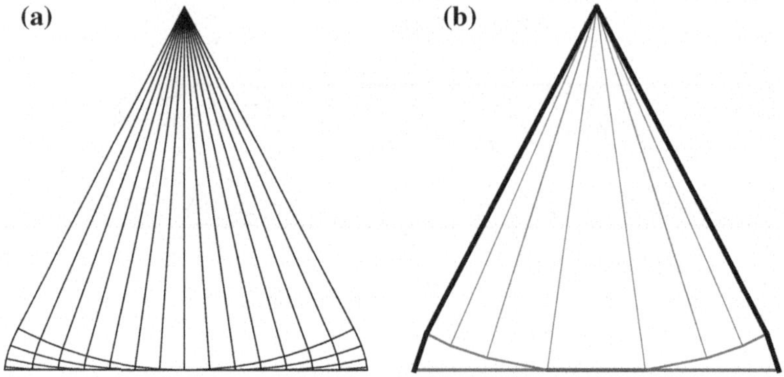

Fig. 4.132 The analytical (**a**) and numerical (**b**) solutions to the problem of Fig. 4.125c. Case of $l/h = 1$

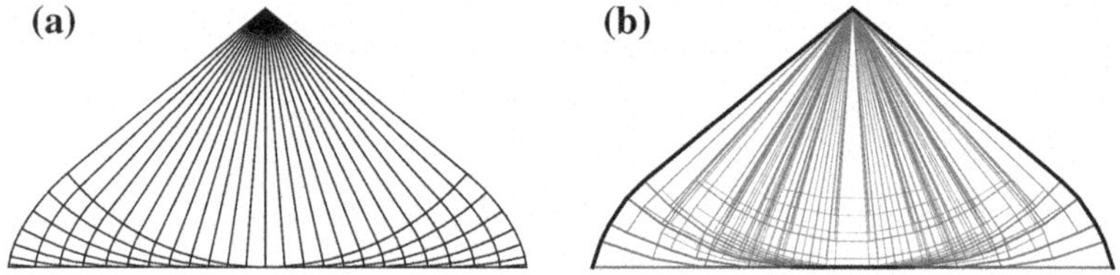

Fig. 4.133 The analytical (**a**) and numerical (**b**) solutions to the problem of Fig. 4.125c. Case of $l/h = 2$

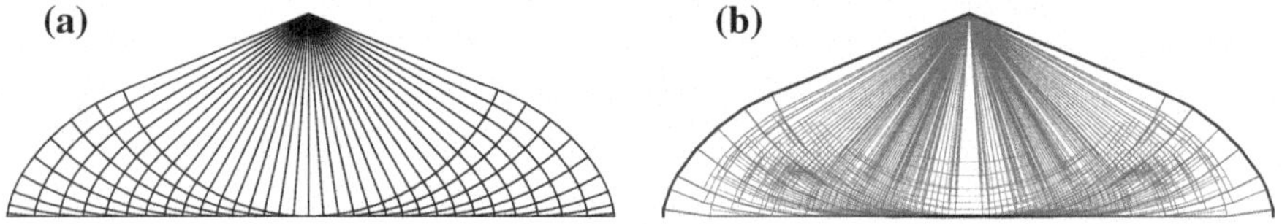

Fig. 4.134 The analytical (**a**) and numerical (**b**) solutions to the problem of Fig. 4.125c. Case of $l/h = 3$

Table 4.5 The volumes of the optimal structures of Fig. 4.125c found analytically and confirmed numerically; the volumes are referred to the reference volume: $V_0 = Ph/\sigma_0$

l/h	V/V_0 analytical	V/V_0 numerical	Relative error (%)
1	2.9804353	2.98196	0.05
2	5.7318376	5.73544	0.06
3	9.8803017	9.88517	0.05
4	15.099504	15.1066	0.05
5	21.164032	21.1733	0.04
6	28.187341	28.1985	0.04
7	36.250713	36.2638	0.04

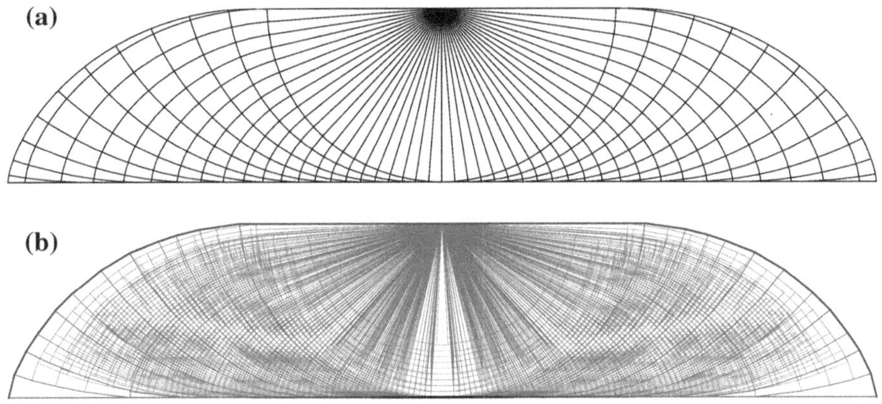

Fig. 4.135 The analytical (**a**) and numerical (**b**) solutions to the problem of Fig. 4.125c. Case of $l/h = 5$

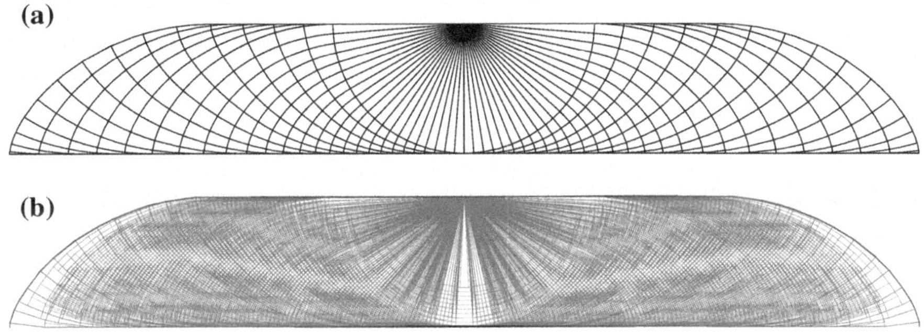

Fig. 4.136 The analytical (**a**) and numerical (**b**) solutions to the problem of Fig. 4.125c. Case of $l/h = 7$

4.14 L-Shaped Domains. Case of $\sigma_C = \sigma_T$

In the first stage of the discussion we shall consider the problem of optimal transmitting the force P to the rolling supports along the AA_1 interval, the design domain being $AA_1BD_1D_2FA$. Note that the design domain is non-convex, with an re-entrant corner B. We predict the volume fraction singularity at this point.

First, consider the case of the point P being close to the line BG_1, within the subdomain which was called CDM in Sect. 4.13, Fig. 4.127. The solution to the problem set in Fig. 4.137a contains two steps. In the first step, we solve the auxiliary problem in which the node B is subject to the additional kinematic constraint: its vertical displacement is set zero, see Fig. 4.137b. This implies vanishing the displacements along the edge AB, see Fig. 4.137c. Upon solving this problem, we add the rigid rotation around A to construct the solution to the initial problem.

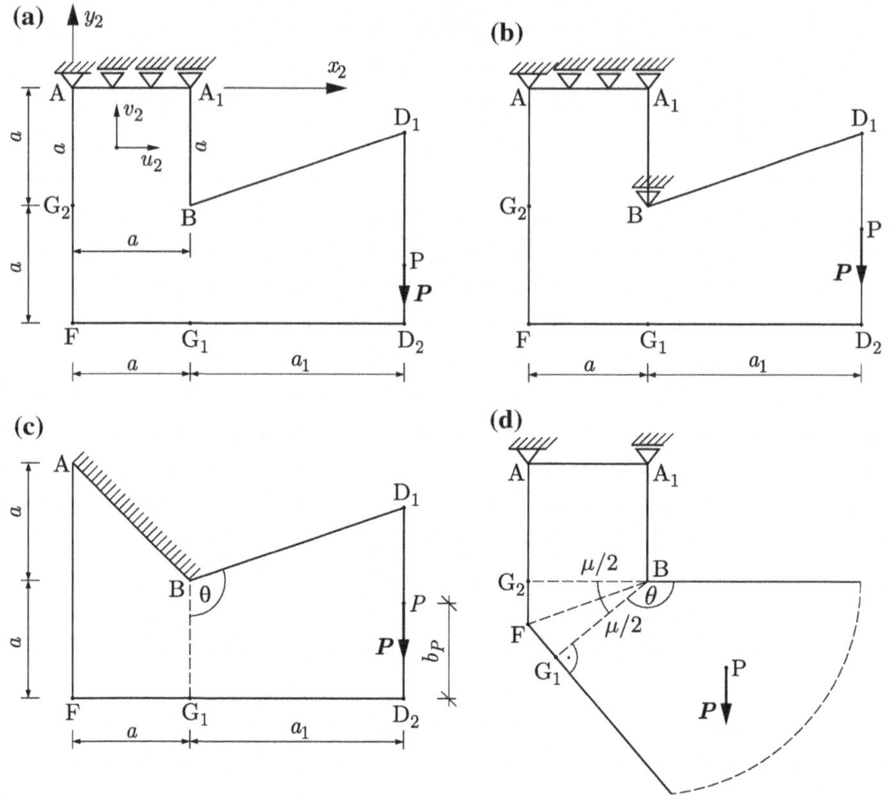

Fig. 4.137 The problem setting (**a**); the auxiliary problem (**b**); the auxiliary problem (**c**); and the case of the line FD_2 being not horizontal (**d**)

4.14.1 Solution of the Auxiliary Problem

Within the design domain of problem of Fig. 4.137b, we construct the Hencky net in the subsequent subdomains, see Fig. 4.138.

The design domain is composed of subdomains. In the domain AG_2B, we introduce the Cartesian parameterization (x_1, y_1) of origin at B; the positive directions of the displacements (u_1, v_1) are shown in Fig. 4.138. In the domain G_2G_1DB, we introduce the polar parameterization (r, β) of origin at B; $-\pi/2 \leq \beta \leq \theta$; $r \leq a$. At arbitrary point T, we define the positive directions of displacements u, v: radial and circumferential.

The domain $G_1E_2D_3$ is adjacent to the parametric line $x = 0$ and is parameterized by the curvilinear orthogonal coordinate system (α, β). Along the line G_1D_3 we have $\alpha = 0$; along G_1E_2 we have $\alpha = \beta$; along D_3E_2 we have $\beta = \alpha_P$. Point E_2 has coordinates (α_P, α_P).

In the domain E_2PDD_3 (which is adjacent to $G_1E_2D_3$), the parametric lines (α, β) are extensions of the lines from the domain $G_1E_2D_3$. The coordinates of the

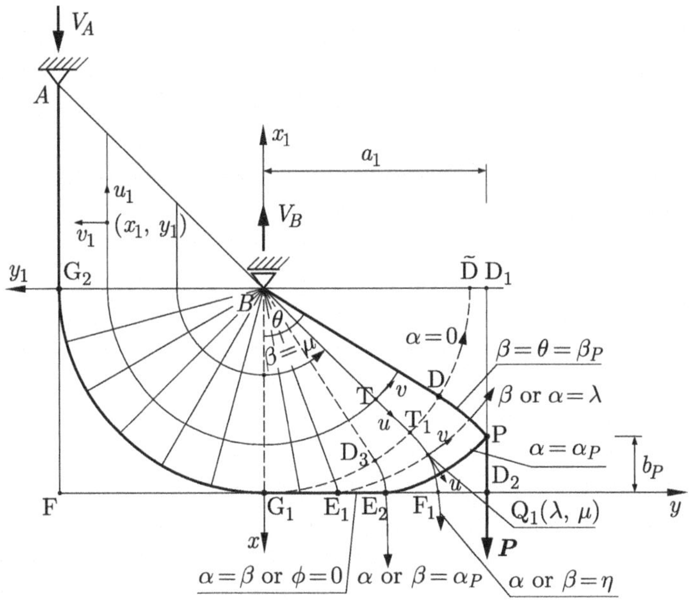

Fig. 4.138 The Hencky net for the problem of Fig. 4.137b

vertices are as follows: $E_1(\lambda, \lambda)$, $D_3(0, \alpha_P)$, $E_2(\alpha_P, \alpha_P)$, $P(\alpha_P, \beta_P)$, $D(0, \beta_P)$. Since β_P is the angle between BD and the vertical line, this coordinate is one of characteristics of the optimal layout. It is helpful to assume $\beta_P = \theta$. An arbitrary point of this domain is denoted by $Q_1(\lambda, \mu)$. The parametric line α going through Q_1 crosses the axis y at the point $F_1(\mu, \mu)$. The line β going through Q_1 crosses the axis y at the point $E_1(\lambda, \lambda)$. This point lies between $G_1(0, 0)$ and $E_2(\alpha_P, \alpha_P)$.

The parametric lines in the domains E_2PDD_3 and $G_1E_2D_3$ are given by (4.698), (4.699), where $r_0 = a$ and by (4.96). Along $\alpha = 0$ the parametric line β is a circle:

$$x = -a(1 - \cos\beta), \qquad y = a\sin\beta \tag{4.772}$$

The points G_1, D_3, T_1, D lie on this circle.

Let us proceed to computing displacements. In the domain G_2BA, we predict

$$u_1 = -x_1 + y_1, \qquad v_1 = u_1 \tag{4.773}$$

while in the domain G_2G_1DB we assume

$$u(r, \beta) = r, \qquad v(r, \beta) = -r(2\beta + 1 + \pi) \tag{4.774}$$

which is compatible with (4.209) if $C_1 = 0$; $d_1 = -1 - \pi$, $\epsilon_1 = -1$, $\epsilon_2 = -1$. Thus, in this domain we have

$$\varepsilon_r = 1, \qquad \varepsilon_\vartheta = -1, \qquad \omega = -2\beta - (1+\pi) \tag{4.775}$$

In particular, $\omega(G_1) = -(1+\pi)$ while the point G_1 has coordinates $(0,0)$ in the (α, β) parameterization of the domain G_1E_2PD. We have found the value $\omega(0,0)$ in the domain G_1E_2PD.

The displacement field (4.774) is now extended to the domain G_1E_2PD by the equations (4.107) which here assume the form

$$u(\lambda,\mu) = \hat{u}_0(\lambda,\mu) + \int_0^\lambda [\cos(\lambda-\alpha) + (-\omega_0 + 2\mu + 2\alpha)\sin(\lambda-\alpha)]A(\alpha,\mu)d\alpha \tag{4.776}$$

$$v(\lambda,\mu) = \hat{v}_0(\lambda,\mu) + \int_0^\lambda [\sin(\lambda-\alpha) - (-\omega_0 + 2\mu + 2\alpha)\cos(\lambda-\alpha)]A(\alpha,\mu)d\alpha \tag{4.777}$$

since $\omega(\alpha,\mu)$ is given by (4.92) with $\kappa = 1$ and $\omega_0 = -(1+\pi)$ (which assures continuity of the field ω, see Sect. 3.1) and $A(\alpha,\mu)$ is given by (4.701), where $r_0 = a$; moreover, $\hat{u}_0$, $\hat{v}_0$ are given by (4.108) with $\theta = 0$, and according to (4.774) we find

$$u(0,\mu) = a, \qquad v(0,\mu) = -a(2\beta + 1 + \pi) \tag{4.778}$$

Integration in (4.776) and in (4.777) can be performed analytically using the rules (4.150) and (4.151). We find

$$\frac{u(\lambda,\mu)}{a} = (1+k)[F_1(\lambda,\mu) - F_3(\lambda,\mu)] + 2(\mu-\lambda)G_1(\lambda,\mu) + G_0(\lambda,\mu) + 2F_2(\lambda,\mu) \tag{4.779}$$

$$\frac{v(\lambda,\mu)}{a} = (1+k)[F_2(\lambda,\mu) - F_0(\lambda,\mu)] - 2(\mu-\lambda)G_0(\lambda,\mu) + G_1(\lambda,\mu) - 2F_1(\lambda,\mu) \tag{4.780}$$

with $k = \pi$.

The vertical displacements (along the axis x) of the point $Q_1(\lambda,\mu)$ is given by

$$u_x(\lambda,\mu) = u(\lambda,\mu)\cos(\mu-\lambda) - v(\lambda,\mu)\sin(\mu-\lambda) \tag{4.781}$$

We predict that the displacement field (u, v) is the maximizer of the problem (3.18). Having this maximizer, one can compute the volume of the optimal truss of Fig. 4.138:

$$V_1 = \frac{P}{\sigma_0} u_x(\alpha_P, \beta_P) \tag{4.782}$$

If point P lies on the line y, then the formula for the volume simplifies to

$$V_1 = \frac{Pa}{\sigma_0}\{(1+\pi)[F_1(\lambda,\lambda) - F_3(\lambda,\lambda)] + G_0(\lambda,\lambda) + 2F_2(\lambda,\lambda)\} \tag{4.783}$$

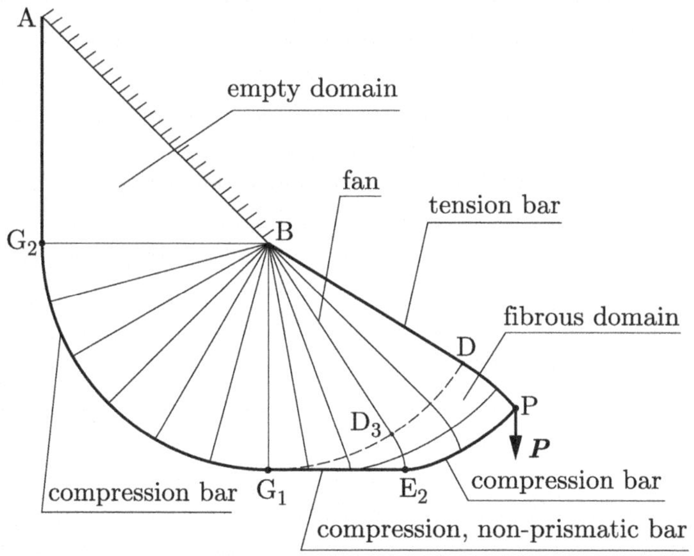

Fig. 4.139 The solution to the auxiliary problem of Fig. 4.137c

If point $P = G_1$, the structure reduces to single bar BG_1 of the volume Pa/σ_0.

The layout of the material is depicted in Fig. 4.139.

The optimal structure is reinforced by ribs. The ribs AG_2, G_2G_1, G_1E_2, E_2P are in compression and the ribs BD, DP are in tension. The rib G_1E_2 is not prismatic: its area of the cross section increases from E_2 to G_1; the other ribs are prismatic. The interior of the domain AG_2B is empty. In the domain $BG_2G_1D_3D$, the fibrous microstructure is radial. In the domain $G_1E_2PDD_3$, the fibres form a net of two families of orthogonal curves. The layout of the fibres is given by the distribution of the Hemp forces T_1, T_2; one can find them by the methods of Sect. 4.12.

4.14.2 *The Solution to the Initial Problem (the Elbow Problem)*

We come back to the problem of Fig. 4.137a. In the domain ABA_1, we assume the virtual displacement field in the form

$$u = -x_2, \qquad v = y_2 \tag{4.784}$$

where the coordinate system (x_2, y_2) has its origin at A, see Fig. 4.140.

Then, $\varepsilon_I = 1$, $\varepsilon_{II} = -1$ and $\varepsilon_I = 1$ along the line y_2, or along BA_1. The displacement of point B, the vector $\boldsymbol{v}_B$ has in the frame (x_2, y_2) the components $(-a, -a)$, see Fig. 4.140.

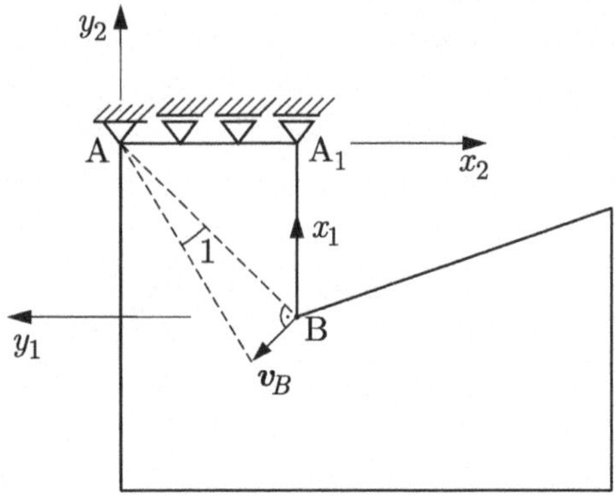

Fig. 4.140 A rigid rotation of the design domain

The field (4.773) has been chosen such that point B does not displace. Now, we rotate the structure of Fig. 4.139 or $AG_2G_1E_2PDB$ around the point A by angle 1, clockwise. Then, the point B changes its position by the vector $\boldsymbol{v}_B$. This rigid motion is described by

$$\tilde{u} = y_2, \qquad \tilde{v} = -x_2 \tag{4.785}$$

where the components of the displacement vector refer to the frame (x_2, y_2). The volume of the optimal structure is given by

$$V = \frac{1}{\sigma_0} P(v_P + \tilde{v}_P) \tag{4.786}$$

where v_P represents the displacement of point P in the direction of the force given by (4.781) and, according to (4.785) the displacement $\tilde{v}_P$ equals $\tilde{v}_P = a + a_1$. Thus,

$$V = V_1 + \underline{\frac{P}{\sigma_0}(a + a_1)} \tag{4.787}$$

where V_1 is given by (4.783). The optimal structure is depicted in Fig. 4.141.

It is easy to note that the component underlined in (4.787) is equal to the volume of the bar BA_1 which transmits the force $P_2 = (a + a_1)P/a$ and is fully stressed.

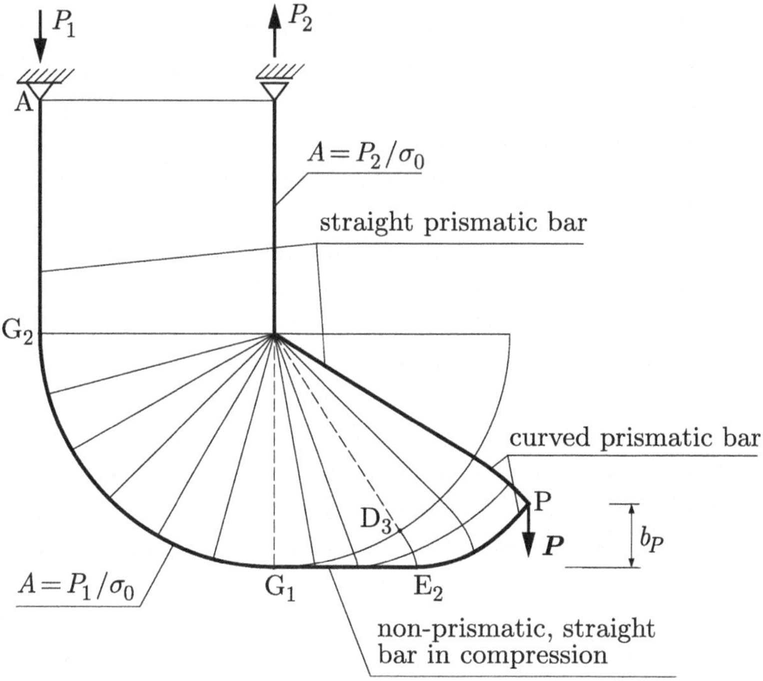

Fig. 4.141 The optimal layout

The presented algorithm of solving the problem of Fig. 4.137 extends to the case of the point P being in a greater distance from the line AF if $a_1 > a$. However, the force P must be parallel to AF since only then the domain AG_2BA_1 is empty. The optimal solution is cut out along the Hencky net in the trapezoidal domain $BG_1D_2D_1$, see Fig. 4.142.

The Hencky net in Fig. 4.142 has been constructed in Sect. 4.13. The problems discussed till now referred to the case of point P lying within the domain BDD_6MG_1B. The most important is the line DD_6D_7M. The next essential line is MN, or $\alpha = \theta$. Fig. 4.142 refers to the case of $0 \le \alpha_P \le \theta$, $\theta \le \beta_P \le 2\theta$. Position of point P introduces the division of the structure into subdomains of vertices at D, D_i, $i = 3, 4, 5, 6, 7, \ldots$, P, E_3, E_2, G_1, B. An exact analysis requires construction of the force fields T_1, T_2 in all the subdomains, which can be done with using the methods of Sect. 4.12. Having these force fields one can compute the density of fibres by (4.170), (4.171), (4.188).

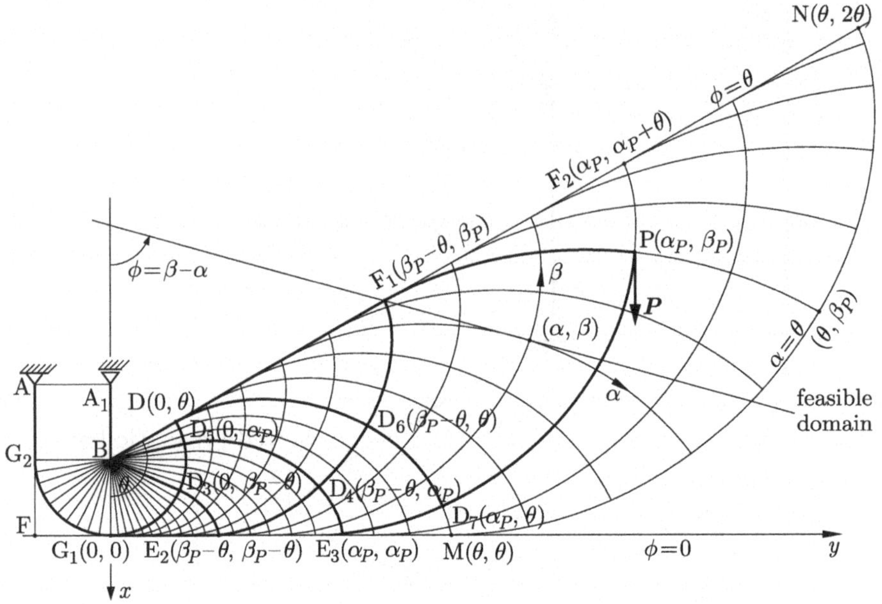

Fig. 4.142 The optimal layout for the case of the point P being in greater distance from AF

4.14.3 Illustrative Examples

In case of $b_P = a, a_1 = a$, the solution is the simplest, see Fig. 4.143a; it is composed of a half of a circle and of two bars. Other, selected solutions are shown in Fig. 4.143b–f. The characteristic analytical and numerical data are set up in Table 4.6; there $\bar{V} = V/V_0$, $V_0 = Pa/\sigma_0$.

The problem of Fig. 4.143e has been numerically solved, see Fig. 4.144.

4.14.4 Case of an Inclined Lower Boundary and the Vertical Point Load

The results found in the previous sections can be extended to the case of the design domain having a non-horizontal lower edge, as in Fig. 4.137d, provided that the point load is parallel to the lines AG_2, A_1B, see Fig. 4.145. Now, the angle $\mu = \angle G_2BG_1 \neq \frac{\pi}{2}$, but is fixed such that $0 < \mu < \frac{\pi}{2}$.

Now, the constant $d_1 = -1 - 2\mu$ and Eqs. (4.774) are generalized to the form

$$u(r, \beta) = r, \qquad v(r, \beta) = -r(2\beta + 1 + 2\mu) \tag{4.788}$$

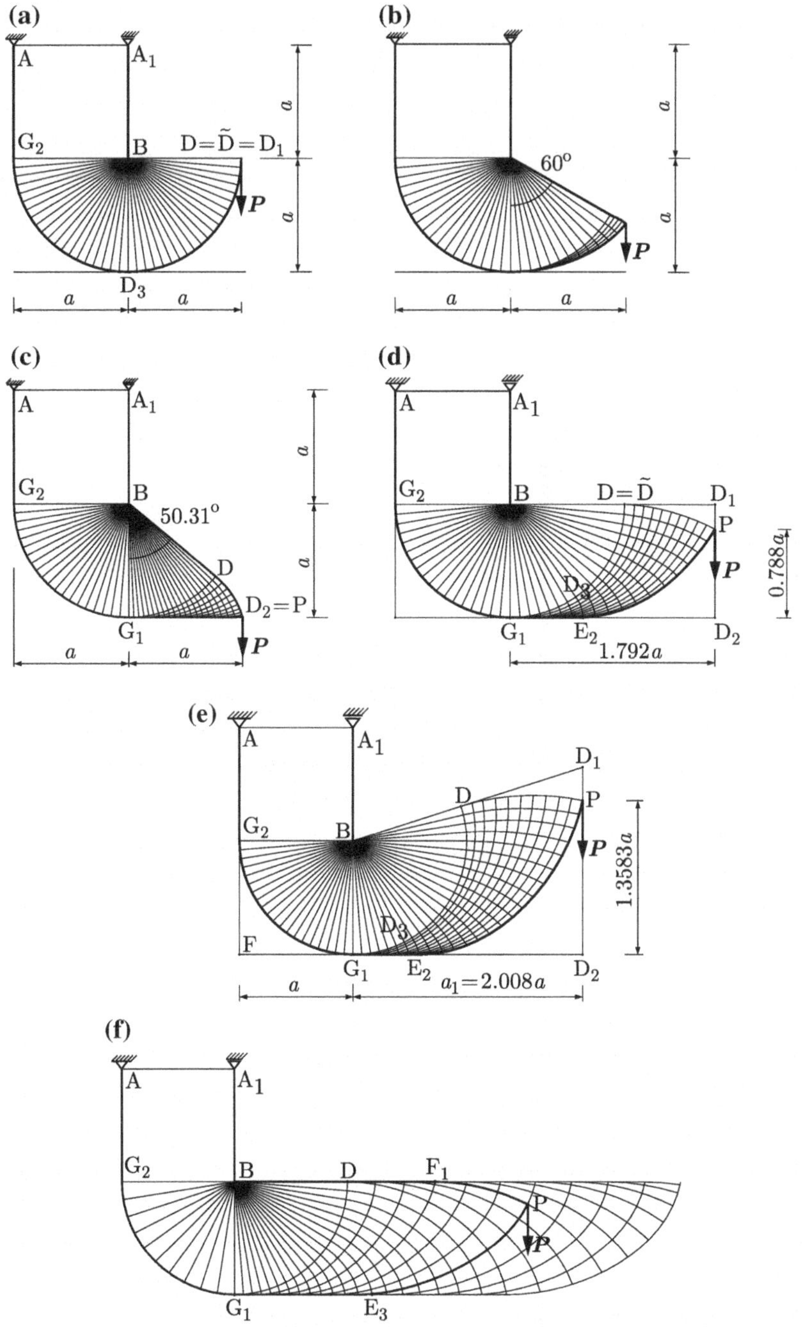

Fig. 4.143 The illustrative examples corresponding to various positions of the point P

Table 4.6 The characteristics of the trusses of Figs. 4.143a–e

Figure	α_P	θ	b_P/a	a_1/a	$\bar{V}$	$\bar{V}_{\text{num}}$	r.err. %
4.143a	0	$90\pi/180$	1	1	9.283	9.28755	0.049
	0.011	$80\pi/180$	0.824	1	9.110106	9.11776	0.084
	0.027	$75\pi/180$	0.732	1	9.02813	9.03460	0.072
	0.09482	$65\pi/180$	0.528	1	8.89096	8.89753	0.074
4.143b	0.162	$60\pi/180$	0.408	1	8.849866	8.85448	0.052
	0.323	$54\pi/180$	0.222	1	8.861026	8.86247	0.016
4.143c	0.8781	$50.31\pi/180$	0	1	9.007675	9.01246	0.053
4.143d	$\pi/6$	$90\pi/180$	0.788	1.792	16.45864	16.46997	0.069
4.143e	$\pi/6$	$108\pi/180$	1.358	2.008	19.1096	19.12494	0.080

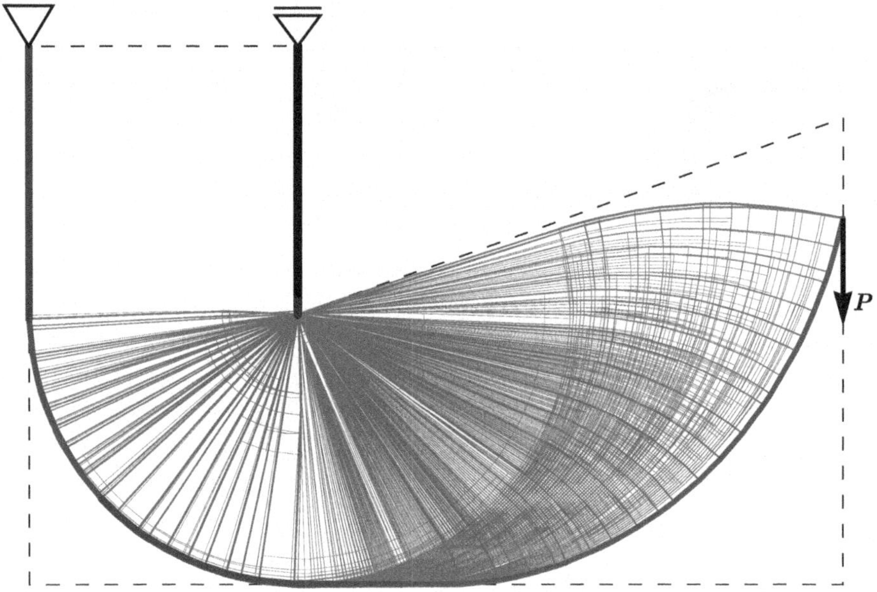

Fig. 4.144 The numerical solution of the problems of Fig. 4.143e found using 150 million of potential bars

The formula (4.787) for the minimal volume holds, but now

$$V_1 = \frac{P}{\sigma_0} u_P(\alpha_P, \beta_P) \tag{4.789}$$

where u_P stands for the displacement along the force P:

$$u_P = u_x(\alpha_P, \beta_P) \sin\mu + u_y(\alpha_P, \beta_P) \cos\mu \tag{4.790}$$

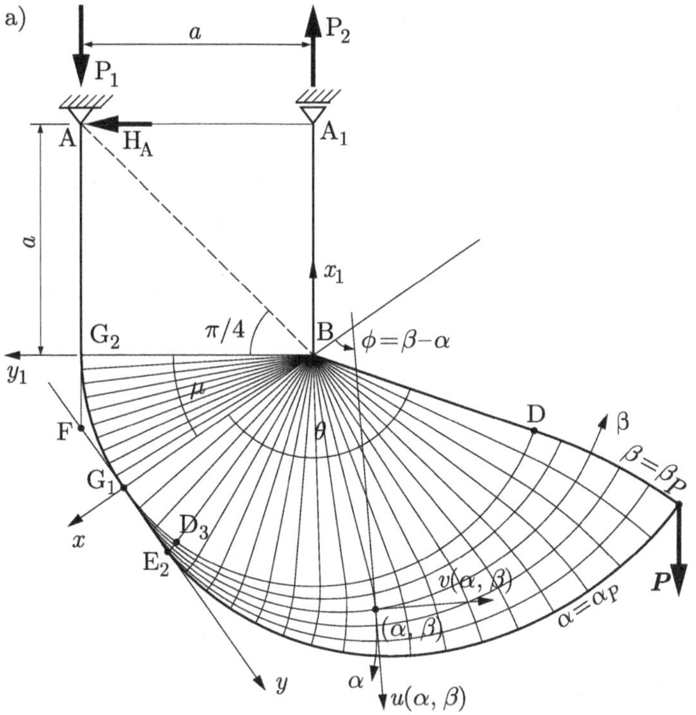

Fig. 4.145 Case of the inclined lower boundary G_1E_2; here $\mu = \pi/5$

with

$$u_x(\alpha, \beta) = u(\alpha, \beta)\cos(\beta - \alpha) - v(\alpha, \beta)\sin(\beta - \alpha) \qquad (4.791)$$
$$u_y(\alpha, \beta) = u(\alpha, \beta)\sin(\beta - \alpha) + v(\alpha, \beta)\cos(\beta - \alpha)$$

Substitution of (4.791) into (4.790) gives

$$u_P(\alpha, \beta) = u(\alpha, \beta)\sin(\beta - \alpha + \mu) + v(\alpha, \beta)\cos(\beta - \alpha + \mu) \qquad (4.792)$$

where $u(\alpha, \beta)$, $v(\alpha, \beta)$ are given by (4.779), (4.780) for $k = 2\mu$. These results are obtained by assuming $\omega_0 = -(1 + 2\mu)$ and performing integration in (4.776), (4.777). The illustrative solution to the problem for the case of $\mu = \pi/5$ is shown in Fig. 4.145.

4.14.5 Cases of Non-vertical Point Load

If the force P deviates from the direction AG_2, then a horizontal reaction at point A appears, contributing to a new substructure within the square domain AG_2BA_1, see Figs. 4.146 and 4.147.

Other examples are set up in Figs. 4.148 and 4.149. The analytical solution to the problem of Fig. 4.148a has been found in Lewiński et al. (2013); its numerical counterpart in Fig. 4.148b has been constructed in the same paper.

The layout of fibres changes drastically if $\frac{1}{2}\mu + \theta < \frac{\pi}{2}$, cf. Fig. 4.149a. In case of the greater design domain, the optimal layout assumes the form shown in Fig. 4.149b.

4.15 The Three-Force Problem. Subclasses of a General Solution

Assume that three forces applied at three different points in the plane are self-equilibrated. Our aim is to find Michell structures in the whole plane equilibrating these forces. Even in the case of $\sigma_C = \sigma_T = \sigma_0$, to be dealt with in this section, this problem is only partly solved till now. Hitherto existing subclasses of analytical solutions will be shown below, all of them confirmed by the ground structure method.

4.15.1 Three-Force Problem in the Whole Plane

4.15.1.1 Formulation and the Main Assumptions

Assume that the forces $\boldsymbol{P} = (P_x, P_y)$, $\boldsymbol{Q} = (Q_x, Q_y)$, $\boldsymbol{F} = (F_x, F_y)$ are applied at the nodes R, N, D, respectively; the forces components are referred to the Cartesian system (x, y) with origin at R, see Fig. 4.150a.

The node R is fixed; hence, P_x, P_y will be viewed as reactions. Point N lies on the axis x. The main subclass of the Michell solutions consists of two circular fans: RBA, NAC and the domain $ABDC$ composed of two families of curvilinear fibres, see Fig. 4.150b. This subclass has been discovered by Chan (1966). As clearly noted by Chan the angles of the fans are bounded by $\pi/2$. Wider angles do not occur, which means that the optimal topologies may differ from that of Fig. 4.150b.

We shall discuss the solutions designed within the full plane or within the half-plane: right to the line RN.

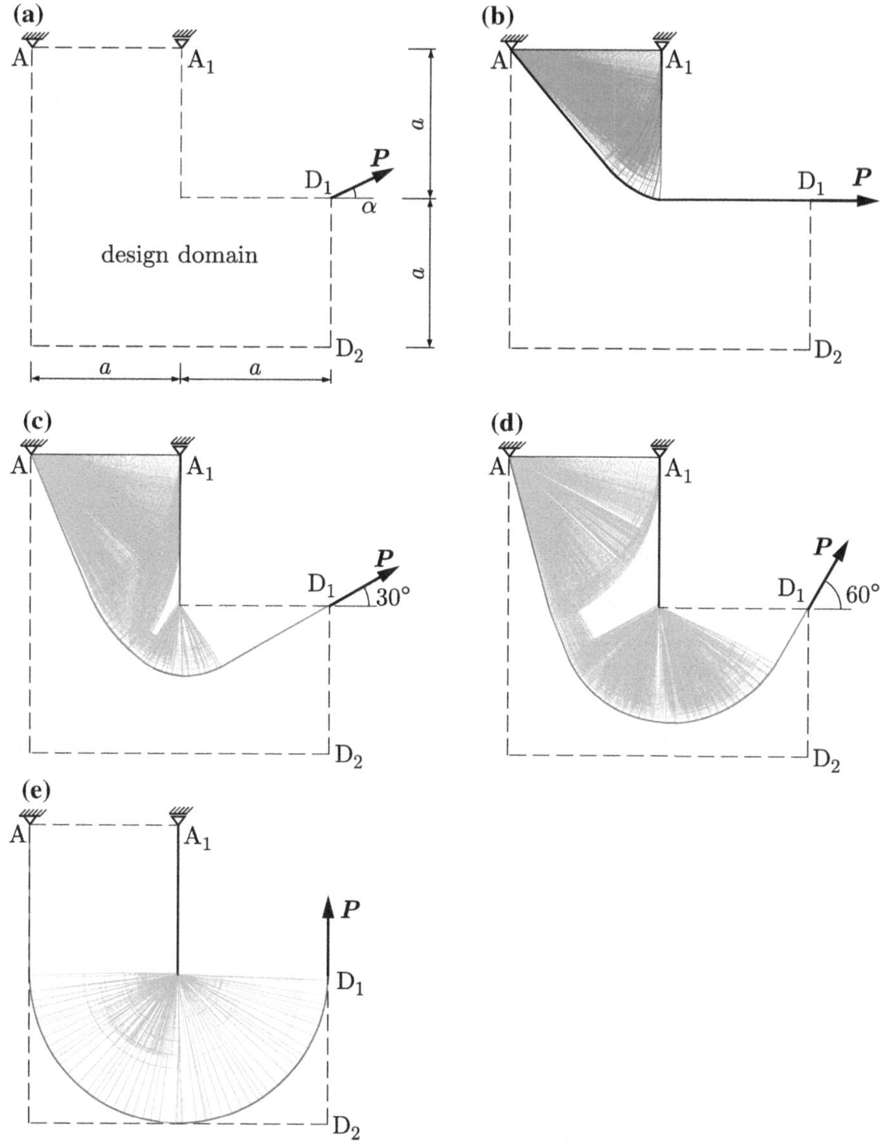

Fig. 4.146 The problem setting concerning the case of application of the force at point D_1 (**a**); the numerical results (the layouts of bars and the volume) corresponding to the angle of inclination of the force P: $\alpha = 0°$, $V = 3.86774853V_0$ (**b**); $\alpha = 30°$, $V = 6.448491646V_0$ (**c**); $\alpha = 60°$, $V = 8.59838176V_0$ (**d**); $\alpha = 90°$, $V = 9.28580985V_0$ (**e**); $V_0 = Pa/\sigma_0$

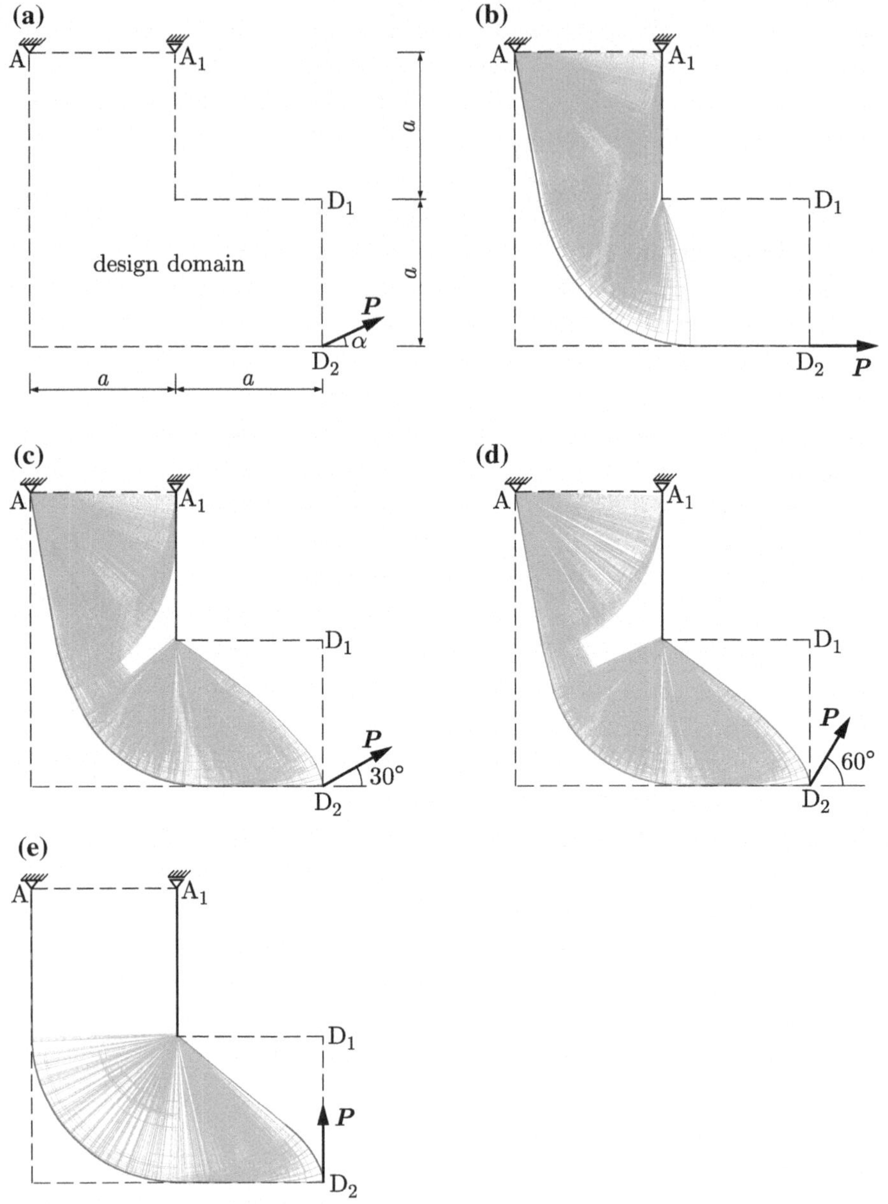

Fig. 4.147 The problem setting concerning the case of application of the force at point D_2 (**a**); the numerical results (the layouts of bars and the volume) corresponding to the angle of inclination of the force P: $\alpha = 0°$, $V = 8.53440250V_0$ (**b**); $\alpha = 30°$, $V = 11.7107173V_0$ (**c**); $\alpha = 60°$, $V = 11.8302059V_0$ (**d**); $\alpha = 90°$, $V = 9.01095103V_0$ (**e**); $V_0 = Pa/\sigma_0$

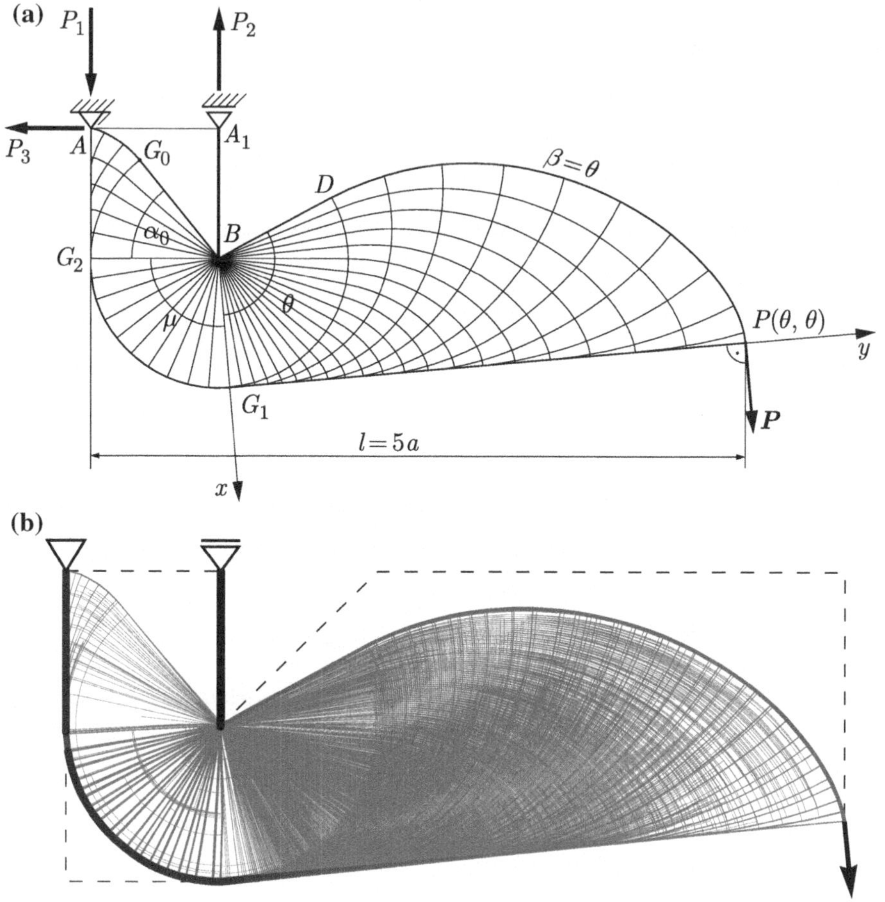

Fig. 4.148 The solution to the illustrative example for the data: $\mu = 95.7106°$, $\theta = 113.4456°$; the analytical solution (**a**); the numerical solution using 474 million of potential bars (**b**)

4.15.1.2 Parameterization

Let us introduce a parameterization of the subdomains RBA, NAC, $ABDC$, see Fig. 4.151.

Within the fan RBA, we introduce a polar system (r, β), $r \leq r_2 = |RA|$, $0 \leq \beta \leq \theta_2$, $\theta_2 = \angle ARB$. Within the fan NAC, we introduce the polar system (r, α), $r \leq r_1 = |NA|$, $0 \leq \alpha \leq \theta_1$, $\theta_1 = \angle ANC$. Moreover, $\angle RAN = \pi/2$; $d = |RN|$. The radii r_1, r_2 can be expressed by the angles γ_1, γ_2; $r_1 = d\cos\gamma_1$, $r_2 = d\cos\gamma_2$. The sides RA, NA of the right-angled triangle RAN determine the axes x_0, y_0 of the Cartesian system with origin at A. The parameterizations of the fans determine the Hencky net (α, β) within $ABDC$, see Sect. 4.8.2 and Fig. 4.54. The coordinates (α, β) of the vertices of this domain are: $A(0, 0)$, $C(\theta_1, 0)$, $B(0, \theta_2)$, $D(\theta_1, \theta_2)$.

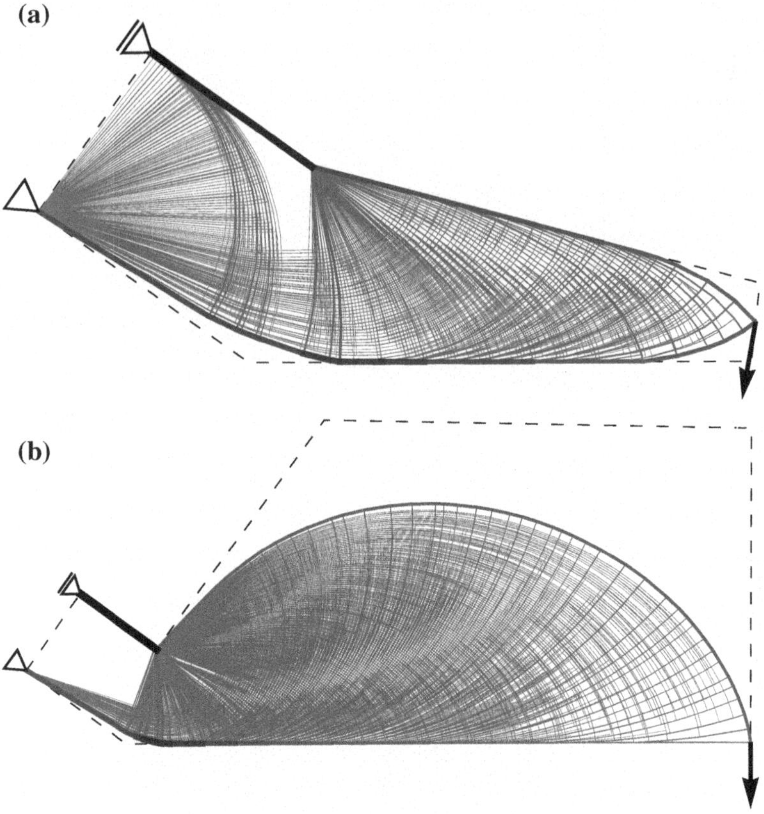

Fig. 4.149 Other numerical solutions

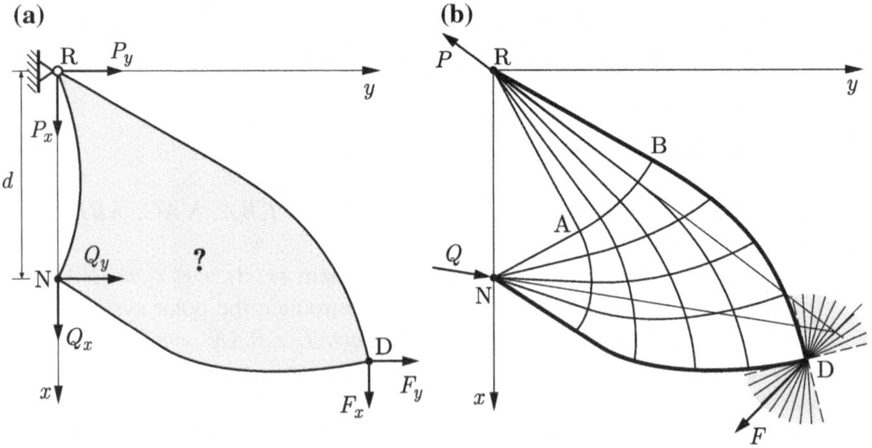

Fig. 4.150 The problem setting (**a**); the layout assumed (**b**)

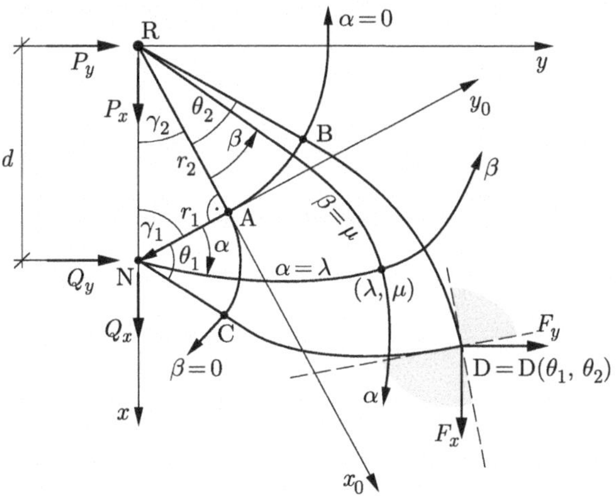

Fig. 4.151 Parameterization of the assumed net of fibres

The Lamé coefficients within $ABDC$ are given by Eqs. (4.401), (4.402). We define $\phi(\alpha, \beta) = \beta - \alpha$ as in Sect. 4.8. The Cartesian coordinates (x_0, y_0) of a point (λ, μ) are given by the right-hand side of (4.410) with $\bar{x}$, $\bar{y}$ given by (4.409). The coordinates (x, y) of a point (λ, μ) are given by

$$
\begin{aligned}
x(\lambda, \mu) &= [x_0(\lambda, \mu) + r_2] \cos \gamma_2 - y_0(\lambda, \mu) \sin \gamma_2 \\
y(\lambda, \mu) &= [x_0(\lambda, \mu) + r_2] \sin \gamma_2 + y_0(\lambda, \mu) \cos \gamma_2
\end{aligned}
\tag{4.793}
$$

By inserting the expressions (4.410), (4.409), one can rearrange (4.793) to the form

$$
\begin{aligned}
\frac{x(\lambda, \mu)}{d} &= \sin \gamma_2 k_1(\lambda, \mu) - \cos \gamma_2 k_2(\lambda, \mu) \\
\frac{y(\lambda, \mu)}{d} &= -\sin \gamma_2 k_3(\lambda, \mu) + \cos \gamma_2 k_4(\lambda, \mu)
\end{aligned}
\tag{4.794}
$$

where

$$
\begin{aligned}
k_1(\lambda, \mu) &= -\sin(\gamma_2 + \mu - \lambda) F_2(\lambda, \mu) + \cos(\gamma_2 + \mu - \lambda) F_1(\lambda, \mu) \\
k_2(\lambda, \mu) &= \sin(\gamma_2 + \mu - \lambda) F_1(\lambda, \mu) - \cos(\gamma_2 + \mu - \lambda) F_0(\lambda, \mu) \\
k_3(\lambda, \mu) &= -\cos(\gamma_2 + \mu - \lambda) F_2(\lambda, \mu) - \sin(\gamma_2 + \mu - \lambda) F_1(\lambda, \mu) \\
k_4(\lambda, \mu) &= \cos(\gamma_2 + \mu - \lambda) F_1(\lambda, \mu) + \sin(\gamma_2 + \mu - \lambda) F_0(\lambda, \mu)
\end{aligned}
\tag{4.795}
$$

The coordinates $x^D = x(\theta_1, \theta_2)$, $y^D = y(\theta_1, \theta_2)$ of point D play an important role in the sequel; they are expressed by

$$\frac{x^D}{d} = \sin\gamma_2 k_1(\theta_1, \theta_2) - \cos\gamma_2 k_2(\theta_1, \theta_2) \tag{4.796}$$
$$\frac{y^D}{d} = -\sin\gamma_2 k_3(\theta_1, \theta_2) + \cos\gamma_2 k_4(\theta_1, \theta_2)$$

4.15.1.3 Computation of the Reactions at the Node N

The condition of vanishing of the moment of all the forces with respect to the hinge R reads

$$-dQ_y - x^D F_y + y^D F_x = 0 \tag{4.797}$$

To find the second equilibrium equation, we apply the kinematic method proposed by Chan (1966). The idea is to consider a strain-free virtual field and demand that the virtual work of the given forces vanishes. More precisely, we construct the virtual field such that the left-hand side of the virtual work (4.175) vanishes. Moreover, we require that $\varepsilon_{r\theta} = 0$ in the fans. The point R is viewed as fixed.

One can check that the representations

$$u^{RAB} = 0, \qquad v^{RAB} = -\frac{C_2^0}{r_2} r \tag{4.798}$$

$$u^{NAC} = C_1^0 \sin\alpha - C_2^0 \cos\alpha, \tag{4.799}$$
$$v^{NAC} = -C_1^0\left(\cos\alpha - \frac{r}{r_1}\right) - C_2^0 \sin\alpha$$

assure that all the components of strain, see (4.206), vanish, and that they are compatible at point A; the sign conventions are shown in Fig. 4.152.

The fields (u, v) in RAB, NAC and $ABDC$ are compatible along AB and AC, hence

$$u^{ABDC}(0, \beta) = 0, \qquad v^{ABDC}(0, \beta) = -C_2^0 \tag{4.800}$$

$$u^{ABDC}(\alpha, 0) = C_1^0(\cos\alpha - 1) + C_2^0 \sin\alpha \tag{4.801}$$
$$v^{ABDC}(\alpha, 0) = C_1^0 \sin\alpha - C_2^0 \cos\alpha$$

Now, we can construct the fields (u, v) within $ABDC$. Since $\epsilon_{11} = 0$, $\epsilon_{22} = 0$, by (4.114) we note that (u, v) satisfy: $\mathscr{L}u = 0$ and $\mathscr{L}v = 0$, where $\mathscr{L}$ is the differential operator given by (4.112). The fields u, v may be constructed using the Riemann formula (4.132), where now $\theta_1 = 0$, $\theta_2 = 0$, $f = 0$. Substitution of (4.800), (4.801) gives

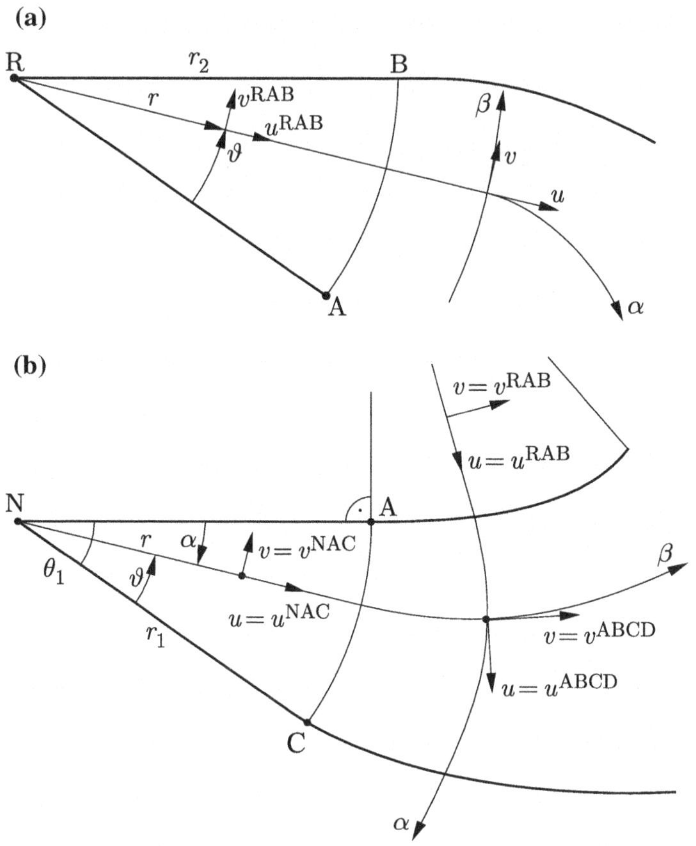

Fig. 4.152 Parameterization of the fans and the adjacent domain

$$u^{ABDC}(\lambda, \mu) = -C_1^0 \int_0^{\lambda} G_0(\lambda - \alpha, \mu) \sin \alpha d\alpha + C_2^0 \int_0^{\mu} G_0(\lambda - \alpha, \mu) \cos \alpha d\alpha \tag{4.802}$$

By performing integration, using (4.137), (4.138), one finds

$$u^{ABDC}(\lambda, \mu) = -C_1^0 F_2(\lambda, \mu) + C_2^0 F_1(\lambda, \mu) \tag{4.803}$$

which gives

$$v^{ABDC}(\lambda, \mu) = C_1^0 F_1(\lambda, \mu) - C_2^0 F_0(\lambda, \mu) \tag{4.804}$$

Now, we can compute displacements of point D: $u^D = u(\theta_1, \theta_2)$, $v^D = v(\theta_1, \theta_2)$, where the superscript $ABDC$ is omitted. The displacements of point D in the directions x and y are computed by the rules

$$\begin{aligned} w_x^D &= u^D \cos(\gamma_2 + \phi^D) - v^D \sin(\gamma_2 + \phi^D) \\ w_y^D &= u^D \sin(\gamma_2 + \phi^D) + v^D \cos(\gamma_2 + \phi^D) \end{aligned} \tag{4.805}$$

where $\phi^D = \theta_2 - \theta_1$. Substitution of (4.803), (4.804) leads to

$$\begin{aligned} w_x^D &= C_1^0 k_3(\theta_1, \theta_2) + C_2^0 k_4(\theta_1, \theta_2) \\ w_y^D &= C_1^0 k_1(\theta_1, \theta_2) + C_2^0 k_2(\theta_1, \theta_2) \end{aligned} \tag{4.806}$$

where k_i are given by (4.795).

Now, we shall find the displacements of node N in the directions x and y. Using (4.799), one finds

$$w_x^N = C_2^0 \cos\gamma_1 + C_1^0 \sin\gamma_1, \qquad w_y^N = -C_2^0 \sin\gamma_1 + C_1^0 \cos\gamma_1 \tag{4.807}$$

We are prepared to write down the equation expressing the virtual work being zero

$$F_x w_x^D + F_y w_y^D + Q_x w_x^N + Q_y w_y^N = 0 \tag{4.808}$$

Let us choose $C_1^o = C_2^o \tan\gamma_1$ to make ${w_y}^N = 0$. Equation (4.808) implies

$$\begin{aligned} Q_x = &-[\cos\gamma_2 k_3(\theta_1, \theta_2) + \sin\gamma_2 k_4(\theta_1, \theta_2)]F_x - \\ &[\cos\gamma_2 k_1(\theta_1, \theta_2) + \sin\gamma_2 k_2(\theta_1, \theta_2)]F_y \end{aligned} \tag{4.809}$$

This is the Chan formula for Q_x, crucial for the further study.

4.15.1.4 The Volume of the Optimal Structure

We shall construct a virtual field (u, v) vanishing at R and satisfying the Michell conditions (4.116) for $\kappa = 1$ within the structure. We start with the triangular domain RAN and propose

$$u^{RAN} = x_0 - Cy_0 + r_2, \qquad v^{RAN} = -y_0 + Cx_0 + Cr_2 \tag{4.810}$$

see Fig. 4.153; C is a constant.

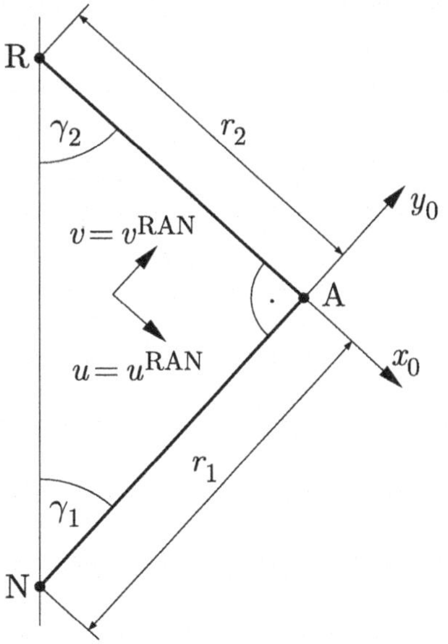

Fig. 4.153 Parameterization of the triangle RAN

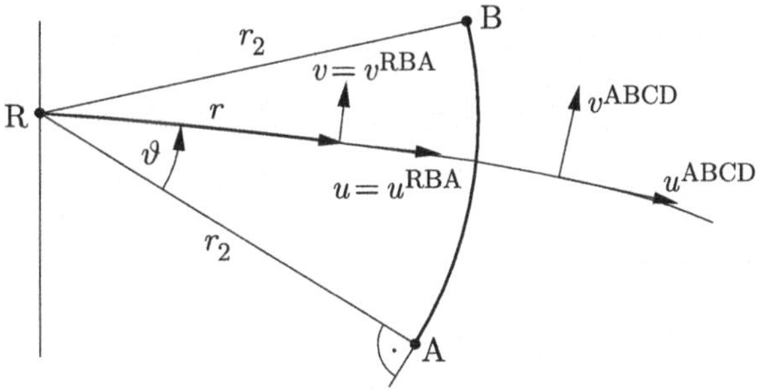

Fig. 4.154 The domain RBA. Polar parameterization

Consider now the fan RBA, see Fig. 4.154. According to the representation (4.209), we put $\epsilon_1 = -1$, $\epsilon_2 = -1$, $C_1 = 0$, $d_1 = \psi$:

$$u^{RBA} = r, \qquad v^{RBA} = -2r\vartheta + \psi r \tag{4.811}$$

To make the fields (4.810), (4.811) compatible, we must take $C = \psi$.

Consider the fan NAC. According to (4.209), we put $\epsilon_1 = 1$, $\epsilon_2 = 1$, $d_1 = \tilde{\psi}$ and find

$$u^{NAC} = -r + C_1 \cos(\vartheta - \varphi), \qquad v^{NAC} = 2r\vartheta + \tilde{\psi} r - C_1 \sin(\vartheta - \varphi) \tag{4.812}$$

To make the fields (4.812), (4.810) compatible along NA, we put $\tilde{\psi} = \psi - 2\theta_1$ and

$$C_1 \cos(\theta_1 - \varphi) = \psi r_2 + r_1, \qquad C_1 \sin(\theta_1 - \varphi) = \psi r_1 + r_2 \tag{4.813}$$

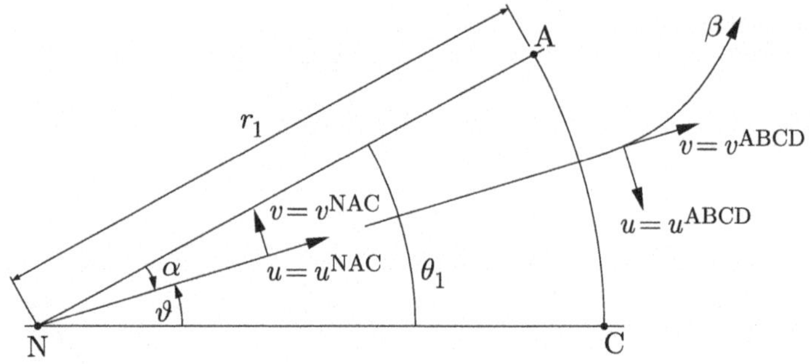

Fig. 4.155 Polar parameterization of the domain NAC

Thus, the above conditions rearrange (4.812) to the form

$$\begin{aligned} u^{NAC} &= -r + (\psi r_2 + r_1)\cos\alpha + (\psi r_1 + r_2)\sin\alpha \\ v^{NAC} &= -2r\alpha + \psi r - (\psi r_1 + r_2)\cos\alpha + (\psi r_2 + r_1)\sin\alpha \end{aligned} \tag{4.814}$$

where $\alpha = \theta_1 - \vartheta$, see Fig. 4.155.

The components of displacements of node N (see Fig. 4.155) defined as in the domain RAN are given by

$$u^{RAN} = \psi r_1 + r_2, \qquad v^{RAN} = r_1 + \psi r_2 \tag{4.815}$$

where we have made use of $x_0 = 0$, $y_0 = -r_1$, $C = \psi$.

The displacements of N in the directions x and y are

$$\begin{aligned} w_x^N &= u^{RAN}(N)\sin\gamma_1 - v^{RAN}(N)\cos\gamma_1 \\ w_y^N &= u^{RAN}(N)\cos\gamma_1 + v^{RAN}(N)\sin\gamma_1 \end{aligned} \tag{4.816}$$

or

$$w_x^N = \frac{1}{d}[(r_2)^2 - (r_1)^2], \qquad w_y^N = \frac{1}{d}[\psi((r_1)^2 + (r_2)^2) + 2r_1 r_2] \tag{4.817}$$

We see now that ψ represents a small angle of rigid rotation of the whole structure around point R. By putting $\psi = -1$, we rearrange (4.817) to the form

$$w_x^N = d\cos(2\gamma_2), \qquad w_y^N = -d(1 - \sin(2\gamma_2)) \tag{4.818}$$

Let us pass now to the construction of u, v within $ABDC$. One should find the values of these fields along the arcs AB, AC. Due to continuity of displacements along AB, we have

$$u^{ABDC}(\alpha = 0, \beta) = u^{RBA}(r = r_2, \vartheta = \beta),$$
$$v^{ABDC}(\alpha = 0, \beta) = v^{RBA}(r = r_2, \vartheta = \beta) \tag{4.819}$$

or

$$u^{ABDC}(0, \beta) = r_2, \qquad v^{ABDC}(0, \beta) = -2r_2\beta - r_2 \tag{4.820}$$

since $\psi = -1$. By continuity along AC, we find

$$u^{ABDC}(\alpha, 0) = 2r_1\alpha + r_1 + (r_2 - r_1)(\cos\alpha + \sin\alpha)$$
$$v^{ABDC}(\alpha, 0) = -r_1 + (r_2 - r_1)(\sin\alpha - \cos\alpha) \tag{4.821}$$

Note that

$$A(0, \beta) = r_1 + \beta r_2, \qquad A(\alpha, 0) = r_1 \tag{4.822}$$

Our unknown will be $u^0(\alpha, \beta)$ given by (4.118a) with $\kappa = 1$. Having found (4.409) and (4.410), one fixes the boundary conditions for $u^0(\alpha, \beta)$

$$u^0(0, \beta) = r_2, \qquad u^0(\alpha, 0) = r_1 + (r_2 - r_1)(\cos\alpha + \sin\alpha) \tag{4.823}$$

These conditions allow to use Riemann's formula (4.132) for $\theta_1 = \theta_2 = 0$, $f = 0$; thus, we find

$$u^0(\lambda, \mu) = r_2 G_0(\lambda, \mu) + (r_2 - r_1)\int_0^\lambda G_0(\lambda - \alpha, \mu)\cos\alpha d\alpha +$$
$$(r_1 - r_2)\int_0^\lambda G_0(\lambda - \alpha, \mu)\sin\alpha d\alpha \tag{4.824}$$

The integration rules (4.137), (4.138) give

$$u^0(\lambda, \mu) = r_2 G_0(\lambda, \mu) + (r_2 - r_1)[F_1(\lambda, \mu) - F_2(\lambda, \mu)] \tag{4.825}$$

hence

$$u^{ABDC}(\lambda, \mu) = 2r_1\lambda G_0(\lambda, \mu) + r_2[G_0(\lambda, \mu) + 2\mu G_1(\lambda, \mu)] +$$
$$(r_2 - r_1)[F_1(\lambda, \mu) - F_2(\lambda, \mu)] \tag{4.826}$$

and, using the first equation of (4.116), we arrive at

$$v^{ABDC}(\lambda, \mu) = -2r_1\mu G_1(\lambda, \mu) - r_2(1 + 2\mu)G_0(\lambda, \mu) +$$
$$(r_2 - r_1)[F_1(\lambda, \mu) + F_2(\lambda, \mu)] \tag{4.827}$$

The above results make it possible to find the fields w_x^D, w_y^D by (4.805) with u^D, v^D given by (4.826), (4.827) by putting $\lambda = \theta_1$, $\mu = \theta_2$.

Having found w_x^N, w_y^N (see 4.818) and w_x^D, w_y^D, as well as the values of Q_x, Q_y by (4.797), (4.809) one can compute the volume of the structure by (3.81) or by

$$V = \frac{1}{\sigma_0}[F_x w_x^D + F_y w_y^D + Q_x w_x^N + Q_y w_y^N] \tag{4.828}$$

The above result is valid only if the force $\boldsymbol{F}$ is directed such that the bar RBD is in tension and the bar NCD is in compression. The force $\boldsymbol{F}$ must lie within the shaded domain in Figs. 4.150, 4.151.

Let us put below the algorithm of constructing the solution. Given are the positions of points: R, N, D. The given three forces are self-equilibrated; hence, their components satisfy

$$\begin{aligned} P_x + Q_x + F_x &= 0 \\ P_y + Q_y + F_y &= 0 \\ dQ_y + x^D F_y - y^D F_x &= 0 \end{aligned} \tag{4.829}$$

This means that at least three force components should be given. Assume that F_x, F_y, Q_x are chosen as the data of the problem. Thus, one can assume that the following quantities σ_0, d, x^D, y^D, F_x, F_y, Q_x are given (the data of the problem). The above quantities should not be viewed as fully free, since the final layout should satisfy some additional conditions of correctness. This will be cleared up later. These bounds of applicability cannot be explicitly written in advance.

The geometry of the optimal layout is determined by the three angles: θ_1, θ_2, γ_2 which are governed by three equations: (4.796) and (4.809). This system of equations is transcendental; the solutions can be constructed only numerically. One can note that a unique solution can be found for a very wide range of the input data. Let us stress that the algorithm is completed only due to Eq. (4.809). Just this Chan's equation paves the way to construct the solutions.

4.15.1.5 When does the Triangle of the Forces Coincide with the Triangle of the Points of Forces Applications? Henry Chan's Theorem

Assume that

$$F_y = 0, \qquad \theta_1 = \theta_2 = \theta \tag{4.830}$$

Then, the lines of the reactions $\boldsymbol{P}$ and $\boldsymbol{Q}$ intersect at point D where the force $\boldsymbol{F}$ is applied, see Fig. 4.156. This property has been noted and proved by Chan (1966).

To prove this property, it is sufficient to show that (4.829) implies the equality

$$\frac{x^D - d}{y^D} = \frac{Q_x}{Q_y} \tag{4.831}$$

if $y^D \neq 0$ and $Q_y \neq 0$. Let us put $F_y = 0$ into (4.797) to find

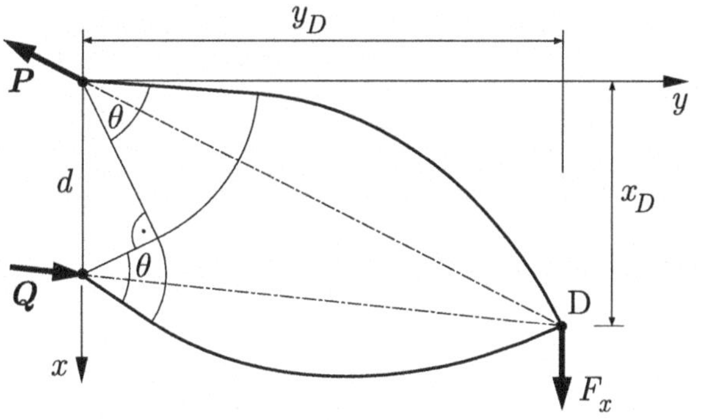

Fig. 4.156 Illustration of the theorem by Henry Chan

$$Q_y = \frac{y^D}{d} F_x \tag{4.832}$$

and now Eq. (4.809) reduces to

$$Q_x = -[\sin \gamma_1 k_3(\theta, \theta) + \cos \gamma_1 k_4(\theta, \theta)] F_x \tag{4.833}$$

On combining the last two equations we find

$$y^D \frac{Q_x}{Q_y} = d[\cos^2 \gamma_2 F_2(\theta, \theta) - \sin^2 \gamma_2 F_0(\theta, \theta)] \tag{4.834}$$

while (4.794) implies

$$x^D - d = d[\cos^2 \gamma_2 F_2(\theta, \theta) - \sin^2 \gamma_2 F_0(\theta, \theta)] \tag{4.835}$$

which ends the proof.

4.15.1.6 Examples Illustrating Chan's Theorem

We refer to the subclass of exact solutions discussed in Sect. 29.5. Assume that σ_0, F_x, $F_y = 0$, d, x^D, y^D are given and that $\theta_1 = \theta_2 = \theta$. Equations (4.796) assume the form

$$2\left(\frac{x^D}{d}\right) = 1 + \cos(2\gamma_2) G_0(\theta, \theta) \tag{4.836}$$
$$2\left(\frac{y^D}{d}\right) = 2F_1(\theta, \theta) + \sin(2\gamma_2) G_0(\theta, \theta)$$

Table 4.7 The numerical data concerning selected solutions to the three-force problem

	F_x	F_y	x_D/d	y_D	θ_1	θ_2	γ_2	$\bar{V}$
Figure 4.157a	1.0	0.0	0.5	2.081	50°	50°	45°	7.451
Figure 4.157b	1.0	0.0	0.1766	1.134	30°	30°	60°	3.094
Figure 4.157c	1.0	0.0	0.0	2.0	51°	51°	60.31°	7.273
Figure 4.157d	1.0	0.0	1.0	1.0	30.42°	30.42°	19.92°	2.835
Figure 4.158a	1.0	−0.5	1.5	1.5	66.77°	29.81°	38.61°	6.831
Figure 4.158b	1.0	−1.0	1.5	0.5	62.89°	23.96°	7.453°	4.750

For fixed d, x^D, y^D, this system can be solved numerically with arbitrary accuracy. The volume of the optimal structure is given by (4.828). The non-dimensional volume will be defined as $\bar{V} = V/V_0$, $V_0 = dF_x/\sigma_0$.

Four different results are shown in Fig. 4.157 for $F_x = 1$. All of them confirm Chan's theorem of Sect. 4.15.1.5; the results are set up in Table 4.7. These results refer to the full plane as the design domain.

4.15.1.7 Two Further Examples

The data for the results in Fig. 4.158a, b are set up in Table 4.7. Now, $\theta_1 \neq \theta_2$ and the point of intersection of $\boldsymbol{P}$ and $\boldsymbol{Q}$ does not coincide with point D. In all the results shown the force $\boldsymbol{F}$ lies in the shaded regions denoted in Fig. 4.150b, hence causes stresses of different signs in the ribs connecting at D. This is the criterion of correctness of the solutions.

4.15.1.8 The Case of the Force F Being Parallel to RN, with a Sliding Support at N

Assume now: $F_y = 0$, $Q_x = 0$, see Fig. 4.159. Chan's equation (4.809) assumes the form

$$\cos\gamma_2 k_3(\theta_1, \theta_2) + \sin\gamma_2 k_4(\theta_1, \theta_2) = 0 \tag{4.837}$$

hence is a purely geometric equation.

Having x^D, y^D, one can determine γ_2, θ_1, θ_2 by solving (4.796) and (4.837). The volume of the optimal structure is expressed by (4.828), which reduces to

$$V = \frac{1}{\sigma_0} F_x w_x^D \tag{4.838}$$

The displacement w_x^D is given by (4.805) with u^D, v^D given by (4.826), (4.827) for $(\lambda, \mu) = (\theta_1, \theta_2)$. A detailed study of the mentioned equations for $(\gamma_2, \theta_1, \theta_2)$ leads

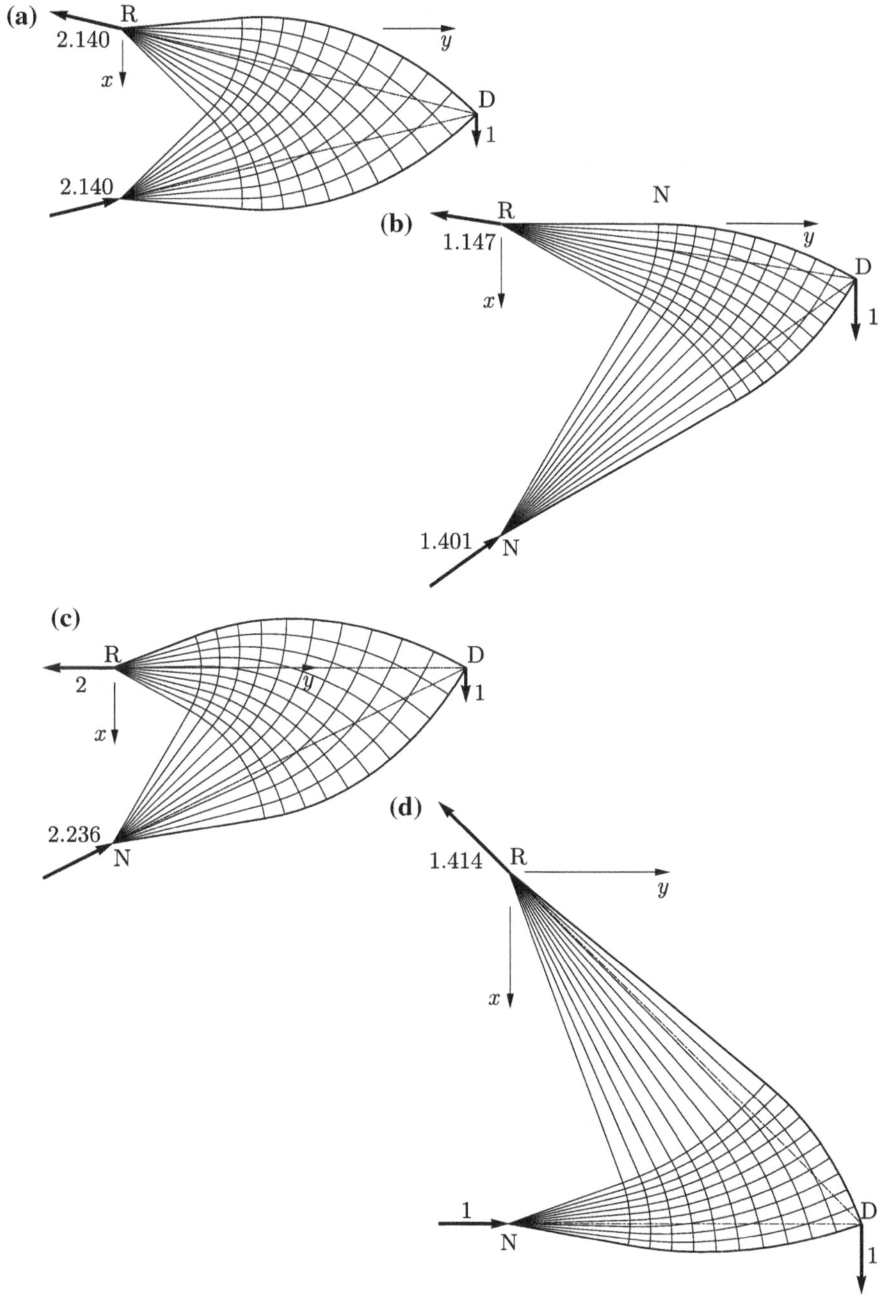

Fig. 4.157 Optimal layouts illustrating Chan's theorem

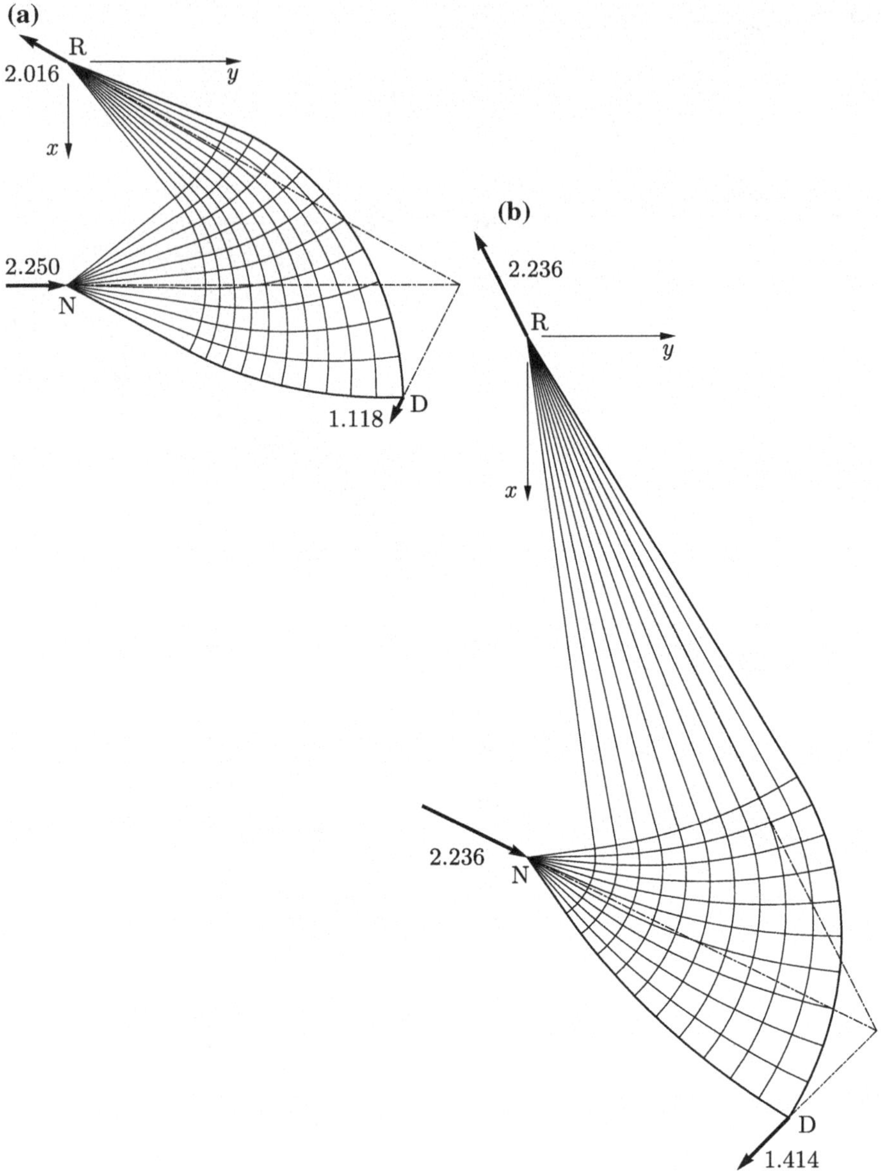

Fig. 4.158 Selected exact solutions to the three-force problem in the plane

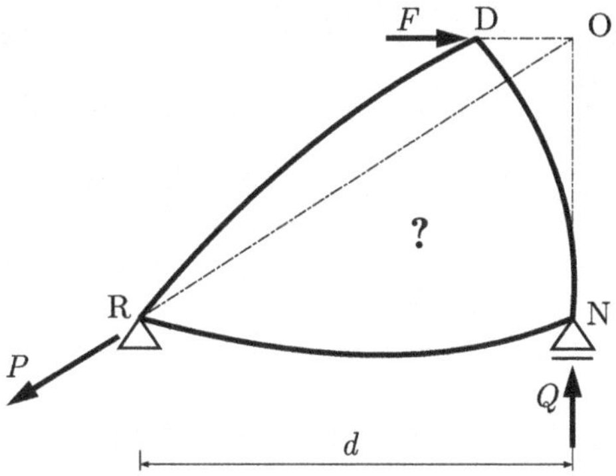

Fig. 4.159 Setting the problem of transmitting a horizontal force to the supports, one of them being a roller

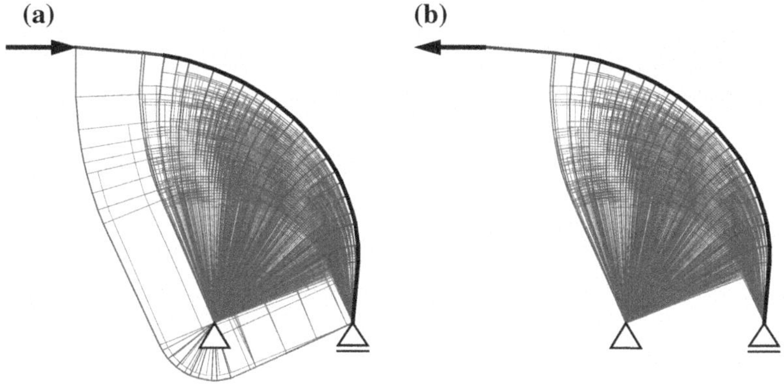

Fig. 4.160 An illustrative full-plane design for $x^D = -d$, $y^D = 2d$ (**a**); a half-plane design showing a change of topology (**b**)

to a conclusion that this system delivers correct results for the full plane as the design domain provided that $\theta_2 + \gamma_2 - \frac{\pi}{2} \leq \theta_1 \leq \theta_2 + \gamma_2$.

In case of $\theta_1 = \theta_2 + \gamma_2$, the force F is tangent to the rib at point D. In case of a full-plane design, a new surrounding structure appears, consisting of the regimes $\mathbb{R}_1$, $\mathbb{R}_2$, see Fig. 4.160a.

In case of a half-plane design, a new straight bar appears transmitting the force to the main part of the structure, see Fig. 4.160b.

The other types of layouts are observed if point D lies closer to the line RN. There appears a fan of curved lines, see Fig. 4.161.

The numerical solution of Fig. 4.161a is confirmed by the analytical result of Fig. 4.161b. The details of constructing the analytical solution are left to the reader.

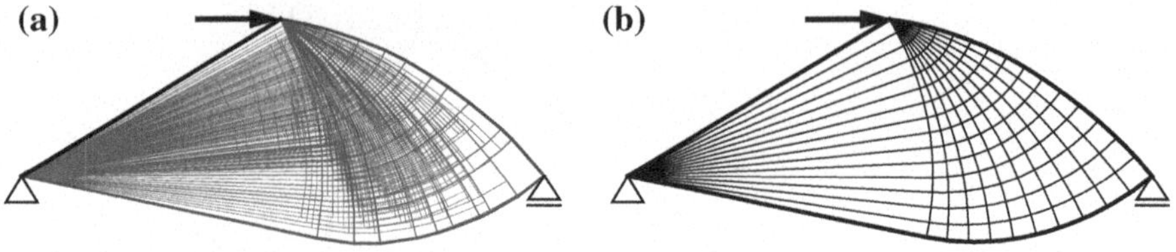

Fig. 4.161 The numerical (**a**) and analytical (**b**) solution for the problem of Fig. 4.159 if the point D lies close to the line RN

4.15.2 Simply Supported Michell Trusses Generated by a Point Load. Half-Plane Solutions

Consider the problem as set in Fig. 4.162: transmit the force F applied at the point D of coordinates (x_D, y_D) to the pin support at R and to the roller at N. The force makes an angle α with the line RN. The half-plane over the line RN is the design domain. Assume that $\sigma_T = \sigma_C = \sigma_0$.

The solution depends on the position of the point D and on the angle α.

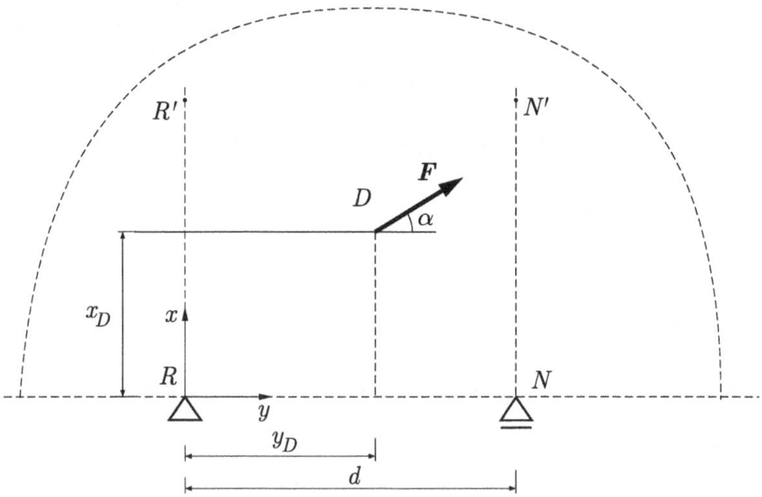

Fig. 4.162 The problem of optimal transmitting the force F to the pin support R and to the roller at N

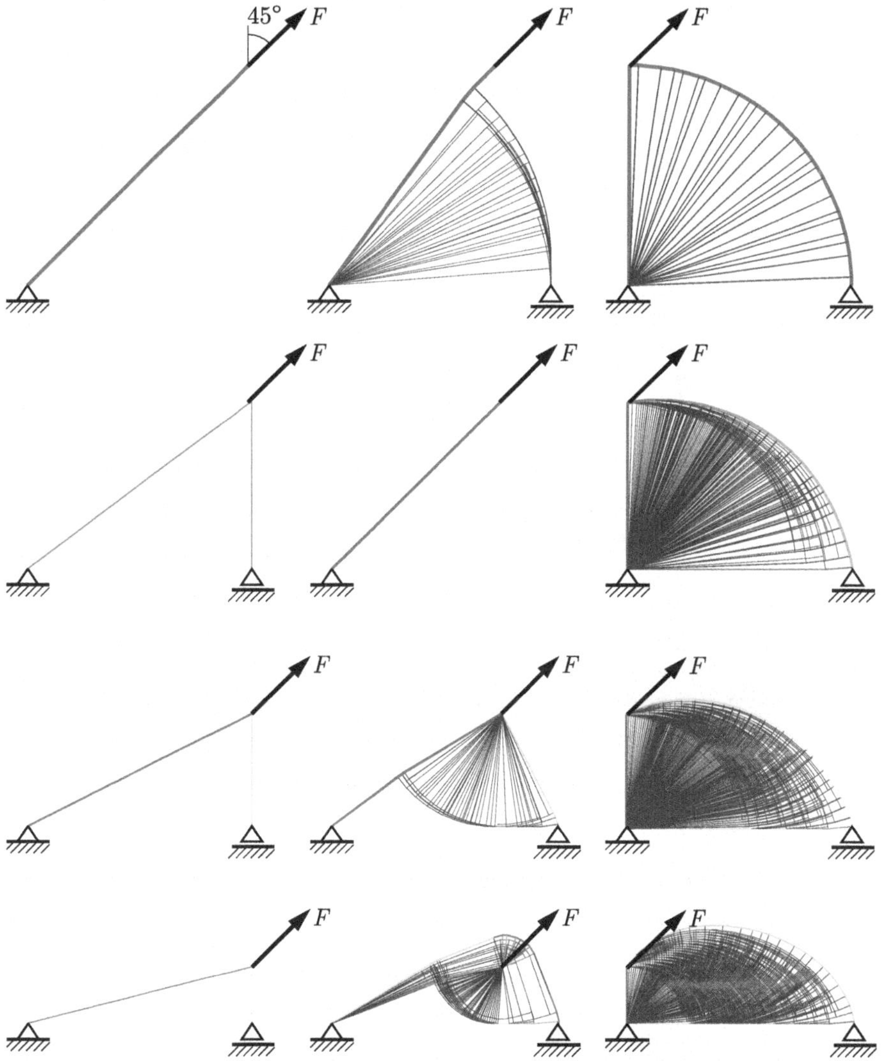

Fig. 4.163 The problem of transmitting the force F, inclined by the angle $\alpha = 45°$, to the pin support at R and to the roller at N, see Fig. 4.162. The numerical solutions by the ground structure method

Consider the case of $\alpha = \pi/4$. If the point D lies on the line NN' perpendicular to RN, then the solution is composed of two bars and can degenerate to a single bar, see Fig. 4.163, the first column of the set of figures. If the point D lies over R, or on the line RR' perpendicular to RN, then a circular fan around the support R occurs, see the last column of the figures of Fig. 4.163. The solution is elementary if $|DR| = d$, see the upper figure in the last column of the figures in Fig. 4.163. In general, the solutions are complex, see the other solutions in this figure. Let us note that the bars

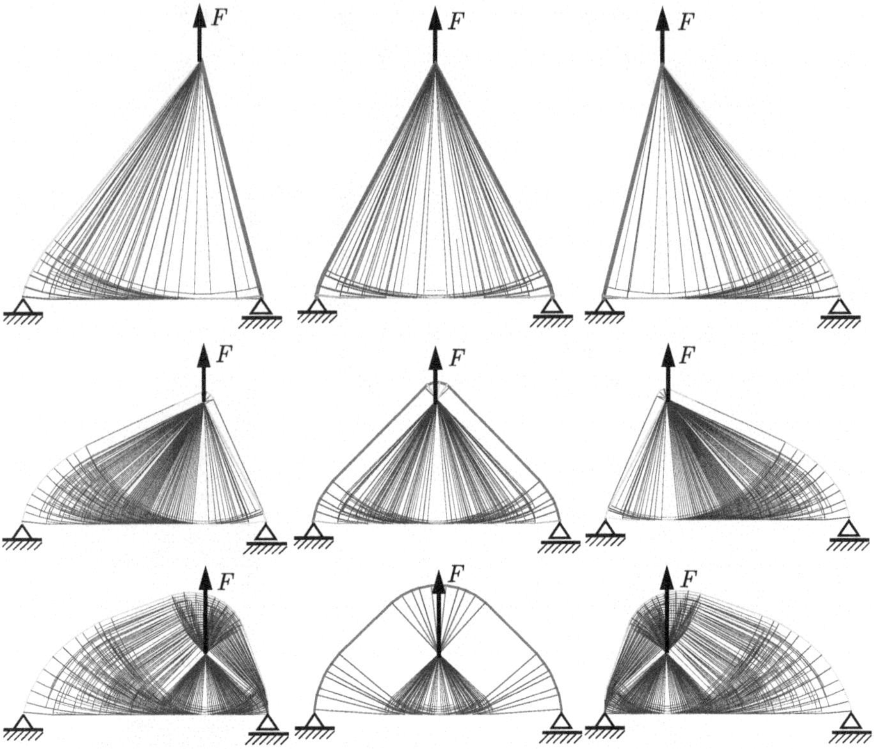

Fig. 4.164 The numerical solutions to the problem of Fig. 4.162 for $\alpha = 90°$

of finite cross sections emerge within the structures (see the second figure in the first raw).

Consider now the case of $\alpha = \pi/2$, i.e. the force is orthogonal to RN, see Fig. 4.164. If the point D lies in a distance d to the line RN, then the structure is composed of a fan and two Chan's domains, see the figures in the upper row of Fig. 4.164. In general, the solutions are more complicated, but the bars of finite cross sections do not appear, see Fig. 4.164.

Selected numerical solutions for the case of $\alpha = \dfrac{3}{4}\pi$ are given in Fig. 4.165. In contrast to the case of $\alpha = \dfrac{\pi}{4}$ the two-bar trusses do not appear.

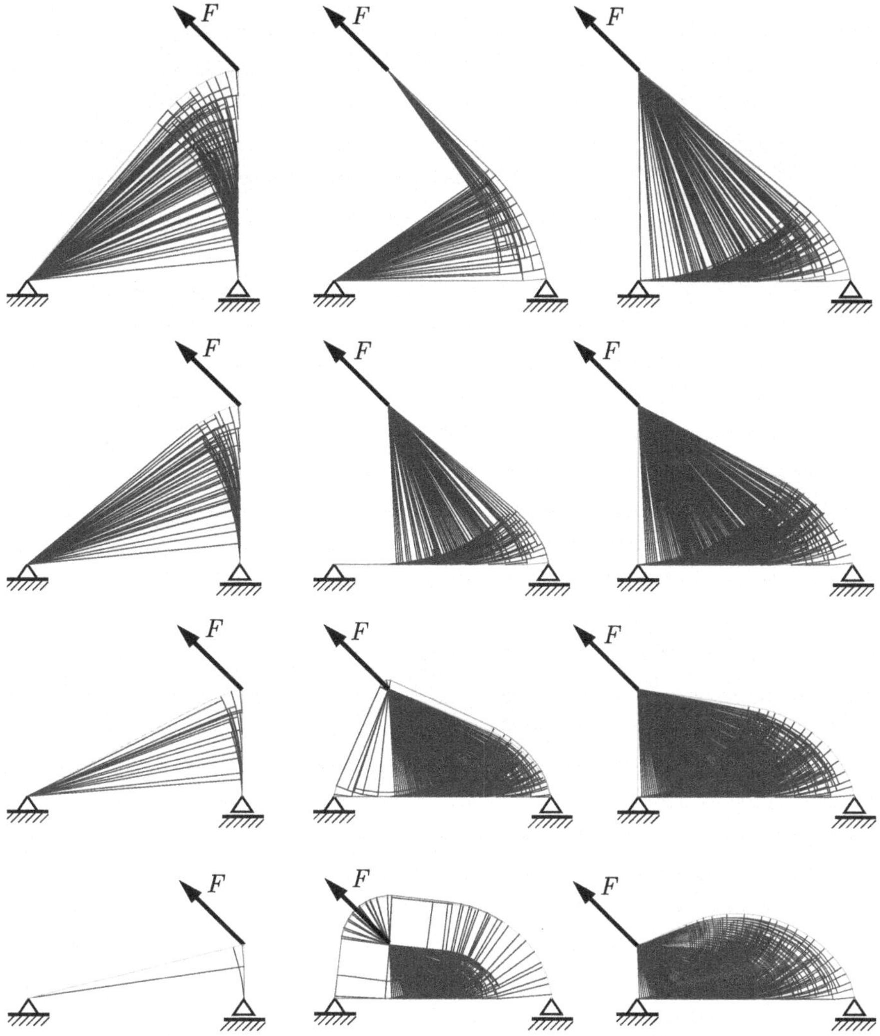

Fig. 4.165 The numerical solutions to the problem of Fig. 4.162 for $\alpha = 135°$

Consider the case of $\alpha = 0°$ (called Rozvany's problem, see Sokół and Lewiński 2016a, b). If point D lies far from the line RN, then the analytical solutions are known, see Fig. 4.166. If point D lies close to RN, then the solutions become complex, and, additionally, a curved bar of finite cross section appears linking D with the support R, see Fig. 4.167.

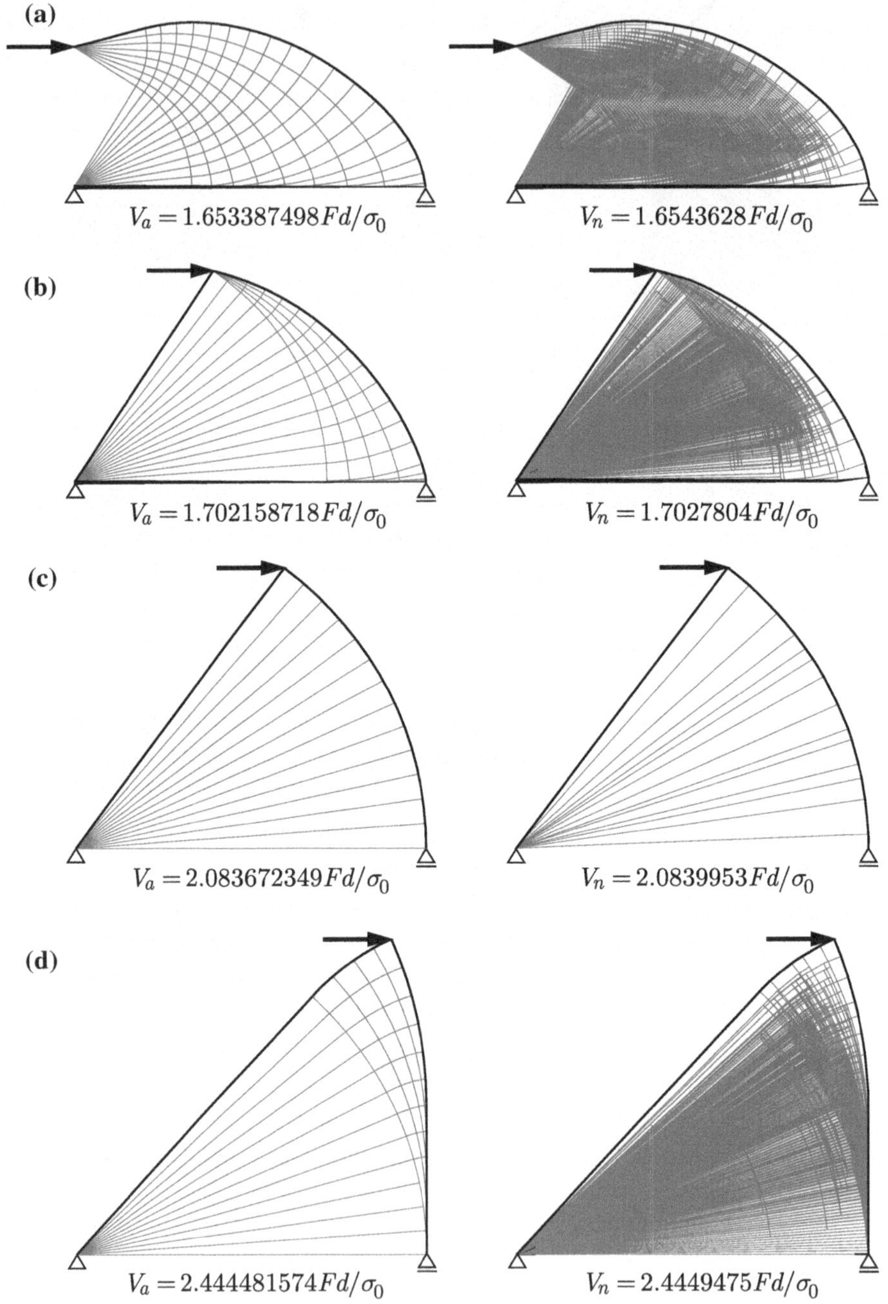

Fig. 4.166 The solutions to the problem of Fig. 4.162 for $\alpha = 0^o$. The analytical solutions (left) and the numerical solutions (right) for **a** $x_D/d = 0.4$, $y_D = 0$; **b** $x_D/d = 0.6$, $y_D/d = 0.4$; **c** $x_D/d = 0.8$, $y_D/d = 0.6$; and **d** $x_D/d = 0.9$, $y_D/d = 0.9$

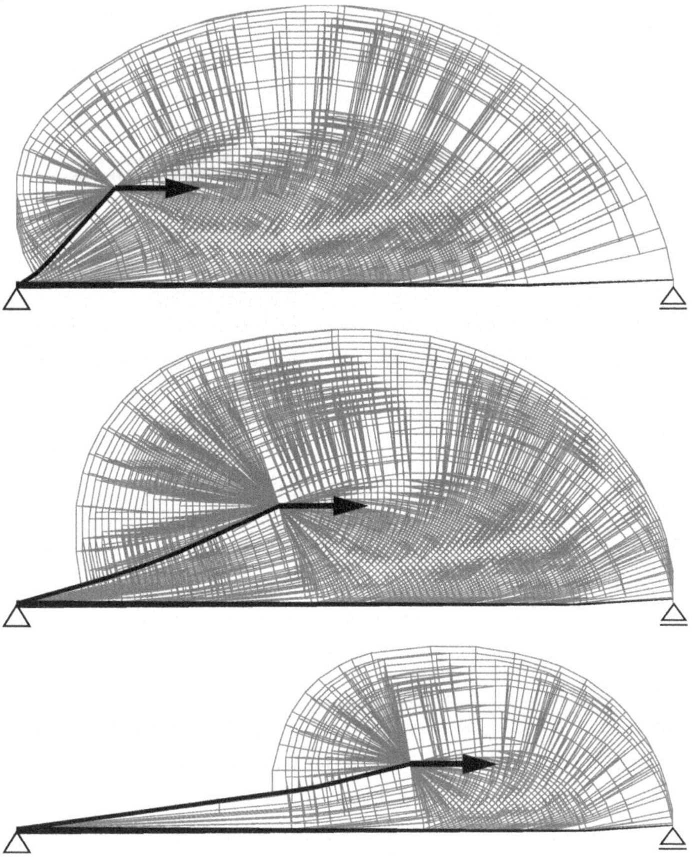

Fig. 4.167 Selected numerical solutions for which analytical solutions are not available

Lastly, compare the solutions to the problem of Fig. 4.162 for $\alpha = 0$, $x_D/d = 2$, $y_D/d = 1/2$ with the solution of the cantilever supported at two pin supports at R and N. Due to the different boundary conditions at N, the solutions do not coincide, see Fig. 4.168.

The numerical prediction of Fig. 4.168b has made it possible to construct the analytical solution using the methods of Sect. 4.12. The details are omitted.

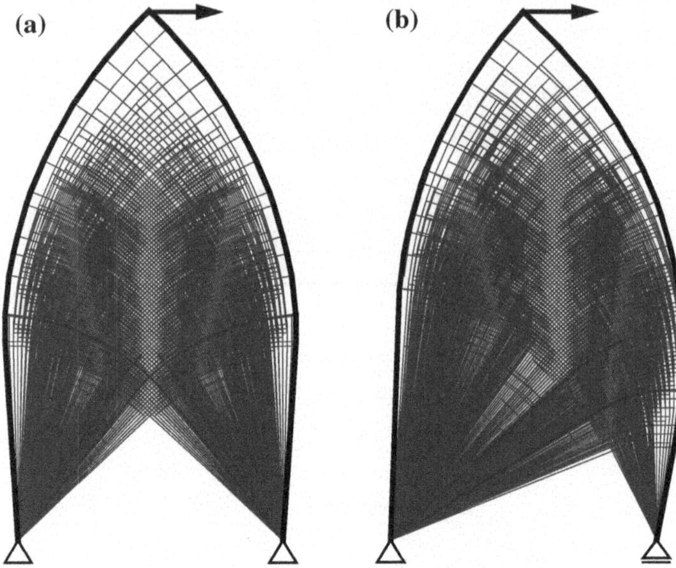

Fig. 4.168 Michell's cantilever (**a**) versus the optimal structure of Fig. 4.162 (**b**) for $\alpha = 0$, $x_D/d = 2$, $y_D/d = 1/2$

4.16 Optimal Pin-Supported Girders Subjected to Symmetrically Located Vertical Forces

The subject of consideration will be structures transmitting systems of vertical forces to two hinges at the supports, both of them unmovable. The design domain is either the half-plane over the line of supports, or the full plane.

4.16.1 Case of Two Forces. Half-Plane Solutions for the Case of $\sigma_C = \sigma_T$

The analytical solutions of the three-force problem can be used to construct an optimal girder transmitting two vertical forces of magnitude P to two pin supports at R and R'. The forces are symmetric with respect to the mid-perpendicular of RR', see Fig. 4.169. The design domain is the half-plane over the line RR'.

Assume that $\sigma_C = \sigma_T = \sigma_0$. Given are σ_0, d, P, L, with $|RR'| = 2L$. Let us define $\xi = d/L$ and $V_0 = PL/\sigma_0$. Our aim is to find the characteristic dimensions of the optimal structure as well as its volume. We predict the layout given in Fig. 4.170.

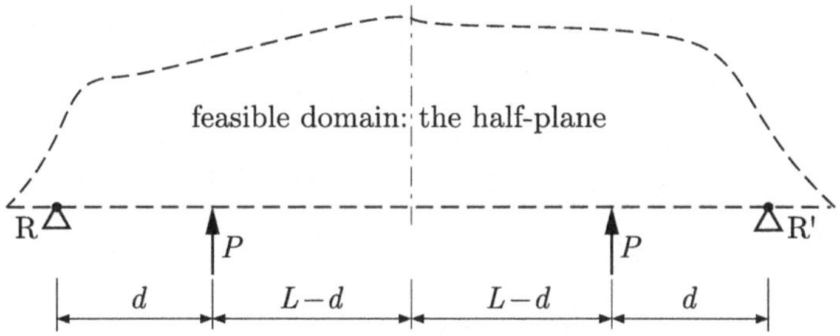

Fig. 4.169 The problem setting. The design domain is the half-plane over the line of supports

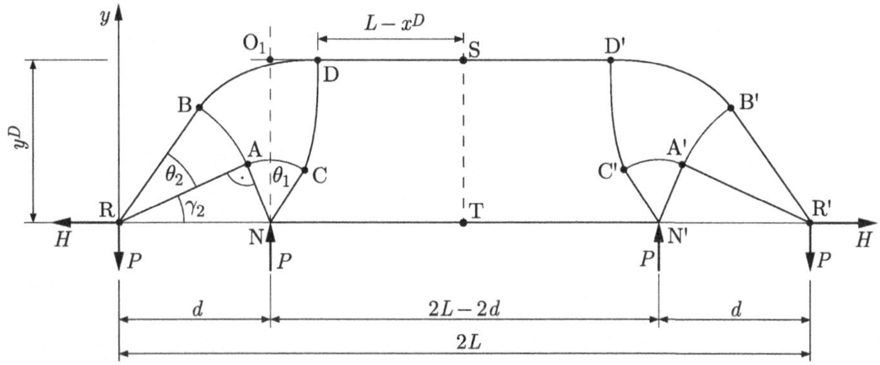

Fig. 4.170 The layout of the optimal solution

Note that the frameworks $RBDCNAR$ and their symmetric image are similar to the structure constructed in Sect. 29.6.3, but now the force at D is horizontal, since it is tangent to the rib BD at D. It turns out that the lower bar NN' appears or not depending on the value of the ratio ξ.

A. The Layouts Without the Bar NN'

A layout is determined by γ_2, θ_2, while θ_1 is directly linked with these angles; indeed, since $\phi_D = \theta_2 - \theta_1$ is an angle between the tangent to the line BD at D and the axis x_0, and since $\theta_1 > \theta_2$, the condition $|\phi_D| = \gamma_2$ implies $\theta_1 = \gamma_2 + \theta_2$, see Fig. 4.171.

As in Sect. 29.6.3, we have here: $F_y = 0$, $Q_x = 0$ while $Q_y = P$. Thus, Eq. (4.837) holds and is expressed by

$$\cos\gamma_2 k_3(\theta_2 + \gamma_2, \theta_2) + \sin\gamma_2 k_4(\theta_2 + \gamma_2, \theta_2) = 0 \tag{4.839}$$

By (4.795), we compute

$$\begin{aligned} k_3(\theta_2 + \gamma_2, \theta_2) &= -F_2(\gamma_2 + \theta_2, \theta_2), \\ k_4(\theta_2 + \gamma_2, \theta_2) &= F_1(\gamma_2 + \theta_2, \theta_2) \end{aligned} \tag{4.840}$$

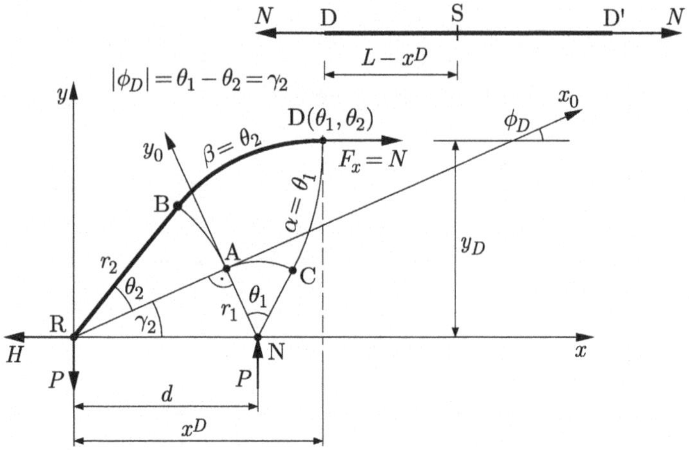

Fig. 4.171 The case of the force at D being horizontal and the force at N being vertical

which reduces (4.839) to the form

$$z(\theta_2, \gamma_2) = 0 \tag{4.841}$$

with

$$z(\theta_2, \gamma_2) = -\cos\gamma_2 F_2(\gamma_2 + \theta_2, \theta_2) + \sin\gamma_2 F_1(\gamma_2 + \theta_2, \theta_2) \tag{4.842}$$

Let us fix θ_2 for a time being. The value of γ_2 is computed by (4.841) and $\theta_1 = \gamma_2 + \theta_2$. Then, we can compute all other characteristics, including coordinates of point D and displacements of points D and N. Then, the force F_x can be found, while $Q_y = P$ is known. The volume of the bar DS equals

$$V_2 = \frac{F_x}{\sigma_0}(L - x^D) \tag{4.843}$$

The volume of the part of the structure: $RBDCNA$ is expressed by

$$V_1 = \frac{1}{\sigma_0}(F_x w_x^D + P w_y^N) \tag{4.844}$$

hence the total volume equals

$$V_A = 2V_0\left(\xi\frac{L - x^D + w_x^D}{y^D} + \frac{w_y^N}{L}\right) \tag{4.845}$$

The volume V_A can be viewed as a function: $V_A(\theta_2, \gamma_2)$, where (θ_2, γ_2) are linked by (4.841) with $0 \leq \gamma_2 \leq \frac{\pi}{2}$.

Assume that $\breve{\theta}_2, \breve{\gamma}_2$ refer to the optimum structure; its volume equals $V_A(\breve{\theta}_2, \breve{\gamma}_2)$ and $z(\breve{\theta}_2, \breve{\gamma}_2) = 0$. Thus, $\breve{\theta}_2, \breve{\gamma}_2$ correspond to an extremum of V_A. By applying the Karush–Kuhn–Tucker conditions, we arrive at

$$z(\theta_2, \gamma_2) = 0, \qquad 2\xi s(\theta_2, \gamma_2) + 1 = 0 \tag{4.846}$$

where

$$s(\theta_2, \gamma_2) = (\gamma_2 + 2\theta_2)s_1(\theta_2, \gamma_2) + \theta_2 s_2(\theta_2, \gamma_2) - \sin\gamma_2 F_1(\gamma_2 + \theta_2, \theta_2) \tag{4.847}$$

and

$$\begin{aligned} s_1(\theta_2, \gamma_2) &= \sin\gamma_2 F_0(\gamma_2 + \theta_2, \theta_2) - \cos\gamma_2 F_1(\gamma_2 + \theta_2, \theta_2) \\ s_2(\theta_2, \gamma_2) &= -\sin\gamma_2 G_0(\gamma_2 + \theta_2, \theta_2) + \cos\gamma_2 G_1(\gamma_2 + \theta_2, \theta_2) \end{aligned} \tag{4.848}$$

The solution of (4.846) is valid only if $\xi_0 \leq \xi \leq \xi_1$; the bounds ξ_0, ξ_1 will be explained in the sequel. Let us note only that for $\xi \geq \xi_1$ the structure in Fig. 4.171 must be stiffened by the bar NN'.

B. The Layouts with the Bar NN'

The layout without the bar NN' ceases to be stable if ξ is greater than a certain threshold value ξ_1. We predict the layout of Fig. 4.170.

The formula $\theta_1 = \gamma_2 + \theta_2$ is still valid, while the formula (4.841) is now wrong. The force $Q_x \neq 0$, see Fig. 4.172.

The volume of the subdomain $RBDCNAR$ is given by (4.828) as in case A, but now $Q_x \neq 0$ and the virtual work $Q_x w_x^N$ must be included. The volume of the bar DS is given by (4.843). The volume of the bar NT is expressed by

$$V_3 = -\frac{Q_x}{\sigma_0}(L - d) \tag{4.849}$$

since we expect that $Q_x < 0$.

The volume of the whole structure equals

$$V_B = V_A + 2V_0\frac{\xi}{y^D}z(\theta_2, \gamma_2)(\xi L - L + w_x^N) \tag{4.850}$$

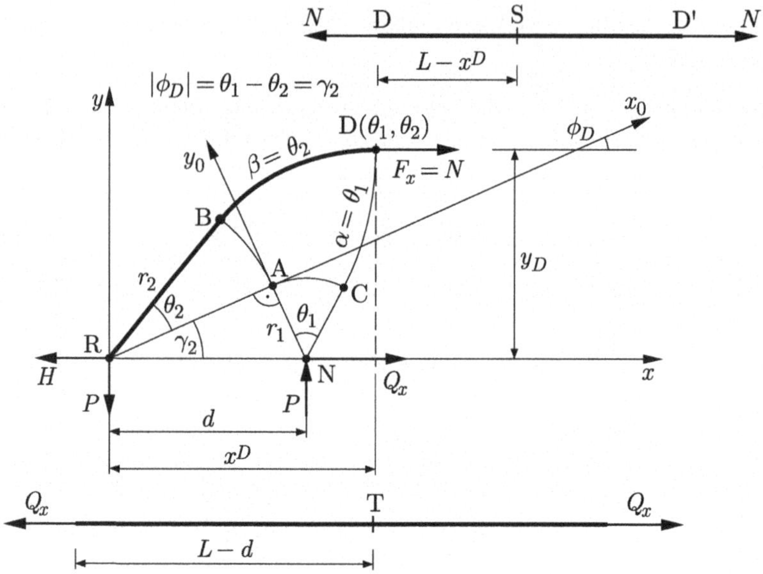

Fig. 4.172 The layout with two horizontal bars

Our task is to find θ_2, γ_2 such that the volume is the smallest; here θ_2, γ_2 are independent (not linked by (4.841)). To simplify the optimization process, we note that $w_x(T) = 0$ or

$$d - L + w_x^N = 0 \tag{4.851}$$

Using (4.817), we rearrange this equation to the form

$$2\xi \cos^2 \gamma_2 = 1 \tag{4.852}$$

hence $\xi \geq \frac{1}{2}$. The stationarity conditions of V_B with respect to θ_2 and γ_2 lead to (4.852) and the second of Eqs. (4.846). This system determines the solution of type B.

The parameter $\xi = \xi_1$, separating two cases: A and B of the optimal layouts, has to satisfy simultaneously: (4.846) and (4.852). This is the system for three unknowns: $\xi = \xi_1, \gamma_2, \theta_2$. The analysis of this system gives

$$\xi_1 = 0.598426 \tag{4.853}$$

In the case of $\xi \leq \xi_1$ in the optimal structure the bar NN' is absent; for $\xi \geq \xi_1$ this bar becomes indispensable for the equilibrium. The selected analytical solutions are set up in Fig. 4.173.

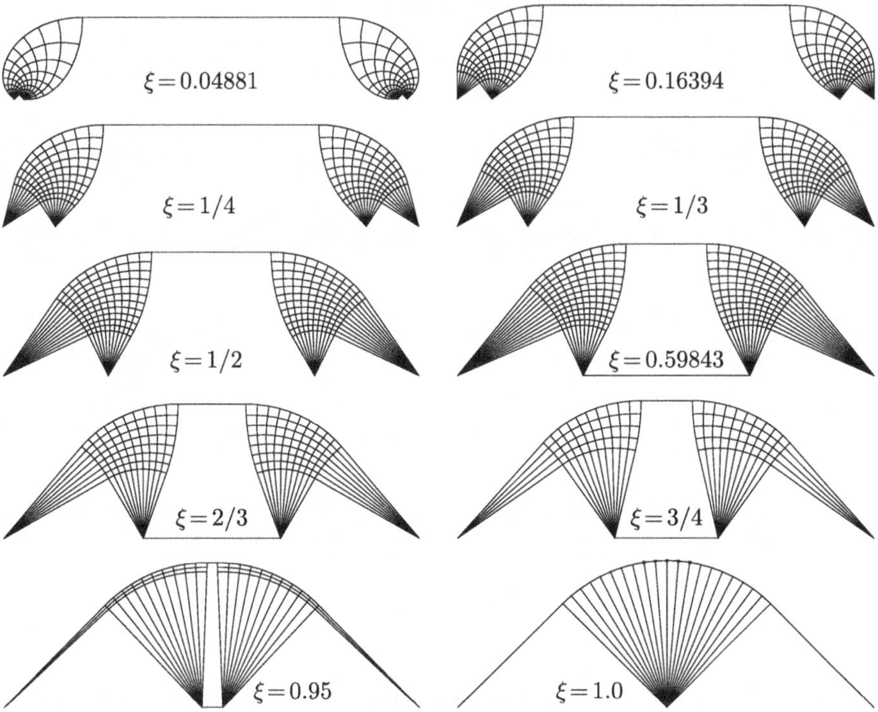

Fig. 4.173 Selected optimal structures for the problem of Fig. 4.169

The height of the girder is increasing if the points N, N' are closer and closer. For $\xi < \xi_1$, these points are not linked by a bar; for $\xi > \xi_1$, this bar appears, and finally, for $\xi = 1$, both the nodes N, N' become a one node. Five of these layouts found by the ground structure method are shown in Fig. 4.174. In each case, the relative error of accuracy of the volume assessment was smaller than 0.52%.

The main characteristics of the optimal layouts of Fig. 4.173 are set up in Table 4.8.

The volume increases along with increase of ξ, see Fig. 4.175; the smallest $\xi = \xi_0 = 0.04881$ refers to the case when the structure becomes tangent to the line $y = 0$.

The angular characteristics γ_2, θ_2, θ_1 depend on ξ. It turns out that θ_1 is smooth in the whole range of $\xi_0 \leq \xi \leq 1$, while θ_2 and γ_2 change their shapes at $\xi = \xi_1$, see Fig. 4.176.

An approximate solution to the problem considered in the present section has been recently proposed by Rojas et al. (2015).

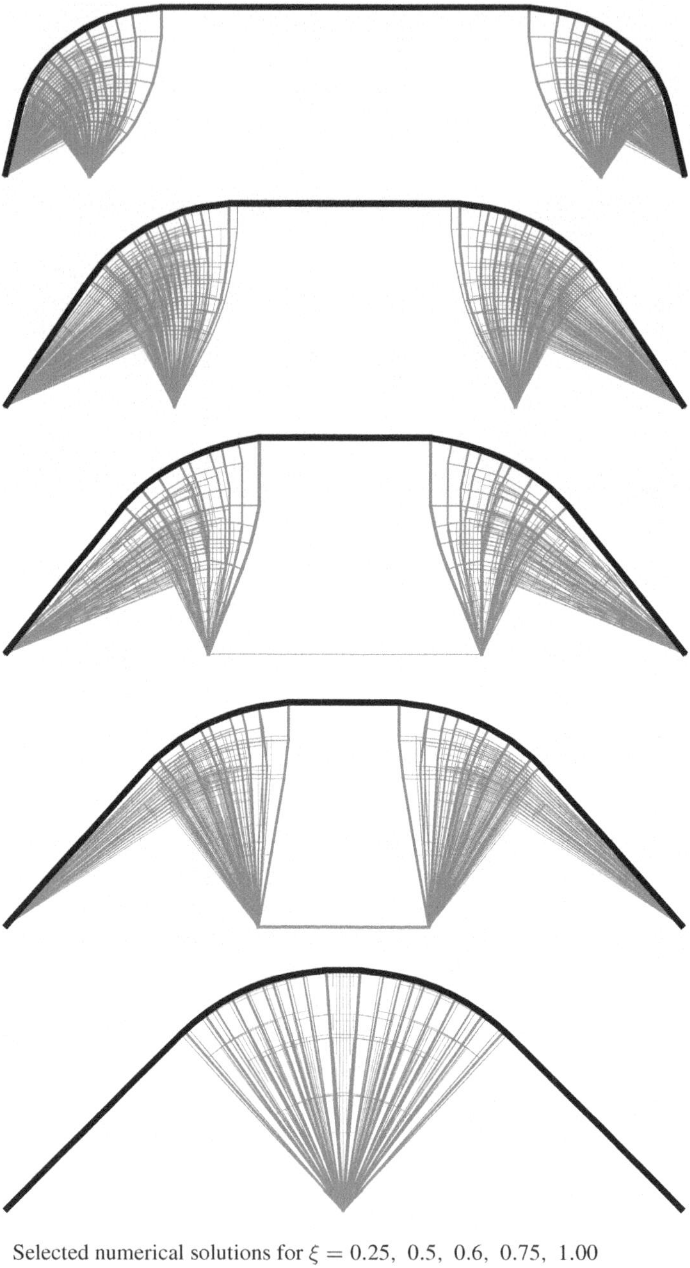

Fig. 4.174 Selected numerical solutions for $\xi = 0.25,\ 0.5,\ 0.6,\ 0.75,\ 1.00$

Table 4.8 The solutions of problem of Fig. 4.169 for selected values of ξ; $2V$ represents the volume of the whole structure

ξ	γ_2	θ_2	θ_1	x^D/L	y^D/L	Q_x/P	F_x/P	$2V/V_0$
0.04881	0.75994	1.57080	2.33074	0.34450	0.41529	0	0.11752	0.85915
0.05	0.75857	1.55696	2.31553	0.34463	0.41538	0	0.12037	0.87436
0.1	0.70214	1.17523	1.87738	0.36392	0.42718	0	0.23410	1.41828
0.16394	0.63881	0.93198	1.57080	0.40739	0.45210	0	0.36262	1.95512
0.2	0.60821	0.84399	1.45221	0.43602	0.46769	0	0.42763	2.21211
0.25	0.57120	0.75291	1.32411	0.47840	0.48983	0	0.51038	2.53184
0.3	0.53950	0.68473	1.22423	0.52280	0.51197	0	0.58598	2.81964
1/3	0.52083	0.64788	1.16871	0.55317	0.52656	0	0.63303	2.99762
0.4	0.48828	0.58848	1.07676	0.61515	0.55517	0	0.72050	3.32723
0.5	0.44871	0.52294	0.97166	0.71019	0.59638	0	0.83840	3.77092
0.59843	0.41759	0.47554	0.89313	0.80519	0.63491	$-2.2\cdot 10^{-10}$	0.94254	4.16355
0.6	0.42053	0.47205	0.89258	0.80588	0.63520	−0.00299	0.94458	4.16953
2/3	0.52360	0.34689	0.87049	0.83617	0.64768	−0.10942	1.02932	4.40104
0.7	0.56394	0.29626	0.86021	0.85191	0.65389	−0.15240	1.07051	4.50303
0.75	0.61548	0.23017	0.84565	0.87599	0.66313	−0.20885	1.13100	4.64170
0.8	0.65906	0.17299	0.83205	0.90046	0.67223	−0.25830	1.19006	4.76534
0.9	0.72973	0.07760	0.80733	0.95005	0.68999	−0.34281	1.30437	4.97460
0.95	0.75907	0.03698	0.79605	0.97501	0.69863	−0.37984	1.35981	5.06287
0.99	0.78035	0.00713	0.78748	0.99500	0.70542	−0.40753	1.40341	5.12657
1	0.78540	0.00000	0.78540	1.00000	0.70711	−0.41421	1.41421	5.14159

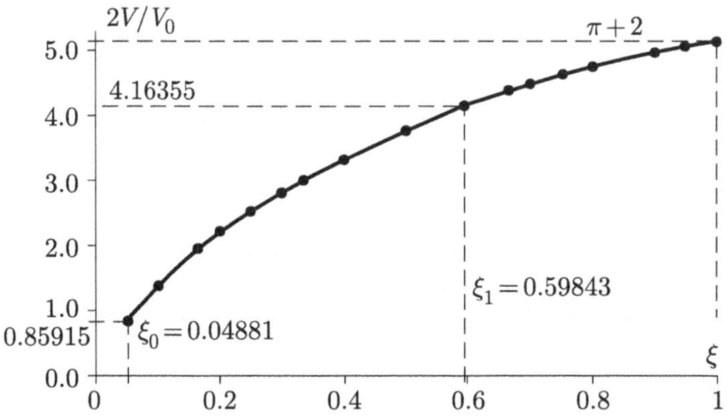

Fig. 4.175 The chart of optimal volume depending on ξ

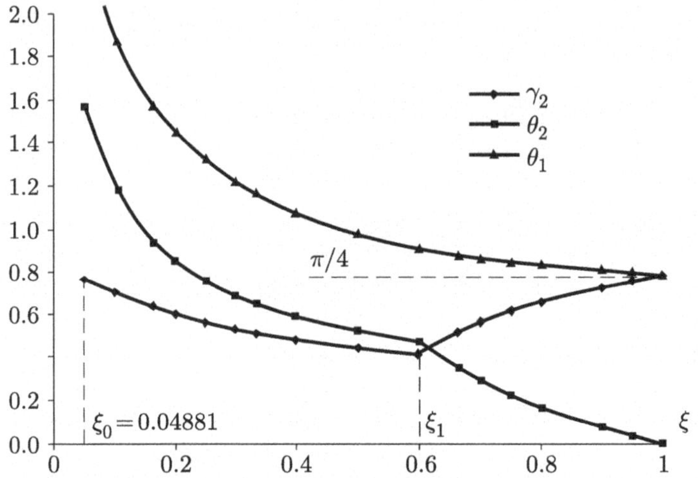

Fig. 4.176 The chart of γ_2, θ_2, θ_1 (in radians) with respect to ξ

4.16.2 Case of Three, Seven and 199 Forces. Half-Plane Solutions for the Case of $\sigma_C = \sigma_T$

Consider the problem of transmitting three forces to the two pin supports. The forces are applied symmetrically, their points of application divide the whole span into four equal parts, see Fig. 4.177.

The numerical solution concerning three forces shows that the horizontal bars along the line of the supports do not appear, see Fig. 4.178a.

In the case of seven forces of equal magnitude, such a horizontal bar is necessary to assure the equilibrium, see Fig. 4.178b. If a great number of forces are applied, then such a horizontal bar appears; it has a variable cross section in these places where the hangers are deviated from the vertical direction. The analytical solution to the problem of a uniform load applied along the line of supports has not been constructed till now. We have at our disposal only semi-inverse solutions referring to a specifically variable load intensity, see Hemp (1975), Chan (1975), Darwich et al. (2010) and Pichugin et al. (2012). The presented numerical solution (Fig. 4.178c, d) concerning 199 point loads approximates the case of q = const.

Fig. 4.177 The problem setting: find a structure transmitting three point loads to two pin supports

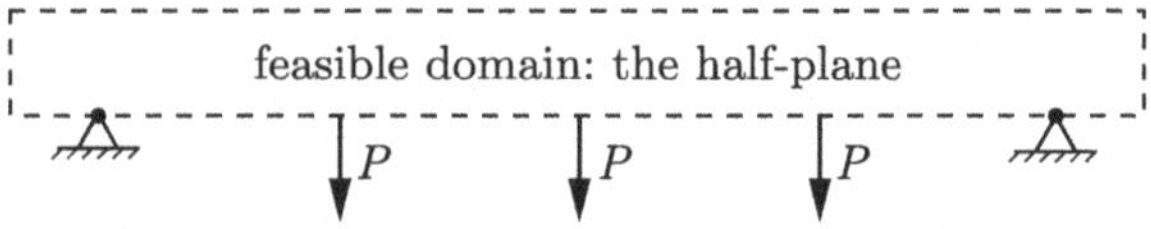

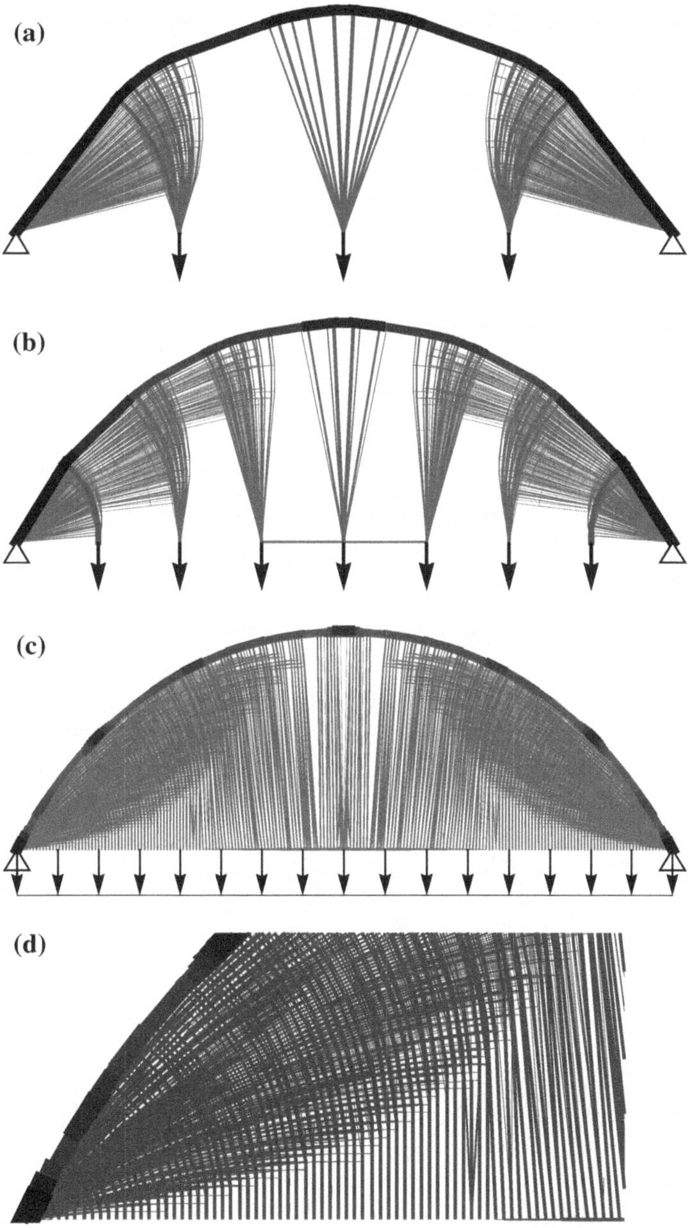

Fig. 4.178 The numerical solutions corresponding to three, seven and 199 forces of equal magnitude. The design domain is the half-plane

4.16.3 Full-Plane Solutions for the Case of $\sigma_C = \sigma_T$

Consider the problem of Fig. 4.169, but now the design domain is the whole plane. Irrespective of the value of $\frac{d}{L}$ a horizontal bar along the line of supports is not necessary, see Fig. 4.179 repeating the results of Sokół and Rozvany (2012).

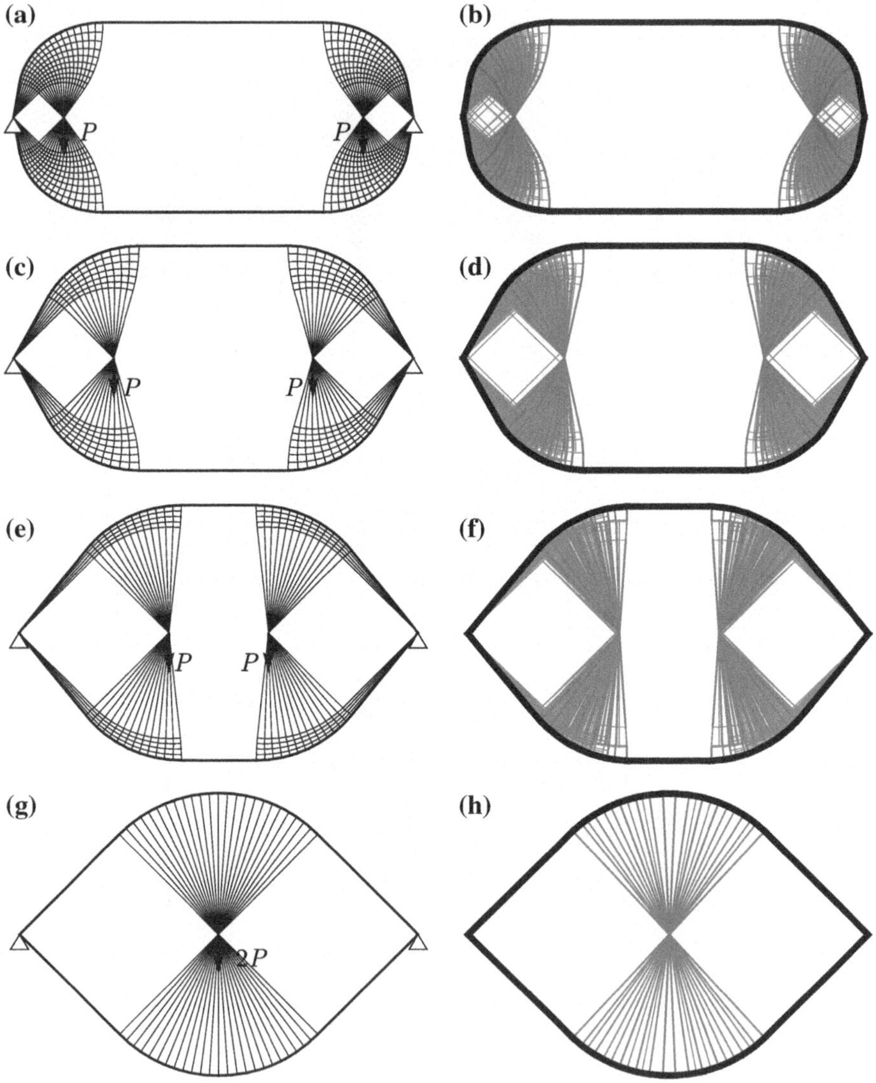

Fig. 4.179 The full-plane solutions. The illustrative examples

Note that if $d = L$, two forces become a single force of magnitude $2P$ and the solution coincides exactly with that of Michell, see Sect. 4.5.

Remark 4.16.1 The optimal designs of Fig. 4.179 hold if one of the supports is a roller, which will be discussed in Sect. 4.18.

4.16.4 Optimal Layouts Constrained by the Allowable Height of the Girder

Consider the problem of optimal transmitting the two vertical forces to the two pin supports (hinges). Now, the design domain is bounded; it is a rectangle of dimensions $2L \times 2h$, see Fig. 4.180. The permissible stresses in tension and compression are equal σ_0.

Then, the exact solutions and the numerical solutions are of the form of Fig. 4.181; the layouts depend on the ratio $\frac{h}{L}$. Note that the layout (e) consists only of straight fibres and circular arcs, hence belongs to the class of the simplest Michell structures, cf. Sect. 4.5. If the height of the domain is made smaller and smaller, the layout changes substantially, see Fig. 4.182.

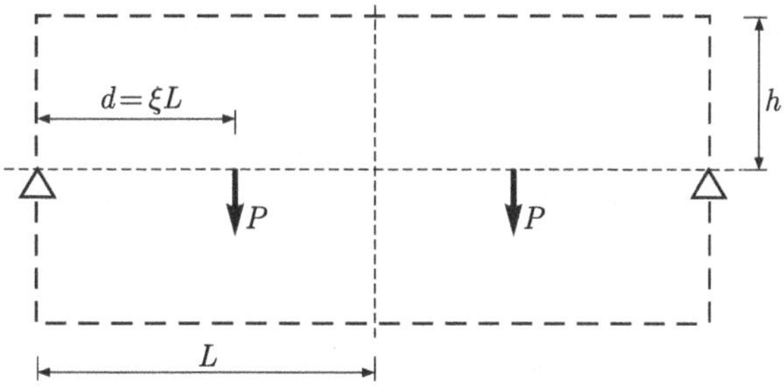

Fig. 4.180 The problem of optimal transmitting two vertical forces to two pin supports; the design domain is a given rectangle

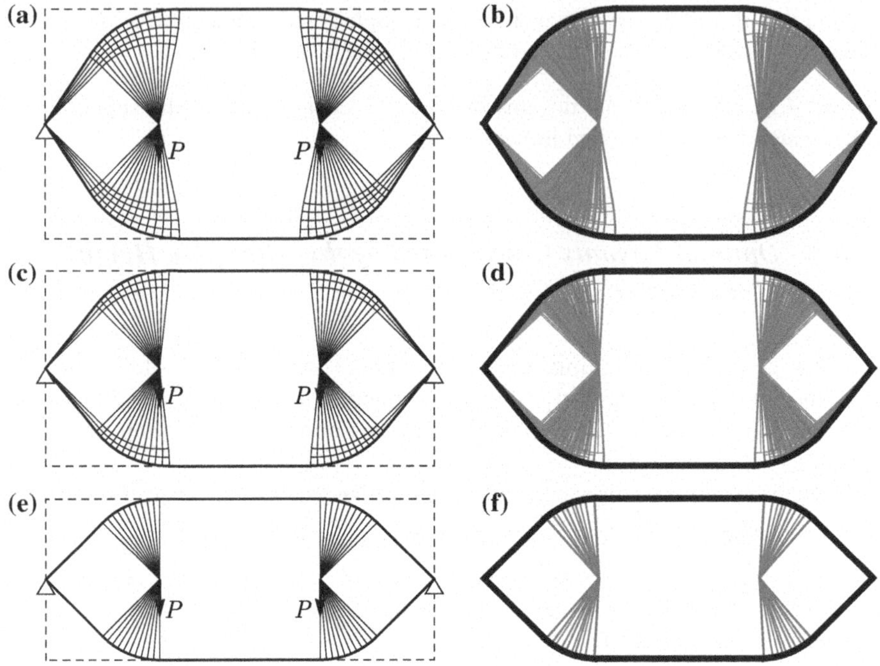

Fig. 4.181 The optimal layouts constrained by the allowable height: $h = h_1$ (**a**); its numerical equivalence (**b**); $h = (h_1 + \frac{d}{\sqrt{2}})/2$ (**c**); its numerical equivalence (**d**); $h = \frac{d}{\sqrt{2}}$ (**e**); and its numerical equivalence (**f**)

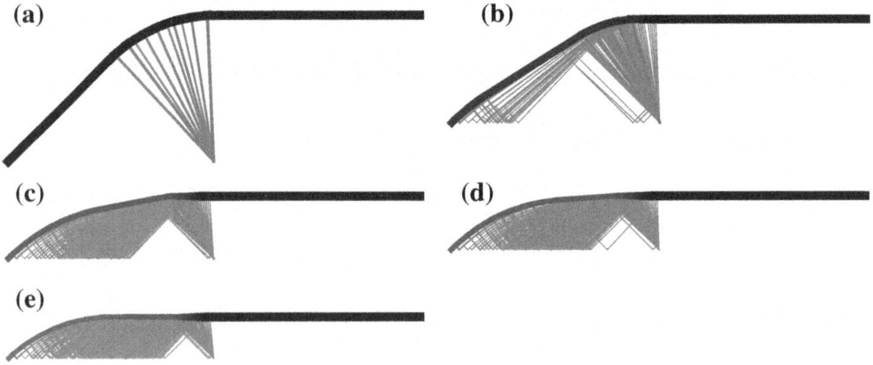

Fig. 4.182 Selected numerical layouts of the upper-left quarter for the problem constrained by allowable heights: $h = \frac{d}{\sqrt{2}}$ (**a**); $h = d/2$ (**b**); $h = 0.3d$ (**c**); $h = 0.26d$ (**d**); and $h = 0.2d$ (**e**)

4.17 Optimal Simply Supported Girders Subjected to Symmetrically Located Vertical Loads. Half-Plane Designs

This section is aimed at constructing the Michell structures transmitting systems of n vertical forces to two pin supports, one of them being a roller. Since the load is vertical one can assume that both supports are rollers. The design domain is the half-plane over the line of supports. The solutions for the case of $n = 2$ and $\sigma_C = \sigma_T$ are found using the available analytical solutions of the three-force problem.

4.17.1 Case of Two Symmetrically Located Point Loads

Let us consider the problem similar to that of Sect. 4.16.1, but now the supports are rollers, see Fig. 4.183.

The design domain is a half-plane over the line of the supports. Because all forces (the loads and the reactions) are vertical, the results will be the same if only one support is a roller. The problem is a subclass of the four-force problem concerning the case of all the forces being parallel. The position of the forces is given by the ratio $\xi = d/L$. Let us start from the numerical solutions, found by the ground structure method, see Fig. 4.184.

These results refer to the cases: $\xi = 0.25,\ 0.5,\ 0.75,\ 1$, the last one referring to the known Michell's solution. In each case of $\xi < 1$ the points of application of the forces become nodes connected by a horizontal bar NN'. The numerical results suggest that the optimal designs are similar to that of Fig. 4.170, but now the horizontal reactions are absent, see Fig. 4.185.

Let us consider the left-hand side of the structure, see Fig. 4.186. Similarly as in the problem of Sect. 4.16, we have now $\theta_1 = \gamma_2 + \theta_2$. We insert this equation into the formulae (4.795) and get

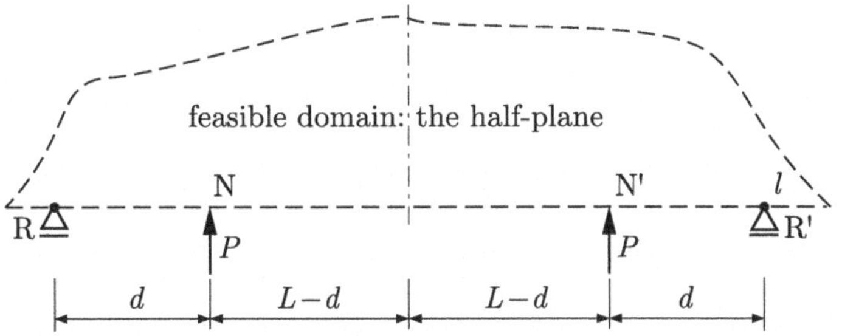

Fig. 4.183 The problem setting: find a structure transmitting two point loads to two roller supports

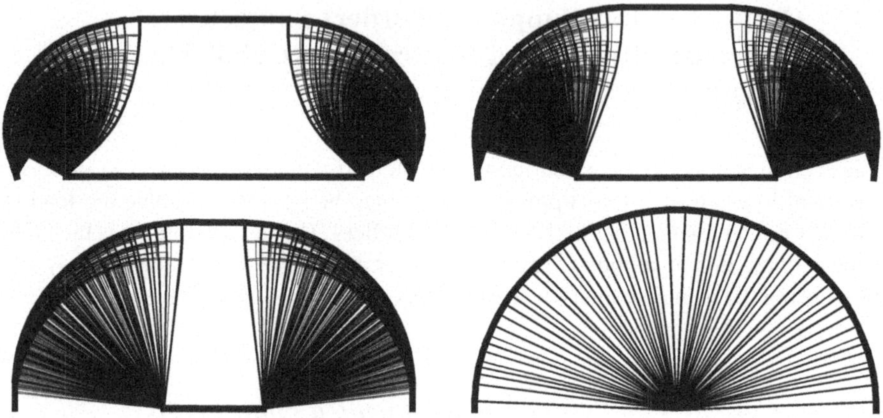

Fig. 4.184 Numerical solutions corresponding to the cases $\xi = 0.25,\ 0.5,\ 0.75,\ 1$

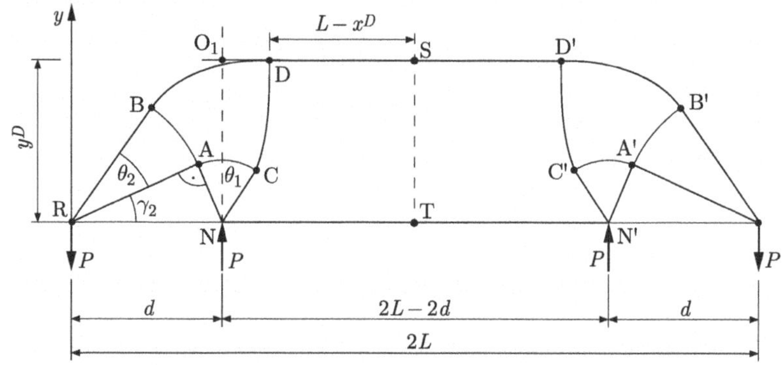

Fig. 4.185 The predicted layout of the optimal girder

Fig. 4.186 The extended three-force diagram with two horizontal bars

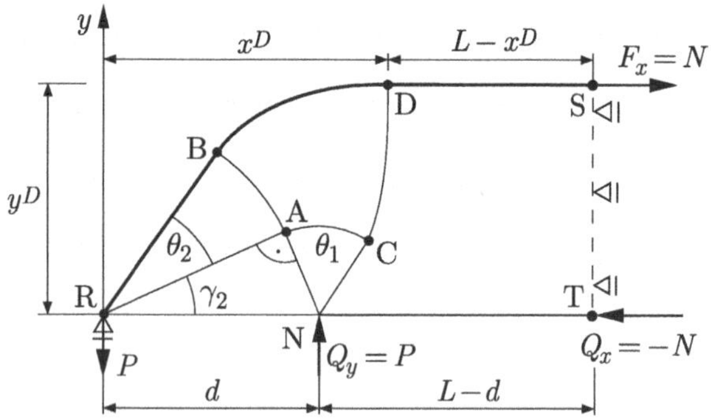

$$k_1(\gamma_2+\theta_2,\theta_2) = F_1(\gamma_2+\theta_2,\theta_2),$$
$$k_2(\gamma_2+\theta_2,\theta_2) = -F_0(\gamma_2+\theta_2,\theta_2) \tag{4.854}$$

and we obtain Eqs. (4.840). The above formulae reduce Eqs. (4.796) setting the position of point D to the form

$$\frac{x^D}{d} = \sin\gamma_2 F_1(\gamma_2+\theta_2,\theta_2) + \cos\gamma_2 F_0(\gamma_2+\theta_2,\theta_2)$$
$$\frac{y^D}{d} = \sin\gamma_2 F_2(\gamma_2+\theta_2,\theta_2) + \cos\gamma_2 F_1(\gamma_2+\theta_2,\theta_2) \tag{4.855}$$

The forces assumed in Fig. 4.151 have now the following values:

$$P_x = 0, \quad P_y = -P, \quad Q_x = -N, \quad Q_y = P,$$
$$F_x = N, \quad F_y = 0 \tag{4.856}$$

while the force N equals

$$N = \frac{d}{y^D} P \tag{4.857}$$

The Chan equation (4.809) assumes the form

$$z(\theta_2,\gamma_2) = 1 \tag{4.858}$$

with $z(\theta_2,\gamma_2)$ being given by (4.842). The volume of the structure of Fig. 4.186 (i.e the half of the girder) equals

$$V = V_1 + V_2 + V_3 \tag{4.859}$$

where V_1 is given by (4.828) and

$$V_2 = \frac{N}{\sigma_0}(L - x^D), \qquad V_3 = \frac{N}{\sigma_0}(L - d) \tag{4.860}$$

Thus, we have assumed that the upper bar is in tension and the lower bar is in compression ($N > 0$). Upon substitution (4.860) into (4.859) and taking into account (4.857), we obtain

$$\frac{V}{V_0} = \frac{\xi}{y^D}(2L - \xi L - x^D + w_x^D - w_x^N) + \frac{1}{L}w_y^N \tag{4.861}$$

where $V_0 = \dfrac{PL}{\sigma_0}$. This formula can also be confirmed by the kinematic method.

The volume V will be treated as a function $V(\theta_2, \gamma_2)$, where $0 \le \gamma_2 \le \frac{\pi}{2}$, $0 \le \theta_2 \le \frac{\pi}{2}$. These bounds on γ_2, θ_2 follow from the condition of the structure being placed in the plane $y \ge 0$.

The problem reduces to

$$\min\left\{V(\theta_2, \gamma_2) \mid z(\theta_2, \gamma_2) = 1, \quad 0 \le \gamma_2 \le \frac{\pi}{2}, \quad 0 \le \theta_2 \le \frac{\pi}{2}\right\} \tag{4.862}$$

The extremality conditions for the problem above may be written in the form of the two equations

$$z(\theta_2, \gamma_2) = 1, \qquad \xi s(\theta_2, \gamma_2) = 1 \tag{4.863}$$

where

$$s(\theta_2, \gamma_2) = -(\gamma_2 + 2\theta_2)s_1(\theta_2, \gamma_2) - \theta_2 s_2(\theta_2, \gamma_2) + s_3(\theta_2, \gamma_2) \tag{4.864}$$

The functions $s_1(\theta_2, \gamma_2)$, $s_2(\theta_2, \gamma_2)$ are given by (4.848) and

$$s_3(\theta_2, \gamma_2) = \sin\gamma_2 F_1(\gamma_2 + \theta_2, \theta_2) + \cos^2\gamma_2 \tag{4.865}$$

The parameter ξ must satisfy the conditions: $\xi_0 \le \xi \le 1$, where $\xi_0 = 0.084776$. The above conditions assure that the structure is contained in the half-plane $y \ge 0$. The condition $\xi = \xi_0$ refers to the case of $\theta_2 = \frac{\pi}{2}$. Thus, ξ_0 is the solution to the system of equations

$$z\left(\frac{\pi}{2}, \gamma_2\right) = 1, \qquad \xi_0 s\left(\frac{\pi}{2}, \gamma_2\right) = 1 \tag{4.866}$$

Now, we can construct the family of solutions indexed by ξ. The illustrative examples referring to $\xi = \xi_0$, $\xi = 0.25,\ 0.5,\ 0.75,\ 0.95,\ 1.00$ are shown in Fig. 4.187.

If $\xi = 1$, we come back to the solution discussed in Sect. 4.5.3. The characteristic parameters of the exact solutions for greater number of the values of ξ from ξ_0 to 1.00 are set up in Table 4.9.

As expected, the volume of the optimal structures increases monotonically for growing ξ. The height y^D as well as x^D and the magnitude of the force $N = F_x = -Q_x$ also increase monotonically. The last solution for $\xi = 1$ coincides with the exact solution of Sect. 4.5.3; then $\gamma_2 = \frac{\pi}{2}$, $\theta_2 = 0$, $\theta_1 = \frac{\pi}{2}$, $x^D = d$, $y^D = d$, $F_x = P$, $2V = 2\pi V_0$.

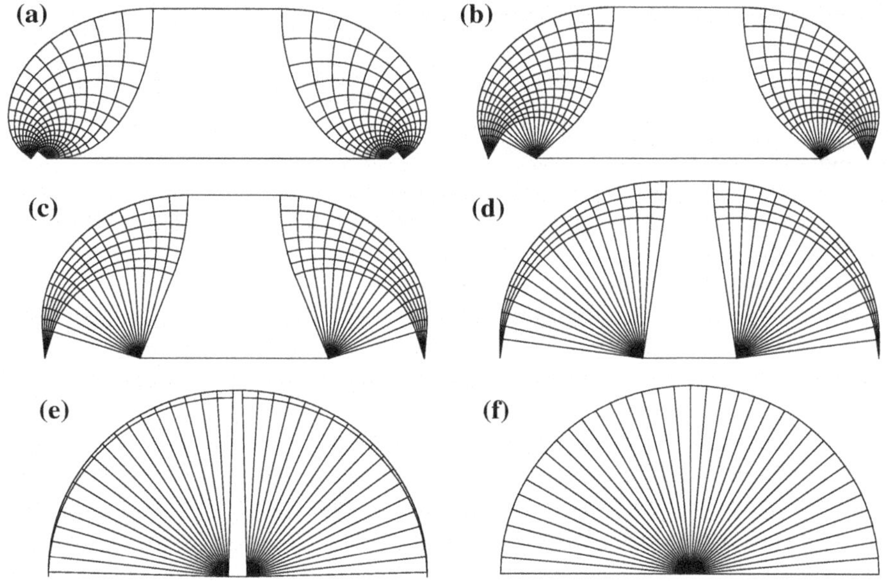

Fig. 4.187 The selected analytical solutions corresponding to $\xi = 0.0847757$, 0.25, 0.5, 0.75, 0.95 and 1

Table 4.9 The main characteristics of the selected optimal solutions of problem posed in Fig. 4.183

ξ	γ_2	θ_2	θ_1	x^D/d	y^D/d	N/P	$2V/V_0$
0.084776	0.89313	1.57080	2.46393	0.65258	0.79928	0.10606	1.53382
0.1	0.90882	1.46115	2.36997	0.64719	0.79509	0.12577	1.72639
0.2	1.01125	0.98932	2.00057	0.64617	0.79200	0.25252	2.75180
0.25	1.06052	0.83619	1.89672	0.65751	0.79915	0.31283	3.15911
0.3	1.10772	0.71246	1.82018	0.67295	0.80935	0.37067	3.51877
0.4	1.19509	0.52276	1.71785	0.71129	0.83513	0.47897	4.13112
0.5	1.27327	0.38278	1.65605	0.75530	0.86433	0.57848	4.63821
0.6	1.34342	0.27430	1.61772	0.80226	0.89435	0.67088	5.06793
0.7	1.40699	0.18706	1.59404	0.85081	0.92370	0.75782	5.43758
0.75	1.43673	0.14933	1.58606	0.87544	0.93783	0.79972	5.60359
0.8	1.46531	0.11476	1.58008	0.90021	0.95148	0.84079	5.75853
0.9	1.51958	0.05334	1.57291	0.95002	0.97707	0.92112	6.03857
0.95	1.54551	0.02580	1.57131	0.97500	0.98889	0.96067	6.16503
0.99	1.56579	0.00503	1.57082	0.99501	0.99785	0.99213	6.26020
1	1.57080	0.00000	1.57080	1.00000	1.00000	1.00000	6.28319

4.17.2 Case of Three or More Forces

Let us consider the problem of optimal transmitting of n vertical forces of equal magnitude P to the movable supports (rollers); the points of application of the forces lie along the line linking the supports; the feasible domain is the half-plane over the line of supports, see Fig. 4.188. We assume $\sigma_T = \sigma_C = \sigma_0$. In each case to be discussed the points of application of the forces divide the span of length l into $n + 1$ equal sections of length $a = l/(n + 1)$. The numerical results given below have been obtained by the ground structure method.

If $n = 3$, the points of applications of the forces introduce the division into four sections of equal length $a = l/4$. We assume $P = qa$ and $V_0 = ql^2/\sigma_0$, q being a given intensity of a uniform vertical load. The optimal structure is similar to that of Fig. 4.187 but now a circular fan over the central force appears, see Fig. 4.189a. The volume of the optimal structure equals $V = 0.9230371\,V_0$.

If $n = 5$, the points of applications of the forces introduce the division into six sections of equal length $a = l/6$. We assume $P = qa$ and $V_0 = ql^2/\sigma_0$. The optimal structure is shown in Fig. 4.189b. The volume of the optimal structure equals $V = 0.95079024\,V_0$.

If $n = 299$, the points of applications of the forces introduce the division into 300 sections of equal length $a = l/300$. We assume $P = qa$ and $V_0 = ql^2/\sigma_0$. The optimal structure is shown in Fig. 4.189c. The volume of the optimal structure equals $V = 0.9766335\,V_0$.

The ribs in compression stiffening the upper edges of the structures of Fig. 4.189a, b, c are prismatic, while the ribs in tension going along the line of supports are of varying cross sections. In case of $q =$ const , this variation becomes continuous. The analytical solution of the latter problem is still unknown.

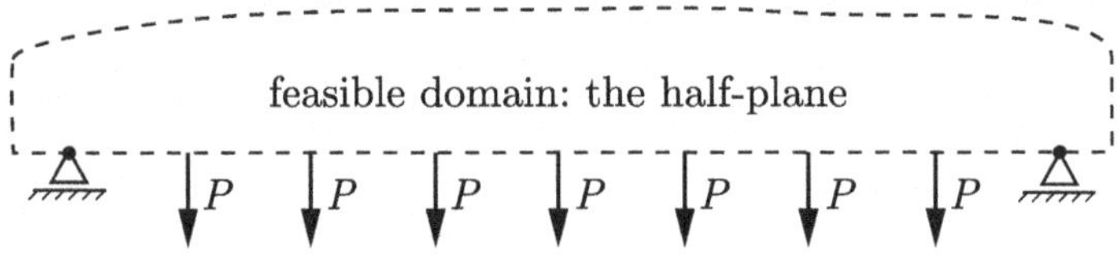

Fig. 4.188 The problem setting for $n = 7$

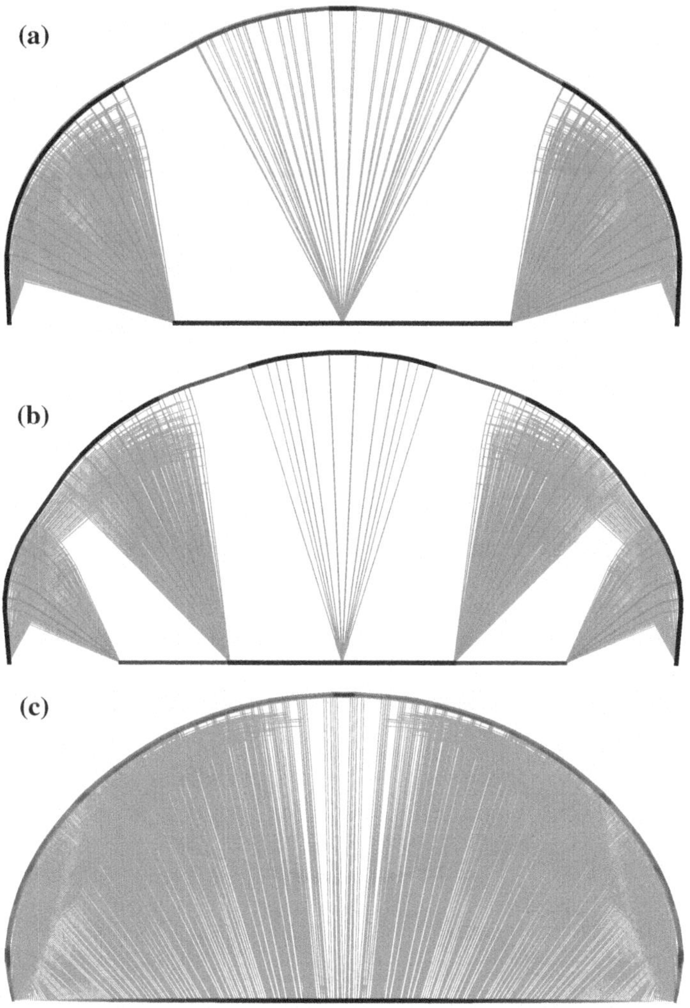

Fig. 4.189 Numerical solutions to the problem of Fig. 4.188 for (a) $n = 3$, (b) $n = 5$ and (c) $n = 299$

4.18 Optimal Simply Supported Girders Subjected to Symmetrically Located Vertical Loads. Full-Plane Designs

One can note that the full-plane solutions to the optimum design problems discussed in Sects. 4.17.1 and 4.17.2 of transmitting the forces to the roller supports coincide with those for girders designed in the whole plane and supported on fixed supports, as already mentioned in Remark 4.16.1. Thus, in particular, the solutions reported in Fig. 4.179 are correct in case of one support (or two of them) being a roller, as in Fig. 4.188.

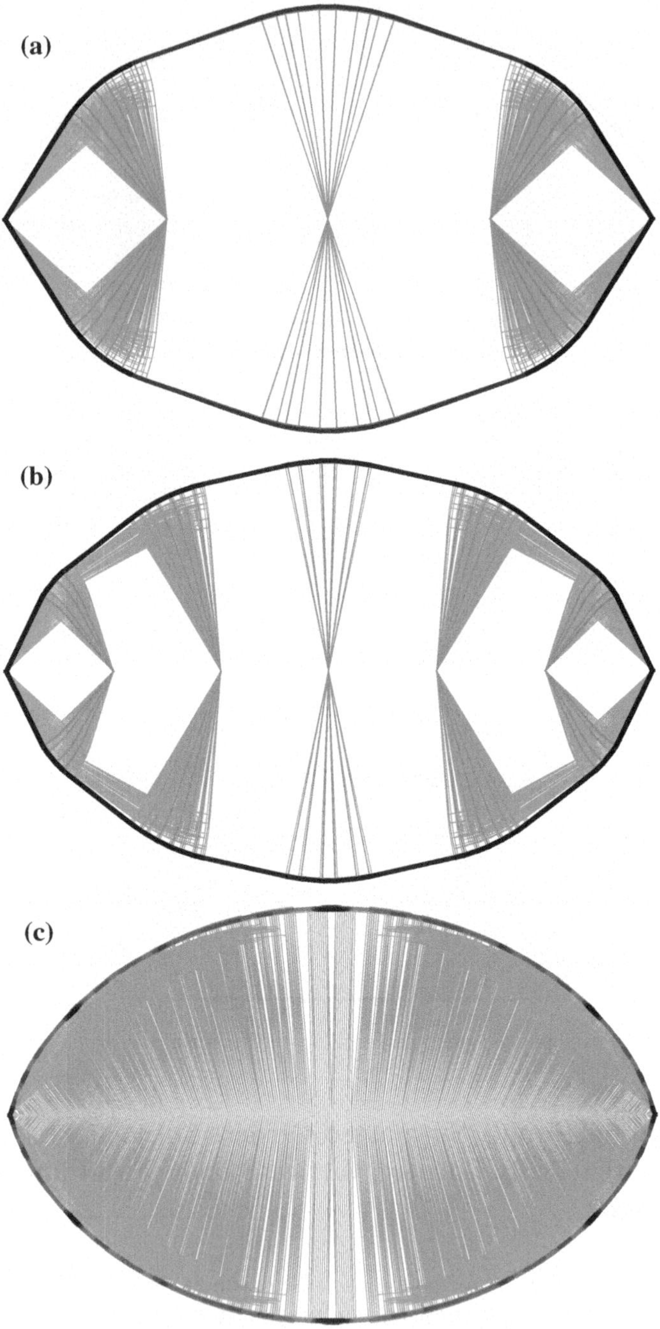

Fig. 4.190 Numerical solutions to the problem of Fig. 4.188, but the feasible domain is the full plane. Cases of **a** $n = 3$; **b** $n = 5$ and **c** $n = 299$

Let us consider the problem of Fig. 4.188, but now the feasible domain is the full plane. If $n = 3$, the points of applications of the forces introduce the division into four sections of equal length $a = l/4$. We assume $P = qa$ and $V_0 = ql^2/\sigma_0$, q being a given intensity of a uniform vertical load. The optimal structure is shown in Fig. 4.190a. The volume of the optimal structure equals $V = 0.73012465\,V_0$.

If $n = 5$, the points of applications of the forces introduce the division into six sections of equal length $a = l/6$. We assume $P = qa$ and $V_0 = ql^2/\sigma_0$. The optimal structure is shown in Fig. 4.190b. The volume of the optimal structure equals $V = 0.745159165\,V_0$, with $V_0 = ql^2/\sigma_0$.

If $n = 299$, the points of applications of the forces introduce the division into 300 sections of equal length $a = l/300$. We assume $P = qa$ and $V_0 = ql^2/\sigma_0$. The optimal structure is shown in Fig. 4.190c. The volume of the optimal structure equals $V = 0.757188131\,V_0$.

4.19 Bibliographical Comments

Curvilinear parameterizations of deformable bodies are discussed in Green and Zerna (1968) and in Fung (1965), while the orthogonal parameterization of shells is cleared up in Novozhilov (1970). The formulae for Christoffel symbols (4.11) and for Gauss curvature (4.18) are repeated after Mazurkiewicz (2004); the formulae (4.17) can be found in Hemp (1973) and Lewiński et al. (1994a, b).

Section 4.1 is inspired by the lecture given in Hemp (1973). In particular, the internal forces (4.45) measured per unit angles have been for the first time introduced by Hemp (1973). That is why we call them Hemp forces.

The theory of Hencky nets has already been elaborated in the classical papers on plasticity by Geiringer (1937) and Hill (1950), cf. Chakrabarty (2006). The original paper on Hencky nets has been written by Carathéodory and Schmidt (1923).

The mathematical methods of constructing the Michell nets along with the adjoint displacement fields have been put forward by A.S.L. Chan (1960), H.S.Y. Chan (1963, 1964, 1966, 1967, 1975), Hemp (1973, 1975), Melchers (2005) and Dewhurst et al. (2009).

The auxiliary displacement fields u^0, v^0 given by (4.118a) satisfy Eqs. (4.119). This paves the way for the systematic construction of the displacement fields u, v, by solving (4.119). This method of construction of the displacement fields u, v has been for the first time put forward in Lewiński et al. (1994a, b).

The functions $G_n(\alpha, \beta)$, $F_n(\alpha, \beta)$ defined in Sect. 4.2 are expressed by the Bessel functions I_n and by the Lommel functions U_n, see Watson (1966). They have been introduced in papers by A.S.L. Chan and H.S.Y. Chan, see Chan (1967). The notation G_n, F_n has been proposed in Lewiński et al. (1994a).

The Riemann formula (4.132) can be found in Geiringer (1937) and Hemp (1973), where also its derivation is outlined. The integration results (4.148)–(4.157) have been reported in Chan (1967), Lewiński et al. (1994a, b) and Graczykowski and Lewiński (2006a, b, 2007a, b).

The equations of equilibrium derived in Sect. 4.3 is compatible with the static analysis performed by Hemp (1973) and with Graczykowski and Lewiński (2007a, b).

The kinematics of the nets formed by one family of straight lines (Sect. 4.4) has been elaborated by Hemp (1973).

The simplest and the most important optimum designs composed of straight lines and circular arcs have been discovered by Michell (1904) and then extended by Hemp (1973).

The Hencky net for the exterior of a circle has been found in the same year 1923 by Prandtl and Carathéodory and Schmidt. The optimal cantilever supported on the circular edge for the case of $\sigma_C = \sigma_T$ has been constructed by Michell (1904) using the kinematic method and extended by Hu and Shield (1961) for the case of arbitrary force direction, cf. Prager (1985). The statics problem of this cantilever, discussed in Sect. 4.6.4, has been solved by Graczykowski and Lewiński (2005), where the proof of the duality gap being zero is given i.e. the possible gap between the solutions to the kinematic and static problems. Dewhurst and Srithongchai (2005) discussed the problem of extension of these results to the case of $\sigma_C \neq \sigma_T$. The results in Graczykowski and Lewiński (2005) concerning the interior of the circle are, in general, not correct; cf. Sect. 4.7.

The problem of Sect. 4.8 concerning the optimal transmitting a point load to an interval of a straight support has been the subject of the paper by Graczykowski and Lewiński (2006a, b). Unfortunately, in this paper, some nets are shown characterized by the angles θ_1, θ_2 (see Fig. 4.54) greater than $\pi/2$. This is impossible; only these layouts are correct for which these angles are smaller or equal $\pi/2$. Therefore, the solutions discussed in Sect. 4.8 refer to the design domains of maximal dimensions shown in Fig. 4.57.

The Michell structures supported on the edge of a polygonal shape have been found in Lewiński and Rozvany (2007); these constructions have been discussed in Sect. 4.9.

The cantilevers supported on the boundary of a square, discussed in Sect. 4.10, were proposed by G. Rozvany and then analytically constructed in Lewiński and Rozvany (2008b). The numerical solution reported in Sect. 4.10.4 confirming these results is published for the first time.

Bounding the feasible domain by a straight line changes the Michell solutions substantially. It is H.S.Y. Chan who first showed how to incorporate such a bound into the kinematic construction of the Michell truss, see Chan (1967). In this paper, a Hencky net tangent to the straight boundary has been mathematically constructed. This method is not straightforward; it requires solving first an integral equation of convolution type. This solution is constructed by using the Laplace transform. The method turns out to be effective due to the exact formulae for the Laplace transform of the special functions F_n, G_n being at our disposal, see Eqs. (1.56), (1.57) and (1.58), (1.59). H.S.Y. Chan has not only proposed the method but also has delivered these exact formulae, thus making the method effective and elegant. In Sect. 4.11.1, all the known formulae of this method are collected. This method is based on the Riemann formula, but in modified version, see Sect. 4.11.2.

The constructions of Michell nets put forward in Sect. 4.12, referring to the trapezoidal design domains, follow the derivations proposed in papers by Graczykowski and Lewiński (2006a, b, 2007a, b). Unfortunately, the results of these papers are valid only if the angles of fans are smaller than the right angle: $\theta_1 \leq \frac{\pi}{2}, \theta_2 \leq \frac{\pi}{2}$, cf. Fig. 4.79.

The problems of Sect. 4.13 have been solved in Lewiński et al. (1994a, b).

The Michell structures designed within the L-shaped domains have been found for the first time in Lewiński and Rozvany (2008a), but, unfortunately, some results of this paper turned out to include errors. All these errors have been detected; the corrections have been published in Lewiński et al. (2013). The constructions of Sect. 4.14 follow the lines of derivations of the mentioned papers.

Section 4.15.1 refers to the three-force problem in the plane, the forces being self-equilibrated. The first paper on this topic was the report by H.S.Y. Chan (1966). This report was kept as confidential till 2006. Professor H.S.Y. Chan has generously sent this report to the first author of the present book on 23rd November 2006, forty years after it was written. This report was the basis of the publication by Sokół and Lewiński (2010); the results of this work are reported in Sect. 4.15.1. Unfortunately, the hitherto available literature on Michell trusses as well as this latter paper do not exhaust the topic and also this book does not deliver all possible layouts of the three-force problem. This problem involves four parameters; hence, it is difficult to cover all cases of optimal designs. For instance, in some designs an additional bar appears, linking the point of application of the load with the main part of the structure, see Fig. 4.160a, repeated after Sokół and Lewiński (2017).

Considering the problem of Sect. 4.15.2 has been suggested to the authors by George Rozvany; the solutions reported there were found in Sokół and Lewiński (2016a, b). The feasible domain may be the half-plane or the whole plane. Professor Rozvany suggested to concentrate attention on the half-plane solutions. Some of them have been constructed analytically, but some of them were very complicated and were only found numerically.

The problem of transmitting vertical forces to two pin supports (or unmovable hinges) within the half-plane over the line of supports is called in the present book the optimal girder problem, cf. Sect. 4.16.1. This problem was first analysed (mainly numerically) by McConnel (1974) and just there the correct optimal shapes of the girders have been predicted. The most important is the simplest case of two symmetrically applied forces. The analytical solution to this problem has been found in Sokół and Lewiński (2010); this theoretical construction is presented in Sect. 4.16.1. An approximate solution is put forward in Rojas et al. (2015). The numerical solution to the bridge problem of transmitting a uniform load to two pinned supports (see Fig. 4.178c) has been found by the second author of the present book. This layout has been discovered for the first time by Tyas et al. (2011). For certain values of σ_T/σ_C, the analytical solutions have been constructed by Pichugin and Gilbert (2011) and Pichugin et al. (2011, 2012).

The problem of optimal transmitting the vertical forces to two rollers (i.e. the hinge supports with a possible horizontal displacement) by a girder lying within the half-plane over the line of supports has been analytically solved in Sokół and Lewiński (2011a, b); these solutions have been numerically confirmed in Sect. 4.17.1.

The numerical solutions corresponding to other vertical force systems are presented in Sect. 4.17.2 for the first time.

The full-plane designs discussed in Sects. 4.16.3 and 4.18 are the same, irrespective of the supports being fixed or movable, which has been first noted by Sokół and Rozvany (2012).

The problem of the optimal girders has been studied in many books on structural mechanics in the past, but nowadays this problem comes back, discussed in the papers inspired by Michell's theory; the remarkable geometric properties of the girders made of finite number of bars have been discovered by Mazurek (2012), Mazurek et al. (2011, 2015).

A natural modification of the problem of the optimal girder is based on the assumption of the load being uniformly distributed in both directions, the girder being put on the infinite number of pin supports, equally spaced. This problem has been set and solved by Pichugin et al. (2015) for various ratios $\frac{\sigma_T}{\sigma_C}$. These new spectacular solutions are shown in Fig. 5 therein.

References

Carathéodory, C., & Schmidt, E. (1923). Über die Hencky-Prandtlschen Kurven. *Zeitschrift für Angewandte Mathematik und Mechanik*, *3*, 468–475.

Chakrabarty, J. (2006). *Theory of plasticity* (3rd ed.). Oxford, UK: Butterworth-Heinemann.

Chan, A. S. L. (1960). *The design of Michell optimum structures*. Cranfield: The College of Aeronautics (Report 142).

Chan, H. S. Y. (1963). *Optimum Michell frameworks for three parallel forces*. Cranfield: The College of Aeronautics (Report AERO 167).

Chan, H. S. Y. (1964). *Tabulation of some layouts and virtual displacement fields in the theory of Michell optimum structures*. Cranfield: The College of Aeronautics (CoA Note Aero 161).

Chan, H. S. Y. (1966). *Minimum weight cantilever frames with specified reactions*. University of Oxford, Department of Engineering Science Laboratory, Parks Road, Oxford, June 1966, No. 1, 010.66, p. 11 with 4 figures. (the paper cited for the first time in Sokół and Lewiński (2010)).

Chan, H. S. Y. (1967). Half-plane slip-line fields and Michell structures. *The Quarterly Journal of Mechanics and Applied Mathematics*, *20*, 453–469.

Chan, H. S. Y. (1975). Symmetric plane frameworks of least weight. In A. Sawczuk & Z. Mróz (Eds.), *Optimization in structural design* (pp. 313–326). Berlin: Springer.

Cox, H. L. (1965). *The design of structures of least weight*. Oxford: Pergamon Press.

Darwich, W., Gilbert, M., & Tyas, A. (2010). Optimum structure to carry a uniform load between pinned supports. *Structural and Multidisciplinary Optimization*, *42*, 33–42.

Dewhurst, P., & Srithongchai, S. (2005). An investigaton of minimum-weight dual-material symmetrically loaded wheels and torsion arms. *Journal of Applied Mechanics, Transactions ASME*, *72*, 196–202.

Dewhurst, P., Fang, N., & Srithongchai, S. (2009). A general boundary approach to the construction of Michell truss structures. *Structural and Multidisciplinary Optimization*, *39*, 373–384.

Fung, Y. C. (1965). *Foundations of solid mechanics*. New Jersey: Prentice Hall.

Geiringer, H. (1937). *Fondements mathématiques de la theorie des corps plastiques isotropes* (Vol. 86). Mémorial des sciences mathématiques. Paris: Gauthier-Villard.

Graczykowski, C., & Lewiński, T. (2005). The lightest plane structures of a bounded stress level transmitting a point load to a circular support. *Control and Cybernetics*, *34*(1), 227–253.

Graczykowski, C., & Lewiński, T. (2006a). Michell cantilevers constructed within trapezoidal domains - Part I: Geometry of Hencky nets. *Structural and Multidisciplinary Optimization*, *32*(5), 347–368.

Graczykowski, C., & Lewiński, T. (2006b). Michell cantilevers constructed within trapezoidal domains - Part II: Virtual displacement fields. *Structural and Multidisciplinary Optimization, 32*(6), 463–471.

Graczykowski, C., & Lewiński, T. (2007a). Michell cantilevers constructed within trapezoidal domains - Part III: Force fields. *Structural and Multidisciplinary Optimization, 33*(1), 27–46.

Graczykowski, C., & Lewiński, T. (2007b). Michell cantilevers constructed within trapezoidal domains - Part IV: Complete exact solutions of selected optimal designs and their approximations by trusses of finite number of joints. *Structural and Multidisciplinary Optimization, 33*(2), 113–129.

Graczykowski, C., & Lewiński, T. (2010). Michell cantilevers constructed within a halfstrip. Tabulation of selected benchmark results. *Structural and Multidisciplinary Optimization, 42*(6), 869–877.

Green, A. E., & Zerna, W. (1968). *Theoretical elasticity*. Oxford: Clarendon Press.

Hemp, W. S. (1973). *Optimum structures*. Oxford: Clarendon Press.

Hemp, W. S. (1975). Michell framework for uniform load between fixed supports. *Engineering Optimization, 1*, 61–69.

Hill, R. (1950). *The mathematical theory of plasticity*. Oxford: Clarendon Press.

Hu, T.-C., & Shield, T. T. (1961). Minimum weight design of discs. *Zeitschrift für angewandte Mathematik und Physik, 12*, 414–433.

Jacot, B. P., & Mueller, C. T. (2017). A strain tensor method for three-dimensional Michell structures. *Structural and Multidisciplinary Optimization, 55*, 1819–1829.

Lewiński, T. (2004). Michell structures formed on surfaces of revolution. *Structural and Multidisciplinary Optimization, 28*(1), 20–30.

Lewiński, T., & Rozvany, G. I. N. (2007). Exact analytical solutions for some popular benchmark problems in topology optimization II: Three-sided polygonal supports. *Structural and Multidisciplinary Optimization, 33*(4–5), 337–349.

Lewiński, T., & Rozvany, G. I. N. (2008a). Exact analytical solutions for some popular benchmark problems in topology optimization III: L-shaped domains. *Structural and Multidisciplinary Optimization, 35*(2), 165–174.

Lewiński, T., & Rozvany, G. I. N. (2008b). Analytical benchmarks for topological optimization IV: Square-shaped line support. *Structural and Multidisciplinary Optimization, 36*(2), 143–158.

Lewiński, T., Zhou, M., & Rozvany, G. I. N. (1994a). Extended exact solutions for least-weight truss layouts - Part I: Cantilever with a horizontal axis of symmetry. *International Journal of Mechanical Sciences, 36*(5), 375–398.

Lewiński, T., Zhou, M., & Rozvany, G. I. N. (1994b). Extended exact solutions for least-weight truss layouts - Part II: Unsymmetric cantilevers. *International Journal of Mechanical Sciences, 36*(5), 399–419.

Lewiński, T., Rozvany, G. I. N., Sokół, T., & Bołbotowski, K. (2013). Exact analytical solutions for some popular benchmark problems in topology optimization III: L-shaped domains revisited. *Structural and Multidisciplinary Optimization, 47*, 937–942.

Melchers, R. E. (2005). On extending the range of Michell-like optimal topology structures. *Structural and Multidisciplinary Optimization, 29*, 85–92.

Mazurek, A. (2012). Geometrical aspects of optimum truss like structures for three-force problem. *Structural and Multidisciplinary Optimization, 45*, 21–32.

Mazurek, A., Baker, W., & Tort, C. (2011). Geometrical aspects of optimum truss like structures. *Structural and Multidisciplinary Optimization, 43*, 231–242.

Mazurek, A., Carrion, J., Beghini, A., & Baker, W. F. (2015). Minimum weight single span bridge obtained using graphic statics. In *Proceedings of International Association for Shell and Spatial Structures (IASS) Symposium, 17–20 August 2015* (11 pp.). Amsterdam, The Netherlands: Future Visions.

Mazurkiewicz, Z. E. (2004). *Thin elastic shells. Linear theory*. Warszawa (in Polish): Oficyna Wydawnicza PW.

McConnel, R. E. (1974). Least-weight frameworks for loads across span. *Journal of the Engineering Mechanics Division, 100*(5), 885–901.

Michell, A. G. M. (1904). The limits of economy of material in frame structures. *Philosophical Magazine*, *8*(47), 589–597.

Novozhilov, V. V. (1970). *Thin shell theory*. Groningen: Walters-Nordhoff.

Pichugin, A. V., & Gilbert, M. (2011). Optimum structure to carry a uniform load between pinned supports: Exact analytical solution. *Proceedings of the Royal Society A*, *467*(2128), 1101–1120.

Pichugin, A. V., Tyas, A., & Gilbert, M. (2011). Michell structure for a uniform load over multiple spans. In *9th World Congress on Structural and Multidisciplinary Optimization, CD ROM*.

Pichugin, A. V., Tyas, A., & Gilbert, M. (2012). On the optimality of Hemp's arch with vertical hangers. *Structural and Multidisciplinary Optimization*, *46*(1), 17–25.

Pichugin, A. V., Tyas, A., Gilbert, M., & He, L. (2015). Optimum structure for a uniform load over multiple spans. *Structural and Multidisciplinary Optimization*, *52*, 1041–1050.

Prager, W. (1978a). Optimal layout of trusses of finite number of joints. *Journal of the Mechanics and Physics of Solids*, *26*, 241–250.

Prager, W. (1978b). Nearly optimal design of trusses. *Computers and Structures*, *8*, 451–454.

Prager, W. (1985). Optimal design of trusses. In M. Save & W. Prager (Eds.), *Structural optimization* (Vol. 1, pp. 121–151). Optimality criteria. New York and London: Plenum Press.

Rojas, C. O., Bravo, J. C., & Espi, M. V. (2015). Near-optimal solutions for two point loads between two supports. *Structural and Multidisciplinary Optimization*, *52*, 663–675.

Rozvany, G. I. N. (2010). On symmetry and non-uniqueness in exact topology optimization. *Structural and Multidisciplinary Optimization*, *43*, 297–317.

Rozvany, G. I. N., & Sokół, T. (2012). Exact truss topology optimization: Allowance for support costs and different permissible stresses in tension and compression - Extensions of a classical solution by Michell. *Structural and Multidisciplinary Optimization*, *45*(3), 367–376.

Sokół, T. (2011). A 99 line code for discretized Michell truss optimization written in Mathematica. *Structural and Multidisciplinary Optimization*, *43*(2), 181–190.

Sokół, T. (2014). Multi-load truss topology optimization using the adaptive ground structure approach. In T. Łodygowski, J. Rakowski, & P. Litewka (Eds.), *Recent advances in computational mechanics* (pp. 9–16). London: CRC Press.

Sokół, T. (2016). A new adaptive ground structure method for multi-load spatial Michell structures. In M. Kleiber, T. Burczyski, K. Wilde, J. Gorski, K. Winkelmann, & T. Smakosz (Eds.), *Advances in Mechanics: Theoretical, Computational and Interdisciplinary Issues* (pp. 525–528). London: CRC Press.

Sokół, T. (2017). On the numerical approximation of Michell trusses and the improved ground structure method. In *Proceedings of 12th World Congress on Structural and Multidisciplinary Optimisation, 05–09 June 2017, Braunschweig, Germany*.

Sokół, T., & Lewiński, T. (2010). On the solution of the three forces problem and its application to optimal designing of a class of symmetric plane frameworks of least weight. *Structural and Multidisciplinary Optimization*, *42*(6), 835–853.

Sokół, T., & Lewiński, T. (2011a). Optimal design of a class of symmetric plane frameworks of least weight. *Structural and Multidisciplinary Optimization*, *44*(5), 729–734.

Sokół, T., & Lewiński, T. (2011b). On the three forces problem in truss topology optimization. Analytical and numerical solutions. In *9th World Congress on Structural and Multidisciplinary Optimization, Book of Abstracts and CD-ROM* (p. 76).

Sokół, T., & Rozvany, G. I. N. (2012). New analytical benchmarks for topology optimization and their implications. Part I: Bi-symmetric trusses with two point loads between supports. *Structural and Multidisciplinary Optimization*, *46*(4), 477–486.

Sokół, T., & Lewiński, T. (2016a). Simply supported Michell trusses generated by a lateral point load. *Structural and Multidisciplinary Optimization*, *54*(5), 1209–1224.

Sokół, T., & Lewiński, T. (2016b). Solution of the three forces problem in a case of two forces being mutually orthogonal. *Engineering Transactions*, *64*(4), 485–491.

Tyas, A., Pichugin, A.V., & Gilbert, M. (2011). Optimum structure to carry a uniform load between pinned supports: exact analytical solution. *Proceedings of the Royal Society A: Mathematical, Physical & Engineering Sciences*, *467*(2128), 1101–1120.

Watson, G. (1966). *Theory of Bessel functions*. Cambridge: Cambridge University Press.

Chapter 5
Selected Classes of Spatial Michell's Structures

The spatial Michell structures are solutions to the problem (3.85) in the kinematic setting and to the dual problem (3.95) in the stress-based setting. The optimality condition (3.84) referring to the spatial setting ($n = 3$) differs from the plane setting ($n = 2$), since in the plane problem, the number of the bounds is equal to the number of the adjoint principal strains and in some regimes the following equalities can be fulfilled $\varepsilon_{II} = -\frac{\sigma_0}{\sigma_C}$, $\varepsilon_I = \frac{\sigma_0}{\sigma_T}$, while in the spatial problem we may have $\varepsilon_{III} = -\frac{\sigma_0}{\sigma_C}$, $\varepsilon_I = \frac{\sigma_0}{\sigma_T}$, with ε_{II} attaining neither of the bounds. Thus, we expect that in many spatial problems the solutions will be either composed of fibrous membrane shells, or will be collections of planar structures not linked in the direction transverse to their planes. The well-known Michell's sphere as well as other optimal shells of revolution subject to the pure torsion load belong to the first class mentioned, see Sects. 5.2 and 5.3. The rotationally symmetric Hemp's structures (Sect. 5.1) belong to the second class.

The continuum grillages are treated in Sect. 5.4 as spatial Michell structures, carrying the out of plane loads. The minimum compliance problem reduces to the kinematic formulation (5.56) and to the dual problem (5.57) expressed in terms of the bending moments. These problems are in analogy with the problems (3.20) and (3.33) inferred from the minimum compliance problem of trusses. In contrast to spatial Michell's structures, the theory of continuum grillages is better developed, but the theoretical work is still in progress. The grillages discussed here have been selected from a rich family of available solutions. The continuum grillages are called Prager–Rozvany grillages to emphasize the essential contribution of these authors towards this topic, see Bibliographical remarks in Sect. 5.5.

T. Lewiński et al., *Michell Structures*,
https://doi.org/10.1007/978-3-319-95180-5_5

5.1 Girders Carrying the Vertical Loading- Hemp's Structures

The simplest Michell structures, i.e. those composed of straight bars and circular bars (cf. Sect. 4.5), upon rotation around their symmetry axes form a family of spatial structures. Interestingly enough, they are optimal despite the absence of a circumferential stiffening, even along the edges. These structures have been discovered by Hemp (1973). This section is aimed at recalling their construction.

5.1.1 Transmitting Vertical Forces to a Supporting Column

Consider the problem of transmitting several vertical forces to a vertical thin and rigid bar. The solution is the collection of planar two-bar trusses of geometry as in Fig. 4.15, see Fig. 5.1.

The angles γ_1, γ_2 are given by (4.228), where $\kappa = \sigma_T/\sigma_C$. If $\kappa = 3$, the angles are $\gamma_1 = \pi/6, \gamma_2 = \pi/3$. Each force P_i generates a two-bar truss in the plane determined by the position of the force, see Fig. 5.1a. If a uniform load of intensity p_z is applied around a circumference, then the optimal structure is composed of two conical shells of fibres lying along their generating lines.

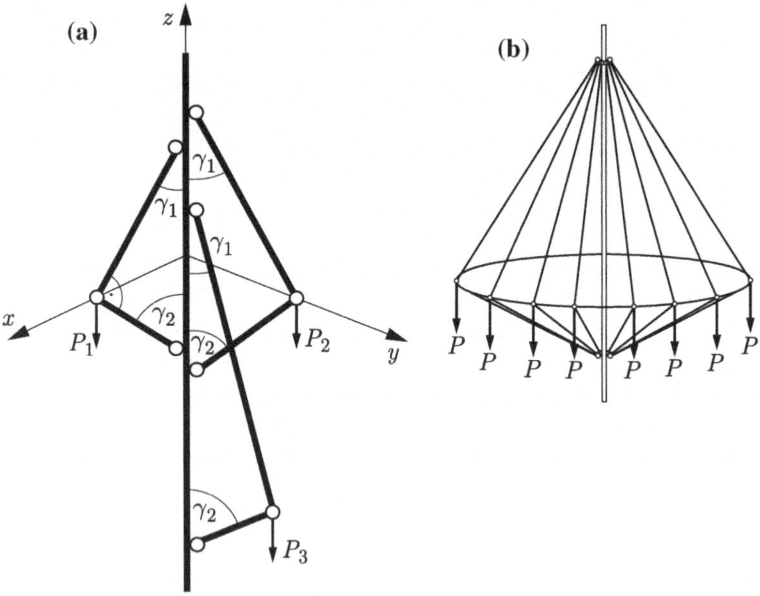

Fig. 5.1 The optimal structure transmitting the vertical forces P_1, P_2, P_3 to the vertical rigid bar; case of $\sigma_T = 3\sigma_C$ (**a**). The optimal structure carrying a uniform vertical load around a circle (**b**)

Let us check the necessary conditions of correctness of the above solutions. It is sufficient to check the condition (4.84). Let us rotate the field $\boldsymbol{v}$ given by (4.217), (4.220) around the axis $y_0 = x_2$, see Fig. 4.15. The circumferential strain at points outside the axis $x_1 = 0$ equals

$$\epsilon_{33} = \frac{v_1}{x_1} \tag{5.1}$$

while according to (4.217), (4.220) we have $v_1 = (1-\kappa)x_1$, where $\kappa = \sigma_T/\sigma_C$ and $\sigma_0 = \sigma_T$. Thus, $\epsilon_{33} = 1-\kappa$. It is seen that this strain component is bounded from both sides

$$-\frac{\sigma_0}{\sigma_C} < \epsilon_{33} < \frac{\sigma_0}{\sigma_T} \tag{5.2}$$

and now $\frac{\sigma_0}{\sigma_T} = 1$ and $\frac{\sigma_0}{\sigma_C} = -\kappa$. Moreover, $\epsilon_{13} = 0$, $\epsilon_{23} = 0$. Therefore, $\epsilon_{33} = \lambda_3(\epsilon)$ and the conditions (3.84) are fulfilled. Moreover, the inequalities (5.2) are not sharp, which means that the optimal structures (e.g. those of Fig. 5.1a, b) do not necessitate a circumferential stiffening.

Due to rotational symmetry, the structure of Fig. 5.1b does not require stiffening of the vertical column; its bending stiffness may be negligible. This structure resembles a hanging lamp, a chandelier, uniformly loaded around its ring.

5.1.2 *Transmitting a Vertical Force to Supports Around the Vertical Axis*

A. Rotating the Half-Plane Solution of Sect. 5.3

We consider the problem of optimal transmitting a vertical force of magnitude $2P$ to the roller supports placed at equal distances to the point of application of the force, see Fig. 5.2a. The solution is not unique since the force $2P$ can be decomposed into two forces: $2\alpha P$ and $2\beta P$, $\alpha + \beta = 1$; the solution is one parameter. We shall show that the solution of Fig. 5.2 is admissible.

To this end, we check the necessary conditions of optimality (3.84). We rotate the field (4.235), (4.238) around the axis z and construct the spatial strain field. The circumferential strain ε_{33} equals

$$\epsilon_{33} = \frac{u_{y_0}}{y_0}, \qquad y_0 = r\sin\theta \tag{5.3}$$

and

$$u_{y_0} = u_r \sin\theta + u_\theta \cos\theta \tag{5.4}$$

where u_r, u_θ are given by (4.235), (4.238). Thus, assuming $\sigma_0 = \sigma_T$, $\frac{\sigma_0}{\sigma_C} = \kappa$ we have

$$\epsilon_{33} = 1 - (1+\kappa)\theta\cot\theta \tag{5.5}$$

Let us consider the expressions

$$\epsilon_{33} + \frac{\sigma_0}{\sigma_C} = (1+\kappa)(1-\theta\cot\theta), \qquad \epsilon_{33} - \frac{\sigma_0}{\sigma_T} = -(1+\kappa)\theta\cot\theta \tag{5.6}$$

By virtue of the estimates

$$0 \le 1-\theta\cot\theta \le 1 \quad \text{if} \quad 0 \le \theta \le \frac{\pi}{2} \tag{5.7}$$

we have

$$-\frac{\sigma_0}{\sigma_C} \le \epsilon_{33} \le \frac{\sigma_0}{\sigma_T} \tag{5.8}$$

which means that the conditions (3.84) are fulfilled. This proves that the structure of Fig. 5.2a is one of optimal structures solving the problem. Its optimality can be confirmed by statical analysis. The equilibrium can be assured without additional stiffening.

If the whole circumference of the circle is a line support sliding in the radial direction, then the solution is a spherical dome linked with the point P by radial fibres, see Fig. 5.2b. The volume of the structure of Fig. 5.2b is the same as the planar structure of Fig. 4.19, hence equals V given by (4.246). The volume of the spherical dome equals V^F given by (4.243). It is worth noting that this result can be confirmed by a direct computation. We shall make use of the membrane theory of shells, see Sect. 2.11.2. We predict here $N^{11} = 0$, $N^{12} = 0$, $N^{22} = N^{22}(\varphi)$. In the spherical shell considered, we have $R_1 = R_2 = l$; $R_0 = l\sin\varphi$. Thus, Eqs. (1.151) reduce to

$$\frac{d}{d\varphi}(l\sin\varphi N^{22}) = 0, \qquad N^{22} + Rp_3 = 0 \tag{5.9}$$

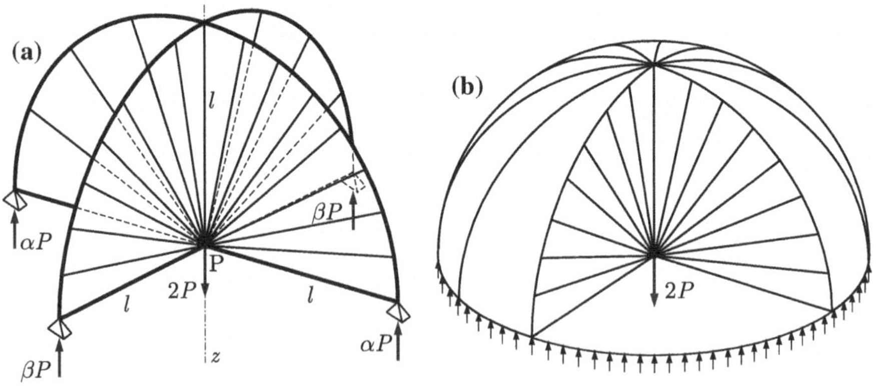

Fig. 5.2 The optimal girders transmitting a force to surrounding supports: **a** case of four forces; **b** case of a uniform load

The boundary condition has the form $N^{22}\left(\frac{\pi}{2}\right) = -p$, where $p = \frac{P}{\pi l}$. Hence

$$p_3(\varphi) = \frac{P}{\pi l^2 \sin\varphi}, \qquad N^{22}(\varphi) = \frac{-P}{\pi l \sin\varphi} \tag{5.10}$$

The volume fraction of the fibres in the spherical dome is

$$\upsilon_F(\varphi) = \frac{1}{\sigma_C}|N^{22}(\varphi)| \tag{5.11}$$

The volume of the fibres in the shell is expressed by the surface integral

$$V^F = \int_0^{2\pi}\int_0^{\frac{\pi}{2}} \upsilon_F(\varphi) l^2 \sin\varphi d\varphi d\vartheta \tag{5.12}$$

and is equal to $V^F = \frac{\pi}{\sigma_C} P l$. Thus, we have obtained the same formula for the volume, as in the planar case, see (4.243).

In the straight fibres, the axial forces appear of value $T = p_3(\varphi)$; the fibres are in tension. The volume fraction of straight fibres is expressed by

$$\upsilon_T(r,\varphi) = \frac{1}{\sigma_T}|T(\varphi)| \tag{5.13}$$

and their volume is expressed by the volume integral

$$V^T = \int_0^{2\pi}\int_0^{\frac{\pi}{2}}\int_0^{l} \upsilon_T(r,\varphi) l^2 \sin\varphi dr d\varphi d\vartheta \tag{5.14}$$

or $V^T = \frac{\pi}{\sigma_T} P l$; this result coincides with the result found in Eq. (4.245). The volume of the whole structure is given by (4.246).

B. Rotating the Full-Plane Solutions of Sects. 4.5.4 **and** 4.5.5

In the present section, we shall show that by rotating the structures of Sects. 4.5.4 and 4.5.5 one obtains the optimal structures referring to the design domain being the whole space.

The planar field of virtual displacements (u, v) given by (4.279)–(4.282) are subject to a rotation around the symmetry axis. This field determines the in-plane components of strain $\epsilon_{11}, \epsilon_{12}, \epsilon_{22}$ as well as out of plane components: $\epsilon_{13}, \epsilon_{23}, \epsilon_{33}$. The components $\epsilon_{13}, \epsilon_{23}$ vanish while $\epsilon_{33} = \frac{u_{y_0}}{y_0}$. We shall prove that in each subdomain the estimates (5.8) hold. We shall put $\sigma_0 = \sigma_T$, $\frac{\sigma_0}{\sigma_C} = \kappa$.

Consider the upper fan. Then, Eqs. (5.3), (5.4) hold but now u_r, u_θ are given by (4.279). We shall find ϵ_{33} in the form (5.5), which proves that (5.8) holds.

Consider the left quadrant of Fig. 4.23. Then, Eqs. (4.253) are satisfied and

$$u_{y_0} = u_y \cos\theta_0 + u_x \sin\theta_0, \qquad y_0 = \frac{x}{\cos\gamma}\sin(\theta_0+\gamma) \tag{5.15}$$

where $y = x\tan\gamma$. We obtain

$$\epsilon_{33} = \frac{1}{\sin(\gamma+\theta_0)}[\sin\theta_0\cos\gamma - \kappa\cos\theta_0\sin\gamma - \theta_0(1+\kappa)\cos(\gamma+\theta_0)] \tag{5.16}$$

Now, we examine the equality

$$(\epsilon_{33}-1)\sin(\gamma+\theta_0) = -(1+\kappa)[\sin\gamma\cos\theta_0(1-\theta_0\tan\theta_0)+\theta_0\cos\gamma\cos\theta_0] \tag{5.17}$$

We note that $\gamma+\theta_0<\pi$, $\theta_0\le\pi/4$, $0<\gamma\le\pi/2$. These inequalities imply that the r.h.s of (5.17) is negative, which proves that $\epsilon_{33}\le 1$.

Now, we examine the equality

$$(\epsilon_{33}+\kappa)\sin(\gamma+\theta_0) = (1+\kappa)\sin\theta_0[\cos\gamma(1-\theta_0\cot\theta_0)+\theta_0\sin\gamma] \tag{5.18}$$

Note that the r.h.s is nonnegative. Thus, $\epsilon_{33}\ge-\kappa$. We see that in the considered domain the condition (5.8) is satisfied.

Consider the left domain of hydrostatic compression, see Eq. (4.281). Then

$$u_{y_0} = u_\rho\cos(\theta_0+\psi) - u_\psi\sin(\theta_0+\psi), \qquad y_0 = \rho\cos(\theta_0+\psi) \tag{5.19}$$

We compute

$$\epsilon_{33} = -\kappa + (1+\kappa)\theta_0\tan(\theta_0+\psi) \tag{5.20}$$

and examine the equality

$$\epsilon_{33}+\kappa = (1+\kappa)\theta_0\tan(\theta_0+\psi) \tag{5.21}$$

We note that the r.h.s. is nonnegative, since $\theta_0+\psi\le\frac{\pi}{2}$, $0\le\theta_0\le\pi/4$. Consequently, $\epsilon_{33}\ge-\kappa$. We examine the equality

$$\epsilon_{33}-1 = (1+\kappa)[\theta_0\tan(\theta_0+\psi)-1] \tag{5.22}$$

and note that the r.h.s. is nonnegative. Hence, $\epsilon_{33}<1$, which confirms (5.8).

Last, consider the lower fan, see (4.282). We compute u_{y_0}, y_0 by Eqs. (5.3), (5.4), where now u_r, u_θ are given by (4.282). We find

$$\epsilon_{33} = -\kappa + (1+\kappa)\theta\cot\theta \tag{5.23}$$

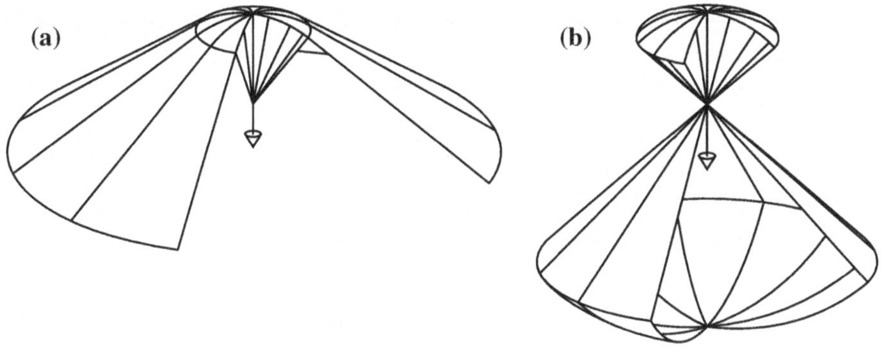

Fig. 5.3 Three-dimensional counterparts of Hemp's planar girders

Hence

$$\epsilon_{33} + \kappa > 0. \qquad \epsilon_{33} - 1 = (1 + \kappa)(\theta \cot \theta - 1) < 0 \tag{5.24}$$

or the strain ϵ_{33} satisfies (5.8).

Concluding, the spatial fields of displacements formed by rotation of the field (u, v) around the symmetry axis satisfies all the optimality conditions (3.84). The conditions of equilibrium do not demand any circumferential stiffening. We conclude that the 3D structures created by rotation of the considered 2D structures are optimal, see Fig. 5.3.

5.2 Michell's Sphere

5.2.1 Shells of Revolution Subjected to Torsion. Case of Arbitrary Meridian

We consider Michell's problem of minimization of the volume of a shell of revolution subject to a rotationally symmetric torsional load induced by uniformly distributed tangent tractions along the circumferential edges, see Fig. 1.11. We assume that the permissible stresses in tension and compression are equal $\sigma_C = \sigma_T = \sigma_0$. The volume of the optimal shell is given by (3.53), where now $d\Omega = dS$ is the elementary area of the middle surface of the shell and the volume fraction of fibres is expressed by (3.54); here $\breve{\tau} = \breve{N}$, $\breve{N}$ being the solution to the problem

$$\mathcal{Z}_2 = \min \left\{ \int_S ||\boldsymbol{N}||^{\star}_{spec} dS \mid \boldsymbol{N} \in \Sigma(S) \right\} \tag{5.25}$$

$\Sigma(S)$ being the set of statically admissible stress resultants $\boldsymbol{N} = (N^{ij})$.

In the considered pure torsion problem, the set $\Sigma(S)$ is a one element set, since only one tensor field $\tilde{N}^{ij}$ of physical components N^{ij} referred to the basis $\boldsymbol{e}_i$ shown in Fig. 1.9

$$N^{11} = 0, \qquad N^{12} = \frac{M}{2\pi[R_0(\varphi)]^2}, \qquad N^{22} = 0 \tag{5.26}$$

satisfies all equilibrium equations as well as the natural boundary conditions, see (1.163). Taking into account $dS = R_0(\varphi)R_2(\varphi)d\vartheta d\varphi$, we express the volume by

$$V_{\min} = \frac{1}{\sigma_0}\int_0^{2\pi}\int_{\varphi_U}^{\varphi_L} 2|N^{12}(\varphi)|R_0(\varphi)R_2(\varphi)d\varphi d\vartheta \tag{5.27}$$

since $||\boldsymbol{N}||^{\star}_{spec} = 2|N^{12}|$ and the volume fraction of fibres is equal to

$$v = \frac{2}{\sigma_0}|N^{12}(\varphi)| \tag{5.28}$$

Substitution of (5.26) into (5.27) and integration over ϑ gives

$$V_{\min} = \frac{2M}{\sigma_0}\int_{\varphi_U}^{\varphi_L}\frac{R_2(\varphi)}{R_0(\varphi)}d\varphi \tag{5.29}$$

and

$$v = \frac{M}{\pi\sigma_0[R_0(\varphi)]^2} \tag{5.30}$$

Let us check the result (5.29) by the kinematic method, by solving the problem (3.19) referring to the membrane shell. Our task is construction of the displacement field $\boldsymbol{v}(\vartheta, \varphi)$ such that the tensor $\boldsymbol{\varepsilon}(\boldsymbol{v})$ is collinear with the tensor of physical components (5.26). We guess that $\boldsymbol{v} = (u(\varphi), 0, 0)$ and choose $u(\varphi)$ such that

$$\epsilon_{11} = 0, \qquad \epsilon_{12} = 1, \qquad \epsilon_{22} = 0 \tag{5.31}$$

where ϵ_{ij} are given by (1.148). By using (1.147), we rearrange the Eq. (1.148) to the form

$$2\epsilon_{12} = \frac{1}{R_0}\frac{\partial v}{\partial\vartheta} + \frac{R_0}{R_2}\frac{\partial}{\partial\varphi}\left(\frac{u}{R_0}\right) \tag{5.32}$$

Thus, the field u satisfies

$$\frac{1}{2}\frac{R_0}{R_2}\frac{d}{d\varphi}\left(\frac{u}{R_0}\right) = 1 \tag{5.33}$$

hence

$$u = 2R_0\int\frac{R_2}{R_0}d\varphi \tag{5.34}$$

The angle of rotation around the axis x_3 equals $\psi = \dfrac{u}{R_0}$, hence

$$\psi = 2 \int \frac{R_2}{R_0} d\varphi \tag{5.35}$$

and the mutual twist of both the rings $\varphi = \varphi_U$ and $\varphi = \varphi_L$ is given by

$$\Delta\psi = 2 \int_{\varphi_U}^{\varphi_L} \frac{R_2(\varphi)}{R_0(\varphi)} d\varphi \tag{5.36}$$

and the virtual work equals

$$\mathcal{L}(\boldsymbol{v}) = M\Delta\psi \tag{5.37}$$

This equation confirms that $V_{\min} = \dfrac{\mathcal{L}(\boldsymbol{v})}{\sigma_0}$, which proves that the results (5.29), (5.30) are correct. The optimal shell is built of fibres following trajectories of the stress resultant field (5.26); the volume fraction of fibres is given by (5.30). For surfaces such that $R_2 = \infty$, the quantity $R_2 d\varphi$ should be replaced by ds_2, ds_2 being elementary length of the meridian.

5.2.2 *Optimization of the Meridian*

Let us introduce the set of functions

$$\mathcal{V}_R = \left\{ f \in C^1[\varphi_U, \varphi_L] \mid f(\varphi_U) = r_U,\ f(\varphi_L) = r_L \right\}$$

Note that $R_0 \in \mathcal{V}_R$, since R_0 measures the distance $|A_0 A|$, see Fig. 1.8 in Chap. 1. Let us change the notation

$$x_3 = x, \quad R_0(\varphi(x)) = f(x), \quad x_3(\varphi_L) = a, \quad x_3(\varphi_U) = b$$

and note that the admissible functions f belong to the set

$$\mathcal{V}_1 = \left\{ f \in C^1[a, b] \mid f(a) = r_L,\ f(b) = r_U \right\}$$

We rewrite (5.29) as

$$V_{\min} = \frac{2M}{\sigma_0} \int_a^b \frac{1}{f(x)} \sqrt{1 + \left(\frac{df}{dx}\right)^2} dx \tag{5.38}$$

The volume of the shell is now explicitly expressed in terms of the function f representing the shape of the meridian. Our aim is to find such a function $f \in \mathcal{V}_1$ which minimizes $V_{\min}$. The Euler equation for the functional

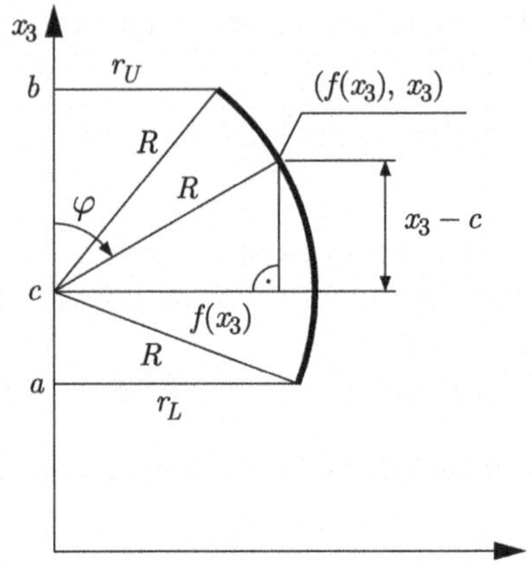

Fig. 5.4 Optimal shape of the meridian: a circular arc

$$f \to \int_a^b \frac{1}{f(x)} \sqrt{1 + \left(\frac{df}{dx}\right)^2} dx$$

assumes the form

$$\frac{d^2}{dx^2} f^2(x) + 2 = 0 \tag{5.39}$$

hence

$$f^2(x_3) + (x_3 - c)^2 = d^2 + c^2 \tag{5.40}$$

and c, d are constants. Let us note that the points $(f(x_3), x_3)$ lie on a circle of the origin $(0, c)$ and radius $R = \sqrt{d^2 + c^2}$, see Fig. 5.4.

We conclude that the lightest shell of revolution has a spherical shape. Thus, $R_1 = R_2 = R$, $R_0 = R \sin \varphi$. By (5.30) we have

$$\upsilon = \frac{M}{\pi \sigma_0 R^2 \sin^2 \varphi} \tag{5.41}$$

and the volume of the spherical shell equals $V_{\min} = V_0$ and by (5.29)

$$V_0 = \frac{2M}{\sigma_0} \int_{\varphi_U}^{\varphi_L} \frac{d\varphi}{\sin \varphi} \tag{5.42}$$

hence

$$V_0 = \frac{2M}{\sigma_0} \ln \left| \frac{\tan\left(\frac{\varphi_L}{2}\right)}{\tan\left(\frac{\varphi_U}{2}\right)} \right| \tag{5.43}$$

If $r_U = r_L$ and $\varphi_L = \pi - \varphi_U$, then

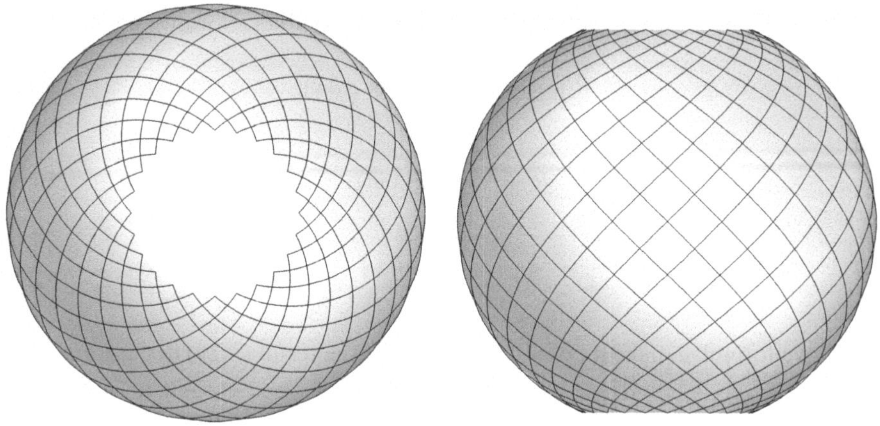

Fig. 5.5 The layout of fibres in the optimal shell

$$V_0 = \frac{4M}{\sigma_0} \ln \left| \cot \left(\frac{1}{2} \varphi_U \right) \right| \tag{5.44}$$

The geometry of fibres of the Michell sphere form rhumb lines (this term was used by Michell) intersecting the meridians at angles $\pm\pi/4$, see Fig. 5.5.

5.3 The Lightest Conical and Cylindrical Shells

5.3.1 The Lightest Conical Shell

We shall specify the result (5.29) to the case of the conical shell linking the given rings of radii r_U and r_L, cf. Fig. 5.6.

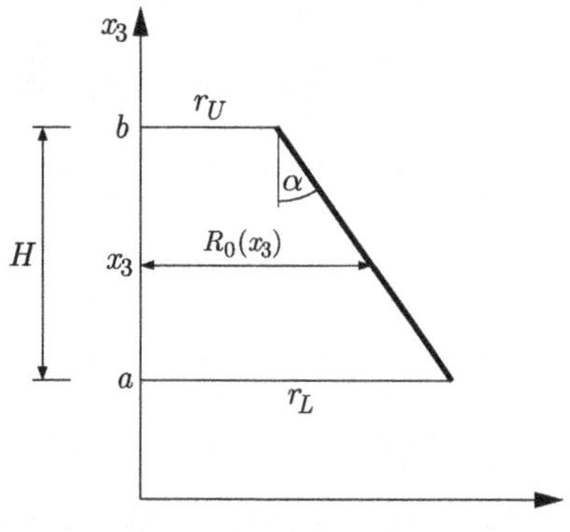

Fig. 5.6 The cross section through the conical shell

Now $\varphi = \text{const}$ and φ cannot parameterize the meridian. Now, we have

$$R_2 d\varphi = ds_2 = -\cos\alpha dx_3 \tag{5.45}$$

and

$$R_0 = r_U + \tilde{x}_3 \tan\alpha \tag{5.46}$$

where

$$\tilde{x}_3 = b - x_3, \qquad H = b - a, \qquad \tan\alpha = \frac{r_L - r_U}{H} \tag{5.47}$$

The above formulae make it possible to rearrange the result (5.29) to the form

$$V_1 = \frac{2M}{\sigma_0 \cos\alpha} \int_0^H \frac{d\tilde{x}_3}{r_U + (\tan\alpha)\tilde{x}_3} \tag{5.48}$$

The volume of the conical shell reads

$$V_1 = \frac{2M}{\sigma_0 \sin\alpha} \ln\left(\frac{r_L}{r_U}\right) \tag{5.49}$$

This result applies to both the cases
$r_L > r_U$, $\sin\alpha > 0$ and $r_L < r_U$, $\sin\alpha < 0$ while ceases to be true for $r_U = r_L$; the latter case will be discussed independently.

One can show that $V_1 > V_0$ or the conical shell is heavier than the spherical one. To be more specific let us fix $\varphi_U = \dfrac{\pi}{30}$. Let us parameterize the volumes by $\varphi_1 = \pi - \varphi_L$, $\varphi_1 \in [\pi/30, 29\pi/30]$. The plot of $\bar{V}_1 = \dfrac{\sigma_0 V_1}{M}$ lies over $\bar{V}_0 = \dfrac{\sigma_0 V_0}{M}$ and both the plots are tangent at $\varphi_1 = \dfrac{29\pi}{30}$, see Fig. 5.7.

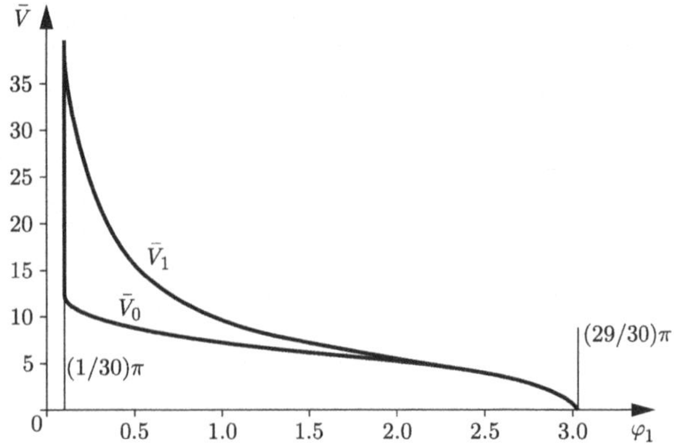

Fig. 5.7 The volumes of the optimal conical and spherical shells versus φ_1. Case of $\varphi_U = \pi/30$

5.3.2 The Lightest Cylindrical Shell

Consider the cylindrical shell of the radius $r_U = r_L = r$ and $\varphi_L = \pi - \varphi_U$, see Fig. 5.8. The result (5.29) assumes the form

$$V_2 = \frac{M}{\sigma_0}\bar{V}_2, \qquad \bar{V}_2 = 4\cot\varphi_U \tag{5.50}$$

The plot of $\bar{V}_2$ lies over $\bar{V}_0$ corresponding to the spherical shell, see Fig. 5.9. Let $\alpha = \pi/2 - \varphi_U$ and express the volumes in terms of α

$$\bar{V}_2 = 4\tan\alpha, \qquad \bar{V}_0 = 4\ln\left|\cos\left(\frac{\pi}{2} - \frac{\alpha}{2}\right)\right| \tag{5.51}$$

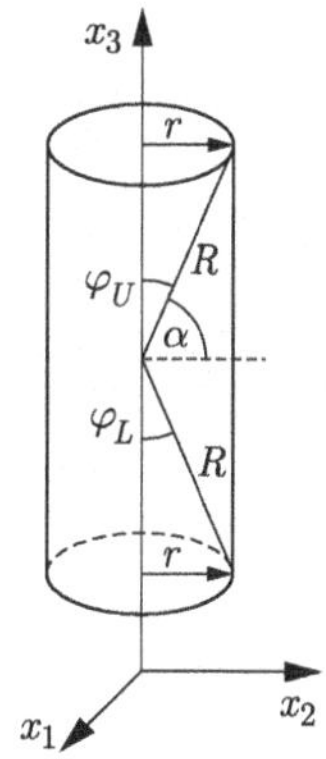

Fig. 5.8 Parameterization of a cylindrical shell

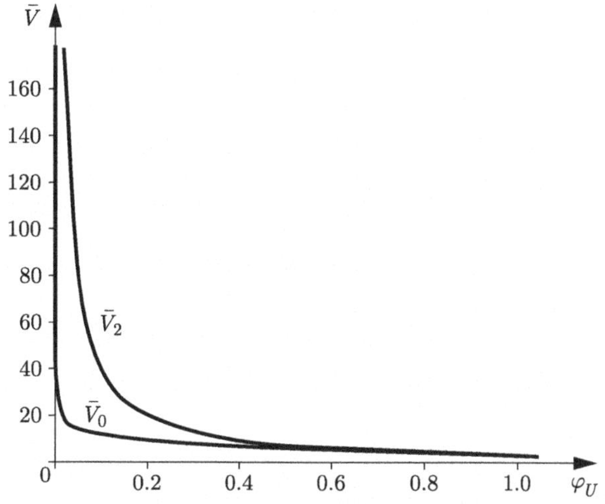

Fig. 5.9 The volumes of the spherical and cylindrical shells versus φ_U

Note that

$$\bar{V}_2 - \bar{V}_0 = \frac{2}{3}\alpha^3 + \frac{11}{30}\alpha^5 + \frac{211}{1260}\alpha^7 + \cdots \tag{5.52}$$

We see that the difference is of order $0(\alpha^3)$ for small α.

5.4 The Least Compliant Systems of Beams Forming a Continuum Layout. Prager–Rozvany Grillages

The problem of constructing the least compliant system of beams with the cost condition (2.73) involving the bending stiffnesses leads to the formula (2.89). This formula expresses the compliance of the system of beams of the given cost (2.73). The problem is reduced to the two mutually dual problems—the static problem (2.90) and the kinematic problem (2.95). If the load is concentrated at the nodes, these problems may be rearranged to the algebraic settings (2.115) and (2.122). This approach applies to the systems of a given and finite number of beams. However, adding new beams contributes to the decrease of the compliance and this process has no natural stop. Thus, there is a need to have a model of infinite number of beams, being infinitely thin and densely packed within the design domain. Such a model has been proposed in 1970s; in this book, we shall call it Prager–Rozvany grillage to honour William Prager's and George Rozvany's landmark contribution towards constituting a firm theoretical base for the problem of optimum design of flexural systems. The Prager–Rozvany approach is based on the two mutually dual auxiliary problems being continuum counterparts of the problems (2.115) and (2.122). The aim of this section is to show a natural construction of this continuum formulation. The point of departure is the kinematic setting (2.95). The quantities $\bar{q}_j$ involved represent displacements or angles of rotation at nodes where the beams are connected or at nodes, where the point loads are applied. Assume that the beams are densely packed and can be viewed as fibres; their deflection is modelled by the scalar function $v = v(x_1, x_2)$ in the given plane design domain Ω. We shall assume that the locking locus condition in (2.95) is satisfied along arbitrary direction, since the fibres may be placed in all possible directions. Let $\boldsymbol{n} = (n_1, n_2)$ be a unit vector in the plane. The change of the curvature in the direction $\boldsymbol{n}$ is expressed by

$$\varkappa_n = \boldsymbol{n} \cdot (\boldsymbol{\varkappa}(v)\boldsymbol{n}) \tag{5.53}$$

where $\boldsymbol{\varkappa}(v)$ is defined as in the thin plate theory, see (1.111). The locking locus condition in (2.95) imposed on the deflection of a system of beams is now replaced by the locking locus condition of the continuum grillage

$$\max_{\|\boldsymbol{n}\|=1} |\varkappa_n| \leq 1 \quad \text{a.e. in } \Omega$$

The above condition is equivalent to

$$\max_{1 \le i \le 2} |\lambda_i(\boldsymbol{\varkappa})| \le 1 \tag{5.54}$$

The left-hand side of this condition is just expressed by the spectral norm of the matrix $\boldsymbol{\varkappa}$, see (3.7). Thus, the condition (5.54) can be written in the form

$$||\boldsymbol{\varkappa}(v)||_{\text{spec}} \le 1 \quad \text{a.e. in } \Omega \tag{5.55}$$

Let $\mathcal{L}(v)$ be the virtual work of the load done on the deflection v. Let $\mathcal{U}_2(\Omega)$ be the set of C^1 class kinematically admissible deflections such that the condition (5.55) is satisfied at almost each point of the domain Ω. The continuum counterpart of the problem (2.95) has the form

$$\mathcal{Z}_1 = \sup\{\mathcal{L}(v) \mid v \in \mathcal{U}_2(\Omega)\} \tag{5.56}$$

The problem is well posed on the basis of the same arguments as those used to justify the problem (3.20). However, the solutions are not unique.

Proceeding as in Sect. 3.2 we may construct the problem dual to (5.56)

$$\mathcal{Z}_2 = \inf\left\{\int_\Omega ||\boldsymbol{M}||^\star_{\text{spec}} dx \mid \boldsymbol{M} \in \Sigma_p(\Omega)\right\} \tag{5.57}$$

where now $\Sigma_p(\Omega)$ is the set of the bending moment fields $\boldsymbol{M} = (M^{ij})$ satisfying the equation of equilibrium

$$\int_\Omega \boldsymbol{M} \cdot \boldsymbol{\varkappa}(v) dx = \mathcal{L}(v) \quad \forall\, v \in \mathcal{V}(\Omega) \tag{5.58}$$

where $\mathcal{V}(\Omega)$ represents the set of kinematically admissible virtual deflections being the scalar fields v in the given domain. In the flexural problem considered the norm involved in Eq. (5.57), dual to the spectral norm, is expressed as below

$$||\boldsymbol{M}||^\star_{\text{spec}} = |\lambda_1(\boldsymbol{M})| + |\lambda_2(\boldsymbol{M})| \tag{5.59}$$

cf. (3.27). Let us note that the problem (5.57) cannot be easy derived from the problems (2.90) and (2.115) concerning the system of finite number of beams.

The components of the solution $\breve{w}$, $\breve{\boldsymbol{\varkappa}} = \boldsymbol{\varkappa}(\breve{w})$ and $\breve{\boldsymbol{M}}$ are linked by the equations similar to (51_a) or

$$\begin{aligned}
&\lambda_i(\breve{\boldsymbol{M}}) > 0 \Rightarrow \lambda_i(\breve{\boldsymbol{\varkappa}}) = 1 \\
&\lambda_i(\breve{\boldsymbol{M}}) = 0 \Rightarrow |\lambda_i(\breve{\boldsymbol{\varkappa}})| \le 1 \\
&\lambda_i(\breve{\boldsymbol{M}}) < 0 \Rightarrow \lambda_i(\breve{\boldsymbol{\varkappa}}) = -1
\end{aligned} \tag{5.60}$$

Note that the alternatives (3.52) apply here, one should replace $(\boldsymbol{\tau}, \boldsymbol{\varepsilon})$ with $(\boldsymbol{M}, \boldsymbol{\varkappa})$. The trajectories of the fields $(\check{\boldsymbol{M}}, \check{\boldsymbol{\varkappa}})$ coincide.

There is a substantial difference between optimal layouts of truss-like Michell structures (being in-plane loaded) and grillage-like optimal structures (being transversely loaded) considered here: the optimal layouts of the beams (of finite stiffness or infinitely thin) consist always of straight beams, while the optimal fibres of in-plane loaded Michell structures may be curved. This property of the optimal grillages has been discovered by Prager (1985); the proof is based on the analysis of the compatibility equations linking the components of the curvature tensor $\boldsymbol{\varkappa}$. Thus, in all the examples discussed below, the optimal layouts will be characterized by families of beams lying along straight lines.

Having found $\check{\boldsymbol{M}}$ one can compute the optimal volume fractions of fibres in bending

$$v_i = \frac{\Lambda}{\mathcal{Z}_2} |\lambda_i(\check{\boldsymbol{M}})| \tag{5.61}$$

in such a manner that the condition (2.73) is fulfilled, or

$$\int_{\Omega} (v_1 + v_2) dx = \Lambda \tag{5.62}$$

The formulation of the problems (5.56), (5.57) comprises the cases of grillages of finite number of beams. Then, the beams should be straight and lie along the trajectories of principal curvatures, or along the principal directions of the tensor field $\check{\boldsymbol{\varkappa}}$. It turns out that such systems of straight beams are statically determinate; the bending moments may be directly computed from the equilibrium conditions, without performing minimization in (5.57).

Remark 5.4.1 The problems (5.56), (5.57) are also continuum counterparts of the auxiliary problems of the minimum compliance problem of grillages of rigid joints, see Sect. 2.5.3, where: (a) the torsion effect has been neglected, (b) the bending moment diagrams do not change signs on given bars.

Consider now two examples. In the first example, no free edges are present, while the second example concerns optimum design in domain with free edges, which will lead to a new phenomena the occurrence of beam-weaves.

Example 5.4.1 The problem is to construct the optimal, least compliant, system of beams, within the square domain Ω of dimensions $2a \times 2a$, transmitting a given vertical force to the boundary. We assume that the beams touching the edge are clamped at the edge; a simple support condition is not allowed. The cost of the design is given by (2.73). The layout of beams and distribution of the bending stiffnesses along the beams are the unknowns of the problem.

This is a classical benchmark problem of the Prager–Rozvany layout theory.

The problem is reduced to the construction of the scalar field $v = v(x_1, x_2)$ within the given domain $\Omega = (0, 2a) \times (0, 2a)$ parameterized by the system (x_1, x_2), as in Fig. 5.10.

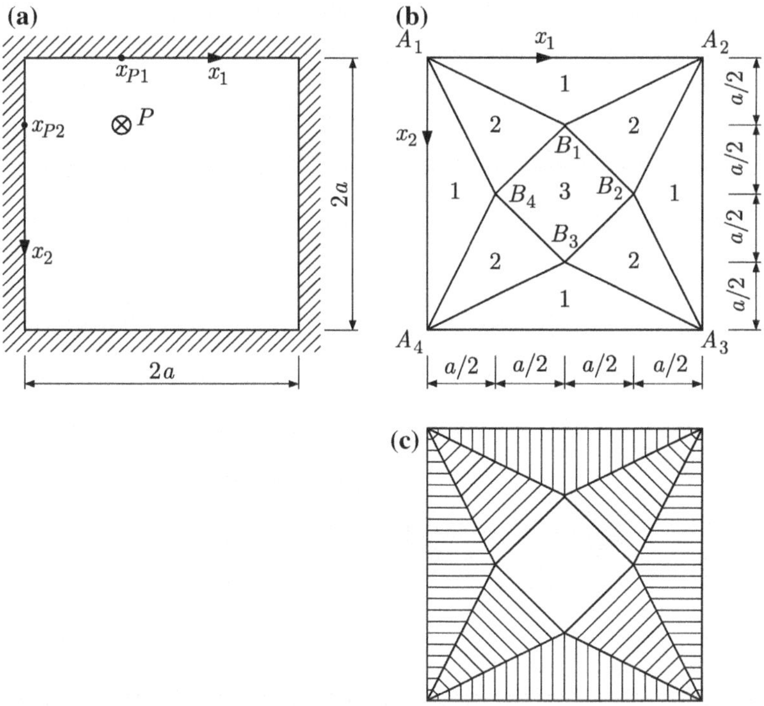

Fig. 5.10 The problem setting: transmit the vertical force P applied at point (x_{P_1}, x_{P_2}) to the clamped edge of a square (**a**), division of the design domains into regimes (**b**), optimal layout of beams (**c**)

Let us introduce the scalar field within the given domain,

$$v(x) = \begin{cases} \frac{1}{2}(x_1)^2 & \text{if } x = (x_1, x_2) \in \Delta A_1 B_4 A_4 \\ \frac{1}{2}(x_2)^2 & \text{if } x \in \Delta A_1 A_2 B_1 \\ -\frac{1}{6}[(x_1)^2 - 4x_1 x_2 + 4(x_2)^2] & \text{if } x \in A_1 B_1 B_4 \\ -\frac{1}{2}[(x_1)^2 + (x_2)^2] + a(x_1 + x_2) - \frac{3}{4}a^2 & \text{if } x \in B_1 B_2 B_3 B_4 \end{cases} \tag{5.63}$$

see Fig. 5.10, where the points A_i, B_i are shown. The values of this field for other points of the domain can be found using symmetry conditions along the lines: $A_1 A_3$, $A_2 A_4$.

The function v thus defined is of C^1 class. Indeed, along the lines $A_1 B_1$, $A_1 B_4$, $B_4 B_1$ the functions v, $\dfrac{\partial v}{\partial x_1}$, $\dfrac{\partial v}{\partial x_2}$, are continuous. Due to the symmetries mentioned, the function v is of C^1 class along the contours of the subdomains depicted in

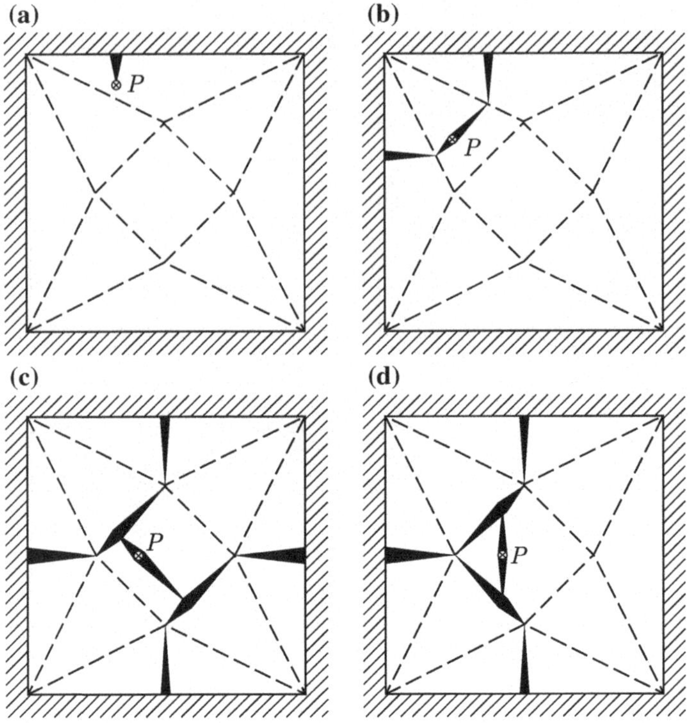

Fig. 5.11 The optimal systems of beams for the force P lying in : domain Ω_1 (**a**); domain Ω_2 (**b**); domain Ω_3: solution no 1 (**c**); solution no 2 (**d**)

Fig. 5.10b. Consequently, the field v is of C^1 class within the whole design domain. Moreover, it is easy to check that $v = 0$ and $\dfrac{\partial v}{\partial \boldsymbol{n}} = 0$ along the contour.

The domain Ω is a sum of the domains Ω_1, Ω_2, Ω_3 indicated in Fig. 5.10b by numbers: 1, 2, 3. The virtual principal strains are:
$\varkappa_I = -1$, $\varkappa_{II} = 0$ in the domain Ω_1
$\varkappa_I = 1$, $\varkappa_{II} = -1/3$ in the domain Ω_2
$\kappa_I = 1$, $\kappa_{II} = 1$ in the domain Ω_3
The principal direction of $\boldsymbol{\varkappa}$ refers to $\varkappa_I$. The lines of principal directions are shown in Fig. 5.10c. In domain Ω_3 these directions are not defined.

Assume that the force is applied at the point $x_P = (x_{P_1}, x_{P_2})$ of the domain Ω_1. Then the solution consists of a single cantilever (a beam clamped at the edge) of the stiffness varying according to the rule (2.96). Assuming that the width $b(x)$ of the beam is the design variable, while its height is constant, then the optimal width $b(x)$ is proportional to $|\breve{M}(x)|$ and here $\breve{M}$ grows linearly from the edge, see Fig. 5.11a.

If the point x_P lies in the domain Ω_2, the optimal system of beams is composed of three beams, see Fig. 5.11b. If point x_P lies in the domain Ω_3, one can construct many systems of beams of the same minimal compliance; two illustrative solutions are depicted in Fig. 5.11c, d.

If the system of beams is designed to carry a system of forces $P_1, \ldots, P_n$ directed downwards, then the optimal system of beams is simply a superposition of the structures corresponding to the subsequent point loads. The field v given by (5.63) is the maximizer of the problem (5.56) for each choice of the loads directed downwards. If the load $q(x)$ is distributed and directed downwards ($q > 0$ at each point), then the optimal structure is a fibrous plate, the fibre directions are shown in Fig. 5.10c; in the domain Ω_3 these directions are not defined, hence one can choose, for instance, the beams along one of the diagonals of the given square domain.

Example 5.4.2 Having the theoretical tools of the present section, we can solve now the problem of Fig. 2.16 of optimally transmitting the force P at C to the two sides OA, OB where the grillage can be simply supported. The design domain is the rhombus $OACB$ of angle $\alpha \leq \pi/4$; the grillage cannot be supported along AC, CB these edges are free. Let us introduce the virtual deflection within the design domain

$$v = \begin{cases} \frac{1}{2}[\tan^2\alpha(x_1)^2 - (x_2)^2] & \text{for } 0 \leq x_1 \leq a \\ \frac{1}{2}[(x_1)^2 - (2ax_1 - a^2)(1 - \tan^2\alpha) - (x_2)^2] & \text{for } a \leq x_1 \leq 2a \end{cases} \tag{5.64}$$

This field vanishes along OA and OB and is continuous along the line AB, where $x_1 = a$. Moreover, the derivative $\dfrac{\partial v}{\partial x_1}$ is continuous along this line. Thus, $v \in C^1(\Omega)$ and $v = 0$ on OA and OB. The tensor of curvature is expressed by

$$\varkappa_{ij} = \begin{cases} \varkappa_{ij}^{(1)} & 0 \leq x_1 \leq a \\ \varkappa_{ij}^{(2)} & a \leq x_1 \leq 2a \end{cases} \tag{5.65}$$

where

$$(\varkappa_{ij}^{(1)}) = \begin{bmatrix} -\tan^2\alpha & 0 \\ 0 & 1 \end{bmatrix} \quad (\varkappa_{ij}^{(2)}) = \begin{bmatrix} -1 & 0 \\ 0 & 1 \end{bmatrix} \tag{5.66}$$

We see that the axes x_1, x_2 are principal directions of the tensors $\boldsymbol{\varkappa}^{(1)}$, $\boldsymbol{\varkappa}^{(2)}$. Due to the assumption of $\alpha \leq \pi/4$ we have

$$\begin{aligned} \lambda_1(\boldsymbol{\varkappa}^{(1)}) &= -\tan^2\alpha, & \lambda_2(\boldsymbol{\varkappa}^{(1)}) &= 1 \\ \lambda_1(\boldsymbol{\varkappa}^{(2)}) &= -1, & \lambda_2(\boldsymbol{\varkappa}^{(2)}) &= 1 \end{aligned} \tag{5.67}$$

Hence, $v \in \mathcal{U}_2(\Omega)$ and $||\boldsymbol{\varkappa}(v)||_{\text{spec}} = 1$ a.e. in Ω. One can compute

$$v_C = v(2a, 0) = \frac{1}{2}a^2(1 + 3\tan^2\alpha) \tag{5.68}$$

The virtual work done by the force P at C equals

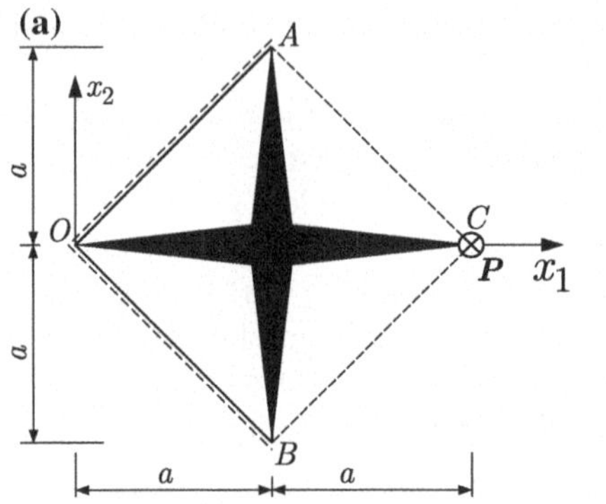

Fig. 5.12 The statically determinate grillages composed of two beams (**a**) and of five beams (**b**) transmitting the point load P to the simply supported edges OA, OB

$$\mathcal{L}(v) = \frac{1}{2} P a^2 (1 + 3 \tan^2 \alpha) \tag{5.69}$$

Case of $\alpha = \pi/4$
In this case, the absolute values of the principal values of the curvature tensor (5.65) are equal to 1 at all the points of the design domain, hence the beams of the optimal grillage should run along the direction of the Cartesian axes x_1, x_2. The grillages of the simplest layouts, with beams going in these directions, both being statically determinate, are shown in Fig. 5.12a, b.

It is easy to check that the integral

$$\int |M| ds = 2Pa^2 \tag{5.70}$$

is the same for both the grillages, the computation could be performed due to the statical determinacy. Thus, for the case of $\alpha = \pi/4$ the following identity holds

$$\mathcal{L}(v) = \int |M| ds \tag{5.71}$$

The left-hand side can be interpreted as $\mathcal{Z}_1$ given by (5.56), see (2.95), while the right-hand side can be viewed as $\mathcal{Z}_2$ given by (5.57), see (2.90). Thus, both the grillages are optimal.

Let us note that the grillages of Fig. 5.12 can be viewed as two first grillages of a family of grillages of similar layout of horizontal and vertical beams. The grillage of Fig. 5.12a introduces the division of OC into $n = 2$ equal parts, while the grillage of Fig. 5.12b introduces the division into $n = 4$ parts. Let us consider a grillage corresponding to the division into $n = 16$ parts, see Fig. 5.13.

Note that within the domain OAB only two beams are placed E_1E_2, E_1', E_2'. If n tends to infinity, the beams within OAB vanish while the free edges AC, BC

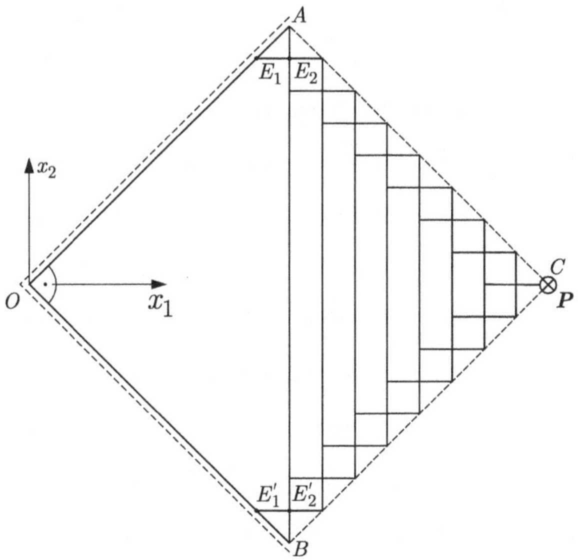

Fig. 5.13 The statically determinate grillage with horizontal 15 beams transmitting the point load P to the simply supported edges OA, OB

become reinforced by so-called beam-weaves, or the rods of a torsional stiffness. These beam-weaves introduce elastic supports along the free edges for the vertical beams. In this example, there is no need for passing to infinity with n, since all the grillages of this family are characterized by the same compliance.

Case of $\alpha < \pi/4$

Consider the grillages of Fig. 5.14a, b. These grillages, characterized by the indices $n = 2$ and $n = 4$, have already been automatically generated by the ground structure method, cf. Fig. 2.17.

One can easily check that these grillages are characterized by different values $\mathcal{Z}_2(n)$, $n = 2$, $n = 4$. This sequence is decreasing and tends to the value (2.150) for n tending to infinity. Let us note that this limit value is equal to the value of the virtual work (5.69) of the load for the virtual field given by (5.64). Thus, the following identity holds

$$\int_{\Omega} ||M||^{\star}_{\text{spec}} = \mathcal{L}(v) \tag{5.72}$$

which proves that the structure of Fig. 5.14c is optimal. This layout has been previously predicted numerically with high accuracy, see Fig. 2.17. The optimal structure of Fig. 5.14c lies beyond the family of grillages, since its free edges are reinforced by beam-weaves capable of transmitting torsion, while the grillages are composed of beams capable only of transmitting bending and transverse shearing, the latter being eliminated by the optimization process. The appearance of reinforcing beam-weaves differs the theory of grillages from the theory of trusses, since the optimal Michell

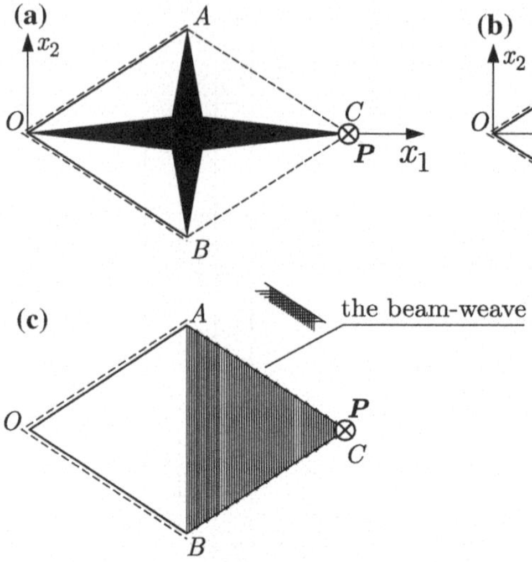

Fig. 5.14 The statically determinate grillages composed of two beams (**a**) and of five beams (**b**) transmitting the point load P to the simply supported edges OA, OB. Case of $\alpha \leq \pi/2$. The limit structure with the beam-weaves along the free edges (**c**)

structures loaded in the plane are reinforced by bars and the bars are included in the original formulation.

5.5 Bibliographical Comments

The spatial Michell structures discussed in Sects. 5.1.1 and 5.1.2 are formed by rotating the planar Michell structures considered in Sect. 4.5 around their vertical symmetry axes. Optimality of such structures has been noted by Hemp (1973); the proofs given in this chapter follow the arguments of this book. These proofs are important, since it is really surprising that these structures do not need a circumferential stiffening.

The problem of torsion discussed in Sect. 5.2.1 was solved by Michell; his formula

$$V_0 = 2M\left(\frac{1}{\sigma_T} + \frac{1}{\sigma_C}\right)\ln(|\cot(\phi_U/2)|) \tag{5.73}$$

can be derived from the formula (5.44) derived in this chapter concerning the case of $\sigma_T = \sigma_C = \sigma_0$. The derivation is standard, as based on Maxwell's and Michell's theorems (see Sect. 2.3) in their versions referring to the case of structures composed of infinite number of infinitely thin members. These theorems apply to the torsion problem considered, since in this problem no kinematic conditions are imposed. The

Michell's counterpart of the formula (5.73) has been published without derivation (see the last equation in Michell 1904); a derivation has been performed by Hemp (1973). The Michell sphere is the lightest among all rotationally symmetric fully stressed membrane shells subject to the torsion load shown in Fig. 1.11. This property has been proved by Lewiński (2004) by standard methods of calculus. The same paper includes the derivation of the formulae for the weight of the optimal shells formed on conical and cylindrical surfaces; these results are repeated in Sects. 5.3.1 and 5.3.2.

A numerical construction of approximants of the Michell sphere is difficult and that is why these constructions have been shown only recently, see Zegard and Paulino (2015), Sokół (2016, 2017). Recently, Jacot and Mueller (2017) put forward a new numerical method, called there MSTM, of constructing Michell structures satisfying at each point the optimality conditions of the regime $\mathbb{T}$, for the case of $\sigma_T = \sigma_C$.

The continuum theory of grillages has a mathematical structure similar to that of the optimal layout of the reinforcement of thin plates of constant thickness. For this latter, problem the first essential theoretical results have been discovered by Morley (1966) and Lowe and Melchers (1972a, b). The theory of optimal grillages has been put forward by Prager and Rozvany (1977), see Rozvany (1976), Prager (1985). The optimal grillage of Fig. 37.1 has been taken from Rozvany (1976). The characteristic feature of the optimal layouts of grillages being composed of straight beams has been noted by Prager (1985). The proof is based on the analysis of the compatibility equations interrelating the strain measures of the theory of thin plates in bending. The optimal solution of the Example 5.4.2 is taken from the same article. The beam-weaves along the free edges have been discovered by Prager and Rozvany (1977). The theory of Prager–Rozvany grillages is based on the problems (5.56), (5.57) which are continuum counterparts of the discretized problems (2.146), (2.142). The problem of correctness of the settings (5.56), (5.57) is encompassed by the results of Bouchitté and Fragala (2007). One can find there the necessary assumptions for their solvability. An exact, very subtle passage from the formulations (2.146), (2.142) to the continuum formulations has been recently performed by Bołbotowski (2018).

The numerical approaches to solve the grillage layouts problems have been proposed by Sigmund et al. (1993), Zhou (2009), Bołbotowski et al. (2018).

References

Bołbotowski, K. (2018). *Theory of grillage optimization-a discrete setting* (In preparation).

Bołbotowski, K., He, L., & Gilbert, M. (2018). Design of optimum grillages using layout optimization. *Structural and Multidisciplinary Optimization, 58*, 851–868.

Bouchitté, G., & Fragalá, I. (2007). Optimality conditions for mass design problems and applications to thin plates. *Archive for Rational Mechanics and Analysis, 184*, 257–284.

Hemp, W. S. (1973). *Optimum structures*. Oxford: Clarendon Press.

Jacot, B. P., & Mueller, C. T. (2017). A strain tensor method for three-dimensional Michell structures. *Structural and Multidisciplinary Optimization, 55*, 1819–1829.

Lewiński, T. (2004). Michell structures formed on surfaces of revolution. *Structural and Multidisciplinary Optimization, 28*(1), 20–30.

Lewiński, T., & Sokół, T. (2014). On basic properties of Michell's structures. In G. I. N. Rozvany & T. Lewiński (Eds.), *Topology optimization in structural and continuum mechanics* (Vol. 549). CISM international centre for mechanical sciences. Courses and lectures. Wien: Springer.

Lowe, P. G., & Melchers, R. E. (1972a). On the theory of optimal constant thickness, fibrereinforced plates. I. *International Journal of Mechanical Sciences*, *14*, 311–324.

Lowe, P. G., & Melchers, R. E. (1972b). On the theory of optimal constant thickness, fibrereinforced plates. II. *International Journal of Mechanical Sciences*, *15*, 157–170.

Michell, A. G. M. (1904). The limits of economy of material in frame structures. *Philosophical Magazine*, *8*(47), 589–597.

Morley, C. T. (1966). The minimum reinforcement of concrete slabs. *International Journal of Mechanical Sciences*, *8*, 305–319.

Prager, W. (1985). Optimal design of grillages. In M. Save & W. Prager (Eds.), *Structural optimization. Volume 1. Optimality criteria* (Vol. 1, pp. 153–200 and 308–317). New York and London: Plenum Press.

Prager, W., & Rozvany, G. I. N. (1977). Optimal layout of grillages. *Journal of Engineering Mechanics*, *5*, 1–18.

Rozvany, G. I. N. (1976). *Optimal design of flexural systems*. London: Pergamon Press.

Sigmund, O., Zhou, M., & Rozvany, G. I. N. (1993). Layout optimization of large FE systems by new optimality criteria methods: Application to beam systems. *Concurrent engineering: Tools and technologies for mechanical system design* (pp. 803–819). Berlin: Springer.

Sokół, T. (2016). A new adaptive ground structure method for multi-load spatial Michell structures. In M. Kleiber, T. Burczyński, K. Wilde, J. Górski, K. Winkelmann, & Ł. Smakosz (Eds.), *Advances in mechanics: Theoretical, computational and interdisciplinary issues* (pp. 525–528). Boca Raton: CRC Press.

Sokół, T. (2017). On the numerical approximation of Michell trusses and the improved ground structure method. In *Proceedings of the 12th World Congress on Structural and Multidisciplinary Optimisation, 5th–9th June 2017, Braunschweig, Germany*.

Zegard, T., & Paulino, G. H. (2015). GRAND3 - Ground structure based topology optimization for arbitrary 3D domains using MATLAB. *Structural and Multidisciplinary Optimization*, *52*, 1161–1184.

Zhou, K. (2009). Optimization of least-weight grillages by finite element method. *Structural and Multidisciplinary Optimization*, *38*, 525–532.

Chapter 6
Optimum Design of Funiculars and Archgrids

Michell structures are designed for a fixed load, applied at fixed places. However, in the engineering practice the load is usually linked with the structure and usually follows its current position, as e.g. the self-weight, or the weight of snow. Thus, majority of loads we face with are design-dependent. A narrow class of such loads are transmissible loads acting along vertical lines, coinciding with the gravity field. This chapter is aimed at putting forward the theory of such optimal structures.

The chapter is organized as follows. First, in Sect. 6.1, we put forward an elementary explanation of peculiar properties of funiculars—the idealized structures perfectly designed to carry transmissible loads. Then, the same subject is discussed in Sect. 6.2 within the Michell-like theory proposed by Rozvany and Prager (1979). The optimal funiculars are called also *Prager-structures*, see Rozvany and Wang (1983).

Section 6.3 concerns spatial counterparts of the funiculars: the archgrids modelling the optimal canopies and roofs designed for transmissible vertical loads.

6.1 Optimal Funiculars—An Elementary Approach

6.1.1 Introduction

We shall put forward the exact solutions to the problem of optimum design of planar structures of minimal weight subject to the condition of the stress being of equal sign and not exceeding a given permissible threshold of compression σ_C. The solutions will refer to the vertical transmissible loads. This is the load the position of which is not fixed; the load can change its application points along the vertical lines, while its intensity per unit length of the horizontal axis x may depend on the ordinate x but is kept independent of the position of its application.

T. Lewiński et al., *Michell Structures*,
https://doi.org/10.1007/978-3-319-95180-5_6

The solutions to the minimum weight design problem corresponding to the transmissible loads are based on the known result of the theory of structures, which links the shape of the funiculars, or the structures which do not undergo bending, with the diagram of bending moments of a simply supported beam subject to the same system of loads.

Let us recall this old theorem, mentioned in many classical textbooks on structural mechanics. Consider a simply supported beam subject to a system of vertical point loads P_1, P_2, P_3 at points of the ordinates x_1, x_2, x_3, see Fig. 6.1.

This load causes the bending moments $M(x)$, the ordinates of the diagram M at points x_1, x_2, x_3 being M_1, M_2, M_3, see Fig. 6.1a. Let H be a force and let $f_i = M_i/H$ be ordinates of the vertices of the three-hinge frame ADB; the hinge at D is assumed at an arbitrary section, see Fig. 6.1b. This frame will be subjected to the same system of loads, as applied to the beam of Fig. 6.1a. One can prove that the stress state in the bars of this frame is uniform at each transverse section, hence the bending moments and the transverse forces vanish: $M = 0$, $Q = Q_M = 0$. Thus, by appropriate choice of the areas $A(x)$ of the cross sections (proportionally to the values of the axial force $T(x)$), one can make the stress distribution uniform in the whole frame. Moreover, the number of hinges can be increased, leading to geometrically

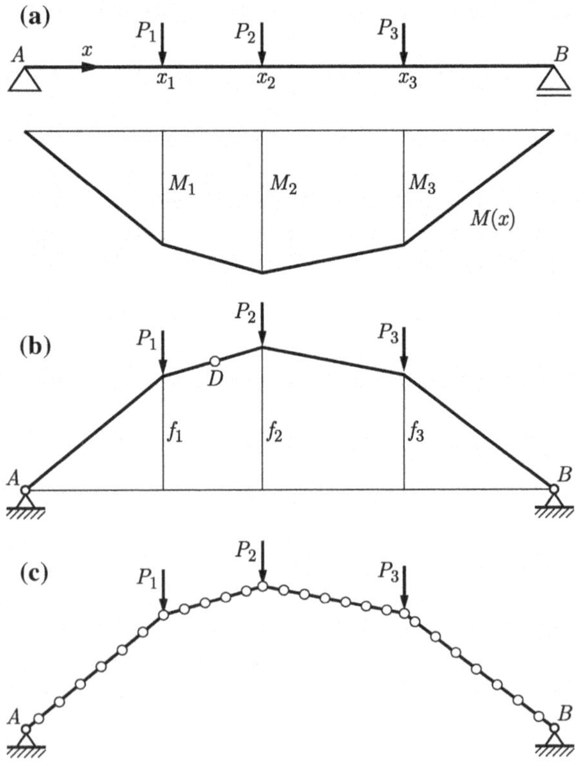

Fig. 6.1 The theorem on funiculars: the system of loads (**a**); a three-hinge frame of the shape of the diagram of the bending moment (**b**); the corresponding funicular (**c**)

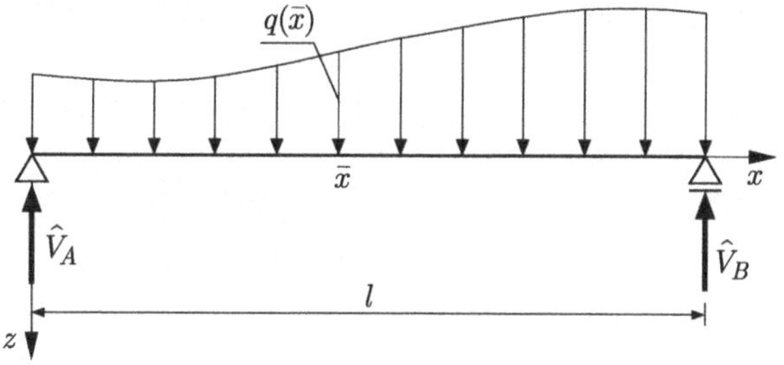

Fig. 6.2 A simply supported beam—the auxiliary problem

variable structures, yet still capable of transmitting the given load to the supports. The number of these additional hinges is arbitrary. In this manner, we build structures with arbitrary number of hinges, ideally suited for a given load and incapable of carrying other loads. These idealized structures are called *funiculars*. In the next sections we shall prove the mentioned property of the funiculars and by appropriate choice of the magnitude of the force H we shall construct the lightest structure. Then we proceed to outline the abstract theory which clears up the existence of the funiculars and delivers new tools of optimum design. This abstract theory has been put forward by George Rozvany and William Prager in 1970s. Formally, this theory is encompassed by the general framework of Michell's theory, as will be discussed later.

6.1.2 The Beam Analogy and the Funicular Equation

Let us consider a simply supported beam under the transverse load of intensity $q(x)$, see Fig. 6.2. The vertical reactions are denoted by $\hat{V}_A$, $\hat{V}_B$. The bending moment at the cross section of the ordinate x is expressed by

$$\hat{M}(x) = \hat{V}_A \cdot x - \int_0^x (x - \bar{x})q(\bar{x})d\bar{x} \tag{6.1}$$

while the transverse force equals

$$\hat{Q}(x) = \hat{V}_A - \int_0^x q(\bar{x})d\bar{x} \tag{6.2}$$

Let us recall the equilibrium equations, cf. (1.3)

$$\frac{d\hat{M}}{dx} = \hat{Q}, \qquad \frac{d\hat{Q}}{dx} + q(x) = 0 \tag{6.3}$$

The condition $\hat{M}(l) = 0$ gives

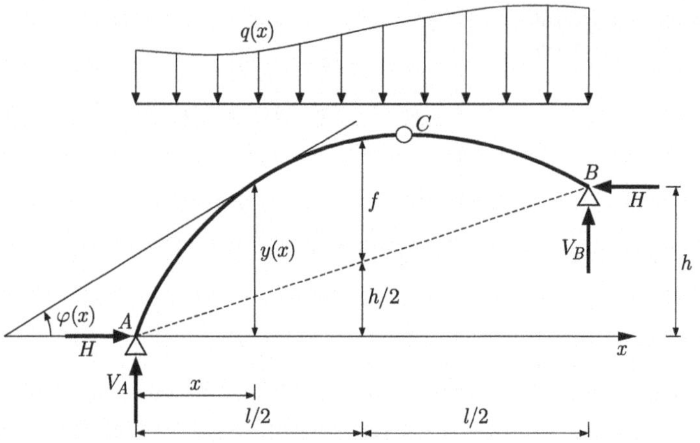

Fig. 6.3 A three-hinges arch under a transverse load of intensity $q(x)$

$$\hat{V}_A = \frac{1}{l}\int_0^l (l - \bar{x})q(\bar{x})d\bar{x} \tag{6.4}$$

Let us consider a three-hinges arch pin-supported at A and B, the line AB being at different levels, see Fig. 6.3. The shape of the arch is given by the function $y(x)$. Let

$$h = y(l), \qquad f = y\left(\frac{l}{2}\right) - \frac{h}{2}, \qquad \tan\beta = \frac{h}{l} \tag{6.5}$$

The bending moment at the section x equals

$$M(x) = V_A x - Hy(x) - \int_0^x (x - \bar{x})q(\bar{x})d\bar{x} \tag{6.6}$$

The condition $M(l) = 0$ leads to

$$V_A = H\tan\beta + \hat{V}_A \tag{6.7}$$

where $\hat{V}_A$ is given by (6.4). Let us substitute (6.7) into (6.6) and introduce the condition of the arch being not bent. The condition $M(x) = 0$ determines the ordinates of the arch:

$$y(x) = (\tan\beta)x + \frac{1}{H}\hat{M}(x) \tag{6.8}$$

The condition $y(l/2) = f + h/2$ gives $\hat{M}(l/2) = Hf$. Equation (6.8) can be expressed as

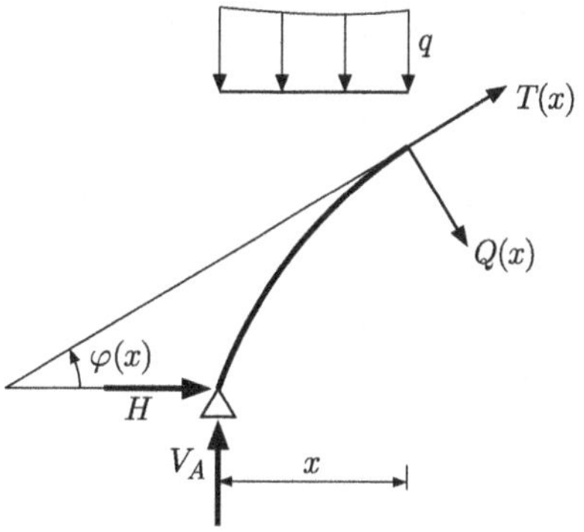

Fig. 6.4 The state of forces acting on a segment of the arch

$$y(x) = (\tan\beta)x + \frac{f}{\hat{M}\left(\frac{l}{2}\right)}\hat{M}(x) \tag{6.9}$$

hence

$$\frac{dy}{dx} = \tan\beta + \frac{f}{\hat{M}\left(\frac{l}{2}\right)}\hat{Q}(x) \tag{6.10}$$

which determines the slope of the tangent, see Fig. 6.3. One can prove that the arch of the shape given by (6.9) is not subject to transverse forces: $Q(x) = 0$; a proof can be found in Hetmański and Lewiński (2007). Due to the transverse force being zero, one can compute the axial force by equilibrating a segment of the arch, see Fig. 6.4 to obtain

$$T(x) = -\frac{H}{\cos\varphi(x)} \tag{6.11}$$

By differentiating (6.10) and using (6.3) one obtains

$$H\frac{d^2y}{dx^2} + q(x) = 0 \tag{6.12}$$

called in the literature: *the funicular equation*. The arch of Fig. 6.3 is a funicular, since $M = 0$, $Q = 0$ and the equilibrium is satisfied if a new hinge is introduced; then arbitrary number of hinges can be introduced, leading to a funicular.

6.1.3 Construction of an Optimal Frame

Let $A = A(x)$ be an area of a cross section of the arch at the ordinate x. Let us assume that the normal stresses within the whole arch are made uniform, or

$$\frac{|T(x)|}{A(x)} = \sigma_C = \text{const}\,.$$

Thus, we take

$$A(x) = \frac{H}{\sigma_C \cos\varphi(x)} \tag{6.13}$$

and assume that $\cos\varphi(x) > 0$. The volume of the arch of the area of the cross section chosen as above equals

$$V = \frac{H}{\sigma_C}\int_0^l \frac{dx}{\cos^2\varphi(x)} \tag{6.14}$$

or

$$V = \frac{H}{\sigma_C}\int_0^l \left[1 + \left(\frac{dy}{dx}\right)^2\right] dx. \tag{6.15}$$

Substitution of (6.10) gives

$$V = \frac{H}{\sigma_C}\int_0^l \left[1 + \left(\tan\beta + \frac{f}{\hat{M}\left(\frac{l}{2}\right)}\hat{Q}(x)\right)^2\right] dx \tag{6.16}$$

Let us introduce a norm of the function $\hat{Q}$ by

$$||\hat{Q}|| = \sqrt{\int_0^l (\hat{Q}(x))^2 dx} \tag{6.17}$$

and let

$$\hat{a} = \frac{l}{\left(\hat{M}\left(\frac{l}{2}\right)\right)^2}||\hat{Q}||^2 \tag{6.18}$$

The formula (6.16) may be written as a function of argument $\eta = f/l$

$$V = \frac{\hat{M}\left(\frac{l}{2}\right)}{\sigma_C}\bar{V}(\eta) \tag{6.19}$$

where

$$\bar{V}(\eta) = \frac{1 + \tan^2\beta}{\eta} + \hat{a}\eta \tag{6.20}$$

The function $\eta \to \bar{V}(\eta)$ is convex and attains a minimum at $\eta = \eta_0$

$$\eta_0 = \frac{\left|\hat{M}\left(\frac{l}{2}\right)\right|}{\sqrt{l}||\hat{Q}||\cos\beta} \tag{6.21}$$

Thus, $f_0 = l\eta_0$ is a rise of the lightest arch of constant stress. The horizontal force $H = H_0$ of this arch equals

$$H_0 = \frac{||\hat{Q}||\cos\beta}{\sqrt{l}} \tag{6.22}$$

Substitution into (6.8) gives the shape of the optimal arch

$$y_0(x) = (\tan\beta)x + \frac{\sqrt{l}}{||\hat{Q}||\cos\beta}\hat{M}(x) \tag{6.23}$$

By differentiation of both sides, we find

$$\frac{1}{||\hat{Q}||}\hat{Q}(x) = \frac{\cos\beta}{\sqrt{l}}\left(\frac{dy_0}{dx} - \tan\beta\right) \tag{6.24}$$

The norm of the square of this function is equal to 1, which results in the optimality condition

$$\frac{1}{l}\int_0^l \left(\frac{dy_0}{dx}\right)^2 dx = 1 + 2\tan^2\beta \tag{6.25}$$

discovered by Rozvany and Prager in 1970s. This condition determines the rise of the optimal arch by imposing a specific value on the average slope of the tangents to the arch axis. It is called: *mean square slope condition.*

Let us substitute (6.21) into (6.20). The volume of the lightest arch equals

$$V_0 = \frac{2\sqrt{l}}{\sigma_C\cos\beta}||\hat{Q}|| \tag{6.26}$$

or

$$V_0 = \frac{2l}{\sigma_C}\frac{1}{\cos^2\beta}H_0 \tag{6.27}$$

The areas of the cross sections vary according to the rule

$$A_0(x) = \frac{H_0}{\sigma_C}\sqrt{1 + \left(\frac{dy_0}{dx}\right)^2} \tag{6.28}$$

in which H_0 is given by (6.22) and

$$\frac{dy_0}{dx} = \tan\beta + \frac{\sqrt{l}}{||\hat{Q}||\cos\beta}\hat{Q}(x) \tag{6.29}$$

Lastly, note that

$$\frac{1}{2}V_0 = \frac{1}{\sigma_C}\int_0^l q(x)\tilde{y}_0(x)dx \tag{6.30}$$

where $\tilde{y}_0(x) = y_0(x) - (\tan\beta)x$. Indeed, by (6.23) we have

$$\int_0^l q(x)\tilde{y}_0(x)dx = \frac{\sqrt{l}}{||\hat{Q}||\cos\beta}\int_0^l \hat{M}(x)q(x)dx. \tag{6.31a}$$

We recall

$$q(x) = -\frac{d\hat{Q}}{dx} \quad \text{and} \quad \hat{Q} = \frac{d\hat{M}}{dx},$$

hence

$$\int_0^l \hat{M}(x)q(x)dx = -\int_0^l \hat{M}(x)\frac{d\hat{Q}}{dx}dx = \tag{6.31b}$$

$$-\int_0^l \left[\frac{d}{dx}(\hat{M}\hat{Q}) - \frac{d\hat{M}}{dx}\hat{Q}\right]dx = \int_0^l (\hat{Q}(x))^2 dx = ||\hat{Q}||^2$$

Thus,

$$\frac{1}{\sigma_C}\int_0^l q(x)\tilde{y}_0(x)dx = \frac{\sqrt{l}||\hat{Q}||}{\sigma_C\cos\beta} \tag{6.32}$$

which confirms (6.30), see (6.26). If the load is in the form of the vertical point loads $P_1, \ldots, P_m$ applied at points of ordinates $x_1, \ldots, x_m$, then Eq. (6.30) assumes the form

$$\frac{1}{2}V_0 = \frac{1}{\sigma_C}\sum_{j=1}^{m} P_j\tilde{y}_0(x_j) \tag{6.33}$$

The formulae (6.30), (6.33) link the optimal volume with the potential energy of the load.

6.1.4 Examples

Example 6.1.1 Consider the case of points A and B being at the same level, i.e. $h = 0$ and assume the transmissible load of intensity $q = \text{const}$. Let us compute

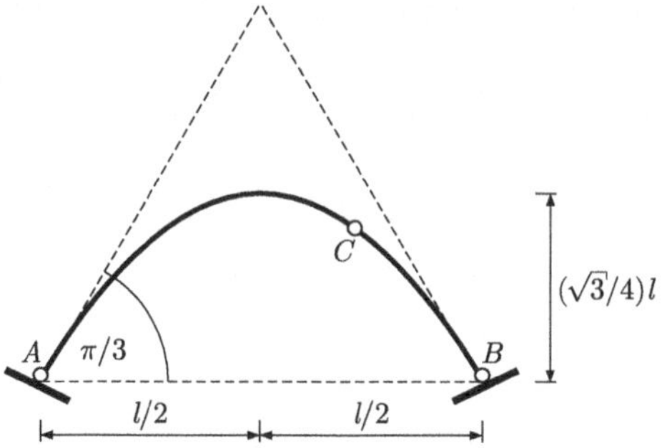

Fig. 6.5 The arch of the least volume assumes the shape of a parabola inscribed into an equilateral triangle

$$\begin{aligned} \hat{Q}(x) &= \frac{1}{2}ql - qx, \qquad ||\hat{Q}||^2 = \frac{1}{12}q^2l^3 \\ \hat{M}(x) &= \frac{1}{2}qlx - \frac{1}{2}qx^2, \qquad \hat{M}\left(\frac{l}{2}\right) = \frac{1}{8}ql^2 \\ H_0 &= \frac{\sqrt{3}}{6}ql, \quad \hat{a} = \frac{16}{3}, \quad \eta_0 = \frac{\sqrt{3}}{4}, \quad \beta = 0 \end{aligned} \tag{6.34}$$

Hence

$$y_0(x) = \sqrt{3}l\left(\frac{x}{l}\right)\left(1 - \frac{x}{l}\right), \qquad y_0\left(\frac{l}{2}\right) = \frac{\sqrt{3}}{4}l \tag{6.35}$$

We conclude that the lightest and fully stressed arch has the shape of a parabola inscribed into an equilateral triangle, see Fig. 6.5.

We see that $\max|\tan\varphi(x)| = \sqrt{3}$. The volume of the lightest arch equals

$$V_0 = \frac{\sqrt{3}}{3}\frac{ql^2}{\sigma_C} \tag{6.36}$$

For the case of $\beta = 0$, the optimality criterion (6.25) simplifies to the form

$$\frac{1}{l}\int_0^l \left(\frac{dy_0}{dx}\right)^2 dx = 1 \tag{6.37}$$

It is easy to check that this optimality criterion is fulfilled. Moreover, the formula (6.30) holds.

Example 6.1.2 Consider the case of $\beta = 0$ and the transmissible load point load of magnitude P along the vertical line at the distance a from the support A, see Fig. 6.6.

The diagram of moments has the shape of a triangle with a vertex for $x = a$. Thus, the three-hinges triangular frames shown in Fig. 6.6, or the frames: AC_1B, AC_2B satisfy the conditions: $M = 0$, $Q = 0$. Among such frames, we should select the lightest. To this end, we shall make use of the criterion (6.25) for $\beta = 0$, or the

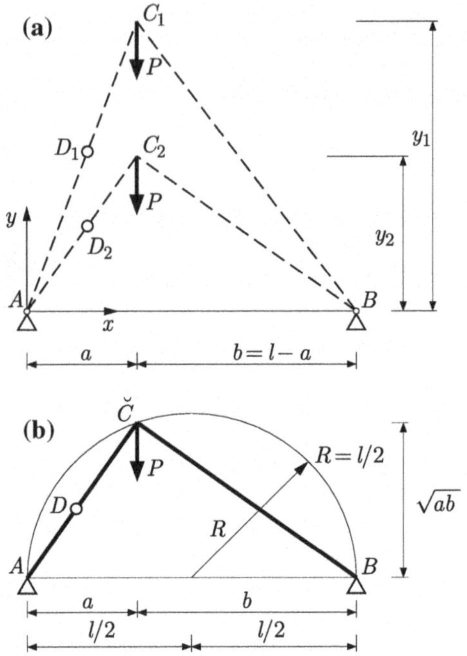

Fig. 6.6 The problem setting and the predicted two-bar solution (**a**); the optimal frame (**b**)

criterion (6.37). The only unknown is the position of point $\breve{C}$ given by the ordinate $y_{\breve{C}}$. The condition (6.37) reduces to the equation

$$\frac{1}{l}\left[\left(\frac{y_{\breve{C}}}{a}\right)^2 a + \left(-\frac{y_{\breve{C}}}{b}\right)^2 b\right] = 1 \tag{6.38}$$

hence

$$y_{\breve{C}} = \sqrt{ab}, \qquad |A\breve{C}| = \sqrt{a^2 + ab}, \qquad |B\breve{C}| = \sqrt{b^2 + ab}$$

Thus, the angle $\angle A\breve{C}B = \frac{\pi}{2}$ and the point $\breve{C}$ lies on the circle of origin at $\left(\frac{l}{2}, 0\right)$ and of radius $R = l/2$, see Fig. 6.6b. Lastly, let us compute

$$||\hat{Q}||^2 = \frac{P^2 ab}{l}, \qquad \hat{M}(a) = \frac{Pab}{l}, \qquad H_0 = \frac{P\sqrt{ab}}{l} \tag{6.39}$$

and compute the volume of the optimal frame

$$V_0 = \frac{2}{\sigma_C} P\sqrt{ab} \tag{6.40}$$

The same result can be obtained by using (6.33); this formula has the following form:

$$\frac{1}{2}V_0 = \frac{1}{\sigma_C}Py_{\breve{C}} \tag{6.41}$$

Let us note that the volumes of the bars are equal

$$V^{A\breve{C}} = \frac{P}{\sigma_C}\sqrt{ab}, \qquad V^{B\breve{C}} = \frac{P}{\sigma_C}\sqrt{ab} \tag{6.42}$$

Example 6.1.3 The transmissible load is composed of a point load P at $x = a$ and the load of uniform intensity $q = \text{const}$ on the interval $2a < x < 3a$; $l = 4a$, see Fig. 6.7a.

In the first stage, we construct the diagram of bending moments $\hat{M}$ in a simply supported beam under the same load, see Fig. 6.7b. Then we construct the transverse force diagram (Fig. 6.7c) and compute its norm by (6.17) to find

$$||\hat{Q}|| = 1.50693qa$$

By using (6.23) we construct the optimal shape $y_0(x)$, see Fig. 6.7d. We compute the horizontal force in this frame by (6.22) to find

$$H_0 = 0.75346qa$$

and then by using (6.27) we compute the volume of the optimal structure

$$V_0 = 6.02771\frac{qa^2}{\sigma_C}$$

The area of the cross sections vary according to (6.28), see Fig. 6.7e. The greatest values appear along the bars at the supports and the smallest value corresponds to the section where the transverse force $\hat{Q}$ vanishes.

The optimal funicular has the shape $y_0(x) = \dfrac{\hat{M}(x)}{H_0}$; the hinge can be assumed at an arbitrary section, for instance-below the point load, see Fig. 6.7f. This structure is composed of three straight bars and of one arch; the arch is tangent to both the adjacent straight bars. The axial stresses are made uniform. This structure is ideally shaped to carry the given load, the only unknown was its height. The rise cannot be too small and cannot be too high; the ideal rise has been determined by the condition (6.37).

The volume of the optimal structure may also be computed by (6.30) and (6.33) or—by the formula

$$\frac{1}{2}V_0 = \frac{1}{\sigma_C}\left[Py_0(a) + \int_{2a}^{3a} qy_0(x)dx\right] \tag{6.43}$$

and the result coincides with that reported above.

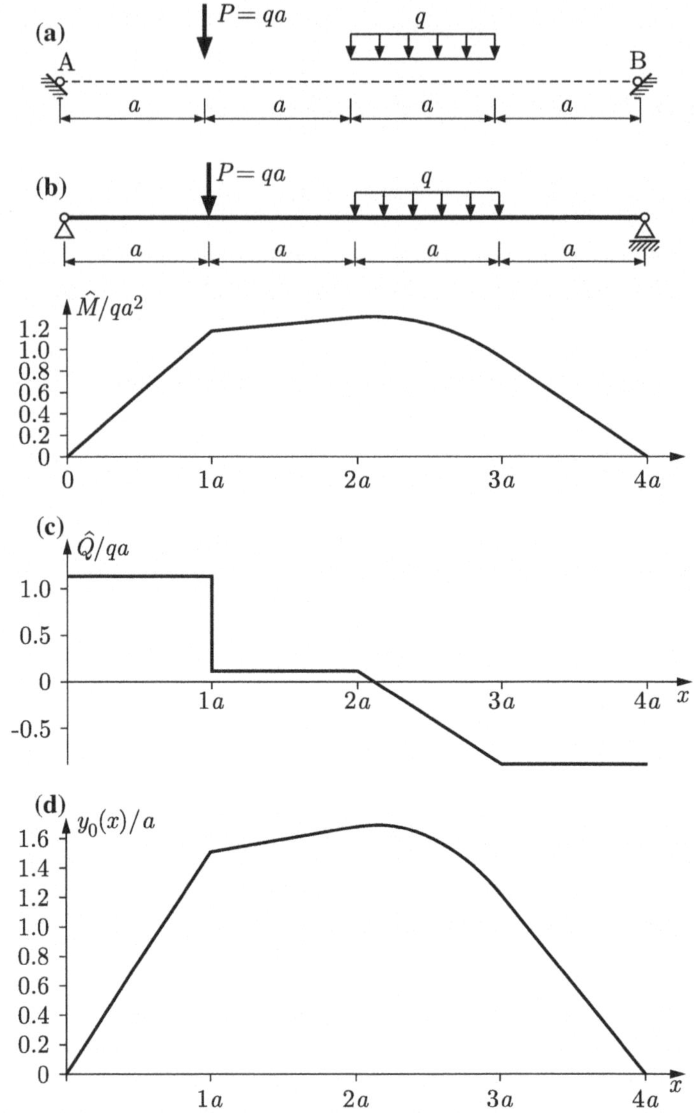

Fig. 6.7 The problem setting (**a**); the auxiliary problem of a beam and the corresponding bending moment diagram $\hat{M}$ (**b**); the transverse force diagram $\hat{Q}$ (**c**) the shape of the optimal frame (**d**); the diagram of variation of the optimal area A_0 of the cross sections (**e**); the final design (**f**)

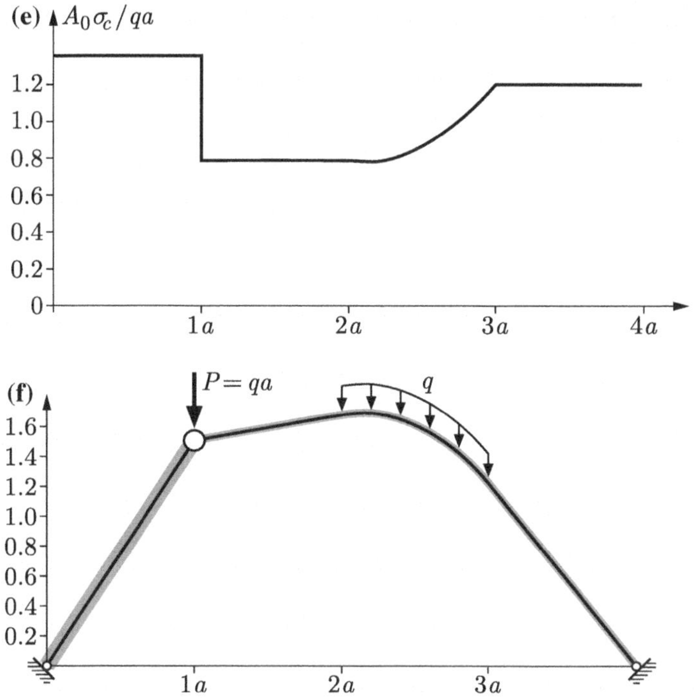

Fig. 6.7 (continued)

6.2 Rozvany–Prager Approach Towards Optimum Design of Planar Funiculars

The present section concerns planar funiculars subject to transmissible loads, i.e. the lightest planar structures in which the stress state is uniform and of equal sign. These structures are either cable networks or archgrids, depending on the sign of the stress: positive or negative. The aim of the present section is to recall the Rozvany–Prager theory of such structures to have a deeper insight into the subject as well as to make it possible to extend the idea of funiculars to spatial girders. This theory is closely linked with Michell's ideas. On the one hand, the Rozvany–Prager process of optimization is more restrictive than Michell's one, since the stress must be of equal sign. On the other hand, the load may change its position along the vertical lines. Despite these differences in the optimum design settings the optimal funiculars turn out to be Michell structures for specially chosen permissible stresses in tension and compression, which will be cleared up at the end of this section.

The structures considered in Sect. 6.1 were archgrids (hence—in compression) but it is now more convenient to set the Rozvany–Prager theory for cable networks; i.e. it is more convenient to assume that the permissible stress is positive and $\sigma_C = 0$ (hence the members are in tension). We shall draw upon the results of Sect. 3.5.

Let us parameterize the design domain Ω by the Cartesian system (x_1, x_2) of basis versors $\boldsymbol{e}_1$, $\boldsymbol{e}_2$. Let us start from the kinematic approach and put $\sigma_C = 0$ in (3.82). The bars cannot appear along the lines $x_1 = \text{const}$, hence we conclude that the strain component along the direction $\boldsymbol{e}_2$ vanishes. Thus, the kinematic problem assumes the form

$$
\begin{aligned}
\tilde{Z}_1 = \sup \{\mathcal{L}(\boldsymbol{v}) \mid \boldsymbol{v} \text{ kinematically admissible},\\
\varepsilon_{\boldsymbol{t}}(\boldsymbol{v}(x)) \le \frac{\sigma_0}{\sigma_T} \quad \forall\, \boldsymbol{t} \text{ such that } \|\boldsymbol{t}\| = 1\\
\varepsilon_{\boldsymbol{e}_2}(\boldsymbol{v}(x)) = 0 \quad \text{for a.e. } x \text{ in } \Omega\}
\end{aligned}
\tag{6.44}
$$

where $\varepsilon_{\boldsymbol{t}}$ is given by (3.79) where $n_i = t_i$ and represents the strain along the direction $\boldsymbol{t}$. Moreover, $\boldsymbol{e}_2$ is the versor of the axis x_2 along which the load is applied.

In the problem discussed later the field $\boldsymbol{v} = \boldsymbol{v}(x_1, x_2)$ will be constructed on the basis of the conditions

$$
\begin{aligned}
&\text{i)} && \max_{\boldsymbol{t}:\ \|\boldsymbol{t}\|=1} \varepsilon_{\boldsymbol{t}}(\boldsymbol{v}(x)) = \frac{\sigma_0}{\sigma_T}\\
&\text{ii)} && \varepsilon_{22}(\boldsymbol{v}(x)) = 0\\
&\text{iii)} && \boldsymbol{v} \ \text{ kinematically admissible}
\end{aligned}
\tag{6.45}
$$

The unit vector $\check{\boldsymbol{t}}$ solving the condition (i) fixes the tangent direction to the frame; this vector is the eigenvector of the tensor $\boldsymbol{\varepsilon}(\boldsymbol{v}(x))$ corresponding to the greater eigenvalue. The condition(ii) assures that along the direction x_2 no bars appear.

Below we shall show the construction of the field $\boldsymbol{v}$ concerning the problem of optimal transmitting three vertical forces P_1, P_2, P_3 to the pin supports A, B; the forces act along the vertical lines: $x_1^0 = a_1$, $x_2^0 = a_1 + a_2$, $x_3^0 = a_1 + a_2 + a_3$, see Fig. 6.8.

The forces P_1, P_2, P_3 cause the bending moment $\hat{M}$ on the simply supported beam; $\hat{M}_i^0 = \hat{M}_i$, see Fig. 6.8b. The optimal structure is predicted according to the rules of Sect. 6.1, or by (6.8). The same shape is obtained by solving the problem (6.44) Let $x_0^0 = 0$, $x_4^0 = l$. We consider the strips: $x_{i-1}^0 \le x_1 \le x_i^0$, $i = 1, 2, 3, 4$ of the widths a_i, which divide the strip $0 \le x_1 \le l$. In the strip $x_{i-1}^0 \le x_1 \le x_i^0$ we introduce the components of the virtual displacements

$$
\begin{aligned}
v_1(x_1, x_2) &= \frac{\sigma_0}{\sigma_T}\left[(1 - \cot^2\theta_i)(x_1 - \mathring{x}_i) + 2\tan\beta(x_2 - f(\mathring{x}_i)) + C_i\right]\\
v_2(x_1, x_2) &= \frac{2\sigma_0}{\sigma_T}\,[f(x_1) - (\tan\beta)x_1]
\end{aligned}
\tag{6.46}
$$

The constants C_i can be chosen such that v_1 be continuous along $x_1 = x_i^0$, $i = 1, 2, 3$ and to assure $v_1 = 0$ for $x_1 = 0$ and $x_1 = l$. The continuity of v_2 follows from the continuity of $f(x_1)$. The conditions $v_2(0, 0) = 0$, $v_2(l, h) = 0$ are satisfied identically, while the condition $v_1(l, h) = 0$ implies

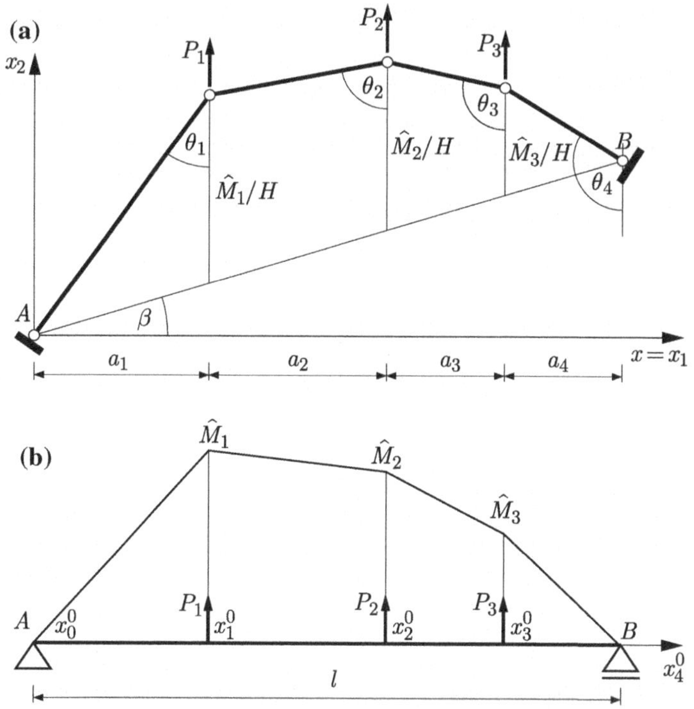

Fig. 6.8 The funicular transmitting the vertical forces to the supports (**a**); the auxiliary problem of a simply supported beam (**b**)

$$\frac{1}{l}\sum_{i=1}^{4}\cot^2\theta_i a_i = 1 + 2\tan^2\beta \tag{6.47}$$

The above condition is equivalent to (6.25) in the case of the considered load. The field $\boldsymbol{v}$ given by (6.46) satisfies the conditions (i), (ii) in (6.45). Indeed, the virtual strains in the strip $x_{i-1}^0 \le x_1 \le x_i^0$ are expressed by:

$$\varepsilon_{11}(\boldsymbol{v}) = \frac{\sigma_0}{\sigma_T}(1 - \cot^2\theta_i), \qquad \varepsilon_{12}(\boldsymbol{v}) = \frac{\sigma_0}{\sigma_T}\cot\theta_i, \qquad \varepsilon_{22}(\boldsymbol{v}) = 0 \tag{6.48}$$

The eigenvalues of this tensor are

$$\varepsilon_I = \frac{\sigma_0}{\sigma_T}, \qquad \varepsilon_{II} = -\frac{\sigma_0}{\sigma_T}\cot^2\theta_i \tag{6.49}$$

Thus, we arrive at the estimates

$$-\frac{\sigma_0}{\sigma_T}\cot^2\theta_i \le \varepsilon_t(\boldsymbol{v}(x)) \le \frac{\sigma_0}{\sigma_T} \tag{6.50}$$

In the subsequent strips (x_{i-1}^0, x_i^0) the vector $\boldsymbol{t}$ coincides with the subsequent bars of the frame. The directions of the first principal strain coincide with the bar directions. To be sure that the solution is correct one should check the equilibrium conditions. Since the solution found is the same as this constructed in Sect. 6.1 by the elementary methods, we are sure that the problem is solved correctly.

Consider now the case of $q = \text{const}$ along the whole interval $0 \le x_1 \le l$ and let q be directed along the axis x_2. We choose the virtual displacement field $\boldsymbol{v} = (v_1, v_2)$ in the form

$$v_1(x_1, x_2) = \frac{4\sigma_0}{\sigma_T}\frac{1}{l^2}x_1\left(x_1 - \frac{l}{2}\right)(l - x_1) \qquad v_2(x_1, x_2) = \frac{2\sqrt{3}\sigma_0}{\sigma_T}\frac{1}{l}x_1(l - x_1) \tag{6.51}$$

The virtual strains are expressed by (6.48), where $\cot\theta_i$ should be replaced by

$$\cot\theta = \sqrt{3}\left(1 - 2\frac{x_1}{l}\right) \tag{6.52}$$

Thus, $\cot\theta = dy_0/dx_1$, where y_0 is given by (6.35). The axis of this arch determines the principal directions of virtual strains. The principal virtual strains are given by (6.49) where $\cot\theta_i$ should be replaced with $\cot\theta$ given by (6.52).

Remark 6.2.1 Assume that θ_1 is smaller than other angles θ_i, as in Fig. 6.8, or, in case of $q = \text{const}$ let θ_1 be equal $\pi/6$. Let us require that the following estimates:

$$-\frac{\sigma_0}{\sigma_C} \le \lambda_i(\varepsilon(\boldsymbol{v})) \le \frac{\sigma_0}{\sigma_T} \tag{6.53}$$

hold, known from the Michell theory, see (3.103). They can be true if

$$-\frac{\sigma_0}{\sigma_C} \le -\frac{\sigma_0}{\sigma_T}(\cot\theta_i)^2$$

for each i or if $\sigma_T \ge \sigma_C(\cot\theta_1)^2$.

Remark 6.2.2 If the loads are directed downwards, the structure is in compression and can be viewed as a Michell structure if $\sigma_C \ge \sigma_T(\cot\theta_1)^2$. In particular, the arch of Fig. 6.5 carrying the transverse load $q = \text{const}$ is a Michell structure if $\sigma_C \ge \sigma_T(\cot(\pi/6))^2$, or if $\sigma_C \ge 3\sigma_T$.

We conclude, that if $\sigma_C < 3\sigma_T$, the Michell structure corresponding to the load of intensity $q = \text{const}$ will not be a parabolic arch.

If $\sigma_C = \sigma_T$, then the condition of compatibility with the Michell theory is of the form $\cot\theta \le 1$ or $\theta \le \pi/4$. This condition is satisfied only in a central part of the arch, hence the arch cannot be optimal. The optimal structure for the transmissible load $q = \text{const}$, for the case of $\sigma_C = \sigma_T$, has been first discovered by Darwich et al. (2010) and thoroughly analytically constructed in Tyas et al. (2011) and Pichugin et al. (2012). This remarkable result will not be displayed here.

The optimal funiculars, or Prager-structures, discussed in this and in the previous section are solutions to the optimum design problems concerning transmissible loads, under an additional assumption on the stress being of given sign. We see now that on having found these solutions one can re-interpret them as Michell structures under given, fixed load with the stress bounded from both the sides characterized by specific values of permissible stresses: σ_C, σ_T. Thus, we confirm here the statement by Rozvany and Wang (1983, p. 522) that 'Prager-structures are actually not modified Michell frames but special cases of Michell frames'. This, by the way, justifies the title of the present book.

6.3 Archgrids of Least Weight

This section discusses optimum design of fully stressed archgrids of least weight transmitting transmissible vertical loads to the edges of a planar domain. The available exact results refer to a single point load, leading to simple trusses of specific geometry. The archgrids assume the form of canopies if the load is distributed. The main theoretical result concerning the distributed loads states that the fibres of archgrids are formed on a single surface.

6.3.1 Transmitting a Vertical Force to the Pin Supports on a Contour of a Circle

Let us consider the optimum design problem posed and solved in the paper by Rozvany, Wang and Dow (1982): construct the least volume structure transmitting a vertical force of magnitude P to pin supports S_i lying on the contour of the circle of origin at O and of radius r. The force is transmissible, its position P given by the vertical ordinate h is an unknown of the problem. The line of the force intersects the plane of the circle at point Q. The least volume structure is assumed to be a spatial truss composed of the bars PS_i, $i = 1, \ldots, n$; $n \geq 2$, see Fig. 6.9a.

If $n = 2$ the truss is planar and then the solution coincides with that of Example 6.1.2 in Sect. 6.1.4. If $n = 3$ and $S_1S_2S_3$ forms a triangle, then the equations of equilibrium of the node P determine uniquely the axial forces T_1, T_2, T_3 in the bars. If $n \geq 3$, the forces T_i (an axial force is positive if the bar is in tension) satisfy the following equilibrium conditions of the node P:

$$\sum_{i=1}^{n} T_i \sin\alpha_i + P = 0 \tag{6.54}$$

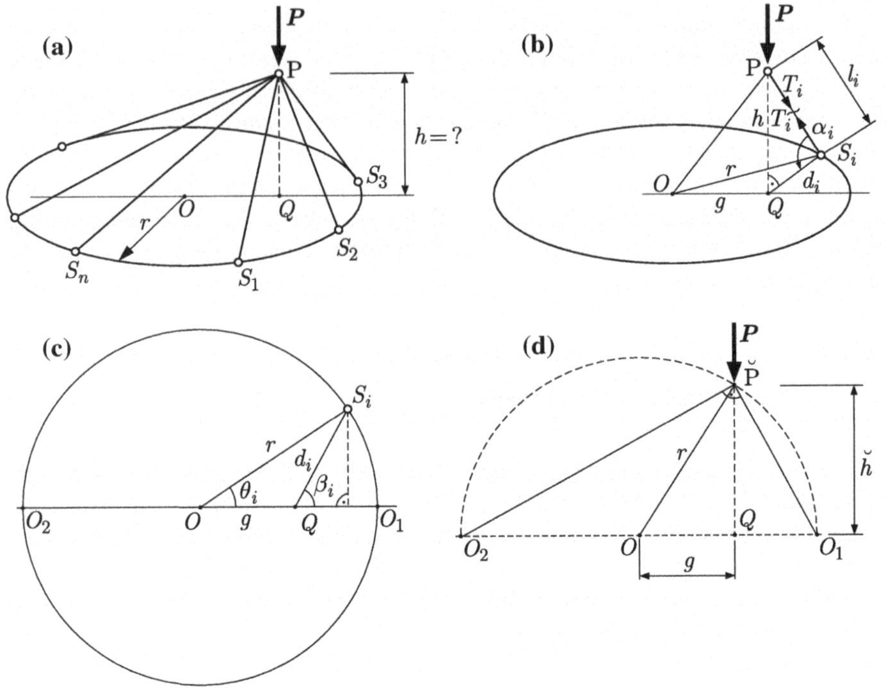

Fig. 6.9 The problem setting (**a**); geometrical consideration (**b**, **c**); the optimal position of the apical node (**d**)

$$\sum_{i=1}^{n} T_i \cos\alpha_i \cos\beta_i = 0 \tag{6.55}$$

$$\sum_{i=1}^{n} T_i \cos\alpha_i \sin\beta_i = 0 \tag{6.56}$$

where $\alpha_i = \angle QS_iP$, $\beta_i = \angle O_1QS_i$, see Fig. 6.9b, c. The angles $\theta_i = \angle QOS_i$ are linked with the angles α_i, β_i by

$$r\sin\theta_i = d_i \sin\beta_i, \qquad r\cos\theta_i = g + d_i\cos\beta_i \tag{6.57}$$

and

$$\sin\alpha_i = \frac{h}{l_i}, \qquad \cos\alpha_i = \frac{d_i}{l_i}, \qquad l_i = \sqrt{h^2 + (d_i)^2}, \\ (d_i)^2 = g^2 + r^2 - 2gr\cos\theta_i \tag{6.58}$$

with $d_i = |QS_i|$, $g = |OQ|$. We predict that all the bars are in compression: $T_i \le 0$. The areas of the cross sections A_i are chosen such that the stresses are made uniform: $A_i = -T_i/\sigma_C$. The volume of the structure is expressed by

$$V = -\frac{1}{\sigma_C} \sum_{i=1}^{n} T_i l_i \tag{6.59}$$

We shall find the height h of the structure for which the volume V is minimal. To solve this problem, we write down the volume in the form

$$V = -\frac{1}{\sigma_C} \sum_{i=1}^{n} (l_i)^2 \frac{T_i}{l_i} \tag{6.60}$$

and make use of the formula, see (6.58)

$$(l_i)^2 = h^2 + g^2 + r^2 - 2gr \cos \theta_i \tag{6.61}$$

We re-write (6.54) as follows

$$\sum_{i=1}^{n} \frac{T_i}{l_i} = -\frac{P}{h} \tag{6.62}$$

Let us substitute (6.61) into (6.60); we find

$$V = -\frac{1}{\sigma_C} \left[(h^2 + g^2 + r^2) \sum_{i=1}^{n} \frac{T_i}{l_i} - 2gr \underline{\sum_{i=1}^{n} \frac{T_i}{l_i} \cos \theta_i} \right] \tag{6.63}$$

Let us compute the term underlined in (6.63)

$$\sum_{i=1}^{n} \frac{T_i}{l_i} \cos \theta_i = \sum_{i=1}^{n} \frac{T_i}{l_i} \frac{1}{r} (g + d_i \cos \beta_i) = \frac{g}{r} \sum_{i=1}^{n} \frac{T_i}{l_i} + \frac{1}{r} \sum_{i=1}^{n} T_i \cos \alpha_i \cos \beta_i =$$
$$\frac{g}{r} \sum_{i=1}^{n} \frac{T_i}{l_i} = -\frac{gP}{rh} \tag{6.64}$$

We have made use of (6.57), (6.55) and (6.62). We compute now the volume by using the results (6.63) and (6.64)

$$V = -\frac{1}{\sigma_C} \left[(h^2 + g^2 + r^2) \left(-\frac{P}{h} \right) - 2gr \left(-\frac{gP}{rh} \right) \right] \tag{6.65}$$

which gives

$$V = \frac{P}{\sigma_C} \left(h + \frac{r^2 - g^2}{h} \right) \tag{6.66}$$

Let us note that $r^2 - g^2 \geq 0$. We compute the optimal height for which the volume given by the above expression is minimal:

$$\check{h} = \sqrt{r^2 - g^2} \tag{6.67}$$

hence, the volume of the optimal structure with the apical point $\check{P}$ equals

$$V = \frac{2}{\sigma_C} P\check{h} \tag{6.68}$$

Let us note that the results above are valid independently of the number of supports n. Note that the apical point $\check{P}$ lies on the circle $O_2\check{P}O_1$ of radius r: $|O\check{P}| = |OO_1|$, see Fig. 6.9d. The result (6.68) is a 3D counterpart of the formula (6.41).

The main result is that the apical point $\check{P}$ lies on the hemisphere over the circle of the contour along which the structure is supported. The number of bars of the optimal structure as well as the position of the hinge supports are not determined (yet must be greater than three).

6.3.2 *Transmitting a Point Load to Four Supports Forming a Cross*

We consider a four-bar truss subject to a single vertical point load of magnitude P at point P, see Fig. 6.10a. The vertical projection of the truss forms a cross.

The aim is to minimize the volume of the truss under the condition of the stress being equal to $-\sigma_C$, hence all the bars are predicted to be in compression. The formulation omits the conditions of deformation compatibility; the setting is the so-called plastic design. We start from decomposing the truss into two-bar trusses subject to the forces: $P/2 - P_1$ and $P/2 + P_1$, see Fig. 6.10b, c. The bars are assumed to be in compression, hence: $-P/2 \le P_1 \le P/2$. The volume of the fully stressed truss of Fig. 6.10b equals

$$V_1 = \frac{1}{\sigma_C}\left(\frac{1}{2}P + P_1\right)\left(h + \frac{ab}{h}\right) \tag{6.69}$$

while the volume of the truss of Fig. 6.10c is

$$V_2 = \frac{1}{\sigma_C}\left(\frac{1}{2}P - P_1\right)\left(h + \frac{cd}{h}\right) \tag{6.70}$$

Let $\eta = P_1/P$ and

$$C(\eta) = \frac{1}{2}(ba + cd) + \eta(ba - cd) \tag{6.71}$$

Due to $|\eta| \le 1/2$ the function $C(\eta)$ is positive. The volume of the structure of Fig. 6.10a is expressed by

$$V = \frac{P}{\sigma_C} f(h, \eta) \tag{6.72}$$

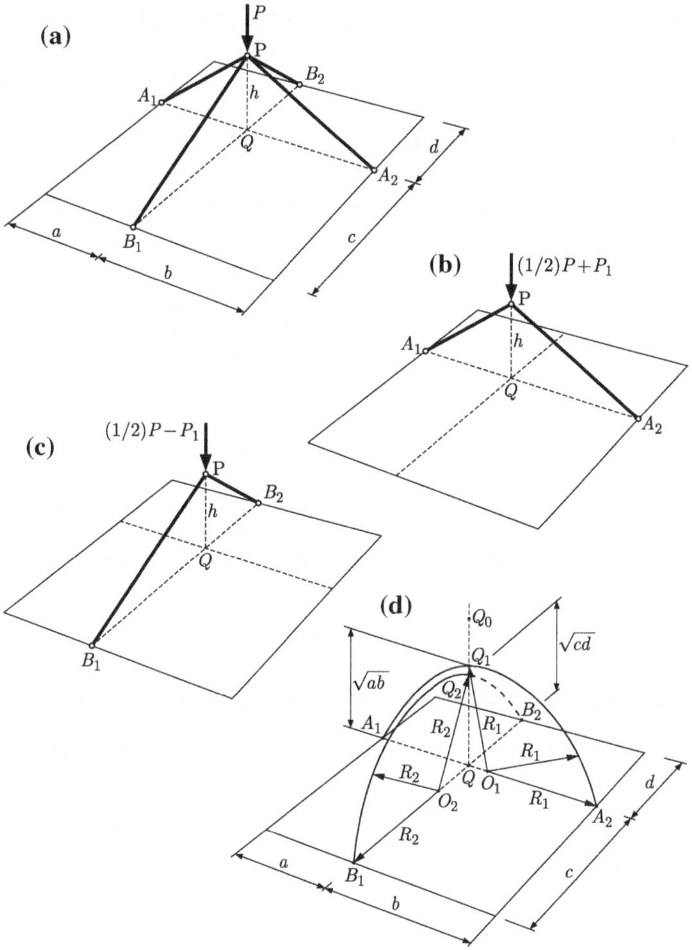

Fig. 6.10 The four-bar truss (**a**); decomposition into the two-bar trusses (**b**), (**c**); finding the lighter truss (**d**)

with

$$f(h, \eta) = h + \frac{C(\eta)}{h} \tag{6.73}$$

The minimizer $h = \breve{h}$ of (6.73) and the optimal value of the volume are expressed by

$$\breve{h} = \sqrt{C(\eta)}, \qquad \breve{V} = \frac{2P}{\sigma_C}\sqrt{C(\eta)} \tag{6.74}$$

In the final step, we perform minimization over η to arrive at

$$\breve{V} = \frac{2P}{\sigma_C} h^*, \qquad h^* = \min\left\{\sqrt{ab}, \sqrt{cd}\right\} \tag{6.75}$$

If $h^\star = \sqrt{ab} < \sqrt{cd}$ then $\eta^\star = 1/2$ and the structure of Fig. 6.10b is subject to the force P, while the structure of Fig. 6.10c is unloaded—its bars have zero cross sections.

If $h^\star = \sqrt{cd} < \sqrt{ab}$ then $\eta^\star = -1/2$ and the structure of Fig. 6.10b is unloaded—its bars have zero cross sections, while the structure of Fig. 6.10c is subject to the force P.

Thus, the minimization process chooses the structure of the smaller rise.

The process of finding the lightest truss is shown in Fig. 6.10d. We compare the circle of origin at O_1 and radius $R_1 = (a + b)/2$ with the circle of origin at O_2 and radius $R_2 = (c + d)/2$. These circles cut along the line QQ_0 orthogonal to the plane $A_1B_1A_2B_2$. The line QQ_0 cuts the contours of the circles at Q_1 and Q_2, their vertical ordinates being equal $\sqrt{ab}$ and $\sqrt{cd}$, respectively. The optimal truss—here the truss: B_1PB_2, with $P = Q_2$—is characterized by the smaller rise.

The final conclusion is that the optimal structure is a two-bar, statically determinate truss. Consequently, the conditions of compatibility of deformation are eventually fulfilled, although they have been initially disregarded.

6.3.3 Transmitting a Point Load to the Edge of a Simply Supported Strip

Let us consider the problem of optimal transmitting the vertical force at a point P to the hinges on the edges of a strip, see Fig. 6.11. The vertical projection of the point P on the plane of the strip is the point Q. Knowing the results of Example 6.1.2 of Sect. 6.1 we predict that the optimal structure will be a truss composed of two bars: $\bar{B}_1P$, $\bar{B}_2P$, see Fig. 6.11.

Due to the result (6.75) minimization of the volume means minimization of the function

$$g(\alpha) = \sqrt{|\bar{B}_2Q| \cdot |\bar{B}_1Q|} \tag{6.76}$$

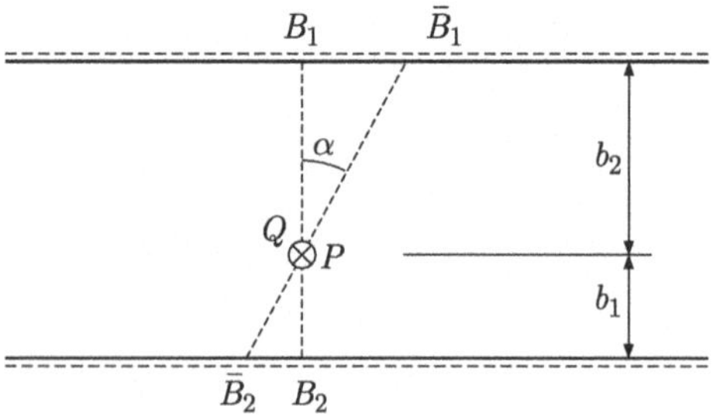

Fig. 6.11 The vertical force is applied over the point Q. The force should be transmitted by a two-bar truss to the edges of the strip

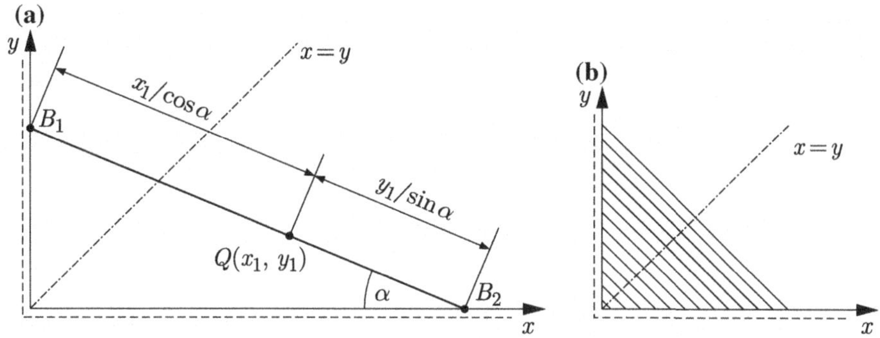

Fig. 6.12 The vertical force is applied at point P over the point Q; the two-bar truss has hinges at points B_1, B_2 (**a**); the optimal directions of the planes of optimal trusses (**b**)

or

$$g(\alpha) = \frac{b_1 b_2}{\cos^2 \alpha} \tag{6.77}$$

Its minimum is attained at $\alpha = 0$. Thus, the truss $B_1 P B_2$ is the lightest. Its plane is orthogonal to the edges of the strip.

6.3.4 *Transmitting a Point Load to the Edges of a Quadrant*

Consider the problem of optimal transmitting a vertical force P to the hinges along the mutually orthogonal edges $x = 0$, $y = 0$ of a quadrant, see Fig. 6.12a

Knowing the result of Example 6.1.2 of Sect. 6.1, we predict the optimal structure to be a two-bar truss $B_1 P B_2$. To minimize the weight of this truss it is sufficient to minimize the function

$$g(\alpha) = \frac{2x_1 y_1}{\sin 2\alpha} \tag{6.78}$$

The minimum is attained at $\alpha = \pi/4$; then $B_1 B_2$ is orthogonal to the line $x = y$, see Fig. 6.12b.

6.3.5 *Transmitting a Point Load to the Boundary of a Square*

We consider an optimal transmission of the vertical force P to the boundary of the square domain of the sides equal b, see Fig. 6.13. A possible truss is composed of the bars: $A_1 P$, $A_2 P$, $B_1 P$, $B_2 P$; the point P lies over the point Q of coordinates (x_1, y_1); let $x_1 + y_1 = a\sqrt{2}$. According to the discussion in Sect. 6.3.4 the interval $A_1 A_2$ should be orthogonal to the diagonal line $O O_1$.

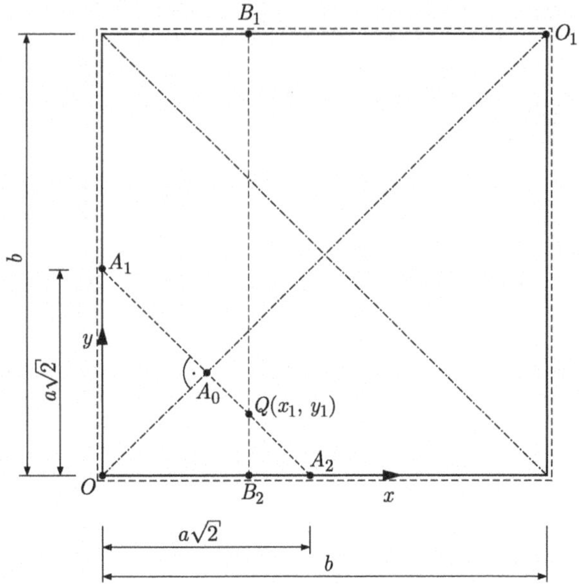

Fig. 6.13 The problem setting: transmit a vertical force at P (its projection is point Q) to the boundary of a square

According to the discussion in Sect. 6.3.3 the line B_1B_2 should be orthogonal to the axis x. We predict that the optimal truss will be either the two-bar truss B_1PB_2 or the truss A_1PA_2, depending on the position of point Q. We apply the condition (6.75). The truss B_1PB_2 is lighter than A_1PA_2, if

$$|B_1Q| \cdot |QB_2| < |A_1Q| \cdot |QA_2| \tag{6.79}$$

or

$$y_1(b - y_1) < \sqrt{(x_1)^2 + (a\sqrt{2} - y_1)^2}\sqrt{(a\sqrt{2} - x_1)^2 + (y_1)^2} \tag{6.80}$$

We assume that Q lies between A_0 and A_2, or $x_1 \geq a\sqrt{2}/2$. The inequality (6.80) simplifies to

$$x_1 > b - a\sqrt{2} \tag{6.81}$$

This bound is the strongest if x_1 is the smallest possible, or for $x_1 = a\sqrt{2}/2$. Hence, we obtain the condition

$$a > \frac{1}{3}b\sqrt{2} \tag{6.82}$$

which introduces the division of the interval OO_1 into three equal intervals: OD_1, D_1D_3, D_3O_1, cf. Fig. 6.14.

If the point Q_1 lies within the square $C_1C_2C_3C_3$, then the optimal structure is a single two-bar truss whose plane is parallel to the sides of the square, like the truss $B_1P_1B_2$, with point P_1 lying over the point Q_1 on the circle of the diameter B_1B_2.

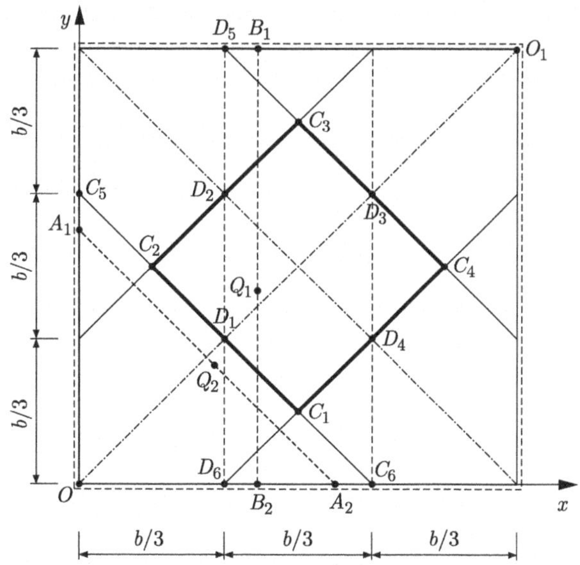

Fig. 6.14 Division of the domain into regimes

If the point Q_2 lies outside the square $C_1C_2C_3C_3$ then the optimal truss is a two-bar truss parallel to a diagonal, like $A_1P_2A_2$, while point P_2 lies over the point Q_2 on a circle of the diameter A_1A_2.

If point Q coincides with one of the vertices C_i, the volume of the discussed two-bar trusses are equal, since e.g.

$$|D_5D_1) \cdot |D_1D_6| = |C_5D_1| \cdot |D_1C_6| \tag{6.83}$$

Remark 6.3.1 Let us recall the optimal layout of beams in the problem of optimal design of a system of beams simply supported along the edges of a square domain and subject to an arbitrary vertical load; this layout is not shown in Sect. 5.4 of the present book, but can be found in the book by Rozvany (1976, Fig. 6.15) and has been recently confirmed numerically in Bołbotowski et al. (2018, Fig. 3). It is seen that this layout is similar but not identical to the layout described in Fig. 6.14. Thus, in general, the grillage solutions do not determine optimal directions of girders carrying vertically transmissible forces.

6.3.6 *Transmitting a Vertical Distributed Load to a Simply Supported Boundary*

The subject of consideration is the optimal transmitting a distributed load of intensity $q(x, y)$, applied over a given planar domain Ω, to its boundary Γ, along which an archgrid is planned to be simply supported. The load is transmissible and is directed

downward along the axis z, orthogonally to the domain Ω, parameterized by the Cartesian system (x, y). The aim is to find the lightest archgrid composed of fibers along x and y directions, fully stressed, with σ_C being the permissible stress in compression. The fibres in x directions form a surface $z_x(x, y)$ and the fibres in y directions form a surface $z_y(x, y)$. The fibres in x directions are subject to the vertical load of intensity $q_x = \frac{1}{2}q(x, y) + p(x, y)$, while the fibres in y directions are subject to $q_y = \frac{1}{2}q(x, y) - p(x, y)$; the load intensity $p(x, y)$ being arbitrary at this stage of the problem formulation, subject only to the condition $|p(x, y)| \leq \frac{1}{2}q(x, y)$ to prevent from tension. The domain Ω is characterized by the sections:
for $y = \text{const}$, $x_1(y) \leq x \leq x_2(y)$ and $c \leq y \leq d$;
for $x = \text{const}$, $y_1(x) \leq y \leq y_2(x)$ and $a \leq x \leq b$;
moreover,

$$L_1(y) = x_2(y) - x_1(y), \qquad L_2(x) = y_2(x) - y_1(x), \tag{6.84}$$

see Fig. 6.15a.

The archgrid is composed of fibres along the x and y directions; the fibres form arches of shapes z_x, z_y, see the section $y = \text{const}$, cf Fig. 6.15c. The arch shown is subject to q_x and to the reactions: $H_x(y)$ and $\hat{V}_A$, $\hat{V}_B$, at the ends $x = x_1(y)$, $x = x_2(y)$. The arch $z_x(x, y)$, for $y = const$, is formed to prevent the bending. We apply now the technique of Sect. 6.1. We introduce a simply supported beam of length $L_1(y)$, subject to $q_x(x, y)$. This load causes reactions $\hat{V}_A$, $\hat{V}_B$, the transverse force $\hat{Q}_x$ and the bending moment $\hat{M}_x$ satisfying

$$\frac{\partial \hat{Q}_x}{\partial x} + q_x = 0, \qquad \hat{Q}_x = \frac{\partial \hat{M}_x}{\partial x} \tag{6.85}$$

and, see (6.8)

$$z_x(x, y) = \frac{1}{H_x(y)} \hat{M}_x(x, y) \tag{6.86}$$

for $x_1(y) \leq x \leq x_2(y)$. In a similar manner we arrive at

$$\frac{\partial \hat{Q}_y}{\partial y} + q_y = 0, \qquad \hat{Q}_y = \frac{\partial \hat{M}_y}{\partial y} \tag{6.87}$$

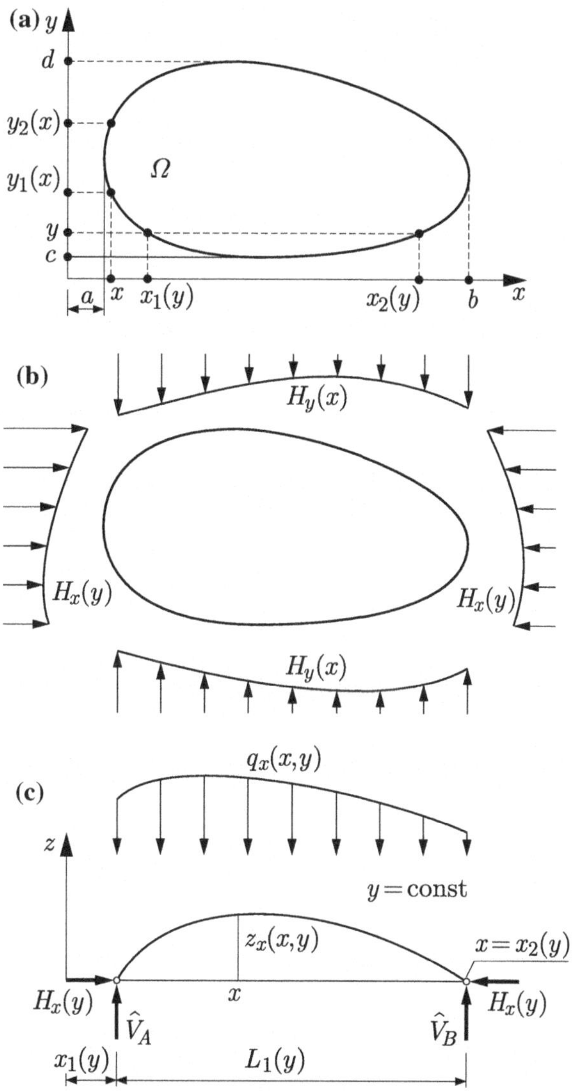

Fig. 6.15 The design domain Ω over which the archgrid is formed (**a**); horizontal forces around the edge (**b**); a transverse section for y = const (**c**)

and

$$z_y(x, y) = \frac{1}{H_y(x)} \hat{M}_y(x, y) \tag{6.88}$$

By adding $(6.85)_1$ and $(6.87)_1$, we find

$$\frac{\partial \hat{Q}_x}{\partial x} + \frac{\partial \hat{Q}_y}{\partial y} + q = 0 \tag{6.89}$$

Let us introduce the angles of slopes φ_x, φ_y such that

$$\tan \varphi_x = \frac{\partial z_x}{\partial x}, \qquad \tan \varphi_y = \frac{\partial z_y}{\partial y} \tag{6.90}$$

The cross sections of fibres are given by analogy with (6.13)

$$A_x = \frac{H_x}{\sigma_C \cos \varphi_x}, \; A_y = \frac{H_y}{\sigma_C \cos \varphi_y} \tag{6.91}$$

The elementary lengths of the arches are

$$ds_x = \frac{dx}{\cos \varphi_x}, \qquad ds_y = \frac{dy}{\cos \varphi_y} \tag{6.92}$$

The volume of all the fibrous arches in the x direction equals

$$V_x = \int_c^d \int_{x_1(y)}^{x_2(y)} A_x ds_x dy \tag{6.93}$$

Hence

$$V_x = \int_c^d \int_{x_1(y)}^{x_2(y)} \frac{H_x(y)}{\sigma_C \cos^2 \varphi_x(x, y)} dx\, dy \tag{6.94}$$

and, similarly, the volume of arches in the y-direction equals

$$V_y = \int_a^b \int_{y_1(x)}^{y_2(x)} \frac{H_y(x)}{\sigma_C \cos^2 \varphi_y(x, y)} dy\, dx \tag{6.95}$$

By using all the equations above, one can rearrange the last two expressions to the forms

$$\sigma_C V_x = \int_c^d L_1(y) H_x(y) dy + \int_c^d \frac{1}{H_x(y)} \int_{x_1(y)}^{x_2(y)} (\hat{Q}_x)^2 dx\, dy \tag{6.96}$$

$$\sigma_C V_y = \int_a^b L_2(x) H_y(x) dx + \int_a^b \frac{1}{H_y(x)} \int_{y_1(x)}^{y_2(x)} (\hat{Q}_y)^2 dy\, dx \tag{6.97}$$

Let us re-write (6.89) in the weak form:

$$G(\tilde{v}) = 0, \qquad G(\tilde{v}) = \int_\Omega q \tilde{v} dx\, dy - \int_\Omega \left(\hat{Q}_x \frac{\partial \tilde{v}}{\partial x} + \hat{Q}_y \frac{\partial \tilde{v}}{\partial y} \right) dx\, dy \tag{6.98}$$

for each appropriately regular $\tilde{v}$ vanishing on Γ.

Let us introduce the Lagrangian

$$\mathcal{L} = \sigma_C(V_x + V_y) + 2\sigma_C G(v) \tag{6.99}$$

involving the fields $\hat{Q}_x(x, y)$, $\hat{Q}_y(x, y)$, $v(x, y)$, $H_x(y)$, $H_y(x)$, while $p(x, y)$ is treated as fixed. The necessary condition for the volume $V_x + V_y$ to be minimal is the stationarity condition

$$\delta\mathcal{L} = 0 \tag{6.100}$$

with respect to the arbitrary variations $\delta\hat{Q}_x(x, y)$, $\delta\hat{Q}_y(x, y)$, $\delta H_x(y)$, $\delta H_y(x)$, δv. We obtain subsequently

$$H_x(y) = \left[\frac{1}{L_1(y)}\int_{x_1(y)}^{x_2(y)} (\hat{Q}_x)^2 dx\right]^{\frac{1}{2}} \tag{6.101}$$

$$H_y(x) = \left[\frac{1}{L_2(x)}\int_{y_1(x)}^{y_2(x)} (\hat{Q}_y)^2 dy\right]^{\frac{1}{2}} \tag{6.102}$$

$$\hat{Q}_x = H_x(y)\frac{\partial v}{\partial x}, \qquad \hat{Q}_y = H_y(x)\frac{\partial v}{\partial y} \tag{6.103}$$

and once again obtain (6.98) and (6.89). Let us note that the formulae (6.101), (6.102) are similar to the formulae (6.22) of the theory of planar funiculars. Substitution of (6.103) into (6.98) gives

$$a_H(v, \tilde{v}) = \int_\Omega q\tilde{v}dx\,dy \tag{6.104}$$

for each $\tilde{v}$ vanishing on Γ and

$$a_H(v, \tilde{v}) = \int_\Omega \left[H_x(y)\frac{\partial v}{\partial x}\frac{\partial \tilde{v}}{\partial x} + H_y(x)\frac{\partial v}{\partial y}\frac{\partial \tilde{v}}{\partial y}\right] dx\,dy. \tag{6.105}$$

This bilinear form is elliptic, since $H_x > 0$, $H_y > 0$. Let us combine (6.103) with (6.86), (6.88) to find

$$H_x\frac{\partial}{\partial x}(z_x - v) = 0, \qquad H_y\frac{\partial}{\partial y}(z_y - v) = 0 \tag{6.106}$$

Since z_x, z_y, v vanish along Γ we conclude

$$z_x(x, y) = v(x, y), \qquad z_y(x, y) = v(x, y) \tag{6.107}$$

We write $z_x = z_y = z(x, y)$ and note that $v = z$ satisfies (6.104). Its local form reads

$$H_x(y)\frac{\partial^2 z}{\partial x^2} + H_y(x)\frac{\partial^2 z}{\partial y^2} + q = 0 \tag{6.108}$$

This equation corresponds to the funicular equation (6.12). Thus, we have proved that the optimal archgrid is a canopy composed on a single surface $z = z(x, y)$. Let us substitute (6.101), (6.102) into the expression for $V = V_x + V_y$. We eventually find

$$\frac{\sigma_C}{2}V = \int_a^b \left[L_2(x)\int_{y_1(x)}^{y_2(x)}(\hat{Q}_y)^2 dy\right]^{\frac{1}{2}} dx + \int_c^d \left[L_1(y)\int_{x_1(y)}^{x_2(y)}(\hat{Q}_x)^2 dx\right]^{\frac{1}{2}} dy \tag{6.109}$$

Let us denote the right hand side of (6.109) by $J(\hat{\boldsymbol{Q}})$. Note that $J(\hat{\boldsymbol{Q}})$ is a positively homogeneous function of degree 1, since $J(\lambda\hat{\boldsymbol{Q}}) = \lambda J(\hat{\boldsymbol{Q}})$ if $\lambda \geq 0$. Moreover,

$$J(\boldsymbol{Q} + \boldsymbol{P}) \leq J(\boldsymbol{Q}) + J(\boldsymbol{P}) \quad \forall \quad \boldsymbol{P} = (P_x, P_y), \quad \boldsymbol{Q} = (Q_x, Q_y) \tag{6.110}$$

and, consequently, J satisfies Jensen's inequality and J is convex. The optimization problem reduces to

$$\min\left\{J(\hat{\boldsymbol{Q}}) \mid \text{div}\ \hat{\boldsymbol{Q}} + q = 0\right\} \tag{6.111}$$

If augmented with regularity assumptions, the problem above may be well posed since the functional is convex and the set of $\hat{\boldsymbol{Q}}$ satisfying (6.89) is non-empty. Assume that $\hat{\boldsymbol{Q}} = (\hat{Q}_x, \hat{Q}_y)$ is a minimizer of (6.111). Then we can compute H_x, H_y by (6.101), (6.102) and construct $z = v$ by solving (6.104). The function z thus constructed satisfies the *mean square slope conditions*

$$\frac{1}{L_1(y)}\int_{x_1(y)}^{x_2(y)}\left(\frac{\partial z}{\partial x}\right)^2 dx = 1 \qquad \frac{1}{L_2(x)}\int_{y_1(x)}^{y_2(x)}\left(\frac{\partial z}{\partial y}\right)^2 dy = 1 \tag{6.112}$$

readily seen from (6.101)–(6.103).

The problem dual to (6.111) has the form

$$\max\left\{\int_\Omega q v d\Omega \mid \nabla v \in B \quad \text{a.e. in } \Omega,\ v = 0 \text{ on } \partial\Omega\right\} \tag{6.113}$$

where B is a locking locus given by

$$B = \left\{\boldsymbol{p} = (p_1, p_2) \;\middle|\; \frac{1}{L_1(y)}\int_{x_1(y)}^{x_2(y)}(p_1)^2\, dx \leq 1, \right.$$
$$\left. \frac{1}{L_2(x)}\int_{y_1(x)}^{y_2(x)}(p_2)^2\, dy \leq 1, \quad (x, y) \in \Omega\right\} \tag{6.114}$$

see Czubacki and Lewiński (2018). Let us note that the function z constructed above satisfies the condition $\nabla z \in B$ since z satisfies the equalities (6.112).

6.4 Bibliographical Comments

While a catenary is a shape of a cable subject to the self-weight, the parabolas are shapes of cables subject to a transverse load of constant intensity q measured with respect to the horizontal axis. The shapes of suspended cables suggest the optimal shapes of arches subject to the vertical loads: the self-weight and the $q =$ const. Already Christiaan Huygens knew that parabolic arches under $q =$ const are uniformly stressed such that the bending moments and the transverse forces vanish, see Darwich et al. (2010) for deep historical remarks. The popular textbooks on structural mechanics published in the 1930s include chapters on funiculars and put emphasis on the analogy between the diagram of the bending moments along a simply supported beam and the shape of the funicular corresponding to the same vertical load. Since this analogy is usually not repeated in the modern textbooks, we have thought appropriate to include Sect. 6.1 with an elementary derivation of the shape of a bending-free arch subject to an arbitrary vertical load; we have followed the paper by Hetmański and Lewiński (2007). Among all such arches stressed up to a certain limit, say $-\sigma_C$ one can search the lightest. The question is well stated, since if an arch is very shallow, the horizontal reactions are very big, which leads to high areas of cross sections to keep the stress at the given level and thus results in designs of great volume. On the other hand, if an arch is higher and higher, its volume increases. Thus, there must exist an optimal rise of an arch corresponding to the smallest volume. This elementary optimization problem is solved in Sect. 6.1 leading to the condition (6.25), the famous '*gradient-dependent mean square slope condition*', as named by Rozvany and Wang (1983). In the case of the supports lying at the same level, the r.h.s. of (6.25) is equal to 1 and the condition assumes a remarkably simple form (6.37). The main conclusion is that optimization introduces a certain slope condition: the optimal arch should be neither too shallow nor too high. The next remarkable result is the formula (6.30), whose right-hand side may be interpreted as the potential energy of the load; also this formula can be found in Rozvany and Wang (1983, Eq. 17).

The optimal arch carrying a constant transverse load assumes the shape of a parabola inscribed into the equilateral triangle, see Fig. 6.5. This property has been noted in Rozvany and Wang (1983), but probably was known earlier.

The optimal two-bar truss carrying a single vertical force, shown in Fig. 6.6, is orthogonal, hence its apical point lies on a circle of the diameter equal to the distance between the supports. This remarkable property has been cleared up in Rozvany et al. (1982).

The optimal funiculars turn out to be global solutions to the topology optimization problem of constructing the lightest structures capable of transmitting a given vertical load to two pin supports and such that the allowable stress assumes either the value $-\sigma_C$—then the optimal structures are arches— or the value σ_T and then the optimal structures are cables. The theory of optimal layout of such structures has been put forward by Prager and Rozvany (1979), Rozvany and Wang (1983) and Wang and Rozvany (1983). The main idea is similar to that of Michell: to embed the statically admissible structures in the field of kinematically admissible virtual displacements undergoing specific locking locus conditions concerning the virtual strains, as explained in Sect. 6.2, see Lewiński and Sokół (2011). The problem is reduced to the kinematic problem (6.44), involving only the upper bound for the stress. Rozvany and Wang (1983) noted that this problem is equivalent to the Michell problem if the load is fixed at the optimal position and the lower bound for the stress is appropriately assumed, see Remark 6.2.1. Note that Michell problems refer to fixed loads while the funicular problem refers to the transmissible loads. Despite this difference, the optimal funiculars may be viewed as Michell structures, as noted in Rozvany and Wang (1983). The term: *transmissible load* appeared later, in the paper by Fuchs and Moses (2000) in which the material interpolation schemes have been applied.

The optimal funiculars in compression are the lightest structures provided that the permissible tensile stress equals zero; similarly, the optimal funiculars in tension are the lightest structures if the permissible compressive stress equals zero. Till 2010 there were no hints in the literature that funiculars may not be optimal if the stress conditions are relaxed to the usual form (2.40). The breakthrough has been done by Wael Darwich, Matthew Gilbert and Andrew Tyas who discovered the true shapes of the optimal structures transmitting a uniform load to two pin supports, see Darwich et al. (2010). The optimal layouts do depend on the ratio σ_T/σ_C; they resemble arches, but their haunch regions are dense nets of bars in compression and tension. These high-precision numerical predictions have been confirmed by Tyas, Pichugin and Gilbert (2011); the exact solution has been found: the optimum virtual displacement field filling up the whole half-plane has been constructed and all the required equilibrium conditions have been checked, thus making the solution complete. This analysis has shed a new light on Hemp's (1974) solutions of the same problem. Aleksey Pichugin, Andrew Tyas and Matthew Gilbert (2012) cleared up that Hemp's solution is fully correct if $\sigma_T/\sigma_C \leq 0.417785$.

The theory of archgrids has been developed in Rozvany and Prager (1979) and Rozvany et al. (1980; 1982). In particular, the problem discussed in Sect. 6.3.1 of optimal transmitting a vertical point load to supports on the contour of a circle has been solved in Rozvany et al. (1982).

The theory of archgrids formed over plane domains, simply supported along its boundary has been put forward by Rozvany and Prager (1979). The main results of this theory have been re-derived in Sect. 6.3.6. The solutions do depend on the assumed directions of the grid on the plane. Note yet, that the contemporary numerical approaches towards the design of gridworks are usually not based on Prager and Rozvany method, but make use of other numerical concepts, see Jiang et al. (2018), Richardson et al. (2013). One of the reason is that the gridworks are in general not formed over a plane domain and in many cases form grid structures composed of three families of bars, see Winslow et al. (2010).

References

Bołbotowski, K., He, L., & Gilbert, M. (2018). Design of optimum grillages using layout optimization. *Structural and Multidisciplinary Optimization*, *58*, 851–868.
Czubacki, R., & Lewiński, T. (2018). On optimal archgrids (submitted).
Darwich, W., Gilbert, M., & Tyas, A. (2010). Optimum structure to carry a uniform load between pinned supports. *Structural and Multidisciplinary Optimization*, *42*, 33–42.
Fuchs, M. B., & Moses, E. (2000). Optimal structural topologies with transmissible loads. *Structural and Multidisciplinary Optimization*, *19*, 263–273.
Hemp, W. S. (1974). Michell framework for uniform load between fixed supports. *Engineering Optimization*, *1*, 61–69.
Hetmański, K., & Lewiński, T. (2007). Forming the frames and arches which do not undergo bending. In W. Szcześniak (Ed.), *Theoretical foundations of civil engineering-XV* (pp. 231–246). Proceedings of the Polish-Ukrainian-Lithuanian transactions. Warszawa (in Polish): Oficyna Wydawnicza PW.
Jiang, Y., Zegard, T., Baker, W. F., & Paulino, G. H. (2018). Form finding of grid-shells using the ground structure and potential energy methods: A comparative study and assessment. *Structural and Multidisciplinary Optimization* (in press).
Lewiński, T., & Sokół, T. (2011). On plane funicular structures. In M. Kamiński & J. B. Obręebski (Eds.), *Lightweight Structures in Civil Engineering: Contemporary Problems : International Seminar of IASS Polish Chapters. Łódź 2 Dec 2011* (pp. 78–82). Naukowe: JBO MP Wydawnictwo.
Pichugin, A. V., Tyas, A., & Gilbert, M. (2012). On the optimality of Hemp's arch with vertical hangers. *Structural and Multidisciplinary Optimization*, *46*, 17–25.
Richardson, J. N., Adriaenssens, S., Coelho, R. F., & Bouillard, Ph. (2013). Coupled form-finding and grid optimization approach for single layer grid shells. *Engineering Structures*, *52*, 230–239.
Rozvany, G. I. N. (1976). *Optimal design of flexural systems*. London: Pergamon Press.
Rozvany, G. I. N., & Prager, W. (1979). A new class of structural optimization problems: Optimal archgrids. *Computer Methods in Applied Mechanics and Engineering*, *19*, 127–150.
Rozvany, G. I. N., & Wang, C.-M. (1983). On plane Prager-structures-I. *International Journal of Mechanical Sciences*, *25*, 519–527.
Rozvany, G. I. N., Nakamura, H., & Kuhnell, B. T. (1980). Optimal archgrids: Allowance for self-weight. *Computer Methods in Applied Mechanics and Engineering*, *24*, 287–304.
Rozvany, G. I. N., Wang, C.-M., & Dow, M. (1982). Prager-structures: Archgrids and cable networks of optimal layout. *Computer Methods in Applied Mechanics and Engineering*, *31*, 91–113.
Tyas, A., Pichugin, A. V., & Gilbert, M. (2011). Optimum structure to carry a uniform load between pinned supports: Exact analytical solution. *Proceedings of the Royal Society A*, *467*, 1101–1120.
Wang, C.-M., & Rozvany, G. I. N. (1983). On plane Prager-structures-II. Non-parallel external loads and allowances for selfweight. *International Journal of Mechanical Sciences*, *25*, 529–541.
Winslow, P., Pellegrino, S., & Sharma, S. B. (2010). Multi-objective optimization of free-form grid structures. *Structural and Multidisciplinary Optimization*, *40*, 257–269.

Chapter 7
Extensions of the Michell Theory

The present chapter concerns the extension of the Michell problems towards the multi-load cases and reveals the impact of the Michell theory on the other problems of topology optimization, especially those concerning the optimum material design.

7.1 Optimum Design of Trusses. Multiple Loads

The aim of the present section is twofold. Section 7.1.1 is aimed at formulating the minimum compliance problem of trusses subject to multiple load cases. Section 7.1.2 concerns a direct extension of Michell's problem to make it possible to encompass multiple loads. Thus, two different methods of designing trusses for multiple load cases are put forward.

The first method is based on constructing the Pareto front of solutions by minimizing a convex combination of compliances corresponding to the subsequent load variants. The solution forms the Pareto set of possible compromising solutions. This set is parameterized by the weights in the convex combinations of compliances. The main disadvantage of this approach is the form of the final problem: it does not reduce to the linear programming problem, hence cannot be easily applied for large-scale settings. However, the resulting optimal solutions are realistic: they are elastic structures capable of carrying the given loads. In many cases, these structures are structurally stable to all possible loads.

The second method assumes the volume as the target function, while the stresses are subject to constraints corresponding to all the subsequent load variants. The optimum designs are composed of layers; these optimum plastic designs suffer kinematic incompatibilities. This optimum design problem can be explicitly solved in case of two load variants by the method developed by Nagtegaal and Prager (1973). For more than three loads, the optimization procedure cannot be decomposed into independent Michell-like problems.

T. Lewiński et al., *Michell Structures*,
https://doi.org/10.1007/978-3-319-95180-5_7

7.1.1 Trusses of Minimal Weighted Compliance and Fixed Volume

The subject of the present section is an extension of the problem of minimum compliance of trusses formulated in Sect. 2.1 to the multiple load cases. We assume as before that the positions of nodes are fixed, while the cost of the design is expressed by the formula (2.1). We consider m variants of loads: $\{P_1^\alpha, \ldots, P_s^\alpha\}$, $\alpha = 1, \ldots, m$. The loads act non-simultaneously. We assume that the αth load causes the state of member forces T_k^α, $k = 1, \ldots, e$ where e is the maximal number of bars linking the given nodes. The member forces $\{T_k^\alpha\}$ are linked with the loads by the equilibrium equations:

$$\boldsymbol{T}^\alpha \cdot \boldsymbol{\Delta}(\bar{\boldsymbol{u}}) = \boldsymbol{P}^\alpha \cdot \bar{\boldsymbol{u}} \,\forall\, \bar{\boldsymbol{u}} \in \mathbb{R}^s \tag{7.1}$$

where $\boldsymbol{\Delta}(\bar{\boldsymbol{u}}) = \boldsymbol{B}\bar{\boldsymbol{u}}$ and $\boldsymbol{B}$ is the geometric matrix. The compliance of the truss subject to the load number α is expressed by

$$\wp^{(\alpha)}(\{EA_k\}) = \min\left\{\sum_{k=1} \frac{(T_k^\alpha)^2 l_k}{EA_k} \mid T_k^\alpha \text{ satisfying (7.1)}\right\} \tag{7.2}$$

Thus, the compliance is viewed as a function of stiffnesses $EA_1, \ldots, EA_e$. Consider a convex combination of the compliances

$$\wp(\{EA_k\}) = \sum_{\alpha=1}^{m} \eta_\alpha \wp^{(\alpha)}(\{EA_k\}) \tag{7.3}$$

where $0 \le \eta_\alpha \le 1$ and $\eta_1 + \ldots + \eta_m = 1$; η_α is a weight reflecting the role of a given load variant. We assume that the weights are given. The subject of the consideration is the problem (2.4) in which now $\wp(EA_k)$ is given by (7.3).

The operation of minimization over the stiffnesses can be performed first, which leads to the formula for the minimal compliance

$$\breve{\wp} = \frac{1}{\Lambda}(Z_2)^2 \tag{7.4}$$

where

$$Z_2 = \min\left\{\sum_{k=1}^{e} \sqrt{\sum_{\alpha=1}^{m} (\sqrt{\eta_\alpha} T_k^\alpha)^2\, l_k} \mid T_k^\alpha \text{ satisfying (7.1)}\right\} \tag{7.5}$$

Let $\breve{T}_k^\alpha$ be the minimizer of (7.5). The optimal stiffnesses are given by

$$\breve{EA}_i = \Lambda \frac{\sqrt{\sum_{\alpha=1}^{m} (\sqrt{\eta_\alpha} \breve{T}_i^\alpha)^2}}{\sum_{k=1}^{e} \sqrt{\sum_{\alpha=1}^{m} (\sqrt{\eta_\alpha} \breve{T}_k^\alpha)^2}\, l_k} \tag{7.6}$$

These results have been found by Bendsøe et al. (1994a) in a different context. The problem (7.5) can be reformulated as below

$$Z_1 = \tag{7.7}$$
$$\max \left\{ \sum_{\alpha=1}^{m} \sqrt{\eta_\alpha} \boldsymbol{P}^\alpha \cdot \boldsymbol{v}^{(\alpha)} \;\middle|\; \boldsymbol{v}^\alpha \text{ such that } \sqrt{\sum_{\alpha=1}^{m} (\varepsilon_k(\boldsymbol{v}^{(\alpha)}))^2} \le 1,\ k = 1, \ldots, e \right\}$$

where $\varepsilon_k(\boldsymbol{v}) = \Delta_k(\boldsymbol{v})/l_k$. The results (7.1)–(7.7) will be given now a new form. Let us introduce the norms in $\mathbb{R}^m$

$$||\boldsymbol{a}||_\eta = \sqrt{\sum_{\alpha=1}^{m} \eta_\alpha (a^{(\alpha)})^2} \qquad ||\boldsymbol{a}||_\eta^\star = \sqrt{\sum_{\alpha=1}^{m} \frac{1}{\eta_\alpha} (a^{(\alpha)})^2} \tag{7.8}$$

where $\boldsymbol{a} = (a^{(1)}, \ldots, a^{(m)}) \in \mathbb{R}^m$. These norms are mutually dual, see (3.23), (3.24) in Chap. 3. Let us define the vectors

$$\boldsymbol{T}_k = ({T_k}^{(1)}, \ldots, {T_k}^{(m)}) \in \mathbb{R}^m \;\; \boldsymbol{\varepsilon}_k = (\varepsilon_k(\boldsymbol{u}^{(1)}), \ldots, \varepsilon_k(\boldsymbol{u}^{(m)})) \in \mathbb{R}^m \tag{7.9}$$

The results (7.5)–(7.7) may be written as follows:

$$Z_2 = \min \left\{ \sum_{k=1}^{e} ||\boldsymbol{T}_k||_\eta l_k \;\middle|\; T_k^\alpha, \alpha = 1, \ldots, m, \text{ satisfying (7.1)} \right\} \tag{7.10}$$

$$\breve{EA}_i = \Lambda \frac{||\breve{\boldsymbol{T}}_i||_\eta}{\sum_{k=1}^{e} l_k ||\breve{\boldsymbol{T}}_k||_\eta} \tag{7.11}$$

$$Z_1 = \max \left\{ \sum_{\alpha=1}^{m} \boldsymbol{P}^{(\alpha)} \cdot \boldsymbol{u}^{(\alpha)} \;\middle|\; \boldsymbol{u}^{(\alpha)} \text{ such that } ||\boldsymbol{\varepsilon}_k||_\eta^\star \le 1,\ k = 1, \ldots, m \right\} \tag{7.12}$$

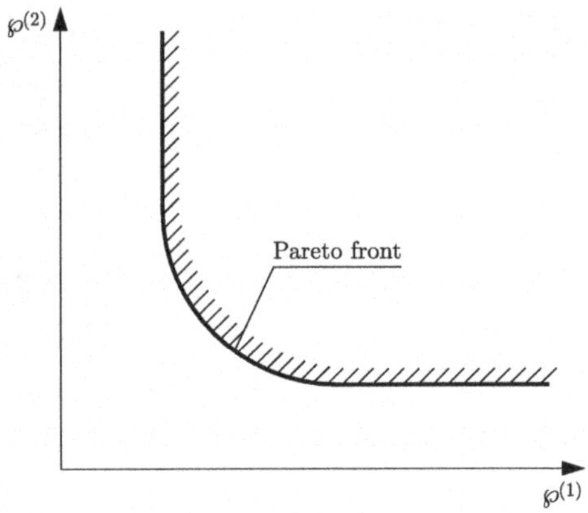

Fig. 7.1 A typical shape of the Pareto front for the two load weighted compliance problem

The above formalism makes it easy to show the equality $Z_1 = Z_2$, or the mutual duality being the problems (7.10) and (7.12). This proof is similar to the proof of (3.41), hence will be omitted.

If $m \geq 2$ the problems (7.10), (7.12) cannot be reduced to a linear programming setting. The numerical methods for (7.10) can be constructed using the non-linear programming methods, like in Czarnecki (2013).

By solving the problems (7.10), (7.12) for subsequent components of the vector $\boldsymbol{\eta} = (\eta_1, \ldots, \eta_m)$, we can build the Pareto front of the compromising solutions. For a selected vector $\boldsymbol{\eta}$, we compute the optimal values of the design variables by using (7.11) and compute the compliances corresponding to the subsequent load cases by (7.1). Then, we treat these compliances as coordinates of points in $\mathbb{R}^m$. For instance, if $m = 2$, $\eta_1 = \eta$, $\eta_2 = 1 - \eta$, the points of coordinates $(\wp^1(E\breve{A}_k), \wp^2(E\breve{A}_k)$ form a contour being a Pareto front. The results by Czarnecki (2013) prove that the Pareto front has the shape like in Fig. 7.1, having two asymptotes. This is the edge of a domain of points satisfying the cost condition (2.1) and not satisfying minimum of (7.3).

Consider now the optimum design problem

$$\min \{\max \{\wp^\alpha(EA_k), \alpha = 1, \ldots, m \mid EA_k \text{ satisfying (7.3)}\}\} \tag{7.13}$$

called the worst case design problem, see Achtziger (1997), Czarnecki (2013) noted that its solution lies on the Pareto front described before.

7.1.2 Trusses of Minimal Volume. Plastic Design

Consider the problem of the volume minimization under the condition of the stress σ in members being bounded: $-\sigma_C \leq \sigma \leq \sigma_T$ for each αth load case, $\alpha = 1, \ldots, m$.

Thus, we extend the problem of Sect. 2.2 to the multiple load case. The problem (2.41) assumes now form:

$$V_{\min} = \min\left\{ \sum_{k=1}^{e} A_k l_k \;\middle|\; T_k^{\alpha} \text{ are linked with } \boldsymbol{P}^{\alpha} \text{ by (7.1) and} \right.$$
$$\left. -A_k\sigma_C \le T_k^{\alpha} \le A_k\sigma_T, A_k \ge 0 \right\} \tag{7.14}$$

By passing to the dual formulation, we obtain

$$V_{\min} = \frac{1}{\sigma_0} Z \tag{7.15}$$

where

$$Z = \max\left\{ \sum_{\alpha=1}^{m} \boldsymbol{P}^{(\alpha)} \cdot \boldsymbol{u}^{(\alpha)} \;\middle|\; \text{over } \boldsymbol{u}^{(\alpha)} \in \mathbb{R}^s,\ \boldsymbol{\Delta}_T^{(\alpha)},\ \boldsymbol{\Delta}_C^{(\alpha)} \in \mathbb{R}^e, \right.$$
$$\boldsymbol{B}\boldsymbol{u}^{(\alpha)} - \boldsymbol{\Delta}_T^{(\alpha)} + \boldsymbol{\Delta}_C^{(\alpha)} = \boldsymbol{0},$$
$$\left. \sum_{\alpha=1}^{m} \left(\frac{\sigma_T}{\sigma_0} \Delta_T^{(\alpha)} + \frac{\sigma_C}{\sigma_0} \Delta_C^{(\alpha)} \right) \le \boldsymbol{l},\quad \boldsymbol{\Delta}_T^{(\alpha)} \ge \boldsymbol{0},\quad \boldsymbol{\Delta}_C^{(\alpha)} \ge \boldsymbol{0} \right\} \tag{7.16}$$

Let us introduce the strain components

$$\varepsilon_k^{(\alpha)C} = \frac{(\Delta_C^{\alpha})_k}{l_k}, \qquad \varepsilon_k^{(\alpha)T} = \frac{(\Delta_T^{\alpha})_k}{l_k}, \tag{7.17}$$

Now the second condition in (7.16) can be expressed by

$$\sum_{\alpha=1}^{m} \left(\frac{\sigma_T}{\sigma_0} \varepsilon_k^{(\alpha)T} + \frac{\sigma_C}{\sigma_0} \varepsilon_k^{(\alpha)C} \right) \le 1 \tag{7.18}$$

where the left-hand side will be called a total normalized strain.

The Decomposition Theorem by Nagtegaal and Prager

Nagtegaal and Prager (1973) noted that in the case of $m = 2$ and $\sigma_C = \sigma_T = \sigma_0$, the problem (7.14) may be composed of two solutions referring to the independent loads:

$$\boldsymbol{P} = \frac{1}{2}(\boldsymbol{P}^{(1)} + \boldsymbol{P}^{(2)}), \qquad \boldsymbol{Q} = \frac{1}{2}(\boldsymbol{P}^{(1)} - \boldsymbol{P}^{(2)}) \tag{7.19}$$

respectively. Thus, the two load problem can be replaced by two problems each of which corresponds to one load case.

Indeed, let the pairs $(T_k^{(1)}, \boldsymbol{P}^{(1)})$, $(T_k^{(2)}, \boldsymbol{P}^{(2)})$ be statically compatible. Then, the areas of the cross sections are estimated as follows:

$$A_k \geq \frac{1}{\sigma_0}|T_k^{(1)}| \quad \text{and} \quad A_k \geq \frac{1}{\sigma_0}|T_k^{(2)}| \tag{7.20}$$

Since the volume is being minimized, at least one of the above constraints has to be active, or

$$\sigma_0 A_k = \max\left\{|T_k^{(1)}|, |T_k^{(2)}|\right\} \tag{7.21}$$

Let us make use of the identity

$$\max\{|a|, |b|\} = \frac{1}{2}|a+b| + \frac{1}{2}|a-b|, a, b \in \mathbb{R} \tag{7.22}$$

to express the areas of the cross sections by

$$A_k = \left|\frac{T_k^{(1)} + T_k^{(2)}}{2\sigma_0}\right| + \left|\frac{T_k^{(1)} - T_k^{(2)}}{2\sigma_0}\right| \tag{7.23}$$

Thus, the volume of the optimal structure equals

$$V = \frac{1}{2\sigma_0}\sum_{k=1}^{e}\left[|T_k^{(1)} + T_k^{(2)}| + |T_k^{(1)} - T_k^{(2)}|\right] l_k \tag{7.24}$$

Let us introduce the component member forces

$$T_k^P = \frac{1}{2}(T_k^{(1)} + T_k^{(2)}), \qquad T_k^Q = \frac{1}{2}(T_k^{(1)} - T_k^{(2)}), \tag{7.25}$$

It is seen that the pairs $(T_k^P, \boldsymbol{P})$, $(T_k^Q, \boldsymbol{Q})$ are statically compatible. The volume of the optimal structure equals

$$V = V^P + V^Q \tag{7.26}$$

where

$$V^P = \min\left\{\sum_{k=1}^{e} A_k^P l_k \;\middle|\; \boldsymbol{B}^\top \boldsymbol{T}^P = \boldsymbol{P},\ |T_k^P| \leq \sigma_0 A_k^P\right\} \tag{7.27}$$

$$V^Q = \min\left\{\sum_{k=1}^{e} A_k^Q l_k \;\middle|\; \boldsymbol{B}^\top \boldsymbol{T}^Q = \boldsymbol{Q},\ |T_k^Q| \leq \sigma_0 A_k^Q\right\} \tag{7.28}$$

Upon solving the above optimum design problems, one eventually computes the areas of the cross sections

$$A_k = A_k^P + A_k^Q \tag{7.29}$$

while the axial forces in the bars are given by

$$T_k^1 = T_k^P + T_k^Q, \qquad T_k^2 = T_k^P - T_k^Q. \tag{7.30}$$

This theorem furnishes a strong tool for solving the two load optimum design problems.

Remark 7.1.1 Let us note that if

$$||\boldsymbol{P}^{(1)}|| = ||\boldsymbol{P}^{(2)}|| \tag{7.31}$$

then the load $\boldsymbol{P}$ is orthogonal to the load $\boldsymbol{Q}$. Indeed, it is easy to check that (7.31) implies

$$\boldsymbol{P} \cdot \boldsymbol{Q} = \frac{1}{4}(||\boldsymbol{P}^{(1)}||^2 - ||\boldsymbol{P}^{(2)}||^2) \tag{7.32}$$

which ends the proof.

Example 7.1.1 (*The spatial cantilever for two alternative loads*) Consider the problem of optimal transmitting two alternate point loads to the base of the parallelepiped of dimensions $d \times d \times 2d$ being the design domain. The forces are applied at the central point of the upper side and are directed parallel to its edges; the magnitudes of the forces are equal: $||\boldsymbol{P}^{(1)}|| = ||\boldsymbol{P}^{(2)}|| = P$. Moreover, we assume that $\sigma_T = \sigma_C = \sigma_0$.

The numerical solution to this problem has been constructed with using the ground structure method with the division: $50 \times 50 \times 100$ of the design domain, which corresponds to the resulting number of potential bars equal 28 701 638 392. The solution of such a large problem can be constructed by using a technique of a selective choice of active bars and nodes developed by Sokół (2016). One can note that the solution is composed of two planar Michell trusses in the two mutually orthogonal planes crossing the design domain along the diagonals of the upper and lower bases. This result confirms the thesis of Nagtegaal and Prager theorem on decomposition of the optimal solution for the two load problems. Moreover, this theorem makes it possible to construct an exact solution to the above problem. To this end, one should consider two independent problems for the component loads given by Eq. (7.19), cf. Fig. 7.2b, c. The optimal planar structures are the Michell–Hemp cantilevers composed of two circular fans and the Hencky nets, see Sect. 4.8. Their volumes can be computed by Eqs. (4.440a), (4.441) by substituting the appropriate value of the angle of the fans; this angle is determined by Eqs. (4.410) specifying the geometrical data. In the case discussed, we obtain $\alpha \approx 0.6193403696 \approx 35.4856°$, which corresponds to the volume $V = 5.808677758\dfrac{Pd}{\sigma_0}$.

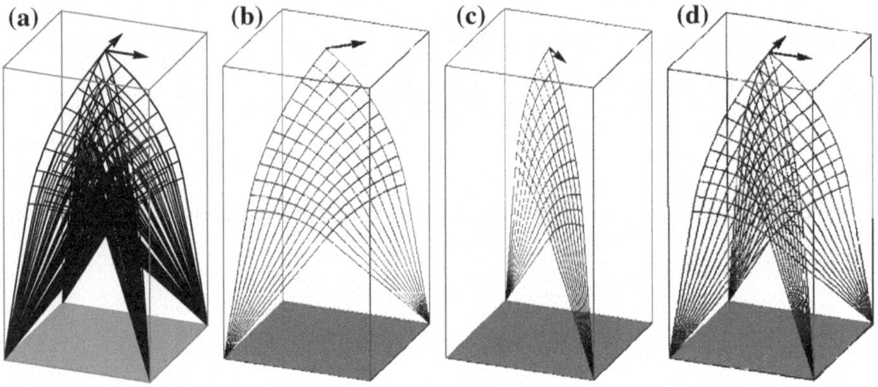

Fig. 7.2 The optimal 3D cantilever for two alternative loads: **a** numerical solution, **b** analytical solution for the first component load $\boldsymbol{P} = \frac{1}{2}(\boldsymbol{P}^{(1)} + \boldsymbol{P}^{(2)})$, **c** analytical solution for the second component load $\boldsymbol{Q} = \frac{1}{2}(\boldsymbol{P}^{(1)} - \boldsymbol{P}^{(2)})$, **d** final solution by superposition of results **b** and **c**

Eventually, the composition of trusses of Fig. 7.2b, c gives the structure of Fig. 7.2d of the total volume $V = 11.61735552\frac{Pd}{\sigma_0}$, while the volume computed numerically equals $V = 11.626442\frac{Pd}{\sigma_0}$, hence the result differs in 0.08%.

Let us note that the spatial truss of Fig. 7.2d is statically determinate. Thus, according to Rozvany et al. (2014a, b, c), the optimal solution within the plastic design setting found here holds for the elastic design setting.

Lastly, let us pay attention to the method of Nowak et al. (2018) which leads also to the optimal designs similar to that of Fig. 7.2a if adopted to the two load cases.

Example 7.1.2 (*Three load case problem*) We consider the McConnel (1974) problem, see Fig. 7.3, of optimal transmitting of three vertical forces to two pin supports. The forces are placed at three points dividing the span l between the supports into four intervals of equal lengths; the magnitudes of the forces are equal, or $||\boldsymbol{P}^{(1)}|| = ||\boldsymbol{P}^{(2)}|| = ||\boldsymbol{P}^{(3)}|| = P$. The design domain is the upper half-plane. As before, we assume that the permissible stresses in tension and compression are equal σ_0. If the forces act simultaneously, which refers to the original McConnel problem, we obtain the optimal truss of the layout already discussed before (Fig. 4.58a). The volume of the truss in Fig. 7.3a equals $V_n = 3.034595\frac{Pl}{\sigma_0}$. This result refers to the relatively small ground structure of division 200×75 comprising about 71 million of potential bars. If symmetry of the problem is taken into account and much denser grid is used, then one arrives at $V_n = 3.032895\frac{Pl}{\sigma_0}$, while the exact analytical solution equals $V_n = 3.031762563\frac{Pl}{\sigma_0}$, see Sokół (2014), where both these results have been obtained.

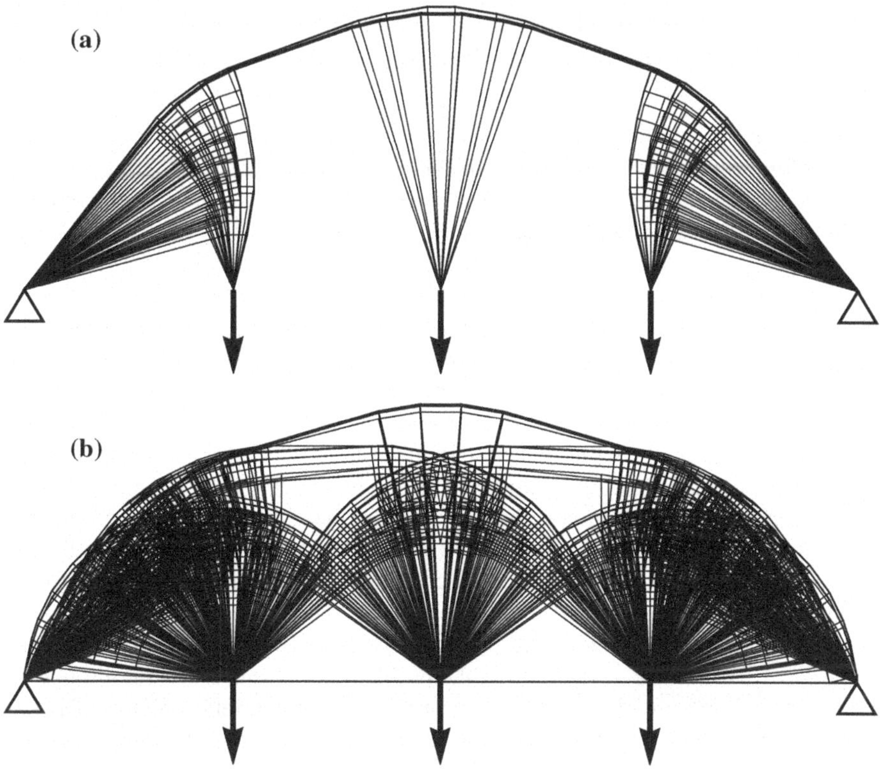

Fig. 7.3 The generalized McConnel problem: **a** optimal layout for three forces acting simultaneously (one load case), **b** optimal layout for three alternative forces

If the forces act non-simultaneously, the optimal layout changes substantially, see Fig. 7.3b. The optimal solution turns out to be a three-layered structure, composed of three Michell frameworks, one for each layer. The stacking is transverse to the given planar design domain.

The form of the solution suggests that the optimal structure may be constructed by appropriate decomposition of the loads, by solving three single load problems independently and, finally, by superimposing the layouts to arrive at the final, layered solution. Unfortunately, such decomposition is unknown, since the available methods of decomposition proposed by Rozvany and Hill (1978) do not apply here. Thus, till now, the exact solution to this problem is not available.

The volume of the structure of Fig. 7.3b has been found numerically: $V_n = 3.894816\frac{Pl}{\sigma_0}$; this volume is greater than the volume of Fig. 7.3a.

7.2 The Methods of Optimum Design of Shape and Material Inspired by the Theory of Michell Structures

Due to availability of the graphically clear and intellectually inspiring exact solutions to Michell's problems of optimum design, this theory has achieved recognition among structural and mechanical engineers. As shown in the previous chapters, the optimal Michell structures emerge from the given design domain Ω; they occupy a material domain Ω_m which occurs to be the support of a measure representing the stress field $\breve{\tau}$ being the minimizer of the functional in (3.95) (in the case of unequal permissible stresses in tension and compression) or of the functional in (3.33) referring to the case when the permissible stresses in tension and compression are equal. Just the specific form of these functionals is the heart of the method: their integrands are expressed by *gauge* functions (see Sect. 15 in Rockafellar 1970) or by norms in $\mathbb{E}^2_s$. This specific feature of the auxiliary problems of the Michell theory determines the exceptional properties of the solutions. The main property is the emergence of a material subdomain Ω_m which is cut out of the design domain Ω. This property is especially well seen if the optimal structure is generated by the set of self-equilibrated point loads and if the design domain is the full plane. Then segments of the boundary of the domain Ω_m are strengthened by ribs, which shows clearly the final shape of the structure. Thus, the theory of Michell structures delivers hints of how to set the problems of a simultaneous solving of the problem of optimum structural topology along with the problem of optimal design of the microstructural properties. The key to success to achieve a synthesis of the two aims is a reduction of the optimization problem to the auxiliary problem of (3.95) type involving the integrand being a gauge or a norm.

The Michell problem reduces to the two mutually dual problems: to the kinematic problem (3.85) and to the static problem (3.95). These problems have the following forms:

$$\mathcal{Z}_1 = \sup\{\mathcal{L}(\boldsymbol{v}) \mid \boldsymbol{v} \text{ kinematically admissible in } \Omega \text{ such that } \rho^\star(\boldsymbol{\varepsilon}(\boldsymbol{v}(x))) \leq 1\} \tag{7.33}$$

and

$$\mathcal{Z}_2 = \inf\left\{\int_\Omega \rho(\boldsymbol{\tau})dx \mid \boldsymbol{\tau} \text{ statically admissible}\right\} \tag{7.34}$$

where $\rho(\cdot)$ is a gauge or a norm in $\mathbb{E}^2_s$, while $\rho^\star(\cdot)$ is the polar gauge or a norm dual to $\rho(\cdot)$ defined by, cf. (3.108)

$$\rho^\star(\boldsymbol{\varepsilon}) = \max_{\boldsymbol{\tau}\in\mathbb{E}^2_s\setminus\{\mathbf{0}\}} \frac{\boldsymbol{\tau}\cdot\boldsymbol{\varepsilon}}{\rho(\boldsymbol{\tau})} \tag{7.35}$$

Recall that the field $\boldsymbol{\tau}$ is statically admissible, if

$$\int_\Omega \boldsymbol{\tau}\cdot\boldsymbol{\varepsilon}(\boldsymbol{v})dx = \mathcal{L}(\boldsymbol{v}) \text{ for each kinematically admissible } \boldsymbol{v} \tag{7.36}$$

By augmenting the problems (7.33), (7.34) with appropriate regularity assumptions, one can prove that $\mathcal{Z}_1 = \mathcal{Z}_2$. The equivalence of the mentioned problems can be read off from the paper by Bouchitté and Fragala (2007). Let us recall that in the case of the permissible stresses in tension and compression being equal, the function $\rho(\cdot)$ is defined by

$$\rho(\boldsymbol{\tau}) = \sum_{i=1}^{n} |\lambda_i(\boldsymbol{\tau})| \tag{7.37}$$

see (3.27), and its dual is expressed by

$$\rho^{\star}(\boldsymbol{\varepsilon}) = \max_{i}\{|\lambda_i(\boldsymbol{\varepsilon})|\} \tag{7.38}$$

The above polar pair gives the following estimate of the scalar product:

$$\boldsymbol{\tau} \cdot \boldsymbol{\varepsilon} \leq \rho(\boldsymbol{\tau})\rho^{\star}(\boldsymbol{\varepsilon}) \tag{7.39}$$

see p. 131, Sec. 15 in Rockafellar (1970). Golay and Seppecher (2001) noted that the problem of optimum design of the thickness $h \geq 0$ of a plate loaded in the plane reduces to the problem (7.33). In Czarnecki and Lewiński (2013), this setting has been complemented with the dual, stress-based problem of type (7.34). The additional assumption: $h \leq h_{\min}$ cancels this simplicity and then the problems like (7.33), (7.34) do not appear.

Bouchitté and Buttazzo (2001) have put forward and solved an abstract problem of optimum design of a support of a mass of an elastic structure. The minimum compliance problem has been led to the pair of problems (7.33), (7.34). These results have been later extended in Bouchitté and Fragala (2007) towards bending problems.

The methods aimed at designing the elastic moduli usually involve the lower and upper local bounds for the moduli, to prevent from degeneracy and not to allow very high values of the moduli. However, such assumptions make the optimum design formulations complicated and make it impossible to reduce the problem to the auxiliary problems (7.33), (7.34). Czarnecki and Lewiński (2012) noted that if these local additional bounds are omitted (the only assumption retained being the positive semi-definiteness of the elastic moduli tensor) then the popular free material design (FMD), proposed in Ringertz (1993) and Bendsøe et al. (1994b), reduces to the pair of problems (7.33), (7.34). By virtue of this reduction, the FMD method becomes the topology optimization method coupled with the optimum material design. This version of FMD will be called: the anisotropic material design (AMD). This result became an inspiration to more realistic versions of FMD: to Young's modulus design (YMD), cubic material design (CMD) and isotropic material design (IMD), put forward by Czarnecki, Czubacki, Dzierżanowski, Lewiński and Wawruch in years 2014–2017. The above methods have been appropriately built in such a manner to make them reducible to the auxiliary problems (7.33), (7.34). Let it be stressed: just the Michell theory was the inspiration for the mentioned methods of material design. These methods will be briefly outlined in the subsequent sections.

7.2.1 In-Plane Loaded Plates of Optimal Thickness: The Variable Thickness Sheet Problem

We consider the problem of optimum choice of the thickness h of an in-plane loaded plate of the middle plane Ω. The plane domain Ω is parameterized by the Cartesian system (x_1, x_2); the axis x_3 is orthogonal to the middle plane. The domain of the plate is given by $\Omega \times [-h(x_1, x_2)/2,\ h(x_1, x_2)/2]$. According to the plane stress assumption, the density of energy is expressed as

$$j^{\star}(\boldsymbol{\tau}) = \frac{1}{2h}[\rho_1(\boldsymbol{\tau})]^2,$$

where

$$\rho_1(\boldsymbol{\tau}) = \sqrt{\frac{1}{E}[(\tau_{11})^2 - 2\nu\tau_{11}\tau_{22} + (\tau_{22})^2 + 2(1+\nu)(\tau_{12})^2]} \tag{7.40}$$

E being Young modulus and ν being Poisson's ratio. Let $\Sigma_p(\Omega)$ be the set of statically admissible stresses in the plate loaded in plane. The compliance of the plate is given by (3.63), where $j^{\star}(\boldsymbol{\tau})$ is defined as above.

The subject of consideration is designing the thickness $h \geq 0$ to minimize the compliance under the condition of the volume being given:

$$\int_{\Omega} h dx = V_0 \tag{7.41}$$

We do not introduce an upper bound on h. The problem thus formulated reduces to the problems (7.33), (7.34) where $\rho = \rho_1$ and the dual norm has the form

$$\rho_1^{\star}(\boldsymbol{\varepsilon}) = \sqrt{\frac{E}{1-\nu^2}[(\varepsilon_{11})^2 + 2\nu\varepsilon_{11}\varepsilon_{22} + (\varepsilon_{22})^2 + 2(1-\nu)(\varepsilon_{12})^2]} \tag{7.42}$$

Let $\breve{\boldsymbol{\tau}}$ be the minimizer of (7.34) with the norm $\rho = \rho_1$. The optimal thickness is given by

$$\breve{h} = V_0 \frac{\rho_1(\breve{\boldsymbol{\tau}})}{\int_{\Omega} \rho_1(\breve{\boldsymbol{\tau}}) dx} \tag{7.43}$$

Thus, $\breve{h} = 0$ there, where $\breve{\boldsymbol{\tau}} = \mathbf{0}$.The solution $\breve{\boldsymbol{\tau}}$ determines the shape of the middle plane Ω of the plate. There, where $\breve{\boldsymbol{\tau}}$ vanishes the domain becomes a void or a segment of the boundary is cut out. The solution of (7.34) determines simultaneously the shape of the middle plane of the plate as well as its thickness.

7.2.2 Optimal Choice of Young's Modulus Distribution: The YMD Method

Consider the problem YMD of optimal choice of Young's modulus $E = E(x)$ within a given spatial design domain Ω in a structure made of an isotropic material of given distribution of Young's modulus $\nu(x)$, $x = (x_1, x_2, x_3)$ is a point of the design domain. The aim is to achieve minimum of the compliance at the given cost. The unit cost E_1 is assumed as equal to the trace of Hooke's tensor. The eigenvalues of the Hooke tensor of an isotropic material are $3k$, 2μ, 2μ, 2μ, 2μ, 2μ, where

$$3k = \frac{E}{1-2\nu}, \qquad 2\mu = \frac{E}{1+\nu} \tag{7.44}$$

with k being the bulk modulus and μ being the shear modulus. Thus, the trace E_1 of Hooke's tensor equals $3k + 10\mu$ or

$$E_1 = Ea, \qquad a = \frac{6-9\nu}{(1+\nu)(1-2\nu)} \tag{7.45}$$

The cost of the design is expressed by

$$\int_\Omega E_1 dx = \Lambda \tag{7.46}$$

The energy density can be expressed as below

$$j^\star(\boldsymbol{\tau}) = \frac{1}{2E_1}[\rho_2(\boldsymbol{\tau})]^2 \tag{7.47}$$

with where

$$\rho_2(\boldsymbol{\tau}) = \sqrt{\frac{6-9\nu}{3(1+\nu)}(\mathrm{tr}\,\boldsymbol{\tau})^2 + \frac{6-9\nu}{1-2\nu}||\mathrm{dev}\,\boldsymbol{\tau}||^2} \tag{7.48}$$

Let $\Sigma_p(\Omega)$ be the set of the stress fields being statically admissible. The surface subjected to tractions will be kept fixed during the optimization process. The compliance is given by (3.63) where $j^\star(\boldsymbol{\tau})$ is given by (7.47). We consider the problem of optimal choice of Young's modulus $E = E(x)$ within the domain Ω to attain minimal compliance under the cost condition (7.46). This problem leads to the two problems (7.33), (7.34) where $\rho = \rho_2$ while the dual norm has the form

$$\rho_2^\star(\boldsymbol{\varepsilon}) = \sqrt{\frac{1+\nu}{3(6-9\nu)}(\mathrm{tr}\,\boldsymbol{\varepsilon}^2) + \frac{1-2\nu}{6-9\nu}||\mathrm{dev}\,\boldsymbol{\varepsilon}||^2} \tag{7.49}$$

Let $\breve{\boldsymbol{\tau}}$ be the minimizer of (7.34). The optimal Young's modulus is given by

$$\breve{E}(x) = \Lambda \frac{(1+\nu(x))(1-2\nu(x))}{6-9\nu(x)} \frac{\rho_2(\breve{\boldsymbol{\tau}}(x))}{\int_\Omega \rho_2(\breve{\boldsymbol{\tau}})dx} \tag{7.50}$$

The material domain Ω_m is the effective domain of $\breve{\boldsymbol{\tau}}$; the openings appear there, where this field vanishes. Thus, by solving the problem (7.34), we solve simultaneously the shape optimization problem—the material domain Ω_m is constructed—and the problem of optimal distribution of Young's modulus. Moreover, one can prove that the state of stress in the optimal structure coincides with the state of stress $\breve{\boldsymbol{\tau}}$ which solves the auxiliary problem (7.34). The state of deformation in the optimal structure occurs to be characterized by a uniform distribution of the norm $\rho_2^\star(\breve{\varepsilon})$ within the whole design domain.

7.2.3 *Optimal Choice of the Elastic Moduli of a Structure Made of Material of Cubic Symmetry: The CMD Method*

At each point of a body made of a material of cubic symmetry, a triplet of mutually orthogonal unit vectors $(\boldsymbol{n}, \boldsymbol{m}, \boldsymbol{p})$, i.e. such that

$$||\boldsymbol{n}|| = ||\boldsymbol{m}|| = ||\boldsymbol{p}|| = 1, \quad \boldsymbol{n}\cdot\boldsymbol{m} = 0, \quad \boldsymbol{n}\cdot\boldsymbol{p} = 0, \quad \boldsymbol{p}\cdot\boldsymbol{m} = 0 \tag{7.51}$$

determines principal directions of the crystal symmetry. These vectors determine the tensor

$$\boldsymbol{S} = \boldsymbol{n}\otimes\boldsymbol{n}\otimes\boldsymbol{n}\otimes\boldsymbol{n} + \boldsymbol{m}\otimes\boldsymbol{m}\otimes\boldsymbol{m}\otimes\boldsymbol{m} + \boldsymbol{p}\otimes\boldsymbol{p}\otimes\boldsymbol{p}\otimes\boldsymbol{p} \tag{7.52}$$

Let $\boldsymbol{I}^2 = (\delta_{ij})$ be the unit tensor in $\mathbb{E}_s^2$ while the tensor of components

$$\overset{4}{I}_{ijkl} = \frac{1}{2}(\delta_{ik}\delta_{jl} + \delta_{il}\delta_{jk}) \tag{7.53}$$

is the unit tensor in the set $\mathbb{E}_s^4$ of 4th rank Hooke's tensors of the known symmetries see (7.76). Let us introduce the 4th rank tensors

$$\boldsymbol{J} = \frac{1}{3}\overset{2}{\boldsymbol{I}}\otimes\overset{2}{\boldsymbol{I}}, \qquad \boldsymbol{L} = \overset{4}{\boldsymbol{I}} - \boldsymbol{S}, \qquad \boldsymbol{M} = \boldsymbol{S} - \boldsymbol{J} \tag{7.54}$$

They are mutually orthogonal; their linear combinations determine Hooke's tensors of a material of cubic symmetry:

$$\boldsymbol{C} = a\boldsymbol{J} + b\boldsymbol{L} + c\boldsymbol{M} \tag{7.55}$$

with a, b, c being non-negative elastic moduli. The eigenvalues of tensor (7.55) are: a, b, b, b, c, c, hence the trace of tensor (7.55) is expressed by

$$\operatorname{tr} \boldsymbol{C} = a + 3b + 2c \tag{7.56}$$

and just this quantity will be the unit cost of the design. The total cost of the design equals

$$\int_{\Omega} (a + 3b + 2c) dx = \Lambda \tag{7.57}$$

The energy density equals

$$j^{\star}(\boldsymbol{\tau}) = \frac{1}{2} \boldsymbol{\tau} \cdot \left(\frac{1}{a} \boldsymbol{J} + \frac{1}{b} \boldsymbol{L} + \frac{1}{c} \boldsymbol{M} \right) \boldsymbol{\tau} \tag{7.58}$$

If the moduli are positive. Let $\Sigma_p(\Omega)$ be the set of statically admissible stresses. The compliance of the structure is given by (3.63). Minimization of the compliance over the unit vectors moduli $(\boldsymbol{n}, \boldsymbol{m}, \boldsymbol{p})$ satisfying (7.51) and over the scalar non-negative fields a, b, c satisfying (7.57) reduces to the problems (7.33), (7.34) with $\rho = \rho_3$ expressed by

$$\rho_3(\boldsymbol{\tau}) = \frac{1}{\sqrt{3}} |\operatorname{tr} \boldsymbol{\tau}| + \sqrt{2} ||\operatorname{dev} \boldsymbol{\tau}|| \tag{7.59}$$

while the dual norm reads

$$\rho_3^{\star}(\boldsymbol{\varepsilon}) = \max \left\{ \frac{1}{\sqrt{3}} |\operatorname{tr} \boldsymbol{\varepsilon}|, \ \frac{1}{\sqrt{2}} ||\operatorname{dev} \boldsymbol{\varepsilon}|| \right\} \tag{7.60}$$

Let $\breve{\boldsymbol{\tau}}$ be the minimizer of the problem (7.53) with the norm (7.59). Let us define

$$Z = \int_{\Omega} \rho_3(\breve{\boldsymbol{\tau}}) dx \tag{7.61}$$

The optimal moduli are given by

$$\breve{a} = \frac{\Lambda}{Z} \frac{1}{\sqrt{3}} |\operatorname{tr} \breve{\boldsymbol{\tau}}|, \qquad \breve{b} = 0 \qquad \breve{c} = \frac{\Lambda}{Z} \frac{1}{\sqrt{2}} ||\operatorname{dev} \breve{\boldsymbol{\tau}}|| \tag{7.62}$$

while the optimal versors $(\breve{\boldsymbol{n}}, \breve{\boldsymbol{m}}, \breve{\boldsymbol{p}})$ follow the trajectories of the stress field $\breve{\boldsymbol{\tau}}$. Thus, the optimal Hooke's tensor is singular. Assume that the Cartesian basis $\boldsymbol{e}_1$, $\boldsymbol{e}_2$, $\boldsymbol{e}_3$ coincides with the directions of the cubic symmetry given by the triplet $(\breve{\boldsymbol{n}}, \breve{\boldsymbol{m}}, \breve{\boldsymbol{p}})$ at a point x. Then the stress components $\breve{\tau}_{12}$, $\breve{\tau}_{13}$, $\breve{\tau}_{23}$, vanish; this means that the corresponding shear moduli are zero. The directions of the principal directions of cubic symmetry are ideally suited to the trajectories of the stress. The CMD method cuts out the material domain by the criterion: $\breve{\boldsymbol{\tau}} = \boldsymbol{0}$ and determines the material domain Ω_m.

7.2.4 Optimal Choice of the Bulk and Shear Moduli of a Structure Made of an Isotropic Material: The IMD Method

We consider the structure made from an isotropic material of the bulk modulus $k(x)$ and of the shear modulus $\mu(x)$. The structure is to be designed to transmit a traction load $\boldsymbol{p}$ to the support. The stress fields τ_{ij} which satisfy the desired equilibrium conditions constitute the set $\Sigma(\Omega)$. The energy density equals

$$j^{\star}(\boldsymbol{\tau}) = \frac{1}{2}\left[\frac{1}{3k}\left(\frac{\mathrm{tr}\,\boldsymbol{\tau}}{\sqrt{3}}\right)^2 + \frac{1}{2\mu}||\mathrm{dev}\,\boldsymbol{\tau}||^2\right] \tag{7.63}$$

while the compliance is given by (3.63). The unit cost is assumed as equal the trace of the Hooke tensor. The eigenvalues of the Hooke tensor of the isotropic material are: $3k$, 2μ, 2μ, 2μ, 2μ, 2μ. Thus, the trace equals $3k + 10\mu$ and the total cost is expressed by

$$\int_{\Omega} (3k + 10\mu)dx = \Lambda \tag{7.64}$$

The moduli: $k(x) \geq 0$, $\mu(x) \geq 0$ are the design variables. The problem of minimization of the compliance with the cost condition (7.63) leads to the two mutually dual problems (7.33), (7.34), where $\rho = \rho_4$ and

$$\rho_4(\boldsymbol{\tau}) = \frac{1}{\sqrt{3}}|\mathrm{tr}\,\boldsymbol{\tau}| + \sqrt{5}||\mathrm{dev}\,\boldsymbol{\tau}|| \tag{7.65}$$

The dual norm reads

$$\rho_4^{\star}(\boldsymbol{\varepsilon}) = \max\left\{\frac{1}{\sqrt{3}}|\mathrm{tr}\,\boldsymbol{\varepsilon}|,\ \frac{1}{\sqrt{5}}||\mathrm{dev}\,\boldsymbol{\varepsilon}||\right\} \tag{7.66}$$

Let $\boldsymbol{\tau} = \breve{\boldsymbol{\tau}}$ be the minimizer of (7.34) with the norm (7.64). The effective domain of $\breve{\boldsymbol{\tau}}$ determines the material domain Ω_m. The condition $\breve{\boldsymbol{\tau}} = \mathbf{0}$ cuts out the structure (or the material domain Ω_m) from the design domain. The optimal moduli are expressed by the field $\breve{\boldsymbol{\tau}}$ as below

$$\breve{k} = \frac{\Lambda}{3\sqrt{3}}\frac{|\mathrm{tr}\,\breve{\boldsymbol{\tau}}|}{Z} \qquad \breve{\mu} = \frac{\sqrt{5}\Lambda}{10}\frac{||\mathrm{dev}\,\breve{\boldsymbol{\tau}}||}{Z} \tag{7.67}$$

with

$$Z = \int_{\Omega} \rho_4(\breve{\boldsymbol{\tau}})dx \tag{7.68}$$

The solution $\breve{\tau}$ divides the material domain into the subdomains

(i) $\breve{k}(x) = 0,\ \breve{\mu}(x) > 0$
(ii) $\breve{k}(x) > 0,\ \breve{\mu}(x) = 0$
(iii) $\breve{k}(x) > 0,\ \breve{\mu}(x) > 0$

Let us recall the formula for the Poisson ratio

$$\breve{\nu} = \frac{3\breve{k} - 2\breve{\mu}}{2(3\breve{k} + \breve{\mu})} \tag{7.69}$$

In case (i), we have $\breve{\nu}(x) = -1$ or the lower bound for the Poisson ratio is attained. In case (ii) $\breve{\nu}(x) = 1/2$ and the Poisson ratio attains the upper bound. In case (iii), we have $-1 < \breve{\nu}(x) < 1/2$. Moreover, if $3\breve{k}(x) < 2\breve{\mu}(x)$ then $\breve{\nu}(x) < 0$ or the material is *auxetic*. In the optimization problem, the design variables are subject only to the conditions of being non-negative, hence the auxetic material may appear in large subdomains of the material domain.

The distribution of the Young modulus introduces two subdomains of the material domain. According to the formula

$$\breve{E} = 3\left(\frac{1}{3\breve{k}} + \frac{1}{2\breve{\mu}} + \frac{1}{2\breve{\mu}}\right)^{-1} \tag{7.70}$$

this modulus vanishes in cases (i) or (ii), while is positive in case of (iii). Indeed, the Young modulus is the harmonic mean of the moduli: $3\breve{k}$, $2\breve{\mu}$, $2\breve{\mu}$.

The isotropic material design (IMD) method described above has its 2D formulation. In the 2D setting, the eigenvalues of Hooke's tensor are $2k$, 2μ, 2μ and the trace of Hooke's tensor equals $2k + 4\mu$. The norm (7.64) assumes the form

$$\rho_4(\boldsymbol{\tau}) = \frac{1}{\sqrt{2}}|\mathrm{tr}\,\boldsymbol{\tau}| + \sqrt{2}||\mathrm{dev}\,\boldsymbol{\tau}|| \tag{7.71}$$

while the formulae (7.67) read

$$\breve{k} = \frac{\Lambda}{2\sqrt{2}}\frac{|\mathrm{tr}\,\breve{\boldsymbol{\tau}}|}{Z} \qquad \breve{k} = \frac{\sqrt{2}\Lambda}{4}\frac{||\mathrm{dev}\,\breve{\boldsymbol{\tau}}||}{Z} \tag{7.72}$$

where Z is given by (7.68) with the norm (7.71). The moduli $\breve{k}$, $\breve{\mu}$ determine the 2D counterparts of Young's modulus and Poisson's ratio:

$$\breve{E} = 2\left(\frac{1}{2\breve{k}} + \frac{1}{2\breve{\mu}}\right)^{-1}, \qquad \breve{\nu} = \frac{\breve{k} - \breve{\mu}}{\breve{k} + \breve{\mu}} \tag{7.73}$$

The Young modulus is expressed by the harmonic mean of the moduli $2\breve{k}$, $2\breve{\mu}$.

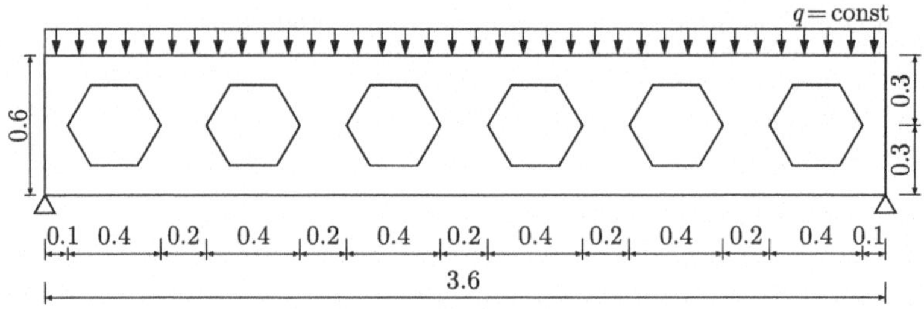

Fig. 7.4 The problem setting

In case (i), we have $\breve{\nu}(x) = -1$; in case (ii) $\breve{\nu}(x) = 1$ and in case (iii): $-1 < \breve{\nu}(x) < 1$. Moreover, if $\breve{k}(x) < \breve{\mu}(x)$ then $\breve{\nu}(x) < 0$ or the material is auxetic in the 2D meaning.

Consider the problem of an optimal transmitting of the load on the upper edge of a deep beam pin-supported at the lower corners, see Fig. 7.4.

We assume that the girder—the deep beam—is of dimensions 3.6 m by 0.6 m. A uniform distributed load of intensity $q = 1\,\mathrm{N/m^2}$ is applied along the upper edge. The design cost is assumed as equal $E_0|\Omega|$ with $E_0 = 1\,\mathrm{N/m^2}$. We consider two problems. In the problem (a), the rectangular domain—the middle plane of the deep beam—is the design domain. In the problem (b), the design domain has six hexagonal openings, see Fig. 7.4.

To solve the problem (7.34) with the norm (7.71), we apply the numerical method developed by Sławomir Czarnecki, basing on a construction of appropriate approximants of the stress fields being statically admissible; the details are published in Czarnecki and Lewiński (2012, 2014a, b), Czarnecki (2015), Czarnecki and Wawruch (2015). The IMD method makes it possible to design the bulk and shear moduli simultaneously, hence the auxetic properties of the optimal material may appear and they do appear. In the problem (a) of Fig. 7.4, the design domain has no openings; the main load carrying part is an arch spreading to the pin supports. In the central part of the design, an auxetic subdomain appears, see Fig. 7.5a.

In problem (b) with openings the optimal distribution of the moduli $\breve{E}$, $\breve{\nu}(x)$ is highly irregular. Yet we observe a strengthening running along the edges of the openings which transmits the load to the pin supports.

7.2.5 *Optimum Design of Anisotropy: The AMD Method*

Let us consider a structure made from an elastic anisotropic material subject to tractions $\boldsymbol{p}$ on a certain part of its boundary; the statically admissible stresses constitute a set $\Sigma_p(\Omega)$. All the components of the Hooke tensor $\boldsymbol{C} = (C_{ijkl})$, referred to a Carte-

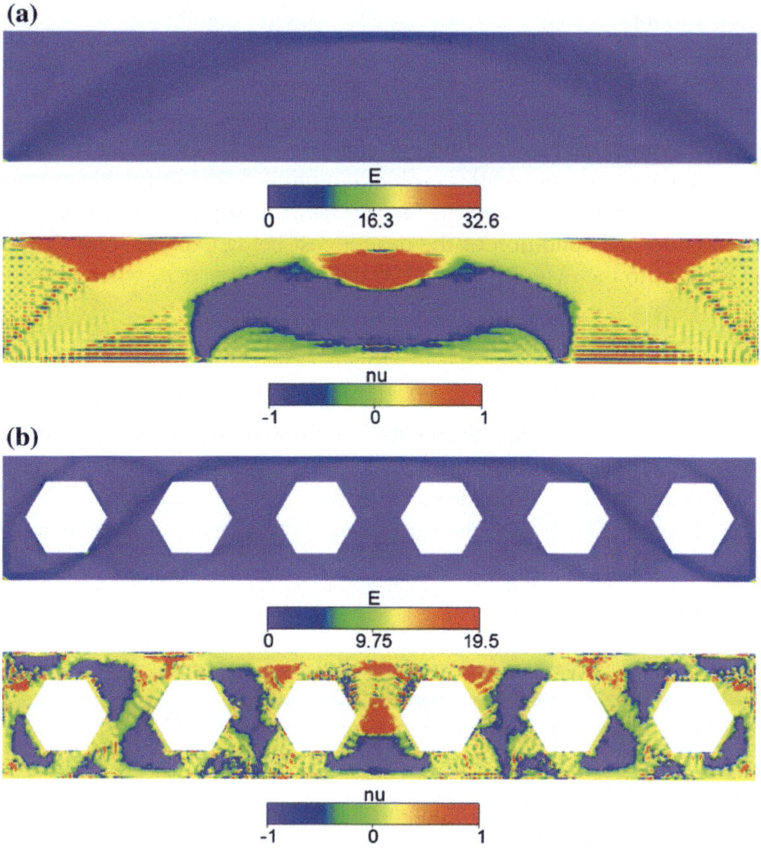

Fig. 7.5 The optimal layouts of the Young modulus $\breve{E}$ and Poisson ratio $\breve{\nu}$ for the rectangular design domain (**a**) and for the domain with hexagonal openings (**b**)

sian coordinate system x_1, x_2, x_3 with the basis $\boldsymbol{e}_1, \boldsymbol{e}_2, \boldsymbol{e}_3$, are design variables. The energy density is expressed by

$$j^{\star}(\boldsymbol{\tau}) = \frac{1}{2}\boldsymbol{\tau} \cdot (\boldsymbol{C}^{-1}\boldsymbol{\tau}) \tag{7.74}$$

while the compliance is given by (3.63). The unit cost of the design is taken as the trace of tensor $\boldsymbol{C}$ or $\operatorname{tr}\boldsymbol{C} = C_{ijij}$; $i, j = 1, 2, 3$. The total cost is given by

$$\int_{\Omega} C_{ijij} dx = \Lambda \tag{7.75}$$

The design variables are subject to the usual symmetry conditions:

$$C^{ijkl} = C^{jikl}, \qquad C^{ijkl} = C^{klij} \tag{7.76}$$

and to the condition of positive semi-definiteness of tensor $\boldsymbol{C}$. We focus on the anisotropic material design (AMD) problem: minimize the compliance (3.63) with the energy density (7.73), subject to the symmetry and positive semi-definiteness conditions, under the condition of the design cost being expressed by (7.75). This problem reduces to the two problems (7.33), (7.34) with $\rho = \rho_5$ where

$$\rho_5(\boldsymbol{\tau}) = ||\boldsymbol{\tau}|| \tag{7.77}$$

and $||\cdot||$ is the Euclidean norm in $\mathbb{E}^2_s$, or: $||\boldsymbol{\tau}|| = \sqrt{\tau_{ij}\tau_{ij}}$. The norm dual to the Euclidean norm is also the Euclidean norm. Unfortunately, the problem formulated above leads to an unrealistic design. Indeed, the problem reduces to the problems (7.33) and (7.34) with the Euclidean norm defining the integrand in (7.34). The minimizer $\boldsymbol{\tau} = \breve{\boldsymbol{\tau}}$ of (7.34) with the Euclidean norm determines the tensor $\breve{\boldsymbol{C}}$ in the form

$$\breve{\boldsymbol{C}} = \breve{\lambda}_1(x)\breve{\boldsymbol{\omega}}_1(x) \otimes \breve{\boldsymbol{\omega}}_1(x) \tag{7.78}$$

with

$$\breve{\lambda}_1(x) = \frac{\Lambda}{\displaystyle\int_\Omega \rho_5(\breve{\boldsymbol{\tau}})dx}\rho_5(\breve{\boldsymbol{\tau}}(x)), \qquad \breve{\boldsymbol{\omega}}_1(x) = \frac{\breve{\boldsymbol{\tau}}(x)}{||\breve{\boldsymbol{\tau}}(x)||} \tag{7.79}$$

Thus, the modulus $\breve{\lambda}_1$ is the only non-zero eigenvalue of tensor $\breve{\boldsymbol{C}}$. The remaining five eigenvalues vanish. The modulus $\breve{\lambda}_1$ is positive there, where $\breve{\boldsymbol{\tau}} \neq \boldsymbol{0}$ and this subdomain Ω_m is a material domain, cut out from the design domain.

Therefore, the AMD problem discussed leads to the solution (7.78) which directs the Hooke tensor along the stress field $\breve{\boldsymbol{\tau}}$. This construction made of this specific material is ideally suited for a given load and is unstable with respect to any other load.

Remark 7.2.1 Due to singularity of tensor (7.78), the AMD method in the form discussed above cannot be helpful in practice. One of the possible remedies is to extend AMD to encompass many load scenarios, e.g. by constructing the Pareto fronts. This extension has been put forward in Czarnecki and Lewiński (2014a, b). There, a proof is provided that the optimal Hooke tensor becomes non-singular if at least six independent load cases are considered.

7.3 Bibliographical Comments

The topic of multi-load truss topology optimization considered in Sect. 7.1 has a rich literature. A special attention is paid to this problem, since the trusses optimized for many load variants are usually structurally stable in contrast to classical Michell's structures optimized only for one load condition. Moreover, many different formulations can be adopted, including plastic or elastic approaches, the latter being usually augmented with additional constraints concerning displacements, the compliance or stresses (or mixture of these constraints). The important preliminary studies in this field have been done by Drucker et al. (1951), Sved (1954), Drucker and Shield (1957), Prager (1978a, b), Drucker et al. (1951), Schmidt (1962), Rozvany (1964), Prager and Shield (1967), Sved and Ginos (1968), Spillers and Lev (1971), Pedersen (1972), Hemp (1973) and McKeown (1974).

The extension of the Michell truss theory to multiple load conditions in the context of plastic layout optimization should be attributed to Hemp and Prager who independently and in different ways obtained similar formulations and conclusions. For example, the decomposition and superposition principles for particular case of two alternative load conditions (see Sect. 7.1.2, see also Spillers and Lev 1971) have been derived by Nagtegaal and Prager (1973) by a static approach. The approach involved virtual displacement and strain fields satisfying appropriate optimality criteria. Hemp (1973, see Sect. 1 there) obtained the same result directly from the linear programming formulation of the problem. Later on, Rozvany and Hill (1978) have extended the superposition principle to more than two load conditions, yet adding strong restrictions on applied loads being hard to satisfy. Thus, the generalization of Nagtegaal–Prager's superposition principle for any number of load condition may be viewed as an open problem. The optimum solution presented in Example 7.1.2 suggests that the decomposition of the applied loads into a proper set of component loads is possible and probably is unique for a given problem.

It should be noted that for one load case problems with the material of the same properties in tension and compression (i.e. $\sigma_T = \sigma_C = \sigma_0$, $E = \text{const}$), the plastic and elastic formulations lead to the same optimal topology due to statical determinacy of the optimal truss. This is not so, in general, for trusses with multiple load conditions. In the plastic design, we require only the existence of a statically admissible stress field, which in trusses is represented by a discrete set of member forces satisfying equilibrium equations (1.64). In elastic design, we must also satisfy elastic compatibility, which requires satisfying all Eqs. (1.65). If the truss is statically determinate then there is no problem with finding a proper vector of nodal displacements which produces compatible elongations of bars (and member forces), and this is true even for multiple load conditions. Unfortunately, in general, multi-load problems yield statically indeterminate structures (see Example 7.1.2) and then elastic and plastic designs differ. It seems that for multi-load case problems, the pure plastic formulation is not enough, but surprisingly even for statically indeterminate optimal solution the topology of elastic and plastic designs can be the same with only slightly different cross-sectional areas of bars (see Rozvany, Sokół and Pomezanski 2014c).

Moreover, the multi-load plastic design can be formulated as a linear programming problem, which enables finding very accurate approximations of exact multi-load Michell trusses. And, finally, if an optimal plastic solution forms a statically determinate truss (like in Example 7.1.1) than it is also an optimal elastic design. For this reason, the multi-load plastic design is attractive as leading to interesting results which would be very hard to obtain directly from the elastic formulation due to its non-linear character and often due to too large size of the optimization problem. Specialized methods of solving large-scale truss topology optimization problems within the plastic design have been proposed by Pritchard et al. (2005), Sokół (2014, 2016, 2018) and Sokół and Rozvany (2013, 2015).

The multi-load elastic design of trusses has been thoroughly studied by Bendsøe (1989, 1995), Bendsøe et al. (1994a), Bendsøe and Sigmund (2004), see also Ben-Tal and Bendsøe (1993). We would like to draw the reader's attention to the seminal papers of Achtziger (1992, 1993, 1998), Achtziger et al. (1992), where the equivalence of many different optimum design formulations is strictly proved. The general properties of optimal topologies of multi-load trusses have been noted by Kirsch (1989), Prager and Rozvany (1977) and Rozvany (1964, 1976, 1989, 1992), see also Rozvany et al. (1995). Some unexpected properties of exact least-weight elastic trusses are discussed in Rozvany, Zhou and Birker (1993), Rozvany and Birker (1994) and Rozvany et al. (1994). These papers show that for multi-load problems, contrary to classical Michell theory, the orthogonal microstructures can be non-optimal and the optimal topology can be singular, which prevents most iterative computational algorithms from reaching the global optimal solution (cf. Sved and Ginos 1968). For having a deeper insight into the exact solutions of multi-load optimum design problems (in elastic and plastic settings), the reader is referred to the papers by Rozvany et al. (2014a, b, c).

The multiple load optimum design needs scalarization. The scalarization (7.3) based on the convex combination of the target functions is standard, see Achtziger (1997) and the references therein. However, it is rarely emphasized that this scalarization is the effective method of constructing a Pareto front. In the context of minimum compliance of trusses, this method of construction of the Pareto front has been put forward by Czarnecki (2013). Just there, the formulation of the auxiliary problem (7.5) has been constructed and numerically solved using the non-linear programming methods. Moreover, one of the solutions lying on the Pareto front is just the solution to the worst case design problem (7.13), cf. Eqs. (63)–(70) in Bendsøe et al. (1994a) concerning the convexification-based formulation. Other reformulation of the worst case problem (7.13) has been derived and discussed in Makrodimopoulos et al. (2010), see also Logo et al. (2018). The problem (7.7) dual to (7.5) as well as the reformulations (7.10)–(7.12) are published for the first time.

The problem of designing the thickness h of a plate loaded in the plane is the classical problem of optimum design. Usually, the thickness is subject to the conditions: $h \geq h_{\min} > 0$. Then, the problem is well posed, see Cea and Malanowski (1970), Litvinov and Panteleev (1980), Bendsøe (Sec. 1.5.1, 1995), Petersson (1999). The case $h \geq 0$ has been considered in Golay and Seppecher (2001) and there the formulation (7.33) with the norm (7.42) has been derived. The formulation (7.34) can be read off from Sec. 5.2.5 in Allaire (2002). Both the problems (7.33) and (7.34) and the result (7.43) have been derived in Czarnecki and Lewiński (2013).

The YMD method has been proposed in Czarnecki and Lewiński (2017a). The results of this paper have much in common with the results of Bouchitté and Buttazzo (2001) concerning optimization of a measure which is a support of the structure designed. The YMD method has been recently used for the 3D printing of the most stiff prototypes, see Lewiński et al. (2018). New results concerning YMD can be found in Czarnecki et al. (2017) and Dzierżanowski and Lewiński (2018).

The CMD method has been put forward in Czubacki and Lewiński (2015); the method has been based on Walpole (1984) representation of the Hooke tensor of a material of a cubic symmetry. The formulae for the optimal moduli (7.61) are compatible with Zohdi's (2003) results.

The IMD method has been proposed by Czarnecki (2015) and Czarnecki and Wawruch (2015), while its extension to the case of multiple loads has been put forward in Czarnecki and Lewiński (2017b). In the IMD method, the bulk and shear moduli are independent design variables; consequently, the auxetic characteristics (see Lim 2015) are predicted as optimal in certain subregions of the optimal structures. The results of Figs. 7.4, 7.5 are taken from Czarnecki et al. (2015).

The AMD method has been proposed for the first time by Ringertz (1993) and Bendsøe et al. (1994b), where it is called the free material design. The result (7.77) can be read off from the latter work, yet there the construction (7.78) basing on the problem (7.34) with the norm (7.76) has not been shown. The above results have been derived in Czarnecki and Lewiński (2012). The extension of AMD towards multiple loads has been proposed in Bendsøe et al. (1994a, b), while the extension of the problem (7.34) for multiple loads has been derived in Czarnecki and Lewiński (2014a, b). The problem of existence of solutions within AMD has been studied by Werner (2000) and Haslinger et al. (2010). The numerical methods for the AMD method have been elaborated by Zowe et al. (1997) and Kočvara and Stingl (2007) for the case of the Hooke tensor being positive definite.

All the material optimization methods discussed in Sect. 7.2 reduce to the two mutually dual problems (7.33), (7.34). The methods differ in the function $\rho(\boldsymbol{\tau})$ involved. In each case, $\rho(\boldsymbol{\tau})$ is a norm in $\mathbb{E}^s_2$, hence all these norms are equivalent. This does not mean that the problems (7.34) are equivalent. In particular, the effective domain of the minimizer $\check{\boldsymbol{\tau}}$ depends on the form of the norm $\rho(\boldsymbol{\tau})$. The problems (7.33), (7.34) are mutually dual, which follow from the general theory of duality, cf. Bouchitté and Fragala (2007).

The definition (7.35) can be extended to the case of functions $\rho(\cdot)$ which assume non-negative values, are positively homogeneous of degree 1, assume the zero value at $\boldsymbol{\tau} = \mathbf{0}$ and are convex. Such functions are called *gauge functions*, cf. Rockafellar (Sec. 15, 1970). In the present book, only the function $\rho(\cdot)$ given by (3.93) is a gauge and is not a norm.

References

Achtziger, W. (1992). *Truss topology design under multiple loadings*. DFG-Report (FSP Applied Optimization and Control), No. 367, Universitat Bayreuth, FRG.

Achtziger, W. (1993). Minimax compliance truss topology subject to multiple loadings. In M. P. Bendsøe & C. A. Mota Soares (Eds.), *Topology optimization of structures* (pp. 43–54). Dordrecht: Kluwer.

Achtziger, W. (1997). Topology optimization of discrete structures: An introduction in view of computational and nonsmooth aspects. In G. I. N. Rozvany (Ed.), *Topology optimization in structural mechanics* (Vol. 374, pp. 57–100), CISM courses and lectures. Wien: Springer.

Achtziger, W. (1998). Multiple load truss topology and sizing optimization: Some properties of minimax compliance. *Journal of Optimization Theory and Applications*, *98*(2), 255–280.

Achtziger, W., Bendsøe, M. P., Ben-Tal, A., & Zowe, J. (1992). Equivalent displacement based formulations for maximum strength truss topology design. *IMPACT of Computing in Science and Engineering*, *4*, 315–345.

Allaire, G. (2002). *Shape optimization by the homogenization method*. New York: Springer.

Bendsøe, M.P., & Sigmund, O. (2004). *Topology Optimization: Theory, Methods, and Applications*. Berlin: Springer.

Bendsøe, M. P. (1989). Optimal shape design as a material distribution problem. *Structural Optimization*, *1*, 193–202.

Bendsøe, M. P. (1995). *Optimization of structural topology, shape and material*. Berlin: Springer.

Bendsøe, M. P., Ben-Tal, A., & Zowe, J. (1994a). Optimization methods for truss geometry and topology design. *Structural Optimization*, *7*, 141–159.

Bendsøe, M. P., Guedes, J. M., Haber, R. B., Pedersen, P., & Taylor, J. E. (1994b). An analytical model to predict optimal material properties in the context of optimal structural design. *Journal of Applied Mechanics, Transaction of ASME*, *61*, 930–937.

Ben-Tal, A., & Bendsøe, M. P. (1993). A new method for optimal truss topology design. *SIAM Journal on Optimization*, *3*, 322–358.

Bouchitté, G., & Buttazzo, G. (2001). Characterization of optimal shapes and masses through Monge–Kantorovich equation. *Journal of European Mathematical Society*, *3*, 139–168.

Bouchitté, G., & Fragala, I. (2007). Optimality conditions for mass design problems and applications to thin plates. *Archive for Rational Mechanics and Analysis*, *184*, 257–284.

Cea, J., & Malanowski, K. (1970). An example of a max-min problem in partial differential equations. *SIAM Journal on Control*, *8*, 305–316.

Czarnecki, S. (2013). Edgeworth-Pareto optimal trusses of least compliance. In S. Jemioło & M. Lutomirska (Eds.), *Mechanics of materials* (pp. 61–75). Oficyna Wydawnicza Politechniki Warszawskiej.

Czarnecki, S. (2015). Isotropic material design. *Computational Methods in Science and Technology*, *21*, 49–64.

Czarnecki, S., & Lewiński, T. (2012). A stress-based formulation of the free material design problem with the trace constraint and single loading condition. *Bulletin of the Polish Academy of Sciences: Technical Sciences*, *60*(2), 191–204.

Czarnecki, S., & Lewiński, T. (2013). On minimum compliance problems of thin elastic plates of varying thickness. *Structural and Multidisciplinary Optimization*, *48*(1), 17–31.

Czarnecki, S., & Lewiński, T. (2014a). A stress-based formulation of the free material design problem with the trace constraint and multiple load conditions. *Structural and Multidisciplinary Optimization*, *49*(5), 707–731.

Czarnecki, S., & Lewiński, T. (2014b). The Free material design in linear elasticity. In G. I. N. Rozvany & T. Lewiński (Eds.), *Topology optimization in structural and continuum mechanics* (Vol. 549, pp. 213–257), CISM International Centre for Mechanical Sciences. Courses and Lectures. Wien, CISM, Udine: Springer.

Czarnecki, S., & Lewiński, T. (2017a). On material design by the optimal choice of Young's modulus distribution. *International Journal of Solids and Structures*, *110–111*, 315–331.

Czarnecki, S., & Lewiński, T. (2017b). Pareto optimal design of non-homogeneous isotropic material properties for the multiple loading conditions. *Physica Status Solidi B: Basic Solids State Physics*, *254*(1600821), 1–14.

Czarnecki, S., & Wawruch, P. (2015). The emergence of auxetic material as a result of optimal isotropic design. *Physica Status Solidi B*, *252*, 1–11.

Czarnecki, S., Łukasiak, T., & Lewiński, T. (2017). The isotropic and cubic material designs. Recovery of the underlying microstructures appearing in the least compliant continuum bodies. *Materials*, *10*(10), 1137. https://doi.org/10.3390/ma10101137.

Czarnecki, S., Czubacki, R. Lewiński, T. & Wawruch, P. (2015). Topology optimization of continuum structures made of non-homogeneous materials of isotropic or cubic symmetry, pp. 83–88. In: Q. Li., G. P. Steven & Zhongpu (Leo) Zhang (Eds.), *Advances in Structural and Multidisciplinary Optimization – Proceedings of the 11th World Congress on Structural and Multidisciplinary Optimization. International Society for Structural and Multidisciplinary Optimization (ISSMO), Sydney, Australia 7–12, June 2015* (pp. 316–321) (ISBN 13: 978-0-646-94394-7); www.aeromech.usyd.edu.au/WCSMO2015/.

Czubacki, R., & Lewiński, T. (2015). Topology optimization of spatial continuum structures made of non-homogeneous material of cubic symmetry. *Journal of Mechanics of Materials and Structures*, *10*(4), 519–535.

Drucker, D. C., Shield, R. T. (1957). Design for minimum weight. In: *Proceedings of the 9th International Congress for Applied Mechanics (held in Brussels 1956)* (Vol. 5, pp. 212–222).

Drucker, D. C., Greenberg, J. H., & Prager, W. (1951). The safety factor of an elastic-plastic body in plane stress. *Journal of Applied Mechanics*, *18*, 371–378.

Dzierżanowski, G., & Lewiński, T. (2018). Young's modulus control in material and topology optimization. In: A. Schumacher, Th. Vietor, S. Fiebig, K.-U. Bletzinger, K. Maute (Eds.), *Advances in Structural and Multidisciplinary Optimization. Proceedings of the 12th World Congress of Structural and Multidisciplinary Optimization (WCSMO12)* (pp 1374–1385). Cham: Springer International Publishing AG.

Golay, F., & Seppecher, P. (2001). Locking materials and the topology of optimal shapes. *European Journal Of Mechanics. A, Solids*, *20*, 631–644.

Haslinger, J., Kočvara, M., Leugering, G., & Stingl, M. (2010). Multidisciplinary free material optimization. *SIAM Journal on Applied Mathematics*, *70*(7), 2709–2728.

Hemp, W. S. (1973). *Optimum structures*. Oxford: Clarendon.

Kirsch, U. (1989). Optimal topologies of truss structures. *Computer Methods in Applied Mechanics and Engineering*, *72*, 15–28.

Kočvara, M., & Stingl, M. (2007). Free material optimization for stress constraints. *Structural and Multidisciplinary Optimization*, *33*, 323–335.

Lewiński, T., Czarnecki, S., Czubacki, R., Łukasiak, T., & Wawruch, P. (2018). Constrained versions of the free material design methods and their applications in 3D printing. In: A. Schumacher, Th. Vietor, S. Fiebig, K.-U. Bletzinger, K. Maute (Eds.), *Advances in Structural and Multidisciplinary Optimization. Proceedings of the 12th World Congress of Structural and Multidisciplinary Optimization (WCSMO12)* (pp. 1317–1332). Cham: Springer International Publishing AG.

Lim, T. C. (2015). *Auxetic materials and structures*. New York: Springer.

Litvinov, V. G., & Panteleev, A. D. (1980). The problem of optimization of plates of varying thickness. *Mekh. Tverdovo Tela*, *2*, 174–181.

Logo, J., Balogh, B., & Pinter, E. (2018). Topology optimization considering multiple loading. *Computers and Structures*. (in press).

Makrodimopoulos, A., Bhaskar, A., & Keane, A. J. (2010). A compliance based design problem of structures under multiple load cases. *Structural and Multidisciplinary Optimization*, *42*, 739–743.

McConnel, R. E. (1974). Least-weight frameworks for loads across span. *Journal of Engineering Mechanics Division*, *100*(5), 885–901.

McKeown, J. J. (1974). A note on the maximum number and density of distribution of members in elastic structures of minimum weight under multiple loading conditions. *International Journal of Solids and Structures*, *10*, 309–312.

Nagtegaal, J. C., & Prager, W. (1973). Optimal layout of a truss for alternative loads. *International Journal of Mechanical Sciences*, *15*, 583–592.

Nowak, M., Sokolowski, J., & Zochowski, A. (2018). Justification of a certain algorithm for shape optimization in 3D elasticity. *Structural and Multidisciplinary Optimization*, *57*,721–734.

Pedersen, P. (1972). On the optimal layout of multi-purpose trusses. *Computers and Structures*, *2*, 695–712.

Petersson, J. (1999). A finite element analysis of optimal variable thickness sheets. *SIAM Journal on Numerical Analysis*, *36*, 1759–1778.

Prager, W. (1958). *On a problem of optimum design*. Brown University, Division of Applied Mechanics, Technical Report No 38.

Prager, W. (1978a). Optimal layout of trusses of finite number of joints. *Journal of the Mechanics and Physics of Solids*, *26*, 241–250.

Prager, W. (1978b). Nearly optimal design of trusses. *Computers and Structures*, *8*, 451–454.

Prager, W., & Rozvany, G. N. (1977). Optimization of the structural geometry. In A. R. Bednarek & L. Cesari (Eds.), *Dynamical systems (Proceedings of International Conference on Gainsville, Florida)* (pp. 265–293). New York: Academic Press.

Prager, W., & Shield, R. T. (1967). A general theory of optimal plastic design. *Journal of Applied Mechanics*, *34*, 184–186. Rozvany, G. J. N. (1976) *Optimal design of flexural systems*. Oxford: Pergamon.

Pritchard, T. J., Gilbert, M., & Tyas, A. (2005). Plastic layout optimization of large-scale frameworks subject to multiple load cases, member self-weight and with joint length penalties. WCSMO-6, Rio de Janeiro, Brazil, 30 May–03 June.

Ringertz, U. T. (1993). On finding the optimal distribution of material properties. *Structural Optimization*, *5*(4), 265–267.

Rockafellar, T. (1970). *Convex analysis*. New Jersey: Princeton University Press.

Rozvany, G. I. N. (1964). Optimal synthesis of prestressed structures. *Journal of the Structural Division ASCE*, *90*, 189–211.

Rozvany, G. I. N. (1976). *Optimal design of flexural systems*. Oxford: Pergamon Press.

Rozvany, G. I. N. (1989). *Structural design via optimality criteria*. Dordrecht: Kluwer.

Rozvany, G. I. N. (1992). Optimal layout theory: analytical solutions for elastic structures with several deflection constraints and load conditions. *Structural and Multidisciplinary Optimization*, *4*, 247–249.

Rozvany, G. I. N., & Birker, T. (1994). On singular topologies in exact layout optimization. *Structural Optimization*, *8*, 228–235.

Rozvany, G. I. N., & Hill, R. H. (1978). Optimal plastic design: superposition principles and bounds on the minimum cost. *Computer Methods in Applied Mechanics and Engineering*, *13*, 151–173.

Rozvany, G. I. N., Zhou, M., & Birker, T. (1993). Why multi-load topology designs based on orthogonal microstructures are in general non-optimal. *Structural Optimization*, *6*, 200–204.

Rozvany, G. I. N., Birker, T., & Lewiński, T. (1994). Some unexpected properties of exact least-weight plane truss layouts with displacement constraints for several alternate loads. *Structural Optimization*, *7*, 76–86.

Rozvany, G. I. N., Bendsøe, M. P., & Kirsch, U. (1995). Layout optimization of structures. *Applied Mechanics Reviews ASME*, *48*, 41–119.

Rozvany, G. I. N., Pomezanski, V., & Sokół, T. (2014a). Exact compliance-based multi-load exact truss topology optimization revisited - also background material for an authors' reply to a discussion by Logo. *Structural and Multidisciplinary Optimization*, *50*, 193–205.

Rozvany, G. I. N., Pomezanski, V., Sokół, T., & Pinter, E. (2014b). Major advances in exact structural topology optimization: stress and displacement based multi-load design. In H. Rodrigues, J. Herskovits, C. M. Soares, J. M. Guedes, A. Araujo, J. Folgado, F. Moleiro, & J. A. Madeira (Eds.), *Engineering Optimization IV* (pp. 835–840). Boca Raton: CRC Press.

Rozvany, G. I. N., Sokół, T., & Pomezanski, V. (2014c). Fundamentals of exact multi-load topology optimization - stress-based least-volume trusses (generalized Michell structures) - Part I: Plastic design. *Structural and Multidisciplinary Optimization*, *50*, 1051–1078.

Schmidt, L. C. (1962). Minimum weight layouts of elastic, statically determinate, triangulated frames under alternative load systems. *Journal of the Mechanics and Physics of Solids*, *10*, 139–149.

Sokół, T. (2014). Multi-load truss topology optimization using the adaptive ground structure approach. In T. Łodygowski, J. Rakowski, & P. Litewka (Eds.), *Recent advances in computational mechanics (Ch. 2)* (pp. 9–16). London: CRC Press.

Sokół, T. (2016). A new adaptive ground structure method for multi-load spatial Michell structures. In M. Kleiber, T. Burczyński, K. Wilde, J. Górski, K. Winkelmann, & L. Smakosz (Eds.), *Advances in mechanics: Theoretical, computational and interdisciplinary issues* (pp. 525–528). Leiden: CRC Press.

Sokół, T. (2018). On the numerical approximation of Michell trusses and the improved ground structure method. In: A. Schumacher, T. Vietor, S. Fiebig, K. U. Bletzinger, K. Maute (Eds.), *Advances in Structural and Multidisciplinary Optimization. WCSMO 2017*. Cham: Springer.

Sokół, T., & Rozvany, G. I. N. (2013). On the adaptive ground structure approach for multi-load truss topology optimization. In: *10th World Congress on Structural and Multidisciplinary Optimization, WCSMO-10, Orlando, Florida, USA, May 19–24* (Vol. 5428) (full text on http://www2.mae.ufl.edu/mdo/papers/5428.pdf).

Sokół, T., & Rozvany, G. I. N. (2015). On the numerical optimization of multi-load spatial Michell trusses using a new adaptive ground structure approach. In: *World Congress on Structural and Multidisciplinary Optimization, WCSMO-11* (Vol. 1181) (full text available on http://www.aeromech.usyd.edu.au/WCSMO2015/papers/1181_paper.pdf).

Spillers, W. R., & Lev, O. (1971). Design for two loading conditions. *International Journal of Solids and Structures*, *7*, 1261–1267.

Sved, G. (1954). The minimum weight of certain redundant structures. *Australian Journal of Applied Science*, *5*, 1–8.

Sved, G., & Ginos, Z. (1968). Structural optimization under multiple loading. *International Journal of Mechanical Sciences*, *10*, 803–805.

Walpole, L. J. (1984). Fourth-rank tensors of the thirty-two crystal classes: Multiplication tables. *Proceedings of the Royal Society of London. A*, *391*, 149–179.

Werner, R. (2000). *Free material optimization. Mathematical analysis and numerical solution*. Ph.D. thesis, Institute of Applied Mathematics II, University of Erlangen-Nuerenberg, p. 147.

Zohdi, T. I. (2003). Constrained inverse formulations in random material design. *Computer Methods in Applied Mechanics and Engineering*, *192*, 3179–3194.

Zowe, J., Kočvara, M., & Bendsøe, M. (1997). Free material optimization via mathematical programming. *Mathematical Programming*, *79*, 445–466.

Chapter 8
Industrial Applications

The theory of Michell structures teaches us how to optimally transmit the given load to the support, hence shows how to make structures ideally suited to the given load. The exact solutions to the Michell theory have been an inspiration for developing new numerical methods of Topology Optimization, which nowadays does contribute to essential changes in the design methods in many domains of engineering. The aim of this chapter is to outline the impact of these ideas on selected methods of designing in civil and mechanical engineering.

8.1 Civil Engineering Structures Inspired by Michell's Solutions

Michell's structures, characterized by the minimal weight and the maximum stiffness have significantly influenced the contemporary architectural forms as well as the structural concepts in the modern civil engineering. These structures had especially fascinated and inspired architects and structural engineers involved in the designing of high-rise buildings, tensegrity structures and long-span bridges. Indeed, the analytical methods of topology optimization furnish structures being light, stiff and beautiful. They possess all features postulated by the Roman architect Marcus Vitruvius Polio – *firmitas* (strength, stiffness, durability), *utilitas* (utility, function), *venustas* (aesthetics, beauty), cf. Vitruvius (ca. 80–15 BC).

T. Lewiński et al., *Michell Structures*,
https://doi.org/10.1007/978-3-319-95180-5_8

8.1.1 Building Designs Based on Michell's Concepts

8.1.1.1 Fully Stressed Design of High-Rise Buildings

One of the main assumptions of Michell's theory is shaping a structure in a way that the entire material is fully stressed. The principle of fully stressed design has a direct impact on the optimal design of civil engineering structures. For instance, it can be successfully applied in the design of buildings, in which the dominating part of loading is caused by gravity forces. In such a case, the principle of fully stressed design is applied at the scale of the entire building, and it is one of the main criteria determining the global building shaping and its construction.

Examples of application of the principle of fully stressed design are observed in contemporary architecture in abandoning the cuboid shapes of high-rise buildings in favour of narrowing upwards obelisk shapes such as cones or prisms. In the case of cuboid-shaped buildings, the distribution of stresses caused by gravity forces is evidently non-uniform and thus utilization of the material is clearly non-optimal. In turn, in case of an obelisk shape the variation of the gravity load is compensated by the variation of the cross section of the building. Therefore, the distribution of stresses tends to be more uniform and utilization of the material is significantly improved. The adequate measure of shape optimality is introduced in civil engineering as the 'critical height of the building', defined as the height for which the stresses at the foundation level is equal to the compressive strength of the material. Let us note that an obelisk-shaped building of the same bottom area and total volume as the cuboid building has exactly three times greater critical height, which results directly from the formulae defining volumes of both the buildings.

Optimal shaping of high-rise buildings subjected to gravity forces has been described by Zalewski and Zabłocki (2002). In order to obtain fully stressed state of material at each level of the building the cross section A has to vary in terms of the building height y and satisfy the equation:

$$\frac{\int_y^{\infty} A(\bar{y})d\bar{y}}{A(y)}\rho g = \sigma_{\max} \tag{8.1}$$

where ρ is the mass density, g is gravitational acceleration and $\sigma_{\max}$ is the compression strength of the material. The solution of Eq. (8.1) reads

$$A(y) = A_0 \exp\left(-\frac{\rho g}{\sigma_{\max}}y\right) \tag{8.2}$$

and it indicates that a hypothetical fully stressed building has a shape of an obelisk in which the cross-section area decreases exponentially along with the height, Fig. 8.1a.

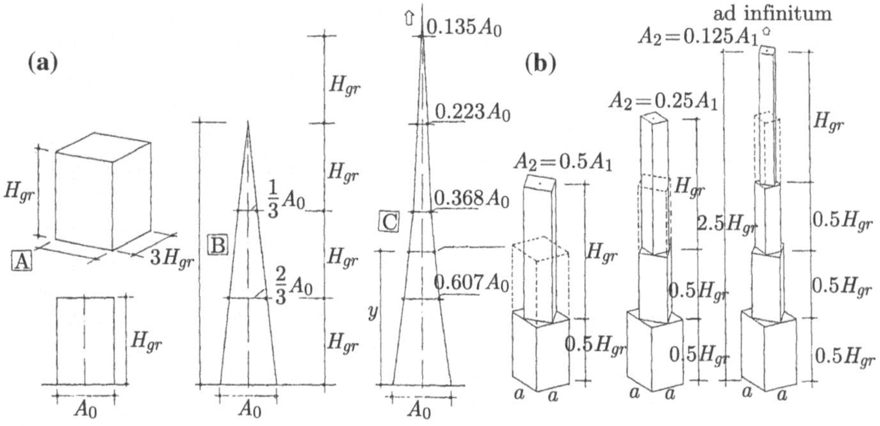

Fig. 8.1 **a** Comparison of critical height of buildings of various shapes, **b** approximation of the optimal fully stressed shape by cuboids of decreasing cross section; Zalewski and Zabłocki (2002)

The critical height of such building equals to infinity. The optimal shape can be also approximated by a system of vertically arranged cuboids, whose cross sections provide that the stress at the bottom of each equals to the compressive strength of the material, Fig. 8.1b.

Nowadays, shaping the high-rise buildings according to the principle of fully stressed design seems to be an important tendency in the contemporary architecture. This trend is reflected in the shapes of the skyscrapers breaking the world height records at the time of their construction. The Willis Tower (formerly Sears Tower) in Chicago designed by Skidmore, Owings and Merill (Fazlur Rahman Khan and Bruce Graham) and erected in 1973 is composed of nine squared tubes bundled together at the bottom of the building and reaching various heights (maximum 442 m), Fig. 8.2a. In turn, the Petronas Towers in Kuala Lumpur designed by Cesar Pélli and Thornton Tomasetti and completed in 1996 have the shape narrowing in its upper part, which achieves the height of 451 m, Fig. 8.2b. Eventually, the Burj Khalifa in Dubai designed by SOM (Adrian Smith and William F. Baker) and constructed in 2009 has an obelisk shape clearly resembling optimal fully stressed exponential obelisk from the Fig. 8.1a which reaches the enormous height of 828 m, Fig. 8.2c.

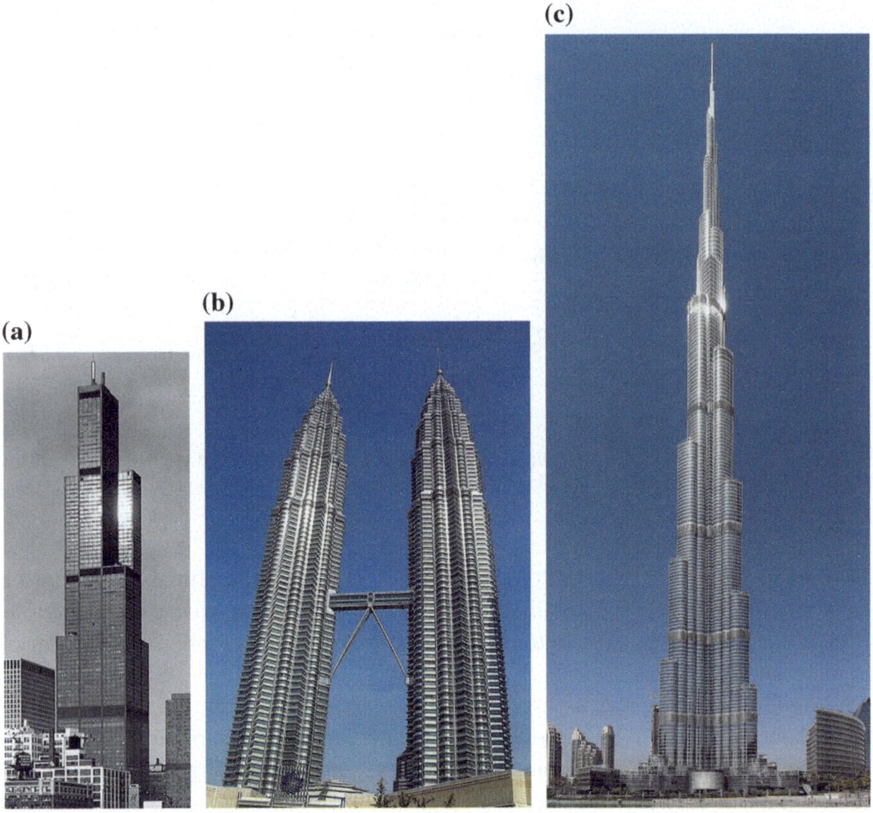

Fig. 8.2 High-rise buildings resembling obelisks: **a** Willis Tower in Chicago, USA (SOM: Fazlur Rahman Khan and Bruce Graham, image: courtesy of SOM © Ezra Stoller | Esto), **b** Petronas Towers in Kuala Lumpur, Malaysia (Cesar Pélli and Thornton Tomasetti, image: courtesy of Wikipedia), **c** Burj Khalifa in Dubai, United Arab Emirates (SOM: Adrian Smith and William F. Baker, image: courtesy of Wikipedia)

8.1.1.2 Michell-Inspired Concepts of Buildings by Zalewski and Zabłocki

The influence of the analytical solutions of Michell's problems on the optimal shaping of high-rise buildings as well as unconventional design concepts are summarized in the papers by the architects (Zalewski and Zabłocki 2002; Zalewski 2000; Zabłocki 2000). The authors reject the classical assumption about the dominating role of the gravity forces and the resulting optimality of the obelisk shape. They consider buildings higher than 200 m, with the basis not larger than one-fifth of the height, subject to the forces generated by the transverse loadings caused by winds or earthquakes. The authors put emphasis on the similarity of a mechanical response of such a building to the response of a large ground-fixed cantilever subjected to the action of bending forces. This constitutes the motivation for the development of the concept of 'wingy'

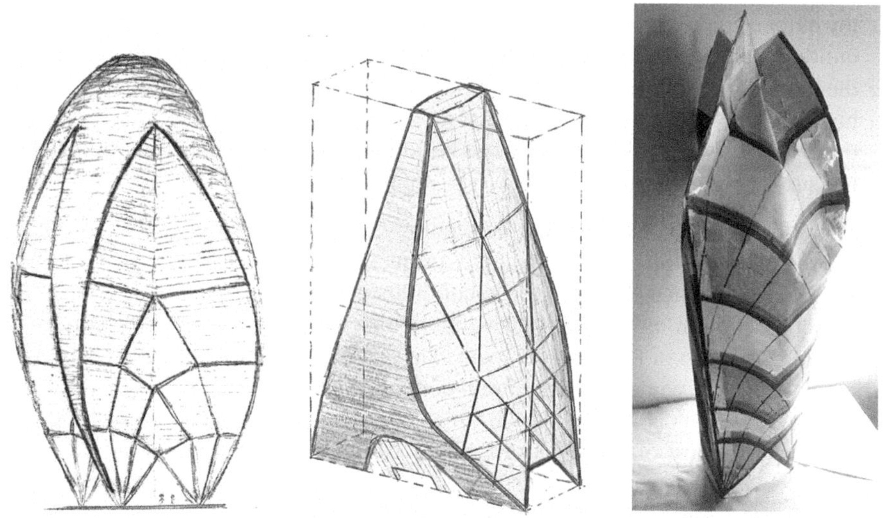

Fig. 8.3 The original sketches and paper model of the buildings based on Michell's cantilever developed by Wacław Zalewski (by courtesy of Wacław Zalewski and www.wz-structure.org)

and 'tulip-like' buildings characterized by untypical bulbous shapes typical for the cantilevers predicted by the Michell theory, Fig. 8.3. Zalewski and Zabłocki focus on the Michell cantilever transmitting a point load to a straight support, corresponding to equal allowable stresses in tension and compression, discussed in Sect. 4.8.2. Moreover, they intuitively apply optimal structure rationalization by reducing it to trusses composed of dozens of elements, cf. Zalewski and Allen (1998).

Zalewski and Zabłocki (2002) have noted the superiority of Michell's cantilever over other structural systems and estimated the profits of its application. They consider various vertical skeletal structures of a given slenderness (the height to the basis width ratio $H/a = 6$) transmitting a given horizontal load of magnitude P to a straight support. It appears that the volume of simple two-bar structures is the greatest, whereas the volume of the Michell cantilever is almost two times smaller. Further, the coefficient of shape effectiveness $\alpha = V\sigma_{\max}/(PH)$ is introduced and applied for the comparison of two-bar structures and Michell's structures of various slenderness. In case of $H/a = \sqrt{2}/2$ both structures have an identical topology and thus both coefficients α are equal. However, when structure slenderness increases the coefficient α for two-bar structure raises significantly faster than the one for Michell's cantilever. The ratio of both coefficients, indicating ineffectiveness of two-bar structure, almost reaches the value of 2 for slenderness $H/a = 6$ and further raises monotonically. This clearly indicates that geometry of a Michell-like cantilever constitutes a proper choice for the structural system of transversally loaded high-rise building.

The comparison of bulbous buildings based on Michell-like cantilevers and classical cuboid buildings allows Zalewski and Zabłocki to distinguish the following characteristic features of their innovative architectural concept:

1. the bulbous building of the same base area and total floor area as the conventional building can be substantially lower;
2. the bulbous building of the same base area and height as the conventional building has a significantly larger total floor area;
3. the characteristic shape of the bulbous building provides its visual separation from the surrounding buildings of a classical shape.

Zalewski and Zabłocki adopted the topology and geometry of optimal Michell's cantilevers to develop several detailed projects of high-rise 'wingy' buildings (Figs. 8.4, 8.5, 8.6 and 8.7). All these buildings possess a characteristic shape and construction combining three or four wings connected to a central service core. The supporting structure of the wings is located at the planes of their exterior walls. The arrangement of the wings and the central core allows to combine a characteristic for Michell's cantilevers ability of transmitting transverse forces and standard for high-rise buildings ability of resisting substantial vertical loads caused by self-weight.

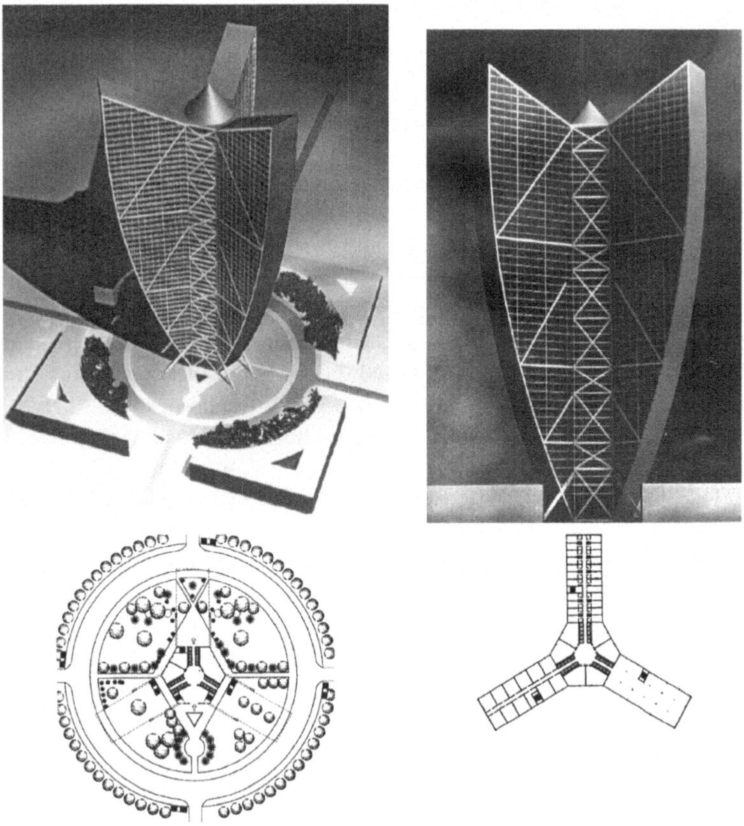

Fig. 8.4 The project of three wings skyscraper resembling Michell's cantilever (by courtesy of W. Zalewski and W. Zabłocki)

Moreover, by such construction, the architects have gained an attractive and innovative shape of the building as well as good lighting of spaces located inside the wings.The additional advantage is a beneficial building aerodynamics, which gives the possibility of acquiring wind energy.

The first proposed building is a three-winged skyscraper in which upwardly expanding wings are connected to a hexagonal vertical service core (Fig. 8.4). The building is 216 m high and has 49 stories. Expanding the wings in the upward direction allows to obtain a visually interesting shape of the building and to significantly increase the total floor area. The building consists of the so-called mega-floors carrying located between them intermediate lighter floors. The supporting structure visible on the facades of the wings is strongly simplified in comparison to optimal Michell's cantilever. Although the orthogonality of the load-bearing bars is not preserved, the bars transferring tensile and compressive forces resulting from the wind loads can be distinguished. On the other hand, the loads due to the self-weight are carried by columns and transmitted both to the central core and to the vertical edges of the wings.

The second building proposed by Zalewski and Zabłocki has 209 m in height and 50 functional floors (Fig. 8.5). Similarly to the building from the previous project, it consists of three wings connected to the central core. However, now the wings narrow in their upper parts and the entire body of the building more closely resembles a theoretical shape of the Michell cantilever. The supporting structure responsible for transferring bending forces resulting from the wind loads is situated on the external surfaces of the wings. The members in compression and tension are located in a similar way as in Michell's structures, i.e. along perpendicular curves indicating directions of principal stresses. In turn, the gravity forces are transmitted by the columns. The curved concrete foundation of the building clearly resembles fan regions of the optimal cantilevers.

The third proposed building reaches a height of about 400 m and has 91 floors (Fig. 8.6). The building consists of four perpendicular wings attached to the central octagonal core. It is designed on a plan of cruciform in order to increase the total stiffness. In addition, the building is equipped with two stiffening horizontal platforms dedicated for recreational purposes, located at the height of 120 and 320 m. Similarly, as in previous buildings, the internal forces caused by the wind loads are transferred by the supporting structures located on the outer facades of the wings, roughly resembling classical Michell cantilevers.

In the design of the fourth building inspired by Michell's theory, the architects had abandoned the concept of several independent wings. The proposed building has the shape of a tulip and it is built on a plan of a clover (Fig. 8.7). The building is created by rotation of a plane Michell structure around the vertical axis of symmetry and by making the design more slender. The geometry of the outer walls is now more complicated than in 'wingy' buildings. Visible on the facade, the double-curvature steel structure that transmits lateral forces resulting from the wind loads surrounds the entire building. In turn, the vertical loads are transferred by the central service core. The building is 164 m high and has 41 floors.

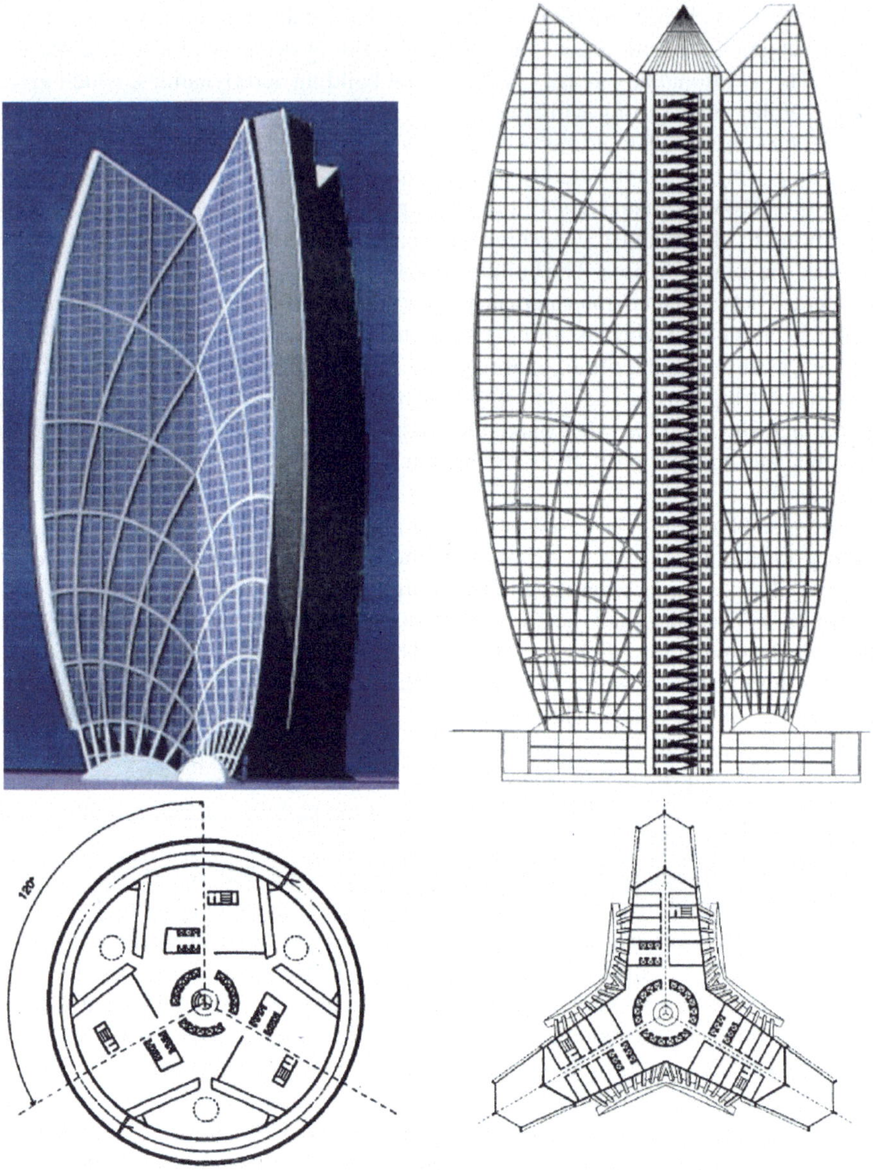

Fig. 8.5 The project of three wings bulbous skyscraper shaped exactly as Michell's cantilever (by courtesy of W. Zalewski and W. Zabłocki)

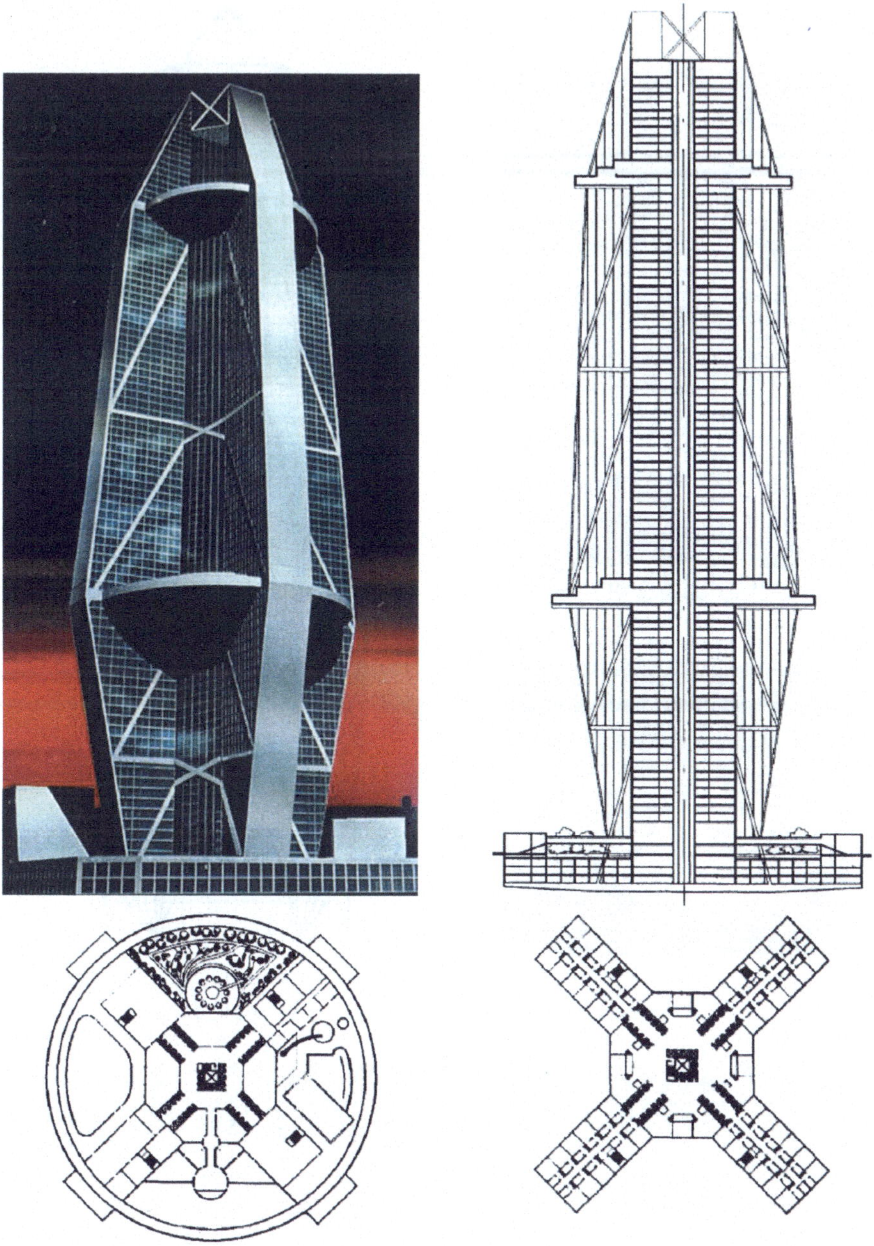

Fig. 8.6 The project of four wings skyscraper (by courtesy of W. Zalewski and W. Zabłocki)

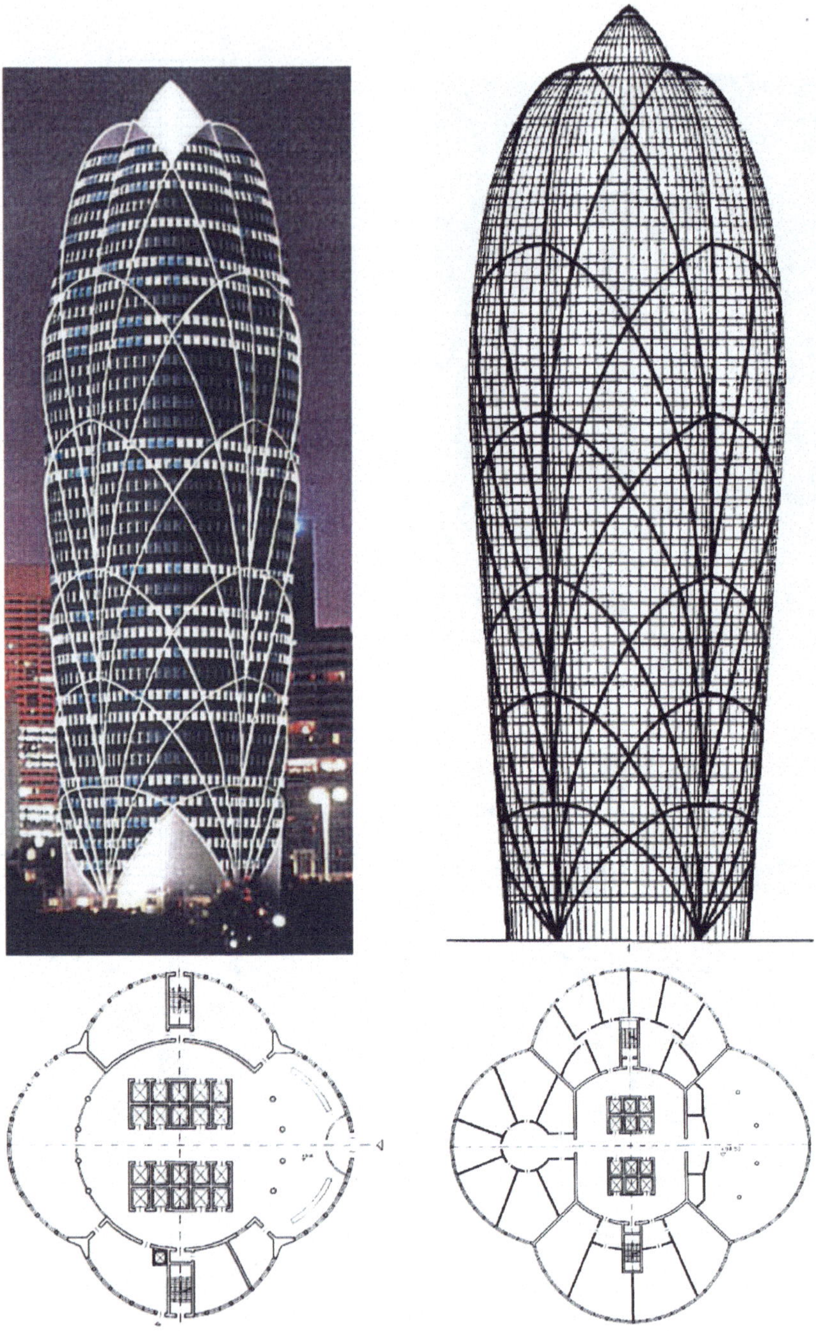

Fig. 8.7 The project of the tulip-shaped building inspired by Michell's theory (by courtesy of W. Zalewski and W. Zabłocki)

8.1.1.3 Optimal Building Design by Skidmore, Owings and Merill

The fundamental principles of topology optimization including the theory of Michell's structures have been effectively utilized by the architectural, urban planning and engineering Skidmore, Owings and Merill (SOM) office. The company had successfully applied the methods of topology optimization in many of their notable projects including world's most recognizable buildings, see the paper by L.L. Stromberg, A. Beghini, W.F. Baker and H. Paulino (2011) and the paper by W.F. Baker, A. Beghini and A. Mazurek (2011). These papers draw upon Maxwell's theorem, the methods of constructing discrete Michell's structures and the methods of graphic statics.

In particular, the optimal layouts of cantilevers subjected to uniform lateral loading (see Sect. 4.12) were inspiration for the optimal design of braced frames of high-rise buildings, cf. Stromberg et al. (2011). The optimal design problem was solved numerically with the use of a material interpolation method, which enabled imposing additional constraints on the design. The mainly considered constraints were the following: (i) the symmetry of the optimal layout with respect to vertical axis of the wall and (ii) the pattern gradation, i.e. the requirement of using repeatable patterns of variable size along the height of the wall. The obtained optimal layout taking into account both these constraints (Fig. 8.8a) was considered as a preliminary design of optimal laterally braced frame of a building.

The illustrative problem presented by Stromberg et al. (2011) includes pattern gradation with division of the considered domain into several separate sections (Fig. 8.8b). The application of topology optimization technique combined with the pattern gradation constraint allows to determine a diagrid-type bracing system, which satisfies the criteria of structural optimality and feasibility. The method reveals optimal angles of the bracing members including sharp angle at the bottom of the structure aimed at resisting the overturning moments and a small angle at the top of the structure aimed at sustaining shear loads. The similarity of the obtained layout to the bracing applied in the (designed by Skidmore, Owings and Merill) John Hancock Centre (Chicago, USA) can be clearly observed.

An innovative, integrated topology optimization technique for designing braced frames has been presented by Stromberg et al. (2012). The method is based on a concurrent use of both continuum four-node quadrilateral finite elements and discrete two-node beams. In contrast to classical material density methods, the proposed technique results in much more transparent and feasible layouts characterized by the presence of distinct longitudinal members. The method has been applied to the conceptual design of the 288 m tall high-rise building including lateral braced frame, see Fig. 8.9a, Beghini (2013). The substantial differences in comparison to a bracing layout applied by SOM in earlier designed buildings can be easy noted (Fig. 8.9b).

Another, more complicated problem is an optimal design of a three-dimensional bracing system for the entire building. The corresponding example presented in Stromberg et al. (2012) concerns an optimal layout of a three-dimensional bracing for (designed by Skidmore, Owings and Merill) Lotte Tower in Seoul, Korea, which has untypical shape with a square base and rectangular top (Fig. 8.10a). The results

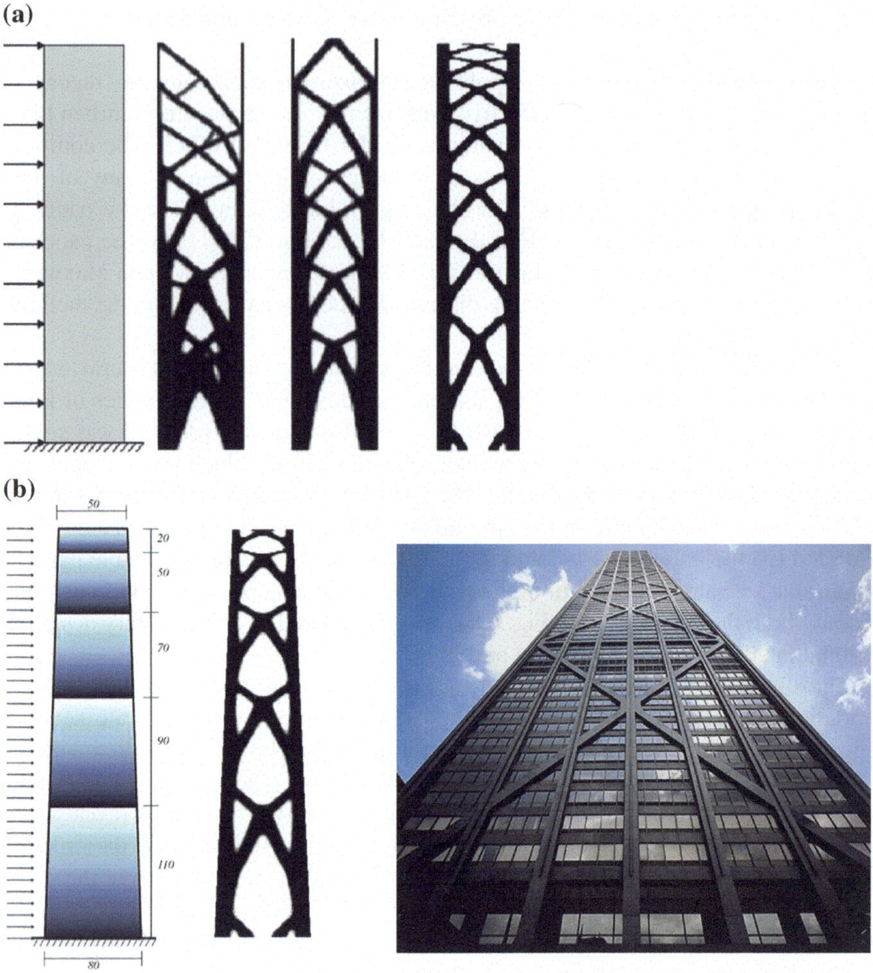

Fig. 8.8 Optimal design of lateral bracing: **a** problem formulation, layouts: without constraints, with symmetry constraints and with pattern gradation, **b** assumed pattern gradation, obtained layout (courtesy of Springer (Stromberg et al. (2011)) and the bracing of 875 North Michigan Avenue (formerly John Hancock Center), (SOM: Fazlur Rahman Khan, image: courtesy of SOM ©Ezra Stoller | Esto)

obtained with the use of the material interpolation method correspond to the cases of wind loading about one axis of symmetry, two axes of symmetry and application of pattern gradation constraints (Fig. 8.10b). The conducted optimization reveals the fundamental aspects of high-rise buildings designs including change of columns sizes and diagrid angles along the height. The structural layout corresponding to unidirectional wind forces resembles the topology of the two-dimensional Michell cantilever located in a bounded rectangular domain. The material tends to concentrate at the extreme sub-domains creating the flanges subjected to tension and compression and

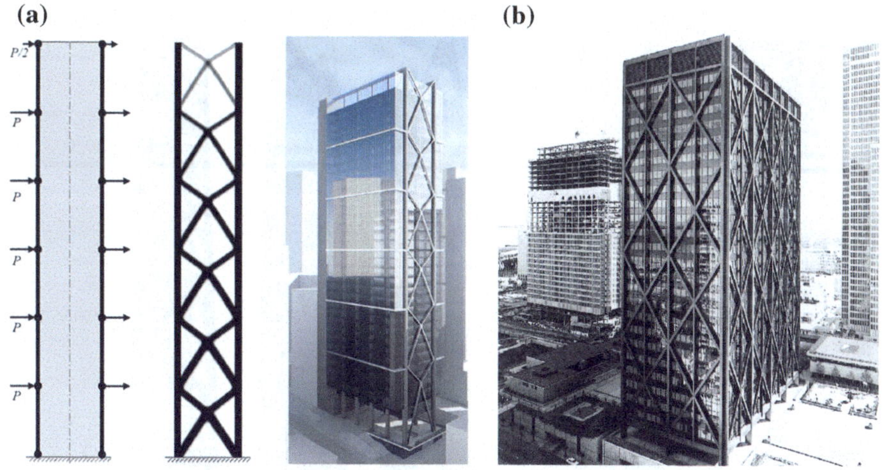

Fig. 8.9 Various designs of lateral bracing: **a** application of integrated topology optimization method (courtesy of Wiley (Stromberg et al. 2012; Beghini 2013)), **b** bracing in the Alcoa Building in San Francisco, USA (SOM, image: courtesy of SOM © Mak Takahashi)

counteracting the bending moment. In turn, the braces located in the perpendicular direction create the web which counteracts shear deformation and resembles an orthogonal fibrous domain of the optimal Michell's cantilever.

The problem of a bracing design for Lotte Tower was independently solved by the unstructured ground structure method by Tomás Zegard (2014). The optimization was performed separately for two loading scenarios assuming unidirectional lateral loading and torsion applied at the top of the tower (Fig. 8.10c). Both obtained solutions reveal the similarities to Michell's structures. For the case of unidirectional lateral loading, the optimal structure contains the straight pillars located in the regions of tension and compression as well as the net of approximately orthogonal members at the sidewalls of the shell. In turn, the structure optimized for torsion contains a diagrid pattern resembling the one known from Michell's sphere. When the high-rise building is subjected to bending and torsion the final optimal design is composed of two aforementioned designs.

The following example of the application of the methods of topology optimization by engineers from Skidmore, Owings and Merill is the design of Broadgate – Exchange House in London, UK (Fig. 8.11a, b). Since the structure spans over 78-m-long rail yard, it combines the function of a building and a bridge and it requires extraordinary structural system. The designers had decided to use two large-scale arches which support and strengthen the entire construction. The weight of the building is transferred to the arches by multiple vertical hangers, while the load from the arches is transferred to the supports. In addition, two oblique members connect the arch and the central bottom part of the structure. The shape of the arch was optimized in order to provide minimal weight and maximal stiffness of the structural system. According to the description presented by Baker et al. (2011) the optimization was based on the Maxwell's load-path theorem. In the considered case, this

Fig. 8.10 Three-dimensional bracing design for Lotte tower: **a** rendering of the Lotte tower and computational mesh (Stromberg et al. 2012), **b** optimal solution obtained with the use of density method (Stromberg et al. 2012), **c** optimal solution obtained with the use of ground structure method (Zegard 2014) (by courtesy of Springer and T. Zegard)

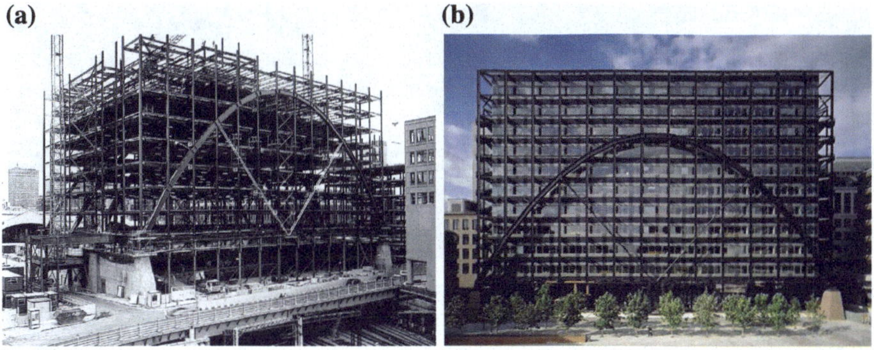

Fig. 8.11 Broadgate Exchange House in London: **a** construction of the building, **b** contemporary view of the building front (SOM: Srinivasa "Hal" Iyengar, William F. Baker, image: courtesy of SOM © SOM)

theorem enables replacing minimization of the total weight of structural system, by minimization of the weight of bars in tension (the hangers and the bottom tie), which significantly simplifies the design process.

The final design clearly resembles the topology of the Michell structures transferring the vertical loading into two pin supports. At first, it contains a compressed arch, which is a typical element of the optimal layout both in case of concentrated and uniformly distributed loading. Secondly, it includes vertical hangers. Since the point loading is not present the fan of straight members characteristic for such type of loading is reduced to two extreme members located at right angle to the arch.

Another example of the application of topology optimization in architectural and structural design is the project prepared by SOM for the Zendai competition aimed at creating a unique and innovative upper 'bridge' connecting several high-rise buildings (Beghini et al. 2013), Fig. 8.12. The preliminary model was subjected to distributed gravity loading and constrained with pin supports at the location of the towers.
The corresponding topology optimization problem was solved with the use of density-based method. The application of the numerical optimization had allowed obtaining an innovative design which shares essential features with Michell's solution. In particular, the proposed design of the bridge is composed of arches and located orthogonally approximately straight members, which is a typical layout of Michell's structures with pin supports and point loads.

The structure designed by SOM, which is probably the most influenced by the theory of Michell's structures, is the CITIC Financial Center in Shenzhen, China (currently under construction), Fig. 8.13. The centre consists of two towers (200–300 m high), which include external bracings strongly inspired by Michell cantilevers.

The lower tower is based on a strongly simplified layout of the optimal cantilever with only two members at each level, which is doubled at the width of the structure (Fig. 8.14). In turn, the higher tower is based on the optimal cantilever layout discretized by using three parametric lines. The towers are designed to resist two different types of lateral forces—seismic loading acting at structure base and wind

Fig. 8.12 Model for the conceptual design of the upper and lower ‘bridges’ spanning several towers for the Zendai competition; Beghini et al. (2013) (by courtesy of SOM and ASME)

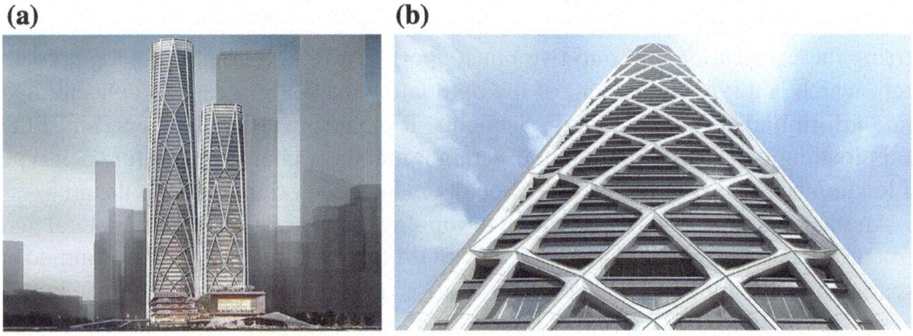

Fig. 8.13 CITIC Financial Centre in Shenzen, China: **a** rendered general view on both towers, **b** detailed view of the bracing of the top tower (SOM: Craig W. Hartman, image: courtesy of SOM © SOM)

loading prevailing at structure top. Therefore, the initially assumed optimal layout is morphed between bottom and top of the structure by substantial stretching in vertical direction. As a result, the orthogonality of the members from the optimal Michell's layout is not preserved, but instead, the angles between members vary over the height of the building. Eventually, the similarity to the original Michell cantilevers design is the most visible in case of higher tower, where we can observe two bottom circular fans of straight members connecting at right angle and inclined members on both sides of the walls located at an acute angle to the edges of the building (Fig. 8.14).

The resistance to wind loading is provided by the tapered form of the entire building, by chamfered corners at its upper part and its porous crown. In contrast, the durability to the seismic loading is assured by ductile steel links located at the corners of the frames, designed to yield during extreme seismic conditions. The links

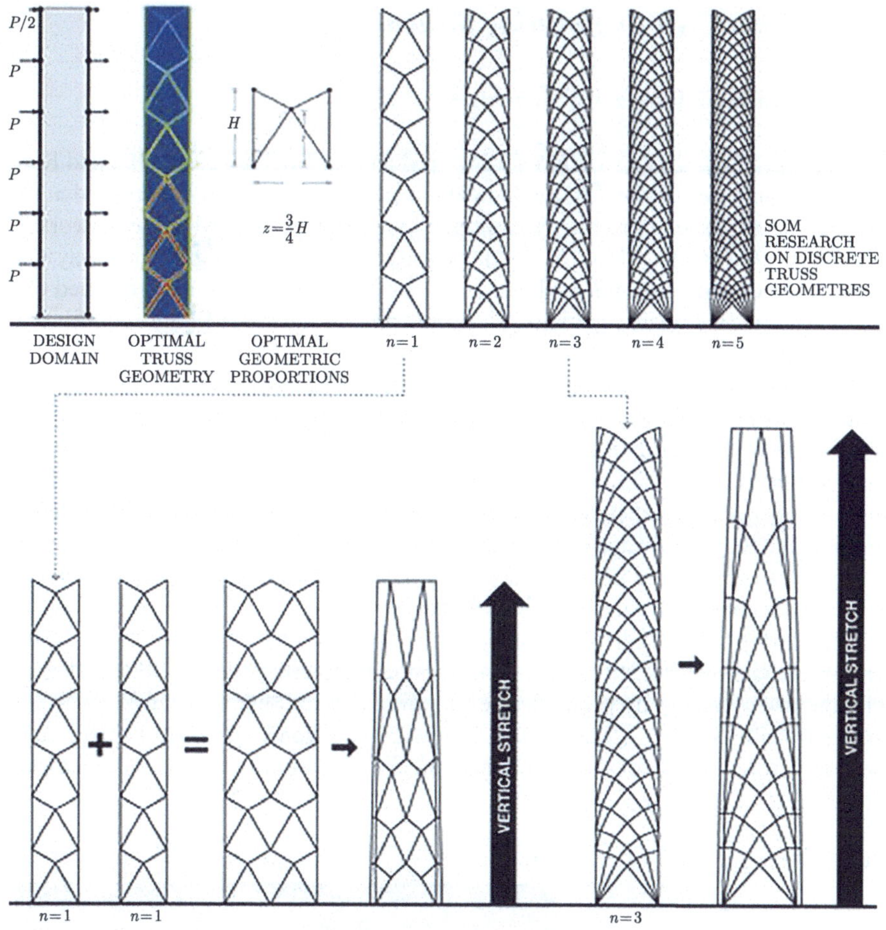

Fig. 8.14 Basic explanation of the design process of the external bracings of both towers (by courtesy of SOM and Spark Awards)

provide dissipation of the energy of the seismic motions and allow damping of the structure vibrations. In addition, the lower part of the structure, which is the most endangered by seismic excitation, contains a larger number of members in order to effectively resist a seismic loading. The above combination of high global stiffness of the entire structure and flexibility of its joints provides optimal design for a presence of wind and seismic loading when maximal displacement has to be controlled and vibration energy has to be dissipated.

8.1.2 *Other Civil Engineering Structures*

8.1.2.1 Optimal Tensegrity Structures

The concept of tensegrity structures was introduced by Buckminster Fuller and Kenneth Snelson in the 1960s. The fundamental principle of tensegrity assumes that the structure is composed of isolated compressed members embedded in the network of tensioned cables (Fig. 8.15a, b). The tensegrity structure is designed in a way that compressed elements are not in direct contact with each other, while tensioned elements delineate the system spatially. The initial prestress is used to assure equilibrium of the entire structure.

Although the concept of tensegrity is not directly based on Michell's theory, it is strongly related to the field of topology optimization. In particular, the crucial stage of design of tensegrity structures is selecting its proper layout, which provides optimal distribution of internal forces and structure stability. Moreover, as a result of the applied construction principle the elements of such structures are never subjected to bending, but they are always working either in tension or compression. Consequently tensegrity structures are characterized by small weight and large stiffness, which makes them apparently similar to the Michell structures.

The concept of tensegrity was adopted by Wacław Zalewski in order to design large-scale engineering structures. Two important structures designed by Zalewski with the use of tensegrity principle are shopping centre 'Supersam' in Warsaw and spectacle hall 'Spodek' ('Saucer') in Katowice in Poland, see Foxe (2018). Both these structures effectively combine the concept of tensegrity and certain elements

(a) **(b)**

Fig. 8.15 The examples of tensegrity structures: **a** Skylon presented in 1951 at Festival of Britain, **b** cantilever-shaped tensegrity structure (by courtesy of www.complexities.com and www.quora.com)

Fig. 8.16 General view of 'Supersam' in Warsaw, photograph taken in 1960s (by courtesy of Warsaw University of Technology Digital Library)

of the theory of Michell's structures in order to obtain an innovative lightweight structural form.

'**Supersam**' was a large self-service store characterized by pioneering construction of its tensegrity roof of dimensions 82 × 42 m covering three separate parts of the interior—the central storage and two lateral commercial spaces. The roof structure supported by steel girders and cables was proposed by architects Jerzy Hryniewiecki, Maciej Krasiński and Maciej Gintowt, while the construction was designed by Wacław Zalewski, Stanisław Kuś and Andrzej Żórawski. This structure, erected in 1962 was one of the landmarks of Warsaw (Fig. 8.16) until its demolition in 2006.

Although the structure with a hanging roof was planned from the beginning of the project, the designers wanted to avoid the standard construction with sloping roof backstays, which would occupy limited external space and could be easily damaged, affecting global structure stability. Therefore, an alternative design of the hanging roof was iteratively developed by Wacław Zalewski. The final proposed solution was tensegrity structure realized with the use of alternating convex arches and concave prestressed cables of parabolic shape, which supports the corrugated roof covering (Figs. 8.17 and 8.18). Both the arches and the cables rest firmly on a massive monolithic construction located at the centre of the structure, while their supports at the outer walls are slender, see Zalewski (1962).

The distinctive feature of the proposed design of the covering is that the interplay of internal forces does not take place in one plane. The design is based on two fundamental assumptions. The first one is constant value of axial forces in steel members along the entire span of the roof, see Zalewski (1963). The second assumption is the exact balance between outward component of arch thrust and the inward component of cable pull at each of the supporting columns, Fig. 8.17. Thus, the columns are

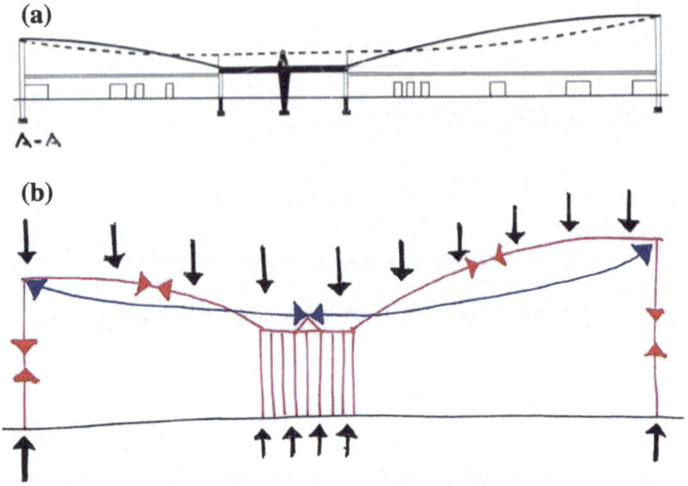

Fig. 8.17 **a** Drawing of general plan of the structure, **b** Force diagram by Edward Allen and David M. Foxe (MIT Exhibit, 2006) (by courtesy of www.wz-structure.org)

Fig. 8.18 Detailed views of arches and cables constituting tensegrity structure of the roof (by courtesy of www.wz-structure.org)

not subjected to any horizontal loading; they transfer exclusively vertical forces, see Zalewski (1964).

Wacław Zalewski describes his structure as follows Zalewski (1964): 'The massive construction of the internal part supports the located every three metres arch ribs of the lateral halls. The outward forces exerted at the columns of external walls are

Fig. 8.19 'Spodek' in Katowice (photo from 1973, by courtesy of www.katowice.fotopolska.eu)

taken over, with the use of horizontal beams, by located between the compressed arches, convex in the bottom direction tendons, which span the entire width of 80 m. Over the central part of the roof, the straight sections of the tendons are captured by directing upwards columns. This provides proper stability of the entire structure. The covering is designed as a steel structure, in which the lower chord is prestressed. The alternately located arches and cables create ridges and valleys of the folded covering. The contours of the edges of folds follow the diagram of the bending moments for a certain case of a three-span beam with two extreme spans subjected to distributed loading and the central span totally free of force'. The above description reveals the uniqueness of the designed structure and deep understanding of its structural behaviour by the main designer.

We note the similarity of the 'Supersam' construction to the Michell structures since its main load-carrying members are exclusively in compression or in tension and, moreover, the internal forces in these members remain constant. In contrast, the dissimilarity to the Michell structures results from the fact that main compressed and tensioned members are located in parallel to each other, not orthogonally.

'Spodek' is a multipurpose arena for cultural and sport events which can hold 11 500 spectators, formed as a 'flying saucer', an iconic shape of UFO (Fig. 8.19). It was the first arena constructed in Poland using the principle of tensegrity and still remains one of the largest objects of this type. The architects involved in design of Spodek were Maciej Gintowt, Jerzy Hryniewiecki and Maciej Krasiński, while the structure was designed by Andrzej Żórawski, Aleksander Włodarz and Wacław Zalewski. The erection of the arena was started in 1964 and, after a temporary stop of works due to suspicion of alleged construction errors, it was successfully accomplished in 1971. Since this time Spodek hosted a large number of international scale events and still remains a landmark of the city of Katowice (Fig. 8.19).

The main challenge in design of 'Spodek' was the requirement of object accommodation to various functions including hosting sport events with preferable surrounding audience, music concerts with directional audience as well as exhibitions or fairs. The second important challenge was location of the object at mining waste dump site characterized by medium mining damage with a possibility of local cave-ins and the corresponding requirement for special type of structure foundation.

The architects and designers had decided to apply untypical asymmetric form of the structure which resembles a bowl or an inverted dome. From the architectural viewpoint, such shape allows to conveniently adapt the interior configuration to various types of events. Even more importantly, from the engineering standpoint it enables reduction of the contact area with the ground and provides that entire structure settles uniformly as a single rigid object. Therefore, the possibility of potential structural damages caused by the ground settlement is minimized.

The entire construction is based on the concept of the balance of tensile and compressive forces. The inverted dome is subjected to circumferential tensile forces caused by its self-weight and the weight of spectators. These forces are balanced by a circumferential prestressed ring located at the top of the bowl and a system of orthogonal tensioned cables spanning to internal steel ring (Fig. 8.20). The cables constitute a support for the inclined surface of the covering. The above construction of the roof is the first known large-scale application of a cable structure based on tensegrity principle. This concept was tested by Wacław Zalewski by creating a number of wire models, which allowed to observe the state of structure equilibrium and to analyze the feasibility of the proposed design (Fig. 8.21).

Wacław Zalewski (1964) describes the structure as follows. 'The aspiration to describe the function of the building by its shape is not always possible to be realized. The project of the multipurpose arena in Katowice is, however, one of the examples of using such possibility. The structure is equipped with the spectators stand located circularly and gradually increasing its diameter in the upward direction. The extending upwards supporting structure being the housing for the stands has the shape of a cone. In this structure the circumferential tension is counterbalanced by prestressing, while the entire weight of the roof and the stands flows down into the foundation ring decreased significantly in comparison to the roof. The spatial work of the structure excludes the possibility of dilatating it into the smaller entities, which in case of expected motions of the ground caused by mining damages imposes large requirements for the structure. For the first location of the structure at the weak and soft ground, the foundation was designed as an inverted bowl closing the cone

(a)

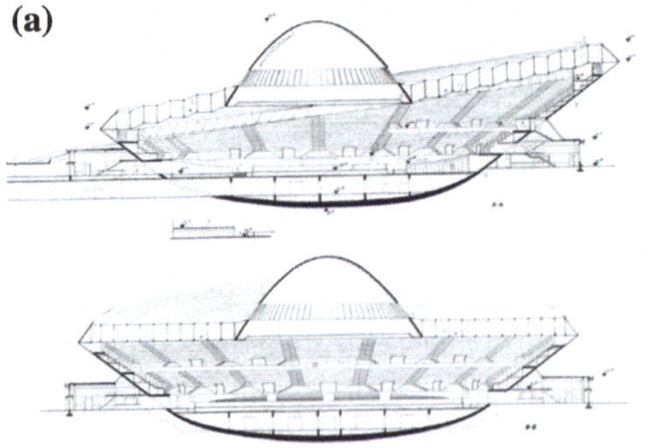

(b)

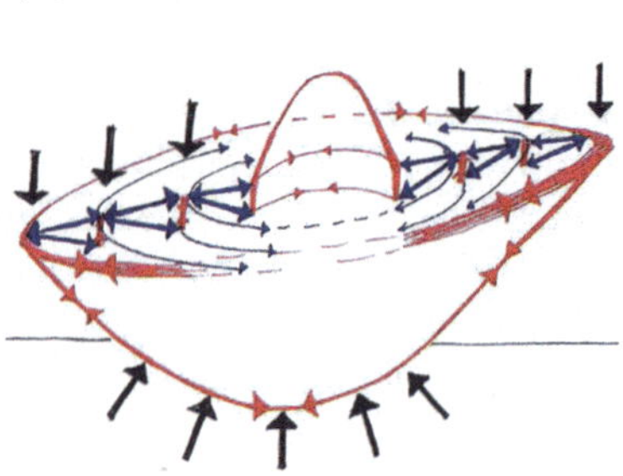

Fig. 8.20 **a** Architectural drawing of the structure, **b** Force diagram by Edward Allen and David M. Foxe (MIT Exhibit, 2006) (by courtesy of Waclaw Zalewski and www.wz-structure.org)

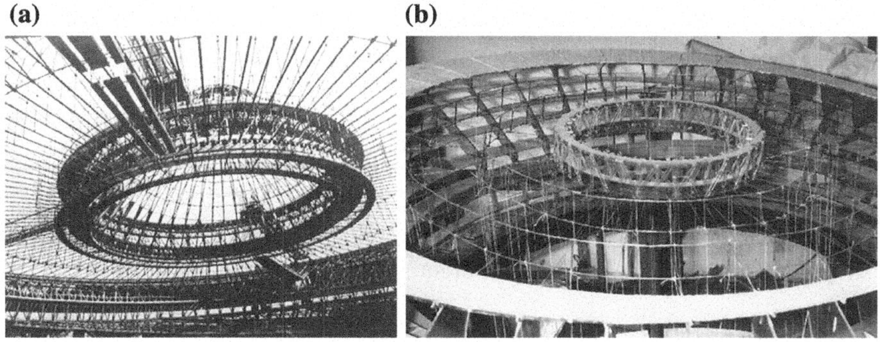

Fig. 8.21 **a** The half-constructed roof viewed from the interior of the building, looking up toward the central tension ring, **b** A scale model in a wire used for early load testing of the concept (by courtesy of www.structure.net and Wacław Zalewski)

from the bottom side and denting into the soft soil. The new, final location is characterized by rather strong grounds, however their global motion might occur. The foundation adjusted to these conditions assumes supporting the cone on the circular ring stiffened by horizontal diaphragm with the possibility of sliding with respect to the creeping ground. The covering of the hall consists of concentric rings in a distance from each other of 3 m. Each internal ring is hanged on larger neighbouring ring with the use of skew hangers. Each hanger consists of two chords, the upper one which is compressed and the lower one which is in tension. The crescent section of the roof, resulting from the asymmetry of the entire hall, is a mixed cantilever-ring structure'.

Let us note that the roof of 'Spodek' satisfies both the tensegrity principle and the fundamental principles of Michell's structures. The tensegrity principle is fulfilled since the compressed ring at the top of the bowl is not connected to other compressed members but exclusively to tensioned cables. Moreover, the cables are approximately perpendicular to the prestressed external and the internal ring, which is characteristic for Michell's structures.

8.1.2.2 Domes of Constant Strength

The concepts known from Michell's structures can be observed in the rational design of shells of revolution. An illustrative example is optimal shaping of a shell dome subjected to a self-weight, see the books written by Flügge (1973), Sect. 2.3 and by Novozhilov (1964), Chap. 28. Herein, we recall the discussion by Flügge and present the sketch of his solution.

It is known that equilibrium of the shell dome of an arbitrary shape under self-weight may be satisfied without the presence of bending. The objective of the formulated optimization problem is to find the constant-strength shell for which membrane stresses have the same magnitude at every point in every direction. The subsequent steps of the problem solution, are as follows:

1. We start from the classical equilibrium equations (1.151), where we introduce the distributed loads $p_2 = \gamma t \sin\varphi$, $p_3 = -\gamma t \cos\varphi$ resulting from the self-weight γ per unit area, as well as the constant values of membrane stresses in the meridional and circumferential directions $N^{11} = N^{22} = -\sigma t$. We obtain the equilibrium equation:
$$\left(\frac{1}{R_1} + \frac{1}{R_2}\right)\sigma t = \gamma t \cos\varphi \tag{8.3}$$
2. By using (1.147), we obtain the differential equation for the angle $\varphi(R_0)$:
$$\frac{d\varphi}{dR_0} = \frac{\gamma}{\sigma} - \frac{\tan\varphi}{R_0} \tag{8.4}$$
This equation can be solved by numerical integration starting at the apex, where we have $R_1 = R_2$ and thus $d\varphi/dR_0 = \gamma/2\sigma$.
3. The meridian of the shell in Cartesian coordinates can be found by a simple quadrature:
$$z = \int \tan\varphi dR_0 \tag{8.5}$$
4. The thickness of the shell can be determined from the equilibrium equation (1.151), where we introduce the assumed distributed loading caused by self-weight and constant values of membrane stresses:
$$\frac{dt}{d\varphi} = \frac{\gamma}{\sigma} t \tan\varphi \frac{dR_0}{d\varphi} \tag{8.6}$$
5. Especially simple form of the above equation is obtained upon transformation into orthogonal coordinates R_0, z:
$$\frac{dt}{dz} = \frac{\gamma}{\sigma} t. \tag{8.7}$$
Thus, the optimal thickness is given by the following formula:
$$t = t_0 \exp\frac{\gamma z}{\sigma}. \tag{8.8}$$

An illustrative constant-strength shell is presented in Fig. 8.22a. The thickness of the shell is the smallest at the apex and it increases exponentially in the direction of the support. The value of angle φ can be further extended, however, the exponential growth of thickness will briefly cause that the shell will no longer remain a thin-walled structure. Important conclusion from the above derivation can be drawn by considering a spherical dome, in which $R_1 = R_2 = a$. In such a dome the stresses caused by self-weight are proportional exclusively to γa so they increase proportionally to the diameter, independently of the wall thickness. Consequently, there exists a

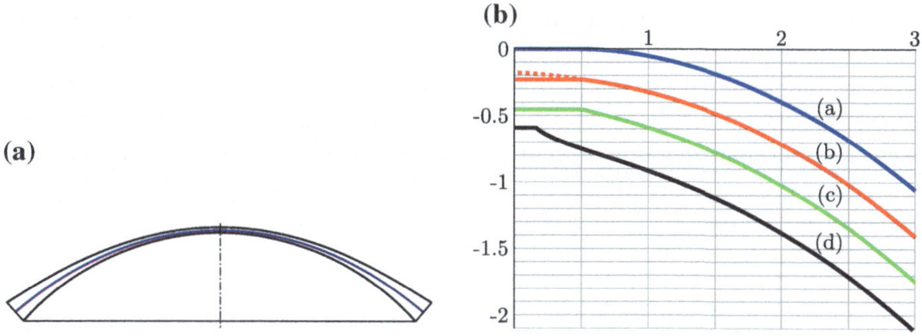

Fig. 8.22 Constant-strength domes determined by Flügge: **a** optimal shape and thickness of closed shell dome, **b** optimal shapes of shell domes with skylights

Fig. 8.23 The dome of the Pantheon resembling the constant-strength dome (by courtesy of www.monolithic.org and www.wikipedia.com)

maximal critical size of the dome, which can be obtained by using material of given ratio σ/γ. The condition of a constant stress provides the size of the largest dome.

The above solution can be generalized to the case of a shell dome with an opening at the top. For this aim, the numerical integration of the equation governing $\varphi(R_0)$ has to be started at the arbitrary edge $R_0 = b$ and arbitrary angle $\varphi(b) = \varphi_0$. Such formulation includes two parameters and thus provides a large variety of possible shapes of domes. In particular, the choice $\varphi_0 = 0$ results in a shape presented in Fig. 8.22b (curve 'a'), which starts with the ring carrying the compressive force σtb. In turn, the assumption $\varphi_0 > 0$ yields the variety of different shapes including the one obtained by starting integration at the apex (Fig. 8.22b, curve 'b') and the one starting with the inflection point (Fig. 8.22b, curve 'c'). Assumption of the larger values of φ_0 results in shapes similar to those presented in Fig. 8.22b, curve 'd'.

Certain aspects of the design of constant-strength shells can be found in the construction of the dome covering Pantheon in Rome, which has the diameter of 43.4 m and the estimated mass of 5000 tons, Fig. 8.23. It was the largest Roman dome at the time when it was built and still remains the largest unreinforced concrete dome in the world, cf. Lancaster (2006). The thickness of the Pantheon dome significantly varies along its meridian—it is the thinnest near the top oculus (1.2 m) and the thickest near the supporting walls of the rotunda (5.9 m), Moore (1995). The bottom part of the dome includes external 'step-rings', which create additional mass counteracting the external lateral thrust. Moreover, the coffering of the internal side of the dome changes in meridian direction in order to achieve variation of dome thickness and create a grid of orthogonal reinforcing ribs. Eventually, the stone of smaller self-weight is applied at the upper part of the dome in order to obtain proper weight distribution in the entire structure.

8.1.2.3 Other Miscellaneous Structures

Besides high-rise buildings, tensegrities and domes many other miscellaneous structures of a various purpose and function had followed the principles of design of Michell's structures. One of such constructions was the sculpture 'Tree of Life' designed for the exhibition Expo 2015 in Milan and located in the front of Italian pavilion, see Fig. 8.24. The project was created by Marco Balich, who was inspired by paving of the Piazza del Campidoglio in Rome designed by Michelangelo Buonarroti.

The entire structure consists of an internal skeleton made of steel and it is entirely covered with wood. The sculpture stands in the middle of the artificial lake. The intertwined roots of a tree resemble intersecting Michell cantilevers supported on a circle, rotated around vertical axis of symmetry. The roots of the tree pass smoothly into its trunk, which is covered with an orthogonal net, typical for the optimal structures subjected to torsion such as Michell's sphere. At the top of the structure, the trunk passes into vertical branches of a tree resembling again rotated Michell's cantilevers supported on a circle. The entire structure has the shape of a canopy of a height of 35 m and span of 45 m. The monument is equipped with illumination and water jets.

Although the structure does not have any practical function and carry exclusively the self-weight, it is aesthetically appealing and impressive. According to the

Fig. 8.24 The 'Tree of life' monument at Milan Expo 2015 (by courtesy of www.wantedinmilan.com and www.vinodadvani.com)

designer, the monument represents the Nature as a source of life, symbolizes many roots and branches of the Italian territory and symbolizes a social and economic recovery of the country.

8.1.3 Cable-Stayed, Arch and Deployable Bridges

8.1.3.1 General Indications for Design of Bridges

One of the most fundamental Michell's problems is finding a structure located in a half-plane, which transfers a single vertical point load, a set of vertical point loads or vertical distributed loading into two pin supports forming a horizontal line. Although the problem is elementary, its solution is non-trivial and it is exceptionally important from the engineering point of view; it gives significant hints on the optimal design of bridges.

The primary problem of the above class is aimed at finding the lightest structure which transfers a single vertical point load located at arbitrary height on the mid-perpendicular of the section connecting two simple supports, see Sect. 4.5.4. The case which particularly refers to the design of bridges involves the point load at the level of the supports. The corresponding optimal layout includes two compressed straight bars starting from the supports and the circular fan composed of straight hangers, spreading outwards from the applied force (Fig. 4.23a, the case of $\chi = 0$, $\theta_2 = \pi/4$). Since the structure is externally statically indeterminate, the optimal layout depends on the ratio of admissible stresses in tension and compression. When allowable stresses in tension and compression are equal the angle of the fan equals $\pi/2$. In turn, when allowable stresses in tension are larger the inclination angle of the fan gradually increases and eventually the optimal layout is a half-circular arch with internal hangers. In contrast, when allowable stresses in compression are larger the inclination angle of the fan decreases and eventually the optimal layout is composed of two straight compressed bars of infinitesimally small inclination angle to the horizontal axis and a single infinitesimally short hanger, see Fig. 8.25.

The solution of the above simple problem suggests the preliminary optimal layout of the bridge. If a material characterized by large tensile strength can be used, it should

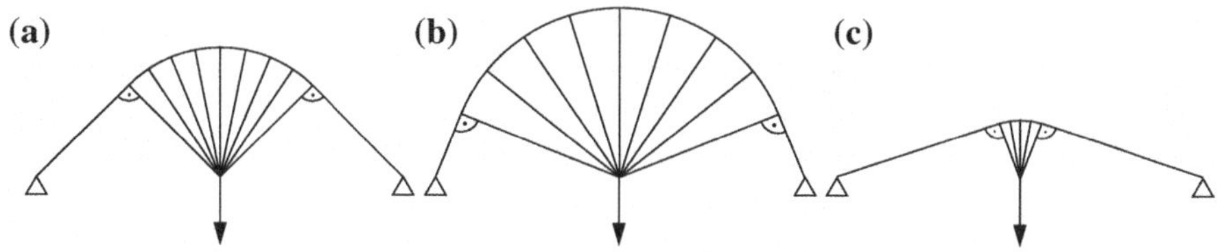

Fig. 8.25 Optimal layout of arch bridge loaded by a point forces at centre of the span: **a** the case of equal allowable stresses, **b** the case of larger allowable tensile stresses, **c** the case of larger allowable compressible stresses

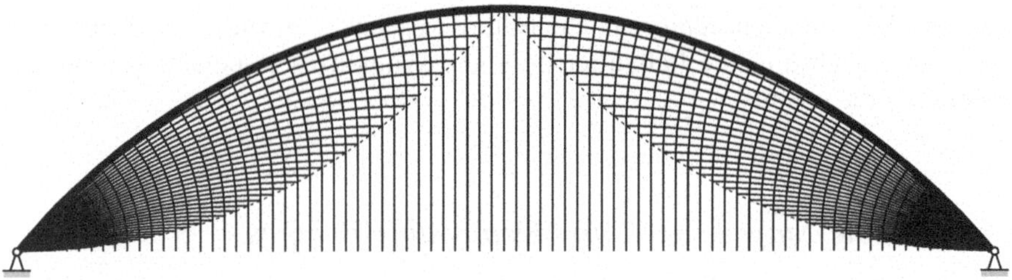

Fig. 8.26 The optimal structure found by Pichugin et al. (2012) transferring a uniform vertical loading for the case of $\sigma_T/\sigma_C = 0.417785$ (by the courtesy of SMO)

be applied extensively. It is appropriate to construct the fan of a large inclination angle, which will not significantly increase the weight of the members in tension, but will result in the favourable shape of the external arch generating small horizontal reactions and decrease of its volume despite increase of length. In contrast, when the designer can use the material characterized by large compressive strength he should design a fan of possibly small inclination angle. Obviously, in case of steel structures, the assumption of larger admissible tensile stresses is more realistic due to occurrence of buckling phenomenon. Thus, the application of the circular fan of a large angle is recommended, see Fig. 8.25b.

The topology optimization problem which refers more directly to design of bridges is aimed at finding the lightest structure transmitting a uniformly distributed transmissible vertical loading to two pin supports; this problem has been partly solved by Hemp (1974). The exact solution to this problem found by Tyas et al. (2011) has been commented in Sect. 6.4. A slightly different topology optimization problem involves the distributed vertical loading of constant intensity applied at the level of the supports. In this case, the optimal layout cannot be directly deducted from the previous solution for transmissible distributed loading since the problem belongs to the subclass when the Hencky net depends on the location of the applied force. The optimal structure consists of the compressed arch with either vertical or curved hangers ranging from the arch to the deck, Fig. 4.178c. The structure includes a horizontal tie bar, located at the level of supports, which does not extend to the entire structure span. The layout of the structure, especially the length of horizontal tie bar depends on the ratio of allowable stresses in tension and compression, see Pichugin et al. (2012).

The indications for bridge design resulting from the solutions in Fig. 4.178 are that the compressed arch should be equipped with hangers of various curvature located at the entire span and supported by lateral members starting at the supports. The range of lateral supporting members should increase along with decrease of ratio of allowable tensile and compressive stresses, see Fig. 8.26.

Finally, the topology optimization problem which gives the best insight into the design of bridges is finding the lightest structure for uniformly distributed loading located over multiple spans. The analytical and numerical solution of this essential problem was found by Pichugin et al. (2015). In case of equal stresses in tension

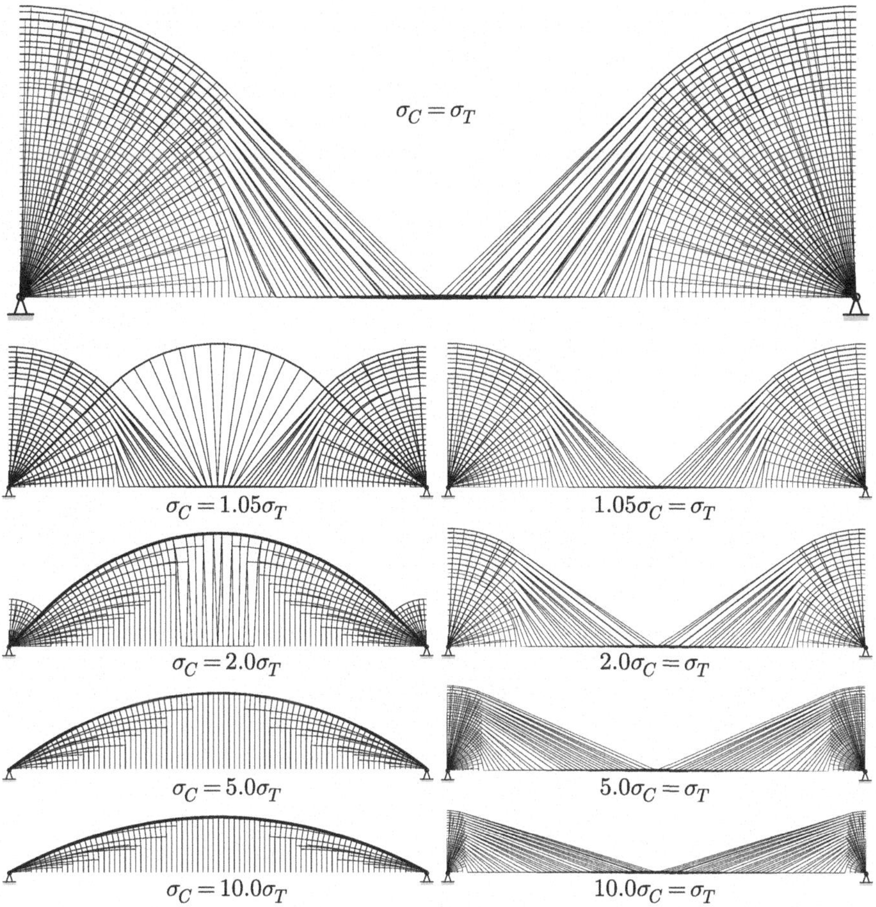

Fig. 8.27 Single section of the optimal structure transmitting a uniform load located over multiple spans determined in Pichugin et al. (2015). The effects of varying the ratio of allowable stresses (by the courtesy of SMO)

and compression, a single section of the optimal structure consists of two lateral fibrous domains composed of straight members in compression and curved members in tension, as well as the central region where only straight members in tension and a horizontal deck are present (Fig. 8.27). The most interesting features of the optimal layout are revealed when various ratios of admissible stresses in tension and compression are considered. When the allowable compressive stress increases an arch spanned between the supports gradually appears, while the size of the lateral fans progressively diminishes. Eventually, the optimal structure is composed of a compressed arch equipped with hangers in tension and reinforcing compressed members starting from both the supports. In contrast, when the allowable tensile stress increases the length of central straight hangers gradually raises, while lateral fans shrink and occupy exclusively small regions near the supports. Finally, the optimal structure is composed of two inclined posts starting at the supports and straight hangers located between the posts and the horizontal deck.

The main conclusion resulting from the above remarkable solution is that the optimal structural form of the bridge strongly depends on the ratio of admissible stresses in tension and compression. In case of a material with a large allowable compressive stress, the optimal structure appears to be the classical arch bridge equipped with vertical hangers and, possibly, additional reinforcing members starting from the supports. In contrast, in case of a material with a large allowable tensile stress, the optimal structure appears to be a classical cable-stayed bridge with pillars above the supports and straight hangers. However, the pillars are not vertical but slightly inclined into the centre of the span. The other, indirect conclusion is that a suspension bridge never appears as an optimal structural form.

8.1.3.2 Optimal Cable-Stayed Bridges

The above rules of optimal design of bridges were either deliberately or intuitively applied by various bridge engineers. In particular, the superiority of cable-stayed bridges over suspension bridges was pointed out by several notable structural engineers including Fritz Leonhardt, who had proved that they are much more economical, stiffer and aerodynamically safer, see Leonhardt and Zellner (1972). The group led by Leonhardt had designed the all-steel cable-stayed bridge for crossing of the Strait of Messina between Calabria in continental Italy and Sicily (Fig. 8.28). The main span of the designed bridge should have had the length of 1800 m and included 6 road lanes and 2 railway tracks. The suspension cables should have been reinforced by curved orthogonal members, created the shape similar to circular fans known from the Michell structures. Unfortunately, the bridge over the Strait of Messina has never been built.

Fig. 8.28 Messina Straits Bridge linking Sicily and Italian peninsula design by Gruppo Lambertini (1982), Leonhardt (1986)

In the review paper, Leonhardt (1986) stresses the requirement of applying high-quality steel for the cables and compares its usage in suspension and cable-stayed bridges. The constructor writes: 'In designing a cable stayed bridge there is the first a problem of finding the required quantity of high strength steel for the cables. For a bridge with a 1800 m main span and 38 m width, a suspension bridge needs 50 700 tons of steel whereas a cable stayed bridge only needs 21 170 tons of steel. In the latter case this represents only about 40% the amount of steel. Furthermore, a suspension bridge needs a stiffening girder with a bending stiffness which must be about ten times larger than that of the cable stayed bridge. The suspension bridge needs additional heavy anchor blocks which can be extremely costly if the navigation clearance is high and foundation conditions are poor. The total cost of such a suspension bridge can easily be 20–30% percent above the costs of a cable stayed bridge'. In his considerations, Leonhardt is directly oriented towards minimization of the structure weight. Moreover, the emphasized requirement of using high-strength steel for the cables is closely correlated with known from Michell's theory fact that cable-stayed bridge is the most efficient structural form in case of larger tensile strength of the material.

The largest cable-stayed bridge designed and erected with participation of Fritz Leonhardt was Helgeland bridge in Nordland county in Norway with the longest span of 425 m and the total length of the deck of 1065 m (Fig. 8.29a). Currently, as of 2018, the cable-stayed bridge with the longest span in the World is the Russky Bridge

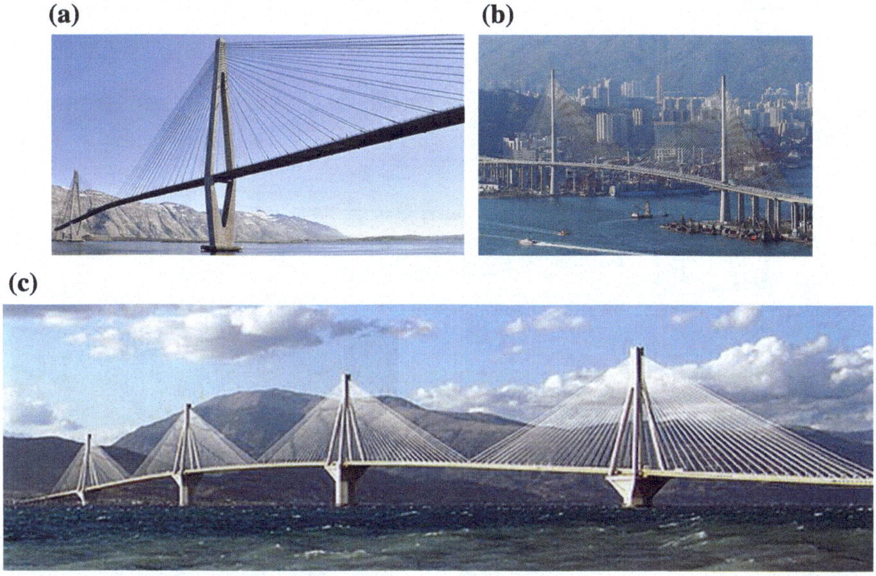

Fig. 8.29 Notable cable-stayed bridges: **a** Helgeland Bridge by Fritz Leonhardt, **b** Stonecutters Bridge in Hong Kong, **c** Rio–Antirrio Bridge over Gulf of Corinth, (by the courtesy of Pinterest and Wikipedia)

built across the Eastern Bosphorus Strait near Vladivostok in Russia (longest span 1104 m, total length 3100 m). It is followed by Sutong Yangtze River Bridge between Nantong and Changshu (China) with the longest span of 1088 m and Stonecutters Bridge between Tsing Yi Island and Stonecutters Island (Hong Kong) with the longest span of 1018 m (Fig. 8.29b). The other classification can be performed with respect to length of cable-stayed deck. According to such criterion the longest cable-stayed bridges in the World are Jiaxing-Shaoxing Sea Bridge across the Hangzhou Bay in Shaoxing, China (2540m), Millau Viaduct over valley of the River Tarn in southern France (2460 m) and Rio-Antirrio Bridge which crosses the Gulf of Corinth and links mainland Greece with Peloponnese peninsula (2252 m), Fig. 8.29c.

A similarity of cable-stayed bridges to layouts resulting from Michell's theory appears if one rejects the classical structural form with the exactly vertical orientation of pylons. Nowadays, it can be observed that more and more designs utilize either straight or curved pylons inclined in the direction of the main span of the bridge. The examples of such structures can be found both among footbridges and road bridges. In particular, the example of the pedestrian overpass with straight pylons inclined to the main span is the footbridge on Dunajec River between Poland and Slovakia in Sromowce Niżne designed by Jan Biliszczuk (Fig. 8.30a). Moreover, an illustrative example of structures with curved pylons is the footbridge over 'Zakopianka' road in

Fig. 8.30 Footbridges with pylons inclined in the direction of the deck, erected in Poland (by the courtesy of http://pieniny-na-weekend.pl; http://www.powiatlive.pl/malopolska/; Andrzej Waligóra http://waligora.net.pl/)

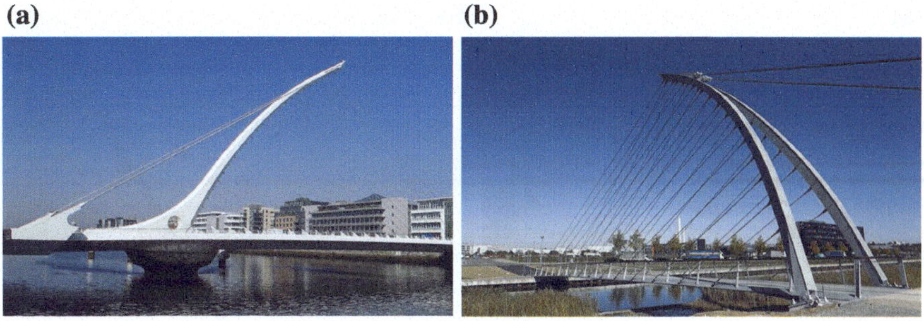

Fig. 8.31 Footbridges with curved and inclined pylons: **a** Samuel Beckett bridge in Dublin **b** footbridge across the lake Langer See (by the courtesy of Wikipedia and www.sbp.de)

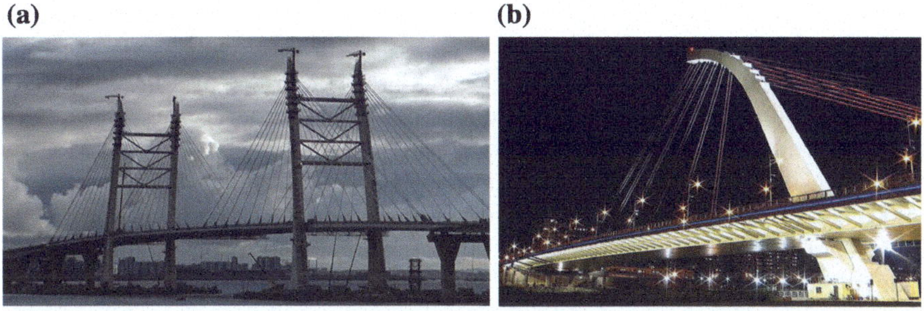

Fig. 8.32 **a** Bridge over Korabelny Fairway in Sankt Petersburg, **b** Dazhi Bridge in Taipei (by the courtesy of www.youtube.com and Wikipedia)

Poland (Fig. 8.30b), while an example of a structure with biforked pylon inclined into both spans is the footbridge over Żwirki and Wigury street in Warsaw (Fig. 8.30c).

The curvature and inclination of pylons in the direction of the deck was applied in well-known cantilever-spar footbridges designed by Santiago Calatrava, including Samuel Becket Bridge in Dublin (Ireland) and Puente de la Mujer in Puerto Madero (Argentina), Fig. 8.31a. Another example is the footbridge across the lake Langer See in Sindelfingen (Germany) designed by 'Schlaich, Bergermann und Partner', which is composed of a curved pylon and straight hangers attached at almost right angle, which corresponds to fan regions of Michell's structures (Fig. 8.31b). Among larger scale structures, it is worth mentioning the Korabelny Fairway Bridge in Sankt Petersburg (currently in the final stage of construction) whose pylons are inclined in the direction of the central span and Dazhi Bridge in Taipei which is supported by single curved and inclined pylon (Fig. 8.32).

An alternative design of the cantilever-spar cable-stayed bridge utilizes a pylon being inclined in the outward direction in order to counteract the torque generated by the weight of the deck. Such a design was popularized by Santiago Calatrava who had used the concept in Alamillo Bridge in Sevilla, Serreria Bridge in Valencia and Sundial Bridge over the Sacramento River in Turtle Bay in California

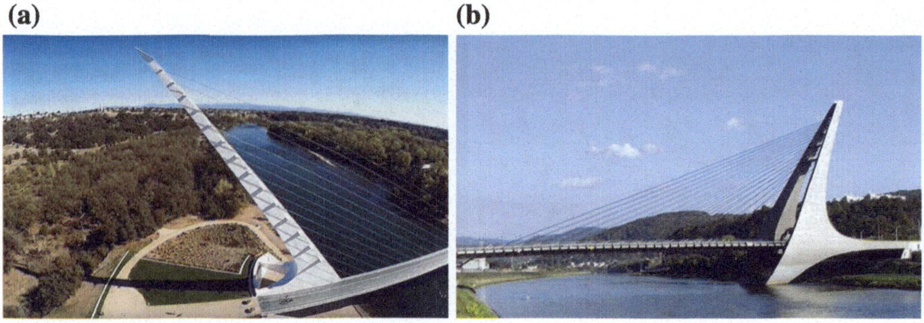

Fig. 8.33 Cable-stayed bridges with pylon inclined outwards from the deck: **a** Sundial Bridge in Turtle Bay in USA, **b** Mariansky Bridge in Czech Republic (by the courtesy of www.youtube.com and www.structurae.net)

(Fig. 8.33a). Another structure of this type is Mariansky Bridge over the Elbe River in Czech Republic (Fig. 8.33b). However, the design of cantilever-spar cable-stayed bridge with an outward-inclined pylon was criticized by Stanisław Kuś and Wacław Zalewski as being non-optimal from mechanical point of view. Kuś and Zalewski (2000) had compared such design against standard cable-stayed bridge with central straight pylon. They had raised the following arguments against cantilever-spar bridges:
(i) maximal force in the outward-inclined pylon is almost 4 times larger than in the straight pylon, (ii) maximal force in the deck generated by tendons is approximately 2 times larger, (iii) tensile forces in the tendons are in both cases identical, however their length in outward-inclined cantilever-spar bridge is always larger, (iv) the cost of the foundation for an inclined pylon substantially exceeds costs of the foundation for the straight one. Thus, according to these authors, the cantilever-spar bridges with pylon inclined in outward direction are chosen to be built exclusively due to their originality, aesthetics and representative function.

8.1.3.3 Optimal Arch Bridges

The similarities to Michell's structures can be also observed in construction of arch bridges. Most long-span arch bridges use the design resulting directly from Michell's theory, which includes a compressed arch and hangers in tension. In particular, the longest arch bridge in the world is Chaotianmen Yangtze River Bridge in Chongqing, China (span 552 m) in which the compressed arch is constructed as a truss structure of variable height, Fig. 8.34a. Moreover, the second longest bridge is Lupu Bridge over the Huangpu River in Shanghai, China (span 550 m) in which the arch is constructed from welded steel box sections, Fig. 8.34b.

Several smaller arch bridges can be considered as particularly interesting due to non-standard shaping of the arch or untypical system of transmitting the force from the deck to the arch. The first noteworthy structure is Zubizuri Bridge in Bilbao

Fig. 8.34 The World longest steel arch bridges: **a** Chaotianmen bridge in Chongqing, **b** Lupu Bridge in Shanghai (by the courtesy of https://i.pinimg.com and www.bridge.info.net)

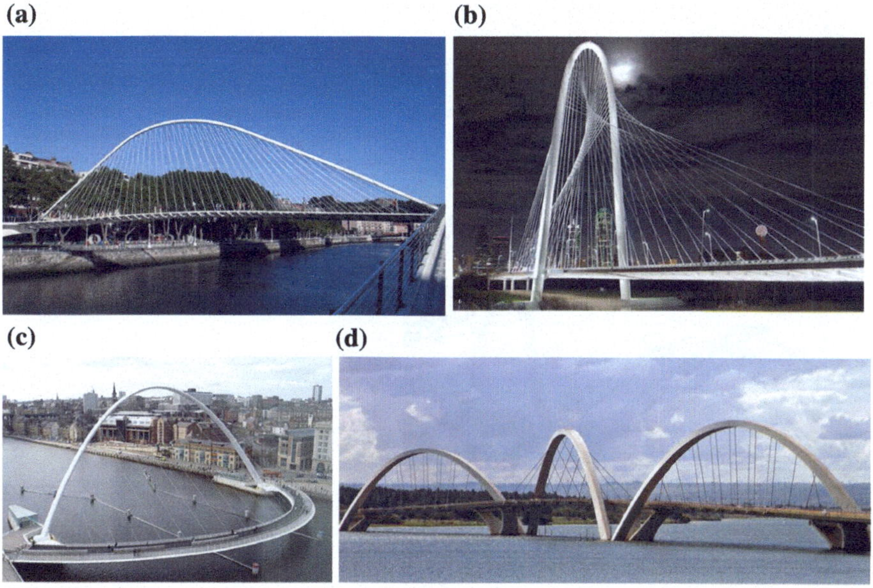

Fig. 8.35 Noteworthy arch bridges: **a** Zubizuri Bridge, **b** Margaret Hunt Hill Bridge, **c** Gateshead Millenium Bridge, **d** Ponte Juscelino Kubitschek (by the courtesy of www.thousandwonders.com, Wikipedia, www.wonderfullengineering.com and Wikipedia)

(Spain) designed by Santiago Calatrava (Fig. 8.35a). In this footbridge, the arch consists of two almost straight segments near the supports and central curved part. The hangers are not vertical but their inclination angle significantly increases near the supports, which partially corresponds to reinforcing members of Michell's structure transmitting uniform distributed loading to two horizontal supports. The second distinguished construction is Margaret Hunt Hill bridge over Trinity River in Dallas (Texas) also designed by Santiago Calatrava (Fig. 8.35b). In this bridge, the arch does not span the length of the bridge but instead it is located perpendicularly to the bridge in its central part. The hangers spread in both sides of the bridge and connect the arch with the central line of the deck. The third extraordinary structure is Gateshead Millenium Bridge located over River Tyne between Gateshead and Newcastle (England) designed by Wilkinson Eyre (Fig. 8.35c). In this case, the deck of the footbridge is curved in horizontal plane; the arch is not located directly above the deck and hangers are not located vertically. In addition, the bridge is equipped with large joints at both ends and can be tilted in order to provide the waterway for the vessels. Finally, the fourth remarkable construction is Ponte Juscelino Kubitschek in Brasilia (Brazil) designed by Alexandre Chan and Mário Vila Verde (Fig. 8.35d). In

(a)

(b)

Fig. 8.36 Bridges combining cable-stayed and arch design: **a** Seri Saujana Bridge in Malaysia, **b** Lianxiang Bridge in China (by the courtesy of Wikipedia Commons)

this bridge, the supports of each of the three arches are located at the opposite side of the deck, while the hangers from each arch are attached to both sides of the deck.

Eventually, several constructed bridges successfully combine the concept of a cable-stayed bridge with pylons and oblique cables with the concept of an arch bridge with vertical hangers. Therefore, their construction corresponds to the layout of the optimal Michell's structure for the case of similar admissible stresses in tension and in compression. The illustrative example of such structure is Seri Saujana Bridge in Putrajaya, Malaysia (Fig. 8.36a), in which the cables attached to the pylons support the middle of the deck, while hangers attached to the arch support its edges. The second example is Lianxiang Bridge in Hunan province in China (Fig. 8.36b), which has straight H-shaped pylons and an arch constructed as a spatial truss.

8.1.3.4 Optimal Deployable Bridges

Topology optimization techniques were recently successfully applied in development of novel deployable lightweight bridges for emergency situations, cf. Ario et al. (2013). The development of the potential structural form of the bridge was based on formulation and solution of the corresponding topology optimization problem. The design domain was a rectangular area located between two supports, see Fig. 8.37.

The optimization process was conducted with the use of the ground structure method utilizing equilibrium equations in the non-linear setting. The design domain Ω was represented by continuum of micro-trusses composed of pin-jointed linear springs. The assumed design variables were the stiffnesses of the micro-truss members. The goal of the optimization was to achieve a fully stressed design in which stresses in all members of micro-truss have equal values. The solution of the problem was obtained by using the adaptive topology optimization method developed by Maute and Ramm (1995a, b). The modification of the actual design of the structure was based on actual local values of stresses. After each iteration, the tangent stiffness matrix was updated and new equilibrium equation was established for the consecu-

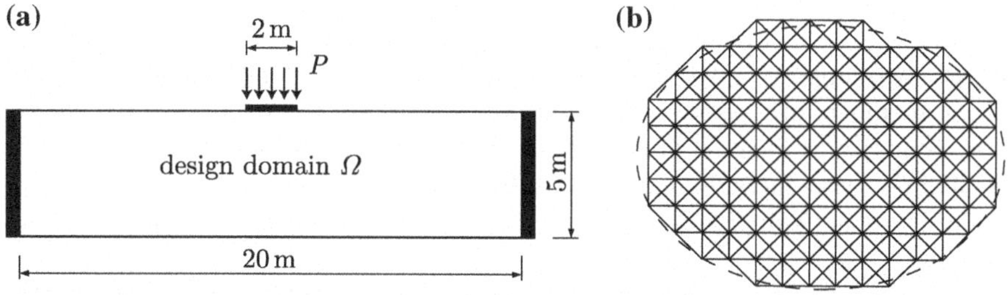

Fig. 8.37 Topology optimization problem aimed at finding optimal topology of deployable bridge: **a** design domain, **b** continuum of micro-trusses; Ario et al. (2011) (by the courtesy of Automation in Construction)

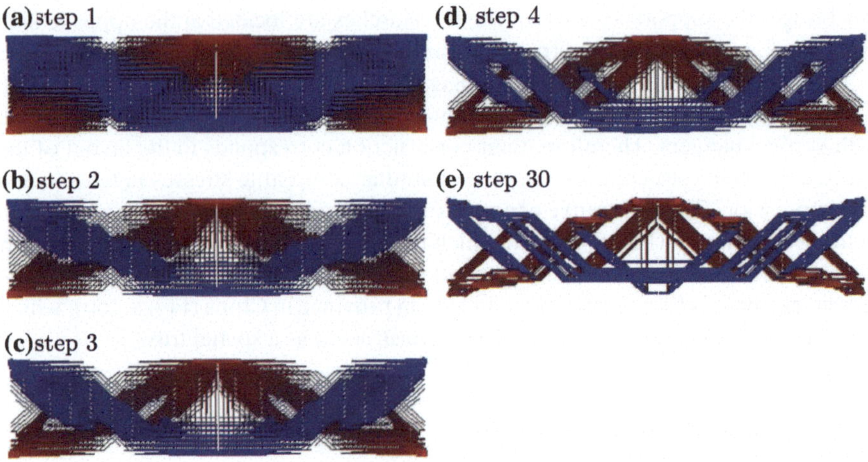

Fig. 8.38 Subsequent steps of solution of topology optimization problem aimed at finding optimal design of the deployable bridge; Ario et al. (2013) (by the courtesy of Automation in Construction)

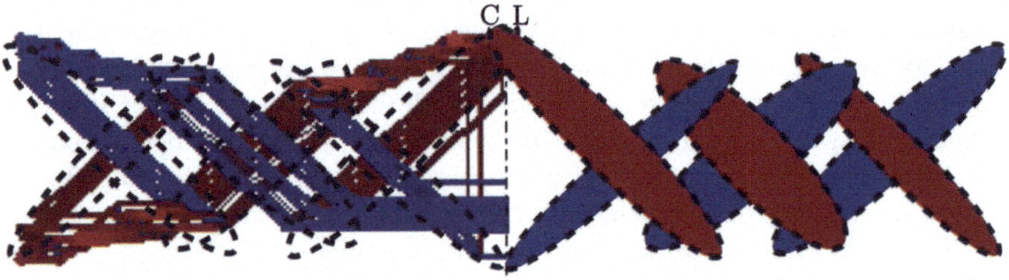

Fig. 8.39 Applied method of design rationalization and simplification of the layout used for final design of deployable bridge; Ario et al. (2013) (by the courtesy of Automation in Construction)

tive load step. The results from subsequent steps of the optimization procedure are presented in Fig. 8.38.

The final solution depicted in Fig. 8.38e contains easily recognizable areas of tension and compression, which can be considered as potential locations of members of a deployable bridge. The compression area has the shape of an arch spanned between two supports and reaching the point of application of the loading. Moreover, several members in tension are located perpendicularly to the arch, which clearly corresponds to the location of cables in a cable-stayed bridge and resembles classical Michell's structures. In addition, the final layout includes horizontal members, which significantly improves the global resistance to bending and provides structure stability.

The requirement of constructing the structure, which is fully deployable enforces the specific methods of layout rationalization. In order to allow rotational motion of the diagonal members, the horizontal members have to be temporarily eliminated. In the left part of Fig. 8.39 the tension and compression members, that can be maintained in the construction of a deployable bridge, are surrounded by dotted lines. In

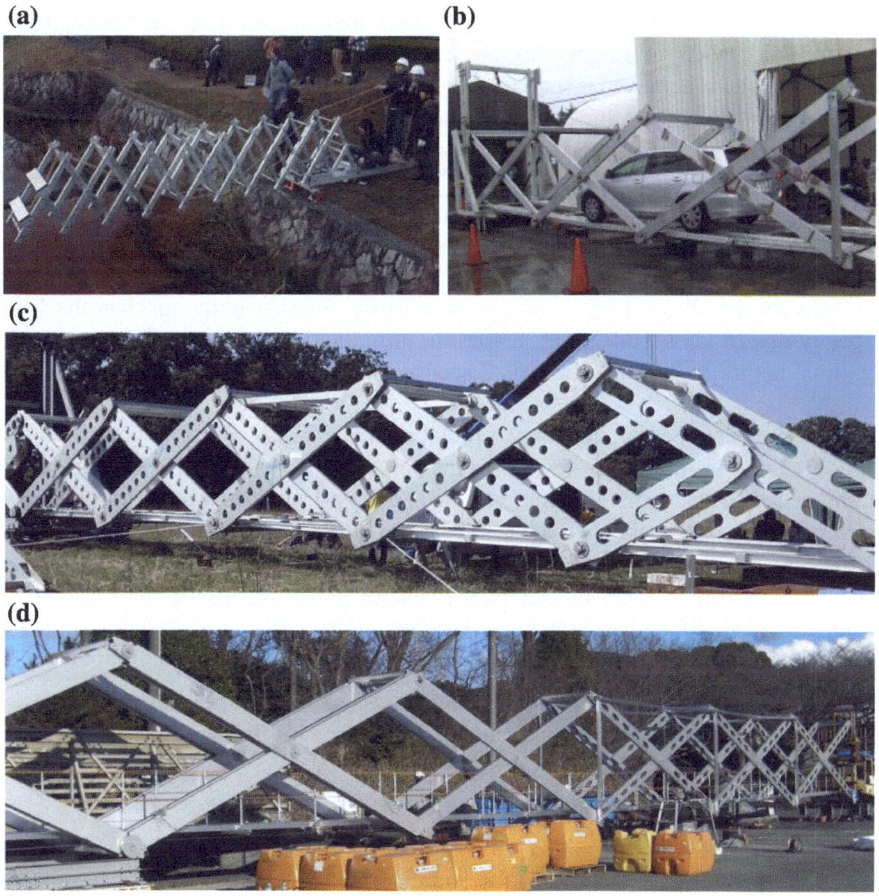

Fig. 8.40 The subsequent stages of the design of mobile bridge: **a** the prototype footbridge (MB0.1), **b** the Mobile Bridge MB1.1, **c** the Mobile Bridge MB4.2, **d** the footbridge Mobile Bridge MB30

addition, in the right part of the figure, the geometry of these members is simplified in order to constitute the basis for the final design. The obtained final layout resembles a scissors structure, which fulfills the requirements of deployment and folding. However, the scissors structure has to be reinforced by the horizontal members in order to form a complete structure of the bridge and provide its adequate bending stiffness corresponding to in-plane and out-of-plane directions. The reinforcement can be achieved by using members in tension and compression at the bottom and top of the structure, respectively. Moreover, the cables can be used as members in tension for reducing the total weight of the bridge.

Let us note that in presented case the topology optimization was successfully applied to find the optimal layout of the structure, which can be easily expanded and folded to the initial state during the construction works as quickly as possible.

The prototype of a deployable bridge was a narrow footbridge for pedestrians (Mobile Bridge MB0.1, Fig. 8.40a). The length of the folded footbridge was 1 m and when expanded it achieved a span of 8.2 m. The footbridge could carry the loading caused by the weight of three persons and it was expanded manually, without any automatic equipment.

The following constructed deployable bridges were dedicated for vehicles, had relatively short length and were composed of two or three scissors sections (MB1.0 and MB1.1, respectively), Fig. 8.40b. The gathered experience allowed to construct final versions of deployable bridges for vehicles and pedestrians. The vehicle bridge achieves the length of 17.5 m, is composed of 6 scissors units and has the load capacity of 120 kN (MB4.2, Fig. 8.40c). In turn, the deployable footbridge has the maximal length of 27.5 m, contains 8 sections and has the load capacity of 30 kN (MB30, Fig. 8.40d).

The bridge is modular, consists of prefabricated elements and can be assembled without applying multiple screws or joint bolts. The construction of the bridge is strengthened by the deck (enabling vehicle and pedestrian loads) as well as by horizontal and vertical members in selected locations. Determination of the cross sections of these members was also formulated as an optimization problem aimed at maximization of bridge stiffness and solved with the use of a differential evolution method (Chikahiro et al. 2018). The constructed bridges are recently commercially available and can be successfully applied in emergency situations, e.g. during natural disasters such as earthquakes, tsunamis and floods.

8.1.4 Optimal Reinforcement of Concrete Structures

The methods of topology optimization and theory of Michell's structures have been successfully applied to design of optimal distribution of reinforcement in concrete structures. According to the recent building codes the concrete reinforcement has to be located along the trajectories of principal tensile strains, which has to be determined by the designer. One of the methods which allow to approximate the complex stress field in concrete elements is the Strut and Tie (ST) method. In this method, the reinforced concrete material is modelled by a truss structure in which members in compression (struts) correspond to maximal compressive stresses in concrete, while members in tension (ties) correspond to the desired location of reinforcement. Although the design codes suggest application of ST models, they do not specify how these models should be constructed.

A new strategy of an automatic generation of Strut and Tie models by truss topology optimization has been proposed by Bołbotowski and Sokół (2016). In the proposed approach, the feasible domain is bounded by a set of polygons and contains a set of truss nodes located inside. Each node is connected with all other nodes by line segments, which create bars of the dense truss forming an initial ground structure. The generated truss is divided into two substructures corresponding to compressed members made of concrete denoted by $()''$ and members in tension made of steel denoted by $()'$. In order to find an optimal distribution of the reinforcement, it is necessary to determine the cross sections of the members, which provides that the limit state of stress under given loading is not exceeded and the volume of the material in tension is minimized. Therefore, the primal formulation of the problem of reinforcement volume minimization takes the following form:

$$\min \left\{ \frac{\sigma_0}{\sigma_T} \sum_k T'_k \tilde{l}'_k \mid (\boldsymbol{B}')^\top \boldsymbol{T}' - (\boldsymbol{B}'')^\top \boldsymbol{T}'' = \boldsymbol{P};\ T'_k \geq 0,\ T''_k \geq 0, \right\} \tag{8.9}$$

where σ_T is the permissible tensile stress in tension and $\tilde{l}'_k$ denote modified lengths of members in tension taking into account the cost of nodes expressed by penalty coefficient: $\tilde{l}'_k = l'_k + p$. Moreover, $\boldsymbol{B}'$ and $\boldsymbol{B}''$ are geometrical matrices of both substructures (see Sect. 1.2.7), $\boldsymbol{T}'$ and $\boldsymbol{T}''$ denote axial forces in their members, while $\boldsymbol{P}$ represents the external loading. The increase of penalty parameter increases the cost of short members and thus simplifies the obtained solution of the optimization problem. The problem dual to the minimization problem posed above takes the form:

$$\max \left\{ \boldsymbol{P}^\top \bar{\boldsymbol{u}} \;\middle|\; \frac{\sigma_0}{\sigma_T} \tilde{l}'_k \geq \Delta'_k(\bar{\boldsymbol{u}}),\ \Delta''_k(\bar{\boldsymbol{u}}) \geq 0 \right\} \tag{8.10}$$

where $\bar{\boldsymbol{u}}$ is a vector of adjoint displacements of the truss nodes, whereas $\boldsymbol{\Delta}'$ and $\boldsymbol{\Delta}''$ are vectors of adjoint elongations of members of both ground structures.

One of the difficulties related to the application of the above methods is boundedness of the feasible domain. Since stresses in compressed members cannot exceed the limit value for concrete, their cross sections are often relatively large. On the other hand, they have to fit in the assumed, bounded feasible domain. Unfortunately, such geometrical condition cannot be introduced directly without violating the linearity of the problem. Thus, the structure with bars located partially beyond the feasible domain is considered as the first iteration of the solution. In the second iteration, the bars intersecting the boundary are eliminated and a new ground structure is created. The procedure is repeated until all elements of the ground structure are located inside feasible domain of the problem. The above procedure is presented in Fig. 8.41.

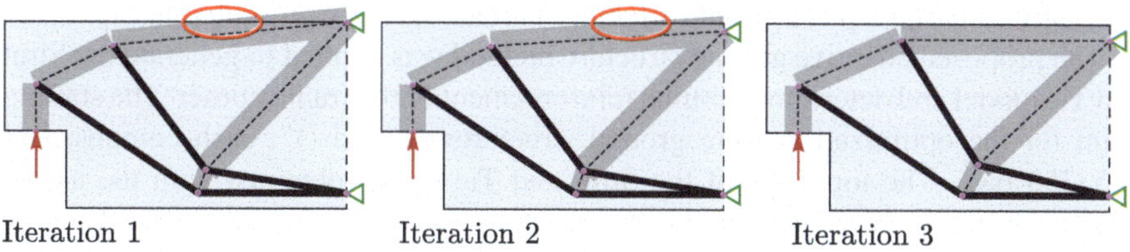

Fig. 8.41 Illustration of the algorithm allowing control of stresses in concrete; after Bołbotowski, and Sokół (2016)

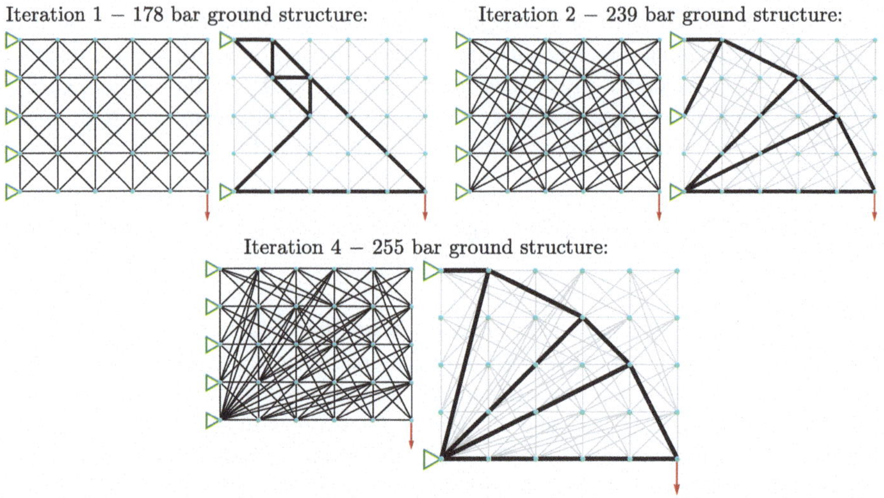

Fig. 8.42 Solution of the illustrative topology optimization problem with the use of the adaptive ground structure method; after Bołbotowski and Sokół (2016)

The other difficulty arising in the proposed method is a long time of computation in case of large number of potential bars. Despite the fact that the considered optimization problem is linear its solution becomes challenging for the number of bars larger than half million. This issue is resolved by the application of the adaptive ground structure method. The optimization problem for fully connected ground structure is not handled straightaway, but instead a sequence of subproblems formulated by smaller ground structures is solved iteratively. The algorithm starts with first subiteration, in which the ground structure with bars of length 1 connecting exclusively the neighbouring nodes is applied. For such ground structure, the solution of primal and dual problem is obtained. In the second subiteration, both ground structures are constructed by adding bars of length 2, whose elongations violate the constraints of the dual problem. These steps are repeated until elongations of all the bars in the ground structures satisfy the constraints of the dual problem. In such a situation, the solution of the dual problem for the current substructure is the same as the solution of the dual problem for the initial structure and the same holds for the primal problems. The algorithm of adaptive ground structure is schematically presented in Fig. 8.42.

The proposed adaptive ground structure method was applied to generate the Strut and Tie model and determine optimal reinforcement of the frame corner. The starting point for the optimization were ground structures $()'$ and $()''$, each composed of 1253353 bars. The topology of the Strut and Tie model obtained with the use of adaptive ground structure method is presented in Fig. 8.42a. Note that the ST model thus constructed has the characteristic features of Michell's structures and it uses 45% less steel than the structure proposed in Eurocode 2. The two following examples concern typical structural elements used in concrete structures, i.e. the cantilever of a medium length (Fig. 8.43b) and the short cantilever fixed in a column (Fig. 8.43c).

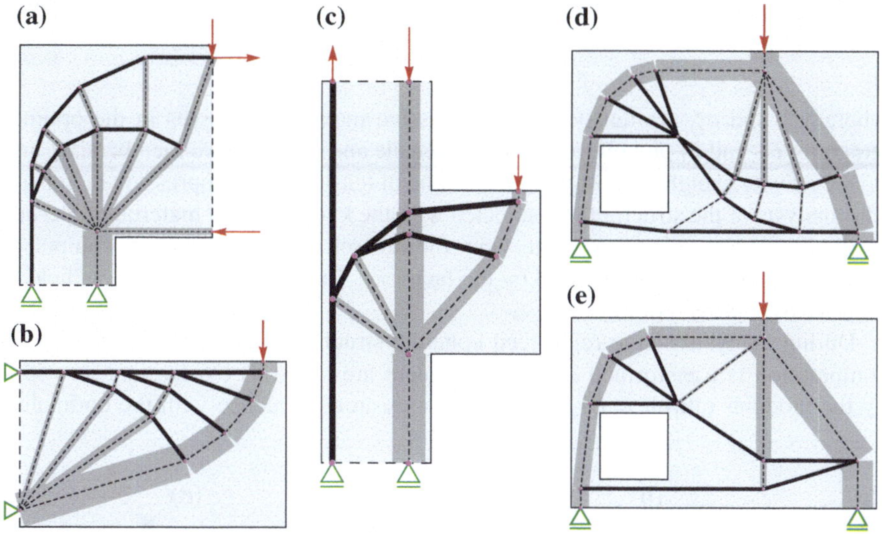

Fig. 8.43 Examples of application of topology optimization for constructing Strut and Tie models: **a** frame corner, **b** cantilever of a medium length, **c** cantilever fixed in a column, **d** deep beam (ground structure method), **e** deep beam (adaptive ground structure method); after Bołbotowski and Sokół (2016)

The ST model for the first cantilever shows a close similarity to the classical Michell cantilever corresponding to a large ratio of tensile and compressive stresses. In both cases, the application of the adaptive ground structure method allows to shorten the computation time approximately 8 times in comparison to the standard method.

The effectiveness of the adaptive ground structure method was also proved by analysis of the optimal reinforcement of a deep beam presented in Fig. 8.43d, e. Initially, the problem was solved with the use of the classical ground structure method, the coarse distribution of nodes and the large value of the penalty coefficient. As a result, a very simplistic, lacking details ST model was obtained (Fig. 8.43d). In the second attempt, the adaptive ground structure method had allowed to use a fine distribution of nodes resulting in the total number of ground structure elements exceeding 16 million and to apply a small value of penalty coefficient. As a result, a more detailed distribution of reinforcement was obtained and the amount of required steel was reduced by 5.5% (Fig. 8.43e).

In contrast, in Victoria et al. (2011), the problem of the optimal location of reinforcement was solved with the use of Isolines Topology Design (ITD) method. The aim of the conducted research was the analysis of the influence of mechanical properties of steel and concrete on the resulting topology of Strut and Tie models.

The optimal layout for maximum stiffness and minimum volume (or weight) is obtained taking into account the criterion introduced in Dewhurst (2005) which states that in the optimum structure all structural members should have the constant fixed stress ratio:

$$k_1 = \frac{\sigma_T^\star}{\sigma_C^\star} = \left(\frac{\rho_T E_T}{\rho_C E_C}\right)^{\frac{1}{2}}$$

where: $\sigma_T^\star$ and $\sigma_C^\star$ are tensile and compressive maximum stresses in the optimal structure; E_T and E_C are elastic moduli of tensile and compressive members; ρ_T and ρ_C are density, weight or cost per unit volume of tension and compression members. The analysis of the structure is conducted with the knowledge of material properties (ρ and E) for the steel and concrete. Once the stresses in each member are computed, all negative values are multiplied by the factor k_1. Then, the resulting stress field is used for the optimization.

During analysis of the reinforced concrete structures, the concrete working in compression is transformed into an equivalent area of steel by dividing its width by the modular ratio $n = E_T/E_C$. This assures an equivalence with the optimality

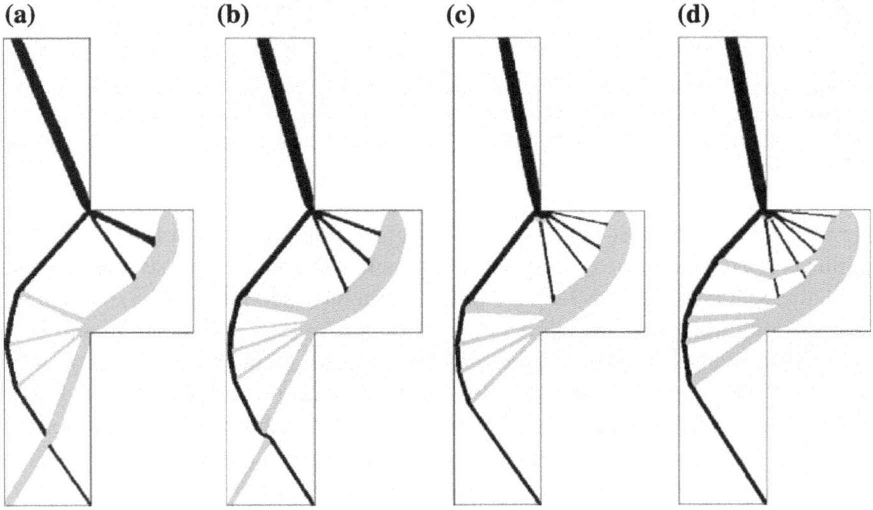

Fig. 8.44 Effect of the k_1 value on the Strut and Tie model for a volume fraction of the 20%: **a** $k_1 = 2.07$, **b** $k_1 = 3.09$, **c** $k_1 = 4.02$, **d** $k_1 = 5.00$; Victoria et al. (2011)

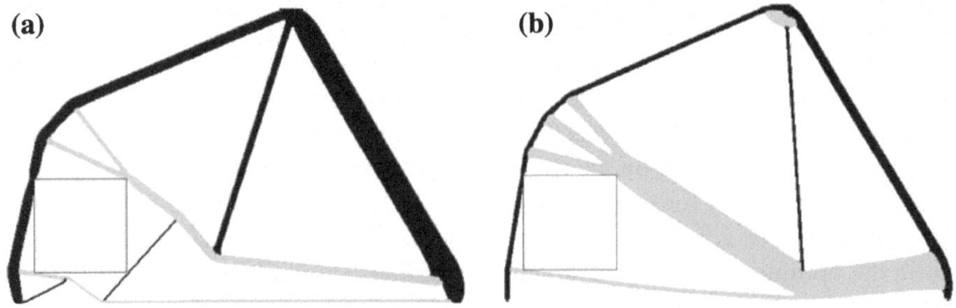

Fig. 8.45 Strut and Tie models for the deep beam with opening and a vertical load applied upwards for a volume fraction of 17%: **a** model assuming the same properties of steel and concrete, **b** the proposed model; Victoria et al. (2011)

criteria by Dewhurst, yet with equivalent mass densities for the two materials. The examples showing the effectiveness of the method and the effect of different mechanical properties of steel and concrete include a single short corbel and a deep beam with an opening (Figs. 8.44 and 8.45, respectively).

8.2 Mechanical Engineering Applications, Design Rationalization, Additive Manufacturing

Not only did Michell's theory inspired the designers of civil structures such as buildings and bridges, but it also affected other branches of engineering. In the field of mechanical engineering, the most influenced disciplines are the design of pressure vessels and the design of structures of improved crashworthiness. On the other hand, the requirement of practical application of complex Michell's solutions had stimulated development of a new discipline aimed at an adequate simplification of the optimal designs, the so-called design rationalization. Moreover, the newly developed additive manufacturing techniques including 3D printing strongly facilitated practical realization of the optimal structures, and thus enabled bridging the field of topology optimization and the manufacturing technology.

8.2.1 Pressure Vessels and Containers

The design of vessels and containers is based on classical theories of shells including membrane theory of shells and bending theory of thin shells, as described in classical books of Novozhilov (1964) and Flügge (1973). Such theories provide a convenient description of shell kinematics, the corresponding equilibrium equations and the methods of stress calculation. Although shell theories are fundamental for the process of optimal design of pressure vessels and containers, they are not sufficient tools for this purpose. The problem of optimal design requires formulation of the optimality criteria related to stiffness or weight of the shell. Such optimality criteria can be reformulated to the elimination of bending and obtaining constant values of membrane stresses, which indicates that problem of optimal design of pressure vessels brings together theory of shells and theory of Michell's structures.

The illustrative problem of this type is the design of constant-strength tank serving for storage of liquids, in which membrane stresses are equal to each other and constant at each point of the shell. The solution of such problem is drop-shaped tanks, whose geometry is strongly affected by pressure at the surface of the liquid. In particular, in the case of large surface pressure, the shape of the tank tends to a sphere, while in case of low surface pressure the shape of the tank becomes flattened and extended in the horizontal direction. This well-known solution was derived by Novozhilov (1964) and it will not be presented herein. Nevertheless, it is worth mentioning that

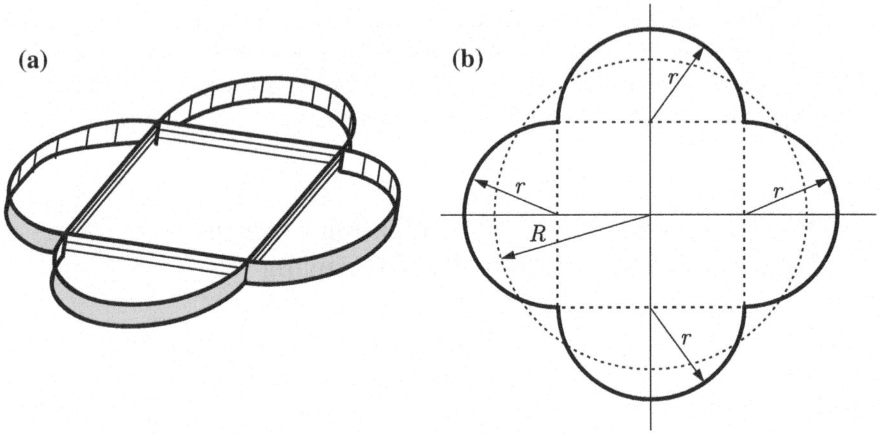

Fig. 8.46 Innovative designs of the ring embracing the tank utilizing internal ties proposed by Zalewski (2005)

constant-strength tanks had found their industrial applications, e.g. as large-scale drinking water reservoirs or smaller scale tanks for transporting various types of fluids.

A different approach to the optimum design of pressure vessels was presented by Zalewski (2005). The author considers two designs of rings embracing the tanks of the same cross-section area. The first ring has a standard shape of a circle of a radius R. The second ring has a shape of a clover; it is composed of four half-circles of radius r and utilizes four ties located inside the tank, see Fig. 8.46. The radii of half-circles are chosen such that the areas of the surfaces inside both rings are exactly the same. Both rings have equal height h and they are subjected to the same hydrostatic pressure p caused by the fluid inside the tank.

A simple comparative analysis of both types of ring designs is based on calculation of mass of steel required assuming the cross-sectional area corresponding to a maximal allowable tensile stress σ_0. In the case of the circular ring, the volume of the material equals

$$V_1 = \frac{phR}{\sigma_0}(2\pi R) = 2\pi hR^2 \frac{p}{\sigma_0} \tag{8.11}$$

In turn, the volume of the ring of clover-shaped tank is composed of volume of n half-rings of allowable stress σ_0 and n internal ties of allowable stress σ_{tie}:

$$V_2 = \frac{n\pi phr^2}{\sigma_0} + \frac{2nphr^2}{\sigma_{tie}} \cot\left(\frac{\pi}{n}\right) \tag{8.12}$$

Thus, the volume of the second ring can be expressed as

$$V_2 = V_1 \left(\frac{r}{R}\right)^2 \left(\frac{n}{2} + \frac{\sigma_0}{\sigma_{tie}} \frac{n}{\pi \tan\left(\frac{\pi}{n}\right)}\right) \tag{8.13}$$

By using geometrical relations between radii of circular parts of both the rings, one obtains

$$V_2 = V_1 \left(\frac{\pi \tan\left(\frac{\pi}{n}\right)}{\pi \tan\left(\frac{\pi}{n}\right) + 2} + \frac{2}{\pi \tan\left(\frac{\pi}{n}\right) + 2} \frac{\sigma_0}{\sigma_{tie}}\right) \tag{8.14}$$

In a special case of the tank shaped as four-leaf clover, we obtain Zalewski's formula:

$$V_2 = V_1 \left(0,611 + 0,389 \frac{\sigma_0}{\sigma_{tie}}\right) \tag{8.15}$$

Equal allowable stresses of both component materials result in identical volume of both structures, which is obviously in agreement with Maxwell's theorem since the internal forces in all elements of both structures are of the same sign. However, application of ties made of steel of larger allowable tensile stress enables decreasing total volume of material. For example, when the limit tensile strength of ties is three times greater than the limit tensile strength of external ring, the proposed innovative design leads to the substantially smaller total volume of the structure $V_2 = 0.74V_1$.

Further, Zalewski considers two cylindrical tanks created by the extrusion of previously analysed rings, i.e. the classical cylindrical tank and clover-shaped tank, Fig. 8.47. Both tanks allow for storage of the same volume of the fluid. Moreover, according to previous considerations, equal tensile strength of material used for curved parts of the ring and the ties results in equal masses of both tanks. However, application of various types of steel reveals several significant advantages of the proposed tank design.

1. In the clover-shaped tank, the amount of material used for external shell is smaller than in the cylindrical tank. This allows to reduce the total cost since steel used for the ties is usually less expensive.
2. The mass and cost of the clover-type tank can be further reduced by applying ties made of high-strength steel.
3. Smaller thickness of the external shell of the clover-shaped tank resulting from its lower values of forces significantly simplifies construction works due to possibility of prefabricating and rolling for transport.
4. The clover-type tanks are usually welded at joints, which creates additional 'beams' transferring the forces to the bottom of the tank.

In the following examples, Zalewski analyses more complex shapes of the tanks, in which the external shell is composed of six or eight semi-spheres, whereas the internal ties connect edges of the shell, Fig. 8.48. The analysis of the fully stressed design allows to reveal decomposition of the total volume of the tank into the part corresponding to the external shell and the part corresponding to the internal ties. For the considered structures, the corresponding formulae take the following form:

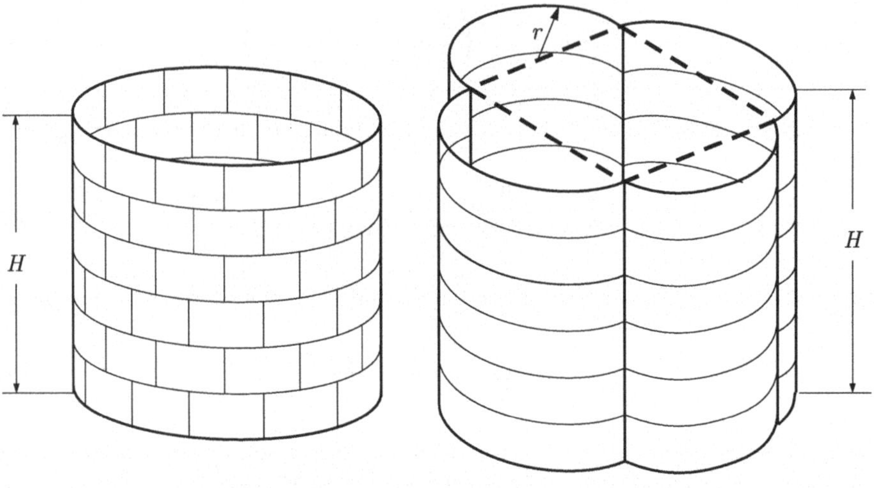

Fig. 8.47 Two designs of the tank of the same volume considered by Zalewski (2005)

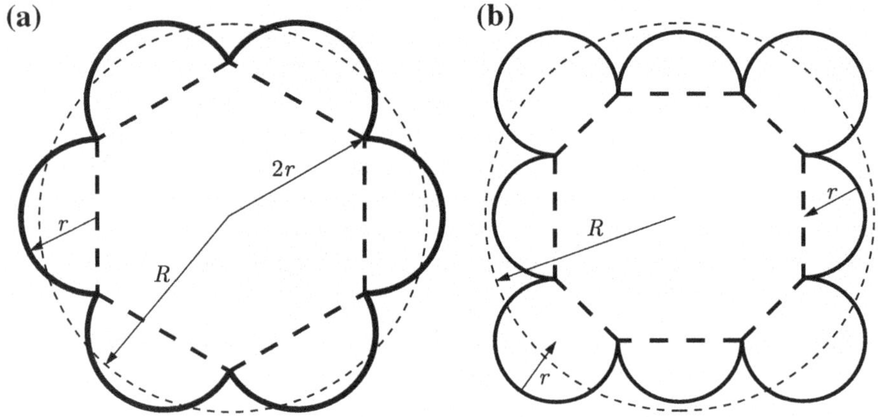

Fig. 8.48 More complex designs of clover-shaped tanks considered by Zalewski (2005)

$$V_2 = V_1 \left(0,476 + 0,524 \frac{\sigma_0}{\sigma_{tie}}\right) \quad \text{and} \quad V_2 = V_1 \left(0,495 + 0,505 \frac{\sigma_0}{\sigma_{tie}}\right) \tag{8.16}$$

For both the considered geometries, the assumption of equal tensile strength of both materials results in similar volumes of external surface and internal ties. In turn, in the case, when tensile strength of the ties tends to infinity, the dominating part of volume is concentrated on external surface and it is approximately half of the volume of circular ring. The example also shows the typical feature of Michell's structure, i.e. the decrease of the volume along with increasing complexity of the structural design.

8.2.2 Applications for Crashworthiness

The methods and techniques of topology optimization were also adapted to optimal design of crashworthy structures or the structures, which serve for dissipation of the impact energy and mitigation of the impact process. In fact, the discipline of crashworthiness was affected by computational methods of topology optimization, not by the Michell's theory. Nevertheless, application of the computational methods for crashworthy design often leads to the layouts at least partially resembling Michell's structures. Thus, some of these results will be presented in the current section.

The rapid development of the computational methods of topological optimization was initiated in 1970s by experts in the control theory, who introduced the homogenization as the method of relaxation of optimum design problems concerning two-material and shape designs. The application of the homogenization method for linear elastic problems was further extended to take into account buckling load criterion (e.g. Neves et al. 1995) and applied for design of structures subjected to large plastic deformations (e.g. Maute et al. 1998). These had paved the way to the application of the topology optimization methods for structural crashworthiness. The problem of compliance minimization was reformulated into the problem of energy absorption maximization by Mayer et al. (1996). The strict relaxation by homogenization has been there replaced by a material density method. The corresponding objective function was based on strain energies weighted at specified time instants. The proposed methods were applied to maximize crash energy absorption of given structural components. Further research efforts were focused on application of developed methods to optimize crashworthy structures for railway industry (Yamakawa et al. 1999) as well as automotive industry (Soto 2002). The development of application of homogenization methods ranging from minimization of compliance in simple elastic structures, through minimization of natural frequencies of vibrating systems to maximization of energy absorption in complex three-dimensional structures subjected to large deformations was described by Soto (2001).

Another method of optimal crashworthy design was proposed by C.B.W. Pedersen (2003a, b). In his approach the topology of the energy absorbing truss structure was optimized in order to achieve desired, preliminarily assumed reaction force generated by the structure or assumed deceleration of its selected nodes during the impact process. In order to solve such topology optimization problems, the author had applied a ground structure composed of 2D beam elements with plastic hinges at the ends, which can undergo large displacements and rotations. A non-linear finite element solution was obtained with implicit backward Euler algorithm, while analytical sensitivities were computed by direct differentiation method. The optimized energy-absorbing structures were further used to improve the crashworthiness of various components transportation vehicles, e.g. bumper of the car or train end-underframe.

The illustrative problem solved by Pedersen is presented in Fig. 8.49. The design domain with the initial ground structure is subjected to the impact of the object of certain mass and initial velocity. Two types of ground structures for topology

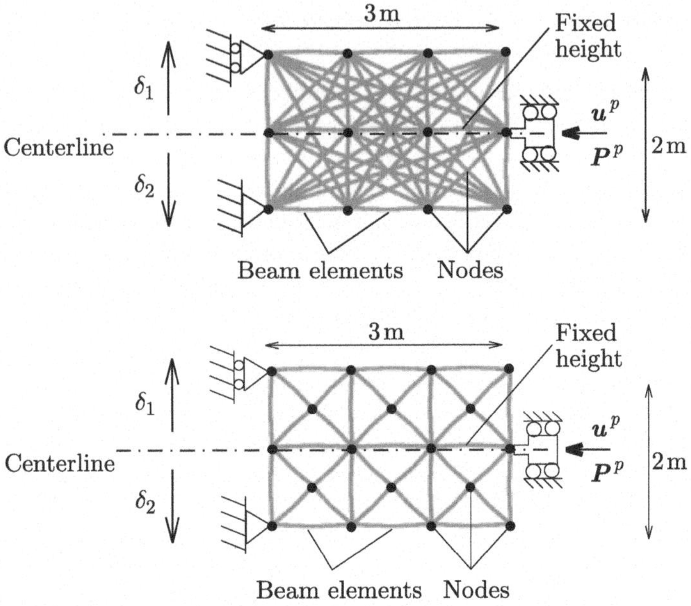

Fig. 8.49 Problem definition and two types of applied ground structures; the imperfections of the beams are adjusted by parameters δ_1 and δ_2 around the centerline

optimization are applied: the ground structure with irregular crossing elements (Fig. 8.49a) and the ground structure with shorter elements aligned in adjacent squared cells (Fig. 8.49b). The objective of topology optimization is to find the structure which allows to obtain predefined force–displacement characteristics, defined, e.g. by several design points marked by circles in Fig. 8.51.

Two structures obtained by using regular ground structure, the same six design points but different types of design imperfections (symmetric and skew-symmetric) are depicted in Fig. 8.50. It can be observed that the type of applied imperfections significantly affects the topology of the obtained crashworthy structure. In turn, Fig. 8.51 presents the response of non-optimized ground structure, which is significantly different than desired response, the intermediate stages of optimization (50 and 200 iterations), as well as the final response of the optimized structure (500 iterations), which reaches the assumed design points with high precision. In addition, the figure shows the change of crashworthy structure response caused by modification of applied imperfections and hardening parameters h'. Although the corresponding force–displacement curves slightly diverge, the global crushing characteristics and the collapse mechanism of the structure remain in all cases similar. This proves the robustness of the obtained optimal design on imperfections and hardening parameters.

Fig. 8.50 Left: undeformed optimal structures; middle: 10% deformation; right: 20% deformation. Structure **a** optimized for symmetric imperfections ($\delta_1 = 0.02\,\text{m}$, $\delta_2 = 0.02\,\text{m}$); structure **b** optimized for skew-symmetric imperfections ($\delta_1 = 0.02\,\text{m}$, $\delta_2 = -0.02\,\text{m}$). In both cases, the hardening parameter $h' = 0.02E$

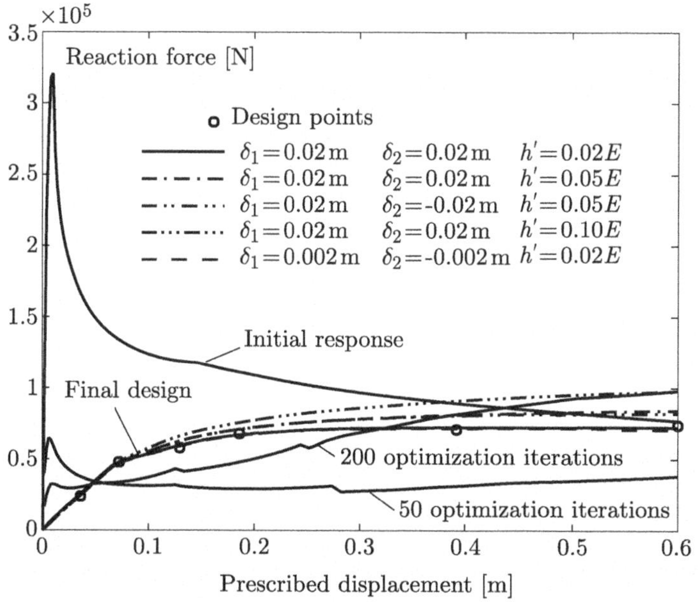

Fig. 8.51 Response for the structure in Fig. 8.49a optimized for $\delta_1 = 0.02\,\text{m}$, $\delta_2 = 0.02\,\text{m}$, $h' = 0.02E$; the optimized structure is analysed for different imperfections and hardening parameters

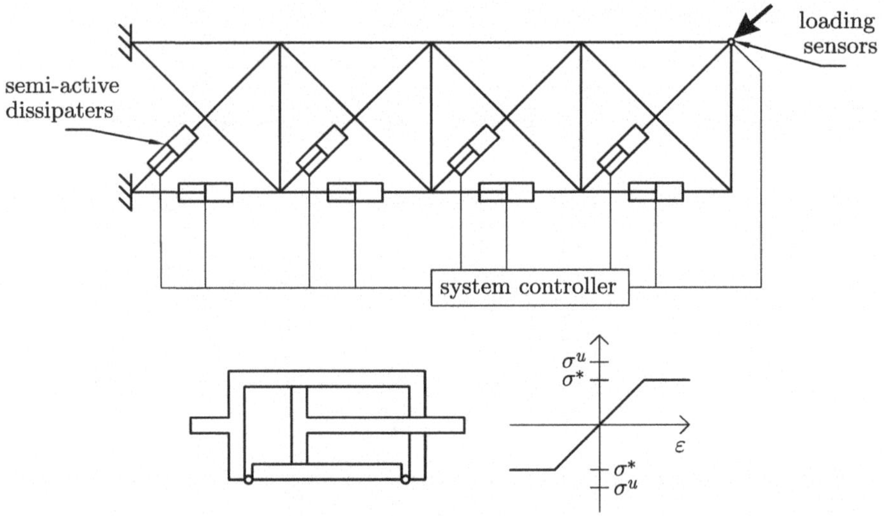

Fig. 8.52 Adaptive impact absorbing skeletal structure equipped with structural fuses as proposed by Holnicki-Szulc et al. (1998), Holnicki-Szulc and Bielecki (2000)

An alternative approach to optimal design of crashworthy truss structures subjected to impact loading was proposed by Holnicki-Szulc et al. (1998). The authors postulated replacing selected passive members of the truss structure with adaptive members ('structural fuses') characterized by a controllable level of the yield stress, Fig. 8.52. The general paradigm of Adaptive Impact Absorption (AIA) was formulated as real-time adaptation of energy-absorbing structure to actual dynamic loading. It was executed either by proper tuning of the mechanical properties of structural fuses just before the impact event or by online modification of their characteristics during the energy absorption process.

The initial papers dedicated to AIA problems were focused on optimal tuning of the yield stress levels in adaptive elements of the truss structures subjected to static loading and small deformations. These assumptions had substantially simplified conducted numerical analysis and enabled application of the Virtual Distortion Method (VDM) as versatile reanalysis and remodelling tool, see Kołakowski et al. (2008). Theoretical development of the AIA concept and solution methods was combined with its practical applications in the design of adaptive barriers providing maximal energy absorption or adaptive car buffers (Marzec and Holnicki-Szulc 1999; Holnicki-Szulc and Bielecki 2000). Further applications concerned optimal design of adaptive railway cars exposed to dynamic loading (Holnicki-Szulc and Knap 2004), which was based on preliminary optimisation of elements cross sections and optimal choice of yield stress levels aimed at maximization of energy dissipation.

A complete methodology of the optimal design of adaptive truss structures was proposed by Wikło and Holnicki-Szulc (2008a, b). In the first paper, the authors had proposed the application of the Virtual Distortion Method for fast remodelling of

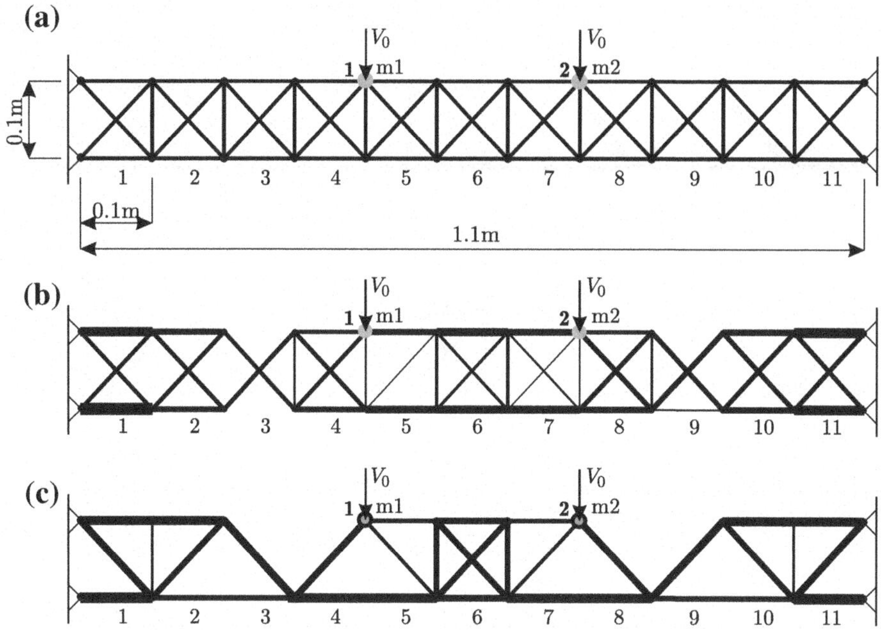

Fig. 8.53 Example of truss remodelling for optimal impact reception (Wikło and Holnicki-Szulc 2008a): **a** initial topology, **b** optimal topology optimal for asymmetric load case, **c** optimal topology for symmetric load case (by the courtesy of SMO)

the truss structure based on modification of the material distribution. The method was aimed at finding the structure of a given maximal volume, which is characterized by the largest global stiffness (smallest averaged deflections) under given dynamic loading. The examples of structures optimally remodelled for asymmetric and symmetric dynamic loading are presented in Fig. 8.53. In the second paper, Wikło and Holnicki-Szulc had formulated and solved the problem of optimal adaptation to impact loading, which is composed of three stages: optimal remodelling, optimal choice of location of adaptive elements, optimal tuning of the yield stress levels for given impact loading. The procedure of optimal adaptation is presented in Fig. 8.54. The method starts with simplification of the topology obtained from the step of remodelling (Fig. 8.54a). Then, elements I and II, which carry the largest internal forces are selected to be replaced by adaptive elements (Fig. 8.54b). Finally, the yield stress levels in these adaptive elements are adjusted such that they maximize global energy dissipation during the action of the impact loading.

Other research in the field of Adaptive Impact Absorption was focused on the analysis of adaptive truss structures subjected to large deformations, contact interactions as well as the processes of snap-through and plastic folding. For example, structural fuses with controllable yield stress levels were used in a quasi one-dimensional multi-folding structure to control sequence of the process of plastic folding and the

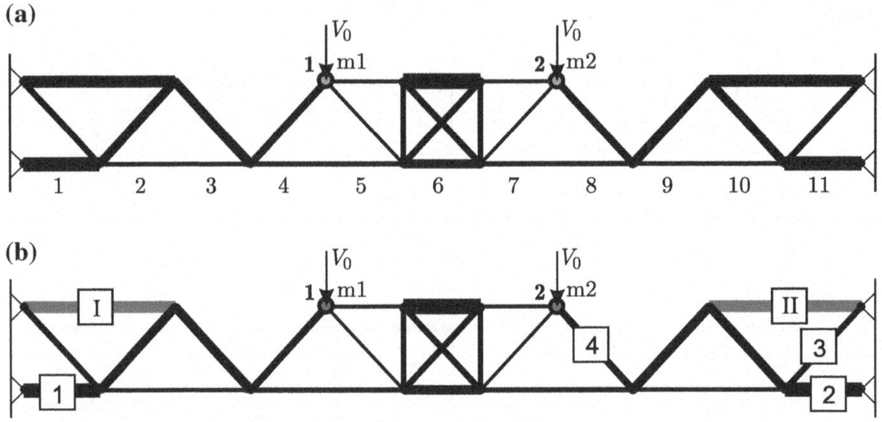

Fig. 8.54 Example of optimal adaptation to impact loading (Wikło and Holnicki-Szulc 2008b): **a** simplification of the topology obtained from the remodelling step, **b** selection of the locations of adaptive elements (by the courtesy of SMO)

corresponding process of energy dissipation (Holnicki-Szulc et al. 2003; Pawłowski and Wikło 2004). Moreover, a novel design of an adaptive skeletal bumper to be applied in offshore tower and optimal adjustment of the yield stress level of its members for various impact scenarios was proposed by Pawłowski and Holnicki-Szulc (2004).

Although the above examples diverge from classical topology optimization problems, they shed a new light on the problem of optimal design of crashworthy structures. They clearly prove that obtaining optimal dissipative characteristics for various types of external loading cannot be achieved by the use of passive structure, but it requires an application of adaptive structure with controllable mechanical parameters. Therefore, in the field of crashworthy design the methods of topology optimization and methods of structural adaptation should be applied jointly since their synergy is required to obtain optimal energy-absorbing structures.

In the context of practical applications, one should also mention the successful use of topology optimization methods for the optimal design of the car body. Herein, we will focus on the novel design concept proposed by Engineering Design Research Laboratory (Indiana University – Purdue University, Indianapolis), which had resulted in a car body including orthogonal curved stiffeners resembling layouts of Michell's structures, see Fig. 8.55. The proposed vehicle has a shape of a water droplet and it is composed of an envelope structure with embedded internal space frame (ribcage). The task of the ribcage is to provide strength and durability of the car during standard operation as well as energy absorption capabilities in case of collisions. The envelope's material is polymer composite, while the ribcage is built of porous material, e.g. functionally graded aluminum alloy foam. The entire structure composed of monocoque envelope and internal spaceframe is intended to be built by using additive manufacturing technology.

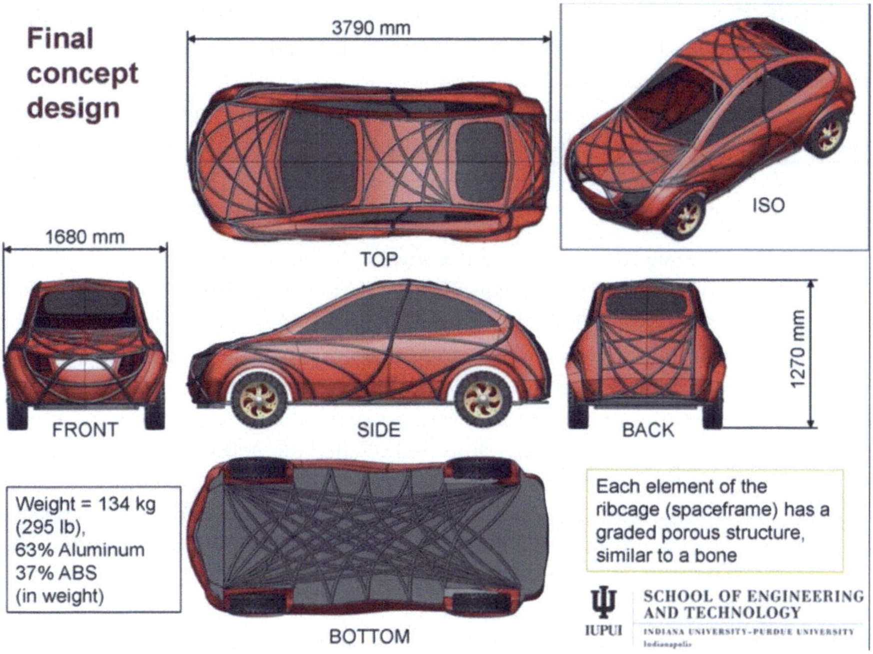

Fig. 8.55 Innovative concept of a car body with internal ribcage resembling a Michell's structure proposed by Engineering Design Research Laboratory (by the courtesy of Andres Tovar)

Both the envelope of the car and the ribcage were designed with the use of dedicated crashworthiness topology optimization algorithms. In particular, the lightweight components were optimized by using a dedicated algorithm based on Hybrid Cellular Automata, which modifies distribution of material in order to maximize internal energy subject to the mass constraint. In turn, the design of lightweight, energy absorbing, progressively folding components was conducted by simultaneous maximization of specific energy absorption, minimization of peak crushing force and maximization of crash load efficiency. The designers had proposed an algorithm utilizing the initial design generated by cellular automata, which is clustered and optimized using sequential metamodel-based genetic programming. According to the proposed concept, the optimal multiscale design of the space frame maximizes internal energy while internal functionally graded cellular structure extremizes other crashworthiness indicators. The optimization of the structure was performed for various loading conditions including the two frontal pole impacts (full frontal and with offset) and three side impacts.

The application of optimization algorithms leads to a novel construction, which is characterized by the following features:

- Multiscale structural layout. At the level of the entire structure, the space frame surrounding the car has a shape of a ribcage which clearly follows the principles of Michell's structures. At the level of the component, each tubular element of the space frame is filled with functionally graded cellular structure resembling structure of the bone.

- Novel lightweight multi-material design. The application of the polymer composite and monocoque design of the envelope provide desirable safety characteristics. In turn, application of metallic, functionally graded and cellular space frame provides a high level of strength, crashworthiness and resulting passengers protection.

Moreover, the paper by Mrzygłód and Kuczek (2014) presents a new optimization-based methodology of crashworthiness design, particularly suitable for application in design of a car body for high-speed trains. The proposed method includes three separate stages. In the first stage, the constant criterion surface algorithm (CCSA) makes use of the finite element model composed of solid elements in order to find a preliminary quasi-optimal solution. In the second step, the obtained solid model is replaced by an equivalent model composed of shell elements, whose thicknesses are parameterized and optimized by using a genetic algorithm. Eventually, in the last stage, the CCSA optimization algorithm is applied again to remove excessive material from the optimized structure. The CCSA algorithm used in the first and third step of optimization is a 'hard-kill'-type method, in which the solution is generated by iterative elimination of elements of a low value of criterion function. The additional procedure within CCSA algorithm is an introduction of design domain constraints, which are aimed at elimination of the infeasible solutions resulting from fine discretization of the solid model.

A general goal of the crashworthy design is to obtain rigid structure of a passenger car for a high-speed train, which will effectively protect the passengers in case of collision or derailment. In order to achieve high impact resistance, the car body structure is divided into several sub-systems, which define passengers' survival cells. The static loading is applied in three orthogonal directions in order to obtain the crashworthy characteristics independent of impact scenario. The corresponding design domain being a quarter of passengers' cell, the applied boundary conditions and loading are presented in Fig. 8.56a–e.

The topology optimization problem is to minimize the mass of the structure taking into account stress constraints. In the first stage of optimization, the design domain is discretized by solid elements. The solid elements with low value of criterion function are removed until the solution approaches a limit of constraint (maximal assumed value of Huber–Mises stress). The obtained quasi-optimal solution (Fig. 8.56f) is considered as a result of the first stage of optimization. In the second stage of solution the above quasi-optimal model composed of solid elements is transformed into a model of thin-walled structure composed of shell elements (Fig. 8.57). The elements are divided into 20 groups of 20 thickness parameters which are visualized by different colours and numbers in Fig. 8.57a. The parametric FE model is subjected to optimization with the use of genetic algorithm of the Evolutionary Optimization System software. The optimization problem is to find 20 thickness parameters, which minimize the mass of the structure and does not violate imposed stress constraints.

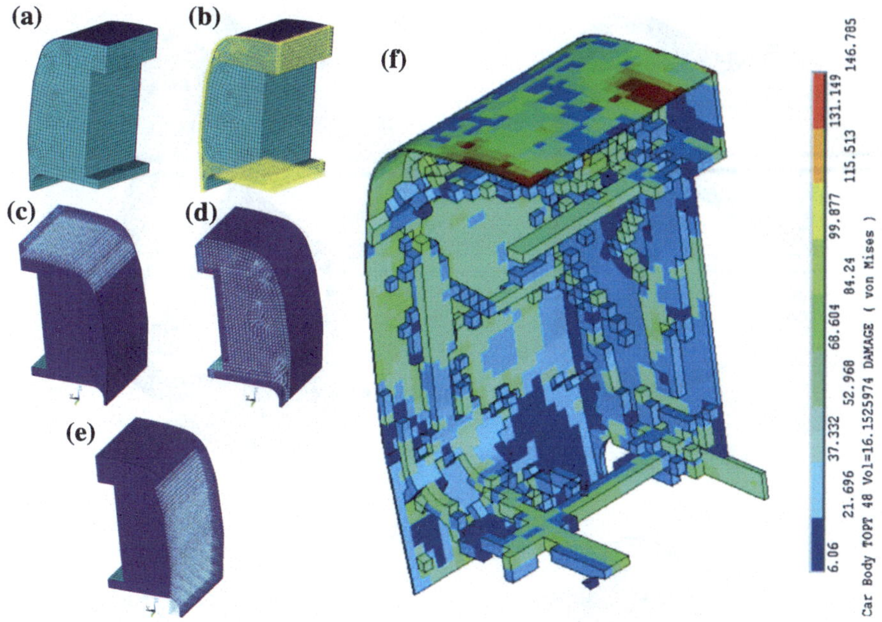

Fig. 8.56 The design domain (**a**), boundary conditions of the FE model (**b**) and three load directions of stage I: *Z* (**c**), *Y* (**d**), *X* (**e**) and final result of the first stage of optimization (**f**); by the courtesy of SMO

Finally, in the last stage of optimization, the obtained FE model of a thin-walled structure is again subjected to topology optimization with the use of CCSA algorithm aimed at minimization of structure mass subject to stress constraints. The design domain is defined by the FE model from the second stage of optimization with 20 frozen thicknesses of the shell structure. The topology optimization algorithm improves the solution by removing the excessive material in selected locations of the structure. The first quasi-optimum solution which approaches a limit of Huber–Mises stress is accepted as the final solution of the optimization problem. The final optimized FE model is presented in Fig. 8.57c, while the complete design of the optimal passengers' survival cell in form of a CAD model is depicted in Fig. 8.57d.

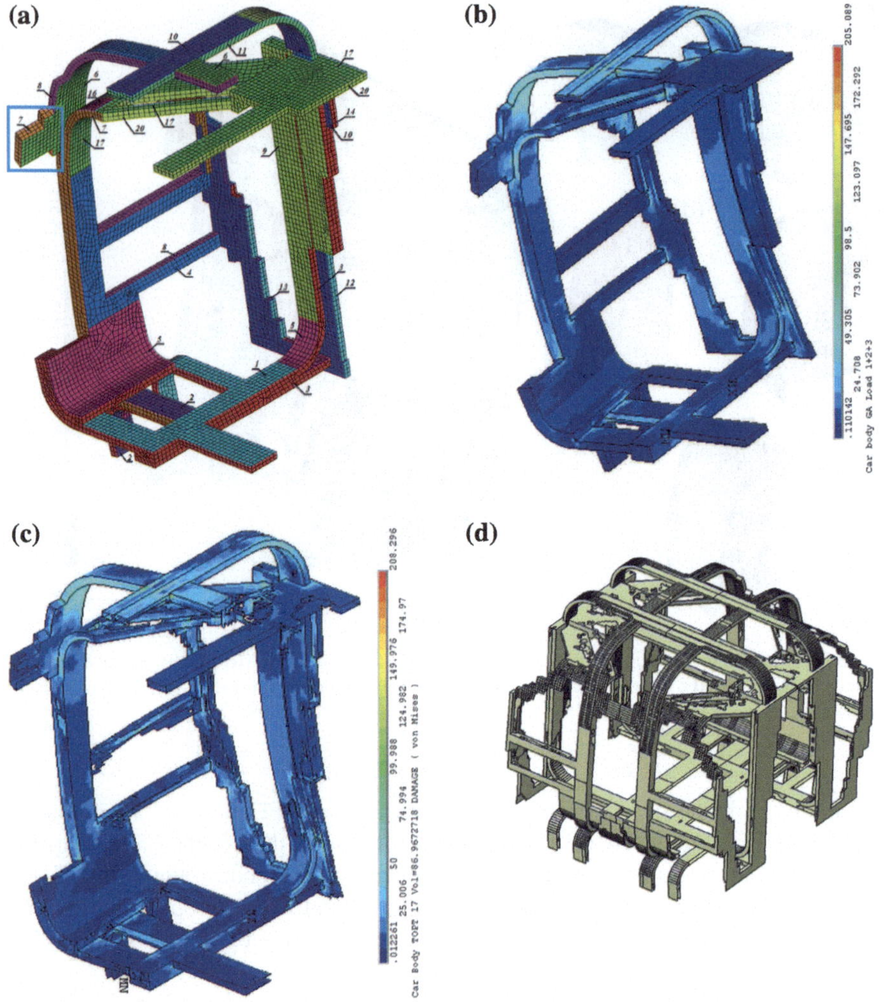

Fig. 8.57 The FE shell model of thin-walled structure (the thickness parameters are visualized by colours and numbers) (**a**), stress map of the optimized FE model (**b**), final result of topology of the third stage of optimization (**c**), the CAD model of passengers' survival cell (**d**)

8.2.3 Design Rationalization

Due to complex internal structure of Michell's layouts composed of infinitely dense regions of orthogonal fibres and reinforcing bars their manufacturing and direct application in engineering practice is impossible. Therefore, the process of simplification of Michell's layouts and their adjustment for the practical requirements (the so-called design rationalization) has to be performed prior to their introduction into a mechanical system and application as load-bearing components.

The most straightforward from the theoretical point of view and simultaneously the simplest numerically method of the Michell designs' rationalization is based on the application of arbitrarily coarser net of parametric lines. In such an approach, the intersections of the parametric lines are considered as nodes of a truss structure which are connected with each other by straight bars. The method leads to approximation of Michell's structures by a sequence of trusses with finite number of joints. The weight of each truss obviously exceeds the weight of a Michell's structure. However, it decreases along with the increase of the number of joints and approaches optimal value as the number of joints tends to infinity.

The parametric lines chosen for construction of approximating truss structures should be distributed regularly with an offset of a certain angle. However, their layout should be determined by the assumed number of nodes and bars of the truss or, alternatively, the assumed increase of weight in comparison to the original Michell structure. Let us note that the each of the obtained truss is not the lightest in the class of trusses of given number of joints and bars. The illustrative sequence of trusses of finite number of joints approximating classical Michell's cantilever supported on a circle is presented in Fig. 4.43, while the corresponding numbers of joints and bars as well as normalized weights are juxtaposed in Table 4.2.

The method which can be classified as design rationalization was proposed by Mazurek et al. (2011). The authors had proposed graphical rules which allow constructing discrete optimum truss structures of finite number of joints, transmitting a single point load to two pin supports (the three-point problem), see Fig. 2.2. The advantage of the graphical approach is that construction of optimal structures does not require any analytical or numerical procedures. In the above-mentioned paper, the topology optimization was aimed either at minimization of theoretical weight of the structure or minimization of deflection of the loaded point. Formulated optimization problems were initially solved with the use of a gradient-based optimization procedure. The authors had disclosed that optimum structures transmitting a single point load to two supports are characterized by the same values of four angles between nearly orthogonal families of members. Consequently, they had concluded that optimal structure of given number of bars corresponding to a given location of the point load can be constructed exclusively by using simple geometrical rules. In the following paper (Mazurek 2012) the above method was extended from the three-point problem to three-force problem.

In contrast to previously described discrete trusses based on coarser net of parametric lines, the discrete trusses based on graphical rules of optimality are the lightest in the class of trusses of finite number of joints. Moreover, when the number of bars tends to infinity, the obtained layout tends to the layout of the Michell's structure. Thus, the method creates a sequence of discrete optimum trusses which approximate optimal Michell's structure in a best possible way.

The most comprehensive analysis of the methods of rationalization was recently presented by He and Gilbert (2015). The authors had considered the problem of design

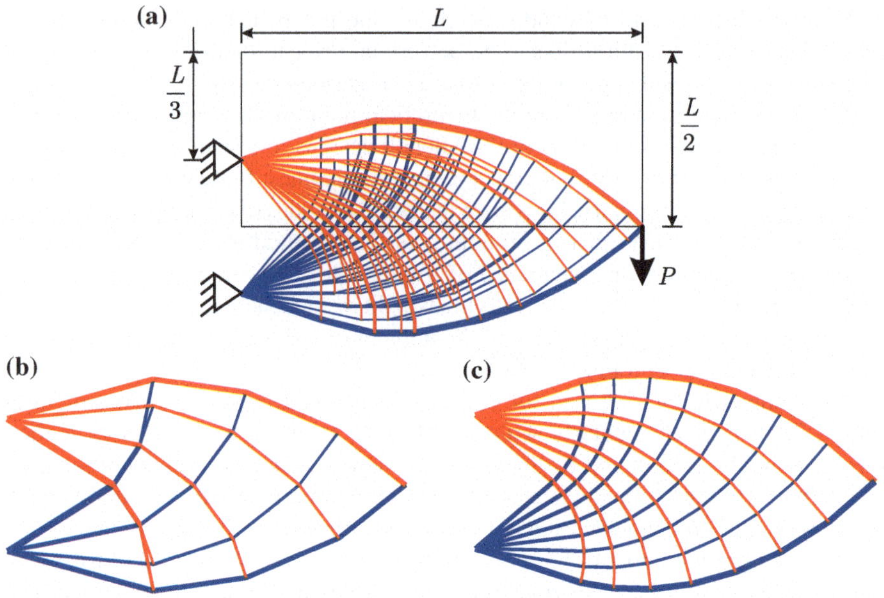

Fig. 8.58 Hemp's cantilever: **a** problem definition and layout optimization solution obtained using 30 × 15 nodal divisions, $V = 4.3541PL/\sigma_0$; **b** rationalized solution obtained using joint length method, $V = 4.3863PL/\sigma_0$; **c** rationalized solution obtained using geometry optimization, $V = 4.3318PL/\sigma_0$ (by the courtesy of SMO; He and Gilbert 2015)

rationalization for the structures obtained with the use of the ground structure method and linear programming. In order to solve the problem of design rationalization, they consider two groups of techniques. The first group is based on the modification of the initially posed optimization problem in order to eliminate excessive complexity of the resulting optimal structure. In turn, techniques from the second group introduce simplifications of the optimal structure obtained as a solution of the classical topology optimization problem during the post-processing step.

The classical method which belongs to the first group is introduction of the 'joint lengths' representing additional volume and cost of the joints. The introduction of the joint lengths leads to a modification of general formulation of the layout optimization problem based on the ground structure method. However, it does not violate linearity of the optimization problem, which still can be efficiently solved with the use of linear programming techniques. The proposed method introduces constraints on the optimal design and it leads to layouts characterized by a smaller complexity but a greater structural volume. In the final step, the real volume of the structure is computed by diminishing the total obtained volume by the volume corresponding to joints lengths. Let us note that the 'joint lengths' method can be applied in the ground structure method, however, it cannot be directly applied in the process of finding the solutions to Michell's problems by standard analytical methods.

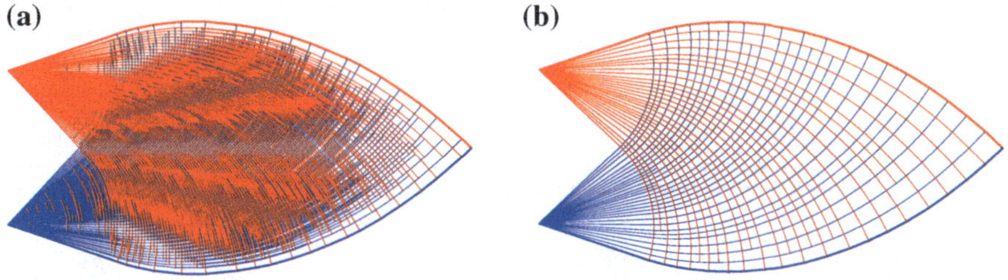

Fig. 8.59 Hemp's cantilever: **a** layout optimization solution obtained using 150 × 75 nodal divisions, $V = 4.3258PL/\sigma_0$; **b** rationalized solution obtained using geometry optimization, $V = 4.3228PL/\sigma_0$; (by the courtesy of SMO, He and Gilbert 2015)

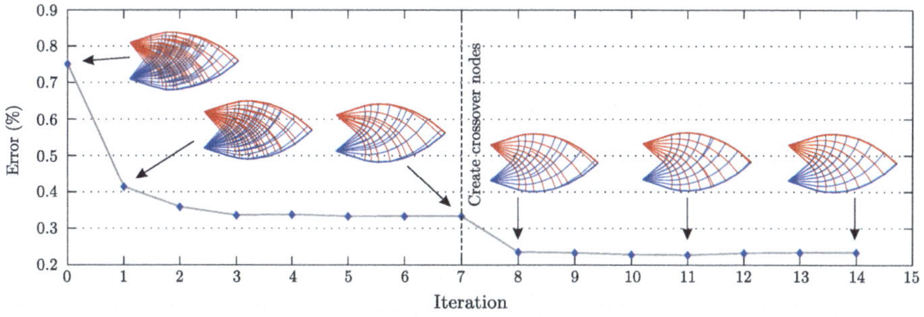

Fig. 8.60 Hemp's cantilever: geometry optimization solutions obtained during the iterative solution procedure; (by the courtesy of SMO; He and Gilbert 2015)

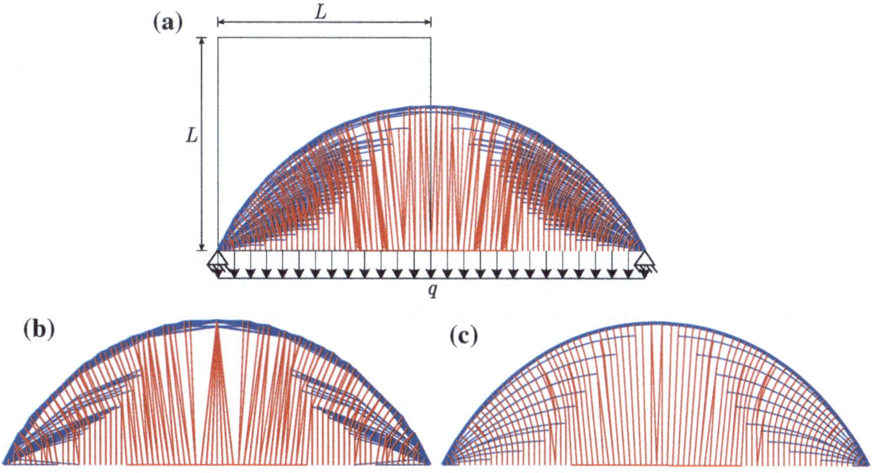

Fig. 8.61 Hemp's arch with distributed load: **a** problem definition and layout optimization solution obtained using 40 × 40 nodal divisions, $V = 3.1679qL^2/\sigma_0$; **b** rationalized solution obtained using joint length method $V = 3.2044qL^2/\sigma_0$; **c** rationalized solution obtained using geometry optimization, $V = 3.1550qL^2/\sigma_0$; (by the courtesy of SMO; He and Gilbert 2015)

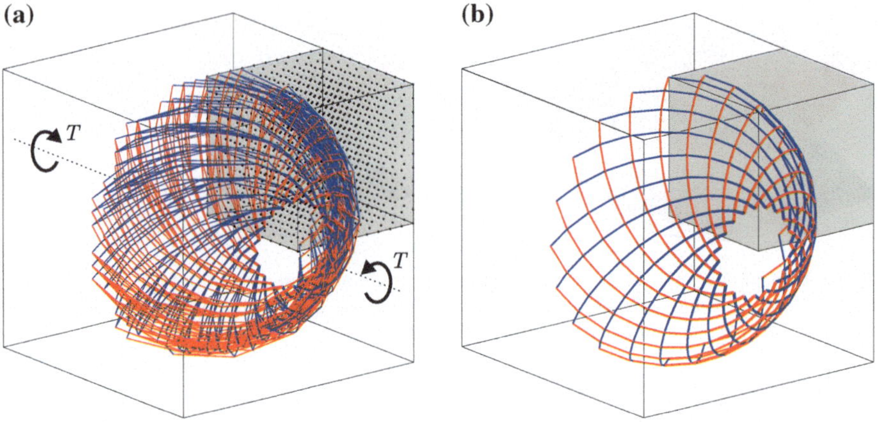

Fig. 8.62 Michell's sphere ($R = 50L$, $T = 100PL$, $\varphi = 18°$): **a** layout optimization solution obtained using $10 \times 10 \times 10$ nodal divisions, $V = 768.34PL/\sigma_0$; **b** rationalized solution obtained using geometry optimization, $V = 740.26PL/\sigma_0$ (half of the full structure shown in both cases); (by the courtesy of SMO; He and Gilbert 2015)

The methods from the second group do not interact within the procedure of constructing the standard solutions of the topology optimization problem, but simplify the obtained layouts at the post-processing step. The rationalization is based on the subsequent optimization of the layout from the applied topology optimization method. The optimization is performed with respect to locations of the nodes and cross sections of the bars, while the constraints are imposed on the structure equilibrium as well as on the maximal tensile and compressive stresses in the bars (He and Gilbert 2015). The above additional optimization problem solved during the post-processing step is by definition non-linear and thus there is no guarantee that the optimal solution will be found. Moreover, the method utilizes special procedures for merging bars and joints, which are executed before each iteration of geometry optimization. The method has substantial advantages in rationalizing the layouts of the optimal structures. In case of application to layouts obtained from the ground structure method, it allows to obtain designs of substantially smaller complexity and reduced total volumes. Moreover, its simplified version (deprived of the initial procedure of merging bars and joints) can be directly applied to discretized layouts of Michell's structures.

The above described two design rationalization techniques were implemented, thoroughly analysed and compared by He and Gilbert (2015). The methods were applied to various optimal structures obtained with the use of ground structure method, corresponding to point and distributed loads, self-weight and multiple load cases. It was shown that 'joint length' method leads to simplified structural layouts, however with increased total volume. On the other hand, it was proved that the method based on geometry optimization is related to considerable numerical cost but it results in structural layouts with decreased complexity and structural volume.

Several examples of the application of rationalization techniques are presented in Figs. 8.58, 8.59, 8.60, 8.61 and 8.62.

The above examples clearly show that design rationalization by geometry optimization allows to transform complex and chaotic structures obtained from the ground structure methods to much simpler and harmonic structures resembling optimal Michell's designs. On the other hand, it is expected that application of the rationalization method for discretized Michell's layouts leads to optimal structures of finite number of joints as described by Mazurek et al. (2011).

8.2.4 Additive Manufacturing

Additive manufacturing is a common name for all non-standard manufacturing techniques in which the material is joined or solidified layer by layer, the process being controlled by the computer to create a three-dimensional object. The additive manufacturing overcomes the limits of classical manufacturing techniques and it becomes more available in the small industry and for private users. The examples of popular and widely available techniques are Fused Decomposition Modelling, Stereolithography, Selective Laser Sintering and Selective Laser Melting.

Additive manufacturing is considered as a discipline which bridges the gap between topology optimization and its applications. The additive manufacturing techniques can be applied to manufacture optimal structures determined by using the ground structure method, the density-based methods or the theory of Michell's structures. However, in each case, the optimal layout obtained from selected topology optimization method requires a dedicated post-processing. This has to be performed very carefully in order not to disturb inherent features of the optimal design.

An original program which utilizes solutions obtained from various topology optimization methods and prepares adequate outputs for additive manufacturing process was proposed by Zegard and Paulino (2016). In general, the form of the output depends on applied topology optimization method. When three-dimensional ground structure method is applied the output is generated by using cylinders of proper radii for all members and spheres of radii equal to the radii of the largest of connected elements. In the case of two-dimensional ground structure method, the output is simply an extrusion of obtained two-dimensional representation. The resulting surface is further tessellated, typically by triangular elements. The illustrative 3D and 2D models, which are ready for additive manufacturing are presented in Fig. 8.63a.

In turn, when density-based methods, such as Solid Isotropic Material with Penalization, are applied for finding optimal topology then the solution is characterized by non-uniform, continuous distribution of density within the considered domain. Such a solution is not directly applicable for 3D printing and thus it requires special interpretation. The simplest method is based on the assumption of the arbitrary cut-off ratio and definition of the solid material in the domains where density exceeds this ratio. Unfortunately, the obtained layout is no longer optimal and, moreover, the

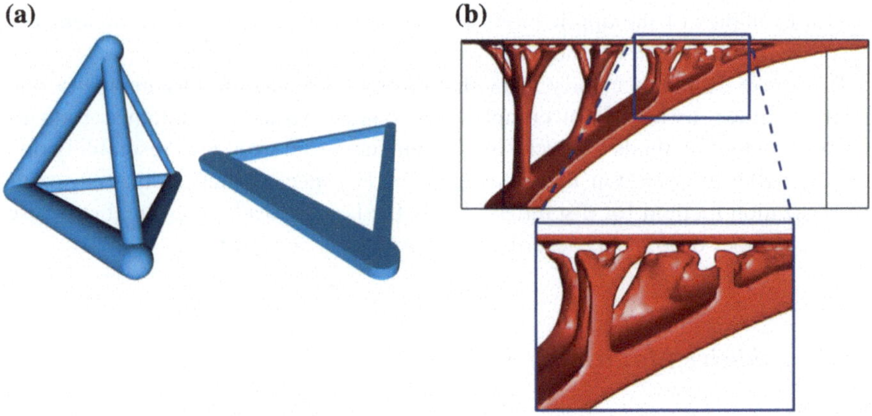

Fig. 8.63 **a** Ground structures optimization: 3D model composed of 6 members and 2D model composed of 3 members, **b** Density-based optimization: symmetric quarter-domain result for constant penalization, with detail of the disconnected members; (by the courtesy of SMO; Zegard and Paulino 2016)

volume constraint may be violated. In addition, certain crucial parts of the structure may be mistakenly eliminated, see Fig. 8.63b.

As proposed by Zegard and Paulino (2016) the three-dimensional isosurface of the model can be determined by interpolating four-dimensional data including three coordinates of each material point and the corresponding value of material density. The surface is further tessellated into triangles. The visualization and inspection of the three-dimensional density-based topologies prior to manufacturing can be performed by using a specialized software such as TOPslicer graphical tool by Zegard and Paulino. The software enables slicing of the structure at arbitrary plane in order to observe distribution of the densities, as well as selecting different cut-off ratios and visualization of the resulting solid body. Thus the problem such as presented in Fig. 8.63b can be identified and possibly avoided.

The plots of the layouts obtained from both ground structures method and density-based method are presented and compared in Fig. 8.64. It can be noticed that the structure obtained by using the ground structure method is composed of a larger number of smaller members with clearly noticeable orthogonality. Thus, it clearly resembles Michell's structures. On the other hand, the structure obtained by using density-based method is composed of thicker non-orthogonal members and its similarity to optimal Michell's structures is hardly recognizable.

Additive manufacturing of the layouts determined directly from the theory of Michell's structures is related to other challenges than manufacturing of the layout determined by ground structures or density-based methods. The main difficulty results from the fact that the layout has the form of infinitely dense net of fibres and it has to be rationalized in order to obtain a feasible engineering solution. The design rationalization can be performed by one of the methods described in the Sect. 8.2.3.

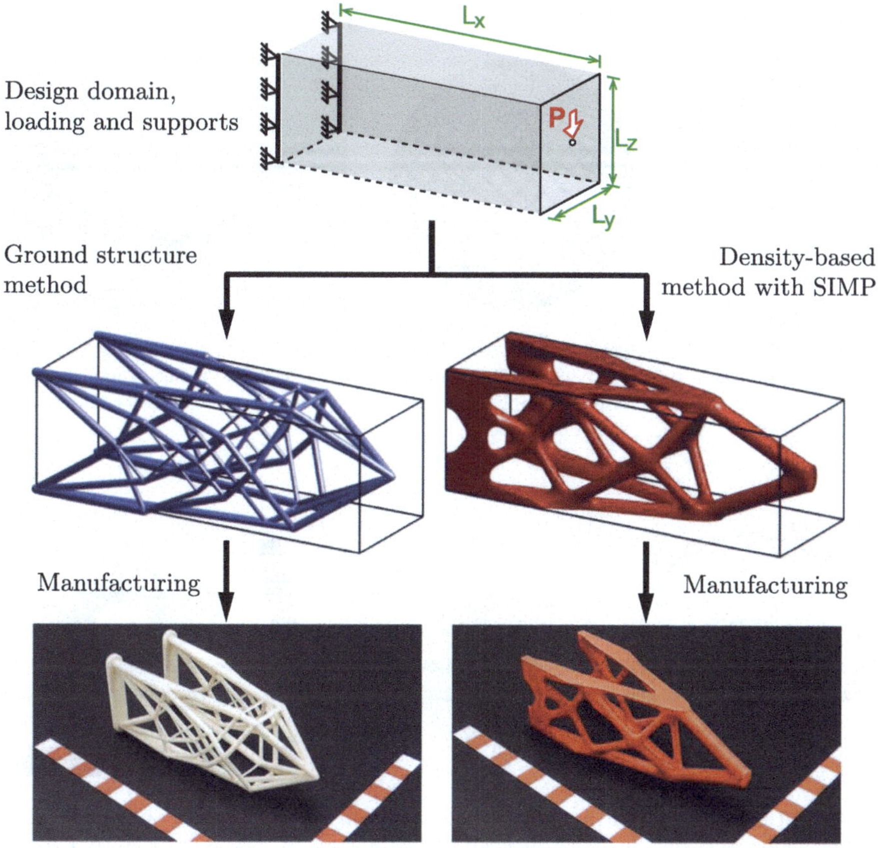

Fig. 8.64 Comparison of the solutions obtained by ground structure and density-based methods and the corresponding manufactured models (by the courtesy of SMO; Zegard and Paulino 2016)

The range of simplification depends on the precision of the applied additive manufacturing method, which is usually in the range of millimetres.

The illustrative examples of the manufacturing of classical Michell's layouts are the three-dimensional Michell's sphere (Fig. 8.65) and the two-dimensional flower problem (Fig. 8.66). In both cases the approximate solutions were found by using the ground structure method. As it can be seen the additive manufacturing technology enables the production of structures of relatively dense distribution of fibres, which resemble theoretical layouts obtained from strict mathematical formulations.

The complete workflow for the automatic preparation of the model for additive manufacturing was proposed by Smith et al. (2016). Such workflow starts from the layout from the optimization procedure (typically – ground structure method) and ends up with the CAD model ready for application in the manufacturing process. The workflow consists of multiple steps related to prerequisite operations

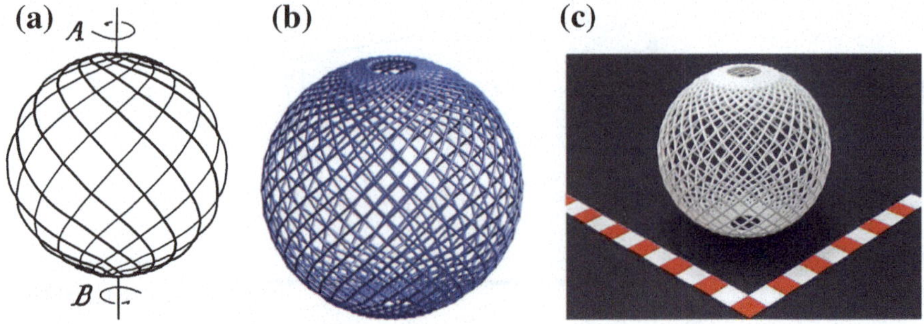

Fig. 8.65 Torsion ball problem: **a** optimal structure to balance a moment pair with no geometrical restrictions, **b** ground structure method solution obtained by Zegard and Paulino, **c** model manufactured using an Selective Laser Sintering (SLS) process; (by the courtesy of SMO; Zegard and Paulino 2016)

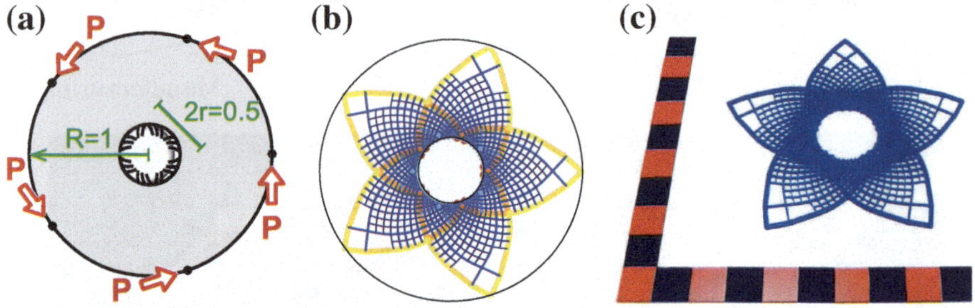

Fig. 8.66 The flower problem: **a** donut-shaped domain supported at the inner ring and loaded at 5 points in the outer ring, **b** ground structure method solution by Zegard and Paulino; **c** model manufactured using an Fused Deposition Modelling (FDM) process; (by the courtesy of SMO; Zegard and Paulino 2016)

(steps 1–2), obtaining a viable solution (steps 3–8) and transformation of the structure to continuum (steps 9–11):

1. Problem is defined by assigning the design domain, the load cases and the supports.
2. Layout optimization by ground structure method with fine node layout is performed in order to determine reference optimal design composed of large number of members and reference volume of the structure being close to the optimal one.
3. More practical, realizable layout is obtained either by rationalization of previously determined nearly optimal design or by performing new layout optimization by ground structure method with coarser nodes distribution.
4. The obtained layout is modified by performing the following operations: (i) merging collinear members, (ii) removing overlapping segments of collinear members, (iii) merging members that diverge from each other at small angle.

5. The locations of elements crossovers are detected, nodes are created at these locations and the adjoining elements are split into two parts.
6. Secondary sizing optimization is conducted in order to update the member areas in the feasible (coarser) layout.
7. The areas of the compression members, which violate the Euler buckling criterion are resized. The areas of all members are increased in order to fulfill the fabrication limit.
8. The condition of total structure volume being within the prescribed limit in comparison to the optimal volume is checked.
9. The volume of material at each node is increased depending on the number and direction of the connecting members in order to avoid high localized stresses.
10. Solid model is generated by using Non-Uniform Rational B-Splines (NURBS). Each member is created by extruding circular cross section along the length of the line element. The sphere is created at each node.
11. Standard Tessellation Language (TSL) model for additive manufacturing and standard STEP/IGES model for commercial FE packages are generated.

The above methodology was applied to optimize, manufacture and experimentally test two structures: (i) three-dimensional, simply supported beam subjected to a point load and (ii) air-brake hinge for supersonic car. The electron beam melting (EBM) technique was selected as the additive manufacturing method.

The definition of the topology optimization problem of the beam (Fig. 8.67a) had allowed to define the benchmark solution (Fig. 8.67b) and two candidate solutions based on two different nodal spacing (Fig. 8.67c, d). The procedure of members resizing had resulted in larger volume of the finer candidate solution and thus the coarser candidate solution was selected for preliminary study.

After constructing the elastic–plastic finite element models of various designs of the beam (Fig. 8.68a, b) the corresponding simulations of the response to static loading were conducted. The simulations had revealed that stress concentration is successfully mitigated by an increase of the material volume at nodes. Further, various designs of the beam were manufactured by EBM technique with the use of Ti-6Al-4V powder and subjected to point load in the middle of the span (Fig. 8.68c). The detailed comparison of obtained weights of the beams and their ultimate loading was conducted. It was found that optimized design is 7% lighter than the initial one and could successfully carry the central loading of assumed magnitude.

A similar analysis was conducted for the air-brake hinge for supersonic car. The initial design of the hinge and formulation of the layout optimization problem are presented in Fig. 8.69a, b. In turn, the final design obtained by using the ground structure method and the above-outlined model preparation methodology as well as the manufactured component are presented in Fig. 8.69c, d. The manufactured component has the typical features of Michell's structure including the reinforcing bar with orthogonally connected thin radial members. The component was tested experimentally in order to determine load–displacement curve and check the ultimate loading. In this case, the component had appeared to be 69% lighter than the initial design and to satisfy assumed load capacity requirements. The obtained huge

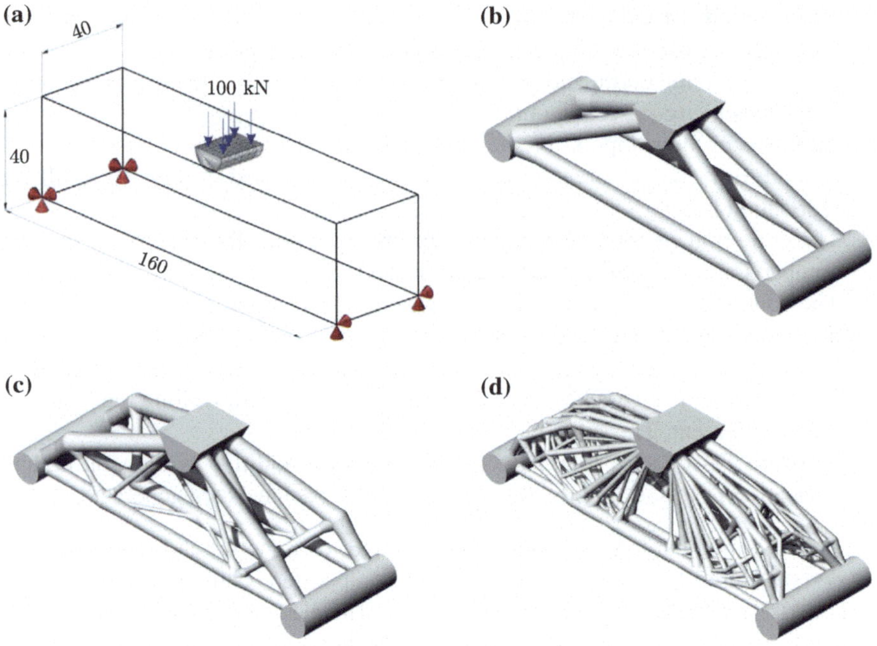

Fig. 8.67 The beam example: **a** design domain, loading and boundary conditions, **b** benchmark solution, **c** candidate solution obtained using 89096 potential members, **d** candidate solution obtained using 499213 potential members; (by the courtesy of SMO; Smith et al. 2016)

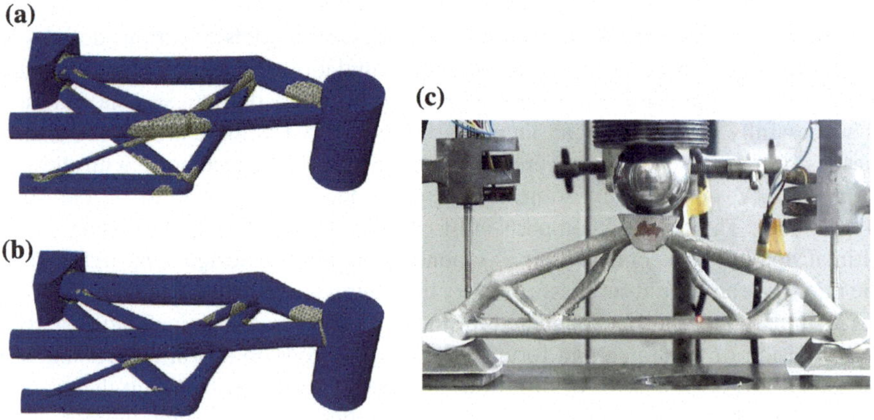

Fig. 8.68 The beam example: **a**, **b** elasto-plastic FE analysis of the optimized specimen without and with joint expansion (highlighted are the regions of plastic strain at 80% of the design load of 80 kN), **c** specimen ultimate load obtained through the buckling of the four central members in compression (by the courtesy of SMO; Smith et al. 2016)

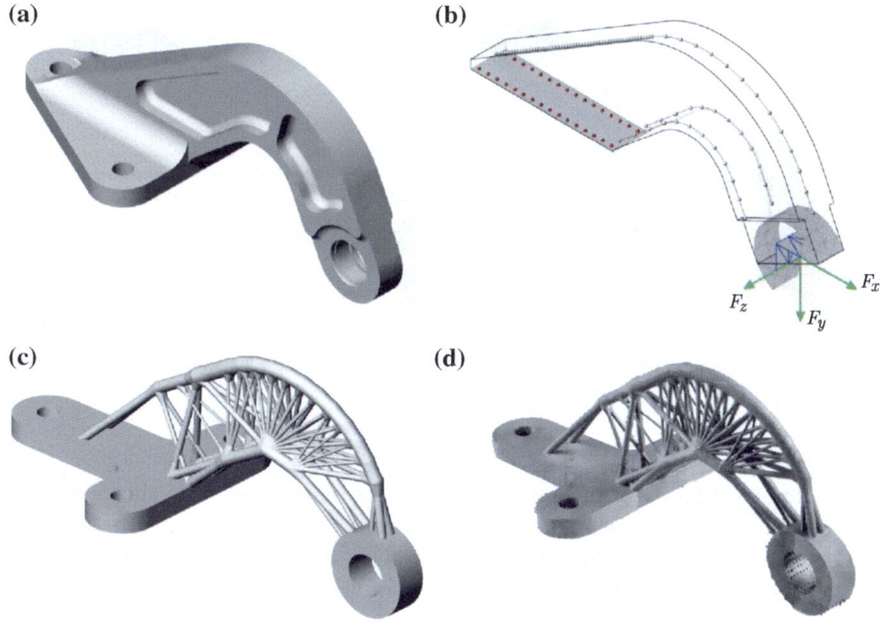

Fig. 8.69 Air-brake hinge example: **a** original air-brake hinge design, **b** definition of the topology optimization problem, **c** final design for air-brake hinge, **d** final design additively manufactured from Ti-6Al-4V using the EBM process; (by the courtesy of SMO; Smith et al. 2016)

difference in weight corresponds to the huge difference in efficiency between the continuous structures and Michell's structures occupying the same spatial domain.

In turn, He et al. (2018) had proposed the optimization framework, which includes human interaction as one of the steps amid other automated procedures (the so-called 'human-in-the-loop' optimization framework). The proposed framework consists of the following steps:

1. layout optimization using a ground structure approach,
2. rationalization by using geometry optimization (moving nodes, merging nodes, elimination of collinear members as described previously),
3. determination of the candidate truss layout,
4. human interaction (manual modification of the structural layout by moving, adding or deleting members, application of the additional constraints),
5. re-optimization of the structure by geometry optimization and sizing,
6. determination of the final structure layout and repeating steps 4 to 5, if needed.

The interaction of a human may have various objectives including removal of too slender members which can be prone to buckling or hard for manufacturing, removal of the cases with too many members intersecting at one node or improvement of the structure aesthetics. The re-use of the geometry optimization and sizing is aimed

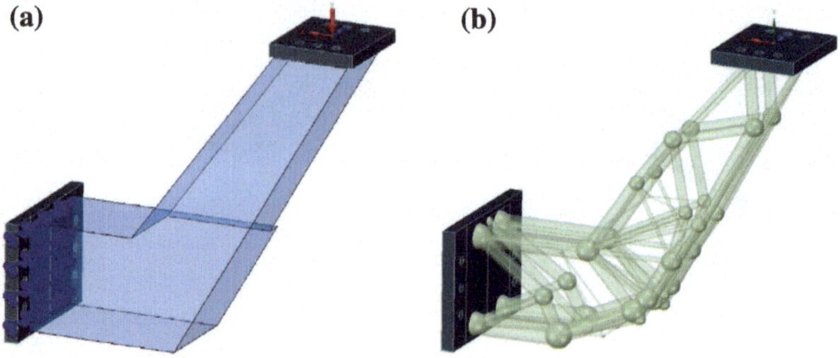

Fig. 8.70 Satellite antenna mount: **a** design problem, **b** initial design generated by layout and geometry optimization (by the courtesy of SMO; He et al. 2018)

at checking the validity of the human-modified layout. In particular, this applies to the situation when human interaction substantially deteriorates the structure, e.g. causes loss of stability by removal of the crucial elements. The applied optimization algorithm is intended to detect such cases and propose the alternative layout which is ready for additive manufacturing.

The presented example concerns optimization of the satellite antenna mount. The problem formulation including the design domain, boundary conditions and loading is depicted in Fig. 8.70a, whereas the initial design created with the use of geometry optimization is presented in Fig. 8.70b. The improvement of the layout feasibility and applicability is obtained in several steps. Initially, all thin members are removed and nodal positions are automatically adjusted in order to restore structure equilibrium (Fig. 8.71a). Then collinear members are removed (Fig. 8.71b) and nodes located close to each other are merged (Fig. 8.71c). The obtained design of the antenna mount (Fig. 8.71d) is only 3% heavier than the initial design. It is characterized by relatively simple, applicable shape; the design is ready for the additive manufacturing process.

The Young modulus design method (YMD), see Sect. 7.2.2, turns out to be a convenient starting point for the manufacturing of structures specifically formed to transmit a given set of in-plane loads; see Fig. 8.72a concerning the design of a stiffest bracket (Lewiński 2018). In the first step the optimum layout of Young's modulus is constructed by solving the stress-based problem (7.34) with the integrand being the 2D counterpart of (7.48). The optimal structure emerges from the design domain, see Fig. 8.72b. Next, the non-homogeneous isotropy is modelled by a hexagonal gridwork of varying thicknesses of ligaments. Such gridworks are characterized by the effective Poisson ratio equal to 0.33 almost independently of the thickness of ligaments, which justifies using YMD with a predefined Poisson's ratio. The virtual model of the structure is imported to AutoCAD, saved as a set of primitives connected to form one planar set, extruded and smoothed, see Fig. 8.73a. The final model is exported to the STL file and the structure is manufactured with the use of 3D printing technology, see Fig. 8.73b. The numerical analysis indicates that the stress distribution in the optimized bracket is more uniform than stresses distributions in

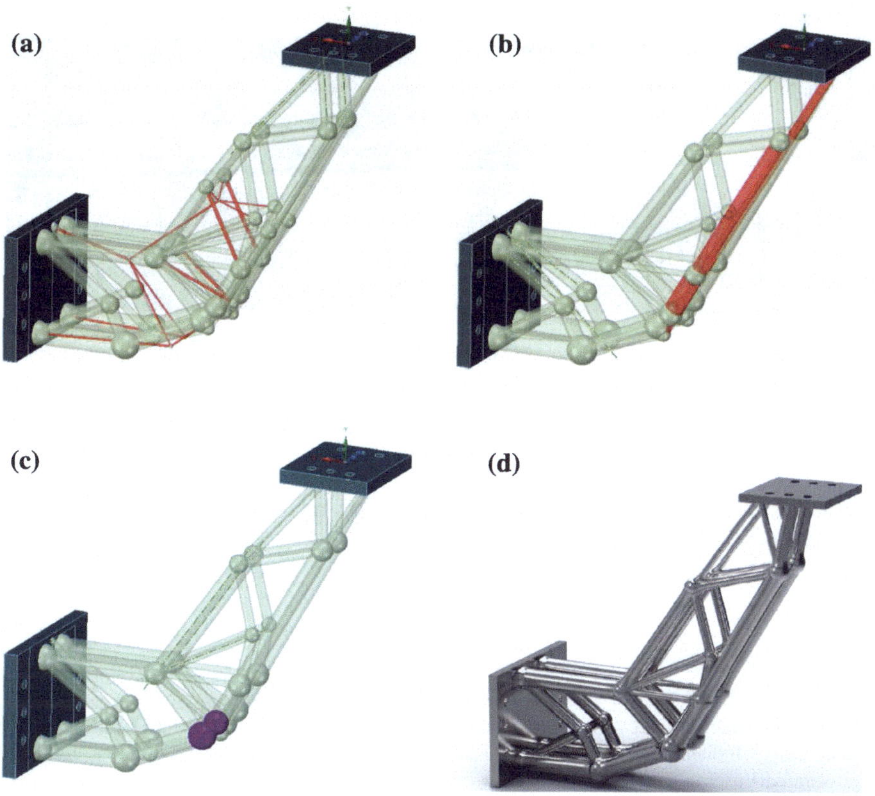

Fig. 8.71 Satellite antenna mount refinement steps: **a** removal of slender bars, **b** removal of collinear members, **c** merging nodes, **d** final 3D printed design; (by the courtesy of SMO; He et al. 2018)

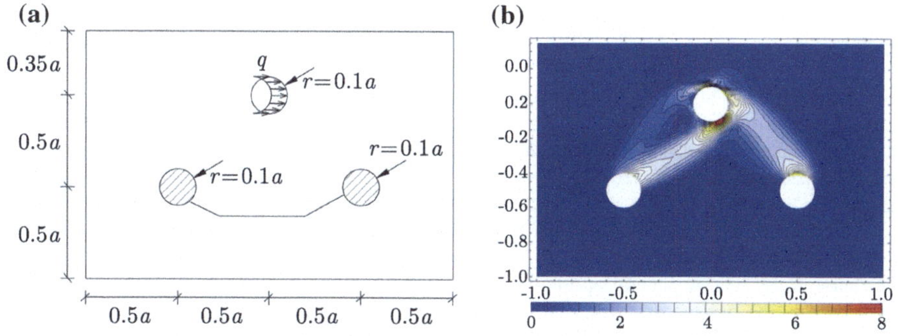

Fig. 8.72 The problem setting: the horizontal load q applied along the boundary of the upper circular aperture should be transmitted to the external boundaries of the two circular apertures below the horizontal line of the resultant of the load (**a**). The optimal distribution of the modulus E found by YMD with given Poisson's ratio equal 0.33 (**b**). Published by courtesy of Springer Int. Publ. AG, Cham, Switzerland; reprinted from Lewiński et al. (2018)

(a) (b)

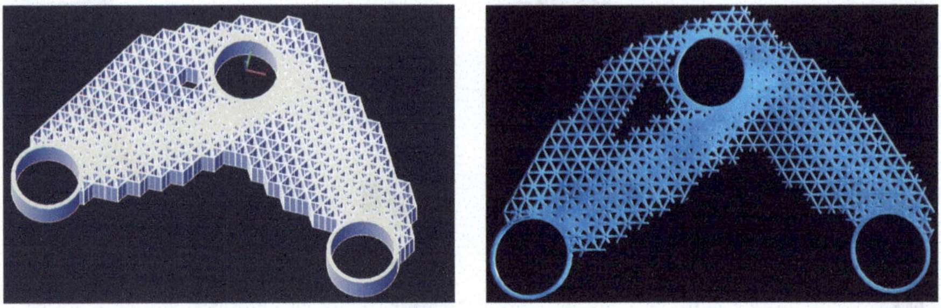

Fig. 8.73 3D-view of the final AutoCAD model (**a**), Photograph of the 3D-printed prototype (**b**). Published by courtesy of Springer Int. Publ. AG, Cham, Switzerland; reprinted from Lewiński et al. (2018)

alternative designs, not supported by optimization methods; for more details, see Lewiński et al. (2018).

References

Aerodynamic water droplet with strong lightweight bone structure, Engineering Design Research Laboratory, Indiana University-Purdue University Indianapolis. http://www.engr.iupui.edu/~tovara/edrl/projects/Litecar.

Ario, I., Nakazawa, M., Tanaka, Y., Tanikura, I., & Ono, S. (2013). Development of a prototype deployable bridge based on origami skill. *Automation in Construction*, *32*, 104–111.

Baker, W. F., Beghini, A., & Mazurek, A. (2011). Applications of structural optimization in architectural design, ASCE structures congress.

Beghini, L. L. (2013). *Building science through topology optimization*. Doctoral dissertation, University of Illinois at Urbana-Champaign.

Beghini, A., Beghini, L. L., & Baker, W. F. (2013). Applications of structural optimization in architectural design, ASCE structures congress.

Bołbotowski, K., & Sokół, T. (2016). New method of generating Strut and Tie models using truss topology optimization. In M. Kleiber, T. Burczyński, K. Wilde, J. Górski, K. Winkelmann, & Ł. Smakosz (Eds.), *Advances in mechanics: Theoretical, computational and interdisciplinary issues*. London UK: CRC Press.

Chikahiro, Y., Ario, I., Pawłowski, P., Graczykowski, C., Holnicki-Szulc, J. (2018). Optimization of the reinforcement layout in scissor type bridge by combination of FEM and Differential Evolution Algorithm (Submitted).

Dewhurst, P. (2005). A general optimality criterion for combined strength and stiffness of dual-material-property structures. *International Journal of Mechanical Sciences*, *47*, 293–302.

Flügge, W. (1973). *Stresses in Shells* (2nd ed.). Berlin: Springer.

Foxe, D. M. (2018). Reassessing Zalewski's shells designs, and surfaces. In C. Mueller, S. Adriaenssens (Eds.), *Creativity in Structural Design, Proceedings of the IASS Symposium 2018, July 16–20, 2018, MIT, Boston, USA*.

He, L., & Gilbert, M. (2015). Rationalization of trusses generated via layout optimization. *Structural and Multidisciplinary Optimization*, *52*, 677–694.

He, L., Gilbert, M., Johnson, T., Smith, C. (2018). Human-in-the-Loop Layout and Geometry. In A. Schumacher, T. Vietor, S. Fiebig, K. U. Bletzinger, K. Maute (Eds.), *Advances in structural and multidisciplinary optimization. WCSMO 2017. Optimization of structures and components*. Cham: Springer.

Hemp, W. S. (1974). Michell framework for uniform load between fixed supports. *Engineering Optimization*, *1*(1), 61–69.

Holnicki-Szulc, J., & Bielecki, T. (2000). Structures with highest ability of adaptation to overloading. In *Proceedings of IUTAM Symposium on Smart Structures and Structronic Systems. 26–29 September 2000. Magdeburg, Germany*.

Holnicki-Szulc, J., & Knap, L. (2004). Adaptive crashworthiness concept. *International Journal of Impact Engineering*, *30*(6), 639–663.

Holnicki-Szulc, J., Maćkiewicz, A., & Kołakowski, P. (1998). Design of adaptive structures for improved load capacity. *AIAA Journal*, *36*(3), 471–476.

Holnicki-Szulc, J., Pawłowski, P., & Wikło, M. (2003). High-performance impact absorbing materials - The concept, design tools and applications. *Smart Materials and Structures*, *12*, 461–467.

Kołakowski, P., Wikło, M., & Holnicki-Szulc, J. (2008). The virtual distortion method - A versatile reanalysis tool for structures and systems. *Structural and Multidisciplinary Optimization*, *36*(3), 217–234.

Kuś, S., & Zalewski, W. (2000). Shaping structures. In *Symposium, Conceptual Designing-Structural Shaping. Corrugated Iron Structures. Cable Structures* (pp. 11–26) Rzeszów: Rzeszów University of Technology (in Polish).

Lancaster, L. C. (2006). Materials and Construction of the Pantheon in Relation to the Developments in Vaulting in Antiquity. In G. Grasshof, M. Wäfler & M. Heinzelmann (Eds.), *The Pantheon in Rome. Contributions to the Conference. Bern, November 9–12, 2006*.

Leonhardt, F., & Zellner, W. (1972). *Vergleiche zwischen Hange - And Schragkabel brücken für Spann weiten über 600 m* (Vol. 32, p. 1972). Zürich: der IVBH-Abhandlugen.

Leonhardt, F. (1986). *1986*. PCI Journal (September - October: Cable stayed bridges with prestressed concrete.

Lewiński, T., Czarnecki, S., Czubacki, R., Łukasiak, T., & Wawruch, P. (2018). Constrained versions of the free material design methods and their applications in 3D printing. In A. Schumacher, Th. Vietor, S. Fiebig, K.-U. Bletzinger, K. Maute (Eds.), *Advances in Structural and Multidisciplinary Optimization. Proceedings of the 12th World Congress of Structural and Multidisciplinary Optimization (WCSMO12)* (pp. 1317–1332). Cham, Switzerland: Springer International Publishing.

Marzec, Z., & Holnicki-Szulc, J. (1999) Adaptive barriers with maximal impact energy absorption. In *Proceedings of 3rd World Congress on Structural and Multidisciplinary Optimization. 17–21 May 1999, Buffalo, NY*.

Maute, K., & Ramm, E. (1995a). Adaptive topology optimization. *Structural Optimization*, *10*, 100–112.

Maute, K., & Ramm, E. (1995b). General shape optimization – an integrated model for topology and shape optimization. In *Proceedings of the First World Congress of Structural and Mutidisciplinary Optimization, 28 May–2 June 1995, Goslar, Germany, Elsevier, Oxford, 1995* (pp. 299–306).

Maute, K., Schwartz, S., & Ramm, E. (1998). Adaptive topology optimization of elasto-plastic structures. *Structural and Multidisciplinary Optimization*, *5*, 81–89.

Mayer, R. R., Kikuchi, N., & Scott, R. A. (1996). Applications of topology optimization techniques to structural crashworthiness. *International Journal for Numerical Methods in Engineering*, *39*, 1383–1403.

Mazurek, A. (2012). Geometrical aspects of optimum truss like structures for three-force problem. *Structural and Multidisciplinary Optimization*, *45*, 21–32.

Mazurek, A., Baker, W. F., & Tort, C. (2011). Geometrical aspects of optimum truss like structures. *Structural and Multidisciplinary Optimization*, *43*, 231–242.

Moore, D. (1995). The Pantheon. www.romanconcrete.com.

Mrzygłód, M., & Kuczek, T. (2014). *Structural and Multidisciplinary Optimization*, *49*(2), 327–336. https://doi.org/10.1007/s00158-013-0972-z.

Neves, M. M., Rodrigues, H., & Guedes, J. M. (1995). Generalized topology design of structures with a buckling load criterion. *Structural and Multidisciplinary Optimization, 10*, 71–78.
Novozhilov, V. V. (1964). *Thin Shell Theory*. Groningen, The Netherlands: Nordhoff.
Pawłowski, P., & Holnicki-Szulc, J. (2004). Adaptive structures under extreme loads – Impact detection, self-adaptation, self-repairing. In *Proceedings of 3rd European Conference on Structural Control, 12–15 July 2004, Vienna, Austria.*
Pawłowski, P., & Wikło, M. (2004). Design of adaptive structures under random impact conditions. In J. Holnicki-Szulc & C. A. Mota-Soares (Eds.), *Advances in smart technologies in structural engineering*. Berlin: Springer.
Pedersen, C. B. W. (2003a). Topology optimization design of crushed 2D-frames for desired energy absorption history. *Structural and Multidisciplinary Optimization, 25*, 368–382.
Pedersen, C. B. W. (2003b). Topology optimization for crashworthiness of frame structures. *International Journal of Crashworthiness, 8*, 29–39.
Pichugin, A. V., Tyas, A., & Gilbert, M. (2012). On the optimality of Hemp's arch with vertical hangers. *Structural and Multidisciplinary Optimization, 46*, 17–25.
Pichugin, A. V., Tyas, A., Gilbert, M., & He, L. (2015). Optimum structure for a uniform load over multiple spans. *Structural and Multidisciplinary Optimization, 52*, 1041–1050.
Smith, Ch J, Gilbert, M., Todd, I., & Derguti, F. (2016). Application of layout optimization to the design of additively manufactured metallic components. *Structural and Multidisciplinary Optimization, 54*, 297–1313.
Soto, C. A. (2001). Structural topology optimization: From minimizing compliance to maximizing energy absorption. *International Journal of Vehicle Design, 25*(1), 142–163.
Soto, C. A. (2002). Application of structural topology optimization in the automotive industry: Past, present and future. In *Proceedings of 5th World Congress of Computational Mechanics. 9–12 July 2002, Vienna, Austria.*
Stromberg, L. L., Beghini, A., Baker, W. F., & Paulino, G. H. (2011). Application of layout and topology optimization using pattern gradation for the conceptual design of buildings. *Structural and Multidisciplinary Optimization, 43*, 165–180.
Stromberg, L. L., Beghini, A., Baker, W. F., & Paulino, G. H. (2012). Topology optimization for braced frames: Combining continuum and beam/column elements. *Engineering Structures, 37*, 106–124.
Tyas, A., Pichugin, A. V., & Gilbert, M. (2011). Optimum structure to carry a uniform load between pinned supports: Exact analytical solution. *Proceedings of the Royal Society A: Mathematical, Physical and Engineering Science, 467*(2128), 1101–1120.
Victoria, M., Querin, O. M., & Marte, P. (2011). Generation of strut-and-tie models by topology design using different material properties in tension and compression. *Structural and Multidisciplinary Optimization, 44*, 247–258.
Wikło, M., & Holnicki-Szulc, J. (2008a). Optimal design of adaptive structures: Part I. Remodeling for impact reception. *Structural and Multidisciplinary Optimization, 37*(3), 305–318.
Wikło, M., & Holnicki-Szulc, J. (2008b). Optimal design of adaptive structures: Part II. Adaptation to impact loads. *Structural and Multidisciplinary Optimization, 37*(4), 351–366.
Yamakawa, H., Tsutsui, Z., Takemae, K., Ujita, Y., & Suzuki, Y. (1999). Structural optimization for improvement of train crashworthiness in conceptual and preliminary design. In *Proceedings of 3rd World Congress on Structural and Multidisciplinary Optimization, 17–21 May 1999, Buffalo, NY.*
Zabłocki, W. (2000). Optimization of structures and new forms of tall buildings. *Architektura, 74*(11), 96–98. (in Polish).
Zalewski, W. (1962). Construction de la toiture du supermarket a Varsovie. In N. Esquillan & Y. Saillard (Eds.), *Hanging roofs: Proceedings of the IASS Colloquium on Hanging Roofs, Continuous Metallic Shell Roofs and Superficial Lattice Roofs, Paris 9–11 July, 1962.*
Zalewski, W. (1963). Constance de la force comme critere de la forme ationelle d'une construction.
Zalewski, W. (1964). Some new structural forms created in the period 1950–60 (in Polish).
Zalewski, W., & Allen, E. (1998). *Shaping Structures. Statics*. New York: Wiley.

Zalewski, W., & Zabłocki, W. (2002). Engineering inspirations in shaping tall buildings. In *Lightweight Structures in Civil Engineering, Proceedings of the International Symposium, Warsaw, Poland, 24–28 June 2002* (pp. 109–118).
Zalewski, W. (2000). Strength and lightness - The muses of a structural designer. *Architektura*, *74*(11), 94–95. (in Polish).
Zalewski, W. (2005). Shaping structures. Introduction and examples. In *Proceedings of the 7th Symposium: New Achievements of Science in Civil Engineering. Shaping Structures, Cable Structures, Corrugated Iron Structures* (pp. 49–72), Rzeszów (in Polish).
Zegard, T. (2014). *Structural optimization: From continuum and ground structures to additive manufacturing*. Doctoral dissertation, University of Illinois at Urbana-Champaign.
Zegard, T., & Paulino, G. H. (2016). Bridging topology optimization and additive manufacturing. *Structural and Multidisciplinary Optimization*, *53*(1), 175–192.

The manufacturer's authorised representative in the EU is Springer Nature Customer Service Centre GmbH, Europaplatz 3, 69115 Heidelberg, Germany. If you have any concerns regarding our products, please contact ProductSafety@springernature.com

Printed and bound by CPI Group (UK) Ltd, Croydon, CR0 4YY
05/10/2026
02228139-0019